INTERMEDIATE ALGEBRA

Third Edition

INTERMEDIATE ALGEBRA

Third Edition

John Tobey

Jeffrey Slater

North Shore Community College
Danvers, Massachusetts

Prentice Hall, Upper Saddle River, NJ 07458

Library of Congress Cataloging-in-Publication Data

Tobey, John (date)
   Intermediate algebra / John Tobey, Jeffrey Slater.—3rd ed.
        p. cm.
   Includes index.
   ISBN 0-13-850884-4
   1. Algebra. I. Slater, Jeffrey (date). II. Title.
QA154.2.T634 1998
512.9—dc21   97-41712
                                     CIP

Editor-in-chief: *Jerome Grant*
Editorial Director: *Tim Bozik*
Acquisitions Editor: *Karin E. Wagner*
Editorial Assistant: *April Thrower/Joanne Wendelken*
Assistant Vice President of Production and Manufacturing: *David W. Riccardi*
Production Editor: *York Production Services*
Senior Managing Editor: *Linda Mihatov Behrens*
Executive Managing Editor: *Kathleen Schiaparelli*
Marketing Manager: *Jolene Howard*
Marketing Assistant: *Jennifer Pan*
Interior Design: *Amy Rosen/Lisa Jones*
Cover Design: *Amy Rosen*
Creative Director: *Paula Maylahn*
Art Director: *Joseph Sengotta*
Art Manager: *Gus Vibal*
Manufacturing Buyer: *Alan Fischer*
Manufacturing Manager: *Trudy Pisciotti*
Photo Researcher: *Rona Tuccillo*
Photo Editor: *Lori Morris-Nantz*
Supplements Editor: *Audra J. Walsh*
Cover Photo: *Burton Pritzker/Photonica*

Photo credits appear on page P–1, which
constitutes a continuation of the copyright page.

© 1998, 1995, 1991 by Prentice-Hall, Inc.
Simon & Schuster/A Viacom Company
Upper Saddle River, New Jersey 07458

Printed in the United States of America

10 9 8 7 6 5 4 3 2 1

ISBN 0-13-850884-4

Prentice-Hall International (UK) Limited, *London*
Prentice-Hall of Australia Pty. Limited, *Sydney*
Prentice-Hall Canada Inc., *Toronto*
Prentice-Hall Hispanoamericana, S.A., *Mexico City*
Prentice-Hall of India Private Limited, *New Delhi*
Prentice-Hall of Japan, Inc., *Tokyo*
Simon & Schuster Asia Pte. Ltd., *Singapore*
Editora Prentice-Hall do Brasil, Ltda., *Rio de Janeiro*

This book is dedicated to
John Tobey III, Marcia Yvonne Tobey, and Melissa Tobey LaBelle.
They are three college graduates who would make any parent proud.

Contents

CHAPTER 5

Polynomials   269

CHAPTER 6

Rational Algebraic Expressions and Equations   333

CHAPTER 9

The Conic Sections 503

CHAPTER 10

Functions 557

CHAPTER 11

Logarithmic and Exponential Functions 609

This book was written with your needs and interests in mind. The original manuscript and the first editions of this book have been class tested with students all across the country. Based on the suggestions of many students, the book has been refined and improved to maximize your learning while using this text.

We realize that students who enter college have sometimes never enjoyed mathematics or never done well in a mathematics course. You may find that you are anxious about taking this course. We want you to know that this book has been written to help you overcome those difficulties. Literally thousands of students across the country have found an amazing ability to learn mathematics as they have used previous editions of this book. We have incorporated several learning tools and various types of exercises and examples to assist you in learning this material.

It helps to know that learning mathematics is going to help you in life. Perhaps you have entered this course feeling that little or no mathematics will be necessary for you in your future job. However, elementary school teachers, bus drivers, laboratory technicians, nurses, secretaries, telephone operators, cable TV repair personnel, photographers, pharmacists, salespeople, doctors, architects, inspectors, counselors, and custodians who once believed that they needed little if any mathematics are finding that the mathematical skills presented in this course can help them. Mathematics, you will find, can help you too. In this book, a great number of examples and problems come from everyday life. You will be amazed at the number of ways mathematics is used in the world each day.

Our greatest wish is that you will find success and personal satisfaction in the mathematics course.

SUGGESTIONS FOR STUDENTS

1. Be sure to take the time to read the boxes marked **Developing Your Study Skills.** These ideas come from faculty and students throughout the United States. They have found ways to succeed in mathematics and they want to share those ideas with you.

2. **Read** through each section of the book that is assigned by your instructor. You will be amazed at how much information you will learn as you read. It will "fill in the pieces" of mathematical knowledge. Reading a math book is a key part of learning.

3. Study carefully each **sample example.** Then work out the related practice problem. Direct involvement in doing problems and not just thinking about them is one of the greatest guarantees of success in this course.

4. Work out all assigned **homework problems.** Verify your answers in the back of the book. Ask questions when you don't understand or when you are not sure of an answer.

5. Be sure to take advantage of **end-of-chapter helps.** Study the Chapter Organizers. Do the Chapter Review problems and the Practice Test problems.

6. If you need help, remember there are teachers and tutors who can assist you. There is a **Student Solutions Manual** for this textbook that you will find most valuable. This manual shows worked out solutions for all the odd-numbered exercises as well as odd-numbered Diagnostic Pretest, Chapter Review, Chapter Test, and Cumulative Test Problems in this book. (If your college bookstore does not carry the Student Solutions Manual, ask them to order a copy for you.)

7. Watch the **videotapes.** Every section of this text is explained in detail on a videotape. These tapes were prepared by Michael Mayne and John (Biff) Pietro of Riverside Community College. These popular mathematics professors were selected by students from a large group of college mathematics teachers who have done videotapes as being the most effective and helpful in teaching mathematics by video-tape. Thousands of students from across the country have been helped by the humor-ous, interesting, mathematically accurate video presentation that is carefully keyed to each section of the text. The work is solved using exactly the same methods as explained in the text. These professors show the step-by-step worked out solutions for many of the even-numbered problems in the homework exercises.

8. Use **MathPro Explorer,** a tutorial software package that allows you to be tutored over any section in the book and also allows you to test yourself on your mastery of any section of any chapter of the text. This software package will be made available in IBM PC or Macintosh versions at each college campus that adopts this textbook. *MathPro Explorer* will allow you to receive help in textbook topics or to explore new mathematical ideas and procedures.

9. Remember, learning mathematics takes time. However, the time spent is well worth the effort. **Take the time** to study, do homework, review, ask questions, and just reflect over what you have learned. The time you invest in learning mathematics will reap dividends in your future courses and your future life.

We encourage you to look over your textbook carefully. Many important features have been designed into the book to make learning mathematics a more enjoyable activity. There are some special features, unique to this textbook, that students from all over this country have told us that they found especially helpful.

FOUR KEY TEXTBOOK FEATURES TO HELP YOU

1. **Practice problems with worked-out solutions.** Immediately following every sample example is a similar problem called the Practice Problem. If you can work it out correctly by following the sample example as a general point of reference, then you will likely be able to do the homework exercises. If you encounter some difficulty, then you will find helpful the completely worked out solutions to the Practice Prob-lems that appear at the end of the text at the section marked with yellow page edges.

2. **Student-friendly application problems.** You will find in every chapter of this book application problems that sound realistic and interesting. Many of the problems were actually written or suggested by students. As you develop your problem-solving and reasoning skills in this course, you will encounter a number of real-world situations that help you to see how very helpful mathematical skills are in today's complex world.

3. **Putting Your Skills to Work problems.** In each chapter are unique problem sets that ask you to analyze in depth some mathematical aspect of daily life. You will be asked to extend your knowledge, do some creative thinking, and to work in small groups in a **Cooperative Learning Situation.** You will even have a chance to explore some web sites and see how the Internet can assist you in learning. These sections will awaken some new interests and help you to develop the critical thinking skills so necessary both in college and after you graduate.

4. The Chapter Organizer. Everything you need to learn in any one chapter of this book is readily available at your fingertips in the Chapter Organizer. This very popular chart summarizes all methods covered in the chapter, gives page references, and shows a sample example completely worked out for every major topic covered. Students have found this tool a most helpful way to master the content of any chapter of the book.

We have spent more than 25 years teaching mathematics. Each teaching day we find our greatest joy is helping students learn. We take a personal interest that each student has a good learning experience in taking this course. If you have some personal comments, suggestions, or ideas for future editions of this textbook, please write to us at:

> Prof. John Tobey and Prof. Jeffrey Slater
> Prentice Hall Publishing
> Office of the College Mathematics Editor
> One Lake Street
> Upper Saddle River, NJ 07458

We wish you success in this course and in your future life!

> John Tobey
> Jeffrey Slater

GUIDE FOR STUDENTS

How to Use **Intermediate Algebra, Third Edition** to enrich
your class experience and prepare for tests.

CHAPTER 3
Equations, Inequalities, and Functions

Each chapter opens with an application.
These applications relate to the material you find
in the chapter as well as an extended discovery
called *Putting Your Skills to Work* which you will
encounter later in the chapter.

uppose you developed superior new graphing calculators and wanted to manufacture and sell them. How would you determine what the demand will be for the new graphing calculator that you will soon introduce? How could you determine the profitability of this new product? If you would like to find out, then turn to the Putting Your Skills To Work on page 158.

page 129

Putting Your Skills to Work

DETERMINING THE DEMAND FOR A PRODUCT

When a manufacturing company is interested in introducing a new product, one of the variables that will affect the demand for the product will be the price. Consider the situation where an electronics company is planning to introduce a new graphing calculator. The market research department has given the company management the price versus demand table shown at the right.

Price	Estimated Demand (in thousands)
$ 40	128
$100	80
$120	64
$200	0

Put your new skills to work.
These multi-part projects are relevant
to chapter concepts as well to problems
you encounter from day to day.

page 158

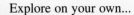

PROBLEMS FOR INDIVIDUAL INVESTIGATION AND ANALYSIS

1. Plot the points and draw the graph. Write the equation of the line in the form $y = mx + b$, where x represents the price in dollars and y represents the estimated demand for the calculator measured in thousands.

2. Find the slope. What does the slope indicate? Find the y-intercept. What does the y-intercept indicate?

3. What is the effect on the demand for the calculator by lowering the price by \$10? by \$100?

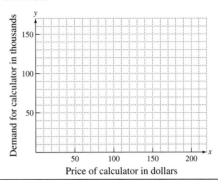

page 158

PROBLEMS FOR GROUP INVESTIGATION AND COOPERATIVE ACTIVITY

Together with other members of your class see if you can determine the following.

4. How many items can the manufacturer expect to sell if the calculator is priced at \$110?

5. If the cost equation for this product is $C = 210{,}000 + 70n$, where n is the number of calculators sold, what would be the cost to manufacture this number of calculators?

Work cooperatively with fellow students to solve more involved problems. You will find that problems are often solved collaboratively in the workplace.

page 158

 INTERNET CONNECTION: Go to `http://www.prenhall.com/tobey` to be connected.

Site: Supply and Demand

This site displays a sample supply and demand graph. The demand graph is the one you have studied, whereas the supply graph shows the quantity of product that the manufacturer is willing to produce for a given price. (Note that this graph shows price on the vertical axis instead of the horizontal axis.)

8. As you move from the left of the graph to the right, the demand decreases but the supply increases. Explain why this is reasonable.

9. The intersection of the line showing supply and the lines showing demand is clearly marked. What is the significance of this point? (*Hint:* What will happen if the price is above or below the price indicated by this point?)

Surf the 'Net for real data that relates to the *Putting Your Skills to Work* explorations. Extend the Concepts!

page 158

These features have been included in this text to help you make connections to mathematics. Use them to explore, connect, and discover.

PREPARE YOURSELF

Chapter Pretests will familiarize you with the objectives of each chapter before you begin.

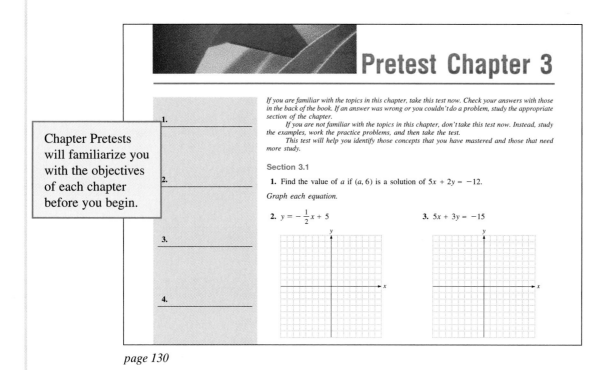

Pretest Chapter 3

If you are familiar with the topics in this chapter, take this test now. Check your answers with those in the back of the book. If an answer was wrong or you couldn't do a problem, study the appropriate section of the chapter.

If you are not familiar with the topics in this chapter, don't take this test now. Instead, study the examples, work the practice problems, and then take the test.

This test will help you identify those concepts that you have mastered and those that need more study.

Section 3.1

1. Find the value of a if $(a, 6)$ is a solution of $5x + 2y = -12$.

Graph each equation.

2. $y = -\frac{1}{2}x + 5$

3. $5x + 3y = -15$

1. _____

2. _____

3. _____

4. _____

page 130

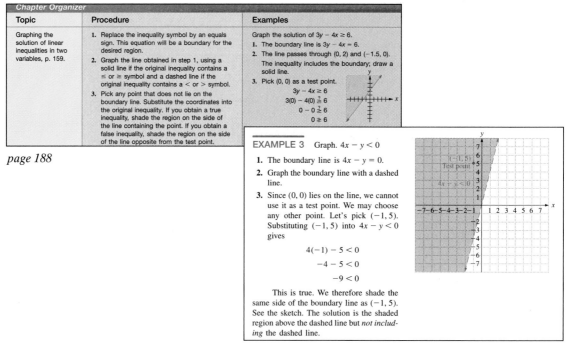

Chapter Organizer		
Topic	**Procedure**	**Examples**
Graphing the solution of linear inequalities in two variables, p. 159.	1. Replace the inequality symbol by an equals sign. This equation will be a boundary for the desired region. 2. Graph the line obtained in step 1, using a solid line if the original inequality contains a $\leq$ or $\geq$ symbol and a dashed line if the original inequality contains a $<$ or $>$ symbol. 3. Pick any point that does not lie on the boundary line. Substitute the coordinates into the original inequality. If you obtain a true inequality, shade the region on the side of the line containing the point. If you obtain a false inequality, shade the region on the side of the line opposite from the test point.	Graph the solution of $3y - 4x \geq 6$. 1. The boundary line is $3y - 4x = 6$. 2. The line passes through $(0, 2)$ and $(-1.5, 0)$. The inequality includes the boundary; draw a solid line. 3. Pick $(0, 0)$ as a test point. $\quad 3y - 4x \geq 6$ $\quad 3(0) - 4(0) \overset{?}{\geq} 6$ $\quad 0 - 0 \overset{?}{\geq} 6$ $\quad 0 \ngeq 6$

page 188

EXAMPLE 3   Graph. $4x - y < 0$.

1. The boundary line is $4x - y = 0$.

2. Graph the boundary line with a dashed line.

3. Since $(0, 0)$ lies on the line, we cannot use it as a test point. We may choose any other point. Let's pick $(-1, 5)$. Substituting $(-1, 5)$ into $4x - y < 0$ gives

$$4(-1) - 5 < 0$$

$$-4 - 5 < 0$$

$$-9 < 0$$

This is true. We therefore shade the same side of the boundary line as $(-1, 5)$. See the sketch. The solution is the shaded region above the dashed line but *not including* the dashed line.

page 160

Have a plan—Develop a problem-solving strategy.

CHECK YOUR UNDERSTANDING

Intermediate Algebra, Third Edition includes many different types of exercises—Apply yourself and *expand* your understanding.

GRAPHING CALCULATOR

Exploring y-intercepts

Using a graphing calculator, on the same set of axes graph

$$y_1 = 2x$$
$$y_2 = 2x + 1$$
$$y_3 = 2x - 1$$
$$y_4 = 2x + 2$$

Where does each graph cross the y-axis? What affect does b have on the graph of $y = mx + b$? What would the graph of the line $y = 2x - 5$ look like? Use your graphing calculator to verify.

page 149

3.1 Exercises

Verbal and Writing Skills

1. Graphs are used to show the relationship between the _____ in an equation.

2. The x-axis and the y-axis intersect at the _____.

3. Explain in your own words why (a, b) is an ordered pair. Give an example.

4. $(5, 1)$ is a solution to the equation $2x - 3y = 7$. What does this mean?

page 137

To Think About

31. A bottled iced-tea manufacturer saw profits increase 65% last year. This year, due to so much new competition, profits fell 40%. Profits this year were $17,820,000.00. What was the realized profit for the iced-tea company two years ago?

32. Northwest Airlines gives out peanuts to all passengers who want them. On several recent flights to Boston 72% of the passengers asked accepted the peanuts and 23% rejected the peanuts. A total of 36 passengers were asleep and could not be asked. How many passengers were on these flights?

33. The area of the shaded triangle is 6 cm². Find the area of parallelogram *ACDF*.

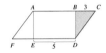

34. The volume of the sphere is 38,808 cm³. Find the volume of the cylinder (use $\pi \approx \frac{22}{7}$).

page 95

Chapter 3 Test

1. Graph the line with equation $2x + 3y = -10$. Plot at least three points.

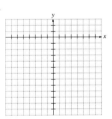

1. _____

2. _____

Use these features as guides to make yourself a better problem-solver.

2. Find the slope of the line passing through $(2, -3)$ and $\left(\frac{1}{2}, -6\right)$.

page 193

Cumulative Review Problems

Solve for x.

39. $2(3ax - 4y) = 5(ax + 3)$

40. $\frac{1}{2}(3x + 5y) = 2(x - 3y)$

41. $\frac{1}{3}x + \frac{1}{5}y = \frac{1}{10} + \frac{1}{2}x$

42. $0.12(x - 4) = 1.16x - 8.02$

page 186

ENHANCE YOUR LEARNING

Intermediate Algebra, Third Edition is more than a textbook; it is an integrated package of instruction. Ask your professor about these supplements, which are a part of the **Intermediate Algebra** suite of learning materials.

Most items are keyed specifically to this text.

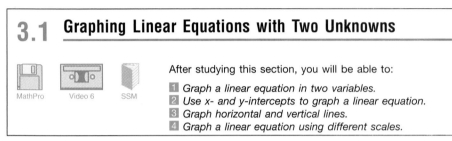

3.1   Graphing Linear Equations with Two Unknowns

MathPro   Video 6   SSM

After studying this section, you will be able to:

1. *Graph a linear equation in two variables.*
2. *Use x- and y-intercepts to graph a linear equation.*
3. *Graph horizontal and vertical lines.*
4. *Graph a linear equation using different scales.*

page 132

• Each section of the text begins with a reminder of the additional companion tools that have been designed to enhance your learning experience.

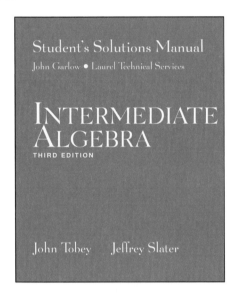

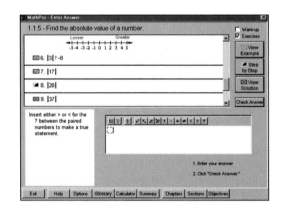

• Includes complete step-by-step solutions for every odd-numbered exercise, and every odd-numbered exercise in the Chapter Pretests, Chapter Reviews, Chapter Tests, and Cumulative Tests.

• MathPro Explorer: Interactive and Tutorial Software
• For Windows and Power Macintosh
• Generates unlimited practice exercises and instant feedback

New York Times/Themes of the Times
Newspaper-format supplement–
*ask your professor about
this free supplement*

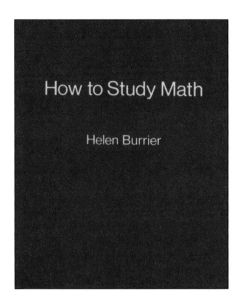

- Tips include how to prepare for class, how to study for and take tests, and how to improve your grades

- Team taught video instruction covers each section of the text

- A guide to navigation strategies through the Internet as well as practice exercises and lists of resources

- Visit the companion website for related and extended applications

www.prenhall.com/tobey

Acknowledgments

This book and the other two books that are part of this series are the products of many years of work and many contributions from faculty and students across the country. We would like to thank the many reviewers and participants in focus groups and special meetings with the authors. Your written reports, suggestions, ideas, and proposals were most helpful in the process of refining and improving this textbook.

Our deep appreciation to each of the following:

George J. Apostolopoulos, DeVry Institute of Technology

Katherine Barringer, Central Virginia Community College

Rita Beaver, Valencia Community College

Jamie Blair, Orange Coast College

Larry Blevins, Tyler Junior College

Brenda Callis, Rappahannock Community College

Joan P. Capps, Raritan Valley Community College

Judy Carter, North Shore Community College

Robert Christie, Miami-Dade Community College

Mike Contino, California State University at Heyward

Judy Dechene, Fitchburg State University

Floyd L. Downs, Arizona State University

Barbara Edwards, Portland State University

Janice F. Gahan-Rech, University of Nebraska at Omaha

Colin Godfrey, University of Massachusetts, Boston

Carl Mancuso, William Paterson College

Janet McLaughlin, Montclair State College

Gloria Mills, Tarrant County Junior College

Norman Mittman, Northeastern Illinois University

Elizabeth A. Polen, County College of Morris

Ronald Ruemmler, Middlesex County College

Sally Search, Tallahassee Community College

Ara B. Sullenberger, Tarrant County Junior College

Michael Trappuzanno, Arizona State University

Cora S. West, Florida Community College at Jacksonville

Jerry Wisnieski, Des Moines Community College

We have been greatly helped by a supportive group of colleagues who not only teach at North Shore Community College but who have provided a number of ideas as well as extensive help on all of our mathematics books. We particularly give a special word of thanks to Hank Harmeling, Tom Rourke, Wally Hersey, Bob McDonald, Judy Carter, Bob Campbell, Rich Ponticelli, Russ Sullivan, Kathy LeBlanc, Lora Connelly, Sharyn Sharaf, Donna Stefano, and Nancy Tufo. Joan Peabody has done an excellent job of typing various materials for the manuscript and her help is gratefully acknowledged.

As a new edition of a book gets finalized new and fresh ideas are always helpful. We want to thank Judy Carter for contributing several new exercise problems in every section of the book and for providing a number of excellent new problems for the graphing calculator. We also want to thank Louise Elton for providing several new applied problems and suggested applications. Error checking is a challenging task and few can do it well. So we especially want to thank Cindy Trimble and her colleagues for accuracy checking the content of the book at different stages of text preparation.

Each textbook is a combination of ideas, writing, and revisions from the authors and wise editorial direction and assistance from the editors. We want to thank our Prentice Hall editor, Karin Wagner, for her administrative support and encouragement, and for her helpful insight and perspective on each phase of the revision of the textbook. Her patience, her willingness to listen, and her flexibility to adapt to changing publishing decisions have been invaluable to the production of this book. We also express our thanks to Jerome Grant and Melissa Acuña, who have continued to support our writing projects and given wise direction and focus to our work. We also thank Dennis Krebs and Barbara Mack for their work, suggestions, and assistance in getting this book into production.

Book writing is impossible for us without the loyal support of our families. Our deepest thanks and love to Nancy, Johnny, Melissa, Marcia, Shelley, Rusty, and Abby. Your understanding, your love and help, and your patience have been a source of great encouragement. Finally, we thank God for the strength and energy to write and the opportunity to help others through this textbook.

We have each spent more than 25 years teaching mathematics. Each teaching day we talk to our colleagues and share ideas of how math instruction can be improved. If you have some personal comments, suggestions, or ideas for future editions of this textbook, please write to us at:

Prof. John Tobey and Prof. Jeffrey Slater
Prentice Hall Publishing
Office of the College Mathematics Editor
One Lake Street
Upper Saddle River, NJ 07458

You may also reach us via e mail at:
johntobey@aol.com or jeffslater@aol.com

John Tobey
Jeffrey Slater

Diagnostic Pretest: Intermediate Algebra

Chapter 1

1. Evaluate. $3 - (-4)^2 + 16 \div (-2)$

2. Simplify. $(3xy^{-2})^3(2x^2y)$

3. Simplify. $3x - 4x[x - 2(3 - x)]$

4. Evaluate $F = \dfrac{9}{5}C + 32$ when $C = -35°$.

Chapter 2

Solve the following.

5. $-8 + 2(3x + 1) = -3(x - 4)$

6. When ice floats in water approximately $\frac{8}{9}$ of the height of the ice lies under water. If the tip of an iceberg is 23 feet above the water, how deep is the iceberg below the water line?

7. $4 + 5(x + 3) \geq x - 1$

8. $\left|3\left(\dfrac{2}{3}x - 4\right)\right| \leq 12$

Chapter 3

9. Find the slope and the y-intercept of $3x - 5y = -7$.

10. Find the equation of a line that passes through $(-5, 6)$ and $(2, -3)$.

11. Is this relation also a function? $\{(5, 6), (-6, 5), (6, 5), (-5, 6)\}$

12. If $f(x) = -2x^2 - 6x + 1$, find $f(-3)$.

Chapter 4

Solve the following.

13. $3x + 5y = 30$
$5x + 3y = 34$

14. $2x - y + 2z = 8$
$x + y + \ z = 7$
$4x + y - 3z = -6$

15. A speed boat can travel 90 miles with the current in two hours. It can travel upstream 105 miles against the current in three hours. How fast is the boat in still water? How fast is the current?

16. Graph the system.
$x - y \leq -4$
$2x + y \leq \ \ 0$

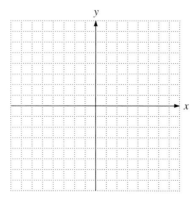

1.	
2.	
3.	
4.	
5.	
6.	
7.	
8.	
9.	
10.	
11.	
12.	
13.	
14.	
15.	
16.	

17. _____

17. Multiply.
$(3x - 4)(2x^2 - x + 3)$

18. Divide.
$(2x^3 + 7x^2 - 4x - 21) \div (x + 3)$

18. _____

19. Factor. $125x^3 - 8y^3$

20. Solve. $2x^2 - 7x - 4 = 0$

19. _____

Chapter 6

21. Simplify.
$$\frac{10x - 5y}{12x + 36y} \cdot \frac{8x + 24y}{20x - 10y}$$

22. Combine. $2x - 1 + \dfrac{2}{x + 2}$

20. _____

21. _____

23. Simplify.

$$\frac{\dfrac{1}{x + h} - \dfrac{1}{x}}{h}$$

24. Solve for x.

$$\frac{x}{x - 2} + \frac{3x}{x + 4} = \frac{6}{x^2 + 2x - 8}$$

22. _____

Chapter 7

23. _____

Assume all expressions under radicals represent nonnegative numbers.

25. Multiply and simplify.
$(\sqrt{3} + \sqrt{2x})(\sqrt{7} - \sqrt{2x^3})$

26. Rationalize the denominator.
$$\frac{3\sqrt{x} + \sqrt{y}}{\sqrt{x} - \sqrt{y}}$$

24. _____

25. _____

27. Solve and check your solutions.
$2\sqrt{x - 1} = x - 4$

26. _____

Chapter 8

28. Solve. $x^2 - 2x - 4 = 0$

29. $x^4 - 12x^2 + 20 = 0$

27. _____

30. Graph $f(x) = (x - 2)^2 + 3$. Label the vertex.

28. _____

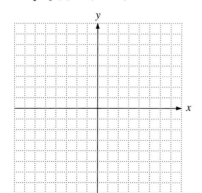

29. _____

30. _____

Chapter 9

31. Write in standard form the equation of a circle with a center at $(5, -2)$ and a radius of 6.

32. Write in standard form the equation of an ellipse whose center is at $(0, 0)$ and whose intercepts are at $(3, 0)$, $(-3, 0)$, $(0, 4)$, and $(0, -4)$.

33. Solve the following nonlinear system of equations.
$$x^2 + 4y^2 = 9$$
$$x + 2y = 3$$

Chapter 10

34. If $f(x) = 2x^2 - 3x + 4$ find $f(a + 2)$.

35. Graph on one axis $f(x) = |x + 3|$ and $g(x) = |x + 3| - 3$.

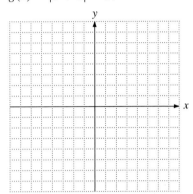

36. If $f(x) = \dfrac{3}{x + 2}$ and $g(x) = 3x^2 - 1$, find $g[f(x)]$.

37. If $f(x) = -\dfrac{1}{2}x - 5$ find $f^{-1}(x)$.

Chapter 11

38. Find y if $\log_5 125 = y$.

39. Find b if $\log_b 4 = \dfrac{2}{3}$.

40. What is $\log 10{,}000$?

41. Solve for x. $\log_6 (5 + x) + \log_6 x = 2$

31. _____	
32. _____	
33. _____	
34. _____	
35. _____	
36. _____	
37. _____	
38. _____	
39. _____	
40. _____	
41. _____	

Basic Concepts

Can you imagine how much Earth weighs compared to Saturn? Do you know which planet in our solar system has the greatest density? What is the escape velocity of Mercury compared to Earth? How can these types of quantities be predicted? The answers may surprise you. To investigate these turn to the Putting Your Skills to Work problems on page 50.

If you are familiar with the topics in this chapter, take this test now. Check your answers with those in the back of the book. If an answer was wrong or you couldn't do a problem, study the appropriate section of the chapter.

If you are not familiar with the topics in this chapter, don't take this test now. Instead, study the examples, work the practice problems, and then take the test.

This test will help you identify those concepts that you have mastered and those that need more study.

Section 1.1

For the set $\{\pi, \sqrt{9}, \sqrt{7}, -5, 3, \frac{6}{2}, 0, \frac{1}{2}, 0.666, \dots \}$

1. List the irrational real numbers.

2. List the whole numbers.

3. What is the property of real numbers that justifies the following equation?

$$(x + y) + z = x + (y + z)$$

Section 1.2

Simplify.

4. $30 \div (-6) + 3 - 2(-5)$

5. $\dfrac{\frac{1}{4} - \frac{1}{3}}{-\frac{1}{2}}$

Section 1.3

Evaluate.

6. $3^3 - \sqrt{4(-5) + 29}$

7. $(-4)^3 + 2[3^2 - 2^2]$

Section 1.4

Simplify each expression.

8. $(-7a^2b)(-2a^0b^3c^2)(-a^{-3})$

9. $\dfrac{(3x^{-1}y^2)^3}{(4x^2y^{-2})^2}$

10. $\dfrac{4x^3y^2}{-16x^2y^{-3}}$

11. $\left(\dfrac{3a^{-5}b^0}{2a^{-2}b^3}\right)^2$

12. $(-2x^3y^{-2})^{-2}$

13. Write 0.000058 in scientific notation.

14. Write 8.95×10^7 in decimal form.

Section 1.5

Simplify.

15. $-5x + 3x^2 + 2x^3 + x - 8x^2$

16. $3ab^2(-2a^2 + 3ab^2 - 1)$

17. $-2\{x + 3[y - 5(x + y)]\}$

Section 1.6

18. Evaluate $5a^2 - 3ab + 2b$ when $a = -3$ and $b = -2$.

19. Use the formula $A = \pi r^2$ to find the area of a circle with a radius of 4 meters. (Use $\pi \approx 3.14$.)

20. Use the formula $T = 2\pi\sqrt{L/g}$ to find the period of a pendulum in seconds where the length of the pendulum $L = 512$ and gravity $g = 32$. (Use $\pi \approx 3.14$.)

1. _____

2. _____

3. _____

4. _____

5. _____

6. _____

7. _____

8. _____

9. _____

10. _____

11. _____

12. _____

13. _____

14. _____

15. _____

16. _____

17. _____

18. _____

19. _____

20. _____

1.1 The Real Number System

After studying this section, you will be able to:

1 *Identify the subsets of the real numbers to which a given number belongs.*
2 *Identify which property of the real numbers is used in a given equation.*

1 Identifying Subsets of the Real Numbers

A **set** is a collection of objects called **elements**. A set of numbers is simply a listing, within braces { }, of the numbers (elements) in the set. For example,

$$S = \{1, 3, 5, \ldots\}$$

is the set of positive odd numbers. The three dots (called an *ellipsis*) mean that the set is *infinite;* in other words, we haven't written all the possible elements of the set.

Some important sets of numbers that we will study are the following:

- Natural numbers
- Whole numbers
- Integers
- Rational numbers
- Irrational numbers
- Real numbers

Definition

The **natural numbers** N (also called the *positive integers*) are the counting numbers:

$$N = \{1, 2, 3, \ldots\}$$

The **whole numbers** W are the natural numbers plus 0:

$$W = \{0, 1, 2, 3, \ldots\}$$

The **integers** I are the whole numbers plus the *negative* of all natural numbers:

$$I = \{\ldots, -3, -2, -1, 0, 1, 2, 3, \ldots\}$$

The **rational numbers** Q include the integers and all *quotients* of integers (but division by zero is not allowed):

$$Q = \left\{\frac{a}{b} \,\middle|\, a \text{ and } b \text{ are integers but } b \neq 0\right\}$$

The last expression means "the set of all fractions a divided by b, such that a and b are integers but b is not equal to zero." (The | is read "such that.") The letters a and b are **variables;** that is, they can represent different numbers.

A rational number can be written as a terminating decimal, $\frac{1}{8} = 0.125$, or as a repeating decimal, $\frac{2}{3} = 0.6666\ldots$. For repeating decimals we often use a bar over the repeating digits. Thus, we write $0.232323\ldots$ as $0.\overline{23}$ to show that the digits 23 repeat indefinitely. A terminating decimal has a definite number of digits. A repeating decimal goes on forever, but the digits repeat in a definite pattern.

Some numbers in decimal notation are nonterminating and nonrepeating; in other words, we can't write them as quotients of integers. Such numbers are called irrational numbers.

The **irrational numbers** are numbers whose decimal forms are nonterminating and nonrepeating.

For example, $\sqrt{3} = 1.7320508\ldots$ can be carried out to an infinite number of decimal places with no repeating pattern of digits. In Chapter 7 we study irrational numbers extensively.

We can now describe the set of real numbers.

> The set R of **real numbers** is the set of all numbers that are rational or irrational.

The following figure will help you see the relationship among all the sets of numbers that we have described.

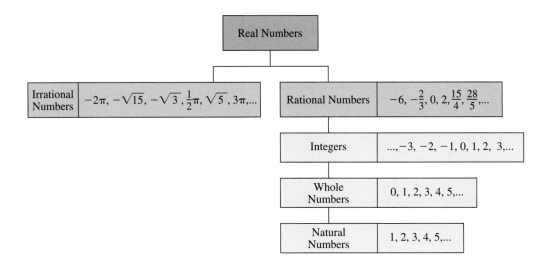

The figure shows that the natural numbers are contained within the set of whole numbers (or we could say that the set of whole numbers contains the set of natural numbers). If all the members of one set are also members of another set, then the first set is a **subset** of the second set. The natural numbers are a subset of the whole numbers. The natural numbers are also a subset of the integers, rational numbers, and real numbers.

EXAMPLE 1   Name the sets to which each of the following numbers belongs.

(a) 5   **(b)** 0.2666 . . .   **(c)** 1.4371826138526 . . .

(d) 0   **(e)** $\dfrac{1}{9}$

(a) 5 is a natural number. Thus we can say that it is also a whole number, an integer, a rational number, and a real number.

(b) 0.2666 . . . is a repeating decimal. Thus it is a rational number and a real number.

(c) 1.4371826138526 . . . doesn't have any repeating pattern. Therefore, it is an irrational number and a real number.

(d) 0 is a whole number, an integer, a rational number, and a real number.

(e) $\dfrac{1}{9}$ can be written as 0.111 . . . (or $0.\overline{1}$), so it is a rational number and a real number.

Practice Problem 1   Name the sets to which the following numbers belong.

(a) 1.26   **(b)** 3   **(c)** $\dfrac{3}{7}$ **(d)** -2 **(e)** 5.182671 . . . ∎

2 Identifying the Properties of Real Numbers

A property of something is a characteristic that we know to be true. For example, one property of water is that it freezes at 32°F.

Real numbers also have properties. There are some characteristics of real numbers that we have found to be true. For example, we know that we can add any two real numbers, and the order in which we add the numbers does not affect the sum: $6 + 8 = 8 + 6$. Mathematicians call this property the **commutative property.** Since all real numbers have this characteristic, we can write the property using letters:

$$a + b = b + a$$

where a and b are any real numbers.

We will look at another property with which you may be familiar, the **associative property.** The way in which you group numbers does not affect the sum:

$$(3 + 8) + 4 = 3 + (8 + 4)$$

$$11 + 4 = 3 + 12$$

$$15 = 15$$

We can write this property using letters:

$$(a + b) + c = a + (b + c)$$

The **identity property** states that, if we add a unique real number (called an *identity element*) to any other real number, that number is not changed. Unique means only one such number has this property. For addition, zero (0) is the identity element. Thus

$$99 + 0 = 99$$

and, in general

$$a + 0 = a$$

The last property we will describe is the **inverse property.** For any real number a there is a unique real number $-a$ such that if we add them we obtain the identity element. For example,

$$8 + (-8) = 0, \qquad -1.5 + 1.5 = 0$$

In general,

$$a + (-a) = 0$$

These are the properties of real numbers for addition. These properties apply to multiplication as well. Multiplication of real numbers is commutative: $3 \cdot 8 = 8 \cdot 3$. It is also associative: $3 \cdot (5 \cdot 6) = (3 \cdot 5) \cdot 6$. Note that the dot $\cdot$ indicates multiplication. Thus $3 \cdot 5$ means multiply 3 and 5; $a \cdot b$ means multiply a and b. The properties are summarized in the following table.

PROPERTIES OF REAL NUMBERS		
Addition	Property	Multiplication
(for all real numbers a, b, c):		
$a + b = b + a$	Commutative properties	$a \cdot b = b \cdot a$
$a + (b + c) = (a + b) + c$	Associative properties	$a \cdot (b \cdot c) = (a \cdot b) \cdot c$
$a + 0 = a$	Identity properties	$a \cdot 1 = a$
$a + (-a) = 0$	Inverse properties	$a \cdot \dfrac{1}{a} = 1$ when $a \neq 0$
Distributive property of multiplication over addition		
$a(b + c) = a \cdot b + a \cdot c$		

To Think About What is the identity element of multiplication? Why? Give a numerical example of the inverse property of multiplication. What is $\dfrac{1}{a}$ if a is $\dfrac{2}{5}$?

We will use each of these properties to simplify expressions and to solve equations. It will be helpful for you to become familiar with these properties and to recognize what these properties can do.

EXAMPLE 2 State the name of the property that justifies each statement.

(a) $5 + (7 + 1) = (5 + 7) + 1$ **(b)** $x + 0.6 = 0.6 + x$

(c) $\dfrac{1}{5} + \left(-\dfrac{1}{5}\right) = 0$ **(d)** $\pi + 0 = \pi$

(a) Associative property of addition **(b)** Commutative property of addition
(c) Inverse property of addition **(d)** Identity property of addition

Practice Problem 2 State the name of the property that justifies each statement.

(a) $9 + 8 = 8 + 9$ **(b)** $17 + 0 = 17$
(c) $-4 + 4 = 0$ **(d)** $7 + (3 + x) = (7 + 3) + x$ ∎

EXAMPLE 3 State the name of the property that justifies each statement.

(a) $5 \cdot y = y \cdot 5$ **(b)** $6 \cdot \dfrac{1}{6} = 1$

(c) $56 \cdot 1 = 56$ **(d)** $12 \cdot (3 \cdot 5) = (12 \cdot 3) \cdot 5$

(a) Commutative property of multiplication **(b)** Inverse property of multiplication
(c) Identity property of multiplication **(d)** Associative property of multiplication

Practice Problem 3 State the name of the property that justifies each statement.

(a) $6 \cdot (2 \cdot w) = (6 \cdot 2) \cdot w$ **(b)** $4 \cdot \dfrac{1}{4} = 1$

(c) $4 \cdot 15 = 15 \cdot 4$ **(d)** $76 \cdot 1 = 76$ ∎

The distributive property links multiplication with addition. It states that

$$a(b + c) = (a \cdot b) + (a \cdot c)$$

We can verify this property by providing a numerical example.

$$5(2 + 7) = (5 \cdot 2) + (5 \cdot 7)$$
$$5(9) = \quad 10 \quad + \quad 35$$
$$45 = 45$$

EXAMPLE 4 Use the distributive property to rewrite each expression and evaluate.

(a) $12(10 + 5)$ **(b)** $(18 \cdot 3) + (18 \cdot 7)$

(a) $12(10 + 5) = (12 \cdot 10) + (12 \cdot 5)$ **(b)** $(18 \cdot 3) + (18 \cdot 7) = 18(3 + 7)$
$\qquad\qquad\qquad = 120 + 60$ $\qquad\qquad\qquad\qquad = 18(10)$
$\qquad\qquad\qquad = 180$ $\qquad\qquad\qquad\qquad = 180$

Practice Problem 4 Use the distributive property to rewrite each expression and evaluate.

(a) $(20 + 9)3$ **(b)** $(29 \cdot 9) + (29 \cdot 1)$ ∎

1.1 Exercises

Verbal and Writing Skills

1. How do integers differ from whole numbers?

2. How do rational numbers differ from integers?

3. What is a terminating decimal?

4. What is a repeating decimal?

	Check the column(s) to which the number belongs.	Natural Numbers	Whole Numbers	Integers	Rational Numbers	Irrational Numbers	Real Numbers
5.	2046						
6.	0						
7.	-17						
8.	$\dfrac{105}{3}$						
9.	$0.\overline{34}$						
10.	$2.713713\ldots$						
11.	$-\dfrac{8}{7}$						
12.	$-2\dfrac{1}{3}$						
13.	$\dfrac{\pi}{5}$						
14.	$\sqrt{2}$						
15.	0.91						
16.	$1.314278619\ldots$ (no discernible pattern)						
17.	$7.040040004\ldots$ (pattern of increasing number of zeros between the 4's						
18.	$54.989898\ldots$						

Problems 19–27 refer to the set $\left\{ -25, -\dfrac{28}{7}, -\dfrac{18}{5}, -\pi, -0.763, -0.333\ldots, 0, \dfrac{1}{10}, \dfrac{2}{7}, \dfrac{\pi}{4}, \sqrt{3}, 9, \dfrac{283}{5}, 52.8 \right\}.$

19. List the rational numbers.

20. List the negative integers.

21. List the counting numbers.

22. List the whole numbers.

23. List the negative real numbers.

24. List the irrational numbers.

25. List the positive integers.

26. List the negative rational numbers.

27. List the terminating decimals that are not integers.

List all the elements of each set.

★ **28.** $\{a \mid a$ is a counting number less than 8$\}$

★ **29.** $\{b \mid b$ is an even integer$\}$

★ **30.** $\{x \mid x$ is a negative integer between -5 and $-1\}$

★ **31.** $\{x \mid x$ is a whole number less than 10$\}$

Indicate whether the statement is true or false.

32. All integers are rational numbers.

33. All counting numbers are whole numbers.

34. All integers are whole numbers.

35. Every integer is an irrational number.

Name the property that justifies each statement. All variables represent real numbers.

36. $(6 + 1) + 3 = 6 + (1 + 3)$

37. $(-6) + 2 = 2 + (-6)$

38. $\dfrac{1}{8} \cdot 8 = 1$

39. $3\left(7 + \dfrac{1}{2}\right) = 3 \cdot 7 + 3 \cdot \dfrac{1}{2}$

40. $3.2 + 0 = 3.2$

41. $5.6 + (-5.6) = 0$

42. $\dfrac{5}{3} \cdot \dfrac{7}{8} = \dfrac{7}{8} \cdot \dfrac{5}{3}$

43. $\dfrac{5}{6} \cdot 1 = \dfrac{5}{6}$

44. $4(1.2 + 6) = 4(1.2) + 4(6)$

45. $6 \cdot \left(\dfrac{2}{5} \cdot 3\right) = \left(6 \cdot \dfrac{2}{5}\right) \cdot 3$

46. $7 + 3 = 3 + 7$

47. $0 + \pi = \pi$

48. $x \cdot \dfrac{1}{x} = 1$

49. $\sqrt{2} + (1 + \sqrt{3}) = (\sqrt{2} + 1) + \sqrt{3}$

50. $4 \cdot (3 \cdot x) = (4 \cdot 3) \cdot x$

51. $2(x + 5) = 2 \cdot x + 2 \cdot 5$

52. $(a + b) + 3 = a + (b + 3)$

53. $x \cdot \dfrac{2}{3} = \dfrac{2}{3} \cdot x$

54. $x + \sqrt{3} = \sqrt{3} + x$

55. $y + 0 = y$

56. $x(y + 4) = x \cdot y + x \cdot 4$

57. $x \cdot 1 = x$

58. $(-x) + x = 0$

59. $\sqrt{3}(x + 2) = \sqrt{3} \cdot x + \sqrt{3} \cdot 2$

Applications

60. The Siberian tiger is the largest living cat in the world, an endangered subspecies of the tiger family. It is estimated that there are no more than 200 of these animals left in the wild, and approximately the same number in captivity. Of the 8 subspecies of tiger, 3 are extinct and the remaining 5 are endangered.
(a) What is the percentage of subspecies of tiger that have become extinct?
(b) What is the percentage of subspecies of tiger that are endangered?

61. According to the world's geologists, the tallest mountain in the world, measured from its base, is the island of Hawaii, which is *all mountain*. The tallest peak is Mauna Kea, 13,784 feet above sea level, rising from a sea bottom that is 18,000 feet below sea level. Mount Everest's peak is 29,022 feet above sea level, but it rises from a plateau 12,000 feet high. How high is each mountain measured from its base to its peak?

1.2 Operations with Real Numbers

MathPro

Video 1

SSM

After studying this section, you will be able to:

1 *Find the absolute value of any real number.*
2 *Add, subtract, multiply, and divide real numbers.*
3 *Perform mixed operations of addition, subtraction, multiplication, and division in the proper order.*

1 Finding the Absolute Value of a Real Number

We can think of real numbers as points on a line, called the *real number line,* where positive numbers lie to the right of zero and negative numbers lie to the left of zero.

The number line helps us to understand the important concept of absolute value and lets us see how to add and subtract signed numbers.

The **absolute value** of a number x, written $|x|$, is its distance from 0 on the number line.

For example, the absolute value of 5 is 5, because 5 is located 5 units from 0 on the number line. The absolute value of -5 is also 5 because -5 is located 5 units from 0 on the number line. We illustrate this concept on the number line below.

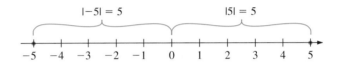

Even though -5 is located in the opposite direction from 5, we are concerned only with distance, which is always positive. (We don't say, for instance, that we traveled -10 miles.) Thus we see that the absolute value of any number is always positive or zero. We formally define absolute value next.

Definition

Absolute value of x: $\quad |x| = \begin{cases} x, & \text{if } x \geq 0 \\ -x, & \text{if } x < 0 \end{cases}$

The $\geq$ symbol means "is greater than or equal to." The $<$ symbol means "is less than."

EXAMPLE 1 Evaluate.

(a) $|6|$ **(b)** $|-8|$ **(c)** $|0|$ **(d)** $|5 - 3|$

(a) $|6| = 6$ **(b)** $|-8| = 8$
(c) $|0| = 0$ **(d)** $|5 - 3| = |2| = 2$

Practice Problem 1 Evaluate.

(a) $|-4|$ **(b)** $|3.16|$ **(c)** $|0|$ **(d)** $|12 - 7|$ ■

2 Addition, Subtraction, Multiplication, and Division of Real Numbers

Addition of Real Numbers

Addition of real numbers can be pictured on the number line. For example, to add +6 and +5, we start at 6 on the number line and move 5 units in the positive direction (to the right).

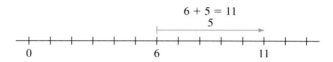

$$6 + 5 = 11$$

To add two negative numbers, say −4 and −3, we start at −4 and move 3 units in the negative direction (to the left). In other words, we add their absolute values. We state this procedure as a rule.

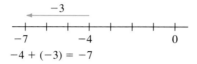

$$-4 + (-3) = -7$$

Rule 1.1

To add two real numbers with the *same* sign, add their absolute values. The sum takes the common sign.

EXAMPLE 2   Add.

(a) $-5 + (-0.6)$ **(b)** $-\dfrac{1}{2} + \left(-\dfrac{1}{3}\right)$ **(c)** $\dfrac{2}{5} + \dfrac{3}{7}$ **(d)** $1.5 + 9.02$

We apply Rule 1.1 to all three problems.

(a) $-5 + (-0.6) = -5.6$

(b) $-\dfrac{1}{2} + \left(-\dfrac{1}{3}\right) = -\dfrac{3}{6} + \left(-\dfrac{2}{6}\right) = -\dfrac{5}{6}$

> Obtain a common denominator before adding.

(c) $\dfrac{2}{5} + \dfrac{3}{7} = \dfrac{14}{35} + \dfrac{15}{35} = \dfrac{29}{35}$

(d) $1.5 + 9.02 = 1.50 + 9.02 = 10.52$

Be sure to align the decimal points when adding or subtracting.

Practice Problem 2   Add.

(a) $3.4 + 2.6$ **(b)** $\left(-\dfrac{3}{4}\right) + \left(-\dfrac{1}{6}\right)$ **(c)** $(-5) + (-37)$ ■

What do we do if the numbers have different signs? Let's add -4 and 3. Again using our number line, we start at -4 and move 3 units in the positive direction.

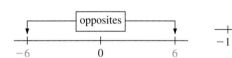

$$-4 + 3 = -1$$

Rule 1.2

To add two real numbers with different signs, find the difference of their absolute values. The answer takes the sign of the number with the larger absolute value.

EXAMPLE 3   Add.

(a) $12 + (-5)$ **(b)** $-12 + 5$ **(c)** $-\dfrac{1}{3} + \dfrac{1}{4}$ **(d)** $8 + (-8)$

(a) $12 + (-5) = 7$ **(b)** $-12 + 5 = -7$

(c) $-\dfrac{1}{3} + \dfrac{1}{4} = -\dfrac{4}{12} + \dfrac{3}{12} = -\dfrac{1}{12}$ **(d)** $8 + (-8) = 0$

Practice Problem 3   Add.

(a) $26 + (-18)$ **(b)** $24 + (-30)$ **(c)** $-\dfrac{1}{5} + \dfrac{2}{3}$ **(d)** $-8.6 + 8.6$ ∎

Notice in part (d) that if you add two numbers that are opposites the answer will *always* be zero. The **opposite** of a number is the number with the same absolute value but a different sign. The opposite of -6 is 6. The opposite of $\frac{2}{3}$ is $-\frac{2}{3}$.

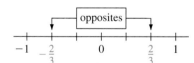

The opposite of a number is also called the **additive inverse.**

To Think About Why is the opposite of a number called the additive inverse of the number?

Subtraction of Real Numbers

We define subtraction in terms of addition and use our rules for adding real numbers. So, actually, you already *know* how to subtract if you understood how to add real numbers.

Rule 1.3

To subtract b from a, add the opposite (inverse) of b to a. Thus $a - b = a + (-b)$.

EXAMPLE 4   Use Rule 1.3 to subtract.

(a) $5 - 7$ **(b)** $-8 - 2$ **(c)** $8 - 6$

(d) $-3 - (-4)$ **(e)** $-12 - (-3)$ **(f)** $-0.06 - 0.55$

(a) $5 - 7 = 5 + (-7) = -2$ **(b)** $-8 - 2 = -8 + (-2) = -10$

(c) $8 - 6 = 8 + (-6) = 2$ **(d)** $-3 - (-4) = -3 + 4 = 1$

(e) $-12 - (-3) = -12 + 3 = -9$ **(f)** $-0.06 - 0.55 = -0.06 + (-0.55) = -0.61$

Practice Problem 4   Use Rule 1.3 to subtract.

(a) $9 - 2$ **(b)** $-8 - (-3)$ **(c)** $-5 - (-14)$

(d) $18 - 26$ **(e)** $\dfrac{1}{2} - \left(-\dfrac{1}{4}\right)$ **(f)** $-0.35 - 0.67$ ∎

Multiplication and Division of Real Numbers

Recall that the product or quotient of two positive numbers is positive.

$$(+2)(+5) = (2)(5) = 2 + 2 + 2 + 2 + 2 = 10$$

Thus

$$+10 \div +5 = +2 \quad \text{because} \quad (+2)(+5) = +10$$

Suppose that the signs of the numbers we want to multiply or divide are different. What will be the sign of the answer? We again use repeated addition for multiplication.

$$(-2)(5) = (-2) + (-2) + (-2) + (-2) + (-2) = -10$$

Similarly,

$$(-5)(2) = (-5) + (-5) = -10$$

Thus

$$-10 \div 5 = -2 \quad \text{and} \quad -10 \div 2 = -5$$

Notice that multiplying or dividing a negative number by a positive number gives a negative answer.

Rule 1.4

When you multiply or divide two real numbers with different signs, the answer is a negative number.

EXAMPLE 5   Evaluate.

(a) $\dfrac{12}{-6}$ **(b)** $5(-8)$ **(c)** $(-13)(2)$ **(d)** $(-5)\left(\dfrac{2}{17}\right)$ **(e)** $\dfrac{-\dfrac{1}{3}}{7}$ **(f)** $\dfrac{1.6}{-0.08}$

(a) $\dfrac{12}{-6} = -2$ **(b)** $5(-8) = -40$

(c) $(-13)(2) = -26$ **(d)** $(-5)\left(\dfrac{2}{17}\right) = \left(-\dfrac{5}{1}\right)\left(\dfrac{2}{17}\right) = -\dfrac{10}{17}$

(e) $\dfrac{-\dfrac{1}{3}}{7} = -\dfrac{1}{3} \div 7 = \left(-\dfrac{1}{3}\right)\dfrac{1}{7} = -\dfrac{1}{21}$ **(f)** $\dfrac{1.6}{-0.08} = -20$

Practice Problem 5 Evaluate.

(a) $(-3)(7)$ **(b)** $\left(-\dfrac{2}{5}\right)\left(\dfrac{3}{4}\right)$ **(c)** $\dfrac{150}{-30}$

(d) $\dfrac{-4}{\dfrac{2}{5}}$ **(e)** $-36 \div (18)$ **(f)** $\dfrac{0.27}{-0.003}$ ∎

Now let us look at multiplication and division of two negative numbers. We begin by stating

$-a(0) = 0$	*Zero property.*
$-a(-b + b) = 0$	*Rewrite 0 as $-b + b$ using the inverse property.*
$(-a)(-b) + (-a)(b) = 0$	*Use the distributive property.*
$ab \quad + \quad -ab \;\; = 0$	*By the inverse property.*

We know that $(-a)(b) = -ab$, a negative number. For the third statement above to be true, $(-a)(-b)$ must be a positive number, ab. Thus we state that $(-a)(-b) = ab$.

Rule 1.5
When you multiply or divide two real numbers with like signs, the answer is a positive number.

EXAMPLE 6   Evaluate.

(a) $(-8)(-2)$ **(b)** $\dfrac{-3}{5} \cdot \dfrac{-2}{11}$ **(c)** 1.3×0.05

(d) $(-6) \div (-12)$ **(e)** $\dfrac{-15}{-3}$ **(f)** $\dfrac{2}{3} \div \dfrac{5}{7}$

(a) $(-8)(-2) = 16$ **(b)** $\dfrac{-3}{5} \cdot \dfrac{-2}{11} = \dfrac{(-3)(-2)}{(5)(11)} = \dfrac{6}{55}$

(c) $1.3 \times 0.05 = 0.065$ **(d)** $(-6) \div (-12) = \dfrac{-6}{-12} = \dfrac{1}{2}$

(e) $\dfrac{-15}{-3} = 5$ **(f)** $\dfrac{2}{3} \div \dfrac{5}{7} = \dfrac{2}{3} \cdot \dfrac{7}{5} = \dfrac{(2)(7)}{(3)(5)} = \dfrac{14}{15}$

Practice Problem 6   Evaluate.

(a) $(-9)(-3)$ **(b)** $(1.23)(0.06)$ **(c)** $\dfrac{-20}{-2}$

(d) $-60 \div (-5)$ **(e)** $\left(\dfrac{3}{7}\right) \div \left(\dfrac{1}{5}\right)$ ∎

Before we leave this topic, we take one more look at division by zero.

$$-\dfrac{5}{3} \div 0 = ? \quad \text{means} \quad -\dfrac{5}{3} = ? \times 0$$

In words, "What times 0 is equal to $-\frac{5}{3}$?" By the zero property, any number times zero is zero. Thus our division problem is impossible to solve. Division by 0 is not defined.

Division by 0 is impossible.

---

3 Performing Mixed Operations

It is very important to perform all mathematical operations in the proper order. Otherwise, you will get wrong answers.

If addition, subtraction, multiplication, and division are written horizontally, do the operations in the following order:

1. Do all multiplications and divisions from left to right.
2. Do all additions and subtractions from left to right.

EXAMPLE 7   Evaluate. $20 \div (-4) \times 3 + 2 + 6 \times 5$

1. Beginning at the left, we do any multiplication or division as we encounter it. Here we do division first, then multiplication.

$$20 \div (-4) \times 3 + 2 + 6 \times 5 = -5 \times 3 + 2 + 6 \times 5$$
$$= -15 + 2 + 6 \times 5$$
$$= -15 + 2 + 30$$

2. Next we add or subtract from left to right.

$$-15 + 2 + 30 = -13 + 30$$
$$= 17$$

Practice Problem 7   Evaluate. $5 + 7(-2) - (-3) + 50 \div (-2)$ ■

EXAMPLE 8   Evaluate.

(a) $6(-3) - 5(-7) + (-8)$ **(b)** $(-20) \div (-5)(3) + 6 - 5(-2)$

(c) $\dfrac{13 - (-3)}{5(-2) - 6(-3)}$

(a) $6(-3) - 5(-7) + (-8)$ **(b)** $(-20) \div (-5)(3) + 6 - 5(-2)$
     $= -18 + 35 + (-8)$ $= (4)(3) + 6 - 5(-2)$
     $= 17 + (-8)$ $= 12 + 6 + 10$
     $= 9$ $= 28$

(c) $\dfrac{13 - (-3)}{5(-2) - 6(-3)} = \dfrac{13 + 3}{-10 + 18} = \dfrac{16}{8} = 2$

Practice Problem 8   Evaluate.

(a) $3(-5) + 3(-6) - 2(-1)$ **(b)** $6(-2) + (-20) \div (2)(3)$

(c) $\dfrac{7 + 2 - 12 - (-1)}{(-5)(-6) + 4(-8)}$ ■

1.2 Exercises

Evaluate if a and b are greater than 0.

1. $|8|$

2. $|12|$

3. $|-9|$

4. $|-23|$

5. $|8.3|$

6. $\left|3\dfrac{1}{2}\right|$

7. $|3-5|$

8. $|9-14|$

9. $|a|$

10. $|-b|$

Perform the operations indicated. Write your answer in simplest form.

11. $5 + (-8)$

12. $-6 + (-12)$

13. $-7 - (-2)$

14. $-6 - (+4)$

15. $5 - 6 + 2 - 4$

16. $3 - 7 - 6 + 1$

17. $5\left(-\dfrac{1}{3}\right)$

18. $(-24) \div (4)$

19. $\dfrac{-15}{-3}$

20. $(-16)(-2)$

21. $-6.02 + (-3.05)$

22. $-12.6 + 8.3$

23. $-4.9 + 10.5$

24. $1.4 - (-3.6)$

25. $-1.2 - (+1.9)$

26. $-\dfrac{1}{3} + \dfrac{3}{4}$

27. $\dfrac{1}{7} + \dfrac{3}{4}$

28. $-\dfrac{5}{6} + \dfrac{1}{8}$

29. $\dfrac{-5}{12} + \left(-\dfrac{3}{4}\right)$

30. $-\dfrac{2}{5} - \dfrac{3}{7}$

31. $(-7)(-2)(-1)(3)$

32. $(18)\left(-\dfrac{1}{2}\right)(3)(-1)$

33. $-8 + (6)\left(-\dfrac{1}{3}\right)$

34. $-12 - (-3)(2)$

35. $(-6)(-2) + 3(-4)$

36. $\dfrac{\dfrac{2}{7}}{-\dfrac{3}{5}}$

37. $\dfrac{\dfrac{-26}{13}}{5}$

38. $\dfrac{-\dfrac{3}{4}}{-24}$

39. $-3 + 3 \div (1.5)$

40. $8(0.5) - 9 \div (-0.3)$

Perform each of the following operations, if possible.

41. $8 + (-8)$

42. $-6.2 + 6.2$

43. $\dfrac{-5}{0}$

44. $\dfrac{-12}{0}$

45. $\dfrac{0}{7}$

46. $\dfrac{0}{4}$

47. $\dfrac{3-3}{-8}$

48. $\dfrac{-5+5}{6}$

49. $\dfrac{-7+(-7)}{-14}$

50. $\dfrac{-6+(-6)}{-18}$

Mixed Practice

Perform the following operations in the proper order. Write your answer in simplest form.

51. $5 + 6 - (-3) - 8 + 4 - 3$

52. $12 - 3 - (-4) + 6 - 5 - 8$

53. $\dfrac{12 - 4(3)}{2 - 6}$

54. $\dfrac{8(-2) + 6}{1 - 6}$

55. $9(-1) - 4(-3) + 6 - 2$

56. $-8(-4) + 3(-2) + 5 - 7$

57. $7 + 30 \div 10 - 15(-3)$

58. $100 \div 2 \div 5 + 3(-2) - (-6)$

59. $\dfrac{1 + 49 \div (-7) - (-3)}{-1 - 2}$

60. $\dfrac{72 \div (-4) + 3(-4)}{5 - (-5)}$

61. $4\left(-\dfrac{1}{2}\right) + \left(\dfrac{2}{3}\right)(9)$

62. $(-2.5)(4) - (3.2)(2)$

63. $\dfrac{1.63482 - 2.48561}{(16.05436)(0.07814)}$

64. $(1.783)(2.5725) - (1.0526)(-5.9812)$

To Think About

65. Three numbers are multiplied: $a \cdot b \cdot c = d$. The value of d is a negative number. What are the possibilities of being positive or negative for a, b, and c?

66. Three numbers are multiplied: $a \cdot b \cdot c = d$. The value of d is a positive number. What are the possibilities of being positive or negative for a, b, and c?

Cumulative Review Problems

Name the property illustrated by each equation.

67. $5 + 17 = 17 + 5$

68. $4 \cdot (3 \cdot 6) = (4 \cdot 3) \cdot 6$

Refer to the set of numbers $\left\{-16, -\frac{1}{2}\pi, 0, \sqrt{3}, \frac{19}{2}, 9.36, 10.\overline{5}\right\}$

69. List the irrational numbers.

70. List the rational numbers.

Applications

71. 4,891,500,000,000,000,000,000,000 equals the number of atoms found in a pound of iron. How many atoms in a ton of iron?

72. During the 1800s, all postal rates were determined by the distance (measured to the nearest mile) the mail had to travel. In 1831, the rates were 6 cents for a trip up to 30 miles, 10 cents for 31–80 miles, $12\frac{1}{2}$ cents for 81–150 miles, $18\frac{3}{4}$ cents for 151–400 miles, and 25 cents over 400 miles. If John Quincy Adams mailed one letter 95 miles away, two letters 20 miles away, 5 letters 200 miles away, 7 letters 40 miles away, and 3 letters 500 miles away, how much postage did he pay?

1.3 Powers, Square Roots, and the Order of Operations

MathPro

Video 2

SSM

After studying this section, you will be able to:

1 *Raise a number to a positive integer power.*
2 *Find square roots of numbers that are perfect squares.*
3 *Evaluate expressions by using the proper order of operations.*

1 Raising a Number to a Power

Exponents or powers are used to indicate repeated multiplication. For example, we can write $6 \cdot 6 \cdot 6 \cdot 6$ as 6^4. 4 is the exponent or power that tells us how many times 6 appears as a factor. In the expression 6^4, 6 is the **base** and 4 is the **exponent.** This is called exponential notation.

$$6^4 = \overbrace{6 \cdot 6 \cdot 6 \cdot 6}^{4 \text{ factors}}$$
$$\underset{\text{base of } 6}{\uparrow}$$

Exponential Notation

If x is a real number and n is a positive integer, then

$$x^n = \underbrace{x \cdot x \cdot x \cdot x \cdots}_{n \text{ factors}}$$

EXAMPLE 1   Write in exponential notation.

(a) $3 \cdot 3 \cdot 3 \cdot 3 \cdot 3 \cdot 3 \cdot 3$ **(b)** $(-2)(-2)(-2)(-2)(-2)$

(c) $x \cdot x \cdot x \cdot x \cdot x \cdot x \cdot x \cdot x$ **(d)** $(a + b)(a + b)(a + b)$

(a) $3 \cdot 3 \cdot 3 \cdot 3 \cdot 3 \cdot 3 \cdot 3 = 3^7$

(b) $(-2)(-2)(-2)(-2)(-2) = (-2)^5$

(c) $x \cdot x \cdot x \cdot x \cdot x \cdot x \cdot x \cdot x = x^8$

(d) $(a + b)(a + b)(a + b) = (a + b)^3$ (Note that the base is $a + b$)

Practice Problem 1   Write in exponential notation.

(a) $(-4)(-4)(-4)(-4)$ **(b)** $z \cdot z \cdot z \cdot z \cdot z \cdot z \cdot z$ ■

EXAMPLE 2   Evaluate.

(a) 4^3 **(b)** $(-2)^4$ **(c)** 3^5 **(d)** $(-5)^3$ **(e)** -2^4 **(f)** $\left(\dfrac{1}{3}\right)^3$

(a) $4^3 = 4 \cdot 4 \cdot 4 = 64$

(b) $(-2)^4 = (-2)(-2)(-2)(-2) = (4)(-2)(-2) = (-8)(-2) = 16$
Notice that we are raising -2 to the fourth power. That is, the base is -2. We use parentheses to clearly indicate a negative base.

(c) $3^5 = 3 \cdot 3 \cdot 3 \cdot 3 \cdot 3 = 243$

(d) $(-5)^3 = (-5)(-5)(-5) = (25)(-5) = -125$

(e) $-2^4 = -(2 \cdot 2 \cdot 2 \cdot 2) = -16$
Here the base is 2. We wish to find the negative of 2 raised to the fourth power.

(f) $\left(\dfrac{1}{3}\right)^3 = \left(\dfrac{1}{3}\right)\left(\dfrac{1}{3}\right)\left(\dfrac{1}{3}\right) = \dfrac{1}{27}$

Practice Problem 2 Evaluate.

(a) $(-3)^5$ **(b)** $(-3)^6$ **(c)** 2^5 **(d)** $(-4)^4$ **(e)** -4^4 **(f)** $\left(\frac{1}{5}\right)^2$ ■

To Think About Look at Practice Problem 2. What do you notice about raising a negative number to an even power? to an odd power? Will this always be true? Why?

2 Finding Square Roots

We say that the square root of 16 is 4 because $4 \cdot 4 = 16$. You will note that, since $(-4)(-4) = 16$, the square root of 16 is also -4. For practical purposes, we are usually interested in the positive square root. We call this the **principal square root.** $\sqrt{}$ is the square root symbol and is called the **radical.** Thus

$$\text{radical} \rightarrow \sqrt{9} = 3$$
$$\text{radicand} \overset{\uparrow}{\underset{}{\rule{0pt}{0pt}}} \quad \overset{\uparrow}{\underset{}{\rule{0pt}{0pt}}} \text{principal square root}$$

The number or expression under the radical sign is called the **radicand.** Both 16 and 9 are called *perfect squares* because their square root is an integer.

> If x is an integer and a is a positive real number such that $a = x^2$, then x is a **square root** of a, and a is a **perfect square.**

EXAMPLE 3 Find the square roots of 25. What is the principal square root?

Since $(-5)^2 = 25$ and $5^2 = 25$, the square roots of 25 and 5 are -5. The principal square root is 5.

Practice Problem 3 What are the square roots of 49? What is the principal square root of 49? ■

EXAMPLE 4 Find the principal square root.

(a) $\sqrt{36}$ **(b)** $\sqrt{81}$ **(c)** $\sqrt{0}$ **(d)** $-\sqrt{25}$ **(e)** $-\sqrt{49}$

(a) $\sqrt{36} = 6$ because $6^2 = 36$ **(b)** $\sqrt{81} = 9$ because $9^2 = 81$
(c) $\sqrt{0} = 0$ because $0^2 = 0$ **(d)** $-\sqrt{25} = -(\sqrt{25}) = -(5) = -5$
(e) $-\sqrt{49} = -(\sqrt{49}) = -(7) = -7$

Remember: A principal square root is *positive*.

Practice Problem 4 Find the principal square root.

(a) $\sqrt{100}$ **(b)** $\sqrt{1}$ **(c)** $-\sqrt{100}$ **(d)** $-\sqrt{36}$ **(e)** $\sqrt{0}$ ■

SCIENTIFIC CALCULATOR

Exponents

You can use a scientific calculator to evaluate $(-5)^3$.

Press these keys:

5 $\boxed{+/-}$ $\boxed{y^x}$ 3 $\boxed{=}$

The display should read:

$$\boxed{-125}$$

On a graphing calculator, use the $\boxed{\wedge}$ key to raise to a power. Also it is necessary to use the parentheses around a negative in order to raise the entire number to the power. Compare raising a negative number to an even power with and without the parentheses. Try.

(a) 4^7 **(b)** $(-8)^3$
(c) 0.3^6 **(d)** -5^3
(e) 1^{15} **(e)** 0.4^4

Note: When you evaluate 0.3^6, your calculator may display the answer in scientific notation. See page 32 for a discussion on scientific notation.

We can also find square roots of rational numbers.

EXAMPLE 5   Evaluate.   **(a)** $\sqrt{0.04}$ **(b)** $\sqrt{\dfrac{25}{36}}$

(a) $(0.2)^2 = (0.2)(0.2) = 0.04$. Therefore, $\sqrt{0.04} = 0.2$.

(b) We can write $\sqrt{\dfrac{25}{36}}$ as $\dfrac{\sqrt{25}}{\sqrt{36}}$. $\dfrac{\sqrt{25}}{\sqrt{36}} = \dfrac{5}{6}$. Thus $\sqrt{\dfrac{25}{36}} = \dfrac{5}{6}$.

Practice Problem 5   Evaluate.   **(a)** $\sqrt{0.09}$ **(b)** $\sqrt{\dfrac{4}{81}}$ ∎

The square roots of some numbers are irrational numbers. To illustrate, $\sqrt{3}$ and $\sqrt{7}$ are irrational numbers. The decimal values of such numbers can never be found exactly since irrational numbers are nonterminating, nonrepeating decimal numbers. However, we often need rational numbers to *approximate* square roots that are irrational. They can be found on a calculator with a square-root key. To find $\sqrt{3}$ on most calculators, we enter the 3 and then enter the $\boxed{\sqrt{}}$ key. Using a calculator, we might get 1.7320508 as our approximation. If you do not have a calculator, you can use the square root table at the back of the book. From the table we have $\sqrt{3} \approx 1.732$. We will use the approximate values of square roots in later chapters.

3 *Order of Operations of Real Numbers*

Parentheses are used in numerical and algebraic expressions to group numbers and variables. When evaluating an expression containing parentheses, evaluate the numbers inside the parentheses first. When we need more than one set of parentheses, we may also use brackets. To evaluate such an expression, work from the inside out.

EXAMPLE 6   Evaluate.

(a) $8 + 5(2 + 3 + 6)$ **(b)** $2(8 + 7) - 12$ **(c)** $12 - 3[7 + 5(6 - 9)]$

(a) $8 + 5(2 + 3 + 6) = 8 + 5(11)$ *Work inside the parentheses first.*
$\qquad\qquad\qquad\quad = 8 + 55$ *Multiply.*
$\qquad\qquad\qquad\quad = 63$ *Add.*

(b) $2(8 + 7) - 12 = 2(15) - 12$ *Work inside the parentheses first.*
$\qquad\qquad\qquad\ = 30 - 12$ *Multiply.*
$\qquad\qquad\qquad\ = 18$ *Add.*

(c) $12 - 3[7 + 5(6 - 9)] = 12 - 3[7 + 5(-3)]$ *Begin with the innermost grouping symbol.*
$\qquad\qquad\qquad\qquad\quad = 12 - 3[7 + (-15)]$ *Multiply.*
$\qquad\qquad\qquad\qquad\quad = 12 - 3[-8]$ *Work inside the grouping symbol.*
$\qquad\qquad\qquad\qquad\quad = 12 + 24$ *Multiply.*
$\qquad\qquad\qquad\qquad\quad = 36$ *Add.*

Practice Problem 6   Evaluate.

(a) $7 - 3(8 + 5 - 6)$ **(b)** $6(12 - 8) + 4$ **(c)** $5[6 - 3(7 - 9)] - 8$ ∎

To Think About Rewrite each expression in Example 6 without grouping symbols. Then evaluate each expression. Remember to first multiply from left to right and then add or subtract from left to right. Explain why the answers may differ.

The fraction bar acts like a grouping symbol. We must evaluate the expressions above and below the fraction bar before we divide.

EXAMPLE 7 Evaluate. $\dfrac{(5)(-2)(-3)}{6-8+4}$

$$\frac{(5)(-2)(-3)}{6-8+4} = \frac{30}{2} = 15$$

Practice Problem 7 Evaluate. $\dfrac{5+2(-3)-10+1}{(1)(-2)(-3)}$ ■

A radical or absolute value symbol groups the quantities within it. Thus we evaluate the terms within the grouping symbol before we find the square root or the absolute value.

EXAMPLE 8 Evaluate.

(a) $\sqrt{(-3)^2 + (4)^2}$ **(b)** $|5 - 8 + 7 - 13|$

(a) $\sqrt{(-3)^2 + (4)^2} = \sqrt{9 + 16} = \sqrt{25} = 5$

(b) $|5 - 8 + 7 - 13| = |-9| = 9$

Practice Problem 8 Evaluate.

(a) $\sqrt{(-5)^2 + 12^2}$ **(b)** $|-3 - 7 + 2 - (-4)|$ ■

When many arithmetic operations or grouping symbols are used, we use the following order of operations:

Order of Operations for Calculations

1. Combine numbers inside grouping symbols.
2. Raise numbers to their indicated powers and take any indicated roots.
3. Multiply and divide numbers from left to right.
4. Add and subtract numbers from left to right.

EXAMPLE 9 Evaluate. $(4 - 6)^3 + 5(-4) + 3$

$(4 - 6)^3 + 5(-4) + 3 = (-2)^3 + 5(-4) + 3$	*Combine the $4 - 6$ in parentheses.*
$= -8 + 5(-4) + 3$	*Cube -2. $(-2)^3 = -8$.*
$= -8 - 20 + 3$	*Multiply $5(-4)$.*
$= -25$	*Combine $-8 - 20 + 3$.*

Practice Problem 9 Evaluate. $-7 - 2(-3) + 4^3$ ■

SCIENTIFIC CALCULATOR

Order of Operations

You can use a scientific calculator to evaluate $3 + 4 \times 5$. Enter

3 $+$ 4 $\times$ 5 $=$

If the display is $\boxed{23}$, the correct order of operations is built in. If the display is not 23, you will need to modify the way you enter the problem. You should enter

4 $\times$ 5 $+$ 3 $=$

Try.

$6 + 3 \cdot 4 - 8 \div 2$

On a graphing calculator, enter the expression as written. Graphing calculators function according to the order of operations.

EXAMPLE 10 Evaluate. $2 + 66 \div 11 \cdot 3 + 2\sqrt{36}$

$$
\begin{aligned}
2 + 66 \div 11 \cdot 3 + 2\sqrt{36} &= 2 + 66 \div 11 \cdot 3 + 2 \cdot 6 \qquad \text{\textit{Evaluate} } \sqrt{36}. \\
&= 2 + 6 \cdot 3 + 2 \cdot 6 \qquad \text{\textit{Divide} } 66 \div 11. \\
&= 2 + 18 + 12 \qquad \text{\textit{Multiply} } 6 \cdot 3 \text{ \textit{and} } 2 \cdot 6. \\
&= 32 \qquad \text{\textit{Add} } 2, 18, \text{ \textit{and} } 12.
\end{aligned}
$$

Students often find that they make errors in problems like Example 10 when they try to omit steps. Therefore we recommend that when doing problems in this section it is best to make a separate step for each of the four priorities listed in the box on page 21.

Practice Problem 10 Evaluate. $5 + 6 \cdot 2 - 12 \div (-2) + 3\sqrt{4}$ ∎

EXAMPLE 11 Evaluate. $\dfrac{2 \cdot 6^2 - 12 \div 3}{4 - 8}$

We evaluate the numerator first.

$$
\begin{aligned}
2 \cdot 6^2 - 12 \div 3 &= 2 \cdot 36 - 12 \div 3 \qquad \text{\textit{Raise to a power.}} \\
&= 72 - 4 \qquad \text{\textit{Multiply and divide from left to right.}} \\
&= 68 \qquad \text{\textit{Subtract.}}
\end{aligned}
$$

Next we evaluate the denominator.

$$
4 - 8 = -4
$$

Thus

$$
\frac{2 \cdot 6^2 - 12 \div 3}{4 - 8} = \frac{68}{-4} = -17
$$

Practice Problem 11 Evaluate. $\dfrac{2(3) + 5(-2)}{1 + 2 \cdot 3^2 + 5(-3)}$ ∎

As we become a more technologically oriented society, it is becoming more necessary to understand in what order a computer or a scientific calculator performs arithmetic operations. Be sure you take the time to master this procedure. If you have a scientific calculator, be sure to become familiar with how it works for these types of problems.

1.3 Exercises

Verbal and Writing Skills

1. In the expression a^3, identify the base and the exponent.

2. When a negative number is raised to an odd power, is the result positive or negative?

3. When a negative number is raised to an even power, is the result positive or negative?

4. Will $-a^n$ always be negative? Why or why not?

5. What are the square roots of 121? Why are there two answers?

6. What is the principle square root?

Write in exponential form.

7. $2 \cdot 2 \cdot 2 \cdot 2 \cdot 2 \cdot 2 \cdot 2 \cdot 2$

8. $5 \cdot 5 \cdot 5 \cdot 5 \cdot 5 \cdot 5 \cdot 5 \cdot 5 \cdot 5$

9. $x \cdot x \cdot x \cdot x \cdot x \cdot x$

10. $y \cdot y \cdot y \cdot y \cdot y \cdot y \cdot y$

11. $(-6)(-6)(-6)(-6)(-6)$

12. $(-8)(-8)(-8)$

Evaluate.

13. 3^5

14. 2^5

15. $(-4)^3$

16. $(-5)^2$

17. -8^2

18. -2^4

19. $(-1)^3$

20. $(-3)^2$

21. -1^4

22. -3^2

23. $\left(\dfrac{2}{3}\right)^2$

24. $\left(\dfrac{3}{4}\right)^2$

25. $\left(-\dfrac{1}{5}\right)^3$

26. $\left(-\dfrac{1}{6}\right)^3$

27. $(0.6)^2$

28. $(0.7)^2$

29. $(-0.2)^2$

30. $(0.05)^3$

31. $(-1.1)^4$

32. $(-1.2)^4$

Find each principal square root.

33. $\sqrt{25}$

34. $\sqrt{36}$

35. $\sqrt{81}$

36. $\sqrt{49}$

37. $-\sqrt{16}$

38. $-\sqrt{64}$

39. $\sqrt{\dfrac{25}{49}}$

40. $\sqrt{\dfrac{1}{100}}$

41. $\sqrt{0.09}$

42. $\sqrt{0.25}$

43. $\sqrt{0.0004}$

44. $\sqrt{0.0016}$

45. $\sqrt{4901 - 1}$

46. $\sqrt{8101 - 1}$

47. $\sqrt{4 + 12}$

48. $\sqrt{5 + 4}$

49. $\sqrt{\dfrac{5}{36} + \dfrac{31}{36}}$

50. $\sqrt{\dfrac{1}{9} + \dfrac{3}{9}}$

★ 51. $-\sqrt{-0.36}$

★ 52. $-\sqrt{-0.49}$

Perform the proper order of operations to evaluate each of the following.

53. $8 \div (4 \cdot 2) - 6$

54. $6(2 - 4) + 5$

55. $\sqrt{16} + 8(-3)$

56. $27 \div 3 \cdot 2 - (-1)$

57. $(5 + 2 - 8)^3 - (-7)$

58. $\sqrt{(-3)^2 + (2)^2} + 3$

59. $2(-4)^2 + 25 \div 5 + \sqrt{9}$

60. $(8 - 6 - 7)^2 \div 5 - 6$

61. $(12 - 7)(3 - 2)(1 - 8)$

62. $2\sqrt{16} + (-4)^2 - 3$

63. $(4^3)(0) + 7(-2)$

64. $3\sqrt{25} + 2(0) - 8$

65. $16 - 2[3 + (5 - 9)]$

66. $8 + 5[6 - (3 + 4)]$

67. $7[(5 - 8) + 3]$

68. $9[(7 - 2) + 6]$

69. $\dfrac{7 + 2(-4) + 5}{8 - 6}$

70. $\dfrac{2^3 + 3^2 - 5}{4}$

71. $\dfrac{-3(2^3 - 1)}{-7}$

72. $\dfrac{4 + 2(3^2 - 12)}{-2}$

73. $\dfrac{|2^2 - 5| - 3^2}{-5 + 3}$

74. $\dfrac{\sqrt{(-5)^2 - 3} + 14}{|19 - 6 + 3 - 25|}$

75. $\dfrac{|3 - 2^3| - 5}{2 + 3}$

76. $\dfrac{|6 - 2^5| - 4}{19 + 3}$

77. $(5.986)^5$

78. $(0.325)^4$

79. $\sqrt{22{,}934{,}521}$

80. $\sqrt{13{,}593{,}969}$

Cumulative Review Problems

State the property illustrated by each equation.

81. $9 + (8.6 + 2.0) = (9 + 8.6) + 2.0$

82. $5(x + 4) = 5(x) + 5(4)$

83. Due to conservation projects and regulated hunting, the polar bear population has risen from 5,000 to 40,000. What is the percentage of increase in the polar bear population?

84. The pressure at the center of the planet Jupiter is 81,000 tons per square inch, and the pressure at the center of the earth is 27,000 tons per square inch. How many times greater is the pressure at the center of Jupiter than the pressure at the center of the earth?

1.4 Integer Exponents and Scientific Notation

After studying this section, you will be able to:

1. Express negative exponents in a form requiring only positive exponents.
2. Evaluate products involving exponents.
3. Use the quotient rule of exponents.
4. Raise a power to a power.
5. Express numbers in scientific notation.

1 Expressing Negative Exponents in a Form Requiring Only Positive Exponents

We now extend the definition of an exponent to include negative integer exponents.

Definition of Negative Exponents

If x is any nonzero real number and n is an integer,

$$x^{-n} = \frac{1}{x^n}$$

EXAMPLE 1   Simplify. Do not leave negative exponents in your answer.

(a) 4^{-1} **(b)** 2^{-5} **(c)** w^{-6}

(a) $4^{-1} = \frac{1}{4^1} = \frac{1}{4}$ **(b)** $2^{-5} = \frac{1}{2^5} = \frac{1}{32}$ **(c)** $w^{-6} = \frac{1}{w^6}$

Practice Problem 1   Simplify. Do not leave negative exponents in your answer.

(a) 3^{-2} **(b)** 7^{-1} **(c)** z^{-8} ∎

EXAMPLE 2   Simplify.   **(a)** $\left(\frac{2}{3}\right)^{-4}$ **(b)** $\left(\frac{1}{5}\right)^{-2}$

(a) $\left(\frac{2}{3}\right)^{-4} = \frac{1}{\left(\frac{2}{3}\right)^4} = \frac{1}{\frac{16}{81}} = (1)\left(\frac{81}{16}\right) = \frac{81}{16}$

(b) $\left(\frac{1}{5}\right)^{-2} = \frac{1}{\left(\frac{1}{5}\right)^2} = \frac{1}{\frac{1}{25}} = (1)\left(\frac{25}{1}\right) = 25$

Practice Problem 2   Simplify.   **(a)** $\left(\frac{3}{4}\right)^{-2}$ **(b)** $\left(\frac{1}{3}\right)^{-3}$ ∎

2 The Product Rule of Exponents

Numbers and variables in exponential notation can be multiplied quite simply if *the base is the same*. For example, we know that

$$(x^3)(x^2) = (x \cdot x \cdot x)(x \cdot x)$$

Since the factor x appears five times, it must be true that

$$x^3 \cdot x^2 = x^5$$

Hence we can state a general rule.

Rule 1.6 Product Rule of Exponents

If x is a real number and n, m are integers, then

$$x^m \cdot x^n = x^{m+n}$$

Remember that we don't usually write an exponent of 1. Thus $3 = 3^1$ and $x = x^1$.

EXAMPLE 3 Multiply. Leave your answer in exponential form.

(a) $x^5 \cdot x^8$ (b) $4^3 \cdot 4^{10}$

(c) $y \cdot y^6 \cdot y^3$ (d) $(a + b)^2(a + b)^3$

(a) $x^5 \cdot x^8 = x^{5+8} = x^{13}$ (b) $4^3 \cdot 4^{10} = 4^{3+10} = 4^{13}$

(c) $y \cdot y^6 \cdot y^3 = y^{1+6+3} = y^{10}$

(d) $(a + b)^2(a + b)^3 = (a + b)^{2+3} = (a + b)^5$ (The base is $a + b$.)

Practice Problem 3 Multiply. Leave your answer in exponential form.

(a) $w^5 \cdot w$ (b) $2^8 \cdot 2^{15}$

(c) $x^2 \cdot x^8 \cdot x^6$ (d) $(x + 2y)^4(x + 2y)^{10}$ ■

EXAMPLE 4 Multiply.

(a) $(3x^2)(5x^6)$ (b) $(2x)(3y^2)$ (c) $(5x^2y)(-2xy^3)$

(a) $(3x^2)(5x^6) = (3 \cdot 5)(x^2 \cdot x^6) = 15x^8$

(b) $(2x)(3y^2) = 6xy^2$ *Because x and y are different bases, we cannot combine their exponents.*

(c) $(5x^2y)(-2xy^3) = (5)(-2)(x^2 \cdot x^1)(y^1 \cdot y^3) = -10x^3y^4$

Practice Problem 4 Multiply.

(a) $(7w^3)(2w)$ (b) $(-5xy)(-2x^2y^3)$ (c) $(x)(2x)(3xy)$ ■

Using the Product Rule with Negative Exponents

Rule 1.6 says that the exponents are integers. Thus they can be negative. Hence

$$x^m \cdot x^{-n} = x^{m+(-n)} = x^{m-n}$$

$$x^{-m} \cdot x^{-n} = x^{-m+(-n)} = x^{-(m+n)}$$

EXAMPLE 5 Multiply. Then simplify your answer.

(a) $(-3x^4y^{-3})(2x^{-5}y^{-6})$ **(b)** $(8a^{-3}b^{-8})(2a^5b^5)$

(a) $(-3x^4y^{-3})(2x^{-5}y^{-6}) = -6x^{4-5}y^{-3-6}$
$= -6x^{-1}y^{-9}$ *To simplify means positive exponents.*
$= -6\left(\dfrac{1}{x}\right)\left(\dfrac{1}{y^9}\right)$ *The coefficient -6 does not appear in the denominator.*
$= \dfrac{-6}{xy^9}$

(b) $(8a^{-3}b^{-8})(2a^5b^5) = 16a^{-3+5}b^{-8+5}$
$= 16a^2b^{-3}$
$= 16a^2\left(\dfrac{1}{b^3}\right)$
$= \dfrac{16a^2}{b^3}$

Practice Problem 5 Multiply. Then simplify your answer.

(a) $(2x^4y^{-5})(3x^2y^{-4})$ **(b)** $(7xy^{-2})(2x^{-5}y^{-6})$ ■

3 The Quotient Rule of Exponents

We now develop the rule for dividing numbers with exponents. We know that

$$\frac{x^5}{x^3} = \frac{x \cdot x \cdot \cancel{x} \cdot \cancel{x} \cdot \cancel{x}}{\cancel{x} \cdot \cancel{x} \cdot \cancel{x}} = x \cdot x = x^2$$

Note that $x^{5-3} = x^2$. This leads us to the following general rule.

Rule 1.7 Quotient Rule of Exponents

If x is a nonzero real number and m and n are integers,

$$\frac{x^m}{x^n} = x^{m-n}$$

EXAMPLE 6 Divide. Leave your answer in exponential form.

(a) $\dfrac{x^{12}}{x^3}$ **(b)** $\dfrac{5^{16}}{5^7}$ **(c)** $\dfrac{y^3}{y^{20}}$ **(d)** $\dfrac{2^{20}}{2^{30}}$

(a) $\dfrac{x^{12}}{x^3} = x^{12-3} = x^9$ **(b)** $\dfrac{5^{16}}{5^7} = 5^{16-7} = 5^9$

(c) $\dfrac{y^3}{y^{20}} = y^{3-20} = y^{-17} = \dfrac{1}{y^{17}}$ **(d)** $\dfrac{2^{20}}{2^{30}} = 2^{20-30} = 2^{-10} = \dfrac{1}{2^{10}}$

Practice Problem 6 Divide. Leave your answer in exponential form.

(a) $\dfrac{w^8}{w^6}$ **(b)** $\dfrac{3^7}{3^3}$ **(c)** $\dfrac{x^5}{x^{16}}$ **(d)** $\dfrac{4^5}{4^8}$ ■

We can use our definition of negative exponents to find another important rule of exponents. Since

$$x^{-n} = \frac{1}{x^n}$$

it must be true that

$$x^{-n} \cdot x^n = 1$$
$$x^{-n+n} = 1 \qquad \textit{Using the product rule.}$$
$$x^0 = 1 \qquad \textit{Since } -n + n = 0.$$

Rule 1.8 Raising a Number to the Zero Power

For any nonzero real number x: $x^0 = 1$

EXAMPLE 7   Simplify.

(a) $3x^0$ **(b)** $(3x)^0$ **(c)** $(-2x^{-5})(y^3)^0$ **(d)** $\left(\dfrac{xyz}{abc}\right)^0$

(a) $3x^0 = 3(1)$ *Since $x^0 = 1$.*   **(b)** $(3x)^0 = 1$ *Note that the entire expression*
$\quad = 3$ *is raised to the zero power.*

(c) $(-2x^{-5})(y^3)^0 = (-2x^{-5})(1)$ **(d)** $\left(\dfrac{xyz}{abc}\right)^0 = 1$

$\qquad\qquad\qquad = (-2)\left(\dfrac{1}{x^5}\right)$

$\qquad\qquad\qquad = \dfrac{-2}{x^5}$

Practice Problem 7   Simplify.

(a) $6y^0$ **(b)** $5x^2y^0$ **(c)** $(3xy)^0$ **(d)** $(5^{-3})(2a)^0$ ■

EXAMPLE 8   Divide. Then simplify your answer.

(a) $\dfrac{26x^3y^4}{-13xy^8}$ **(b)** $\dfrac{-150a^3b^4c^2}{-300abc^2}$

(a) $\dfrac{26x^3y^4}{-13xy^8} = \dfrac{26}{-13} \cdot \dfrac{x^3}{x} \cdot \dfrac{y^4}{y^8} = -2x^2y^{-4} = -\dfrac{2x^2}{y^4}$

(b) $\dfrac{-150a^3b^4c^2}{-300abc^2} = \dfrac{-150}{-300} \cdot \dfrac{a^3}{a} \cdot \dfrac{b^4}{b} \cdot \dfrac{c^2}{c^2} = \dfrac{1}{2} \cdot a^2 \cdot b^3 \cdot c^0 = \dfrac{a^2b^3}{2}$

Practice Problem 8   Divide. Then simplify your answer.

(a) $\dfrac{30x^6y^5}{20x^3y^2}$ **(b)** $\dfrac{-15a^3b^4c^4}{3a^5b^4c^2}$ ■

EXAMPLE 9   Divide. Then simplify your answer.

(a) $\dfrac{3x^{-5}y^{-6}}{27x^2y^{-8}}$ **(b)** $\dfrac{2^{-4}a^3b^2c^{-4}}{2^{-5}a^{-1}b^3c^{-5}}$

(a) $\dfrac{3x^{-5}y^{-6}}{27x^2y^{-8}} = \dfrac{1}{9}x^{-5-2}y^{-6-(-8)} = \dfrac{1}{9}x^{-5-2}y^{-6+8} = \dfrac{1}{9}x^{-7}y^2 = \dfrac{y^2}{9x^7}$

(b) $\dfrac{2^{-4}a^3b^2c^{-4}}{2^{-5}a^{-1}b^3c^{-5}} = 2^{-4-(-5)}a^{3-(-1)}b^{2-3}c^{-4-(-5)} = 2^{-4+5}a^{3+1}b^{-1}c^{-4+5}$

$$= 2a^4b^{-1}c = \dfrac{2a^4c}{b}$$

Practice Problem 9   Divide. Then simplify your answer.

(a) $\dfrac{2x^{-3}y}{4x^{-2}y^5}$ **(b)** $\dfrac{5^{-3}xy^{-2}}{5^{-4}x^{-6}y^{-7}}$ ∎

4 The Power Rule for Exponents

The expression $(x^4)^3$ means that $(x^4)^3 = x^4 \cdot x^4 \cdot x^4 = x^{4+4+4} = x^{4\cdot3} = x^{12}$. In the same way (try it, using the definitions and rules already proved) we can show that

$$(xy)^3 = x^3y^3$$

and

$$\left(\frac{x}{y}\right)^3 = \frac{x^3}{y^3} \qquad (y \neq 0)$$

Therefore, we have

Rule 1.9 Power Rules of Exponents

If x and y are any real numbers and n and m are integers,

$$(x^m)^n = x^{mn}, \qquad (xy)^n = x^ny^n$$

$$\left(\frac{x}{y}\right)^n = \frac{x^n}{y^n}, \qquad \text{if } y \neq 0$$

EXAMPLE 10   Use the power rules of exponents to simplify.

(a) $(x^6)^5$ **(b)** $(2^8)^4$ **(c)** $[(a+b)^2]^4$

(a) $(x^6)^5 = x^{6\cdot5} = x^{30}$

(b) $(2^8)^4 = 2^{32}$ *Careful. Don't change the base of 2.*

(c) $[(a+b)^2]^4 = (a+b)^8$ *The base is $a+b$.*

Practice Problem 10   Use the power rules of exponents to simplify.

(a) $(w^3)^8$ **(b)** $(5^2)^5$ **(c)** $[(x-2y)^3]^3$ ∎

EXAMPLE 11 Simplify.

(a) $(3xy^2)^4$ **(b)** $\left(\dfrac{2a^2b^3}{3ab^4}\right)^3$ **(c)** $(2a^2b^{-3}c^0)^{-4}$

(a) $(3xy^2)^4 = 3^4x^4y^8 = 81x^4y^8$

(b) $\left(\dfrac{2a^2b^3}{3ab^4}\right)^3 = \dfrac{2^3a^6b^9}{3^3a^3b^{12}} = \dfrac{8a^3}{27b^3}$

(c) $(2a^2b^{-3}c^0)^{-4} = 2^{-4}a^{-8}b^{12} = \dfrac{b^{12}}{2^4a^8} = \dfrac{b^{12}}{16a^8}$

Practice Problem 11 Simplify.

(a) $(4x^3y^4)^2$ **(b)** $\left(\dfrac{4xy}{3x^5y^6}\right)^3$ **(c)** $(3xy^2)^{-2}$ ■

We need to derive one more rule. You should be able to follow the steps.

$$\frac{x^{-m}}{y^{-n}} = \frac{\dfrac{1}{x^m}}{\dfrac{1}{y^n}} = \frac{1}{x^m} \cdot \frac{y^n}{1} = \frac{y^n}{x^m}$$

Rule 1.10 Rule of Negative Exponents

If n and m are positive integers and x and y are nonzero real numbers, then

$$\frac{x^{-m}}{y^{-n}} = \frac{y^n}{x^m}$$

For example,

$$\frac{x^{-5}}{y^{-6}} = \frac{y^6}{x^5} \quad \text{and} \quad \frac{2^{-3}}{x^{-4}} = \frac{x^4}{2^3} = \frac{x^4}{8}$$

EXAMPLE 12 Simplify. **(a)** $\dfrac{3x^{-2}y^3z^{-1}}{4x^3y^{-5}z^{-2}}$ **(b)** $\left(\dfrac{5xy^{-3}}{2x^{-4}yz^{-3}}\right)^{-2}$

(a) $\dfrac{3x^{-2}y^3z^{-1}}{4x^3y^{-5}z^{-2}} = \dfrac{3y^3y^5z^2}{4x^3x^2z^1}$ *Only variables with negative exponents will change their position.*

$\qquad\qquad = \dfrac{3y^8z^2}{4x^5z^1}$

$\qquad\qquad = \dfrac{3y^8z}{4x^5}$

(b) $\left(\dfrac{5xy^{-3}}{2x^{-4}yz^{-3}}\right)^{-2} = \dfrac{5^{-2}x^{-2}y^6}{2^{-2}x^8y^{-2}z^6}$

$\qquad\qquad = \dfrac{2^2y^6y^2}{5^2x^8x^2z^6}$

$\qquad\qquad = \dfrac{4y^8}{25x^{10}z^6}$

Practice Problem 12 Simplify. **(a)** $\dfrac{7x^2y^{-4}z^{-3}}{8x^{-5}y^{-6}z^2}$ **(b)** $\left(\dfrac{4x^2y^{-2}}{x^{-4}y^{-3}}\right)^{-3}$ ■

EXAMPLE 13 Simplify. $(-3x^2)^{-2}(2x^3y^{-2})^3$ Express your answer with positive exponents only.

$$(-3x^2)^{-2}(2x^3y^{-2})^3 = (-3)^{-2}x^{-4} \cdot 2^3x^9y^{-6}$$
$$= \frac{2^3x^9}{(-3)^2x^4y^6} = \frac{8x^9}{9x^4y^6} = \frac{8x^5}{9y^6}$$

Practice Problem 13 Simplify. $(2x^{-3})^2(-3xy^{-2})^{-3}$ Express your answer with positive exponents only. ■

⑤ Scientific Notation

Scientific notation is a convenient way to write numbers. For example, we can write 5000 as 5×10^3 since $10^3 = 1000$, and we can write 0.005 as 5×10^{-3} since $10^{-3} = \frac{1}{1000}$ or 0.001. In scientific notation, the first factor is a number between 1 and 10. The second factor is a power of 10.

Scientific Notation

A positive number written in scientific notation has the form $a \times 10^n$, where $1 \leq a < 10$ and n is an integer.

To change a number from decimal notation to scientific notation, follow the steps below. Remember that the first factor must be a number between 1 and 10. This determines where to place the decimal point.

Rule 1.11 Convert from Decimal Notation to Scientific Notation

1. Move the decimal point from its original position to the right of the first nonzero digit.
2. Count the number of places that you moved the decimal point. This number is the power of 10 (that is, the exponent).
3. If you moved the decimal point to the right, the exponent is negative; if you moved it to the left, the exponent is positive.

EXAMPLE 14 Write in scientific notation.

(a) 7816   (b) 15,200,000   (c) 0.0123   (d) 0.00046

(a) $7816 = 7.816 \times 10^3$ *We moved the decimal point three places to the left, so the power of 10 is 3.*

(b) $15,200,000 = 1.52 \times 10^7$

(c) $0.0123 = 1.23 \times 10^{-2}$ *We moved the decimal point two places to the right, so the power of 10 is -2.*

(d) $0.00046 = 4.6 \times 10^{-4}$

Practice Problem 14 Write in scientific notation.

(a) 128,320   (b) 476   (c) 0.0123   (d) 0.007 ■

We can also change from scientific notation to decimal notation. We simply move the decimal point to the right or left the number of places indicated by the power of 10.

EXAMPLE 15   Write in decimal form.

(a) 1.28×10^2 **(b)** 8.8632×10^4 **(c)** 6.6×10^5
(d) 6.032×10^{-2} **(e)** 4.4861×10^{-5}

$\qquad\qquad$ ⌐— Move the decimal point two places to the right.

(a) $1.28 \times 10^2 = 128. = 128$
(b) $8.8632 \times 10^4 = 88{,}632$
(c) $6.6 \times 10^5 = 660{,}000$

$\qquad\qquad$ ⌐— Move the decimal point two places to the left.

(d) $6.032 \times 10^{-2} = 0.06032$
(e) $4.4861 \times 10^{-5} = 0.000044861$

Practice Problem 15   Write in decimal form.

(a) 3×10^4 **(b)** 4.62×10^6 **(c)** 1.973×10^{-3}
(d) 6×10^{-8} **(e)** 4.931×10^{-1} ■

Using scientific notation and the laws of exponents greatly simplifies calculations.

EXAMPLE 16   Evaluate using scientific notation. $\dfrac{(0.000000036)(0.002)}{0.000012}$

Rewrite the expression in scientific notation.

$$\frac{(3.6 \times 10^{-8})(2.0 \times 10^{-3})}{1.2 \times 10^{-5}}$$

Now rewrite using the commutative property.

$$\frac{\overset{3}{\cancel{(3.6)}}(2.0)(10^{-8})(10^{-3})}{\underset{1}{\cancel{(1.2)}}(10^{-5})} = \frac{6.0}{1} \times \frac{10^{-11}}{10^{-5}} \qquad \textit{Simplify and use the laws of exponents.}$$

$$= 6.0 \times 10^{-11-(-5)}$$
$$= 6.0 \times 10^{-6}$$

Practice Problem 16   Evaluate using scientific notation. $\dfrac{(55{,}000)(3{,}000{,}000)}{5{,}500{,}000}$ ■

EXAMPLE 17   In a scientific experiment, a scientist stated that a proton is theoretically traveling at 3.36×10^5 meters per second. If that is the correct speed, how far would the proton travel in 2×10^4 seconds?

Here we use the idea that the rate times the time equals the distance. So we multiply 3.36×10^5 by 2×10^4. Using the commutative and associative properties, we can write this as

$$3.36 \times 2 \times 10^5 \times 10^4 = 6.72 \times 10^9$$

The proton would travel 6.72×10^9 meters.

Practice Problem 17   The mass of Earth is considered to be 6.0×10^{24} kilograms. A scientist is studying a star whose mass is 3.4×10^5 times larger than the mass of the earth. If the scientist is correct, what is the mass of this star? ■

1.4 Exercises

Simplify. Do not leave negative exponents in your answer.

1. 3^{-2}

2. 4^{-3}

3. x^{-5}

4. y^{-4}

5. $\left(\dfrac{3}{4}\right)^{-3}$

6. $\left(\dfrac{4}{5}\right)^{-2}$

7. $\left(-\dfrac{1}{9}\right)^{-1}$

8. $\left(-\dfrac{1}{5}\right)^{-3}$

9. $\left(\dfrac{1}{x}\right)^{-2}$

10. $\left(\dfrac{1}{y}\right)^{-5}$

11. $x^4 \cdot x^8$

12. $y^{10} \cdot y$

13. $b \cdot b^4 \cdot b^6 \cdot b^0$

14. $y^5 \cdot y \cdot y^0 \cdot y^3$

15. $3^{12} \cdot 3^5$

16. $2^{18} \cdot 2^{10}$

17. $(3x)(-2x^5)$

18. $(5y^2)(3y)$

19. $(-12x^3y)(-3x^5y^2)$

20. $(-20a^3b^2)(5ab)$

21. $(2x^0y^5z)(-5xy^0z^8)$

22. $(4^0x^2y^3)(-3x^0y^6)$

23. $(-5a^{-3}b)(-4a^{-2}b^{-2})$

24. $(-4a^4b^{-3})(a^{-4}b)$

25. $(6x^{-3}y^{-5})(-2x^3y^{-6})$

26. $(-5x^{-8}y^{-2})(3x^{-5}y^2)$

27. $\dfrac{x^{12}}{x^7}$

28. $\dfrac{y^{18}}{y^{20}}$

29. $\dfrac{a^{20}}{a^{25}}$

30. $\dfrac{x^4}{x^7}$

31. $\dfrac{2^8}{2^5}$

32. $\dfrac{3^{16}}{3^{18}}$

33. $\dfrac{2x^3}{x^8}$

34. $\dfrac{4y^3}{8y}$

35. $\dfrac{10ab^5c}{-2ab^2}$

36. $\dfrac{-64x^2y}{4x^2}$

37. $\dfrac{12a^{-5}b^{-4}c}{18a^{10}b^{-3}c}$

38. $\dfrac{-15x^8yz^{-4}}{-35x^0y^5z^{-2}}$

39. $\dfrac{-14a^{-12}b^{-10}}{8a^{-15}b^{-20}}$

40. $\dfrac{21x^{-12}y^{-3}}{-14x^{-16}y^{-8}}$

41. $(x^2)^8$

42. $(a^5)^7$

43. $(3a^5b)^4$

44. $(2xy^6)^5$

45. $\left(\dfrac{x^2y^3}{z}\right)^6$

46. $\left(\dfrac{x^3}{y^5z^8}\right)^4$

47. $\left(\dfrac{-2x^0y^6z}{y^4}\right)^3$

48. $\left(\dfrac{-5x^4yz^3}{z^2}\right)^2$

49. $\left(\dfrac{3ab^{-2}c^3}{4a^0b^4}\right)^2$

50. $\left(\dfrac{5a^3bc^0}{-3a^{-2}b^5}\right)^3$

51. $\left(\dfrac{2xy^2}{x^{-3}y^{-4}}\right)^{-3}$

52. $\left(\dfrac{3x^{-4}y}{x^{-3}y^2}\right)^{-2}$

53. $(x^{-1}y^3)^{-2}(2xy^4)^2$

54. $(x^3y^{-2})^{-2}(5x^{-5}y)^2$

55. $\dfrac{(-4a^3b^{-5})^3}{(-2a^{-6}b^{-1})^{-2}}$

56. $\dfrac{(-5ab^3)^{-2}}{(-2a^8b^{-3})^3}$

Mixed Practice

Simplify. Express your answer with positive exponents only.

57. $\dfrac{2^{-3}a^2b^{-4}}{2^{-4}a^{-2}b^3}$

58. $\dfrac{3^4a^{-3}b^2c^{-4}}{3^3a^4b^{-2}c^0}$

59. $(3x^4y^{-2}z^0)^{-2}$

60. $(2x^{-3}y^2)^{-3}$

61. $\left(\dfrac{x^{-3}}{y^{-4}}\right)^{-2}$

62. $\left(\dfrac{2^{-1}y^{-5}}{x^{-6}}\right)^{-2}$

63. $\dfrac{a^{-2}b^0c}{ab^{-5}c}$

64. $\dfrac{b^{-2}c^4d^0}{b^{-3}c^4d^{-3}}$

65. $\left(\dfrac{14x^{-3}y^{-3}}{7x^{-4}y^{-3}}\right)^{-2}$

66. $\left(\dfrac{25x^{-1}y^{-6}}{5x^{-4}y^{-6}}\right)^{-2}$

67. $\dfrac{7^{-8}\cdot 5^{-6}}{7^{-9}\cdot 5^{-5}\cdot 6^0}$

68. $\dfrac{9^{-2}\cdot 8^{-10}\cdot 4^0}{9^{-1}\cdot 8^{-9}}$

69. $(3x^{-4}y^5)(-2x^5y^{-6})$

70. $(-4x^3y^{-5})(2x^{-8}y^{-5})$

71. $\dfrac{(3ab^2)^2}{(2a^4b^3)^{-3}}$

72. $\dfrac{(4a^2b^3)^2}{(2ab)^{-2}}$

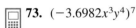 **73.** $(-3.6982x^3y^4)^7$

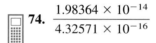 **74.** $\dfrac{1.98364\times 10^{-14}}{4.32571\times 10^{-16}}$

To Think About

75. The Amazon River is famous for sending forth one fifth of all the moving fresh water on Earth, amounting to 7,200,000 cubic *feet* per second. How many cubic *meters* per second pour out of the mouth of the Amazon? (Use 1 foot ≈ 0.305 meter)

76. Certain moths and butterflies can detect sweetness in a solution when the ratio is 1 : 300,000. Humans, on the other hand, are considered very sensitive, if they are able to detect sweetness in a solution of 1 part sugar to 200 parts water. How much more sensitive are moths and butterflies than the most sensitive humans?

Write in scientific notation.

77. 470

78. 1230

79. 1,730,000

80. 5,318,000,000

81. 0.017

82. 0.093

83. 0.000008346

84. 0.000007116

Write in decimal notation.

85. 7.13×10^5

86. 2.75×10^6

87. 1.863×10^{-2}

88. 7.07×10^{-3}

89. 9.01×10^{-7}

90. 6.668×10^{-9}

Perform the calculations indicated. Express your answer in scientific notation.

91. $(7.2 \times 10^{-3})(5.0 \times 10^{-5})$

92. $(3.1 \times 10^{-5})(2.0 \times 10^{8})$

93. $\dfrac{3.6 \times 10^{-5}}{1.2 \times 10^{-6}}$

94. $\dfrac{4.6 \times 10^{-12}}{2.3 \times 10^{5}}$

Applications

95. A bat emits a sound at a very high frequency that humans cannot hear. The frequency is approximately 5.1×10^{4} cycles per second. How many cycles would occur in 2×10^{2} seconds?

96. In one year, light travels 5.87×10^{12} miles. How far would light travel in 5×10^{3} years?

97. The weight of one oxygen molecule is 5.3×10^{-23} gram. How much would 2×10^{4} molecules of oxygen weigh?

98. The average distance of Earth to the sun is 4.90×10^{11} feet. If a solar probe is launched from Earth traveling at 2×10^{4} feet per second, how long would it take to reach the sun?

Cumulative Review Problems

Evaluate.

99. $(-3)(-2)(-1)(-4)$

100. $5 + 2(-3) + 12 \div (-6)$

101. $5 - 7 + 3 - 4(2)^{2}$

102. $\dfrac{5 + 3 - 4}{32 \div (-8)}$

Developing Your Study Skills

WHY STUDY MATHEMATICS?

Students often question the value of mathematics, particularly algebra. They see little real use for algebra in their everyday lives. This is understandable at the beginning or intermediate levels of algebra, because applications of algebra may not be obvious.

The extensive usefulness of mathematics becomes clear as you take higher-level courses, such as college algebra, statistics, trigonometry, and calculus. You may not be planning to take these higher-level courses now, but your college major may require you to do so.

In our present-day, technological world, it is easy to see mathematics at work. Many vocational and professional areas—such as the fields of business, statistics, economics, psychology, finance, computer science, chemistry, physics, engineering, electronics, nuclear energy, banking, quality control, and teaching—require a certain level of expertise in mathematics. Those who want to work in these fields must be able to function at a given mathematical level. Those who cannot will not make it.

So if your field of study requires you to take higher-level mathematics courses, be sure to realize the importance of mastering the basics of this course. Then you will be ready to advance with your career plans.

1.5 Operations with Variables and Grouping Symbols

MathPro   Video 2   SSM

After studying this section, you will be able to:

1 *Collect like terms in an algebraic expression.*
2 *Multiply algebraic expressions using the distributive property.*
3 *Remove grouping symbols in their proper order to simplify algebraic expressions.*

1 Collecting Like Terms in an Algebraic Expression

A collection of numerical values, variables, and signs of operations is called an **algebraic expression.** An algebraic expression usually contains the sum or difference of several *terms.* A **term** is a real number, or a variable, or a product or quotient of numbers and variables.

> Terms are *always* separated by + or − signs.

EXAMPLE 1   List the terms in each algebraic expression.

(a) $5x + 3y^2$ **(b)** $6y^2 + 3y + 2$ **(c)** $5x^2 - 3xy - 7$

(a) $5x$ is a product of a real number (5) and a variable (x), so $5x$ is a term. $3y^2$ is also a product of a real number and a variable, so $3y^2$ is a term.

(b) $6y^2$ and $3y$ are products, so they are terms. 2 is a real number, so it is also a term.

(c) Note that we can write $5x^2 - 3xy - 7$ in an equivalent form: $5x^2 + (-3xy) + (-7)$. This second form helps us to identify more readily the three terms. The terms are $5x^2$, $-3xy$, and -7.

Practice Problem 1   List the terms in each algebraic expression.

(a) $7x - 2w^3$ **(b)** $5 + 6x + 2y$ ∎

Any factor in a term is the **coefficient** of the product of the remaining factors. For example, $4x^2$ is the coefficient of y in the term $4x^2y$, and $4y$ is the coefficient of x^2 in $4x^2y$. The numerical coefficient of a term is the numerical value multiplied by the variables. The numerical coefficient of $4x^2y$ is 4. However, ''coefficient'' is often used to mean the ''numerical coefficient.'' We will use it this way for the remainder of the book. Thus in the expression $-5x^2y$ the coefficient will be considered to be -5. If no numerical coefficient appears before a variable in a term, the coefficient is understood to be 1. For example, the coefficient of xy is 1. The coefficient of $-x$ is -1.

EXAMPLE 2   Identify the numerical coefficient in each term

(a) $5x^2 - 2x + 3xy$ **(b)** $8x^3 - 12xy^2 + y$
(c) $\frac{1}{2}x^2 + \frac{1}{4}x$ **(d)** $3.4ab - 0.5b$

(a) The coefficient of the x^2 term is 5. The coefficient of the x term is -2. The coefficient of the xy term is 3.

(b) The coefficient of the x^3 term is 8. The coefficient of the xy^2 term is -12. The coefficient of the y term is 1.

(c) The coefficient of the x^2 term is $\frac{1}{2}$. The coefficient of the x term is $\frac{1}{4}$.

(d) The coefficient of the ab term is 3.4. The coefficient of the b term is -0.5.

Practice Problem 2   Identify the numerical coefficient in each term.

(a) $5x^2y - 3.5w$ **(b)** $\frac{3}{4}x^3 - \frac{5}{7}x^2y$ **(c)** $-5.6abc - 0.34ab + 8.56bc$ ■

We can add or subtract terms if they are **like terms;** that is, the terms have the same variables and the same exponents. When we collect like terms, we are using the following form of the distributive property:

$$ba + ca = (b + c)a$$

EXAMPLE 3   Collect like terms by showing the use of the distributive property.

(a) $8x + 2x$ **(b)** $5x^2y + 12x^2y$

(a) $8x + 2x = (8 + 2)x = 10x$

(b) $5x^2y + 12x^2y = (5 + 12)x^2y = 17x^2y$

Practice Problem 3   Collect like terms by showing the use of the distributive property.

(a) $9x - 12x$ **(b)** $4ab^2c + 15ab^2c$ ■

EXAMPLE 4   Collect like terms.

(a) $5a^2b + 3ab - 8a^2b + 2ab$ **(b)** $7x^2 - 2x - 8 + x^2 + 5x - 12$

(c) $\frac{1}{3}x^2 + \frac{1}{4}x - \frac{1}{6}x^2$ **(d)** $2.3x^2 - 5.6x + 5.8x^2 - 7.9x$

(a) $5a^2b + 3ab - 8a^2b + 2ab = -3a^2b + 5ab$

Note that we cannot combine $-3a^2b + 5ab$ since they are not like terms.

(b) $7x^2 - 2x - 8 + x^2 + 5x - 12 = 8x^2 + 3x - 20$

Remember that the coefficient of x^2 is 1, so you are adding $7x^2 + 1x^2$.

(c) $\frac{1}{3}x^2 - \frac{1}{6}x^2 + \frac{1}{4}x = \frac{2}{6}x^2 - \frac{1}{6}x^2 + \frac{1}{4}x$

$$= \frac{1}{6}x^2 + \frac{1}{4}x$$

(d) $2.3x^2 - 5.6x + 5.8x^2 - 7.9x = 8.1x^2 - 13.5x$

Practice Problem 4   Collect like terms.

(a) $12x^3 - 5x^2 + 7x - 3x^3 - 8x^2 + x$

(b) $\frac{1}{3}a^2 - \frac{1}{5}a - \frac{4}{15}a^2 + \frac{1}{2}a + 5$

(c) $4.5x^3 - 0.6xy - 9.3x^3 + 0.8xy$ ■

2 Multiplying Algebraic Expressions Using the Distributive Property

We can use the distributive property $a(b + c) = ab + ac$ to multiply algebraic expressions. The kinds of expressions usually encountered are called *polynomials*. **Polynomials** are variable expressions that contain terms with *nonnegative* integer exponents. Some examples of polynomials are $6x^2 + 2x - 8$, $5a + b$, $16x^3$, and $5x + 8$.

EXAMPLE 5   Use the distributive property to multiply $-2x(x^2 + 5x)$.

The distributive property tells us to multiply each term in the parentheses by the term outside the parentheses.

$$-2x(x^2 + 5x) = (-2x)(x^2) + (-2x)(5x) = -2x^3 - 10x^2$$

Practice Problem 5   Use the distributive property to multiply $-3x^2(2x - 5)$. ■

There is no limit to the number of terms we can multiply. For example, $a(b + c + d + \ldots) = ab + ac + ad + \cdots$.

EXAMPLE 6   Multiply.

(a) $7x(x^2 - 3x - 5)$ **(b)** $5ab(a^2 - ab + 8b^2 + 2)$

(a) $7x(x^2 - 3x - 5) = 7x^3 - 21x^2 - 35x$

(b) $5ab(a^2 - ab + 8b^2 + 2) = 5a^3b - 5a^2b^2 + 40ab^3 + 10ab$

Practice Problem 6   Multiply.

(a) $-5x(2x^2 - 3x - 1)$ **(b)** $3ab(4a^3 + 2b^2 - 6)$ ■

A parenthesis preceded by no sign or a positive sign $(+)$ can be considered to have a numerical coefficient of 1. A parenthesis preceded by a negative sign $(-)$ can be considered to have a numerical coefficient of -1.

EXAMPLE 7   Multiply.

(a) $(3x^2 + 2)$ **(b)** $+(5x - 7)$
(c) $-(2x + 3)$ **(d)** $-4x(x - 2y)$

(e) $\dfrac{2}{3}(6x^2 - 2x + 3)$

(a) $(3x^2 + 2) = 1(3x^2 + 2) = 3x^2 + 2$ **(b)** $+(5x - 7) = +1(5x - 7) = 5x - 7$
(c) $-(2x + 3) = -1(2x + 3) = -2x - 3$ **(d)** $-4x(x - 2y) = -4x^2 + 8xy$
(e) $\dfrac{2}{3}(6x^2 - 2x + 3) = \dfrac{2}{3}(6x^2) - \dfrac{2}{3}(2x) + \dfrac{2}{3}(3)$

$$= 4x^2 - \dfrac{4}{3}x + 2$$

Practice Problem 7   Multiply.

(a) $(7x^2 - 8)$ **(b)** $+(5x^2 + 6)$
(c) $-(3x + 2y - 6)$ **(d)** $-5x^2(x + 2xy)$

(e) $\dfrac{3}{4}(8x^2 + 12x - 3)$ ■

3 *Removing Grouping Symbols to Simplify Algebraic Expressions*

To simplify an expression that contains parentheses that are not placed within other parentheses, multiply and then collect like terms.

EXAMPLE 8 Simplify. $5(x - 2y) - (y + 3x) + (5x - 8y)$

$5(x - 2y) - (y + 3x) + (5x - 8y)$

> Remember that you are really multiplying by -1, so don't forget to change this sign.

$= 5x - 10y - y - 3x + 5x - 8y$

$= 7x - 19y$

Practice Problem 8 Simplify. $-7(a + b) - 8a(2 - 3b) + 5a$ ■

To simplify an expression that contains grouping symbols within grouping symbols, work from the inside out. The symbols [ ] and { } are used as grouping symbols that are equivalent to using parentheses.

EXAMPLE 9 Simplify. $-2\{3 + 2[z - 4(x + y)]\}$

$-2\{3 + 2[z - 4(x + y)]\}$

$= -2\{3 + 2[z - 4x - 4y]\}$ *Remove the parentheses by multiplying each term of $x + y$ by -4.*

$= -2\{3 + 2z - 8x - 8y\}$ *Remove the brackets by multiplying each term of $z - 4x - 4y$ by 2.*

$= -6 - 4z + 16x + 16y$ *Remove the braces by multiplying each term by -2.*

Practice Problem 9 Simplify. $-2\{4x - 3[x - 2x(1 + x)]\}$ ■

EXAMPLE 10 Simplify. $-2\{3x + [x - (3x - 1)]\}$

$-2\{3x + [x - (3x - 1)]\}$

$= -2\{3x + [x - 3x + 1]\}$ *Remove the parentheses by multiplying each term of $3x - 1$ by -1.*

$= -2\{3x + [-2x + 1]\}$ *Collect like terms.*

$= -2\{3x - 2x + 1\}$ *Remove the brackets.*

$= -2\{x + 1\}$ *Collect like terms.*

$= -2x - 2$ *Remove the braces by multiplying by -2.*

It would also be correct to leave our answer in its "factored" form, $-2\{x + 1\}$. By common practice, we would then use parentheses instead of braces and write $-2(x + 1)$.

Practice Problem 10 Simplify. $3\{-2a + [b - (3b - a)]\}$ ■

1.5 Exercises

1. What is the coefficient of y in $-5x^2y$?

2. What is the coefficient of x in $3xy^3z$?

3. What are the terms in the expression $5x^3 - 6x^2 + 4x + 8$?

4. What are the terms in the expression $5x^3 + 3x^2 - 2y - 8$?

In Problems 5–10, list the numerical coefficient of each term.

5. $x^2 + 2x + 3y$

6. $x^2 + 2xy + 8y^2$

7. $5x^3 - 3x^2 + x$

8. $6x^2 - x - 6y$

9. $6.5x^3y^3 - 0.02x^2y + 3.05y$

10. $-\dfrac{1}{2}a^2b^2 - \dfrac{10}{3}a^2b - \dfrac{4}{5}ab$

Collect like terms.

11. $3ab + 8ab$

12. $7ab - 5ab$

13. $5x - 2x + 3x - 8y$

14. $2a - 8a + 3b + 5a$

15. $x^2 - 3x - 4x^2 + 2x$

16. $y^2 + 5y - 4y - 5y^2$

17. $4ab - 3b^3 - 5ab + 3b^3$

18. $2a - ab - 2a - ab$

19. $3y^2 - 9y + 2y^2 + 6$

20. $5x^2 - 8x - 10x^2 + 8$

21. $0.1x^2 + 3x - 0.5x^2$

22. $0.7x^2 - 6x + 0.4x^2$

23. $\dfrac{1}{2}a + \dfrac{1}{3}b - \dfrac{1}{3}a - \dfrac{1}{3}b$

24. $\dfrac{1}{4}x + \dfrac{1}{3}y - \dfrac{1}{12}x + \dfrac{1}{6}y$

25. $\dfrac{1}{5}a^2 - 3b - \dfrac{1}{2}a^2 + 5b$

26. $\dfrac{1}{2}x^2 + 6y - \dfrac{1}{7}x^2 - 8y$

27. $1.2x^2 - 5.6x - 8.9x^2 + 2x$

28. $4y^2 - 2.1y - 8.6y - 2.2y^2$

29. $-8a^2b - 2ab + 3ab^2 + 5a^2b - 12ab - 8ab^2$

30. $12mn + 8m^2n - 6mn^2 + mn - 16m^2n - 4mn^2$

Multiply. Simplify your answer wherever possible.

31. $4x(2x + y)$

32. $5y(3x - 2)$

33. $-x(-x^3 + 3x^2 + 5x)$

34. $-2y(y^2 - 3y + 1)$

35. $-3(3a^2 - a + 7)$

36. $-6(4a + 2ab - 7b^2)$

37. $2xy(x^2 - 3xy + 4y^2)$

38. $4ab(a^2 - 6ab - 2b^2)$

39. $\frac{1}{3}(2x + 6xy - 12)$

40. $\frac{1}{4}(7x^2 - 4x + 8)$

41. $\frac{x}{6}(x^2 + 5x - 9)$

42. $\frac{x}{2}(7x^2 - 4x + 1)$

43. $2ab(a^5 - 3a^4 + a^2 - 2a)$

44. $5xy^2(y^3 - y^2 + 3x + 1)$

45. $1.5x^2(x - 2y + 3y^2)$

46. $8(x + 5y) - 2y(x + 1)$

Remove grouping symbols and simplify.

47. $2y - 3(y + 4z)$

48. $5x + 2(-x - 6)$

49. $2(x - 1) - x(x + 1) + 3(x^2 + 2)$

50. $5(x - 2) + (3x - 8) - (x - 2)$

51. $2[x + (x - y)] - 3[x - (x - y)]$

52. $3[2x + (y - 2x)] - 2[x - (3y - x)]$

53. $2\{3x - 2[x - 4(x + 1)]\}$

54. $-3\{3y + 2[y + 2(y - 4)]\}$

55. $2(a^2 - 5) - 3[4 - a(a + 2)]$

56. $-5(b + a^2) - 6[a + a(-3 - a)]$

Cumulative Review Problems

Evaluate.

57. $2(-3)^2 + 4(-2)$

58. $4(2 - 3 + 6) + \sqrt{36}$

59. $\dfrac{5(-2) - 8}{3 + 4 - (-3)}$

60. $(-3)^5 + 2(-3)$

61. The smallest known organism to contain all the chemicals needed to sustain independent life is a bacterium called the *pleuropneumonia* organism. It would take 1,893,500 of them, touching side by side, to stretch an inch. How many *pleuropneumonia* organisms would be found if you put them in a line one kilometer long. Express your answer in scientific notation. (Use 1 inch = 0.0254 meter.)

62. Efficiency of internal combustion engines is lost at the rate of 2% for every 1,000 feet of altitude above sea level. What percentage of efficiency would be lost by power boats and cars at Lake Titicaca, 4167 *meters* above sea level? (Use 1 foot = 0.305 meter.)

1.6 Evaluating Variable Expressions and Formulas

MathPro   Video 3   SSM

After studying this section, you will be able to:

1 *Evaluate a variable expression when the value of the variable is given.*
2 *Evaluate a formula to determine a given quantity.*

1 *Evaluating an Expression*

We want to learn how to evaluate—compute a numerical value for—variable expressions when we know the values of the variables.

Evaluating a Variable Expression

1. Replace each variable (letter) by its numerical value. Put parentheses around the value (watch out for negative values).
2. Carry out each step, using the correct order of operations.

EXAMPLE 1   Evaluate $5 + 2x$ when $x = -3$.

Replace x by -3 and put parentheses around it.
$$5 + 2(-3) = 5 + (-6) = 5 - 6 = -1$$

Practice Problem 1   Evaluate $-6 + 3x$ when $x = -5$. ∎

EXAMPLE 2   Evaluate $x^2 - 5x - 6$ when $x = -4$.

$(-4)^2 - 5(-4) - 6$ *Replace x by -4 and put parentheses around it.*

$= 16 - 5(-4) - 6$ *Square -4.*

$= 16 - (-20) - 6$ *Multiply $5(-4)$.*

$= 16 + 20 - 6$ *Multiply $-1(-20)$.*

$= 30$ *Combine $16 + 20 - 6$.*

Practice Problem 2   Evaluate $2x^2 + 3x - 8$ when $x = -3$. ∎

EXAMPLE 3   Evaluate $(5 - x)^2 + 3xy$ when $x = -2$ and $y = 3$.

$[5 - (-2)]^2 + 3(-2)(3)$

$= [5 + 2]^2 + 3(-2)(3) = [7]^2 + 3(-2)(3)$

$= 49 + 3(-2)(3) = 49 - 18 = 31$

Practice Problem 3   Evaluate $(x - 3)^2 - 2xy$ when $x = -3$ and $y = 4$. ∎

EXAMPLE 4   Evaluate when $x = -3$.

(a) $(-2x)^2$ **(b)** $-2x^2$

(a) $[-2(-3)]^2 = [6]^2 = 36$ *Multiply $(-2)(-3)$ and then square the result.*

(b) $-2(-3)^2 = -2(9)$ *Square -3.*
$\qquad\qquad = -18$ *Multiply by -2.*

Practice Problem 4   Evaluate when $x = -4$.   **(a)** $(-3x)^2$ **(b)** $-3x^2$ ■

To Think About Why are the answers to part (a) and (b) different in Example 4? What does $(-2x)^2$ mean? What does $-2x^2$ mean? Why are the parentheses so important in this situation?

② *Evaluating Formulas*

A ***formula*** is a rule for finding the value of a variable when the values of other variables in the expression are known. The word ''formula'' is usually applied to some physical situation, much like a recipe. For example, we can determine the Fahrenheit temperature F for any Celsius temperature C from the formula

$$F = \frac{9}{5}C + 32$$

EXAMPLE 5   Find the Fahrenheit temperature when the Celsius temperature is $-30°C$.

$$F = \frac{9}{5}(-30) + 32 \qquad\qquad \text{\textit{Substitute the known value -30 for}}$$
$$\text{\textit{the variable C. Then evaluate.}}$$

$$= \frac{9}{\cancel{5}_{1}}(\cancel{-30}^{-6}) + 32 = 9(-6) + 32$$

$$= -54 + 32 = -22$$

Thus, when the temperature is $-30°C$, the equivalent Fahrenheit temperature is $-22°$.

Practice Problem 5   Find the Fahrenheit temperature when the Celsius temperature is $70°C$. ■

EXAMPLE 6   The period T of a pendulum (the time in seconds for the pendulum to swing back and forth one time) is $T = 2\pi\sqrt{\frac{L}{g}}$, where L is the length of the pendulum in feet and g is the acceleration due to gravity. Find the period when $L = 288$ and $g = 32$. Approximate the value of π by 3.14. Round your answer to the nearest tenth of a second.

$$T = 2(3.14)\sqrt{\frac{288}{32}} = 6.28\sqrt{9} = 6.28(3)$$

$$= 18.8 \text{ seconds}$$

The period for a 288-foot-long pendulum to swing back and forth one time is approximately 18.8 seconds.

Practice Problem 6   Tarzan is swinging back and forth on a 128-foot-long rope. Use the above formula to find out how long it takes him to swing back and forth one time. Round your answer to the nearest tenth of a second. ■

EXAMPLE 7   An amount of money invested or borrowed (not including interest) is called the *principal*. Find the amount A to be repaid on a principal p of $1000.000 borrowed at an interest rate r of 8% for a time t of 2 years. The formula is $A = p(1 + rt)$.

$$A = 1000[1 + (0.08)(2)] \qquad \textit{Change 8\% to 0.08.}$$

$$= 1000[1 + 0.16]$$

$$= 1000(1.16) = 1160$$

The amount to be repaid (principal plus interest) is $1160.

Practice Problem 7   Find the amount to be repaid on a loan of $600.00 at an interest rate of 9% for a time of 3 years. ■

One of the areas of mathematics where a formula is very helpful is in determining the perimeter, area, or volume of common geometric figures.

EXAMPLE 8   Find the perimeter of a school playground in the shape of a rectangle with length 28 meters and width 16.5 meters. Use the formula $P = 2l + 2w$.

We draw a picture to get a better idea of the situation.

Length = 28 meters

Width = 16.5 meters

$$P = 2l + 2w = 2(28) + 2(16.5) = 56 + 33 = 89$$

The perimeter of the playground is 89 meters.

Practice Problem 8   Find the perimeter of a rectangular-shaped computer chip. The length of the chip is 0.76 centimeter and the width is 0.38 centimeter. ■

The most commonly used formulas for geometric figures are listed below.

Area and Perimeter Formulas

In the following formulas, A = area, P = perimeter, and C = circumference. The "squares" in the figures like ⌐ mean that the angle formed by the two lines is 90°.

Rectangle

$$A = lw$$
$$P = 2l + 2w$$

Triangle

$$A = \frac{1}{2}ab$$
$$P = b + c + d$$

Parallelogram

$$A = ab$$
$$P = 2b + 2c$$

Rhombus

$$A = ab$$
$$P = 4b$$

Trapezoid

$$A = \frac{1}{2}a(b + c)$$
$$P = b + c + d + e$$

Circle

$$A = \pi r^2$$
$$C = 2\pi r, \quad \text{where } r \text{ is radius}$$
$$C = \pi d, \quad \text{where } d \text{ is diameter}$$

In circle formulas we need to use π, which is an irrational number. Its value can be approximated to as many decimal places as we like, but we will use 3.14 because that is accurate enough for most calculations.

EXAMPLE 9 Find the area of a trapezoid that has a height of 6 meters and bases of 7 and 11 meters.

The formula is $A = \frac{1}{2}a(b + c)$. We are told that $a = 6$, $b = 7$, and $c = 11$, so we put those values into the formula.

$b = 7$ meters

$a = 6$ meters

$c = 11$ meters

$$A = \frac{1}{2}(6)[7 + 11] = \frac{1}{2}(6)(18) = 54$$

The area is 54 *square meters*.

Practice Problem 9 Find the area of a triangle with an altitude of 12 meters and a base of 14 meters. ■

In the following formulas, V = volume and S = surface area.

Rectangular solid $V = lwh$
$S = 2lw + 2wh + 2lh$

Sphere $V = \frac{4}{3}\pi r^3$
$S = 4\pi r^2$

Right circular cylinder $V = \pi r^2 h$
$S = 2\pi rh + 2\pi r^2$

EXAMPLE 10   Find the volume of a sphere with a radius of 3 centimeters.

The formula is $V = \frac{4}{3}\pi r^3$.

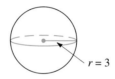

Therefore, we have $V = \dfrac{4}{3}(3.14)(3)^3 \approx \dfrac{4}{3}(3.14)(27)$

$$\approx \frac{4}{\cancel{3}_1}(3.14)(\cancel{27}^9) \approx 113.04$$

The volume is approximately 113.04 cubic centimeters.

Practice Problem 10   Find the volume of a right circular cylinder of height 10 meters and radius 6 meters. ■

Developing Your Study Skills

CLASS ATTENDANCE

A student of mathematics needs to get started in the right direction by choosing to attend class every day, beginning with the first day of class. Statistics show that class attendance and good grades go together. Classroom activities are designed to enhance learning, and therefore you must be in class to benefit from them. Vital information and explanations are given each day that can help you in understanding concepts. Do not be deceived into thinking that you can just find out from a friend what went on in class. There is no good substitute for firsthand experience. Give yourself a push in the right direction by developing the habit of going to class every day.

1.6 Exercises

In this exercise set, round all answers to the nearest hundredth unless otherwise stated. Evaluate each expression for the values given.

1. $4x + 6$; $x = -2$

2. $2x - 6$; $x = 4$

3. $29 - 7x$; $x = 3$

4. $8x + 5$; $x = -6$

5. $x^2 + 5x - 6$; $x = -3$

6. $x^2 + 7x + 12$; $x = -2$

7. $4 - 5x - x^2$; $x = 1$

8. $-3x^2 + 5x + 2$; $x = -1$

9. $-2x^2 + 5x - 3$; $x = -4$

10. $6x^2 - 3x + 5$; $x = 5$

11. $2ax - by - a$; $a = 1$, $b = -2$, $x = \dfrac{1}{2}$, $y = 3$

12. $3ay - 2by + x$; $a = 4$, $x = 1$, $y = -1$, $b = -6$

13. $ax^2 + bxy + y^2$; $a = 4$, $x = -1$, $b = 3$, $y = -2$

14. $x^3 + ax^2 + aby$; $a = 1$, $b = -3$, $x = -2$, $y = 5$

15. $\sqrt{b^2 - 4ac}$; $b = 2$, $a = 1$, $c = -15$

16. $\sqrt{b^2 - 4ac}$; $b = -5$, $a = 2$, $c = -3$

17. Evaluate $2x^2 - 5x + 6$ when $x = -3.52176$. (Round to 5 decimal places.)

18. Evaluate $3x^2 - 7x - 2$ when $x = -0.56736$. (Round answer to 5 decimal places.)

Applications

For Problems 19 and 20, use the formula $F = \frac{9}{5}C + 32$.

19. Find the Fahrenheit temperature when the Celsius temperature is $-60°C$.

20. Find the Fahrenheit temperature when the Celsius temperature is $85°C$.

For Problems 21 and 22, use the formula $C = \frac{5F - 160}{9}$.

21. Find the Celsius temperature if the Fahrenheit temperature is $122°F$.

22. Find the Celsius temperature if the Fahrenheit temperature is $-40°F$.

For problems 23 and 24, use $T = 2\pi\sqrt{\frac{L}{g}}$. Let $\pi = 3.14$ and $g = 32$ feet per second per second.

23. A child is swinging on a rope 32 feet long over a river swimming hole. How long does it take (in seconds) to complete one swing back and forth?

24. A cable acts as a pendulum swinging from a skyscraper under construction. The cable is 512 feet long. How many seconds will it take for the cable to swing back and forth one time?

In Problems 25–28 use the formula $A = p(1 + rt)$.

25. Find A if $p = \$4800$, $r = 0.12$, and $t = 1.5$.

26. Find A if $p = \$3200$, $r = 0.07$, and $t = 2$.

27. Find the amount to be repaid on a loan of $1200 at an interest rate of 11% for 3 years.

28. Find the amount to be repaid on a loan of $1900 at an interest rate of 6% for 3 years.

In Problems 29–32, use the fact that the distance S an object falls in t seconds is given by $S = \frac{1}{2}gt^2$, where $g = 32$ feet per second per second.

29. Find $S = \frac{1}{2}gt^2$ if $g = 32$ and $t = 6$ seconds.

30. Find $S = \frac{1}{2}gt^2$ if $g = 32$ and $t = 3$ seconds.

31. A bolt fell out of a window frame on the twenty-third floor of the John Hancock Tower in Boston. The bolt took 4 seconds to hit the ground. How far did the bolt fall?

32. A piece of the window ledge fell out of a window on the sixty-eighth floor of the Texas Commerce Tower in Houston. It took 7 seconds to hit the ground. How far did the piece of the window ledge fall?

33. Find $z = \dfrac{Rr}{R + r}$ if $R = 36$ and $r = 4$.

34. Find $z = \dfrac{Rr}{R + r}$ if $R = 35$ and $r = 15$.

In Problems 35 and 36, you need the formula $S = \frac{n}{2}[2a + (n - 1)d]$.

35. Find S if $n = 12$, $a = -7$, and $d = 3$.

36. Find S if $n = 16$, $a = 4$, and $d = -3$.

In Problems 37–52, use the geometry formulas on pages 45 and 46 whenever necessary to find the quantities specified.

37. Find the area of a circle with radius of 0.5 inches.

38. Find the circumference of a circle with a diameter of 0.2 meters.

39. Find the area of a triangle with a base of 12 meters and an altitude of 14 meters.

40. Find the area of a triangle with a base of 16 centimeters and a height of 7 centimeters.

41. Find the area of a parallelogram with an altitude of 5 yards and a base of 8 yards.

42. Find the area of a rhombus with a height of 16 centimeters and a base of 5 centimeters.

43. Find the area of a rectangle 0.5 meter long and 0.07 meter wide.

44. Find the surface area of a rectangular solid 12.4 centimeters long, 6.7 centimeters wide, and 1.2 centimeters high.

45. The base of a rhombus is 0.38 meter. Find its perimeter.

46. A trapezoid has sides 5.2, 6.1, 3.5, and 2.2 meters long. Find its perimeter.

47. A right circular cylinder has a height of 7 feet and a radius of 3 feet.
  (a) Find its volume.
  (b) Find its total surface area.

48. A sphere has a radius of 6 meters.
  (a) Find its volume.
  (b) Find its surface area.

49. Find the circumference of a circle when the diameter is 5.78349 meters. (Use $\pi \approx 3.1415927$ or the value of π in your calculator.) Round your answer to five decimal places.

50. Find the area of a circle when the radius is 9.05263 centimeters. (Use $\pi \approx 3.1415927$ or the value of π in your calculator.) (Round your answer to five decimal places.)

51. An engineer must make a stainless steel circle with a radius of 6 centimeters. What is the area of the circle? What is the circumference of the circle? (Use $\pi \approx 3.14$.)

52. A cross section of telephone cable is a circle that has a radius of 8 centimeters. What is the area of the cross section? What is the circumference of this circular cross section? (Use $\pi \approx 3.14$.)

To Think About

53. Human beings existing on 2500 calories per day give off energy equal to the heat of a 120-watt light bulb (104 calories per hour) when they are seated in a room.
  (a) How many watts of power would be comparable, if there were 433 people in a movie theater, without a fan circulating the air?
  (b) How many calories would be expended by the crowd in an hour?

54. The Russian station of Vostok recorded a temperature of $-126.9°F$, $(-88.3°C)$ at the interior of Antarctica, 11,500 feet above sea level. The conditions of high altitude and continuous darkness in winter lead to such frigid levels. If the sun had shone that day, what would be the temperature, *in Fahrenheit and Celsius,* if the temperature has risen $+47.4°F$?

Cumulative Review Problems

Simplify.

55. $-2a - b - 5a + 6b$

56. $3x(x - y) + x(y - 2x)$

57. $\left(\dfrac{-5x^2}{2y^3}\right)^2$

58. $2\{5 - 2[x - 3(2x + 1)]\}$

Putting Your Skills to Work

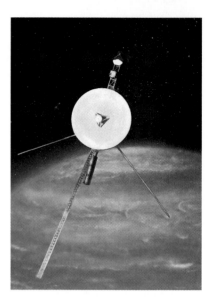

THE MATHEMATICS OF THE PLANETS

With the advent of the Hubble space telescope and the success of space probes such as *Voyager* and *Galileo,* we continue to learn more about the planets that make up our solar system. Some of the mathematical relationships about the planets are quite remarkable. Due to its size, Earth does not weigh anywhere near what some of the larger planets do, yet Earth is very compressed and is in fact the most dense of all the planets in the solar system.

    Consider the following facts:

Planet	Radius in km	Mean Density in kg/m³	Escape Velocity in km/s
Mercury	2439	5430	4.3
Earth	6378	5520	11.2
Saturn	60,268	710	36.0
Neptune	24,764	1670	24.0

PROBLEMS FOR INDIVIDUAL INVESTIGATION AND ANALYSIS

Assume that the planets are spheres. Express your answers in scientific notation. Round your answers to five decimal places. Use $\pi \approx 3.1416$.

1. Find the mass of Earth in kg.

2. Find the mass of Mercury in kg. The radius of Mercury is only 42% of Earth's radius. What percent of the Earth's mass is the mass of Mercury?

PROBLEMS FOR GROUP INVESTIGATION AND COOPERATIVE LEARNING

Together with some other members of your class see if you can answer the following.

3. Find the mass of Saturn in kg. Compare the mass of Earth and the mass of Saturn. How does the mass of the planet seem to influence the escape velocity?

4. The specific gravity is a measure of the power of gravitational force on an object of a given mass if it were placed on different planets. The Earth's specific gravity is defined as 1.00 while Mercury is 0.38, Saturn is 1.07, and Neptune is 1.14. What factor seems to be the best predictor of specific gravity of a planet: the radius of the planet, the mean density, the mass, or the escape velocity?

INTERNET CONNECTION: Go to `http://www.prenhall.com/tobey` to be connected.

Site: The Sun (The Nine Planets)

This site provides a wealth of information about our sun. Note that some of the numbers here are given in an abbreviated form of scientific notation; for example, 1.989e30 means 1.989×10^{30}.

5. Calculate the density of the sun in kg/m³.

6. The earth's density is what percent of the sun's density?

7. Find the mass of the helium in the sun.

8. Find the number of tons of hydrogen that are converted to helium and energy in a century. Express your answer in scientific notation.

Alternate sites (mirrors) are listed here: http://www.nwvoyager.com/nineplanets/nineplanets/mirrors.html

Topic	Procedure	Examples
Commutative property of addition, p. 5.	$a + b = b + a$	$12 + 13 = 13 + 12$
Commutative property of multiplication, p. 5.	$a \cdot b = b \cdot a$	$11 \cdot 19 = 19 \cdot 11$
Associative property of addition, p. 5.	$a + (b + c) = (a + b) + c$	$4 + (3 + 6) = (4 + 3) + 6$
Associative property of multiplication, p. 5.	$a \cdot (b \cdot c) = (a \cdot b) \cdot c$	$7 \cdot (3 \cdot 2) = (7 \cdot 3) \cdot 2$
Identity property of addition, p. 5.	$a + 0 = 0 + a = a$	$9 + 0 = 0 + 9 = 9$
Identity property of multiplication, p. 5.	$a \cdot 1 = 1 \cdot a = a$	$7 \cdot 1 = 1 \cdot 7 = 7$
Inverse property of addition, p. 5.	$a + (-a) = (-a) + a = 0$	$8 + (-8) = -8 + 8 = 0$
Inverse property of multiplication, p. 5.	If $a \neq 0$, $a\left(\dfrac{1}{a}\right) = \dfrac{1}{a}(a) = 1$	$15\left(\dfrac{1}{15}\right) = \dfrac{1}{15}(15) = 1$
Distributive property of multiplication over addition, p. 5.	$a(b + c) = a \cdot b + a \cdot c$	$7(9 + 4) = 7 \cdot 9 + 7 \cdot 4$
Addition of real numbers, p. 11.	*Addition:* To add two real numbers with the *same sign,* add their absolute values and use the common sign. To add two real numbers with *different signs,* take the difference of their absolute values, and the answer takes the sign of the number with the larger absolute value.	$9 + 5 = 14 \qquad -7 + (-3) = -10$ $-\dfrac{1}{5} + \left(\dfrac{3}{5}\right) = \dfrac{2}{5} \qquad -42 + 19 = -23$
Subtraction of real numbers, p. 12.	*Subtraction:* To subtract b from a, *add the opposite* of b to a. $a - b = a + (-b)$	$12 - (-3) = 12 + (+3) = 15$ $-7.2 - (+1.6) = -7.2 + (-1.6) = -8.8$
Multiplication and division of real numbers, p. 13.	*Multiplication and Division:* When you multiply or divide two real numbers with like signs, the answer is a *positive* number. When you multiply or divide two real numbers whose signs are *different,* the answer is a *negative* number.	$(-6)(-3) = 18 \qquad -20 \div (-4) = 5$ $(-8)(5) = -40 \qquad 16 \div (-2) = -8$
Order of operations of real numbers, p. 21.	For involved numerical expressions, use this order of operations. 1. Combine numbers inside grouping symbols. 2. Raise numbers to their indicated powers and take any indicated roots. 3. Multiply and divide numbers from left to right. 4. Add and subtract numbers from left to right.	Simplify: $5 + 2(5 - 8)^3 - 12 \div (-4)$. $5 + 2(-3)^3 - 12 \div (-4)$ $\quad = 5 + 2(-27) - 12 \div (-4)$ $\quad = 5 + (-54) - (-3)$ $\quad = -49 + 3$ $\quad = -46$
Absolute value of a number, p. 10.	$\lvert x \rvert = \begin{cases} x, & \text{if } x \geq 0 \\ -x, & \text{if } x < 0 \end{cases}$	$\lvert 6 \rvert = 6 \qquad \lvert 0 \rvert = 0$ $\lvert -2 \rvert = 2 \quad \lvert -3.6 \rvert = 3.6$

Topic	Procedure	Examples
Rules of Exponents for multiplication and division, pp. 26 and 27.	If x, y are any nonzero real numbers and m, n are integers: Multiplication: $x^m x^n = x^{m+n}$ Division: $\dfrac{x^m}{x^n} = x^{m-n}$	$(2x^5)(3x^6) = 6x^{11}$ $\dfrac{15x^8}{5x^3} = 3x^5$
Negative exponents, pp. 25 and 30.	Transforming negative exponents to positive exponents: $x^{-n} = \dfrac{1}{x^n}$ $\qquad$ $\dfrac{x^{-n}}{y^{-m}} = \dfrac{y^m}{x^n}$	$x^{-6} = \dfrac{1}{x^6}$ $\quad$ $2^{-8} = \dfrac{1}{2^8}$ $\quad$ $\dfrac{x^{-4}}{y^{-5}} = \dfrac{y^5}{x^4}$
Zero exponent, p. 28.	$x^0 = 1$ when $x \neq 0$	$x^0 = 1$ $\quad$ $5^0 = 1$ $\quad$ $(3ab)^0 = 1$
Power rules, p. 29.	$(x^m)^n = x^{mn}$ $(xy)^n = x^n y^n$ $\left(\dfrac{x}{y}\right)^n = \dfrac{x^n}{y^n}$, if $y \neq 0$	$(7^3)^4 = 7^{12}$ $(3x^{-2})^4 = 3^4 x^{-8} = \dfrac{3^4}{x^8}$ $\left(\dfrac{2a^2}{b^3}\right)^{-4} = \dfrac{2^{-4}a^{-8}}{b^{-12}} = \dfrac{b^{12}}{2^4 a^8}$
Scientific notation, p. 31.	A positive number is written in scientific notation if it is in the form $a \times 10^n$, where $1 \leq a < 10$ and n is an integer.	$128 = 1.28 \times 10^2$ $2{,}568{,}000 = 2.568 \times 10^6$ $13{,}200{,}000{,}000 = 1.32 \times 10^{10}$ $0.16 = 1.6 \times 10^{-1}$ $0.00079 = 7.9 \times 10^{-4}$ $0.0000034 = 3.4 \times 10^{-6}$
Combining like terms, p. 37.	Combine terms that have identical letters and exponents.	$7x^2 - 3x + 4y + 2x^2 - 8x - 9y = 9x^2 - 11x - 5y$
Using the distributive property, p. 38.	Use the distributive law to remove parentheses. $a(b + c) = ab + ac$	$3(5x + 2) = 15x + 6$
Removing grouping symbols, p. 39.	1. Remove innermost grouping symbols first. 2. Then remove remaining innermost grouping symbols. 3. Continue until all grouping symbols are removed. 4. Combine like terms.	$5\{3x - 2[4 + 3(x - 1)]\} = 5\{3x - 2[4 + 3x - 3]\}$ $= 5\{3x - 8 - 6x + 6\}$ $= 15x - 40 - 30x + 30$ $= -15x - 10$
Evaluating variable expressions, p. 42.	1. Replace each letter by the numerical value given. 2. Follow the order of operations in evaluating the expression.	Evaluate $2x^3 + 3xy + 4y^2$ for $x = -3$, $y = 2$. $2(-3)^3 + 3(-3)(2) + 4(2)^2$ $= 2(-27) + 3(-3)(2) + 4(4)$ $= -54 - 18 + 16$ $= -56$
Using formulas, p. 43.	1. Replace each variable in the formula by the given values. 2. Evaluate the expression. 3. Label units carefully.	Find the area of a circle with radius of 4 feet. Use $A = \pi r^2$ and π as approximately 3.14. $A = (3.14)(4 \text{ feet})^2$ $= (3.14)(16 \text{ feet}^2)$ $= 50.24 \text{ feet}^2$ The area of the circle is approximately 50.24 square feet.

Chapter 1 Review Problems

Check the columns to which the number belongs.

	Natural Numbers	Whole Numbers	Integers	Rational Numbers	Irrational Numbers	Real Numbers
1. -5						
2. $\dfrac{7}{8}$						
3. 3						
4. $0.\overline{3}$						
5. $2.1652384\ldots$ (no discernible pattern)						

In Problems 6 and 7, name the properties of real numbers that justify each statement.

6. $(2 + 3) + (7 + x) = (2 + 3) + (x + 7)$

7. $5(2 \cdot x) = (5 \cdot 2) \cdot x$

8. Are all rational numbers also real numbers?

Compute, if possible.

9. $-8 - (-3)$

10. $-1.6 + (-5.2)$

11. $-4(-3)$

12. $-12 \div (+3)$

13. $\left(-\dfrac{3}{5}\right)\left(\dfrac{2}{3}\right)$

14. $\left(-\dfrac{5}{7}\right) \div \left(\dfrac{5}{-13}\right)$

15. $(4)(-3)(-10)$

16. $5 + 6 - 2 - 5$

17. $(-3.6)(-1.5)$

18. $0 \div (-6)$

19. $7 \div 0$

20. $-12 + (+12)$

21. $2 - 3[(-4) + 6] \div (-2)$

22. $\dfrac{5 - 8}{2 - 7 - (-2)}$

23. $4\sqrt{16} + 2^3 - 6$

24. $4 - 2 + 6\left(-\dfrac{1}{3}\right)$

25. $3 - |-4| + (-2)^3$

26. $\sqrt{(-3)^2} + (-2)^3$

27. $\sqrt{\dfrac{25}{36}} - 2\left(\dfrac{1}{12}\right)$

28. $2\sqrt{16} + 3(-4)(0)(2) - 2^2$

29. $(-0.4)^3$

Simplify. Do not leave negative exponents in your answer.

30. $(3xy^2)(-2x^0y)(4x^3y^3)$

31. $(2^4ab)(2^{-3}a^{-5}b^6)$

32. $\dfrac{5^{-3}x^{-3}y^6}{5^{-5}x^{-5}y^8}$

33. $\dfrac{27ab^3c}{81a^5bc^0}$

34. $\left(\dfrac{-3x^3y}{2x^4z^2}\right)^4$

35. $\dfrac{(-2a^{-3}b^{-4})^3}{(-3a^{-4}b^2c)^{-2}}$

36. $(2^{-1}a^2b^{-4})^3$

37. $\dfrac{3^{-2}x^5y^{-6}}{3x^{-4}y^{-5}}$

38. $\dfrac{(3^{-1}x^{-2}y)^{-2}}{(4^{-1}xy^{-2})^{-1}}$

39. $\dfrac{(2a^{-2}b)^{-3}}{(3a^{-3}b)^{-1}}$

40. $\left(\dfrac{a^5b^2}{3^{-1}a^{-5}b^{-4}}\right)^3$

41. $\left(\dfrac{x^3y^4}{5x^6y^8}\right)^3$

42. Write in scientific notation. 0.00721

43. Change to scientific notation and multiply. $(5,300,000)(2,000,000,000)$ Express your answer in scientific notation.

44. Collect like terms.
$2ab - 4a^2b - 6b^2 - 3ab + 2a^2b + 5b^3$

45. Multiply. $-5ab^2(a^3 + 2a^2b - 3b - 4)$

In Problems 46 and 47, remove grouping symbols and simplify.

46. $3a[2a - 3(a + 4)]$

47. $2x^2 - \{2 + x[3 - 2(x - 1)]\}$

48. Evaluate $5x^2 - 3xy - 2y^3$ when $x = 2$ and $y = -1$.

49. A right circular cylinder has a height of 2 meters and a radius of 3 meters. Find its volume. (Use $V = \pi r^2 h$.)

Developing Your Study Skills

EXAM TIME: HOW TO REVIEW

Reviewing adequately for an exam enables you to bring together the concepts you have learned over several sections. For your review, you will need to:

1. Reread your textbook. Make a list of any terms, rules, or formulas you need to know for the exam. Be sure you understand them all.

2. Reread your notes. Go over returned homework and quizzes. Redo the problems you missed.

3. Practice some of each type of problem covered in the chapter(s) you are to be tested on.

4. Use the end-of-chapter materials provided in your textbook. Read carefully through the Chapter Organizer. Do the Chapter Review Problems. Take the Chapter Test. When you are finished, check your answers. Redo any problems you missed.

5. Get help if any concepts give you difficulty.

For the set $\{-2, 12, \frac{9}{3}, \frac{25}{25}, 0, 2.585858, \ldots, \pi, \sqrt{4}, 2\sqrt{5}\}$,

1. List the real numbers that are not rational.

2. List the integers.

3. Name the property that justifies this statement: $(8 \cdot x)3 = 3(8 \cdot x)$

In Problems 4–21, do not leave negative exponents in your answer.

4. Evaluate.
$(7 - 5)^3 - 18 \div (-3) + 3\sqrt{10 + 6}$

5. Evaluate. $\dfrac{-4 + 2\sqrt{9} - (-2)^3}{|8 - 13|}$

6. Simplify. $(5x^{-3}y^{-5})(-3xy)(-2x^3y^0)$

7. Simplify. $\dfrac{12a^{-2}b^3c^{-1}}{15a^{-4}b^{-1}c^2}$

8. Simplify. $\left(\dfrac{2x^{-3}y^{-1}}{-8x^2y^{-4}}\right)^{-2}$

9. Simplify.
$8y - 5y^2 - 12y - 2y^2 + 7y$

10. Simplify.
$2a + 3b - 6a^2 + b - 8a - 3a^2$

11. Simplify. $3xy^2(4x - 3y + 2x^2)$

12. Write in scientific notation.
0.000002186

13. Write in decimal form. 2.158×10^9

1. _____

2. _____

3. _____

4. _____

5. _____

6. _____

7. _____

8. _____

9. _____

10. _____

11. _____

12. _____

13. _____

Perform the calculation indicated.

14. $(3.8 \times 10^{-5})(4 \times 10^{-2})$

15. $\dfrac{3.6 \times 10^8}{1.2 \times 10^2}$

16. Simplify. $2\{3x - 2[x - 3(x + 5)]\}$

17. Evaluate $2x^2 - 3x - 6$ when $x = -4$.

18. Evaluate $5x^2 + 3xy - y^2$ when $x = 3$ and $y = -3$.

19. Find the area of a trapezoid with an altitude of 12 meters and bases of 6 and 7 meters.

20. Find the area of a circle with a diameter of 12 meters.

21. Find the amount A to be repaid on a principal of $8000.00 borrowed at an interest rate r of 5% for a time t of 3 years. The formula is $A = p(1 + rt)$.

14. _____

15. _____

16. _____

17. _____

18. _____

19. _____

20. _____

21. _____

CHAPTER 2
Linear Equations and Inequalities

Have you noticed that newer cars have a greater fuel efficiency? Did you know that in 1970 the average U.S. automobile only achieved 13.5 miles per gallon of gas? Did you realize that in 1995 this figure had changed to 21.9 miles per gallon? Can you predict what may happen by the year 2015? Turn to the Putting Your Skills to Work on page 79 to investigate this area in more detail.

If you are familiar with the topics in this chapter, take this test now. Check your answers with those in the back of the book. If an answer was wrong or you couldn't do a problem, study the appropriate section of the chapter.

If you are not familiar with the topics in this chapter, don't take this test now. Instead, study the examples, work the practice problems, and then take the test.

This test will help you identify those concepts that you have mastered and those that need more study.

Section 2.1

Solve for x.

1. $\frac{1}{3}(2x - 1) = 4(x + 3)$

2. $\frac{x - 2}{4} = \frac{1}{2}x + 4$

3. $15x - 2 = -38$

4. $0.7x - 1 = 0.4$

Section 2.2

5. Solve for y. $5x - 8y = 15$

6. Solve for a. $5ab - 2b = 16ab - 3(8 + b)$

7. (a) Solve for r. $A = P + Prt$

   (b) Find r when $P = 100$, $t = 3$, and $A = 118$.

Section 2.3

Solve for x.

8. $|3x - 2| = 7$

9. $|8 - x| - 3 = 1$

10. $\left|\frac{2x + 3}{4}\right| = 2$

11. $|5x - 8| = |3x + 2|$

Section 2.4

Use an algebraic equation to find a solution for the problems.

12. The perimeter of a rectangle is 64 centimeters. Its length is 4 centimeters less than three times its width. Find the rectangle's dimensions.

13. Eastern Bank charges its customers a flat fee of $6.00 per month for a checking account plus 12¢ for each check. The bank charged Jose $9.12 for his checking account last month. How many checks did he use?

Section 2.5

Use an algebraic equation to find a solution for the problems.

14. A technician needs 100 grams of an alloy that is 80% pure copper (Cu). She has one alloy that is 77% pure Cu and another that is 92% pure Cu. How many grams of each alloy should she use to obtain 80% pure Cu?

15. A retired couple earned $4725 in simple interest last year from two investments. They deposited part of their $40,000 life savings in an account bearing 10% interest. They invested the rest of their money in a mutual fund earning 15% interest. How much did they deposit? How much did they invest?

1. _____

2. _____

3. _____

4. _____

5. _____

6. _____

7. (a) _____

(b) _____

8. _____

9. _____

10. _____

11. _____

12. _____

13. _____

14. _____

15. _____

Section 2.6

Solve for x. Graph your solution.

16. $7x + 12 < 9x$

17. $3(x - 5) - 5x \geq 2x + 9$

18. $\dfrac{2}{3}x - \dfrac{5}{6}x - 3 \leq \dfrac{1}{2}x - 5$

Section 2.7

Graph each region.

19. $-2 \leq x < 5$

20. $x < -3$ or $x > 0$

Solve for x. Graph your solution.

21. $-2 \leq x + 1 \leq 4$

22. $2x + 3 < -5$ or $x - 2 > 1$

Section 2.8

Solve for x.

23. $|3x + 2| < 8$

24. $\left| \dfrac{2}{3}x - \dfrac{1}{2} \right| \leq 3$

25. $|2 - 5x - 4| > 13$

16.
17.
18.
19.
20.
21.
22.
23.
24.
25.

2.1 First-degree Equations with One Unknown

MathPro

Video 3

SSM

After studying this section, you will be able to:

1 *Solve first-degree equations with one unknown.*

1 Solving First-degree Equations with One Unknown

An **equation** is a mathematical statement that two quantities are equal. A **first-degree equation with one unknown** is an equation in which only one kind of variable appears and that variable has an exponent of 1. The variable itself may appear more than once. The equation $-6x + 7x + 8 = 2x - 4$ is a first-degree equation with one unknown, because there is one variable, x, and that variable has an exponent of 1. Other examples are $5(6y + 1) = 2y$ and $5z + 3 = 10z$.

To solve a first-degree equation with one unknown, we need to find the value of the variable that makes the equation a true mathematical statement. This value is the **solution** or **root** of the equation.

EXAMPLE 1

(a) Is $x = 4$ a root of the equation $5x - 6 = 14$?

(b) Is $a = \frac{1}{3}$ a solution of the equation $2a + 5 = a + 6$?

(a) We replace x by the value 4 in the equation $5x - 6 = 14$.

$$5(4) - 6 \overset{?}{=} 14$$

$$20 - 6 \overset{?}{=} 14$$

$$14 = 14$$

Since we obtain a true statement, $x = 4$ is a root of $5x - 6 = 14$.

(b) We replace a by $\frac{1}{3}$ in the equation $2a + 5 = a + 6$.

$$2\left(\frac{1}{3}\right) + 5 \overset{?}{=} \frac{1}{3} + 6$$

$$\frac{2}{3} + 5 \overset{?}{=} \frac{1}{3} + 6$$

$$\frac{17}{3} \neq \frac{19}{3}$$

This last statement is not true. Thus, $a = \frac{1}{3}$ is not a solution of $2a + 5 = a + 6$.

Practice Problem 1

(a) Is $x = -5$ the root of the equation $2x + 3 = 3x + 8$?

(b) Is $a = -\frac{3}{2}$ a solution of the equation $6a - 5 = -4a + 10$? ∎

This procedure of verifying that a value is indeed a root of an equation is very valuable. We will use it throughout Chapter 2 to check our answers when solving equations.

Equations that have the same solution are said to be **equivalent.** The equations

$$7x - 2 = 12, \qquad 7x = 14, \qquad x = 2$$

are equivalent because the solution of each equation is $x = 2$.

To solve an equation, we perform algebraic operations on it to obtain a simpler, equivalent equation of the form variable = constant or constant = variable.

Properties of Equivalent Equations

1. If $a = b$, then $a + c = b + c$ and $a - c = b - c$.

   If the same number is added to or subtracted from both sides of an equation, the result is an equivalent equation.

2. If $a = b$ and $c \neq 0$, then $ac = bc$ and $\frac{a}{c} = \frac{b}{c}$.

   If both sides of an equation are multiplied or divided by the same nonzero number, the result is an equivalent equation.

To solve a first-degree equation, we isolate the variable on one side of the equation and the constants on another, using the properties of equivalent equations.

EXAMPLE 2   Solve for x. $x - 8.2 = 5.0$

$$x - 8.2 + 8.2 = 5.0 + 8.2 \qquad \textit{Add 8.2 to each side.}$$

$$x = 13.2$$

Thus the solution is $x = 13.2$.

We must check to see if our answer is a valid root of the equation. Checking is especially important when we solve higher-degree equations. We check the validity of our answer in the same way that we determined the validity of roots on page 60; we substitute the answer into the original equation.

$$\textit{Check:} \qquad x - 8.2 = 5.0$$
$$13.2 - 8.2 \overset{?}{=} 5.0$$
$$5.0 = 5.0 \quad \checkmark$$

The statement is valid, so our answer is correct. $x = 13.2$ is a root of the equation.

Practice Problem 2   Solve for y. $y + 5.2 = -2.8$ ■

EXAMPLE 3   Solve for y. $\frac{1}{3}y = -6$

$$3\left(\frac{1}{3}\right)y = 3(-6) \qquad \textit{Multiply each side by 3 to eliminate the fraction.}$$

$$y = -18$$

The solution is $y = -18$. Check this.

Practice Problem 3   Solve for w. $\frac{1}{5}w = -6$ ■

EXAMPLE 4   Solve. $5x + 2 = 17$

$$5x + 2 - 2 = 17 - 2 \qquad \text{Subtract 2 from (or add } -2 \text{ to) both sides.}$$

$$5x = 15$$

$$\frac{5x}{5} = \frac{15}{5} \qquad \text{Divide each side by 5 to isolate the variable } x.$$

$$x = 3$$

Check: We now want to verify that $x = 3$ is the root for this equation. We replace x by 3 in the original equation in order to check our answer.

$$5(3) + 2 \overset{?}{=} 17$$

$$15 + 2 \overset{?}{=} 17$$

$$17 = 17 \quad \checkmark$$

Thus $x = 3$ is the solution.

Practice Problem 4   Solve. $-7x - 2 = 26$ Check your answer. ■

EXAMPLE 5   Solve. $6x - 2 - 4x = 8x + 3$

$$2x - 2 = 8x + 3 \qquad \text{Collect like terms.}$$

$$2x - 8x - 2 = 8x - 8x + 3 \qquad \text{Subtract } 8x \text{ from (or add } -8x \text{ to) each side.}$$

$$-6x - 2 = 3$$

$$-6x - 2 + 2 = 3 + 2 \qquad \text{Add 2 to each side.}$$

$$-6x = 5$$

$$\frac{-6x}{-6} = \frac{5}{-6} \qquad \text{Divide each side by } -6.$$

$$x = -\frac{5}{6}$$

When checking fractional values like $x = -\frac{5}{6}$, extra care will be required in order to perform the fractional operation correctly.

$$Check: \quad 6\left(-\frac{5}{6}\right) - 2 - 4\left(-\frac{5}{6}\right) \overset{?}{=} 8\left(-\frac{5}{6}\right) + 3 \qquad \begin{array}{l} Replace\ x\ by\ -\frac{5}{6} \\ in\ the\ \textbf{original}\ equation. \end{array}$$

$$-5 - 2 + \frac{10}{3} \overset{?}{=} \frac{-20}{3} + 3$$

$$\frac{-21}{3} + \frac{10}{3} \overset{?}{=} \frac{-20}{3} + \frac{9}{3}$$

$$-\frac{11}{3} = -\frac{11}{3} \quad \checkmark$$

Thus $x = -\frac{5}{6}$ is the solution.

Practice Problem 5   Solve. $8w - 3 = 2w - 7w + 4$ ■

When the equation contains grouping symbols and fractions, use the following procedure.

Procedure for Solving First-degree Equations with One Unknown

1. Remove grouping symbols in the proper order.
2. If fractions exist, multiply both sides of the equation by the least common denominator (LCD) of all the fractions.
3. Collect like terms if possible.
4. Add or subtract a variable term on both sides of the equation to obtain all variable terms on one side of the equation.
5. Add or subtract a numerical value on both sides of the equation to obtain all numerical values on the other side of the equation.
6. Divide each side of the equation by the coefficient of the variable.
7. Simplify the solution (if possible).
8. Check your solution.

EXAMPLE 6 Solve and check. $3(3x + 2) - 4x = -2(x - 3)$

$$9x + 6 - 4x = -2x + 6 \qquad \textit{Remove the parentheses.}$$

$$5x + 6 = -2x + 6 \qquad \textit{Collect like terms.}$$

$$5x + 2x + 6 = -2x + 2x + 6 \qquad \textit{Add 2x to each side.}$$

$$7x + 6 = 6$$

$$7x + 6 - 6 = 6 - 6 \qquad \textit{Subtract 6 from each side.}$$

$$7x = 0$$

$$\frac{7x}{7} = \frac{0}{7} \qquad \textit{Divide each side by 7.}$$

$$x = 0$$

$$\textit{Check:} \quad 3[3(0) + 2] - 4(0) \stackrel{?}{=} -2(0 - 3)$$

$$3(0 + 2) - 0 \stackrel{?}{=} -2(-3)$$

$$6 - 0 \stackrel{?}{=} 6$$

$$6 = 6 \quad \checkmark$$

Thus $x = 0$ is the solution.

Practice Problem 6 Solve and check. $a - 4(2a - 7) = 3(a + 6)$ ∎

It is not necessary to memorize this 8-step procedure. However, you should refer to it often as you work the homework exercises in Section 2.1 until the sequence of steps is "second nature" to you.

EXAMPLE 7 Solve. $\dfrac{x}{5} + \dfrac{1}{2} = \dfrac{4}{5} + \dfrac{x}{2}$

$$10\left(\dfrac{x}{5} + \dfrac{1}{2}\right) = 10\left(\dfrac{4}{5} + \dfrac{x}{2}\right) \qquad \textit{Multiply each term by the LCD} = 10.$$

$$10\left(\dfrac{x}{5}\right) + 10\left(\dfrac{1}{2}\right) = 10\left(\dfrac{4}{5}\right) + 10\left(\dfrac{x}{2}\right) \qquad \textit{Use the distributive property.}$$

$$2x + 5 = 8 + 5x \qquad \textit{Simplify.}$$

$$2x - 2x + 5 = 8 + 5x - 2x \qquad \textit{Subtract } 2x \textit{ from each side.}$$

$$5 = 8 + 3x$$

$$5 - 8 = 8 - 8 + 3x \qquad \textit{Subtract 8 from each side.}$$

$$-3 = 3x$$

$$-\dfrac{3}{3} = \dfrac{3x}{3} \qquad \textit{Divide each side by 3.}$$

$$-1 = x$$

Check: See if you can verify this solution.

Practice Problem 7 Solve and check. $\dfrac{y}{3} + \dfrac{1}{2} = 5 + \dfrac{y-9}{4}$

$\left(\textit{Hint: You can write } \dfrac{y-9}{4} \textit{ as } \dfrac{y}{4} - \dfrac{9}{4}.\right)$ ∎

An equation that contains many decimals can be multiplied by the appropriate power of 10 to clear the equation of decimals.

EXAMPLE 8 Solve and check. $0.9 + 0.2(x + 4) = -3(0.1x - 0.4)$

$$0.9 + 0.2x + 0.8 = -0.3x + 1.2 \qquad \textit{Remove the parentheses.}$$

$$10(0.9) + 10(0.2x) + 10(0.8) = 10(-0.3x) + 10(1.2) \qquad \begin{array}{l}\textit{Multiply each side}\\\textit{of the equation}\\\textit{by 10 and use the}\\\textit{distributive property.}\end{array}$$

$$9 + 2x + 8 = -3x + 12 \qquad \textit{Simplify.}$$

$$2x + 17 = -3x + 12 \qquad \textit{Collect like terms.}$$

$$2x + 3x + 17 = -3x + 3x + 12 \qquad \textit{Add } 3x \textit{ to each side.}$$

$$5x + 17 = 12$$

$$5x + 17 - 17 = 12 - 17 \qquad \textit{Add } -17 \textit{ to each side.}$$

$$5x = -5$$

$$x = -1 \qquad \textit{Divide each side by 5.}$$

Check: $0.9 + 0.2(-1 + 4) \overset{?}{=} -3[(0.1)(-1) - 0.4]$

$$0.9 + 0.2(3) \overset{?}{=} -3[-0.1 - 0.4]$$

$$0.9 + 0.6 \overset{?}{=} -3(-0.5)$$

$$1.5 = 1.5 \quad ✓$$

Thus $x = -1$ is the solution.

Practice Problem 8 Solve and check. $4(0.01x + 0.09) - 0.07(x - 8) = 0.83$ ∎

2.1 Exercises

Verbal and Writing Skills

1. Is $x = -20$ a root of the equation $3x - 15 = 45$? Why or why not?

2. Is $y = \frac{3}{5}$ a solution to the equation $5y + 9 = 12$? Why or why not?

3. What is the first step in solving the equation $\frac{x}{3} + \frac{3}{4} = 2 - \frac{x}{2}$? Why?

4. What is the first step in solving the equation $0.7 + 0.03x = 4$? Why?

5. Would you clear the equation $x + 3.6 = 8$ of decimals by multiplying each term by 10? Why or why not?

6. Would you clear the equation $x - \frac{1}{4} = 3$ of fractions by multiplying each term by 4? Why or why not?

Solve Exercises 7 through 48. Check your solutions.

7. $-16 + x = 43$

8. $17 + x = -24$

9. $-5x = 35$

10. $-12x = -48$

11. $7x - 8 = 20$

12. $5x + 3 = 43$

13. $5y + 8 = 4y - 2$

14. $8y - 3 = 2y + 3$

15. $6x - 12 = -3x - 21$

16. $-10 + 3x = 2 - 3x$

17. $5 - 4x + 15x = 10x$

18. $-3x + 6 - 8x = -5$

19. $3a - 5 - 2a = 2a - 3$

20. $5a - 2 + 4a = 2a + 12$

21. $4(y - 1) = -2(3 + y)$

22. $5(2 - y) = 3(y - 2)$

23. $(4x - 3) - (2x - 7) = -1(x - 6)$

24. $8 - (4x - 5) = x - 7$

25. $5(y + 1) + 2 = y - 3(2y + 1)$

26. $5 - 2(3 - y) = 2(2y + 5) + 1$

27. $\frac{y}{4} + \frac{1}{2} = \frac{2}{3}$

28. $\frac{y}{5} + \frac{1}{3} = \frac{7}{15}$

29. $\frac{y}{3} + 2 = \frac{4}{5}$

30. $\frac{y}{2} + 4 = \frac{1}{6}$

31. $\dfrac{5x}{6} - \dfrac{3}{5} = \dfrac{2x}{3}$

32. $\dfrac{1}{2} + \dfrac{3x}{7} = \dfrac{x}{4}$

33. $\dfrac{x}{3} + \dfrac{3}{4} = \dfrac{11}{3} - \dfrac{x}{4}$

34. $\dfrac{x}{5} - \dfrac{4}{6} = \dfrac{x}{6} - \dfrac{2}{3}$

35. $\dfrac{1}{2}(x + 3) - 2 = 1$

36. $5 - \dfrac{2}{3}(x + 2) = 3$

37. $\dfrac{7x}{3} + 5 = 3x + 5$

38. $\dfrac{5x}{6} - 8 = 2x - 8$

39. $0.3x + 0.4 = 0.5x - 0.8$

40. $0.7x - 0.2 = 0.5x + 0.8$

41. $0.2(x - 4) = 3$

42. $0.6(2x + 1) = 1$

43. $1.5x + 4 = 1.2x - 2$

44. $0.8x + 3 = 0.6x + 2$

45. $0.21 - 0.06x = 0.22 - 0.08x$

46. $0.04x - 0.03 = 0.05 + 0.12x$

47. $0.08 = 0.4x + 2$

48. $0.17 = 0.5x + 3$

Mixed Practice

Solve each of the following.

49. $\dfrac{2}{3}(x + 6) = 1 + \dfrac{4x - 7}{3}$

50. $2y - 5 - \dfrac{4}{3}(2y + 6) = -\dfrac{5}{3}$

51. $\dfrac{1}{5} - \dfrac{x}{3} = \dfrac{x - 2}{5}$

52. $\dfrac{1}{6} - \dfrac{x}{2} = \dfrac{x - 5}{3}$

53. $\dfrac{4y-1}{10} = \dfrac{5y+2}{4} - 4$

54. $\dfrac{y+5}{7} = \dfrac{5}{14} - \dfrac{y-3}{4}$

55. $2(5x-1) + x(2x-4) - (2x^2 + 1) = 0$

56. $3(4x+5) - x(2x+1) + 2x(x-3) = 0$

57. $0.3 + 0.4(2-x) = 6(-0.2 + 0.1x) + 0.3$

58. $3(0.3 + 0.1x) + 0.1 = 0.5(x+2)$

59. $\dfrac{1}{2}(x+2) = \dfrac{2}{3}(x-1) - \dfrac{3}{4}$

60. $x - \dfrac{5}{3}(x-2) = \dfrac{1}{9}(x+2)$

Solve for x. Round your answer to four decimal places.

 61. $9.8615x - 2.3218 = 18.0716x + 4.9862$

62. $2x + \dfrac{36,942}{79,603} = 5x - \dfrac{88,032}{91,264}$

To Think About

63. When he died in 1973, the artist Picasso left 44,740 pieces of art in the south of France. They consisted of the following: 1876 paintings, 2880 ceramics, 11,589 drawings and sketches, and 27,042 etchings, engravings, and lithographs. The rest were sculptures.
   (a) What percentage of the works found were sculptures?
   (b) How many more etchings, engravings and lithographs were there than sculptures?

64. 10,545 people attended a symphony concert. Of the people attending, 6,579 were men and 3966 were women. Of the men, 5491 wore black shoes and the rest wore sneakers or other shoes. Of the women, 1460 wore pants and the rest wore skirts or dresses.
   (a) What percentage of the audience wore skirts or dresses?
   (b) What would be the ratio of men who wore black shoes to those who wore other footwear?

Cumulative Review Problems

Simplify. Leave positive exponents in your answer.

65. $5 - (4-2)^2 + 3(-2)$

66. $\left(\dfrac{3xy^2}{2x^2y}\right)^3$

67. $(-2)^4 - 12 - 6(-2)$

68. $(2x^{-2}y^{-3})^2(4xy^{-2})^{-2}$

2.2 Literal Equations and Formulas

MathPro   Video 3   SSM

After studying this section, you will be able to:

1 *Solve literal equations for the desired unknown.*

1 *Solving Literal Equations for the Desired Unknown*

A first-degree **literal equation** is an equation that contains variables other than the variable that we are solving for. When you solve for an unknown in a literal equation, the final expression will contain these other letters. We use this procedure to deal with formulas in applied problems.

EXAMPLE 1 Solve for x. $5x + 3y = 2$

$$5x = 2 - 3y \qquad \textit{Subtract 3y from each side.}$$

$$\frac{5x}{5} = \frac{2 - 3y}{5} \qquad \textit{Divide each side by 5.}$$

$$x = \frac{2 - 3y}{5} \qquad \textit{The solution is a fractional expression.}$$

Practice Problem 1 Solve for W. $P = 2L + 2W$ ■

Where possible, collect like terms as you solve the equation.

EXAMPLE 2 Solve for y. $3ay + 8 = 5ay - 7$

$$3ay - 5ay + 8 = -7 \qquad \textit{Subtract 5ay from each side.}$$

$$-2ay = -7 - 8 \qquad \textit{Simplify and subtract 8 from each side.}$$

$$-2ay = -15 \qquad \textit{Simplify.}$$

$$\frac{-2ay}{-2a} = \frac{-15}{-2a} \qquad \textit{Divide each side by the coefficient of y, which is }-2a.$$

$$y = \frac{15}{2a} \qquad \textit{Simplify. (Recall that a negative divided by a negative is positive.)}$$

Practice Problem 2 Solve for w. $8 + 12wx = 18 - 7wx$ ■

For longer problems, follow this procedure.

Procedure for Solving First-degree Literal Equations

1. Remove grouping symbols in the proper order.
2. If fractions exist, multiply all terms on both sides by the LCD.
3. Collect like terms if possible.
4. Add or subtract a term with the desired unknown on both sides of the equation to obtain all terms with the desired unknown on one side of the equation.
5. Add or subtract appropriate terms on both sides of the equation to obtain all other terms on the other side of the equation.
6. Divide each side of the equation by the coefficient of the desired unknown.
7. Simplify the solution (if possible).

Some equations appear more difficult to solve because they contain fractions and parentheses. Immediately remove the parentheses. Then multiply each term by the LCD. The equation will now appear less threatening.

EXAMPLE 3 Solve for b. $A = \dfrac{2}{3}(a + b + 3)$

$$A = \frac{2}{3}a + \frac{2}{3}b + 2 \qquad \text{\textit{Remove parentheses.}}$$

$$3A = 3\left(\frac{2}{3}a\right) + 3\left(\frac{2}{3}b\right) + 3(2) \qquad \text{\textit{Multiply all terms by the LCD 3.}}$$

$$3A = 2a + 2b + 6 \qquad \text{\textit{Simplify.}}$$

$$3A - 2a - 6 = 2b \qquad \begin{array}{l}\text{\textit{Subtract 2a from each side.}}\\\text{\textit{Subtract 6 from each side.}}\end{array}$$

$$\frac{3A - 2a - 6}{2} = \frac{2b}{2} \qquad \begin{array}{l}\text{\textit{Divide each side of the equation}}\\\text{\textit{by the coefficient of b.}}\end{array}$$

$$\frac{3A - 2a - 6}{2} = b \qquad \text{\textit{Simplify.}}$$

Practice Problem 3 Solve for a. $3(2ax - y) = \dfrac{1}{2}(ax + 2y)$ ∎

Be sure to collect like terms after removing parentheses. This will simplify the equation and make it much easier to solve.

EXAMPLE 4 Solve for x. $5(2ax + 3y) - 4ax = 2(ax - 5)$

$$10ax + 15y - 4ax = 2ax - 10 \qquad \text{\textit{Remove parentheses.}}$$

$$6ax + 15y = 2ax - 10 \qquad \text{\textit{Collect like terms.}}$$

$$6ax - 2ax + 15y = -10 \qquad \begin{array}{l}\text{\textit{Subtract 2ax from each side to obtain}}\\\text{\textit{terms containing x on one side.}}\end{array}$$

$$4ax = -10 - 15y \qquad \text{\textit{Simplify and subtract 15y from each side.}}$$

$$\frac{4ax}{4a} = \frac{-10 - 15y}{4a} \qquad \text{\textit{Divide each side by the coefficient of x.}}$$

$$x = \frac{-10 - 15y}{4a}$$

Practice Problem 4 Solve for b. $-2(ab - 3x) + 2(8 - ab) = 5x + 4ab$ ∎

EXAMPLE 5

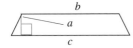

(a) In the formula for the area of a trapezoid $A = \frac{1}{2}a(b + c)$ solve for c.

(b) Find c when $A = 20$ square inches, $a = 3$ inches, and $b = 4$ inches.

(a)

$A = \frac{1}{2}ab + \frac{1}{2}ac$	*Remove the parentheses.*
$2A = 2\left(\frac{1}{2}ab\right) + 2\left(\frac{1}{2}ac\right)$	*Multiply each term by 2.*
$2A = ab + ac$	*Simplify.*
$2A - ab = ac$	*Subtract ab from each side to isolate the ac term.*
$\frac{2A - ab}{a} = c$	*Divide each side by a.*

(b) We use the equation we derived above to find c for the given values.

$$c = \frac{2A - ab}{a}$$

$$= \frac{2(20) - (3)(4)}{3} \qquad \textit{Substitute the given values of A, a, b, to find c.}$$

$$= \frac{40 - 12}{3} = \frac{28}{3} \qquad \textit{Simplify.}$$

Thus side $c = \frac{28}{3}$ inches or $9\frac{1}{3}$ inches.

Practice Problem 5

(a) Solve for h. $A = 2\pi rh + 2\pi r^2$

(b) Find h when $A = 100$, $\pi \approx 3.14$, $r = 2.0$. Round your answer to the nearest hundredth. ∎

Developing Your Study Skills

TAKING NOTES IN CLASS

An important part of mathematics studying is taking notes. To take meaningful notes, you must be an active listener. Keep your mind on what the instructor is saying, and be ready with questions whenever you do not understand something.

If you have previewed the lesson material, you will be prepared to take good notes. The important concepts will seem somewhat familiar. You will have a better idea of what needs to be written down. If you frantically try to write all that the instructor says or copy all the examples done in class, you may find your notes to be nearly worthless when you are home alone. You may find that you are unable to make sense of which you have written.

Write down *important* ideas and examples as the instructor lectures, making sure that you are listening and following the logic. Include any helpful hints or suggestions that your instructor gives you or refers to in your text. You will be amazed at how easily these are forgotten if they are not written down.

Successful note taking requires active listening and processing. Stay alert in class. You will realize the advantages of taking your own notes over copying those of someone else.

2.2 Exercises

Solve for x.

1. $3y - 5x = -8$

2. $-7y + 8x = -13$

3. $5x - 2(x - y) = 3y$

4. $2a - 3(x + a) = 2x$

5. $5abx - 2y - 3abx = 6y$

6. $8d + 8cdx - 3d = 5cdx$

7. $4x + 3y = 8 - x$

8. $7x - 2y + 4 = 8$

9. $y = \dfrac{2}{3}x - 4$

10. $y = -\dfrac{1}{3}x + 2$

11. $2x + a - 3b = 4x - 2a$

12. $3x - 4(x - 2b) = x + 4a$

Solve for the letter specified.

13. $d = rt$; for t

14. $I = prt$; for p

15. $A = \dfrac{h}{2}(B + b)$; for B

16. $C = \dfrac{5}{9}(F - 32)$; for F

17. $\dfrac{2}{3}(x + y) = 2(x - y)$; for y

18. $\dfrac{1}{5}(a + 2b) = 3(a + 2b)$; for b

19. $0.2(a - 3x) = 0.5a - 1.2x$; for x

20. $-6dex + 3y = 5(dex - 3y)$; for d

21. $\dfrac{2}{3} + y = \dfrac{2}{5}b + 3y$; for y

22. $\dfrac{1}{2}A + 3B = \dfrac{1}{3}B + 6$; for B

Solve for x. Round your answer to four decimal places.

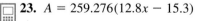 **23.** $A = 259.276(12.8x - 15.3)$

24. $16.932x - 19.832 = 15.428 + 19.3(56x - 12)$

Follow the directions given.

25. (a) Solve for a. $A = \dfrac{1}{2}ab$

(b) Evaluate when $A = 20$ and $b = \dfrac{5}{2}$.

26. (a) Solve for C. $F = \dfrac{9}{5}C + 32$

(b) Evaluate when $F = 23°$.

27. (a) Solve for n. $A = a + d(n - 1)$

(b) Evaluate when $A = 28$, $a = 3$, and $d = 15$.

28. (a) Solve for t. $A = p + prt$

(b) Evaluate when $A = 3400$, $p = 1700$, and $r = 0.06$.

29. (a) Solve for a. $D = Vt + \dfrac{1}{2}at^2$

(b) Evaluate when $D = 46$, $V = 20$, and $t = 2$.

30. (a) Solve for S. $A = \dfrac{\pi r^2 S}{360}$

(b) Evaluate when $A = 0.314$, $r = 2$, and $\pi \approx 3.14$.

Applications

31. The mariner's formula can be written in the form $\dfrac{m}{1.15} = k$, where m is the speed of a ship in miles per hour and k is the speed of the ship in knots (nautical miles per hour).

(a) Solve the formula for m.

(b) Use this result to find the number of miles per hour a ship is traveling if its speed is 29 knots.

32. Some doctors use the formula $NI = 1.08T$ to relate the variables N (the number of patient appointments the doctor schedules in one day), I (the interval of time between each patient appointment), and T (the total number of minutes the doctor can use to see patients in one day).

(a) Solve the formula for N.

(b) Use this result to find the number of patient appointments N a doctor should make if the doctor has 6 hours available for patients and wants to allow 15 minutes between patient appointments. (Round your answer to the nearest whole number.)

In Exercises 33 and 34 the variable C represents the consumption of products in the United States in billions of dollars and D represents disposable income in the United States in billions of dollars.

 33. If economists approximate the economy of the country by the equation $C = 0.6547D + 5.8263$:
(a) Solve the equation for D.

 34. If economists approximate the economy of the country by the equation $C = 0.7649D + 6.1275$:
(a) Solve the equation for D.

(b) Use this result to determine the disposable income D if the consumption C is $9.56 billion. Round your answer to the nearest tenth of a billion.

(b) Use this result to determine the disposable income D if the consumption C is $12.48 billion. Round your answer to the nearest tenth of a billion.

Cumulative Review Problems

Write with positive exponents in simplest form.

35. $(2x^{-3}y)^{-2}$

36. $\left(\dfrac{5x^2y^{-3}}{x^{-4}y^2}\right)^{-3}$

37. Sharon and James want to begin an education fund for their two daughters. They invest $5000 in a certificate of deposit for one year, with an annual return of 5%. $4000 is invested in a more risky venture which has a return of 9%. How much money will they have at the end of one year? (Assume that their risky investment does well.)

38. Drew wants to go to college in Maryland. He and his parents take a long weekend and drive from Kansas City, Missouri. His odometer read 45,711.3 when he started the trip and 46,622.1 when he arrived back home. He started and ended his trip on a full tank of gas. He made gas purchases of 9.9 gallons, 11.7 gallons, 10.6 gallons, 5.8 gallons, and 8 gallons for the car during the trip. How many miles per gallon did the car get on the trip?

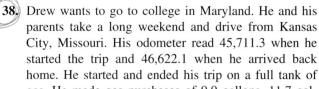

Developing Your Study Skills

PROBLEMS WITH ACCURACY

Strive for accuracy. Mistakes are often made as a result of human error rather than by lack of understanding. Such mistakes are frustrating. A simple arithmetic or sign error can lead to an incorrect answer.

These five steps will help you to cut down on errors.

1. Work carefully, and take your time. Do not rush through a problem just to get it done.

2. Concentrate on one problem at a time. Sometimes problems become mechanical, and your mind begins to wander. You become careless and make a mistake.

3. Check your problem. Be sure that you copied it correctly from the book.

4. Check your computations from step to step. Check the solution in the problem. Does it work? Does it make sense?

5. Keep practicing new skills. Remember the old saying "Practice makes perfect." An increase in practice results in an increase in accuracy. Many errors are due simply to lack of practice.

There is no magic formula for eliminating all errors, but these five steps will be a tremendous help in reducing them.

2.3 Absolute Value Equations

 MathPro Video 4 SSM

After studying this section, you will be able to:

1 *Find the solution to absolute value equations of the form* $|ax + b| = c$.
2 *Find the solution to absolute value equations of the form* $|ax + b| + c = d$.
3 *Find the solution to absolute value equations of the form* $|ax + b| = |cx + d|$.

1 *Solving Absolute Value Equations of the Form* $|ax + b| = c$

From Section 1.2, you know that the absolute value of a number x is the distance between 0 and x on the number line. Let's look at a simple equation $|x| = 4$ and draw a picture.

4 units   4 units
$$-5 \;-4 \;-3 \;-2 \;-1 \quad 0 \quad 1 \quad 2 \quad 3 \quad 4 \quad 5$$
$$x = -4 \qquad\qquad\qquad\qquad x = 4$$

Thus the equation $|x| = 4$ has two solutions $x = 4$ and $x = -4$. Let's look at another example.

$$\text{If} \quad |x| = \frac{2}{3}$$

$$\text{then} \quad x = \frac{2}{3} \quad \text{or} \quad x = -\frac{2}{3}$$

$$\text{because} \quad \left|\frac{2}{3}\right| = \frac{2}{3} \quad \text{and} \quad \left|-\frac{2}{3}\right| = \frac{2}{3}$$

You will be able to solve these relatively simple equations if you recall the definition of absolute value.

$$|x| = \begin{cases} x, & \text{if } x \geq 0 \\ -x, & \text{if } x < 0 \end{cases}$$

Now let's take a look at a more complicated equation: $|ax + b| = c$.

> To solve an equation of the form $|ax + b| = c$, where $a \neq 0$ and c is a positive number, we solve the two equations
>
> $$ax + b = c \quad \text{or} \quad ax + b = -c$$

EXAMPLE 1 Solve $|2x + 5| = 11$, and check your solutions.

By using the procedure established above, we have

$$2x + 5 = 11 \quad \text{or} \quad 2x + 5 = -11$$
$$2x = 6 \qquad\qquad\qquad 2x = -16$$
$$x = 3 \qquad\qquad\qquad x = -8$$

The two solutions are $x = 3$ or $x = -8$.

Check: if $x = 3$ 			if $x = -8$

$$|2x + 5| = 11 \qquad\qquad |2x + 5| = 11$$
$$|2(3) + 5| \overset{?}{=} 11 \qquad\qquad |2(-8) + 5| \overset{?}{=} 11$$
$$|6 + 5| \overset{?}{=} 11 \qquad\qquad |-16 + 5| \overset{?}{=} 11$$
$$|11| \overset{?}{=} 11 \qquad\qquad |-11| \overset{?}{=} 11$$
$$11 = 11 \ \checkmark \qquad\qquad 11 = 11 \ \checkmark$$

Practice Problem 1   Solve $|3x - 4| = 23$ and check your solutions. ■

EXAMPLE 2 Solve $\left|\dfrac{1}{2}x - 1\right| = 5$, and check your solutions.

By using the procedure, we have the two equations

$$\frac{1}{2}x - 1 = 5 \quad \text{or} \quad \frac{1}{2}x - 1 = -5$$

If we multiply each term of each equation by 2, we obtain

$$x - 2 = 10 \quad \text{or} \quad x - 2 = -10$$
$$x = 12 \qquad\qquad x = -8$$

Check: if $x = 12$ $\qquad\qquad$ if $x = -8$

$$\left|\frac{1}{2}(12) - 1\right| \overset{?}{=} 5 \qquad\qquad \left|\frac{1}{2}(-8) - 1\right| \overset{?}{=} 5$$

$$|6 - 1| \overset{?}{=} 5 \qquad\qquad |-4 - 1| \overset{?}{=} 5$$

$$|5| \overset{?}{=} 5 \qquad\qquad |-5| \overset{?}{=} 5$$

$$5 = 5 \checkmark \qquad\qquad 5 = 5 \checkmark$$

Practice Problem 2 Solve and check your solutions.

$$\left|\frac{2}{3}x + 4\right| = 2 \ \blacksquare$$

② Solving Absolute Value Equations of the Form $|ax + b| + c = d$

Notice that in the previous examples the absolute value expression is on one side of the equation and a positive real number is on the other side of the equation. What happens when we encounter an equation of the form $|ax + b| + c = d$?

EXAMPLE 3 Solve. $|3x - 1| + 2 = 5$ Check your solution.

First we will change the equation so that the absolute value expression alone is on one side of the equation.

$$|3x - 1| + 2 - 2 = 5 - 2$$
$$|3x - 1| = 3$$

Now we solve $|3x - 1| = 3$.

$$3x - 1 = 3 \quad \text{or} \quad 3x - 1 = -3$$
$$3x = 4 \qquad\qquad 3x = -2$$
$$x = \frac{4}{3} \qquad\qquad x = -\frac{2}{3}$$

Check: if $x = \dfrac{4}{3}$ $\qquad\qquad$ if $x = -\dfrac{2}{3}$

$$\left|3\left(\frac{4}{3}\right) - 1\right| + 2 \overset{?}{=} 5 \qquad\qquad \left|3\left(-\frac{2}{3}\right) - 1\right| + 2 \overset{?}{=} 5$$

$$|4 - 1| + 2 \overset{?}{=} 5 \qquad\qquad |-2 - 1| + 2 \overset{?}{=} 5$$

$$|3| + 2 \overset{?}{=} 5 \qquad\qquad |-3| + 2 \overset{?}{=} 5$$

$$3 + 2 \overset{?}{=} 5 \qquad\qquad 3 + 2 \overset{?}{=} 5$$

$$5 = 5 \checkmark \qquad\qquad 5 = 5 \checkmark$$

Practice Problem 3 Solve $|2x + 1| + 3 = 8$, and check your solutions. $\blacksquare$

3 Solving Absolute Value Equations of the Form $|ax + b| = |cx + d|$

Let us now consider the possibilities for a and b if $|a| = |b|$.

Suppose $a = 5$; then $b = 5$ or -5.
If $a = -5$, then $b = 5$ or -5.

> To generalize, if $|a| = |b|$, then $a = b$ or $a = -b$.

This allows for all the possibilities described above. We now apply this property to solve more complex equations.

EXAMPLE 4   Solve and check. $|3x - 4| = |x + 6|$

We write the two possible equations.

$$3x - 4 = x + 6 \quad \text{or} \quad 3x - 4 = -(x + 6)$$

Now we solve each equation in the normal fashion.

$$
\begin{array}{ll}
3x - 4 = x + 6 & 3x - 4 = -x - 6 \\
3x - x = 4 + 6 & 3x + x = 4 - 6 \\
2x = 10 & 4x = -2 \\
x = 5 & x = -\dfrac{1}{2}
\end{array}
$$

We will check each solution by substituting them into the *original equation*.

Check: if $x = 5$

$$|3(5) - 4| \stackrel{?}{=} |5 + 6|$$
$$|15 - 4| \stackrel{?}{=} |11|$$
$$|11| \stackrel{?}{=} |11|$$
$$11 = 11 \ \checkmark$$

if $x = -\dfrac{1}{2}$

$$\left|3\left(-\dfrac{1}{2}\right) - 4\right| \stackrel{?}{=} \left|-\dfrac{1}{2} + 6\right|$$
$$\left|-\dfrac{3}{2} - 4\right| \stackrel{?}{=} \left|-\dfrac{1}{2} + 6\right|$$
$$\left|-\dfrac{3}{2} - \dfrac{8}{2}\right| \stackrel{?}{=} \left|-\dfrac{1}{2} + \dfrac{12}{2}\right|$$
$$\left|-\dfrac{11}{2}\right| \stackrel{?}{=} \left|\dfrac{11}{2}\right|$$
$$\dfrac{11}{2} = \dfrac{11}{2} \ \checkmark$$

Practice Problem 4   Solve and check. $|x - 6| = |5x + 8|$ ∎

To Think About Explain how you would solve an absolute value equation of the form $|ax + b| = 0$. Give an example.
Does $|3x + 2| = -4$ have a solution? Why or why not?

2.3 Exercises

Solve each absolute value equation. Check your solutions for Exercises 1 through 18.

1. $|x| = 14$

2. $|x| = 19$

3. $|x + 2| = 7$

4. $|x - 3| = 6$

5. $|2x - 5| = 13$

6. $|2x + 1| = 15$

7. $|5 - 4x| = 11$

8. $|2 - 3x| = 13$

9. $\left|\dfrac{1}{2}x - 3\right| = 2$

10. $\left|\dfrac{1}{4}x + 5\right| = 3$

11. $|0.5 - 0.3x| = 2$

12. $|0.7 - 0.3x| = 2$

13. $|x + 2| - 1 = 7$

14. $|x + 3| - 4 = 8$

15. $|2x + 3| + 3 = 20$

16. $|3x + 5| + 3 = 14$

17. $|4x - 5| - 8 = 3$

18. $|2x - 9| - 1 = 15$

19. $\left|1 - \dfrac{3}{4}x\right| + 4 = 7$

20. $\left|4 - \dfrac{5}{2}x\right| + 3 = 15$

21. $|x + 6| = |2x - 3|$

22. $|x - 4| = |2x + 5|$

23. $|4x + 7| = |5x + 2|$

24. $|6x - 2| = |3x + 1|$

25. $|8 - x| = |4 - 2x|$

26. $|5 + x| = |3 - 4x|$

27. $|1.5x - 2| = |x - 0.5|$

28. $|2.2x + 2| = |1 - 2.8x|$

29. $|3 - x| = |3x + 11|$

Solve for x. Round to the nearest hundredth.

30. $|1.62x + 3.14| = 2.19$

31. $|-0.74x - 8.26| = 5.36$

32. $|9.63x + 1.52| = |-8.61x + 3.76|$

33. $|8.12x - 5.85| + 1.93 = 5.42$

Mixed Practice

Solve each equation, if possible. Check your solutions.

34. $|3(x + 4)| + 2 = 14$

35. $|4(x - 2)| + 1 = 19$

36. $|4x - 20| = 0$

37. $\left|\frac{3}{4}x + 9\right| = 0$

38. $\left|\frac{1}{3}x + \frac{1}{7}\right| = -4$

39. $\left|\frac{3}{4}x - \frac{2}{3}\right| = -8$

40. $\left|\frac{2x - 1}{3}\right| = \frac{5}{6}$

41. $\left|\frac{3x + 1}{2}\right| = \frac{3}{4}$

42. $\left|\frac{1}{2}x + 3\right| = \left|\frac{1}{4}x - 6\right|$

43. $\left|\frac{2}{3}x - 1\right| = \left|\frac{1}{6}x + 2\right|$

44. $\frac{|x + 2|}{-3} = -5$

45. $\frac{|x - 3|}{-4} = -2$

Cumulative Review Problems

Solve for x.

46. $2(3x + 1) - 3(x - 2) = 2x + 5$

47. $\frac{1}{2}(3x + 1) - \frac{1}{6}(7x + 3) = \frac{1}{3}(3 - x)$

Simplify. Do not leave negative exponents in your answer.

48. $\left(\frac{2x^{-2}y}{z^{-1}}\right)^3$

49. A scientist bought 3 new Bunsen burners and 25 new beakers of different sizes last month for $975. This month she bought 3 Bunsen burners and 20 beakers of different sizes for $825. How much did each beaker cost? (All beakers cost the same regardless of size.) How much did each Bunsen burner cost? (All burners cost the same.)

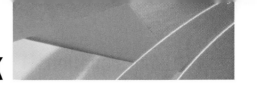

Putting Your Skills to Work

THE MATHEMATICS OF FUEL EFFICIENCY

Cars in the United States are becoming more fuel efficient. In 1970 the average U.S. car consumed 760 gallons of gas per year. In 1994 the average car consumed 551 gallons of gas per year in spite of the fact that in 1994 the average car traveled many more miles per year. This bar graph shows the increasing level of fuel efficiency of cars in the United States over a 25-year period. The graph indicates the number of miles the average car can drive on one gallon of gas.

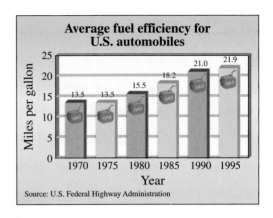

Average fuel efficiency for U.S. automobiles

Source: U.S. Federal Highway Administration

PROBLEMS FOR INDIVIDUAL INVESTIGATION AND ANALYSIS

1. In what five-year period did the greatest increase in automobile fuel efficiency occur?

2. If cars continue to increase in fuel efficiency at the same rate of increase as occurred between 1990 and 1995, how many miles per gallon will the average U.S. car achieve in the year 2005?

PROBLEMS FOR GROUP INVESTIGATION AND COOPERATIVE LEARNING

Together with some other members of your class see if you can answer the following.

3. A number of automotive engineers have predicted that the fuel efficiency of automobiles will continue to increase according to the equation $M = 21.9 + 0.4t$ where M is the average number of miles per gallon and t is the number of years after 1995. If this equation is reliable, what will be the number of miles per gallon achieved by the average car in the year 2015? Theoretically, in what year will automobiles achieve a fuel efficiency of 31.1 miles per gallon?

4. A group of government transportation officials recently predicted that the rate of increase of fuel efficiency will continue until the year 2000 at the same rate of increase that was experienced from 1990 to 1995. Then for the next five years after the year 2000 the rate of increase will be exactly one half of the rate of the previous five years. This pattern will continue, with the rate of each five-year period being half of what it was for the previous five years. Under this scenario, what will be the number of miles achieved by the average car in the year 2015?

INTERNET CONNECTION: Go to http://www.prenhall.com/tobey to be connected.

Site: Zutter Electric Vehicles FAQs

Some of the most fuel-efficient cars in the world do not use gasoline at all—they use electricity. As the graph at this site shows, electric vehicles are expected to become increasingly popular in the coming years.

5. Find the total number of expected electric vehicle sales in 2010 for the United States, Canada, Western Europe, and Japan combined.

6. In what five-year period is the greatest increase in electric vehicle sales anticipated?

7. If sales in Japan continue to increase at the rate of increase expected for 2005–2010, what will be the number of Japanese electric vehicle sales in 2015?

8. If sales in the United States and Canada continue to increase at the rate of increase expected for 2005–2010, what will be the number of U.S. and Canadian electric vehicle sales in 2020?

2.4 Using Equations to Solve Word Problems

MathPro

Video 4

SSM

After studying this section, you will be able to:

1 *Solve applied problems by using equations.*

1 *Solving Applied Problems by Using Equations*

The skills you have developed in solving equations will allow you to solve a variety of applied problems. The following steps may help you to organize your thoughts and provide you with a procedure to solve such problems.

---

1. *Understand the problem.*
   (a) Read the word problem carefully to get an overview.
   (b) Determine what information you will need to solve the problem.
   (c) Draw a sketch. Label it with the known information. Determine what needs to be found.
   (d) Choose a variable to represent one unknown quantity.
   (e) If necessary, represent other unknown quantities in terms of the same variable.

2. *Write an equation.*
   (a) Look for key words to help you translate the words into algebraic symbols.
   (b) Use a given relationship in the problems or an appropriate formula in order to write an equation.

3. *Solve the equation and state the answer.*

4. *Check.*
   (a) Check the solution in the original equation.
   (b) Be sure the solution to the equation answers the question in the word problem. You may need to do some additional calculations if it does not.

The step of writing the equation will be easier if you are familiar with certain English phrases and how to translate these phrases into mathematical symbols.

SYMBOLIC EQUIVALENTS OF ENGLISH PHRASES	
English Phrase	Mathematical Symbol
and, added to, increased by, greater than, plus, more than, sum of	$+$
decreased by, subtracted from, less than, diminished by, minus, difference between	$-$
product of, multiplied by, of, times	$\cdot$ or ( )( ) or $\times$
divided by, quotient of, ratio of	$\div$ or fraction bar
equals, are, is, will be, yields, gives, makes, is the same as, has a value of	$=$

Although we often use *x* to represent the unknown quantity when we write equations, any letter can be used. It is a good idea to use a letter that helps us remember what the variable represents. (For example, we might use *s* for speed or *h* for hours.) We now look at some examples of writing algebraic equations.

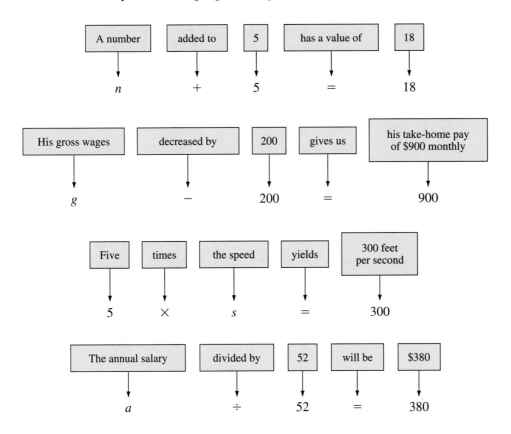

Be careful when translating the expressions "more than" or "less than." The order in which you write the symbols does not follow the order found in the English phrase. For example, "5 less than a number will be 40" is written

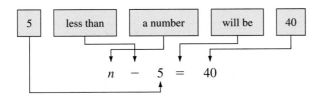

"Two more than a number is −5" is written

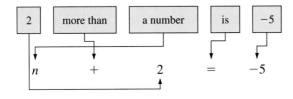

Since addition is commutative, we can write $2 + n = -5$ for "Two more than a number is −5." Be careful, however, with subtraction. "5 less than a number will be 40" must be written as $n - 5 = 40$.

We will now illustrate the use of the four-step procedure in solving a number of word problems.

EXAMPLE 1 Three-fifths of a number is 150. What is the number?

1. *Understand the problem.*

Let x = the unknown number.

When we take $\frac{3}{5}$ of a number, we can represent this by multiplying the number by $\frac{3}{5}$. Let $\frac{3}{5}x = \frac{3}{5}$ of the unknown number.

2. *Write an equation.*

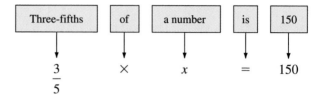

3. *Solve the equation and state the answer.*

$$\frac{3}{5} \cdot x = 150 \qquad \textit{We used the dot just for clarity. It is not necessary to write it.}$$

$$5\left(\frac{3}{5}x\right) = 5(150) \qquad \textit{Multiply each side by 5.}$$

$$3x = 750 \qquad \textit{Simplify.}$$

$$x = 250 \qquad \textit{Divide each side by 3.}$$

Therefore, the number is 250.

4. *Check.*

Is three-fifths of 250 equal to 150?

$$\frac{3}{\cancel{5}_1}(\cancel{250}^{50}) \stackrel{?}{=} 150$$

$$150 = 150 \quad \checkmark$$

Our answer is therefore correct.

Practice Problem 1 When three-fourths of a number is increased by 20, the result is 110. What is the number? ∎

Throughout Chapter 2 we will encounter many applied problems that lead to an equation with fractions. It will be helpful if you remember that these equations are always solved more easily if we first multiply both sides of the equation by the LCD (least common denominator).

EXAMPLE 2 Nancy works for an educational services company that provides computers and software for local public schools. She went to a local truck rental company to rent a truck to deliver some computers to the Newbury Elementary School. The truck rental company charges $40 per day plus 20 cents per mile. Nancy rented a truck for three days and was billed for $177. If the bill was correct, how many miles did she drive?

1. *Understand the problem.*

Let n = the number of miles driven.

Since each mile driven costs 20¢, we multiply the 20¢ (or $0.20) per mile cost by the number of miles n.

Let $0.20n$ = the cost for driving n miles at 20¢ per mile.

2. *Write an equation.*

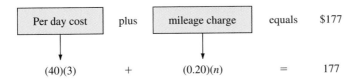

Per day cost	plus	mileage charge	equals	$177
(40)(3)	+	(0.20)(n)	=	177

3. *Solve the equation and state the answer.*

$$120 + 0.20n = 177 \qquad \textit{Multiply } (40)(3).$$

$$0.20n = 57 \qquad \textit{Subtract } 120 \textit{ from each side.}$$

$$\frac{0.20n}{0.20} = \frac{57}{0.20} \qquad \textit{Divide each side by } 0.20.$$

$$n = 285 \qquad \textit{Simplify.}$$

Nancy drove the truck for 285 miles.

4. *Check.*

Does a truck rental for 3 days at $40 per day plus 20¢ per mile for 285 miles come to a total of $177?

We will check our values in the original equation.

$$(40)(3) + (0.20)(n) = 177$$

$$120 + (0.20)(285) \stackrel{?}{=} 177$$

$$120 + 57 \stackrel{?}{=} 177$$

$$177 = 177 \ \checkmark$$

It checks. Our answer is valid.

Practice Problem 2 Western Laboratories rents a computer terminal for $400 per month plus $8 per hour for computer use time. The bill for one year's computer use was $7680. How many hours did Western Laboratories actually use the computer? ■

Some problems involve two or more unknown quantities. We then represent one quantity as n and express the other quantities in terms of n. **Consecutive even integers** are numbers such as $2, 4, 6, \ldots, n, n + 2, \ldots$. Each number in the sequence is larger by 2 than the preceding number. If we let n be the first unknown even integer, then $n + 2$ must be the second even integer, and $n + 4$ must be the third. **Consecutive odd integers** are represented the same way. Thus we can write $1, 3, 5, 7, 9, 11, \ldots, n, n + 2, n + 4, \ldots$.

EXAMPLE 3   The sum of three consecutive even integers is 264. What are the integers?

1. *Understand the problem.*
Let n = the first integer.
Let $n + 2$ = the second consecutive even integer.
Let $n + 4$ = the third consecutive even integer.

2. *Write an equation.*

Since the sum of the three even consecutive integers is 264, we have

$$n + (n + 2) + (n + 4) = 264$$

3. *Solve the equation and state the answer.*

$$n + n + 2 + n + 4 = 264$$
$$3n + 6 = 264$$
$$3n = 258$$
$$n = 86$$

If $n = 86$, then

$$n + 2 = 86 + 2 = 88$$

and

$$n + 4 = 86 + 4 = 90$$

The three consecutive even integers are thus 86, 88, and 90.

4. *Check.*
Are these three numbers consecutive even integers?
Yes. 86, 88, and 90 are all even integers and they are consecutive.   ✓

Does the sum of these three integers total 264?
Yes. $86 + 88 + 90 = 264$.   ✓

Practice Problem 3   The sum of three consecutive odd integers is 195. What are the three integers? ∎

Sometimes we need a simple formula from geometry or some other science to write the original equation.

EXAMPLE 4   An astronaut's space suit contains a small rectangular steel plate that supports the breathing control valve. The length of the rectangle is 3 millimeters more than double its width. Its perimeter is 108 millimeters. Find the width and length.

1. *Understand the problem.*
We draw a sketch to assist us.

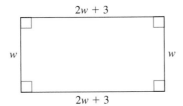

The formula for the perimeter of a rectangle is $P = 2w + 2l$, where $w =$ the width and $l =$ the length. Since the length is compared to the width we will begin with the width.
Let $w =$ the width.
Let $2w + 3 =$ the length.

2. *Write an equation.*

$$P = 2w + 2l$$

$$108 = 2w + 2(2w + 3)$$

3. *Solve the equation and state the answer.*

$$108 = 2w + 4w + 6$$

$$108 = 6w + 6$$

$$102 = 6w$$

$$17 = w$$

For $w = 17$, we have $2w + 3 = 2(17) + 3 = 37$. Thus the rectangle is 17 millimeters wide and 37 millimeters long.

4. *Check.*

$$P = 2s + 2l$$

$$108 \overset{?}{=} 2(17) + 2(37)$$

$$108 \overset{?}{=} 34 + 74$$

$$108 = 108 \ \checkmark$$

Practice Problem 4   The perimeter of a triangular lawn is 162 meters. The length of the first side is twice the length of the second side. The length of the third side is 6 meters shorter than three times the length of the second side. Find the dimensions of the triangle. ∎

2.4 Exercises

Write an algebraic equation and use it to solve each problem.

1. Three-fourths of a number is −96. What is the number?

2. Seven-eighths of a number is −63. What is the number?

3. The population of a city in India doubles. Then it increases by 1.8 million people. The present population is 5.3 million people. What was the original population of the city?

4. The budget for a Midwestern state triples. Then it increases by $2.7 million. The present budget for this state is $14.4 million. What was the original budget for this state?

5. Sarah Chamberlain just moved to Framingham. She can buy a monthly commuter rail pass for $55 for unlimited train travel to Boston. A daily round trip ticket costs $2.70. How many round trip rides per month would Sarah have to make to Boston so that it is less expensive to purchase the pass?

6. The Central Cambridge City Garage charges $5.00 for the first hour and then $3.50 for each additional hour. For how long has the car been parked at this garage when the charge for parking is $82?

7. Roberto and Maria Santanos spend approximately $11.75 per week to wash and dry their family's clothes at a local coin laundry. If a new washer and dryer would cost them a total of $846, how many weeks will it take for the laundry cost to equal the cost to purchase a new washer and dryer?

8. Manchester Community Bank charges its customers $8.00 per month plus 10¢ per check for the use of a standard checking account. Sonja had a checking account there for four months and was charged $39.70 in service charges. How many checks did she write during that period?

9. The sum of three consecutive even integers is 258. What are the integers?

10. The sum of three consecutive odd integers is 237. What are the integers?

11. While commuting to work in Orlando Florida, Marcia drives one half the distance that Melissa drives each day. John drives 17 miles more than Melissa each day. In one day these three people drive 112 miles commuting to work. How far does each drive each day?

12. In a set of three numbers, the first number is one-third of the second number, and the third number is twice the second number. The sum of the three numbers is 70. Find each number.

13. Dave and Jane Wells have a new rectangular-shaped driveway. The perimeter of the driveway is 164 feet. The length is 12 feet longer than four times the width. What are the dimensions of the driveway?

14. A new Youth Opportunity Center is being built in Roxbury. The perimeter of the rectangular shaped playing field is 340 yards. The length of the field is 6 yards less than triple the width. What are the dimensions of the playing field?

15. Russ Camp is repairing the windshield of his 17-foot Bayliner boat. The side window is in the shape of a triangle. The triangular window leaks when Russ takes the boat in the ocean and the spray hits the windshield. Russ purchased 156 centimeters of caulking to go around all three sides of the window and used all but 3 centimeters of it. The first side of the window is 3 centimeters less than double the third side. The second side of the window is 12 centimeters longer than the third side. How long is each side of the window?

16. A technician is making a corner brace for an automobile fuse box. The brace is shaped like a triangle and has a perimeter of 62 millimeters. The length of the second side is two-thirds of the length of the first side. The length of the third side is 10 millimeters less than the length of the first side. Find the length of each side of the triangle.

17. The annual telephone bill at Westmont College is $3000 less than double the annual telephone bill at Golden West College. The total spent each year by these two colleges for telephone services is $91,500. What is the annual bill for each college?

18. The number of employees at Computer Village is 50 less than triple the number of employees at Digital Center. The number of employees at both companies is 470. How many people are employed at each company?

19. Mr. and Mrs. Wong purchased three boxes of Cascade dishwasher detergent. The large box contains triple the number of ounces of the smallest box. The medium box contains 12 less than double the number of ounces of the smallest box. The three boxes together contain 228 ounces of dishwasher detergent.
 (a) How many ounces are in each size of box?
 (b) If the large box sells for $4.90, the medium box for $2.70 and the small box for $1.95, which box is the best buy?

20. An American Airlines M80 left Logan Airport in Boston and climbed to a higher altitude for several minutes. It then stayed at a level altitude of 33,000 feet for three times as long as it had climbed. It took 4 minutes less descending than it did climbing before landing at O'Hare Airport in Chicago. The entire trip took 2 hours and 6 minutes of flying time.
 (a) How many minutes was the aircraft climbing? How many minutes was the aircraft descending?
 (b) What is the climbing rate of the aircraft in terms of feet per minute that was used on the takeoff from Logan Airport?

21. Dr. Price's medical office rents a photocopier. The charge is $129 per month plus $0.03 per copy for all copies in excess of the suggested level of use of 1400 copies per month. Last month the bill from the photocopier center was $520.53. What was the total number of copies made in Dr. Price's office? How many copies in excess of the 1400 copies were made?

22. An electronics company uses a computer chip with a perimeter of 0.05052 centimeter. The length is double the width. Find the length and width.

Cumulative Review Problems

Name the property that justifies each statement.

23. $57 + 0 = 57$

24. $(2 \cdot 3) \cdot 9 = 2 \cdot (3 \cdot 9)$

2.5 Solving More Involved Word Problems

MathPro   Video 5   SSM

After studying this section, you will be able to:

1 *Solve more involved word problems by using an equation.*

1 Solving More Involved Word Problems by Using an Equation

To solve some word problems, we need to understand percents, interest, mixtures, or some other concepts before we can use an algebraic equation as a model.

From arithmetic you know that to find a percent of a number we write the percent as a decimal and multiply the decimal by the number. Thus, to find 36% of 85, we calculate $(0.36)(85) = 30.6$. If the number is not known, we can represent it by a variable.

EXAMPLE 1   The Wildlife Refuge Rangers tagged 144 deer. They estimate that they have tagged 36% of the deer in the refuge. If they are correct, approximately how many deer are in the refuge?

1. *Understand the problem.*
   Let n = the number of the deer in the refuge.
   Let $0.36n$ = 36% of the deer in the refuge.

2. *Write an equation.*

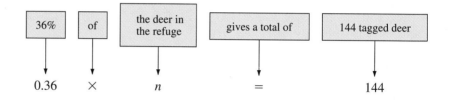

$$0.36 \quad \times \quad n \quad = \quad 144$$

3. *Solve the equation and state the answer.*

$$0.36n = 144$$

$$\frac{0.36n}{0.36} = \frac{144}{0.36} \qquad \textit{Divide each side by } 0.36.$$

$$n = 400 \qquad \textit{Simplify.}$$

There are approximately 400 deer in the refuge.

4. *Check.*
   Is it true that 36% of 400 is 144?

$$(0.36)(400) \overset{?}{=} 144$$

$$144 = 144 \quad \checkmark$$

It checks. Our answer is valid.

Practice Problem 1   Technology Resources, Inc., sold 6900 computer workstations, a 15% increase in sales over the year. How many computer workstations were sold last year? (*Hint:* Let x = the amount of sales last year and let $0.15x$ = the increase in sales over last year.) ■

Adding two numbers yields a total. We can call one of the numbers *x* and the other number (total − *x*). We will use this concept in Example 2.

EXAMPLE 2   Bob's and Marcia's weekly salaries total $265. If they both went from part-time to full-time employment, their combined weekly income would be $655. Bob's salary would double, while Marcia's would triple. How much do they each make now?

1. *Understand the problem.*
Let *b* = Bob's part-time salary.
Since the total of the two part-time weekly salaries is $265, we can
let 265 − *b* = Marcia's part-time salary.

2. *Write an equation.*

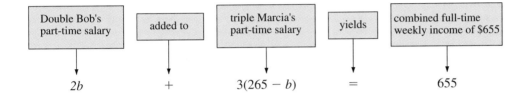

| Double Bob's part-time salary | added to | triple Marcia's part-time salary | yields | combined full-time weekly income of $655 |

$$2b \qquad + \qquad 3(265 - b) \qquad = \qquad 655$$

3. *Solve the equation and state the answer.*

$$2b + 3(265 - b) = 655$$
$$2b + 795 - 3b = 655 \qquad \textit{Remove parentheses.}$$
$$795 - b = 655 \qquad \textit{Simplify.}$$
$$-b = -140 \qquad \textit{Subtract 795 from each side.}$$
$$b = 140 \qquad \textit{Multiply each side by } -1.$$

If *b* = 140, then 265 − *b* = 265 − 140 = 125. Thus Bob's present full-time weekly salary is $140, and Marcia's present full-time weekly salary is $125.

4. *Check.*
Do their present weekly salaries total $265? Yes: 140 + 125 = 265. If Bob's income is doubled and Marcia's is tripled, will their new weekly salaries total $655? Yes: 2(140) + 3(125) = 280 + 375 = 655. ✓

Practice Problem 2   Alicia and Heather each sold a number of cars at Prestige Motors last month. Together they sold 43 cars. If next month Alicia doubles her sales, and Heather triples her sales, they will sell 108 cars. How many cars did each person sell this month? ■

Simple interest is an income from investing money or a charge for borrowing money. It is computed by multiplying the amount of money borrowed or invested (called the principal) by the rate of interest and by the period of time it is borrowed or invested (usually measured in years unless otherwise stated). Hence

$$\text{interest} = \text{principal} \times \text{rate} \times \text{time}$$

$$I = prt$$

All interest problems in this chapter involve simple interest.

EXAMPLE 3

Find the interest on $800 borrowed at an interest rate of 18% for two years.

$$I = prt$$
$$= (800)(0.18)(2)$$
$$= (144)(2)$$
$$= 288$$

The interest charge for borrowing $800 for two years at 18% is $288.

Practice Problem 3 Find the interest on $7000 borrowed at an interest rate of 12% for four years. ■

EXAMPLE 4

Maria has a job as a loan counselor in a bank. She advised a customer to invest part of his money in a 12% interest money market fund and the rest in a 14% interest investment fund. The customer had $6000 to invest. If he earned $772 in interest in one year, how much did he invest in each fund?

1. *Understand the problem.*
   Let x = the amount of money invested at 12% interest.
   The other amount of money is (the total − x).
   Let $6000 − x$ = the amount of money invested at 14% interest.

2. *Write an equation.*

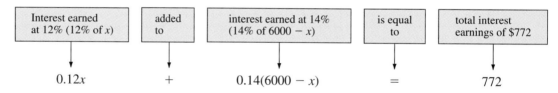

Interest earned at 12% (12% of x)	added to	interest earned at 14% (14% of 6000 − x)	is equal to	total interest earnings of $772
$0.12x$	$+$	$0.14(6000 − x)$	$=$	772

3. *Solve the equation and state the answer.*
$$0.12x + 0.14(6000 − x) = 772$$

$$0.12x + 840 − 0.14x = 772 \qquad \textit{Remove parentheses.}$$

$$840 − 0.02x = 772 \qquad \textit{Add } 0.12x \textit{ to } −0.14x.$$

$$−0.02x = −68 \qquad \textit{Subtract } 840 \textit{ from each side.}$$

$$\frac{−0.02x}{−0.02} = \frac{−68}{−0.02} \qquad \textit{Divide each side by } −0.02.$$

$$x = 3400 \qquad \textit{Simplify.}$$

If $x = 3400$, then $6000 − x = 6000 − 3400 = 2600$. Thus $3400 was invested in the 12% interest money market fund, and $2600 was invested in the 14% interest investment fund.

4. *Check.*
   Do the two amounts of money total $6000?
   Yes: $3400 + $2600 = $6000.
   Does the total interest amount to $772?

$$(0.12)(3400) + (0.14)(2600) \stackrel{?}{=} 772$$

$$408 + 364 \stackrel{?}{=} 772$$

$$772 = 772 \quad \checkmark \quad \textit{Everything checks}$$

Our answers are correct.

Practice Problem 4 Tricia received an inheritance of $5500. She invested part of it at 8% interest and the remainder at 12% interest. At the end of the year she had earned $540. How much did Tricia invest at each amount? ■

Sometimes we encounter a situation where two or more items are combined to form a mixture or solution. These types of problems are called **mixture problems.**

EXAMPLE 5 A small truck has a radiator that holds 20 liters. A mechanic needs to fill the radiator with a solution that is 60% antifreeze. How many liters of a solution that is 70% antifreeze should she mix with a solution that is 30% antifreeze to achieve the desired mix?

1. *Understand the problem.*
   Let x = the number of liters of 70% antifreeze to be used.
   Since the total solution must be 20 liters, we can use $20 - x$ for the other part.
   Let $20 - x$ = the number of liters of 30% antifreeze to be used.

   In this problem a chart or table is very helpful. We will multiply the entry in column (A) by the entry in column (B) to obtain the entry in column (C).

	(A) Number of Liters of the Solution	(B) Percent Pure Antifreeze	(C) Number of Liters of Pure Antifreeze
70% antifreeze solution	x	70%	$0.70x$
30% antifreeze solution	$20 - x$	30%	$0.30(20 - x)$
Final 60% solution	20	60%	$0.60(20)$

2. *Write an equation.*
   Now we form an equation from the entries of column (C).

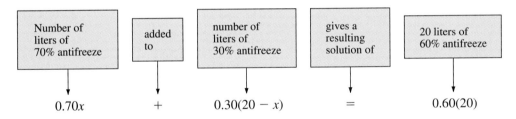

$$0.70x \qquad + \qquad 0.30(20 - x) \qquad = \qquad 0.60(20)$$

3. *Solve the equation and state the answer.*

$$0.70x + 0.30(20 - x) = 0.60(20)$$
$$0.70x + 6 - 0.30x = 12$$
$$6 + 0.40x = 12$$
$$0.40x = 6$$
$$\frac{0.40x}{0.40} = \frac{6}{0.40}$$
$$x = 15$$

   If $x = 15$, then $20 - x = (20 - 15) = 5$. Thus the mechanic needs 15 liters of 70% antifreeze solution and 5 liters of 30% antifreeze solution.

4. *Check.*
   Can you verify that this answer is correct?

Practice Problem 5 A jeweler wishes to prepare 200 grams of 80% pure gold from sources that are 68% pure gold and 83% pure gold. How many grams of each should he use? ■

Some problems involve the relationship distance = rate × time or $d = rt$.

EXAMPLE 6   Frank drove at a steady speed for 3 hours on the turnpike. He then slowed his traveling speed by 15 miles per hour for traveling on secondary roads. The entire trip took 5 hours and covered 245 miles. What was his speed on the turnpike?

1. *Understand the problem.*
Let x = speed on the turnpike in miles per hour.
Let $x - 15$ = speed on the secondary roads in miles per hour.

Again, as in Example 5, a chart or table is very helpful. We will use a chart that records values of $(r)(t) = d$.

	Rate (miles per hour) r	Time (hours) t	Distance (miles) $(r)(t) = d$
Turnpike	x	3	$3x$
Secondary roads	$x - 15$	2	$2(x - 15)$
Entire trip	Not given	5	245

2. *Write an equation.*
Using the distance entries (third column of the table), we have

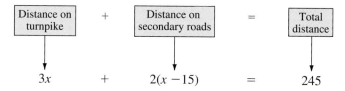

$$3x \quad + \quad 2(x - 15) \quad = \quad 245$$

3. *Solve the equation and state the answer.*

$$3x + 2(x - 15) = 245$$
$$3x + 2x - 30 = 245$$
$$5x - 30 = 245$$
$$5x = 275$$
$$x = 55$$

Thus he traveled at an average speed of 55 miles per hour on the turnpike.

4. *Check.*
Can you verify that this answer is correct?

Practice Problem 6   Wally drove for 4 hours at a steady speed. He slowed his speed by 10 miles per hour for the last part of the trip. The entire trip took 6 hours and covered 352 miles. How fast did he drive on each portion of the trip? ■

Developing Your Study Skills

CLASS PARTICIPATION

People learn mathematics through active participation, not through observation from the sidelines. If you want to do well in this course, be involved in classroom activities. Sit near the front where you can see and hear well, where your focus is on the instruction process and not on those students around you. Ask questions, be ready to contribute toward solutions, and take part in all classroom activities. Your contributions are valuable to the class and to yourself. Class participation requires an investment of yourself in the learning process, which you will find pays huge dividends.

2.5 Exercises

Write an algebraic equation for each problem and solve it.

1. The population of the United States in 1994 was 260.6 million people. This was an increase of 36% from the population in 1964. What was the population of the United States in 1964?

2. The U.S. national debt in 1996 was $5.207 trillion. This was an increase of 62% from the national debt in 1990. What was the national debt in 1990?

3. Jody accepted her first full time teaching job as a starting elementary teacher in Naperville at an annual salary of $27,000 per year. Her friend Julie told her that she is earning 12% more than the starting pay for teachers in Naperville was 3 years ago. What was the starting pay for teachers in Naperville 3 years ago?

4. Eastwing dormitory just purchased a new Sony color television on sale for $340. The sale price was 80% of the original price. What was the original price of the television set?

5. A biologist tagged 242 penguins on an island near Cape Verde, South America. He estimated that approximately 82% of the penguins who live on that island were tagged. If this is the case, approximately how many penguins actually live on this island?

6. The Arbor Society discovered that 95 elm trees were dying of Dutch Elm Disease. The experts determined that the dying trees discovered thus far were only 65% of the trees which were contaminated. If this estimate is correct, approximately how many trees are actually infected?

7. The sum of two numbers is 77. If the first number is multiplied by 4 and the second number is multiplied by 2, the sum is 286. What are the two numbers?

8. The sum of two numbers is 128. If the first number is tripled, and added to two times the second number the result is 311. What are the two numbers?

9. Fred Willianson had some leftover boards that were 16 feet long. He decided to use them for projects in his house. First he cut a 16-foot board into two unequal pieces. He then cut five additional 16-foot boards using exactly the same dimensions. He used all six of the longer pieces to complete a wall. He used four of the shorter pieces to case in a stairwell. In all he used 82 linear feet of boards on the two projects, and the remaining wood was left over. What were the lengths of the boards after he cut them into two pieces?

10. A hospital received a shipment of 8-milligram doses of a medicine. Each 8-milligram package was repacked into two smaller doses of unequal size and labeled packet A and packet B. The hospital then used 17 doses of packet A and 14 doses of packet B in one week. The total use of this medicine at the hospital during that week was 127 milligrams. How many milligrams of the medicine are contained in each packet A? in each packet B?

11. Find the interest on $600 borrowed at an annual interest rate of 16% for 3 years.

12. Find the interest on $900 borrowed at an annual interest rate of 14% for 2 years.

13. Find the interest if $4000 is invested at an annual interest rate of 7.5% for $\frac{1}{2}$ year.

14. Find the interest if $3000 is invested at an annual interest rate of 8% for $\frac{1}{4}$ year.

15. Cynthia is a hard-working single mother on a very limited budget. She invests her savings in only safe investments, because she is not in a position to take risks and lose any of her hard-earned money. She invested $6400 in two types of accounts for one year. The first type earns 5% interest, and the second type earns 8% interest. At the end of this year, Cynthia earned $395 in interest. How much did she invest at each rate?

16. The Johnson's family business did well *this* year due to their investments *last* year. The business earned $6570 from having invested $45,000 in mutual funds. There were two types of funds that Jenna Johnson invested in. The first was a pharmaceutical fund, which paid out interest of 13%. The second was a genetech fund, which paid out interest of 16%. How much did Jenna Johnson invest in each fund last year for her family's business?

17. A retired couple earned $4140 last year from their investments in mutual funds. They originally invested $30,000 in two types of funds. In the science investment fund they earned 12% interest. In the overseas investment fund they earned 15% interest. How much did they invest in each type of fund?

18. Ramon invested $7000 in money market funds. Part was invested at 14% interest and the rest at 11% interest. At the end of one year Ramon had earned $902 in interest. How much did Ramon invest at each amount?

19. A chef has one cheese that contains 45% fat and another cheese that contains 20% fat. How many grams of each cheese should she mix in order to obtain 30 grams of a cheese mixture which is 30% fat?

20. Becky was awarded a college internship at a local pharmaceutical research corporation. She has been given her own corner of the laboratory to try her hand at pharmacology. One of her assignments is to mix two solutions, one containing a solution that is 16% strength and the second containing a solution that is 9% strength. How many milliliters of each should she mix in order to obtain 350 milliliters of a 12% strength solution?

21. A grocer at a specialty store is mixing tea worth $6.00 per pound with tea worth $8.00 per pound. He wants to obtain 144 pounds of tea worth $7.50 per pound. How much of each tea should he mix?

22. The meat department manager at a large food store wishes to mix some hamburger with 30% fat content with some hamburger that has 10% fat content in order to obtain 100 pounds of hamburger with 25% fat content. How much hamburger of each type should she use?

23. A grocer has 200 pounds of candy worth $1.80 per pound. He also has some imported candy worth $3.00 per pound. How much imported candy should he mix with the 200 pounds in order to obtain a mixture worth $2.40 per pound?

24. A pharmacist has 40 liters of a 30% salt solution. He needs to cut the strength to 27%. He has on hand some 15% solution. How much of the 15% solution should the pharmacist add to the 30% solution?

25. Susan drove for 4 hours on secondary roads at a steady speed. She completed her trip by driving 2 hours on an interstate highway. Her total trip was 250 miles. Her average speed was 20 miles per hour faster on the interstate highway portion of the trip. How fast did she travel on the secondary roads?

26. Alice and Wendy flew a small plane for 930 miles. For the first 3 hours they flew at maximum speed. After refueling, they finished the trip at a cruising speed, which was 60 miles per hour slower than maximum speed. The entire trip took 5 hours of flying time. What is the maximum flying speed of the plane?

27. Jung rides the stationary bike in the university health club at 13 kilometers per hour. Cheryl rides for the same amount of time at 9 kilometers per hour. Jung biked exactly 2 kilometers more in her workout. How long was each person's time on the stationary bike?

28. The Clarke family went sailing on the lake. Their boat averaged 6 kilometers per hour. The Rourke family took their outboard runabout for a trip on the lake for the same amount of time. Their boat averaged 14 kilometers per hour. The Rourke family traveled 20 kilometers farther than the Clarke family. How many hours did each family spend on the boat trip?

29. From 1950 to 1970 the town of Spencerville *grew* by 38.6%. From 1970 to 1990 the population *decreased* by 19%. The population in 1990 is 56,133. What was the population in 1950?

30. Tom and Linda invested $18,375 in money market funds. Part was invested at 5.9% interest and part was invested at 7.8% interest. In one year they earned $1243.25 in interest. How much did they invest at each rate?

To Think About

31. A bottled iced-tea manufacturer saw profits increase 65% last year. This year, due to so much new competition, profits fell 40%. Profits this year were $17,820,000.00. What was the realized profit for the iced-tea company two years ago?

32. Northwest Airlines gives out peanuts to all passengers who want them. On several recent flights to Boston 72% of the passengers asked accepted the peanuts and 23% rejected the peanuts. A total of 36 passengers were asleep and could not be asked. How many passengers were on these flights?

33. The area of the shaded triangle is 6 cm². Find the area of parallelogram *ACDF*.

$$A \quad \overset{5}{} \quad B \quad 3 \quad C$$

$$F \quad \overset{3}{} E \quad 5 \quad D$$

34. The volume of the sphere is 38,808 cm³. Find the volume of the cylinder (use $\pi \approx \frac{22}{7}$).

Cumulative Review Problems

Evaluate.

35. $2x^2 - 3x + 1$ when $x = -2$

36. $5a - 2b + c$ when $a = 1$, $b = -3$, $c = -4$

37. $a + d(n - 1)$ when $a = 12$, $d = 2$, $n = 4$

38. $\frac{9}{5}C + 32$ when $C = -25$

2.6 Inequalities

After studying this section, you will be able to:

1 *Determine if one number is less than or greater than another number.*
2 *Graph a linear inequality in one variable.*
3 *Solve a linear inequality in one variable.*

1 *Determining If One Number Is Less Than or Greater Than Another Number*

An inequality means that the two numbers or expressions in the mathematical statement are not necessarily equal. We can use the number line to visualize the concept of inequality.

It is a mathematical property that -1 is less than 3. We write this in the following way:

$$-1 < 3$$

Notice the position of these numbers on the number line. -1 is to the *left* of 3 on the number line. We say that a first number *is less than* a second number if the first number *is to the left of* the second number on the number line.

Looking at these numbers once again, we could also say that 3 is greater than -1. We could read the above symbol from right to left. That is,

$$-1 < 3$$

$\leftarrow$ 3 is greater than -1

The inequality symbol can be read from left to right or from right to left. Notice that the opening of the symbol faces the greater number. Since we generally read from left to right, we will write the symbol in the following way.

$$3 > -1$$

3 is greater than -1 $\rightarrow$

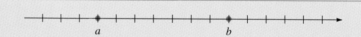

In general, if $a < b$, then it is also true that $b > a$.

EXAMPLE 1   Insert the proper symbol between the numbers.

(a) 8 _____ 6   **(b)** -2 _____ 3   **(c)** -4 _____ -2

(d) $\dfrac{1}{2}$ _____ $\dfrac{1}{3}$ **(e)** 0.56 _____ 0.561   **(f)** -0.033 _____ -0.0329

(a) $8 > 6$ because 8 is to the right of 6 on the number line.

(b) $-2 < 3$ because -2 is to the left of 3 on the number line.

(c) $-4 < -2$ because -4 is to the left of -2 on the number line.

(d) To compare fractions, remember to change to fractions with the same denominator.

$\dfrac{1}{2}$ _____ $\dfrac{1}{3}$ *Change to fractions with the same denominator.*

$\dfrac{3}{6}$ _____ $\dfrac{2}{6}$ *Compare the numerators:* $3 > 2$.

$\dfrac{3}{6} > \dfrac{2}{6}$

Thus, $\frac{1}{2} > \frac{1}{3}$ because $\frac{3}{6} > \frac{2}{6}$.

(e) To compare decimals, annex zeros so that both decimals have the same number of decimal places.

   0.56 _____ 0.561   *Annex a zero to the decimal on the left.*

   0.560 _____ 0.561   *Compare.*

   $0.560 < 0.561$

   Thus $0.56 < 0.561$ because $0.560 < 0.561$.

(f) $-0.033 > -0.0329$ because $-0.0330 > -0.0329$.

Notice that, since both numbers are negative, -0.0330 is to the left of -0.0329 on the number line.

Practice Problem 1 Insert the symbol $<$ or $>$ between the two numbers.

(a) -1 _____ -2 **(b)** $\dfrac{2}{3}$ _____ $\dfrac{3}{4}$ **(c)** -0.561 _____ 0.5555 ■

Numerical or algebraic expressions as well as numbers can be compared.

EXAMPLE 2 Insert the proper symbol between the expressions.

(a) $(5 - 8)$ _____ $(2 - 3)$ **(b)** $|1 - 7|$ _____ $|-4 - 12|$

(c) $(x + 3)$ _____ $(x - 1)$

(a) $(5 - 8)$ _____ $(2 - 3)$ *Evaluate each expression and compare.*
   $-3 < -1$
   Thus $(5 - 8) < (2 - 3)$ because $-3 < -1$.

(b) $|1 - 7|$ _____ $|-4 - 12|$ *Evaluate each expression and compare.*
   $6 < 16$
   Thus $|1 - 7| < |-4 - 12|$ because $6 < 16$.

(c) $(x + 3) > (x - 1)$ *Since x represents the same number in both expressions, $x + 3$ will be greater than $x - 1$.*

Practice Problem 2 Insert $<$ or $>$ between the expressions.

(a) $(-8 - 2)$ _____ $(-3 - 12)$ **(b)** $|-15 + 8|$ _____ $|7 - 13|$

(c) $(x - 3)$ _____ $(x - 4)$ ■

2 Graphing Linear Inequalities in One Variable

Inequality symbols are often used with variables. For example,

$$x > 4$$

means that x can be any number that is greater than 4. x cannot equal 4. We use a number line to graph this inequality.

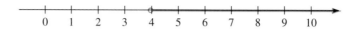

On the number line, we shade the portion that is to the right of 4. Any point in the shaded portion will satisfy the inequality since all points to the right of 4 are greater than 4. The open circle at 4 means that x cannot be 4.

Let's look at the inequality $x < -1$. This means that x can be any number that is less than -1. To graph the inequality, we will shade all points to the left of -1 on the number line.

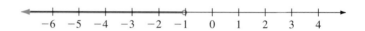

$$x < -1$$

Inequality symbols used with variables can include the equal sign. $x \geq -2$ means all numbers greater than or equal to -2. The graph is

$$x \geq -2$$

The shaded circle at -2 means that the graph includes the point -2. -2 is a solution to the inequality.

Similarly, $x \leq 1$ is graphed as

$$x \leq 1$$

EXAMPLE 3   Graph each inequality.

(a) $x < 0$ **(b)** $x \leq 0$ **(c)** $x > -5$ **(d)** $x \geq -5$ **(e)** $2 < x$

(a) $x < 0$
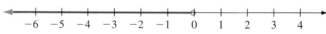

(b) $x \leq 0$

(c) $x > -5$
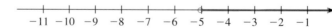

(d) $x \geq -5$
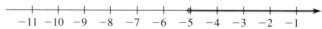

(e) We read an inequality starting with the variable. Thus we read $2 < x$ as x is greater than 2 and graph the expression accordingly.

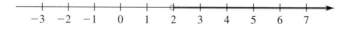

Practice Problem 3   Graph each relationship.

(a) $x > 3.5$

```
  +--+--+--+--+--+--+--+--+-->
 -1  0  1  2  3  4  5  6  7
```

(b) $x \le -10$

```
  +--+--+--+--+--+--+--+--+-->
 -14  -12  -10  -8   -6
```

(c) $x \ge -2$

```
  +--+--+--+--+--+--+--+--+-->
 -5 -4 -3 -2 -1  0  1  2  3
```

(d) $-4 > x$

```
  +--+--+--+--+--+--+--+--+-->
 -8 -7 -6 -5 -4 -3 -2 -1  0
```
■

3 Solving Inequalities in One Variable

Solving a first-degree inequality is similar to solving first-degree equations. We use various properties of real numbers.

Addition and Subtraction Property for Inequalities

If $a < b$, then

$$a + c < b + c \quad \text{and} \quad a - c < b - c$$

for all real numbers a, b, and c.

The same number can be added or subtracted from both sides of an inequality without changing the direction of the inequality. (Remember, any inequality symbol can be used. We use $<$ for convenience.)

EXAMPLE 4   Solve the inequalities. Graph and check their solutions.

(a) $x - 8 < 15$ **(b)** $5x \ge 4x - 2$

(a)
$$x - 8 < 15$$
$$x - 8 + 8 < 15 + 8 \qquad \textit{Add} +8 \textit{ to each side.}$$
$$x < 23 \qquad \textit{Simplify.}$$

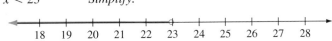

```
 <--+--+--+--+--+--o--+--+--+--+--+-->
   18 19 20 21 22 23 24 25 26 27 28
```

To check, choose any solution in the solution set and see if a true statement results when you substitute the solution into the inequality. We will choose $x = 22.5$.

$$x - 8 < 15 \qquad \textit{Substitute 22.5 for x in the original inequality.}$$

$$22.5 - 8 \overset{?}{<} 15$$

$$14.5 < 15 \quad \checkmark \qquad \textit{True statement.}$$

(b)
$$5x \ge 4x - 2$$
$$5x - 4x \ge 4x - 4x - 2 \qquad \textit{Subtract 4x from each side.}$$
$$x \ge -2 \qquad \textit{Simplify.}$$

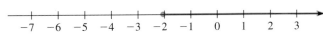

```
  +--+--+--+--+--+--•--+--+--+--+-->
 -7 -6 -5 -4 -3 -2 -1  0  1  2  3
```

To check, we choose $x = 2$, and $x = -1$.

$$5x \ge 4x - 2 \qquad\qquad 5x \ge 4x - 2$$

$$5(2) \overset{?}{\ge} 4(2) - 2 \qquad 5(-1) \overset{?}{\ge} 4(-1) - 2$$

$$10 \ge 6 \quad \checkmark \quad \text{True} \qquad -5 \ge -6 \quad \checkmark \quad \text{True}$$

Practice Problem 4   Solve the inequalities. Graph and check their solutions.

(a) $x + 2 > -12$

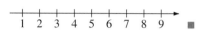

(b) $12x \leq 11x + 5$

Multiplication or Division by a Positive Number

If $a < b$, then

$$ac < bc \quad \text{and} \quad \frac{a}{c} < \frac{b}{c}$$

for all real numbers a, b, c when $c > 0$.

When we multiply or divide both sides of an inequality by a positive number, the direction of the inequality is not changed. That is, if $5x < -15$ and we divide both sides by 5, we obtain $x < -3$.

To check this solution, we choose $x = -4$. $5(-4) < -15$? Yes.

EXAMPLE 5   Solve. $6x + 3 \leq 2x - 5$ Graph your solution.

$6x + 3 - 3 \leq 2x - 5 - 3$ *Subtract 3 from each side.*

$6x \leq 2x - 8$ *Simplify.*

$6x - 2x \leq 2x - 2x - 8$ *Subtract 2x from each side.*

$4x \leq -8$ *Simplify.*

$\dfrac{4x}{4} \leq \dfrac{-8}{4}$ *Divide each side by 4. The inequality is not reversed.*

$x \leq -2$

To check, we choose $x = -2$ and $x = -3$.

$$6x + 3 \leq 2x - 5 \qquad\qquad 6x + 3 \leq 2x - 5$$

$$6(-2) + 3 \overset{?}{\leq} 2(-2) - 5 \qquad 6(-3) + 3 \overset{?}{\leq} 2(-3) - 5$$

$$-9 \leq -9 \ \checkmark \qquad\qquad\qquad -15 \leq -11 \ \checkmark$$

Practice Problem 5   Solve. $8x - 8 \geq 5x + 1$ Graph your solution.

When we multiply or divide both sides of an inequality by a *negative number,* the direction of the inequality is *reversed.* That is, if we divide both sides of $-3x < 21$ by -3, we obtain $x > -7$. If we divide both sides of $-4x \geq -16$ by -4, we obtain $x \leq 4$.

To check this solution, we choose $x = 1$. Is $-3(1) < 21$? Yes. To reverse the inequality symbol is an unusual move. Let's see what would happen if we did not reverse the symbol. If we did not reverse the symbol, the solution would be $x < -7$. To check this solution, we choose $x = -8$. Is $-3(-8) < 21$? No, 24 is not less than 21.

Multiplication or Division by a Negative Number

If $a < b$, then

$$ac > bc \quad \text{and} \quad \frac{a}{c} > \frac{b}{c}$$

for all real numbers a, b, c when $c < 0$.

To Think About Show why you must reverse the inequality symbol when you multiply or divide by a negative number. Provide several examples.

EXAMPLE 6 Solve. $-8x - 12 < -4(x - 4) + 8$.

$-8x - 12 < -4x + 16 + 8$	*Use the distributive property to remove the parentheses.*
$-8x - 12 < -4x + 24$	*Simplify.*
$-8x + 4x - 12 < -4x + 4x + 24$	*Add $4x$ to each side.*
$-4x - 12 < 24$	
$-4x - 12 + 12 < 24 + 12$	*Add 12 to each side.*
$-4x < 36$	
$\dfrac{-4x}{-4} > \dfrac{36}{-4}$	*Divide each side by -4. This reverses the inequality.*
$x > -9$	

Practice Problem 6 Solve. $2 - 12x > 7(1 - x)$ ∎

An inequality that contains all decimal terms is best handled by first multiplying both sides of the inequality by the appropriate power of 10.

EXAMPLE 7 Solve. $-0.3x + 1.0 \le 1.2x - 3.5$

$10(-0.3x + 1.0) \le 10(1.2x - 3.5)$	*Multiply each side by 10.*
$-3x + 10 \le 12x - 35$	
$-3x - 12x + 10 \le 12x - 12x - 35$	*Add $-12x$ to each side.*
$-15x + 10 \le -35$	
$-15x + 10 - 10 \le -35 - 10$	*Add -10 to each side.*
$-15x \le -45$	
$\dfrac{-15x}{-15} \ge \dfrac{-45}{-15}$	*Divide each side by -15. This reverses the direction of the inequality.*
$x \ge 3$	

Practice Problem 7 Solve for x. $0.5x - 0.4 \le -0.8x + 0.9$ ∎

To solve an inequality that contains fractions, multiply both sides of the inequality by the LCD. You may skip this step and multiply each term by the LCD.

EXAMPLE 8 Solve. $\frac{1}{7}(x + 5) > \frac{1}{5}(x + 1)$

$$\frac{x}{7} + \frac{5}{7} > \frac{x}{5} + \frac{1}{5}$$ *Using the distributive property, remove the parentheses.*

$$35\left(\frac{x}{7}\right) + 35\left(\frac{5}{7}\right) > 35\left(\frac{x}{5}\right) + 35\left(\frac{1}{5}\right)$$ *Multiply each term by the LCD.*

$$5x + 25 > 7x + 7$$ *Remove parentheses.*

$$5x + 25 - 25 > 7x + 7 - 25$$ *Subtract 25 from each side.*

$$5x > 7x - 18$$ *Simplify.*

$$5x - 7x > 7x - 7x - 18$$ *Subtract 7x from each side.*

$$-2x > -18$$ *Simplify.*

$$\frac{-2x}{-2} < \frac{-18}{-2}$$ *Divide each side by −2. This reverses the direction of the inequality.*

$$x < 9$$

Practice Problem 8 Solve. $\frac{1}{5}(x - 6) < \frac{1}{3}(x - 2)$ ∎

To Think About Another approach to solving the inequality in Example 8 would be to multiply each side of the inequality by the LCD before removing the parentheses. Try it. Discuss the pros and cons of this approach. Choose the one you like best.

EXAMPLE 9 Solve. $\frac{x}{4} - \frac{3}{4} - 1 \le \frac{x}{2}$

$$4\left(\frac{x}{4}\right) - 4\left(\frac{3}{4}\right) - 4(1) \le 4\left(\frac{x}{2}\right)$$ *Multiply each term by the LCD 4.*

$$x - 3 - 4 \le 2x$$ *Simplify.*

$$x - 7 \le 2x$$ *Combine like terms.*

$$x - x - 7 \le 2x - x$$ *Add −x to each side.*

$$-7 \le 1x$$ *Simplify. Rewrite the expression with the variable on the left.*

$$x \ge -7$$ *Because a < b is equivalent to b > a.*

Practice Problem 9 Solve. $\frac{1}{5} - \frac{1}{10}x \le \frac{3}{10}x + 1$ ∎

2.6 Exercises

Verbal and Writing Skills

True or false?

1. The statement $6 < 8$ conveys the same information as $8 > 6$.

2. Adding $-5x$ to each side of an inequality reverses the direction of the inequality.

3. Dividing each side of an inequality by -4 reverses the direction of the inequality.

4. The graph of $x > -2$ is the set of all points to the right of -2 on the number line.

5. The graph of $x \le 6$ does not include the point at 6 on the number line.

6. To solve the inequality $\frac{2}{3}x + \frac{3}{4} \ge \frac{1}{2}x - 4$, multiply each fraction by the LCD.

Insert the symbol $<$ or $>$ between each pair of numbers.

7. $8 \underline{\hspace{1cm}} -9$

8. $-12 \underline{\hspace{1cm}} 6$

9. $-6 \underline{\hspace{1cm}} -3$

10. $-2 \underline{\hspace{1cm}} -7$

11. $\dfrac{3}{4} \underline{\hspace{1cm}} \dfrac{2}{3}$

12. $\dfrac{5}{6} \underline{\hspace{1cm}} \dfrac{5}{7}$

13. $-3.4 \underline{\hspace{1cm}} -3.41$

14. $-2.69 \underline{\hspace{1cm}} -2.7$

15. $(5 - 8) \underline{\hspace{1cm}} (3 - 2)$

16. $(-12 + 6) \underline{\hspace{1cm}} (18 - 3)$

17. $-\dfrac{10}{3} \underline{\hspace{1cm}} -2$

18. $-3 \underline{\hspace{1cm}} -\dfrac{15}{4}$

Graph each relationship.

19. $x \ge -2$

20. $x \ge -4$

21. $x < 6$

22. $x < 9$

23. $x > \dfrac{5}{4}$

24. $x < \dfrac{2}{3}$

25. $x \le -4.5$

26. $x \ge -7.5$

Solve for x and graph your solution.

27. $3x + 5 \le 20$

28. $2x - 8 \le 10$

29. $5x + 6 > -7x - 6$

30. $2x + 5 > 4x - 5$

31. $0.5x + 0.1 < 1.1x + 0.7$

32. $15 - 3x < -x - 5$

Solve for x.

33. $9x - 16 + 6x \geq 11 + 4x - 5$

34. $7 - 9x - 12 \geq 3x + 5 - 8x$

35. $-2(x + 3) < -9$

36. $2x - 11 + 3(x + 2) < 0$

37. $-4(x + 2) - 3x - 20 > 0$

38. $1 - (x + 3) + 2x > 4$

39. $0.5x - 0.6 \leq 1.4$

40. $-0.3x + 0.4 \leq 2.2$

41. $1.9 > 0.2x + 0.4$

42. $0.5x - 0.2 > 2.8$

43. $0.3(x + 3) \geq 0.5x + 1.4$

44. $-4.5 + 0.4x \geq 0.5(x + 0.8)$

45. $1 + \dfrac{1}{8}x \leq \dfrac{1}{4}x - \dfrac{3}{4} + \dfrac{1}{8}x$

46. $\dfrac{7}{6} - \dfrac{1}{3}x + \dfrac{5}{6}x \leq 1x - \dfrac{4}{3}$

47. $\dfrac{1}{3}(x + 2) \geq 3x - 5(x - 2)$

48. $\dfrac{1}{4}(x + 3) \geq 4x - 2(x - 3)$

49. $\dfrac{1}{2}x - 2 - \dfrac{1}{3}x < \dfrac{1}{4}(x - 8)$

50. $\dfrac{2}{3}x - x + \dfrac{3}{2} < \dfrac{1}{3}(x + 3) + \dfrac{1}{2}$

Solve for x. Round your final answer to nearest hundredth.

51. $156.98x - 32.73 < 181.23x + 77.32$

52. $1.92(6.3x + 4.9) \geq 7.06x - 4.371$

Applications

For Exercises 53 through 56, describe the situation with a linear inequality and then solve the inequality.

53. A phone solicitor selling long-distance services earns $7.75 per hour plus $25 for every new customer she signs up. How many customers must she sign up to earn more than $401.50 during the next 26 working hours?

54. A waitress earns $3 per hour plus an average tip of $4 for every table served. How many tables must she serve to earn more than $52 for a 4-hour shift?

55. A time-share salesperson sells the right to one week's vacation in San Diego, California for a cost of $12,000 for a lifetime's use. The corporation's overall fixed costs are $500,000 per month. The average time-share has a real market value of $5200. The board of directors want the salespersons to sell enough time-shares to realize a profit of $1,100,000 per month. How many time-shares must be sold each month?

56. Charlotte is investing for the first time. She has chosen to invest $3200.00 in a certificate-of-deposit for one year. Since the investment pays simple interest, what interest rate is required if she wants to earn more than $192 in interest?

57. Sandy and Ed Washington live in Chicago. They have been doing their wash at a local laundromat, where they spend $5.75 per week. They are now able to buy a new washing machine for $295. How many weeks will it take for the new washer to become the less expensive choice?

58. A landscaper put in his best price to Morticia and Gomez, who want to put in a new garden. The landscaper will charge Morticia and Gomez $250 to dig up the old garden and then $6.00 per hour for the new planting or $8.00 per hour for the entire job. How many hours could be spent digging up the garden for $250 until it becomes more cost efficient to pay the overall $8.00 per hour?

Cumulative Review Problems

Simplify.

59. $\dfrac{3x^2y^6}{-18x^3y^4}$

60. $(2x^2y^{-3})^2$

61. $\left(\dfrac{3x}{2y^2w^{-4}}\right)^3$

62. $(-4x^{-2}y^4z^{-6})^{-2}$

2.7 Compound Inequalities

MathPro

Video 6

SSM

After studying this section, you will be able to:

1 *Graph a compound inequality using the connective AND.*
2 *Graph a compound inequality using the connective OR.*
3 *Solve a compound inequality and graph the solution.*

1 *Graphing Compound Inequalities Using the Connective AND*

Some inequalities consist of two inequalities connected by the word *and* or the word *or*. They are called **compound inequalities.** The solution of a compound inequality using the connective *and* includes all the numbers that make both parts true at the same time.

EXAMPLE 1   Graph the values of x where $7 < x$ and $x < 12$.

We read the inequality starting with the variable. Thus we graph all values of x, where x is greater than 7 and where x is less than 12. All such values must be between 7 and 12. Numbers that are greater than 7 and less than 12 can be written as $7 < x < 12$.

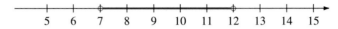

Practice Problem 1   Graph the values of x where $-8 < x < -2$.

■

EXAMPLE 2   Graph the values of x where $-6 \le x \le 2$.

Here we have x is greater than or equal to -6 and x is less than or equal to 2. We remember to include the points -6 and 2 since the symbol contains the equal sign.

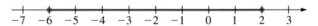

Practice Problem 2   Graph the values of x where $-1 \le x \le 5$.

■

EXAMPLE 3   Graph the values of x where $-8.5 \le x < -1$.

Note the shaded circle at -8.5 and the open circle at -1.

Practice Problem 3   Graph the region $-10 \le x \le -5.5$.

■

EXAMPLE 4   Graph the take-home salary range (*s*) of the full-time employees of Tentron Corporation. Each person earns at least $190 weekly, but not more than $800 weekly in their take-home paycheck.

At least $190 means that the take-home salary of each person is greater than or equal to $190 weekly. We write $s \geq 190$. Not more than means that the salary of each person is less than or equal to $800 weekly. We write $s \leq 800$. Thus *s* may be between 190 and 800 and may include those end points.

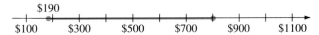

Practice Problem 4   Graph the take-home salary range if each person earns at least $200 per week but never more than $950 per week.

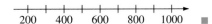

2 *Graphing Compound Inequalities Using the Connective OR*

The solution of a compound inequality using the connective *or* includes all the numbers that belong to either of the two inequalities.

EXAMPLE 5   Graph the region where $x < 3 \ or \ x > 6$.

Notice that the solution to this inequality need not be in both regions at the same time.

Read the inequality as *x* is less than 3 or *x* is greater than 6. *x* can be less than 3 or *x* can be greater than 6. This includes all values to the left of 3 as well as all values to the right of 6 on the number line. We shade these regions.

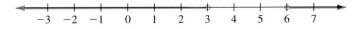

Practice Problem 5   Graph the region where $x < 8 \ or \ x > 12$.

EXAMPLE 6   Graph the region where $x > -2 \ or \ x \leq -5$.

Note the shaded circle at -5 and the open circle at -2.

Practice Problem 6   Graph the region where $x \leq -6 \ or \ x > 3$.

EXAMPLE 7 Male applicants for the state police force in Fred's home state are ineligible for the force if they are shorter than 60 inches or taller than 76 inches. Graph these regions.

The rejected applicant's height h will be less than 60 inches ($h < 60$) *or* will be greater than 76 inches ($h > 76$).

Practice Problem 7 Graph the region if female applicants are ineligible if they are shorter than 56 inches or taller than 70 inches.

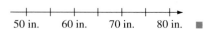

3 Solving Compound Inequalities

When asked to solve for x in two compound inequalities, we normally solve each one separately.

EXAMPLE 8 Solve for x where $3x + 2 > 14$ *or* $2x - 1 < -7$. Graph your solution.

We solve each inequality separately.

$$3x + 2 > 14 \quad or \quad 2x - 1 < -7$$
$$3x > 12 \qquad\qquad 2x < -7 + 1$$
$$x > 4 \qquad\qquad 2x < -6$$
$$\qquad\qquad\qquad x < -3$$

The solution is $x < -3$ *or* $x > 4$.

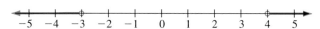

Practice Problem 8 Solve for x and graph where $3x - 4 < -1$ *or* $2x + 3 > 13$.

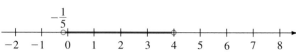

EXAMPLE 9 Solve for x and graph when $5x - 1 > -2$ *and* $3x - 4 < 8$.

We solve each inequality separately.

$$5x - 1 > -2 \qquad and \quad 3x - 4 < 8$$
$$5x > -2 + 1 \qquad\qquad 3x < 8 + 4$$
$$5x > -1 \qquad\qquad\qquad 3x < 12$$
$$x > -\frac{1}{5} \qquad\qquad\qquad x < 4$$

The solution is all the numbers between $-\frac{1}{5}$ and 4, not including these end points.

$$-\frac{1}{5} < x < 4$$

Practice Problem 9 Solve for x when $3x + 6 > -6$ *and* $4x + 5 < 1$.

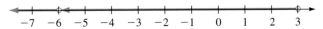

EXAMPLE 10 Solve and graph. $2x + 5 \le 11$ *and* $-3x > 18$

We solve each inequality separately.

$$2x + 5 \le 11 \quad and \quad -3x > 18$$

$$2x \le 6 \qquad\qquad x < -\frac{18}{3}$$

$$x \le 3 \qquad\qquad\qquad x < -6$$

The solution is $x < -6$ *and* at the same time $x \le 3$.

Notice this can be described by *one* region. The only numbers x that satisfy both statements $x \le 3$ *and* $x < -6$ at the same time would be $x < -6$. Thus $x < -6$ is the solution to the compound inequality.

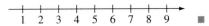

Practice Problem 10 Solve for x when $-2x + 3 < -7$ *and* $7x - 1 > -15$.

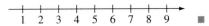

EXAMPLE 11 Solve. $-3x - 2 < -5$ *and* $4x + 6 < -12$

We solve each inequality separately.

$$-3x - 2 < -5 \qquad and \qquad 4x + 6 < -12$$

$$-3x - 2 + 2 < -5 + 2 \qquad 4x + 6 - 6 < -12 - 6$$

$$-3x < -3 \qquad\qquad 4x < -18$$

$$\frac{-3x}{-3} > \frac{-3}{-3} \qquad\qquad \frac{4x}{4} < \frac{-18}{4}$$

$$x > 1 \qquad\qquad\qquad x < -4\frac{1}{2}$$

Now clearly it is impossible for one number to be greater than 1 *and* at the same time less than $-4\frac{1}{2}$.

There is *no solution*. We express this by the notation $\varnothing$, which is an empty set.

Practice Problem 11 Solve for x when $-3x - 11 < -26$ *and* $5x + 4 < 14$. ■

2.7 Exercises

Graph the values of x that satisfy the conditions given.

1. $3 < x$ and $x < 8$

2. $5 < x$ and $x < 10$

3. $-4 < x$ and $x < 2$

4. $-7 < x$ and $x < 1$

5. $7 < x < 9$

6. $3 < x < 5$

7. $-1 < x \le 3$

8. $-1 \le x \le 6$

9. $x > 5$ or $x < 3$

10. $x > 7$ or $x < 2$

11. $x \le -\dfrac{5}{2}$ or $x > 4$

12. $x < -2$ or $x > 4$

13. $x \le -1$ or $x \ge 4$

14. $x \le -6$ or $x \ge 2$

Solve for x and graph your result.

15. $3x + 1 < -2$ and $x \ge -5$

16. $4x - 1 < 7$ and $x \ge -1$

17. $2x - 3 > 0$ or $x - 2 < -7$

18. $x + 1 \ge 5$ or $x + 5 < 2.5$

19. $x < 8$ and $x > 10$

20. $x < 6$ and $x > 9$

Applications

Express as an inequality.

21. The seam on the blue jeans (s) was unacceptable if it was narrower than 10 millimeters or wider than 12 millimeters.

22. The amount of toothpaste in the tube (t) was not properly filled if there were more than 11.2 ounces, or less than 10.9 ounces.

23. The number of campers (c) at the campsite during the independence day weekend was always at least 490 but never more than 2000.

24. The number of cars c driving over Interstate 91 during the evening hours during January was always at least 5000 but never more than 12,000.

Solve the following application problems by using the formula $C = \dfrac{5}{9}(F - 32)$. *Round to the nearest tenth.*

25. The temperature in Mexico City during February can range from 8°C to 23°C. Find an inequality that represents the range in Fahrenheit temperatures.

26. When visiting Montreal this spring, Marcos has been advised that the temperature can range from -20°C to 11°C. Find an inequality that represents the range in Fahrenheit temperatures.

The exchange equation in April 1997 for converting American dollars into Japanese yen was $Y = 129(d - 4)$. *In this equation, d is the number of American dollars, Y is the number of yen, and $4 represents a one-time fee that banks sometimes charge for currency conversion. Use this equation to solve the following problems. (Round answers to the nearest cent.)*

27. Betty is traveling to Japan for two weeks, and has been advised to have between 18,000 yen and 33,000 yen for spending money for each week she is there. Write an inequality that represents the number of American dollars she will need to bring to the bank to exchange money for this two-week period.

28. Paul is traveling to Japan for two weeks, and has been advised to have between 17,000 yen and 29,000 yen for spending money for each week he is there. Write an inequality that represents the number of American dollars he will need to bring to the bank to exchange money for this two-week period.

Solve the compound inequality.

29. $x - 3 > -5$ *and* $2x + 4 < 8$

30. $x + 3 < 7$ *and* $x - 2 < -3$

31. $5x - 7 \geq 3$ *and* $4x - 8 \leq 0$

32. $2x - 5 \geq 1$ *and* $3x - 3 \leq 6$

33. $2x - 5 < -11$ *or* $5x + 1 \geq 6$

34. $3x + 2 < 5$ *or* $5x - 7 > 8$

35. $-3x + 10 > 2x$ *or* $-2x + 5 > 7$

36. $-6x - 8 > 2x$ *or* $-4x + 6 < 8x$

37. $3x - 4 > -1$ *or* $2x + 1 < 15$

38. $4x + 3 < -1$ *or* $2x - 3 > -11$

39. $7x + 2 \geq 2$ *and* $2x + 7 \geq 19$

40. $5x - 6 < 14$ *and* $6x + 5 < -1$

41. $2x + 5 < 3$ *and* $3x - 1 > -1$

42. $6x - 10 < 8$ *and* $2x + 1 > 9$

43. $-4x - 1 \geq -5$ *and* $x + 1 > -1.5$

44. $-7x + 3 \geq -11$ *and* $1 - x < 4.2$

45. $2x - 1 > -9$ *and* $3x + 1 < 13 + x$

46. $-3x - 7 < 2$ *and* $x + 1 > 4x + 7$

47. $6x - 5 \leq -8$ *or* $7x - 5 \leq 16$

48. $x - 4 \geq 1$ *or* $-3x + 1 \geq -5 - x$

Solve for x. Round your answer to two decimal places.

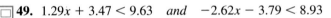

49. $1.29x + 3.47 < 9.63$ *and* $-2.62x - 3.79 < 8.93$

50. $2.35x + 6.62 \geq 5.04x - 1.23$ *or* $9.28x \geq 52.71$

To Think About

Solve the compound inequality.

51. $x \geq 3$ *and* $x > -2$ *and* $x \leq 8$ *and* $x < 12$

52. $x > -2.5$ *and* $x > 6$ *and* $x < 3.5$ *and* $x \leq 14$

53. $\frac{1}{4}(x + 2) + \frac{1}{8}(x - 3) \leq 1$ *and* $\frac{3}{4}(x - 1) > -\frac{1}{4}$

54. $\frac{x - 4}{6} - \frac{x - 2}{9} \leq \frac{5}{18}$ *or* $-\frac{2}{5}(x + 3) < -\frac{6}{5}$

Cumulative Review Problems

Solve for the specified letter.

55. Solve for x: $3y - 5x = 8$

56. Solve for y: $7x + 6y = -12$

57. Solve for a: $3d - 4a = 5x + 2a$

58. Solve for p: $I = prt$

2.8 Absolute Value Inequalities

MathPro   Video 6   SSM

After studying this section, you will be able to:

1 *Solve absolute value inequalities of the form* $|ax + b| < c$.
2 *Solve absolute value inequalities of the form* $|ax + b| > c$.

1 *Solving Absolute Value Inequalities of the Form* $|ax + b| < c$

We begin by looking at $|x| < 3$. What does this mean? $|x| < 3$ means that x is less than 3 units from zero on the number line. We draw a picture.

This picture shows all possible values of x such that $|x| < 3$. We see that this occurs when $-3 < x < 3$. We conclude that $|x| < 3$ and $-3 < x < 3$ are equivalent statements.

Definition

If a is a positive real number, when $|x| < a$, then $-a < x < a$.

EXAMPLE 1   Solve. $|x| \leq 4.5$

$|x| \leq 4.5$ means that x is less than or equal to 4.5 units from zero on the number line. We draw a picture.

Thus the solution is $-4.5 \leq x \leq 4.5$.

Practice Problem 1   Solve. $|x| < 2$

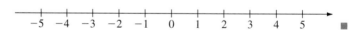 ■

This same technique can be used to solve more complicated inequalities.

EXAMPLE 2   Solve. $|x + 5| \leq 10$ Graph your solution.

$|x + 5| \leq 10$ means that $x + 5$ is less than 10 units from zero on the number line. We draw a picture.

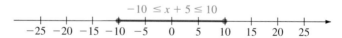

$$-10 \leq x + 5 \leq 10$$

Note that this is not the solution. This does not tell us the values of x that make $-10 \leq x + 5 \leq 10$ a true statement. We need to solve the compound inequality.

To solve this inequality, we add -5 to each part.

$$-10 - 5 \le x + 5 - 5 \le 10 - 5$$

$$-15 \le x \le 5$$

Thus the solution is $-15 \le x \le 5$. We graph this solution.

$$-15 \le x \le 5$$

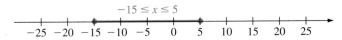

$-15 \le x \le 5$

Practice Problem 2 Solve. $|x - 6| < 15$ Graph your solution. (*Hint:* Choose a scale.)

0

EXAMPLE 3 Solve. $\left| x - \dfrac{2}{3} \right| \le \dfrac{5}{2}$ Graph your solution.

$-\dfrac{5}{2} \le x - \dfrac{2}{3} \le \dfrac{5}{2}$ *If $|x| < a$, then $-a < x < a$.*

$6\left(-\dfrac{5}{2}\right) \le 6(x) - 6\left(\dfrac{2}{3}\right) \le 6\left(\dfrac{5}{2}\right)$ *Multiply each part of the inequality by 6.*

$-15 \le 6x - 4 \le 15$ *Simplify.*

$-15 + 4 \le 6x - 4 + 4 \le 15 + 4$ *Add 4 to each part.*

$-11 \le 6x \le 19$ *Simplify.*

$-\dfrac{11}{6} \le \dfrac{6x}{6} \le \dfrac{19}{6}$ *Divide each part by 6.*

$-1\dfrac{5}{6} \le x \le 3\dfrac{1}{6}$ *Change to mixed numbers to facilitate graphing.*

$-1\dfrac{5}{6} \le x \le 3\dfrac{1}{6}$

$$-1\dfrac{5}{6} \le x \le 3\dfrac{1}{6}$$

Practice Problem 3 Solve. $\left| x + \dfrac{3}{4} \right| \le \dfrac{7}{6}$

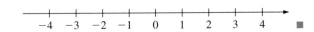

EXAMPLE 4 Solve. $|2(x - 1) + 4| < 8$ Graph your solution.

First we simplify the expression within the absolute value symbol.

$$|2x - 2 + 4| < 8$$

$$|2x + 2| < 8$$

$$-8 < 2x + 2 < 8 \qquad \textit{If } |x| < a, \textit{ then } -a < x < a.$$

$$-8 - 2 < 2x + 2 - 2 < 8 - 2 \qquad \textit{Add } -2 \textit{ to each part.}$$

$$-10 < 2x < 6 \qquad \textit{Simplify.}$$

$$\frac{-10}{2} < \frac{2x}{2} < \frac{6}{2} \qquad \textit{Divide each part by 2.}$$

$$-5 < x < 3$$

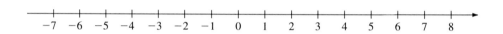

$-5 < x < 3$

Practice Problem 4 Solve and graph. $|2 + 3(x - 1)| < 20$

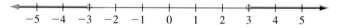

2 Solving Absolute Value Inequalities of the Form $|ax + b| > c$

Now consider $|x| > 3$. What does this mean? $|x| > 3$ means that x is greater than 3 units from zero on the number line. We draw a picture.

This picture shows all possible values of x such that $|x| > 3$. This occurs when $x < -3$ or when $x > 3$. (Note that a solution can either be in the region to the left of -3 on the number line or to the right of 3 on the number line.) We conclude that $|x| > 3$ and $x < -3$ or $x > 3$ are equivalent statements.

> **Definition**
>
> If a is a positive real number when $|x| > a$, then $x < -a$ or $x > a$.

EXAMPLE 5 Solve. $|x| \geq 5\frac{1}{4}$

$|x| \geq 5\frac{1}{4}$ is more than $5\frac{1}{4}$ units from zero on the number line. We draw a picture.

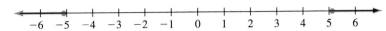

Thus the solution is $x \leq -5\frac{1}{4}$ or $x \geq 5\frac{1}{4}$.

Practice Problem 5   Solve. $|x| > 2.5$

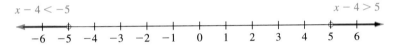

This same technique can be used to solve more complicated inequalities.

EXAMPLE 6   Solve. $|x - 4| > 5$ Graph your solution.

$|x - 4| > 5$ means that $x - 4$ is more than 5 units from zero on the number line. We draw a picture.

$x - 4 < -5$ $x - 4 > 5$

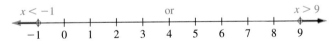

Note this is not the solution. This does not tell us the values of x that make $x - 4 < -5$ or $x - 4 > 5$ a true statement. We need to solve the compound inequality.

We will solve each inequality separately.

$$x - 4 < -5 \qquad or \qquad x - 4 > 5$$
$$x - 4 + 4 < -5 + 4 \qquad x - 4 + 4 > 5 + 4$$
$$x < -1 \qquad\qquad x > 9$$

Thus the solution is $x < -1$ or $x > 9$. We graph the solution on the number line.

$x < -1$ or $x > 9$

Practice Problem 6   Solve. $|x + 6| > 2$ Graph the solution.

EXAMPLE 7   Solve. $|-3x + 6| > 18$

By definition, we have the compound inequality

$$-3x + 6 > 18 \qquad\qquad or \qquad\qquad -3x + 6 < -18$$

$$-3x > 12 \qquad\qquad\qquad\qquad\qquad -3x < -24$$

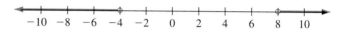 $\dfrac{-3x}{-3} < \dfrac{12}{-3} \longrightarrow$ *Division by negative* $\longrightarrow \dfrac{-3x}{-3} > \dfrac{-24}{-3}$
number reverses inequality

$$x < -4 \qquad\qquad\qquad\qquad\qquad\qquad x > 8$$

The solution is $x < -4$ *or* $x > 8$.

Practice Problem 7   Solve and graph. $|-5x - 2| > 13$

(number line: $-4 -3 -2 -1\ 0\ 1\ 2\ 3\ 4$)   ▪

EXAMPLE 8   Solve. $\left|3 - \dfrac{2}{3}x\right| \geq 5$

By definition, we have the compound inequality.

$$3 - \dfrac{2}{3}x \geq 5 \qquad or \qquad 3 - \dfrac{2}{3}x \leq -5$$

$$3(3) - 3\left(\dfrac{2}{3}x\right) \geq 3(5) \qquad 3(3) - 3\left(\dfrac{2}{3}x\right) \leq 3(-5)$$

$$9 - 2x \geq 15 \qquad\qquad\qquad 9 - 2x \leq -15$$

$$-2x \geq 6 \qquad\qquad\qquad\qquad -2x \leq -24$$

$$\dfrac{-2x}{-2} \leq \dfrac{6}{-2} \qquad\qquad\qquad \dfrac{-2x}{-2} \geq \dfrac{-24}{-2}$$

$$x \leq -3 \qquad\qquad\qquad\qquad x \geq 12$$

The solution is $x \leq -3$ *or* $x \geq 12$.

(number line: $-15 -12 -9 -6 -3\ 0\ 3\ 6\ 9\ 12\ 15$)

Practice Problem 8   Solve and graph. $\left|4 - \dfrac{3}{4}x\right| \geq 5$

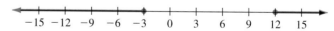

▪

EXAMPLE 9 In building a new car transmission, the diameter d of the transmission must differ from the specified standard s by not more than 0.37 millimeter. A distance of 0.37 mm is less than the thickness of this printed line → | ← . The engineers express this requirement as $|d - s| \leq 0.37$. If the standard s is 216.82 millimeters for a particular car, find the limits of d.

$$|d - s| \leq 0.37$$

$$|d - 216.82| \leq 0.37 \qquad \textit{Substitute the known value of s.}$$

$$-0.37 \leq d - 216.82 \leq 0.37 \qquad \textit{When } |x| \leq a, \textit{ then } -a \leq x \leq a.$$

$$-0.37 + 216.82 \leq d - 216.82 + 216.82 \leq 0.37 + 216.82$$

$$216.45 \leq d \leq 217.19$$

Thus the diameter of the transmission must be at least 216.45 millimeters, but not greater than 217.19 millimeters.

Practice Problem 9 The diameter d of the transmission must differ from the specified standard s by not more than 0.37 millimeter. Solve to find the allowed limits of d for a truck transmission for which the standard s is 276.53 millimeters. ∎

Summary of Absolute Value Equations and Inequalities

It may be helpful to review the key concepts of absolute value equations and inequalities that we have covered in sections 2.3 and 2.8. For real numbers a, b, and c where $a \neq 0$ and $c > 0$:

Form of the equation or inequality	Type of solution obtained	Graphed form of the solution on a number line
$|ax + b| = c$	Two distinct numbers, m and n	
$|ax + b| < c$	The set of numbers between the two numbers m and n, that is $m < x < n$	
$|ax + b| > c$	The set of numbers less than m or the set of numbers greater than n $x < m$ or $x > n$	

2.8 Exercises

Solve and graph the solutions.

1. $|x| \le 8$

2. $|x| < 6$

3. $|x| > 5$

4. $|x| \ge 7$

5. $|x + 4.5| < 5$

$-10 \quad -8 \quad -6 \quad -4 \quad -2 \quad 0 \ 1$

6. $|x + 6| < 3.5$

$-10 \quad -8 \quad -6 \quad -4 \quad -2$

Solve for x.

7. $|x - 6| \le 8$

8. $|x - 7| \le 10$

9. $|2x + 3| \le 17$

10. $|3x + 2| \le 12$

11. $|0.3x - 0.5| \le 0.1$

12. $|0.2x - 0.7| \le 0.3$

13. $\left| x - \dfrac{1}{2} \right| < \dfrac{5}{2}$

14. $\left| x - \dfrac{3}{2} \right| < \dfrac{1}{2}$

15. $\left| \dfrac{1}{4}x + 2 \right| < 6$

16. $\left| \dfrac{1}{5}x + 1 \right| < 5$

17. $|-1 + 3(x + 1)| \le 5$

18. $|-3 + 4(x + 1)| \le 3$

19. $\left| \dfrac{4}{5}(x - 1) \right| < 8$

20. $\left| \dfrac{3}{4}(x - 1) \right| < 6$

21. $\left| \dfrac{2x + 6}{3} \right| < 2$

22. $\left| \dfrac{5x - 3}{2} \right| < 4$

23. $|x + 2| > 5$

24. $|x + 4| > 7$

25. $|x - 2| \ge 3$

26. $|x - 1| \ge 2$

27. $|3x - 8| \ge 7$

28. $|5x - 2| \ge 13$

29. $\left| 3 - \dfrac{3}{4}x \right| > 9$

30. $\left| 3 - \dfrac{2}{3}x \right| > 5$

31. $\left| \dfrac{1}{5}x - \dfrac{1}{10} \right| > 2$

32. $\left| \dfrac{1}{4}x - \dfrac{3}{8} \right| > 1$

33. $|5 - 7x| \ge 9$

34. $|11 - 6x| \ge 7$

35. $\left| \dfrac{1}{3}(x - 2) \right| < 5$

36. $\left| \dfrac{2}{5}(x - 2) \right| \le 4$

Applications

In a certain company the measured thickness m of a helicopter blade must differ from the exact standard s by not more than 0.12 millimeter. The manufacturing engineer expresses this as $|m - s| \leq 0.12$.

37. Find the limits of m if the standard s is 18.65 millimeters.

38. Find the limits of m if the standard s is 17.48 millimeters.

A small computer microchip has dimension requirements. The manufacturing engineer has written the specification that the new length n of the chip can differ from the previous length p by only 0.05 centimeter or less. The equation is $|n - p| \leq 0.05$ *centimeter.*

39. Find the limit of the new length if the previous length is 9.68 centimeters.

40. Find the limits of the new length if the previous length is 7.84 centimeters.

To Think About

41. A student tried to solve the inequality $|4x - 8| > 12$. Instead of writing $4x - 8 > 12$ or $4x - 8 < -12$, which he should have done, he made a mistake. In error he wrote $12 < 4x - 8 < -12$. What is wrong with his approach?

42. A student tried to write a solution to a triple inequality $6 < 4 - 3x < 19$. He made an error and got the wrong answer. He used the following steps:

Step 1 $6 - 4 < 4 - 4 - 3x < 19 - 4$

Step 2 $2 < -3x < 15$

Step 3 $\dfrac{2}{-3} < \dfrac{-3x}{-3} < \dfrac{15}{-3}$

Step 4 $\dfrac{-2}{3} < x < -5$

What *serious error* did he make?

Cumulative Review Problems

Perform the correct order of operations to simplify.

43. $12 \div (-2)(3) - (-5) + 2$

44. $(6 - 4)^3 \div (-4) + 2^2$

In Problems 45 and 46 use $\pi \approx 3.14$. *Round answers to nearest hundredth.*

45. The Outward Bound program in the United States is famous for teaching self-esteem and personal achievement to young people. One of its physical obstacles is for a student to hang on to a rope 19 meters long and swing from one shore to another, then back. The rope swings through a circular arc, measuring $\frac{1}{8}$ of the circumference of a circle. How many seconds does it take for one full *round trip* swing if the student moves at 3 meters per second?

46. The rigging on a sailboat comes loose from the mast. The end of the wire rigging hanging down is 30 feet from the top of the mast. The end of the wire rigging swings through a circular arc, measuring $\frac{1}{6}$ of the circumference of a circle. How many seconds would it take the rigging wire, swinging like a pendulum, to swing *round trip*, if the wind is blowing the end at 8 feet per second?

Topic	Procedure	Examples
Solving first-degree equations, p. 60.	1. Remove any parentheses. 2. If fractions exist, multiply all terms on both sides by the LCD of all the fractions. 3. If decimals exist, multiply all terms on both sides by a power of 10. 4. Collect like terms if possible. 5. Add or subtract terms on both sides of the equation to get all terms with the variable on one side of the equation. 6. Add or subtract a value on both sides of the equation to get all terms not containing the variable on the other side of the equation. 7. Divide both sides of the equation by the coefficient of the variable. 8. Simplify the solution (if possible). 9. Check the solution.	Solve for x. $\dfrac{1}{3}(2x - 3) + \dfrac{1}{2}(x + 1) = 3$ $$\frac{2}{3}x - 1 + \frac{1}{2}x + \frac{1}{2} = 3$$ $$6\left(\frac{2}{3}x\right) - 6(1) + 6\left(\frac{1}{2}x\right) + 6\left(\frac{1}{2}\right) = 6(3)$$ $$4x - 6 + 3x + 3 = 18$$ $$7x - 3 = 18$$ $$7x = 18 + 3$$ $$7x = 21$$ $$x = 3$$ Check. $\dfrac{1}{3}[2(3) - 3] + \dfrac{1}{2}(3 + 1) \overset{?}{=} 3$ $$\frac{1}{3}[3] + \frac{1}{2}(4) \overset{?}{=} 3$$ $$1 + 2 = 3 \quad \checkmark \quad \text{It checks.}$$
Equations and formulas with more than one variable, p. 68.	If an equation or formula has more than one variable, you can solve for a particular variable by using the procedure for solving linear equations. Remember, your goal is to get all terms containing the desired variable on one side of the equation and all other terms on the opposite side of the equation.	Solve for r. $A = P(1 + rt)$ $$A = P + Prt$$ $$A - P = Prt$$ $$\frac{A - P}{Pt} = \frac{Prt}{Pt}$$ $$\frac{A - P}{Pt} = r$$
Absolute value equations, p. 74.	To solve an equation that involves an absolute value, we rewrite the absolute value equation as two separate equations. We solve each equation. If $|ax + b| = c$, then $ax + b = c$ or $ax + b = -c$.	Solve for x. $\|4x - 1\| = 17$ $4x - 1 = 17 \qquad$ or $\quad 4x - 1 = -17$ $4x = 17 + 1 \qquad\qquad 4x = -17 + 1$ $4x = 18 \qquad\qquad\quad 4x = -16$ $x = \dfrac{18}{4} \qquad\qquad\quad x = \dfrac{-16}{4}$ $x = \dfrac{9}{2} \qquad\qquad\qquad\; = -4$

Solving word problems, p. 80.	1. *Understand the problem.*
	  (a) Read the word problem carefully to get an overview.
	  (b) Determine what information you need to solve the problem.
	  (c) Draw a sketch. Label it with the known information. Determine what needs to be found.
	  (d) Choose a variable to present one unknown quantity.
	  (e) If necessary, represent other unknown quantities in terms of that same variable.
	2. *Write an equation.*
	  (a) Look for key words to help you to translate the words into algebraic symbols.
	  (b) Use a given relationship in the problems or an appropriate formula in order to write an equation.
	3. *Solve the equation and state the answer.*
	4. *Check.*
	  (a) Check the solution in the original equation.
	  (b) Be sure the solution to the equation answers the question in the word problem. You may need some additional calculations if it does not.
		Fred and Linda invested $7000 for one year. Part was invested at 3% interest and part was invested at 5% interest. If they earned a total of $320 in one year in interest, how much did they invest at each rate?
		
		1. *Understand the problem.*
		Let x = the amount invested at 3% interest
		Let $7000 - x$ = the amount invested at 5% interest
		2. *Write an equation.*
		$$0.03x + 0.05(7000 - x) = 320$$
		3. *Solve the equation and state the answer.*
		$$0.03x + 350 - 0.05x = 320$$
		$$350 - 0.02x = 320$$
		$$-0.02x = -30$$
		$$x = 1500$$
		$$7000 - x = 5500$$
		Therefore, $1500 was invested at 3% interest and $5500 at 5% interest.
		4. *Check.*
		$$0.03(1500) + 0.05(7000 - 1500) \overset{?}{=} 320$$
		$$45 \quad + \quad 275 \quad \overset{?}{=} 320$$
		$$320 \quad = 320 \quad \checkmark$$
		Yes. It checks.

Topic	Procedure	Examples
Solving linear inequalities, p. 96.	**1.** If $a < b$, then for all real numbers a, b, c, $$a + c < b + c \quad \text{and} \quad a - c < b - c$$ **2.** If $a < b$, then for all real numbers a, b when $c > 0$, $$ac < bc \quad \text{and} \quad \frac{a}{c} < \frac{b}{c}$$ Multiplying or dividing both sides of an inequality by a positive number does **not** reverse the inequality. **3.** If $a < b$, then for all real numbers a, b when $c < 0$, $$ac > bc \quad \text{and} \quad \frac{a}{c} > \frac{b}{c}$$ Multiplying or dividing both sides of an inequality **reverses the direction** of the inequality symbol.	Solve and graph. $3(2x - 4) + 1 \geq 7$ $$6x - 12 + 1 \geq 7$$ $$6x - 11 \geq 7$$ $$6x \geq 18$$ $$x \geq 3$$ Solve and graph. $\frac{1}{4}(x + 3) \leq \frac{1}{3}(x - 2)$ $$\frac{1}{4}x + \frac{3}{4} \leq \frac{1}{3}x - \frac{2}{3}$$ $$3x + 9 \leq 4x - 8$$ $$-1x + 9 \leq -8$$ $$-1x \leq -17$$ $$\frac{-1x}{-1} \geq \frac{-17}{-1}$$ $$x \geq 17$$
Solving compound inequalities containing **and**, p. 106.	The solution is the desired region containing all values of x that meet both conditions.	Graph the values of x satisfying $x + 6 > -3$ and $2x - 1 < -4$. $$x + 6 - 6 > -3 - 6 \quad \text{and} \quad 2x < -3$$ $$x > -9 \qquad\qquad \text{and} \quad x < -1.5$$
Solving compound inequalities containing **or**, p. 107.	The solution is the desired region containing all values of x that meet either of the two conditions.	Graph the values of x satisfying $-3x + 1 \leq 7$ or $3x + 1 \leq -11$. $$-3x + 1 - 1 \leq 7 - 1 \quad \text{or} \quad 3x + 1 - 1 \leq -11 - 1$$ $$-3x \leq 6 \qquad \text{or} \qquad\qquad 3x \leq -12$$ $$\frac{-3x}{-3} \geq \frac{6}{-3} \quad \text{or} \qquad\qquad \frac{3x}{3} \leq \frac{-12}{3}$$ $$x \geq -2 \qquad \text{or} \qquad\qquad x \leq -4$$
Solving absolute value inequalities involving $<$ or $\leq$, p. 112.	When a is a positive real number: If $|x| < a$, then $-a < x < a$. If $|x| \leq a$, then $-a \leq x \leq a$.	Solve and graph. $|3x - 2| < 19$ $$-19 < 3x - 2 < 19$$ $$-19 + 2 < 3x - 2 + 2 < 19 + 2$$ $$-17 < 3x < 21$$ $$-\frac{17}{3} < \frac{3x}{3} < \frac{21}{3}$$ $$-5\frac{2}{3} < x < 7$$

Chapter Organizer

Topic	Procedure	Examples
Solving absolute value inequalities involving > or ≥, p. 114.	When a is a positive real number: If $\|x\| > a$, then $x > a$ or $x < -a$. If $\|x\| \geq a$, then $x \geq a$ or $x \leq -a$.	Solve and graph. $\left\|\dfrac{1}{3}(x-2)\right\| \geq 2$ $\dfrac{1}{3}(x-2) \geq 2 \qquad$ or $\qquad \dfrac{1}{3}(x-2) \leq -2$ $\dfrac{1}{3}x - \dfrac{2}{3} \geq 2 \qquad\qquad \dfrac{1}{3}x - \dfrac{2}{3} \leq -2$ $x - 2 \geq 6 \qquad\qquad x - 2 \leq -6$ $x \geq 6 + 2 \qquad\qquad x \leq -6 + 2$ $x \geq 8 \qquad$ or $\qquad x \leq -4$ $\xleftarrow{\quad\overset{\bullet}{-16}\ \overset{\bullet}{-8}\ \ 0\ \ \ 8\ \ 16\quad}\rightarrow$

Chapter 2 Review Problems

Solve for x.

1. $5x - 1 = -6x - 23$

2. $6 - (3x + 5) = 15 - (x - 6)$

3. $4(x - 1) + 2 = 3x + 8 - 2x$

4. $x - \dfrac{7}{5} = \dfrac{1}{3}x + \dfrac{7}{15}$

5. $\dfrac{x - 4}{2} - \dfrac{1}{5} = \dfrac{7x + 1}{20}$

6. $\dfrac{1}{9}x - 1 = \dfrac{1}{2}\left(x + \dfrac{1}{3}\right)$

7. $4.6x = 2(1.6x - 2.8)$

8. $0.6 - 0.2(x - 3) + 0.5 = 1.5$

9. Solve for y.
 $4x - 8y = 5$

10. Solve for a.
 $2(3ax - 2y) - 6ax = -3(ax + 2y)$

11. **(a)** Solve for W. $P = 2W + 2L$

   (b) Now find W when $P = 100$ meters and $L = 20.5$ meters.

12. **(a)** Solve for F. $C = \dfrac{5F - 160}{9}$

   (b) Now find F when $C = 10°$.

Solve for x.

13. $\|x + 1\| = 8$

14. $\|x - 3\| = 12$

15. $\|4x - 5\| = 7$

16. $\|3x + 2\| = 20$

17. $\|x + 8\| = \|2x - 4\|$

18. $\|3 - x\| = \|5 - 2x\|$

19. $\left\|\dfrac{1}{4}x - 3\right\| = 8$

20. $\|4 - 7x\| = 25$

21. $\|2x - 8\| + 7 = 12$

Solve each problem.

22. Five-eighths of a number is 290. What is the number?

23. The number of men attending Western Tech is 200 less than twice the number of women. The number of students at the school is 280. How many men attend? How many women attend?

24. City Compacts rents cars for $30 per day plus 12¢ per mile. Juanita rented a car for two days and was charged $102. How many miles did she drive the car?

25. The sum of three consecutive even integers is 408. What are the integers?

26. A rectangular piece of copper has a perimeter of 88 mm. The length of the rectangle is 8 mm longer than three times the width. Find the length and width.

27. Alice's employer withholds from her monthly paycheck $102 for federal and state taxes and for retirement. She noticed that the amount withheld for her state income tax is $13 more than that withheld for retirement. The amount withheld for federal income tax is three times the amount withheld for the state tax. How much is withheld monthly for federal tax? state tax? retirement?

28. Hightstown College has 12% fewer students this year than it did last year. There are 2332 students enrolled. How many students were enrolled last year?

29. An auto manufacturer wants to make 260,000 sedans a year. Some will be two-door sedans; the rest will be four-door. He also wants to make three times as many four-door sedans as he does two-door sedans. How many of each type should be manufactured?

30. Lucinda has invested $7000 in mutual funds and bonds. The mutual fund earns 12% simple interest. The bonds earn 8% simple interest. At the end of the year she earned $740 in interest. How much did she invest at each rate?

31. To make a weak solution of 24 liters of 4% acid, a lab technician has some premixed solutions: one is 2% acid; the other, 5% acid. How many liters of each type should he mix to obtain the desired solution?

32. A local coffee specialty shop wants to obtain 30 pounds of a mixture of coffee beans costing $4.40 per pound. They have a mixture costing $4.25 a pound and a mixture costing $4.50 per pound. How much of each should be used?

33. When Eastern Slope Community College opened, the number of students (full-time and part-time) was 380. Since then the number of full-time students has doubled, and the number of part-time students has tripled. There are now 890 students at the school. How many of the present students are full-time? part-time?

Solve for x.

34. $8x + 3 < 5x$

35. $7x + 5 < 2x$

36. $2 - x \geq 3x + 10$

37. $2 - 3x \geq 10 - 7x$

38. $4x - 1 < 3(x + 2)$

39. $3(3x - 2) < 4x - 16$

40. $(x + 6) - (2x + 7) \leq 3x - 9$

41. $(4x - 3) - (2x + 7) \leq 5 - x$

42. $\frac{1}{9}x + \frac{2}{9} > \frac{1}{3}$

43. $\frac{3}{4}x - \frac{1}{4} < 2$

44. $\frac{6}{5} - x \geq \frac{3}{5}x + \frac{14}{5}$

45. $\frac{7}{4} - 2x \geq -\frac{3}{2}x - \frac{5}{4}$

46. $\frac{1}{3}(x - 2) < \frac{1}{4}(x + 5) - \frac{5}{3}$

47. $\frac{1}{3}(x + 2) > 3x - 5(x - 2)$

48. $7x - 6 \leq \frac{1}{3}(-2x + 5)$

Graph the values of x that satisfy the conditions given.

49. $-3 \leq x < 2$

50. $-4 < x \leq 5$

51. $-8 \leq x \leq -4$

52. $-9 \leq x \leq -6$

53. $x < -2 \quad or \quad x \geq 5$

54. $x < -3 \quad or \quad x \geq 6$

55. $x > -5 \quad and \quad x < -1$

56. $x > -8 \quad and \quad x < -3$

57. $x + 3 > 8 \quad or \quad x + 2 < 6$

Solve for x.

58. $x - 2 > 7 \quad or \quad x + 3 < 2$

59. $x + 3 > 8 \quad and \quad x - 4 < -2$

60. $-1 < x + 5 < 8$

61. $0 \leq 5 - 3x \leq 17$

62. $2x - 7 < 3 \quad and \quad 5x - 1 \geq 8$

63. $4x - 2 < 8 \quad or \quad 3x + 1 > 4$

Solve for x.

64. $|x + 3| < 10$

65. $|x + 4| < 3$

66. $|2x - 1| \leq 15$

67. $|3x - 2| \leq 19$

68. $\left|\frac{1}{2}x + 2\right| < \frac{7}{4}$

69. $\left|\frac{1}{5}x + 3\right| < \frac{11}{5}$

70. $|2x - 1| \geq 9$

71. $|3x - 1| \geq 2$

72. $|3(x - 1)| \geq 5$

73. $|2(x - 3)| \geq 4$

74. $|7 + x - 3| > 10$

75. $|8 + x - 7| > 7$

Solve for x.

1. $3x + 2 = -13$

2. $3(7 - 2x) = 14 - 8(x - 1)$

3. $\frac{1}{3}(-x + 1) + 4 = 4(3x - 2)$

4. $0.5x + 1.2 = 4x - 3.05$

5. Solve for *n*. $L = a + d(n - 1)$

6. Solve for *C*. $F = \frac{9}{5}C + 32$

7. Use your answer for Problem 6 to evaluate *C* when $F = -40°$.

8. Solve for *r*. $H = \frac{1}{2}r + 3b - \frac{1}{4}$

Solve for x.

9. $|5x - 2| = 37$

10. $\left|\frac{1}{2}x + 3\right| - 2 = 4$

1. _____

2. _____

3. _____

4. _____

5. _____

6. _____

7. _____

8. _____

9. _____

10. _____

11. _____

12. _____

13. _____

14. _____

15. _____

16. _____

17. _____

18. _____

19. _____

20. _____

Use an algebraic equation to find a solution.

11. A triangle has a perimeter of 69 meters. The second side is twice the length of the first side. The third side is 5 meters longer than the first side. How long is each side?

12. The city hall leased a computer for one year. The city was billed for a one-time $200 installation fee, a monthly rental fee, and a fee for the number of actual hours the computer was used. The contract stated that the monthly rental fee was $280 and the hourly charge was $10 per hour of actual computer use. The first yearly bill was $12,560. How many actual hours was the computer used?

13. Linda needs 10 gallons of solution that is 60% antifreeze. She has a mixture that is 90% antifreeze and one that is 50% antifreeze. How much of each should she use?

14. Lon Triah invested $5000 at a local bank. Part was invested at 6% interest and the remainder at 10%. At the end of one year, Lon earned $428 interest. How much was invested at each rate?

Solve and graph.

15. $5 - 6x < 2x + 21$

16. $-\dfrac{1}{2} + \dfrac{1}{3}(2 - 3x) \geq \dfrac{1}{2}x + \dfrac{5}{3}$

Find the values of x that satisfy the given conditions.

17. $-16 < 2x + 4 < -6$

18. $x - 4 \leq -6$ _or_ $2x + 1 \geq 3$

Solve each absolute value inequality.

19. $|7x - 3| \leq 18$

20. $|3x + 1| > 7$

This test is made up of problems from Chapters 1 and 2.

1. Consider the set of numbers $\left\{-12, -3, 0, \dfrac{1}{4}, 2.16, 2.333..., 2.9614371823,..., -\dfrac{5}{8}, 3\right\}$.
List all the rational numbers.

2. Name the property that justifies $7 + (6 + 3) = (7 + 6) + 3$.

3. Evaluate. $\sqrt{49} + 3(2 - 6)^2 - (-3)$ **4.** Simplify. $(-2x^{-3}y^4)^{-2}$

5. Simplify. $\dfrac{7ab^3}{-14a^5b^{-2}}$ **6.** Find the perimeter of a parallelogram if two sides measure 9 centimeters and 18 centimeters.

7. Find the area of a circle with radius of 7 inches.   **8.** Simplify. $2x(3x - 4) - 5x^2(2 - 6x)$

9. Solve for x.
$\dfrac{1}{4}(x + 5) - \dfrac{5}{3} = \dfrac{1}{3}(x - 2)$ **10.** Solve for b. $h = \dfrac{2}{3}(b + d)$

1.	
2.	
3.	
4.	
5.	
6.	
7.	
8.	
9.	
10.	

11. A triangle has a perimeter of 105 meters. The second side is 10 meters longer than the first. The third side is 5 meters shorter than double the first. How long is each side?

12. A man rented a car for four days at a cost of $19 per day. He was billed for $154.20. He was charged 23¢ for each mile driven. How far did he drive?

13. Wendy invested $6500 for one year. She invested part of 12% interest and part at 10% interest. She earned $690 in interest. How much did she invest at each rate?

14. Hector needs 9 gallons of a solution that is 70% antifreeze. He will combine a mixture of 80% antifreeze with a mixture of 50% antifreeze to obtain the desired 9 gallons. How many gallons of each should he use?

Solve and graph.

15. $-4 - 3x < -2x + 6$

16. $\frac{1}{3}(x + 2) \le \frac{1}{5}(x + 6)$

Find the value of x that satisfies the conditions given.

17. $4 < 3x + 7 < 13$

18. $x + 5 \le -4$ *or* $2 - 7x \le 16$

Solve each absolute value inequality.

19. $\left| \frac{1}{2}x + 2 \right| \le 8$

20. $|3x - 4| > 11$

11. _____

12. _____

13. _____

14. _____

15. _____

16. _____

17. _____

18. _____

19. _____

20. _____

Equations, Inequalities, and Functions

Suppose you developed superior new graphing calculators and wanted to manufacture and sell them. How would you determine what the demand will be for the new graphing calculator that you will soon introduce? How could you determine the profitability of this new product? If you would like to find out, then turn to the Putting Your Skills To Work on page 158.

If you are familiar with the topics in this chapter, take this test now. Check your answers with those in the back of the book. If an answer was wrong or you couldn't do a problem, study the appropriate section of the chapter.

  If you are not familiar with the topics in this chapter, don't take this test now. Instead, study the examples, work the practice problems, and then take the test.

  This test will help you identify those concepts that you have mastered and those that need more study.

Section 3.1

1. Find the value of a if $(a, 6)$ is a solution of $5x + 2y = -12$.

Graph each equation.

2. $y = -\dfrac{1}{2}x + 5$

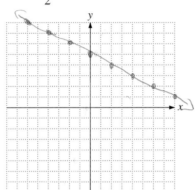

3. $5x + 3y = -15$

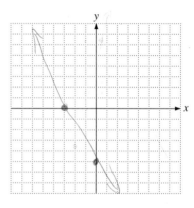

Section 3.2

4. Find the slope of the line passing through the points $(2, 8)$ and $(-1, -6)$.

5. Find the slope of the line *perpendicular* to the line passing through the points $(5, 0)$ and $(-13, -4)$.

Section 3.3

6. Write an equation of the line of slope -2 that passes through $(7, -3)$.

7. Write an equation of the line passing through $(-1, -2)$ and perpendicular to $2x + 4y = -7$.

Section 3.4

Graph the region described by the inequality.

8. $y > -\dfrac{1}{2}x + 3$

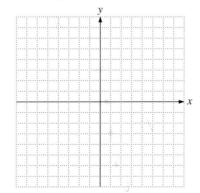

9. $2x - 3y \le -15$

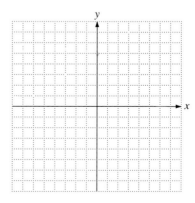

1. _____

2. _____

3. _____

4. _____

5. _____

6. _____

7. _____

8. _____

9. _____

Section 3.5

10. What are the domain and range of this relation?

$$A = \{(3, 4), (5, 6), (0, 1), (3, 2), (4, 7)\}$$

Is the relation a function?

Determine if the graph represents a function.

11.

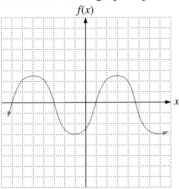

12.

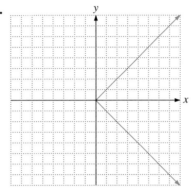

13. If $f(x) = -2x^2 + x - 3$, find $f(-3)$.

14. If $g(x) = |2x - 3|$, find $g(-1)$.

Section 3.6

Graph each function.

15. $p(x) = x^2 + 4$

16. $h(x) = |x - 2|$

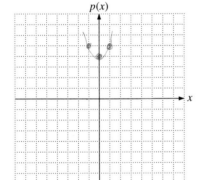

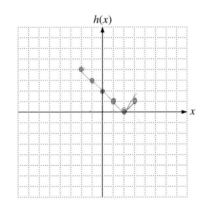

17. Graph the function defined by the table.

x	0	1	4	9	16	25
$f(x)$	0	1	2	3	4	5

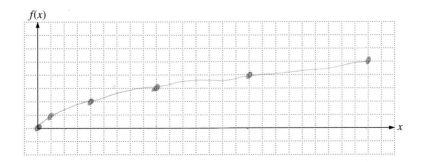

10. _____

11. _____

12. _____

13. _____

14. _____

15. _____

16. _____

17. _____

3.1 Graphing Linear Equations with Two Unknowns

MathPro

Video 6

SSM

After studying this section, you will be able to:

1 *Graph a linear equation in two variables.*
2 *Use x- and y-intercepts to graph a linear equation.*
3 *Graph horizontal and vertical lines.*
4 *Graph a linear equation using different scales.*

Graphs are often used to show the relationship between two sets of data. You may be familiar with graphs that are used in applied mathematics, science, and business.

The following is a graph that could be found in a local newspaper. It shows the daily low temperature in the northeast in January 1994 and compares this to the normal (average) low temperature, as well as to the record low temperature for that month in that region of the country.

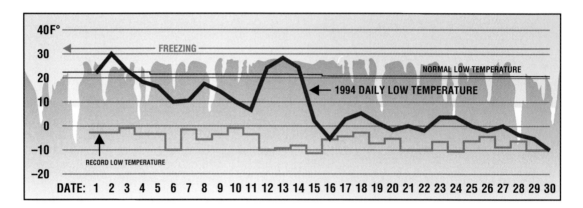

Graphs can also be used in mathematics to show the relationship between the variables in an equation. To do so, we will use two number lines. For convenience we construct two real number lines—one horizontally and one vertically—that intersect to form a rectangular coordinate system. The horizontal line is the **x-axis.** The vertical line is the **y-axis.** They intersect at the **origin.**

To graph an ordered pair of real numbers on a rectangular coordinate system, we begin at the origin. For example, to locate the point (a, b) where $a, b > 0$, we move a units to the right of the origin along the x-axis. Then we move b units up. Note that (a, b) is an ordered pair. Thus order is important. $(a, b) \neq (b, a)$

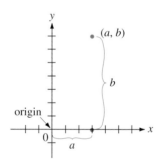

The figure below shows the graphs of (3, 4) and (4, 3).

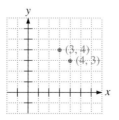

EXAMPLE 1 Graph the ordered pairs $A(3, 2)$, $B(0, -4)$, $C(-2, -1)$, $D(-5, -4)$, and $E(-3, 4)$ on a rectangular coordinate system.

PRACTICE PROBLEM 1

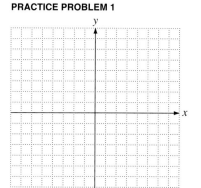

The graph of each ordered pair is shown to the right. Notice that to graph the ordered pair (3, 2) we move 3 units to the *right* and 2 units *up*. In the ordered pair (0, -4) the *x*-coordinate is 0. We do not move either right or left. We move 4 units *down* along the *y*-axis since the *y*-coordinate is -4. To graph the ordered pair (-2, -1), we move 2 units to the *left* and 1 unit *down*. Use your own words to describe how the ordered pairs (-5, -4) and (-3, 4) are graphed.

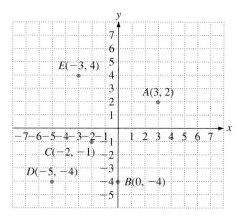

Practice Problem 1 Graph the ordered pairs $A(-4, 1)$, $B(2, -3)$, $C(0, 5)$, $D(-4, -2)$, and $E(4, 0)$ on the rectangular coordinate system in the margin. ∎

1 Graphing Linear Equations

A **solution of an equation** in two variables is an ordered pair of real numbers that *satisfies* the equation. In other words, when we substitute the values of the coordinates into the equation, we get a true statement. For example, (6, 4) is a solution to $4x - 3y = 12$. The ordered pair (6, 4) means that $x = 6$ and $y = 4$.

$$4x - 3y = 12$$
$$4(6) - 3(4) = 12$$
$$24 - 12 = 12$$
$$12 = 12 \checkmark$$

To graph the equation, we could graph all its solutions. However, this would be impractical. It is a mathematical property that an equation of the form $Ax + By = C$, where A, B, and C are constants, is a straight line. Hence to graph the equation we graph three solutions (ordered pairs) and connect them with a straight line.

GRAPHING CALCULATOR

Graphing Ordered Pairs

You can use a graphing calculator to graph ordered pairs. Most graphing calculators can plot statistical data. Use your calculator's statistical plot feature to plot the points in Example 1. The display should be similar to the one below. Be sure to use an appropriate window.

Display:

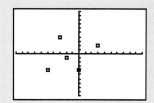

EXAMPLE 2 Graph the equation $y = -3x + 2$.

We choose three values of x and then solve the equation to find the corresponding values of y. Let's choose $x = -1$, $x = 1$, and $x = 2$.

For $x = -1$, $y = -3(-1) + 2 = 5$, so the first point, or solution, is $(-1, 5)$.

For $x = 1$, $y = -3(1) + 2 = -1$, and for $x = 2$, $y = -3(2) + 2 = -4$

We can condense this procedure by using a table to help us construct a graph.

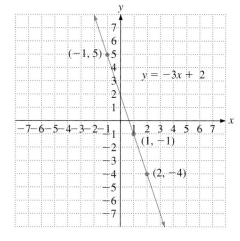

x	y
-1	5
1	-1
2	-4

Practice Problem 2 Graph the equation $y = -4x + 2$. ∎

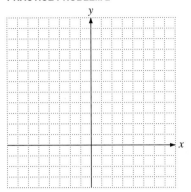

2 Graphing a Linear Equation by Using Intercepts

We can quickly graph a straight line by using the x- and y-intercepts. A straight line that is not vertical or horizontal has these two intercepts. In the figure on the right, the x-intercept is $(a, 0)$ and the y-intercept is $(0, b)$.

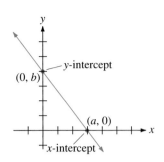

> The **x-intercept** of a line is the point where the line crosses the x-axis (that is, where $y = 0$). The **y-intercept** of a line is the point where the line crosses the y-axis (that is, where $x = 0$).

EXAMPLE 3   Use intercepts to graph the equation $4x - 3y = -12$.

Find the x-intercept by setting $y = 0$:

$$4x - 3(0) = -12$$
$$4x = -12$$
$$x = -3$$

The x-intercept is $(-3, 0)$.

Find the y-intercept by setting $x = 0$:

$$4(0) - 3y = -12$$
$$-3y = -12$$
$$y = 4$$

The y-intercept is $(0, 4)$.

We can now pick any value of x or y to find our third point. Let's pick $y = 2$:

$$4x - 3(2) = -12$$
$$4x - 6 = -12$$
$$4x = -12 + 6$$
$$4x = -6$$
$$x = -\frac{6}{4} = -\frac{3}{2}$$

Hence the third point on the line is $(-\frac{3}{2}, 2)$.

The figure below shows the graph of the equation that corresponds to the following table of values:

x	y
-3	0
0	4
$-\dfrac{3}{2}$	2

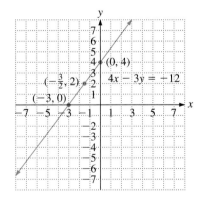

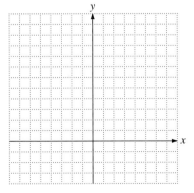

Practice Problem 3   Use intercepts to graph the equation $3x - 2y = -6$. ■

Let's look at the standard form of a linear equation, $Ax + By = C$ when $B = 0$.

$$Ax + (0)y = C$$

$$Ax = C$$

$$x = \frac{C}{A}$$

Notice that when we solve for x we get $x = \frac{C}{A}$, which is a constant. For convenience we will rename it a. The equation then becomes

$$x = a$$

What does this mean? $x = a$ means that, for any value of y, x is a. The graph is a vertical line.

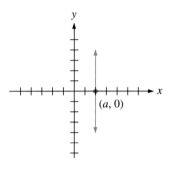

(a, 0)

GRAPHING CALCULATOR

Graphing A Line

You can graph a line given in the form $y = mx + b$ using a graphing calculator. For example, to graph the equation in Example 3, rewrite the equation by solving for y.

$$4x - 3y = -12$$

$$-3y = -4x - 12$$

$$y = \tfrac{4}{3}x + 4$$

Enter the right-hand side of the resulting equation in the Y = editor of your calculator and graph. Choose an appropriate window to show all the intercepts. The following window is $[-10, 10]$ by $[-10, 10]$.

Display:

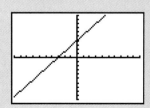

What happens to $Ax + By = C$ when $A = 0$?

$$(0)x + By = C$$

$$By = C$$

$$y = \frac{C}{B}$$

We will rename the constant $\frac{C}{B}$ as b. The equation then becomes

$$y = b$$

What does this mean? $y = b$ means that, for any value of x, y is b. The graph is a horizontal line.

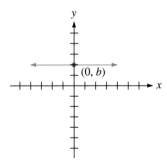

(0, b)

The graph of the equation $x = a$, where a is any real number, is a **vertical line** through the point $(a, 0)$.

The graph of the equation $y = b$, where b is any real number, is a **horizontal line** through the point $(0, b)$.

EXAMPLE 4 Graph each equation.   **(a)** $x = -3$ **(b)** $2y - 4 = 0$

(a) The equation $x = -3$ means that, for any value of y, x is -3. The graph of $x = -3$ is a vertical line 3 units to the left of the origin.

(b) The equation $2y - 4 = 0$ can be simplified.

$$2y - 4 = 0$$
$$2y = 4$$
$$y = 2$$

The equation $y = 2$ means that, for any value of x, y is 2. The graph of $y = 2$ is a horizontal line 2 units above the x-axis.

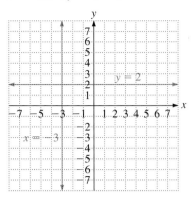

PRACTICE PROBLEM 4

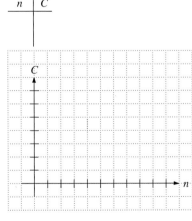

Practice Problem 4   Graph the equation $5x + 8 = 9x$. ∎

4 Graphing a Linear Equation Using Different Scales for the Axes

By common convention, each square on the graph indicates 1 unit, so we don't need to use a marked scale on each axis. But sometimes a different scale is more appropriate. This new scale must then be clearly labeled on each axis.

EXAMPLE 5   A company's finance officer has determined that the monthly cost in dollars for leasing a photocopy machine is $C = 100 + 0.002n$, where n is the number of copies produced in a month in excess of a specified number. Graph the equation using $n = 0$, $n = 30{,}000$, and $n = 60{,}000$. Let the n-axis be the horizontal axis.

For each value of n we obtain C.

When $n = 0$, then $C = 100 + 0.002(0) = 100 + 0 = 100$.

When $n = 30{,}000$, then $C = 100 + 0.002(30{,}000) = 100 + 60 = 160$.

When $n = 60{,}000$, then $C = 100 + 0.002(60{,}000) = 100 + 120 = 220$.

The table of values is

n	C
0	100
30,000	160
60,000	220

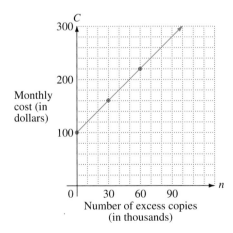

PRACTICE PROBLEM 5

Since n varies from 0 to 60,000 and C varies from 100 to 220, we need a different scale on each axis. We let each square on the horizontal scale represent 10,000 excess copies, and each square on the vertical scale represents $20.

Practice Problem 5   The cost of a product is given by $C = 300 + 0.15n$. Graph the equation using an appropriate scale. Use $n = 0$, $n = 1000$, and $n = 2000$. ∎

3.1 Exercises

Verbal and Writing Skills

1. Graphs are used to show the relationship between the _____ in an equation.

2. The x-axis and the y-axis intersect at the _____.

3. Explain in your own words why (a, b) is an ordered pair. Give an example.

4. $(5, 1)$ is a solution to the equation $2x - 3y = 7$. What does this mean?

Find the missing coordinate.

5. $(-2, \underline{\hspace{0.6cm}})$ is a solution of $y = 3x - 7$.

6. $(-3, \underline{\hspace{0.6cm}})$ is a solution of $y = 4 - 3x$.

7. $(\underline{\hspace{0.6cm}}, \frac{1}{2})$ is a solution of $7x + 14y = -21$.

8. $(\underline{\hspace{0.6cm}}, 2)$ is a solution of $6x - 24y = 12$.

9. $(0, \underline{\hspace{0.6cm}})$ is a solution of $7x - 12y = 24$.

10. $(0, \underline{\hspace{0.6cm}})$ is a solution of $-6x + 3y = 15$.

Graph each equation.

11. $y = 5x - 2$

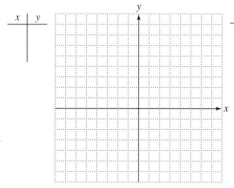

12. $y = -2x + 1$

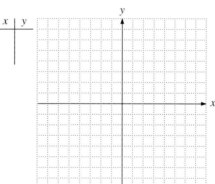

13. $y = 4 - 2x$

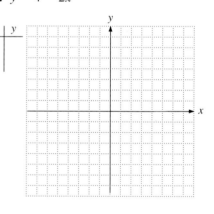

14. $y = -5x - 2$

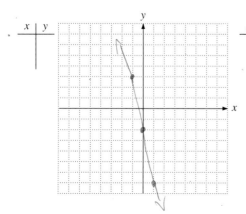

15. $y = \frac{2}{3}x - 4$

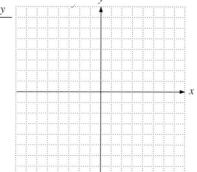

16. $y = \frac{3}{5}x + 2$

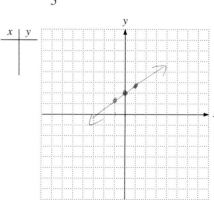

Simplify the equation if possible. Find the x-intercept and the y-intercept. Then find one or two additional ordered pairs that are solutions to the equation. Then graph the equation.

17. $3x - 2y = 6$

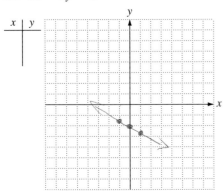

18. $2x + 5y = 10$

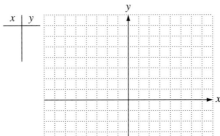

19. $2x - 3y = -9$

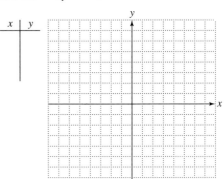

20. $-3x - 4y = 8$

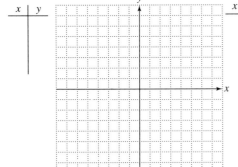

21. $y = 4x$

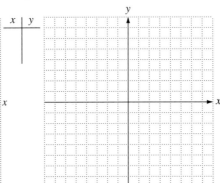

22. $y = -3x$

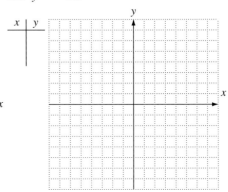

23. $8x + 3 = 4y + 3$

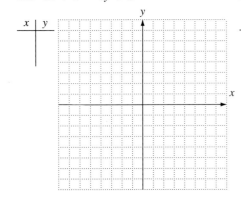

24. $-2x + 5y - 6 = -6$

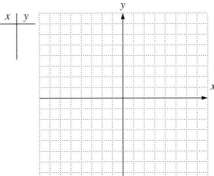

25. $y = -x$

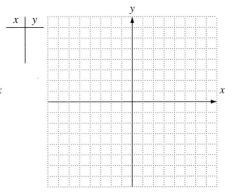

Simplify the equation. Identify the equation as representing a horizontal or a vertical line. Then graph the equation.

26. $x = 6$

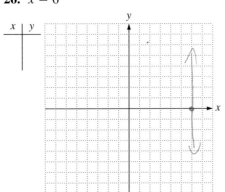

27. $y = -4$

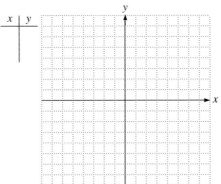

28. $5y + 6 = 2y$

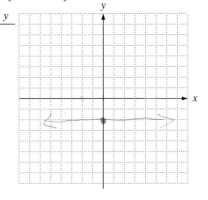

29. $2x - 3 = 3x$

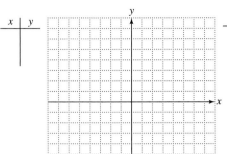

30. $x = 0$

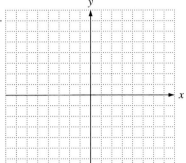

31. $y = 0$

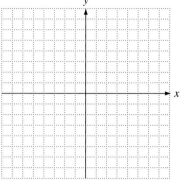

Mixed Practice

Simplify each equation if possible. Then graph the equation by any appropriate method.

32. $y = -1.5x + 2$

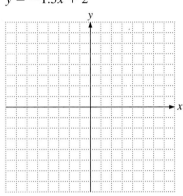

33. $2.5x - 5y = -20$

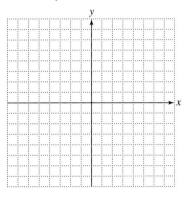

34. $5x + y + 4 = 8x$

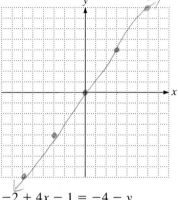

35. $4y + 6x = 5x - 3 + 4y$

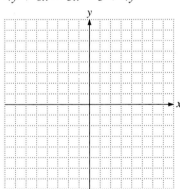

36. $7x + 8y = 2 + 7y + 7x$

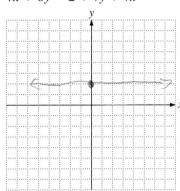

37. $-2 + 4x - 1 = -4 - y$

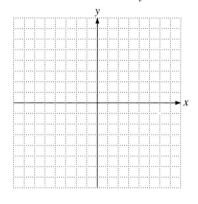

Graph each equation. Use an appropriate scale.

38. $y = 82x + 150$

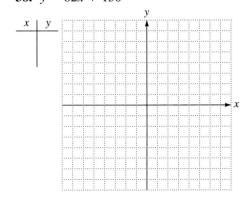

39. $y = 0.06x - 0.04$

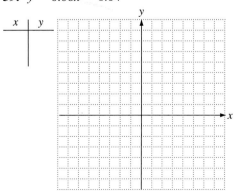

Applications

40. If a baseball is thrown vertically upward by Paul Frydrych when he is standing on the ground, the velocity of the baseball V (feet per second) after T seconds is $V = 120 - 32T$.

   a. Find V for $T = 0, 1, 2, 3,$ and 4.

   b. Graph the equation, using T as the horizontal axis.

   c. What is the significance of the negative value of V when $T = 4$?

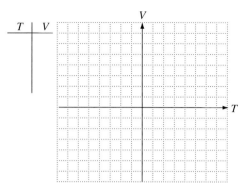

41. A full storage tank on the Robinson family farm contains 900 gallons. Gasoline is then pumped from the tank at a rate of 15 gallons per minute. The equation is $G = 900 - 15m$, where G is the number of gallons of gasoline in the tank and m is the number of minutes since the pumping started.

   a. Find G for $m = 0, 10, 20, 30,$ and 60.

   b. Graph the equation, using m as the horizontal axis.

   c. What would happen if we let $m = 61$?

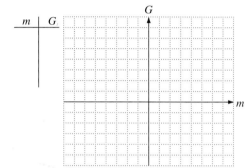

42. Pick an appropriate scale and graph the equation $P = 14.7 - 0.0005d$ for $d = 0, 1000, 2000, 3000, 9000,$ and $15,000$. (This equation predicts the atmospheric pressure in pounds per square inch a person would experience at a height of d feet above sea level.)

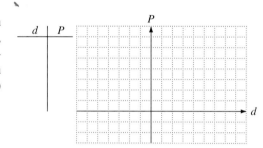

43. Bob and Cheryl Finkelstein belong to a Sky Diving Club. When jumping out of a plane, they try to open parachutes at 5600 feet above the ground. If they accomplish that goal, then their height above the ground measured in feet is given by the equation $h = 5600 - 190t$, where t is the number of seconds since the parachute was opened. Find h for $t = 0, 5, 10, 20,$ and 29. Graph the equation using t as the horizontal axis.

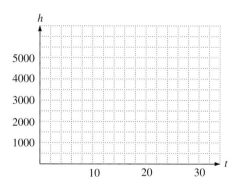

Optional Graphing Calculator Problems

If you have a graphing calculator, use it to graph the following equations. Choose an appropriate window that will allow you to see both the x-intercept and the y-intercept. (Your answers will vary depending on the window selected.)

44. $y = -2.15x + 2.73$ **45.** $y = 1.36x - 1.83$ **46.** $y = 0.713x + 25.82$ **47.** $y = -0.819x - 43.82$

3.2 Slope of a Line

After studying this section, you will be able to:

1 *Find the slope of any nonvertical straight line if two points are known.*
2 *Determine if two lines are parallel or perpendicular by comparing their slopes.*

MathPro   Video 7   SSM

1 *Finding the Slope if Two Points Are Known*

The concept of slope is one of the most useful in mathematics and in practical applications. A carpenter needs to determine the slope (or pitch) of a roof (you may have heard someone say that a roof has a 5:12 pitch). Road engineers must determine the proper slope (or grade) of a roadbed. If the slope is steep, you feel like you're driving almost straight up. Simply put, slope is a variation from the horizontal. That is, slope measures vertical change *(rise)* versus horizontal change *(run)*.

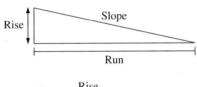

$$\text{Slope} = \frac{\text{Rise}}{\text{Run}}$$

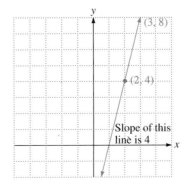

Mathematically, we define slope as follows:

> The **slope of a straight line** with points (x_1, y_1) and (x_2, y_2) is
>
> $$\text{slope} = m = \frac{y_2 - y_1}{x_2 - x_1} \qquad x_2 \neq x_1$$

In this sketch, we see that the rise is $y_2 - y_1$ and the run is $x_2 - x_1$.

EXAMPLE 1   Find the slope of a line passing through

(a) $(2, 4)$ and $(3, 8)$ **(b)** $(-2, -3)$ and $(1, -4)$

(a) First, we need to identify the y-coordinates and the x-coordinates for the points $(2, 4)$ and $(3, 8)$.

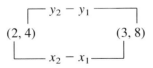

Then we use the formula.

$$\text{slope} = m = \frac{y_2 - y_1}{x_2 - x_1}$$

$$= \frac{8 - 4}{3 - 2}$$

$$= 4$$

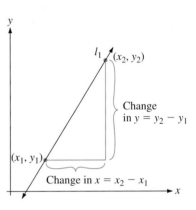

Slope of this line is 4

(b) Identify the *y*-coordinates and the *x*-coordinates for the points $(-2, -3)$ and $(1, -4)$.

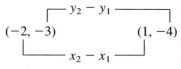

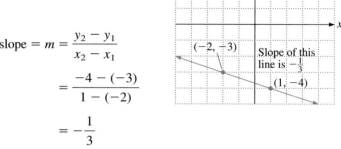

Use the formula.

$$\text{slope} = m = \frac{y_2 - y_1}{x_2 - x_1}$$

$$= \frac{-4 - (-3)}{1 - (-2)}$$

$$= -\frac{1}{3}$$

Practice Problem 1   Find the slope of the line passing through $(-6, 1)$ and $(-5, -2)$. ∎

Notice that it does not matter which coordinates become y_1 and y_2 as long as we subtract the *x*-coordinates in the same order that we subtract the *y*-coordinates. Let's redo part (b) of Example 1.

$$m = \frac{-3 - (-4)}{-2 - 1} = \frac{-3 + 4}{-3} = -\frac{1}{3}$$

To Think About What do you notice about the graph of the lines in part (a) and in part (b) in Example 1? How are the graphs related to the slopes?

The graphs tell us some basic facts about the slope of a line.

> **1.** Lines sloping upward to the right have positive slopes.
> **2.** Lines sloping downward to the right have negative slopes.

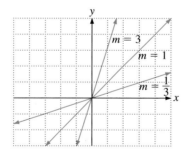

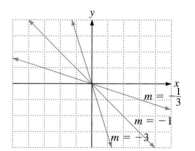

What is the slope of a horizontal line? Let's look at the equation, $y = 4$. $y = 4$ is the horizontal line 4 units above the *y*-axis. It means that, for any *x*, *y* is 4. The slope is

$$m = \frac{y_2 - y_1}{x_2 - x_1} = \frac{4 - 4}{x_2 - x_1} = \frac{0}{x_2 - x_1} = 0$$

In general, for all horizontal lines, $y_2 - y_1 = 0$ and hence the slope is 0.

What is the slope of a vertical line? The equation of the vertical line 4 units to the right of the *y*-axis is $x = 4$. It means that, for any *y*, *x* is 4. The slope is

$$m = \frac{y_2 - y_1}{x_2 - x_1} = \frac{y_2 - y_1}{4 - 4} = \frac{y_2 - y_1}{0}$$

Because division by zero is not defined, we say that a vertical line has no slope or the slope is undefined.

GRAPHING CALCULATOR

Slopes

Using a graphing calculator, on the same set of axes, graph

$$y_1 = 3x + 1$$

$$y_2 = \frac{1}{3}x + 1$$

$$y_3 = -3x + 1$$

$$y_4 = -\frac{1}{3}x + 1$$

How is the coefficient of *x* in each equation related to the slope of the line? Will the graph of the line $y = -2x + 3$ slope upward or downward? How do you know? Verify using your calculator.

The slope of a horizontal line is 0. The slope of a vertical line is undefined.

EXAMPLE 2 Find the slope if possible of a line passing through:

(a) $(1.6, 2.3)$ and $(-6.4, 1.8)$ **(b)** $(\frac{5}{3}, -\frac{1}{2})$ and $(\frac{2}{3}, -\frac{1}{4})$

(c) $(6, -3)$ and $(8, -3)$ **(d)** $(5, -6)$ and $(5, 8)$

(a) $m = \dfrac{1.8 - 2.3}{-6.4 - 1.6} = \dfrac{-0.5}{-8.0} = 0.0625$

(b) $m = \dfrac{-\dfrac{1}{4} - \left(-\dfrac{1}{2}\right)}{\dfrac{2}{3} - \dfrac{5}{3}} = \dfrac{-\dfrac{1}{4} + \dfrac{2}{4}}{-\dfrac{3}{3}} = \dfrac{\dfrac{1}{4}}{-1} = -\dfrac{1}{4}$

(c) $m = \dfrac{-3 - (-3)}{8 - 6} = \dfrac{-3 + 3}{2} = \dfrac{0}{2} = 0$

(d) If we attempt to use the formula, we obtain

$$\frac{8 - (-6)}{5 - 5} = \frac{8 + 6}{0} = \frac{14}{0}$$

This is not defined. Division by zero is not allowed. You cannot divide by zero. These two points represent two points on a vertical line. We cannot find the slope of a vertical line.

Practice Problem 2 Find the slope if possible of a line through:

(a) $(1.8, -6.2)$ and $(-2.2, -3.4)$ **(b)** $(\frac{1}{5}, -\frac{1}{2})$ and $(\frac{4}{15}, -\frac{3}{4})$

(c) $(-3, 12)$ and $(-3, -14)$ **(d)** $(7, -6)$ and $(-5, -6)$ ∎

In dealing with practical situations, such as the grade of a road or the pitch of a roof, we can find the slope directly by using the formula

$$\text{slope} = \frac{\text{rise}}{\text{run}}$$

EXAMPLE 3

(a) Find the slope of a road that rises 34 feet vertically over a horizontal distance of 680 feet.

(b) Find the pitch of a roof as shown in the sketch in the margin.

(a) Slope $= \dfrac{\text{rise}}{\text{run}} = \dfrac{34}{680} = \dfrac{1}{20}$ or 0.05.

When the slope of a road is 0.05 we often say the *grade* of the road is 5%.

(b) Slope $= \dfrac{\text{rise}}{\text{run}} = \dfrac{7.4}{18.5} = 0.4$.

This could also be expressed as the fraction $\frac{2}{5}$. Builders sometimes refer to this as a *pitch* (slope) of 2:5.

Practice Problem 3

Find the slope of a river that drops 25.92 feet vertically over a horizontal distance of 1296 feet. (*Hint:* Use only positive numbers. In everyday use, the slope of a river or road is always considered to be a positive value.) ∎

2 Determining If Two Lines are Parallel or Perpendicular

We can tell a lot about a line by looking at its slope. A positive slope tells us that the line rises from left to right. A negative slope tells us that the line falls from left to right. What might be true of the slopes of parallel lines? Determine the slope of each line in the graphs below and compare the slopes of the parallel lines.

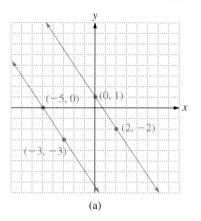

(a)

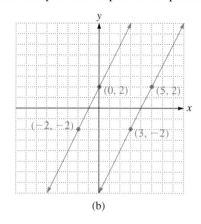

(b)

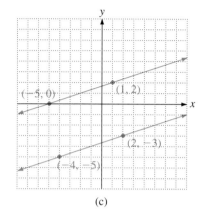

(c)

Since parallel lines are lines that never intersect, their slopes must be equal.

> **Parallel Lines**
>
> Two different lines with slopes m_1 and m_2, respectively, are *parallel* if $m_1 = m_2$.

GRAPHING CALCULATOR

Parallel Lines

Using a graphing calculator, graph both of the following equations on one coordinate plane.

$$y_1 = 7.3457x - 3.5678$$

$$y_2 = 7.2357x + 2.3481$$

Experiment with different graphing windows. For example, first try a window of $[-10, 10]$ by $[-10, 10]$ and then try a window of $[0, 30]$ by $[0, 200]$.

Do the lines appear to be parallel on your graph? Are the lines really parallel? Why or why not?

EXAMPLE 4 A line k passes through the points $(-1, 3)$ and $(2, -1)$. A second line h passes through the points $(-1, -4)$ and $(-4, 0)$. Is line k parallel to line h?

The slope of line k is

$$m_k = \frac{y_2 - y_1}{x_2 - x_1} = \frac{-1 - 3}{2 - (-1)}$$

$$= \frac{-4}{2 + 1} = -\frac{4}{3}$$

The slope of line h is

$$m_h = \frac{y_2 - y_1}{x_2 - x_1} = \frac{0 - (-4)}{-4 - (-1)}$$

$$= \frac{0 + 4}{-4 + 1} = \frac{4}{-3} = -\frac{4}{3}$$

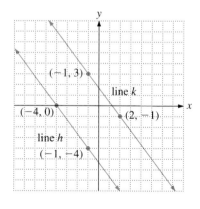

Since both lines are distinct (that is, different) and have the same slope, they are parallel.

Practice Problem 4 Line k passes through $(7, -2)$ and $(-5, -3)$. Find the slope of a line *parallel* to line k. ∎

Now take a look at perpendicular lines. Determine the slope of each line in the following graphs and compare the slopes of the perpendicular lines. What do you notice?

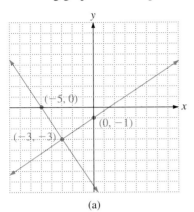
(a)

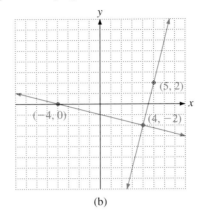
(b)

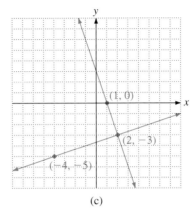
(c)

By definition, two lines are perpendicular if they intersect at right angles (a right angle is an angle of 90°). We would expect that one slope would be positive and the other slope negative. In fact, we could say that, if the slope of a line is 5, then the slope of a line that is perpendicular to it is $-\frac{1}{5}$.

Perpendicular Lines

Two lines with slopes m_1 and m_2, respectively, are *perpendicular* if $m_1 = -\dfrac{1}{m_2}$ ($m_1, m_2 \neq 0$).

EXAMPLE 5   Find the slope of a line that is perpendicular to a line l that passes through $(4, -6)$ and $(-3, -5)$.

The slope of line l is

$$m_l = \frac{-5 - (-6)}{-3 - 4} = \frac{-5 + 6}{-7} = \frac{1}{-7} = -\frac{1}{7}$$

The slope of a line perpendicular to line l must have a slope of 7.

Practice Problem 5   If a line l passes through $(5, 0)$ and $(6, -2)$, what is the slope of a line h that is *perpendicular* to l? ∎

It is helpful to remember that if two lines are perpendicular, their slopes are negative reciprocals. One slope will be positive, the other negative. If one line had a slope of $-\frac{3}{7}$, the slope of a line perpendicular to it would have a slope of $\frac{7}{3}$.

EXAMPLE 6 Without plotting any points, show that the points $A(-5, -1)$, $B(-1, 2)$, and $C(3, 5)$ lie on the same line.

First we find the slope of the line segment from A to B and the slope of the line segment from B to C.

$$m_{AB} = \frac{2 - (-1)}{-1 - (-5)} = \frac{2 + 1}{-1 + 5} = \frac{3}{4}$$

$$m_{BC} = \frac{5 - 2}{3 - (-1)} = \frac{3}{3 + 1} = \frac{3}{4}$$

Since the slopes are equal, we must have one line or two parallel lines. But the line segments have a point (B) in common, so all three points lie on the same line.

Practice Problem 6 Without plotting the points, show that $A(1, 5)$, $B(-1, 1)$, and $C(-2, -1)$ lie on the same line. ∎

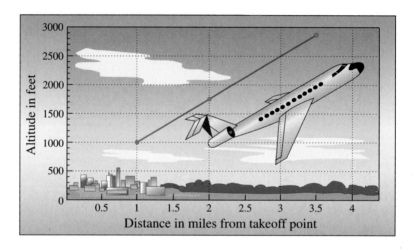

An aircraft taking off climbs into the sky at a certain rate of speed. This measurement of a slope is often called a rate of climb.

EXAMPLE 7 A Gulfstream jet takes off from Orange County Airport in California. At one mile from the takeoff point it is 1000 feet above the airport and begins a specified rate of climb. At two miles from the takeoff point it is 1750 feet above the airport. At 3.5 miles from the takeoff point it is 2865 feet above the airport. Has the jet traveled in a straight line from the one-mile point to the 3.5-mile point?

If we measure the two slopes, from one mile away to two miles away, the slope is 750 feet upward for every one mile horizontal. From two miles away to 3.5 miles away the slope is 1115 feet upward for every 1.5 miles horizontal. The first slope is 750 feet per mile, the second slope is $743\frac{1}{3}$ feet per mile. These slopes are not the same, so the jet has **not** traveled in a straight line.

Practice Problem 7 On the return flight of the jet, it is at an altitude of 4850 feet when it is 4.5 miles from the airport. The altitude is 3650 feet when it is 3.5 miles from the airport. The height is 1010 feet when it is 1.3 miles from the airport. Is the jet descending in a straight line?

3.2 Exercises

Verbal and Writing Skills

1. Slope measures _____ change (rise) versus _____ change (run).

2. A positive slope indicates that the line will be sloping _____ to the right.

3. What is the slope of a horizontal line?

4. Two lines are parallel if their slopes are _____.

5. Does the line passing through $(-3, -7)$ and $(-3, 5)$ have a slope? Give a reason for your answer.

6. Let $(x_1, y_1) = (-6, -3)$ and $(x_2, y_2) = (-4, 5)$. Find $\dfrac{y_2 - y_1}{x_2 - x_1}$ and $\dfrac{y_1 - y_2}{x_1 - x_2}$. Are the results the same. Why or why not?

Find the slope, if possible, of a line passing through each pair of points.

7. $(-7, 2)$ and $(3, -4)$

8. $(-3, 6)$ and $(2, 5)$

9. $(-2, -2)$ and $(6, -6)$

10. $(2, -1)$ and $(-6, 3)$

11. $\left(\dfrac{1}{2}, 3\right)$ and $(1, 5)$

12. $(6, 1)$ and $\left(0, \dfrac{1}{3}\right)$

13. $(-7, 6)$ and $(2, 6)$

14. $(-2, -8)$ and $(5, -8)$

15. $(2.9, -3.5)$ and $(1.5, -2.8)$

16. $(-2, 5.2)$ and $(4.8, -1.6)$

17. $\left(\dfrac{3}{2}, -2\right)$ and $\left(\dfrac{3}{2}, \dfrac{1}{4}\right)$

18. $\left(\dfrac{7}{3}, -6\right)$ and $\left(\dfrac{7}{3}, \dfrac{1}{6}\right)$

Applications

19. Find the slope (grade) of a snowboard "half-pipe" recreation hill that rises 33.6 feet vertically with a horizontal distance of 56 feet.

20. Find the grade of a driveway that rises 6.3 feet vertically with a horizontal distance of 126 feet.

21. Find the slope (pitch) of a perfectly smooth rock formation that rises 35.7 feet vertically with a horizontal distance of 142.8 feet.

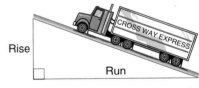

22. Find the slope (pitch) of a roof that rises 3.15 feet vertically with a horizontal distance of 10.50 feet.

23. A river has a slope of 0.16. How many feet does it fall vertically in a horizontal distance of 500 feet?

24. A Maine mountain road used by tractor-trailer trucks has a slope of 0.18. How many feet does it fall vertically in a horizontal distance of 700 feet?

Find the slope of a line parallel to the line that passes through the following points.

25. $(7, 1)$ and $(6.5, 2)$

26. $(3, 5)$ and $(2.8, 6)$

27. $(-7, 1)$ and $(-4, -3)$

28. $(2, -6)$ and $(-5, -8)$

29. $(-6, -3)$ and $(4, -3)$

30. $(2, -5)$ and $(2, 16)$

Find the slope of a line perpendicular *to the line that passes through the following points.*

31. $(8, 12)$ and $(3, 9)$

32. $(-2, 5)$ and $(1, 3)$

33. $(2, -\frac{1}{2})$ and $(1, \frac{5}{2})$

34. $(-\frac{2}{3}, -2)$ and $(-3, \frac{1}{3})$

35. $(-8, 0)$ and $(0, 6)$

36. $(0, -5)$ and $(-2, 0)$

To Think About

37. Show that $ABCD$ is a parallelogram if the four vertices are $A(2, 1)$, $B(-1, -2)$, $C(-7, -1)$, and $D(-4, 2)$. (*Hint:* A parallelogram is a four-sided figure with opposite sides parallel.)

38. Do the points $A(-1, -2)$, $B(2, -1)$, and $C(8, 1)$ lie on a straight line? Explain.

39. Most new buildings are required to have a ramp for the handicapped that has a maximum vertical rise of 5 feet for every 60 feet of horizontal distance.

(a) What is the value of the slope of a handicapped ramp?

(b) If the builder constructs a new building where the handicapped ramp has a horizontal distance of 24 feet, what is the maximum height of the doorway above the level of the parking lot where the handicapped ramp begins?

(c) What will the length of the handicapped ramp become if the architect redesigns the building so that the doorway is 1.7 feet above the parking lot?

40. A small Cessna plane takes off from Hyannis Airport. When the plane is 1 mile from the airport it is flying at 3000 feet altitude. When the plane is 2 miles from the airport it is flying at 4300 feet altitude. Round your answers to the nearest tenth.

(a) If the plane continues flying at the same slope (the same rate of climb), what will be the altitude when the plane is 4.8 miles from the airport?

(b) If the plane continues flying at the same rate of climb how many miles from the airport will it be when it reaches an altitude of 6000 feet?

(c) A Lear jet leaves the airport and has the same altitude (3000 feet) as the Cessna when each plane is 1 mile from the airport. When it is 1.8 miles from the airport it is flying at 4040 feet altitude. Is the plane being flown at the same rate of climb as the Cessna?

Cumulative Review Problems

Evaluate.

41. $2(3 - 6)^3 + 20 \div (-10)$

42. $\dfrac{5 + 3\sqrt{9}}{|2 - 9|}$

Simplify.

43. $(5x^{-3}y^{-4})(-2x^6y^{-8})$

44. $\dfrac{-15x^6y^3}{-3x^{-4}y^6}$

3.3 Graphs and the Equations of a Line

After studying this section, you will be able to:

1. *Use the slope–intercept form of the equation of a line.*
2. *Use the point–slope form of the equation of a line.*
3. *Write the equation of the line passing through a given point that is parallel to or perpendicular to a given line.*

1 Slope–Intercept Form of the Equation of a Line

Recall that the equation of a straight line is $Ax + By = C$. This form of the equation is called the **standard form.** When we look at the equation in standard form, we know the graph is a straight line. However, very little information about the line is revealed. A more useful form of the equation is the **slope–intercept form.** The slope–intercept form immediately reveals the slope of the line and where it intersects the y-axis. This is important information for graphing the line.

We will use the definition of slope to derive the equation. Let a straight line with slope m cross the y-axis at some point $(0, b)$, where b is the y-intercept. Choose any point on the line and label it (x, y). By the definition of slope, $\frac{y_2 - y_1}{x_2 - x_1} = m$. But here $x_1 = 0$ and $y_1 = b$, and $y_2 = y$ and $x_2 = x$, so

$$\frac{y - b}{x} = m$$

Now we solve this equation for y.

$$y - b = mx$$

$$y = mx + b$$

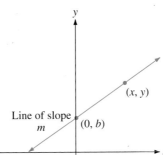

Line of slope m $(0, b)$ (x, y)

> ### Slope–Intercept Form
>
> The **slope–intercept** form of the equation of a line is $y = mx + b$, where m is the slope and b is the y-intercept.

Write the Equation of a Line Given Its Slope and y-Intercept

EXAMPLE 1 Write an equation of the line with slope $-\frac{2}{3}$ and y-intercept 5.

$$y = mx + b$$

$$= \left(-\frac{2}{3}\right)x + (5) \qquad \text{Substitute } -\tfrac{2}{3} \text{ for } m \text{ and 5 for } b.$$

$$= -\frac{2}{3}x + 5$$

Practice Problem 1 Write an equation of the line with slope 4 and y-intercept $-\frac{3}{2}$. ∎

GRAPHING CALCULATOR

Exploring y-intercepts

Using a graphing calculator, on the same set of axes graph

$$y_1 = 2x$$
$$y_2 = 2x + 1$$
$$y_3 = 2x - 1$$
$$y_4 = 2x + 2$$

Where does each graph cross the y-axis? What affect does b have on the graph of $y = mx + b$? What would the graph of the line $y = 2x - 5$ look like? Use your graphing calculator to verify.

GRAPHING CALCULATOR

Exploring Slopes

Using a graphing calculator, on the same set of axes, graph

$$y_1 = x + 1$$
$$y_2 = 2x + 1$$
$$y_3 = 3x + 1$$
$$y_4 = \frac{1}{3}x + 1$$

What affect does m have on the graph of $y = mx + b$? What would the graph of the line $y = \frac{1}{2}x + 1$ look like? Use your graphing calculator to verify.

Write the Equation of a Line Given the Graph

We can write an equation of a line given its graph since we can determine the y-intercept and the slope from the graph itself.

EXAMPLE 2 Write an equation of the line whose graph is shown on the right.

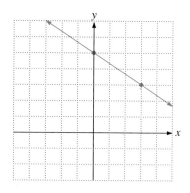

Looking at the graph, we can see that the y-intercept is at $(0, 5)$. That is, $b = 5$. If we can identify the coordinates of another point on the line, we will have two points and we can determine the slope. $(3, 3)$ is another point on the line.

Thus
$$(x_1, y_1) = (0, 5)$$
$$(x_2, y_2) = (3, 3)$$
$$m = \frac{y_2 - y_1}{x_2 - x_1} = \frac{3 - 5}{3 - 0} = -\frac{2}{3}$$

We can now write the equation of the line in slope–intercept form.

$$y = mx + b$$
$$y = -\frac{2}{3}x + 5 \qquad \textit{Substitute } -\tfrac{2}{3} \textit{ for m and 5 for b.}$$

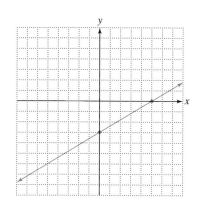

Practice Problem 2 Write an equation of the line whose graph is shown in the margin on the left. ∎

Use the Slope–Intercept Form to Graph an Equation

We have just seen that, given the graph, we can determine an equation of the line if we can identify the y-intercept and determine the slope. We can also draw the graph of an equation without plotting points if we can write the equation in slope–intercept form and then locate the y-intercept on the graph.

EXAMPLE 3 Sketch the graph of the equation $28x - 7y = 21$.

GRAPHING CALCULATOR

Graphing a Line

In order to graph an equation of a line, most graphing calculators require you to have the equation in slope–intercept form. (See box on p. 135.)

First we will change the standard form of the equation into **slope–intercept form.** This is a very important procedure. Be sure that you understand each step.

$$28x - 7y = 21$$
$$-7y = -28x + 21$$
$$\frac{-7y}{-7} = \frac{-28x}{-7} + \frac{21}{-7}$$

$$\overset{\text{slope}}{y = 4x + (\underset{y\text{-intercept}}{-3})} \qquad y = mx + b$$

Thus the slope is 4 and the y-intercept is -3.

To sketch the graph, begin by identifying the point where the graph crosses the y-axis. Since the y-intercept is −3, the point is (0, −3). Plot the point. Now look at the slope. The slope, m, is 4 or $\frac{4}{1}$. This means there is a rise of 4 and a run of 1. From the point (0, −3) go up 4 units and to the right 1 unit to locate a second point on the line. Draw a straight line that contains these two points and you have the graph of the equation $28x - 7y = 21$.

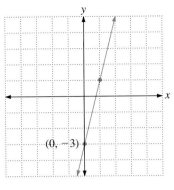

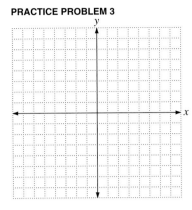

Practice Problem 3   Sketch the graph of the equation $3x - 4y = -8$. ■

2 The Point–Slope Form of the Equation of a Line

What happens if we know the slope of a line and a point on the line that is not the y-intercept? Can we write the equation of the line? By definition of slope,

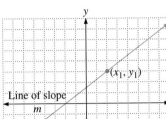

$$m = \frac{y - y_1}{x - x_1}$$

$$m(x - x_1) = y - y_1$$

That is, $y - y_1 = m(x - x_1)$

This is the point–slope form of the equation of a line.

Point–Slope Form

The **point–slope form** of the equation of a line is $y - y_1 = m(x - x_1)$, where m is the slope and (x_1, y_1) are the coordinates of a known point on the line.

Write the Equation of a Line Given Its Slope and One Point on the Line

EXAMPLE 4   Find an equation of the line with slope $-\frac{3}{4}$ that passes through the point $(-6, 1)$. Express your answer in standard form.

Since we don't know the y-intercept, we can't use the slope–intercept form. Therefore, we use the point–slope form.

$$y - y_1 = m(x - x_1)$$

$$y - 1 = -\frac{3}{4}[x - (-6)] \qquad \textit{Substitute the given values.}$$

$$y - 1 = -\frac{3}{4}x - \frac{9}{2} \qquad \textit{Simplify. (Do you see how we did this?)}$$

$$4y - 4(1) = 4\left(-\frac{3}{4}x\right) - 4\left(\frac{9}{2}\right) \qquad \textit{Multiply each term by the LCD 4.}$$

$$4y - 4 = -3x - 18 \qquad \textit{Simplify.}$$

$$3x + 4y = -18 + 4 \qquad \textit{Add } 3x + 4 \textit{ to each side.}$$

$$3x + 4y = -14 \qquad \textit{Add like terms.}$$

The equation in standard form is $3x + 4y = -14$.

Practice Problem 4   Find an equation of the line that passes through $(5, -2)$ and has a slope of $\frac{3}{4}$. Express your answer in slope–intercept form. ■

We can use the point–slope form to find the equation of a line if we are given two points. Carefully study the following example. Be sure you understand each step. This type of problem will be frequently encountered.

EXAMPLE 5   Find the equation of a line that passes through $(3, -2)$ and $(5, 1)$. Express your answer in standard form.

First find the slope.

$$m = \frac{y_2 - y_1}{x_2 - x_1} = \frac{1 - (-2)}{5 - 3} = \frac{1 + 2}{2} = \frac{3}{2}$$

Now we substitute the value of the slope and the coordinates of either point into the point–slope equation. Let's use $(5, 1)$. Then

$y - y_1 = m(x - x_1)$

$y - 1 = \dfrac{3}{2}(x - 5)$ *Substitute $m = \frac{3}{2}$ and $(x_1, y_1) = (5, 1)$.*

$y - 1 = \dfrac{3}{2}x - \dfrac{15}{2}$ *Remove parentheses.*

$2y - 2 = 2\left(\dfrac{3}{2}x\right) - 2\left(\dfrac{15}{2}\right)$ *Multiply each term by 2.*

$2y - 2 = 3x - 15$ *Simplify.*

$-3x + 2y = -15 + 2$ *Add $2 - 3x$ to each side.*

$-3x + 2y = -13$ *Simplify.*

$3x - 2y = 13$ *It is customary, but not necessary to have a positive value for the coefficient of x in the standard form, so we multiply each term by -1.*

Practice Problem 5   Find an equation of the line that passes through $(-4, 1)$ and $(-2, -3)$. Express your answer in standard form. ■

Before we go further, we want to point out that these various forms of the equation of a straight line are just that—*forms* for convenience. We are *not* using different equations each time, nor should you simply try to memorize the different variations without understanding when to use them. They can easily be derived from the definition of slope, as we have seen. And remember, you can *always* use the definition of slope to find the equation of a line. You will find it helpful to review Example 4 and Example 5 for a few minutes before going ahead to Example 6. It is important to see how each example is different in some aspect.

3 Parallel and Perpendicular Lines

Let us now look at parallel and perpendicular lines. Given a line, we can write the equation of a line that is parallel to or perpendicular to the given line because we know that the slopes of parallel lines are equal and that the slopes of perpendicular lines are negative reciprocals of each other.

    In our next example, we consider the case of finding the equation of a line that is parallel to a given line. However we want this new line to pass through a specific point that is not on the given line. This will involve first finding the slope of the given line. Then we will use the point–slope form of the equation of a line. Study carefully each step of this example.

EXAMPLE 6   Find the equation of a line passing through the point $(-2, -4)$ and parallel to the line $2x + 5y = 8$. Express the answer in standard form.

    First we need to find the slope of the line $2x + 5y = 8$. We do this by writing the equation in slope–intercept form.

$$5y = -2x + 8$$

$$y = -\frac{2}{5}x + \frac{8}{5}$$

The slope of the given line is $-\frac{2}{5}$. Since parallel lines have the same slope, the slope of the unknown line is also $-\frac{2}{5}$. Now we substitute $m = -\frac{2}{5}$ and the coordinates of the point $(-2, -4)$ into the point–slope form of an equation of a line.

$$y - y_1 = m(x - x_1)$$

$$y - (-4) = -\frac{2}{5}[x - (-2)] \qquad \text{\textit{Substitute.}}$$

$$y + 4 = -\frac{2}{5}(x + 2) \qquad \text{\textit{Simplify.}}$$

$$y + 4 = -\frac{2}{5}x - \frac{4}{5} \qquad \text{\textit{Remove parentheses.}}$$

$$5y + 5(4) = 5\left(-\frac{2}{5}x\right) - 5\left(\frac{4}{5}\right) \qquad \text{\textit{Multiply each term by LCD}} = 5.$$

$$5y + 20 = -2x - 4 \qquad \text{\textit{Simplify.}}$$

$$2x + 5y = -4 - 20 \qquad \text{\textit{Add }} 2x - 20 \text{ \textit{to each side.}}$$

$$2x + 5y = -24 \qquad \text{\textit{Simplify.}}$$

$2x + 5y = -24$ is the equation of the line passing through the point $(-2, -4)$ and parallel to the line $2x + 5y = 8$.

Practice Problem 6   Find the equation of a line passing through $(4, -5)$ and parallel to the line $5x - 3y = 10$. ∎

Some extra steps are needed if the desired line is to be perpendicular to the given line. Note carefully the approach of Example 7.

EXAMPLE 7 Find the equation of a line that passes through the point $(2, -3)$ and is **perpendicular** to the line $3x - y = -12$. Express the answer in standard form.

Find the slope of the line $3x - y = -12$. Its slope–intercept form is

$$-y = -3x - 12$$
$$y = 3x + 12$$

This line has a slope of 3. Therefore, the slope of a line perpendicular to this line is the negative reciprocal, $-\frac{1}{3}$.

Now substitute the slope $m = -\frac{1}{3}$ and the coordinates of the point $(2, -3)$ into the point–slope form of the equation.

$$y - y_1 = m(x - x_1)$$

$$y - (-3) = -\frac{1}{3}(x - 2) \qquad \textit{Substitute.}$$

$$y + 3 = -\frac{1}{3}(x - 2) \qquad \textit{Simplify.}$$

$$y + 3 = -\frac{1}{3}x + \frac{2}{3} \qquad \textit{Remove parentheses.}$$

$$3y + 3(3) = 3\left(-\frac{1}{3}x\right) + 3\left(\frac{2}{3}\right) \qquad \textit{Multiply each term by the LCD} = 3.$$

$$3y + 9 = -x + 2 \qquad \textit{Simplify.}$$

$$x + 3y = 2 - 9 \qquad \textit{Add } x - 9 \textit{ to each side.}$$

$$x + 3y = -7 \qquad \textit{Simplify.}$$

$x + 3y = -7$ is the equation of a line that passes through the point $(2, -3)$ and is perpendicular to the line $3x - y = -12$.

Practice Problem 7 Find the equation of a line that passes through $(-4, 3)$ and is **perpendicular** to the line $6x + 3y = 7$. Express the answer in standard form. ■

Developing Your Study Skills

MAKING A FRIEND IN THE CLASS

Attempt to make a friend in your class. You may find that you enjoy sitting together and drawing support and encouragement from one another. Exchange phone numbers so you can call each other whenever you get stuck while doing your homework. Set up convenient times to study together on a regular basis, to do homework, and to review for exams.

You must not depend on a friend or fellow student to tutor you, do your work for you, or in any way be responsible for your learning. However, you will learn from one another as you seek to master the course. Studying with a friend and comparing notes, methods, and solutions can be very helpful. And it can make learning mathematics a lot more fun!

3.3 Exercises

Verbal and Writing Skills

1. You are given two points that lie on a line. Explain how you would find the equation of the line.

2. $y = -\frac{2}{7}x + 5$. What can you tell about the graph by looking at the equation?

Write the equation of a line with the given slope and the given y-intercept. Leave the answer in **slope–intercept form.**

3. Slope -3, y-intercept 6

4. Slope 10, y-intercept -2

5. Slope $\frac{3}{4}$, y-intercept -9

6. Slope $-\frac{2}{3}$, y-intercept 5

Write the equation of a line with the given slope and the given y-intercept. Express the answer in **standard form.**

7. Slope-5, y-intercept -8

8. Slope -6, y-intercept -9

9. Slope $\frac{3}{4}$, y-intercept $\frac{1}{2}$

10. Slope $\frac{5}{6}$, y-intercept $\frac{1}{3}$

Write the equation of each line.

11.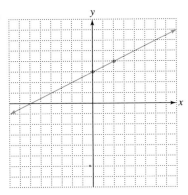

12.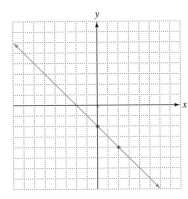

13.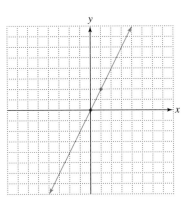

14.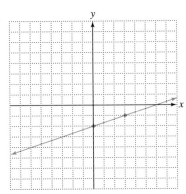

15.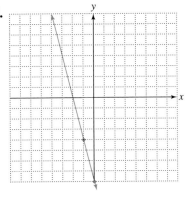

16.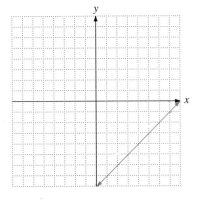

Write each equation in slope–intercept form. Then identify the slope and the y-intercept for each line.

17. $y + 6x = 7$

18. $y - 8x = 12$

19. $2x - 3y = -8$

20. $5x - 4y = -20$

21. $\frac{1}{2}x + 4y = 5$

22. $3x + \frac{2}{3}y = -2$

23. $2x - y = 5$ **24.** $3x - y = 10$

Sketch the graph of each equation.

25. $y = 3x + 4$

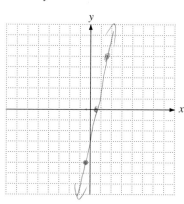

26. $y = \dfrac{1}{2}x - 3$

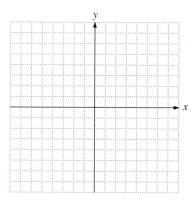

27. $y = -2x$

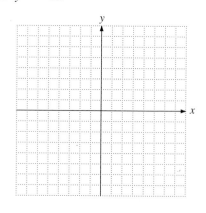

28. $5x - 4y = -20$

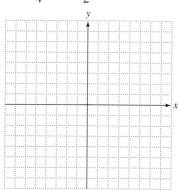

29. $3x + \dfrac{3}{4}y = -\dfrac{3}{2}$

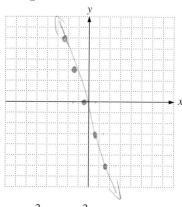

30. $5x + 3y = 18$

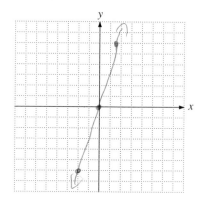

Find the equation of the line that passes through the given point and has the given slope. Express your answer in **slope–intercept form.**

31. $(5, 6)$, $m = 3$ **32.** $(-4, 5)$, $m = 7$ **33.** $(-7, -2)$, $m = 5$

34. $(8, 0)$, $m = -3$ **35.** $(2, -4)$, $m = -\dfrac{2}{3}$ **36.** $(-3, 1)$, $m = -\dfrac{1}{4}$

Find the equation of a line passing through the pair of points. Write the equation in **standard form.**

37. $(-3, 2)$ and $(2, -8)$ **38.** $(6, -1)$ and $(3, 2)$ **39.** $(7, -2)$ and $(-1, -3)$

40. $(-4, -1)$ and $(3, 4)$ **41.** $(0, 6)$ and $(2, -7)$ **42.** $(3, -8)$ and $(-5, 0)$

Find the equation of the line satisfying the conditions given. Express your answer in **standard form.**

43. Parallel to $3x - y = -5$ and passing through $(-1, 0)$

44. Parallel to $5x - y = 4$ and passing through $(-2, 0)$

45. Parallel to $2y + x = 7$ and passing through $(-5, -4)$

46. Parallel to $x = 3y - 8$ and passing through $(5, -1)$

47. Perpendicular to $y = 5x$ and passing through $(4, -2)$

48. Perpendicular to $y = -\frac{2}{3}x$ and passing through $(-3, 1)$

49. Perpendicular to $x - 4y = 2$ and passing through $(3, -1)$

50. Perpendicular to $x + 7y = -12$ and passing through $(-4, -1)$

51. Parallel to $x = -2$ and passing through $(5, 8)$

52. Parallel to $x = -3$ and passing through $(4, 6)$

To Think About

Determine without graphing whether the following pairs of lines are (a) parallel, (b) perpendicular, or (c) neither parallel or perpendicular.

53. $5x - 6y = 19$
$6x + 5y = -30$

54. $-3x + 5y = 40$
$5y + 3x = 17$

55. $y = \frac{2}{3}x + 6$
$-2x - 3y = -12$

56. $y = -\frac{3}{4}x - 2$
$6x + 8y = -5$

57. $y = \frac{3}{7}x - \frac{1}{14}$
$14y + 6x = 3$

58. $y = \frac{5}{6}x - \frac{1}{3}$
$6x + 5y = -12$

Optional Graphing Calculator Problems

If you have a graphing calculator, use it to graph each pair of equations. Do the graphs appear to be parallel?

59. $y = -2.39x + 2.04$ and $y = -2.39x - 0.87$

60. $y = 1.43x - 2.17$ and $y = 1.43x + 0.39$

Putting Your Skills to Work

DETERMINING THE DEMAND FOR A PRODUCT

When a manufacturing company is interested in introducing a new product, one of the variables that will affect the demand for the product will be the price. Consider the situation where an electronics company is planning to introduce a new graphing calculator. The market research department has given the company management the price versus demand table shown at the right.

Price	Estimated Demand (in thousands)
$ 40	128
$100	80
$120	64
$200	0

PROBLEMS FOR INDIVIDUAL INVESTIGATION AND ANALYSIS

1. Plot the points and draw the graph. Write the equation of the line in the form $y = mx + b$, where x represents the price in dollars and y represents the estimated demand for the calculator measured in thousands.

2. Find the slope. What does the slope indicate? Find the y-intercept. What does the y-intercept indicate?

3. What is the effect on the demand for the calculator by lowering the price by $10? by $100?

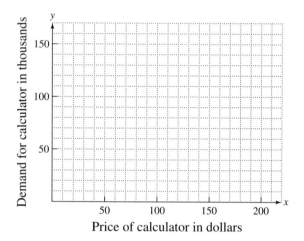

PROBLEMS FOR GROUP INVESTIGATION AND COOPERATIVE ACTIVITY

Together with other members of your class see if you can determine the following.

4. How many items can the manufacturer expect to sell if the calculator is priced at $110?

5. If the cost equation for this product is $C = 210,000 + 70n$, where n is the number of calculators sold, what would be the cost to manufacture this number of calculators?

6. If the revenue for selling these calculators is given by $R = 90n$, where n is the number of calculators sold what would be the profit for manufacturing and selling this number of calculators?

7. If one of the members of your group has a **graphing calculator,** complete the following. Graph the revenue equation $R = 90n$ and the cost equation $C = 210,000 + 70n$ at the same time. For what value of n do the two straight lines intersect? What is the significance of this value?

 INTERNET CONNECTION: Go to http://www.prenhall.com/tobey to be connected.

Site: Supply and Demand

This site displays a sample supply and demand graph. The demand graph is the one you have studied, whereas the supply graph shows the quantity of product that the manufacturer is willing to produce for a given price. (Note that this graph shows price on the vertical axis instead of the horizontal axis.)

8. As you move from the left of the graph to the right, the demand decreases but the supply increases. Explain why this is reasonable.

9. The intersection of the line showing supply and the lines showing demand is clearly marked. What is the significance of this point? (*Hint:* What will happen if the price is above or below the price indicated by this point?)

3.4 Linear Inequalities in Two Variables

After studying this section, you will be able to:

MathPro   Video 8   SSM

1 *Graph a linear inequality in two variables.*
2 *Graph a linear inequality in one variable on the plane.*

1 *Graphing a Linear Inequality in Two Variables*

A linear inequality in two variables is similar to a linear equation in two variables. However, in place of the = sign there appears instead one of the following four inequality symbols: $<, >, \leq, \geq$.

The graph of this type of linear inequality is a half-plane that lies above or below a straight line. It will also include the line if the inequality contains the $\leq$ or the $\geq$ symbols.

Procedure for Graphing Linear Inequalities

1. Replace the inequality symbol by an equals sign. This equation will be a boundary for the desired region.
2. Graph the line obtained in step 1. Use a dashed line if the original inequality contains a $<$ or $>$ symbol. Use a solid line if the original inequality contains a $\leq$ or $\geq$ symbol.
3. Choose a test point that does not lie on the boundary line. Substitute the coordinates into the original inequality. If you obtain an inequality that is true, shade the region on the side of the line containing the test point. If you obtain a false inequality, shade the region on the opposite side of the line from the test point.

If the boundary line does not pass through $(0, 0)$, that is usually a good test point to use.

EXAMPLE 1   Graph $y < 2x + 3$

1. The boundary line is $y = 2x + 3$.
2. Graph $y = 2x + 3$ using a dashed line because the inequality contains $<$.
3. Since the line does not pass through $(0, 0)$, we can use it as a test point. Substituting $(0, 0)$ into $y < 2x + 3$, we have

$$0 < 2(0) + 3$$
$$0 < 3$$

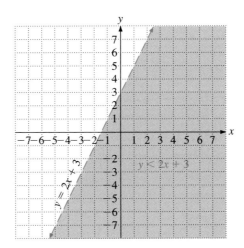

This inequality is true. We therefore shade the region on the same side of the line as $(0, 0)$. See the sketch. The solution is the shaded region *not including* the dashed line.

Practice Problem 1   Graph $y > 3x + 1$ ∎

PRACTICE PROBLEM 1

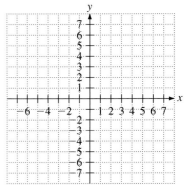

GRAPHING CALCULATOR

Graphing Linear Inequalities

You can graph a linear inequality like $y \leq 3x + 1$ on a graphing calculator. On some calculators you can directly enter the expression for the boundary in the $Y =$ editor of your graphing calculator and then select the appropriate direction for shading. Other calculators may require using a Shade command to draw in the shaded region. In general, most calculator displays do not distinguish between dashed and solid boundaries.

Display:

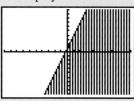

Graph the following:

1. $y < 3.45x - 1.232$
2. $y > -5.346x - 3.678$
3. $3.45y + 4.782x > 6.0238$

PRACTICE PROBLEM 3

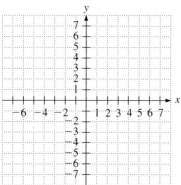

EXAMPLE 2 Graph. $3x + 2y \geq 4$

1. The boundary line is $3x + 2y = 4$.
2. Graph the boundary line with a solid line because the inequality contains $\geq$.
3. Since the line does not pass through $(0, 0)$, we can use it as a test point. Substituting $(0, 0)$ into $3x + 2y \geq 4$ gives

$$3(0) + 2(0) \geq 4$$
$$0 + 0 \geq 4$$
$$0 \geq 4$$

This inequality is false. We therefore shade the region on the opposite side of the line from $(0, 0)$. See the sketch. The solution is the shaded region *including* the boundary line.

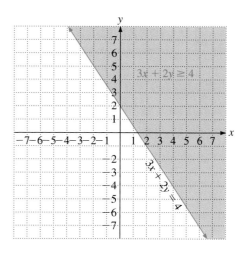

Practice Problem 2 Graph. $-4x + 5y \leq -10$

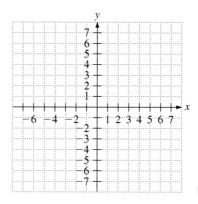

EXAMPLE 3 Graph. $4x - y < 0$

1. The boundary line is $4x - y = 0$.
2. Graph the boundary line with a dashed line.
3. Since $(0, 0)$ lies on the line, we cannot use it as a test point. We may choose any other point. Let's pick $(-1, 5)$. Substituting $(-1, 5)$ into $4x - y < 0$ gives

$$4(-1) - 5 < 0$$
$$-4 - 5 < 0$$
$$-9 < 0$$

This is true. We therefore shade the same side of the boundary line as $(-1, 5)$. See the sketch. The solution is the shaded region above the dashed line but *not including* the dashed line.

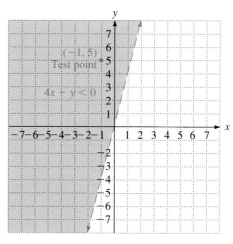

Practice Problem 3 Graph. $3y + x < 0$ ■

EXAMPLE 4 Graph $x < 2$ on a plane.

The variable y is missing in this inequality. If we were to graph the inequality on a line, its graph would be the following:

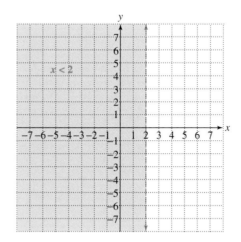

Now if we examine the inequality

$$x + 0y < 2,$$

we find that if $x < 2$ then any value of y would make the inequality true. Thus y can be any value at all and still be included in our shaded region. We therefore draw a vertical dashed line corresponding to $x = 2$. The region we want to shade is the region to the left of the dashed line. Do you see why? Our graph is on the right.

In Example 4 we could have employed the three-step method that was used in Examples 1 through 3. However, most students find that this additional work is not necessary.

Practice Problem 4 Graph $x > -2$ on a plane. ∎

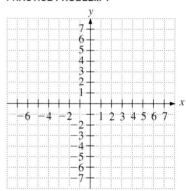

EXAMPLE 5 Graph $3y \geq -12$ on a plane.

First we will need to simplify the original inequality by dividing each side by 3.

$$\frac{3y}{3} \geq \frac{-12}{3}$$

$$y \geq -4$$

This inequality is equivalent to

$$0x + y \geq -4.$$

We find that if $y \geq -4$ any value of x will make the inequality true. Thus x can be any value at all and still be included in your shaded region. We therefore draw a solid horizontal line at $y = -4$. The region we want to shade is the region above the line. The solution to our problem is the line itself and the shaded region above the line.

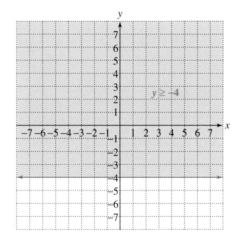

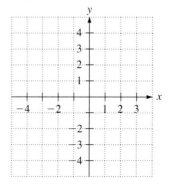

Practice Problem 5 Graph $6y \leq 18$ on a plane. ∎

3.4 Exercises

Graph each region.

1. $y > -2x + 4$

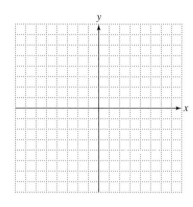

2. $y > -3x + 2$

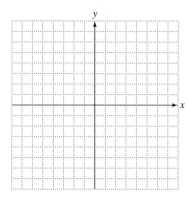

3. $y < \dfrac{2}{3}x - 2$

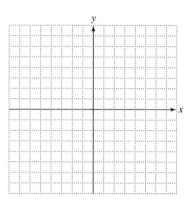

4. $y < \dfrac{3}{4}x - 3$

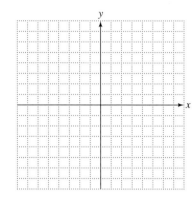

5. $y \le -\dfrac{3}{5}x + 1$

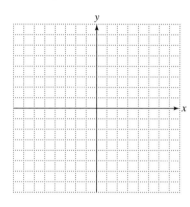

6. $y \le -\dfrac{2}{3}x + 4$

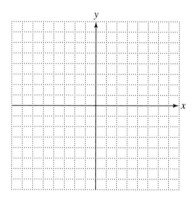

7. $-x + 2y \le 10$

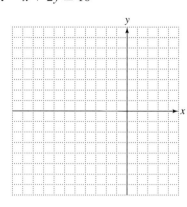

8. $-x + 3y \le 12$

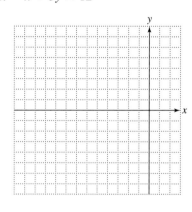

9. $2x + 3y \ge 6$

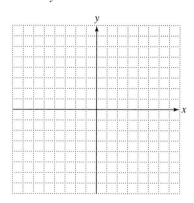

10. $-2x - y > 1$

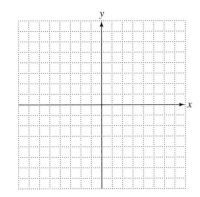

11. $x - y > -2$

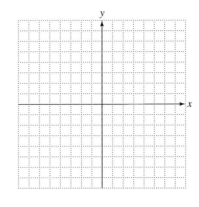

12. $-4x + 2y > -10$

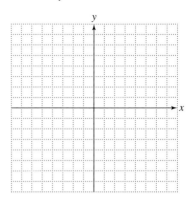

13. $y < -2x$

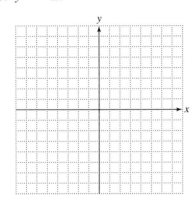

14. $y > -3x$

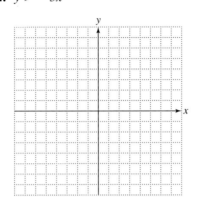

15. $3x - 2y \geq 0$

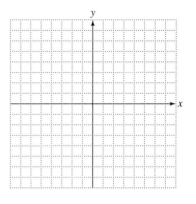

16. $4x - 3y \geq 0$

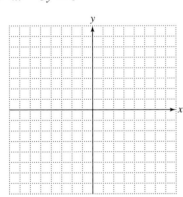

17. When we graph the inequality $3x - 2y \geq 0$ in Exercise 15, why can't we use $(0, 0)$ as a test point?

If we test the point $(-4, 2)$ do we obtain a false statement or a true one?

18. When we graph the inequality $4x - 3y \geq 0$ in Exercise 16, why can't we use $(0, 0)$ as a test point?

If we test the point $(6, -5)$ do we obtain a false statement or a true one?

Graph each inequality in the Cartesian plane.

19. $x > -4$

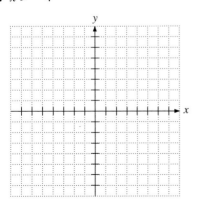

20. $x < 3$

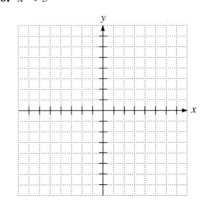

21. $y \leq -1$

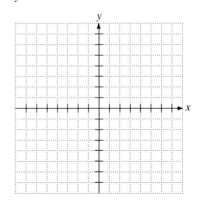

22. $y \geq -1$

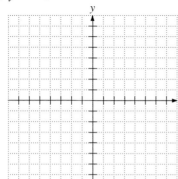

23. $-4x \leq 8$

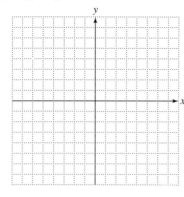

24. $-5x \leq -10$

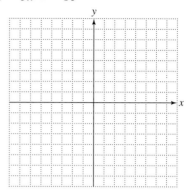

25. $-5y < 0$

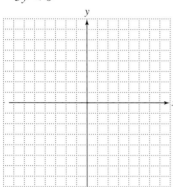

26. $-8y \geq 24$

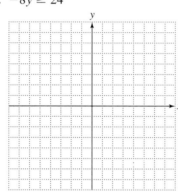

27. $-3x < 0$

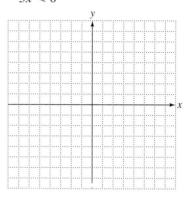

To Think About

28. What *one* region should be shaded to satisfy the inequalities $x + y \leq 3$, $x \geq 0$, and $y \geq 0$?

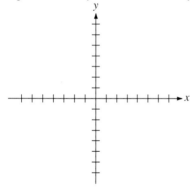

29. What *one* region should be shaded to satisfy the inequalities $x \geq 1$, $y \geq 2$, $x \leq 4$, and $y \leq 5$?

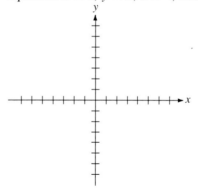

30. An elevator in an office building has a weight limit of 2100 pounds. If x children each weighing 75 pounds and y adults each weighing 175 pounds board the elevator, write a linear inequality using x and y that describes a weight limit of 2100 pounds or less. Graph the inequality.

Cumulative Review Problems

Evaluate each of the following.

31. $3x^2 + 6x - 2$ when $x = -2$

32. $A = \dfrac{1}{2}a(b + c)$ when $a = 6.0$, $b = 2.5$, and $c = 5.5$

3.5 Concept of a Function

After studying this section, you will be able to:

1 *Describe a relation and determine its domain and range.*
2 *Determine if a relation is a function.*
3 *Evaluate a function using function notation.*

MathPro   Video 8   SSM

1 *Describing a Relation*

Whenever we collect and study data, we look for a relationship. If we can determine a relationship between two sets of data, we can make predictions about the data. Let's begin with a simple finite example.

A local pizza parlor offers four different sizes of pizza and prices them according to the table on the right. By looking at the table we can see that the price depends on the size of the pizza. The larger the size of the pie, the more expensive it will be. This appears to be an increasing relation. We can use a set of ordered pairs to show the correspondence between the size of the pizza and the price.

——— *Pizza Plus* ———

Size of Pizza	Price of Pizza
Small	$ 4.50
Medium	$ 7.00
Large	$ 9.25
Party size	$12.75

{(Small, $4.50), (Medium, $7.00), (Large, $9.25), (Party size, $12.75)}

We can graph the ordered pairs to get a better picture of the relationship.

Following convention, we will assign the independent variable to the horizontal axis and the dependent variable to the vertical axis. Note that the dependent variable is price because price *depends* on size.

Just as we suspected, the relation is increasing. That is, the graph goes up as we move from left to right. We draw a line that approximately fits the data. This allows us to analyze the data more easily. Notice that the price of the party-size pizza is more than it should be. We can also see that if Pizza Plus decides to come out with a size between the small pizza and the medium pizza, you would expect it to be priced around $6. Thus we can see that there is a relation between the size of a pizza and its price and that we can express this relation as a table of values, as a set of ordered pairs, or in a graph.

Mathematicians have found it most useful to define a relation in terms of ordered pairs.

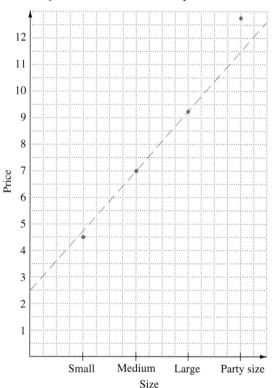

Relation

A relation is any set of ordered pairs.

All the first items of each ordered pair in a relation can be grouped as a set called the **domain** of the relation. The domain of the pizza example consists of {Small, Medium, Large, Party size}. These are all the possible values of the independent variable *size*.

All the second items of each ordered pair in a relation can be grouped as a set called the **range** of the relation. The range of the pizza example consists of {$4.50, $7.00, $9.25, $12.75}. These are the corresponding values of the dependent variable *price*.

EXAMPLE 1 The information in the table below can be found in most almanacs.

Look at the data for the men's 100-meter run. Is there a relation between any two sets of data in this table? If so, describe the relation as a table of values, a set of ordered pairs, and a graph.

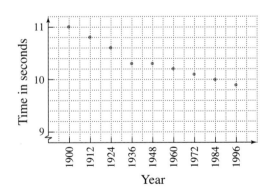

Olympic Games:
100-Meter Run For Men

Year	Winning runner, Country	Time
1900	Francis W. Jarvis, USA	11.0 s
1912	Ralph Craig, USA	10.8 s
1924	Harold Abrahams, Great Britain	10.6 s
1936	Jesse Owens, USA	10.3 s
1948	Harrison Dillard, USA	10.3 s
1960	Armin Harg, Germany	10.2 s
1972	Valery Borzov, USSR	10.14 s
1984	Carl Lewis, USA	9.99 s
1996	Donovan Bailey, Canada	9.84 s

A useful relation might be the correspondence between the year the event occurred and the time in which the race was won. Let's see how this looks in a table, as a set of ordered pairs, and on a graph. We will choose the year as the independent variable.

Table

Year	1900	1912	1924	1936	1948	1960	1972	1984	1996
Time in Seconds	11.0	10.8	10.6	10.3	10.3	10.2	10.14	9.99	9.84

Ordered pairs

{(1900, 11.0), (1912, 10.8), (1924, 10.6), (1936, 10.3), (1948, 10.3), (1960, 10.2), (1972, 10.14), (1984, 9.99), (1996, 9.84)}

Graph

By looking at the graph, we can see that the winning time usually decreases each year. This is a decreasing relation most of the time. What might we expect the winning time to be in 2008? Can we expect the time to decrease indefinitely? Why or why not?

Practice Problem 1   The data for the women's 100-meter run in the Olympic Games for selected years are given in the table on the right. Describe the relation between the year and the time in a table, as a set of ordered pairs, and as a graph.

Olympic Games: 100-Meter Run For Women		
Year	**Winning runner, Country**	**Time**
1936	Helen Stephens, USA	11.5 s
1948	Francina Blankers-Koen, Netherlands	11.9 s
1960	Wilma Rudolph, USA	11.0 s
1972	Renate Stecher, East Germany	11.07 s
1984	Evelyn Ashford, USA	10.97 s
1996	Gail Devers, USA	10.94 s

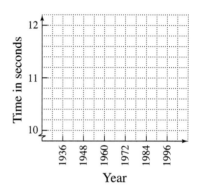

2 Functions

Two of the relations we have just discussed have a special characteristic. Each value in the domain does not have more than one value in the range. This is true of our pizza example. One size of pizza does not have several different prices. This is also true of the relation in the 100-meter run for men, where time corresponds to year. No year has several different winning times.

Looking at this as ordered pairs, we say that no two different ordered pairs have the same first coordinate. We will list each set of ordered pairs.

{(Small, $4.50), (Medium, $7.00), (Large, $9.25), (Party size, $12.75)}

{(1900, 11.0), (1912, 10.8), (1924, 10.6), (1936, 10.3), (1948, 10.3), (1960, 10.2), (1972, 10.14), (1984, 9.99), (1996, 9.84)}

All such relations with this special property are called **functions.**

Function

A function is a relation in which no two different ordered pairs have the same first coordinates.

Notice that for the data we described in Example 1, if we reverse the order of the ordered pairs, the resulting relation is not a function. We can see this readily if we list the ordered pairs.

{(11.0, 1900), (10.8, 1912), (10.6, 1924), (10.3, 1936), (10.3, 1948), (10.2, 1960), (10.14, 1972), (9.99, 1984), (9.84, 1996)}

Two pairs have the same first coordinate. This is not a function.

EXAMPLE 2   Give the domain and range of each relation. Indicate if the relation is a function.

(a) $f = \{(1, 1), (2, 4), (3, 4), (4, 16)\}$ **(b)** $g = \{(2, 8), (2, 3), (3, 7), (5, 12)\}$

(c) $t =$ income tax paid by individuals

Income	14,000	18,000	24,500	33,000	50,000	50,000
Tax	2350	2800	2900	3750	1350	7980

(d) $v =$ average value of a $10,000 whole-life policy

Year	0	5	10	15	20
Value	0	490	1270	2000	2790

(e)

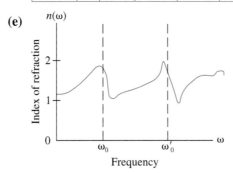

Recall that the domain of a function consists of all the possible values of the independent variable. In a set of ordered pairs, this is the first item in each ordered pair.

The range of a function consists of the corresponding values of the dependent variable. In a set of ordered pairs, this is the second item in each ordered pair.

(a) $\text{Domain}_f = \{1, 2, 3, 4\}$
$\text{Range}_f = \{1, 4, 16\}$
Look at the ordered pairs of the function. No two different ordered pairs have the same first coordinate. f is a function.

(b) $\text{Domain}_g = \{2, 3, 5\}$
$\text{Range}_g = \{8, 3, 7, 12\}$
$(2, 8)$ and $(2, 3)$ have the same first coordinate. g is not a function.

(c) $\text{Domain}_t = \{14{,}000, 18{,}000, 24{,}500, 33{,}000, 50{,}000\}$
$\text{Range}_t = \{2350, 2800, 2900, 3750, 1350, 7980\}$
$(50{,}000, 1350)$ and $(50{,}000, 7980)$ have the same first coordinate. t is not a function.

(d) $\text{Domain}_v = \{0, 5, 10, 15, 20\}$
$\text{Range}_v = \{0, 490, 1270, 2000, 2790\}$
No two different ordered pairs have the same first coordinate. v is a function.

(e) $\text{Domain}_r = \{\text{all positive numbers}\}$
$\text{Range}_r = \{\text{all numbers from 0 to 2}\}$
For every value in the domain there is only one value in the range. No value in the domain has more than one corresponding range value. This is the graph of a function.

Practice Problem 2   Give the domain and range of each relation. Indicate if the relation is a function.

(a) $h = \{(6, 1), (2, 1), (3, 1), (4, 1)\}$

(b) p = car performance

Horsepower	158	161	163	160	161
Top speed (mph)	98.6	89.2	101.4	102.3	94.9

(c)

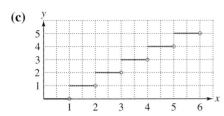

3 Function Notation

Function notation is one way of showing the relationship between the ordered pairs in a function. Looking back at our pizza example, we see that a small pizza costs $4.50. Since size is a function of price, we could write

$$f(\text{Small}) = \$4.50$$

This notation is most useful when working with continuous functions whose domain and range consist of an infinite number of values. We often use equations to describe such functions. You are already familiar with linear equations. Let's look at one such equation.

$$y = 3x - 2$$

In this equation, y is a function of x. That is, for each value of x in the domain there is a unique value of y in the range. We can use function notation to determine the function value y for specific values of x. Let $x = 0$. We write

$$f(x) = 3x - 2$$
$$f(0) = 3(0) - 2 \qquad \textit{Substitute 0 for x.}$$
$$f(0) = -2 \qquad \textit{Evaluate.}$$

We see that, when x is 0, $f(x)$ is -2. That is, the value of the function is -2 when x is 0. *Note:* The notation $f(x)$ does not mean that f is multiplied by x. It means that for any specific value of x there is only one value for y.

It may help you to understand the idea of a function by imagining a "function machine." An item from the domain enters the machine—the function—and a member of the range results.

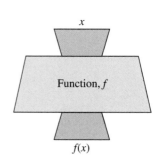

EXAMPLE 3   If $f(x) = 4 - 3x$, find **(a)** $f(2)$ **(b)** $f(-1)$ **(c)** $f\left(\dfrac{1}{3}\right)$

In each case we replace x by the specific value in the domain.

(a) $f(2) = 4 - 3(2)$ **(b)** $f(-1) = 4 - 3(-1)$ **(c)** $f\left(\dfrac{1}{3}\right) = 4 - 3\left(\dfrac{1}{3}\right)$

$\qquad\quad = 4 - 6$ $\qquad\qquad\quad = 4 + 3$ $\qquad\qquad\qquad\quad = 4 - 1$

$\qquad f(2) = -2$ $\qquad\qquad f(-1) = 7$ $\qquad\qquad\qquad f\left(\dfrac{1}{3}\right) = 3$

Practice Problem 3   If $f(x) = 2x^2 - 8$, find

(a) $f(-3)$ **(b)** $f(4)$ **(c)** $f\left(\dfrac{1}{2}\right)$ ∎

3.5 Exercises

Verbal and Writing Skills

1. Explain the difference between a relation and a function.

2. Explain the difference between the domain and the range.

3. What are the three ways you can describe a function.

4. Write the ordered pair for $f(-5) = 8$ and identify the x and y values.

What are the domain and range of each relation? Is the relation a function?

5. $A = \{(10, 1), (12, 2), (14, 3), (16, 4)\}$

6. $B = \{(1, 4), (2, 9), (3, 16), (4, 25)\}$

7. $C = \{(\frac{1}{2}, 5), (-3, 7), (\frac{3}{2}, 5), (\frac{1}{2}, -1)\}$

8. $D = \{(0, 0), (5, 13), (7, 11), (5, 0)\}$

9. $E = \{(3, -2), (4, -2), (5, 7), (8, -6)\}$

10. $F = \{(-7, 2), (-6, -1), (-3, 4), (-7, -6)\}$

11. $G = \{(1, 8), (3, 8), (5, 8), (7, 8)\}$

12. $H = \{(0, 0), (1.3, -2), (4, 8.6), (1.3, 0)\}$

13.
Women's Dress Sizes

USA	6	8	10	12	14
France	38	40	42	44	46

14.
Women's Dress Sizes

France	40	42	44	46	48
Britain	14	15	16	17	18

15.
Some of the World's Longest Rivers

River	Nile	Amazon	Chang Jiang	Ob-Irtysh	Huang He	Congo
Approximate Length in Miles	4160	4000	3964	3362	2903	2900

16.
Normal Monthly Average Fahrenheit Temperature: Pago Pago, Samoa

Month	Jan.	Feb.	Mar.	Apr.	May	June	July	Aug.	Sept.	Oct.	Nov.	Dec.
Temperature	81	81	81	81	80	80	79	79	79	80	80	81

17.

Tallest Buildings, USA

City	Chicago	New York	New York	Chicago	Chicago	New York
Height	1454	1350	1250	1136	1127	1046

18.

Temperature Scales

Fahrenheit	32	41	50	59	68	95
Celsius	0	5	10	15	20	35

19.

Metric Conversion

Miles	1	2	3	4	5
Kilometers	1.61	3.22	4.83	6.44	8.05

Determine if the graph represents the graph of a function.

20.

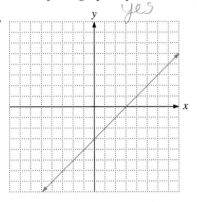

21.

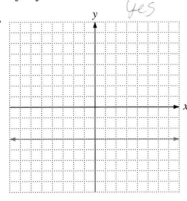

22.

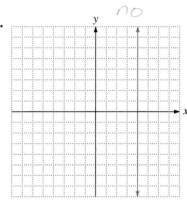

23.

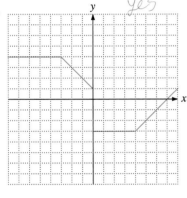

24.

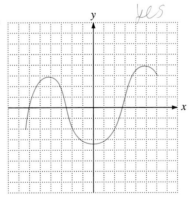

25.

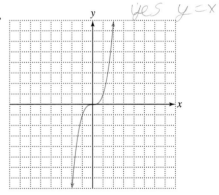

26.

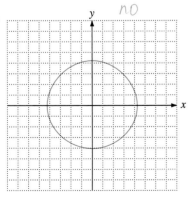

27.

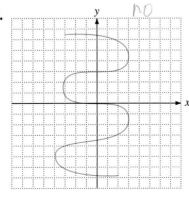

28.

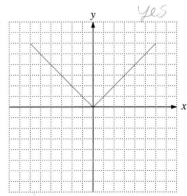

Given the function defined by $f(x) = 3x + 4$, find

29. $f(5)$ **30.** $f(6)$ **31.** $f(-3)$ **32.** $f(-2)$

Given the function defined by $g(x) = 2x - 5$, find

33. $g(-3)$ **34.** $g(-2.4)$ **35.** $g\left(\dfrac{1}{3}\right)$ **36.** $g\left(\dfrac{1}{2}\right)$

Given the function defined by $h(x) = \dfrac{2}{3}x + 2$, find

37. $h(-3)$ **38.** $h(-6)$ **39.** $h(1)$ **40.** $h(2)$

Given the function defined by $r(x) = 2x^2 - 4x + 1$, find

41. $r\left(\dfrac{1}{2}\right)$ **42.** $r\left(\dfrac{1}{3}\right)$ **43.** $r(-2)$ **44.** $r(-3)$

Given the function defined by $t(x) = x^3 - 3x^2 + 2x - 3$, find

45. $t(2)$ **46.** $t(-3)$ **47.** $t(-1)$ **48.** $t(-2)$

To Think About

Find the range of the function for the given domain.

49. $f(x) = x + 3$ Domain $= \{-2, -1, 0, 1, 2\}$

50. $g(x) = x^2 + 3$ Domain $= \{-2, -1, 0, 1, 2\}$

Find the domain of the function for the given range. Be careful to find all possible values.

51. $h(x) = \dfrac{2}{3}x - 4$ Range $= \left\{0, 2, \dfrac{10}{3}, 4\right\}$

52. $d(x) = 3 - \dfrac{1}{4}x$ Range $= \left\{0, \dfrac{1}{4}, \dfrac{3}{4}, 4\right\}$

Cumulative Review Problems

Remove grouping symbols and simplify.

53. $5(3 - x) - 2(3y + x)$

54. $-3\{x + 2[x - 3(2 + x)]\}$

55. Laurie has \$763.21 in her checking account. She writes checks for \$280.55, \$78.91, \$116.01 and \$196.69. She needs to pay a bill of \$424.98 next week, and she wants to keep a minimum deposit of \$100 in her account at all times (so that she is not required to pay a penalty). How much should she deposit into her account?

56. To provide the average modern person living in a developed county with the necessities and luxuries of his or her life, at least twenty tons of raw materials must be dug from the earth every year. If the average life span of a U.S. female is 81 years of age, and the average life span of a U.S. male is 76 years of age, how much raw material would have to be dug for the average couple in a lifetime in the United States?

3.6 Graphing Functions from Equations and Tables

After studying this section, you will be able to:

1 *Graph a function from an equation.*
2 *Graph a function from a table.*

1 Graphing Functions from Equations

Frequently, we are given a function in the form of an equation and are asked to graph the function. Each value of the function $f(x)$, often labeled y, corresponds to a value in the domain, often labeled x. This correspondence is the ordered pair (x, y) or $(x, f(x))$. The graph of the function is the graph of the ordered pairs.

If a function can be written in the form $f(x) = mx + b$, it is called a linear function. The graph of a linear function is a straight line.

EXAMPLE 1   Graph. $f(x) = \dfrac{1}{2}x - 2$

We recognize this as a linear function with a slope of $\frac{1}{2}$ and the y-intercept at -2. We could graph the function using the slope and y-intercept or we could use a table of values. We choose to use a table of values. To avoid obtaining fractions in the range, we will select even integers for domain values.

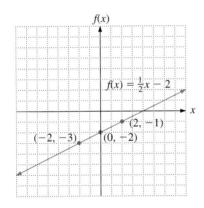

x	$f(x)$
-2	-3
0	-2
2	-1

Practice Problem 1   Graph $f(x) = -3x + 2$ using a table of values.

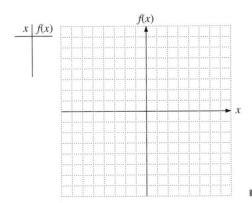

If we can represent real-life situations as functions, we can determine characteristics about the relationship and make predictions based on a table of values or a graph.

EXAMPLE 2   A salesperson earns $15,000 a year plus a 20% commission on her total sales. Express her annual income in dollars as a function of her total sales. Determine values of the function for total sales of $0, $25,000, and $50,000. Graph the function and determine if the function is increasing or decreasing.

Since we want to express income as a function of sales, let's see how income is determined.

income = $15,000 + 20% of total sales

Income depends on total sales. Income is the dependent variable, while total sales is the independent variable. So we will

Let x = the amount of total sales in dollars

$i(x) = 15,000 + 0.20x$ *Change 20% to 0.20.*

$i(0) = 15,000 + 0.20(0) = 15,000$

$i(25,000) = 15,000 + 0.20(25,000) = 20,000$

$i(50,000) = 15,000 + 0.20(50,000) = 25,000$

We can put this information in a table of values.

x	$i(x)$
0	15,000
25,000	20,000
50,000	25,000

To facilitate the graphing, we will use a scale in thousands. Thus we modify the table to make our task of graphing easier.

x (thousands)	$i(x)$ (thousands)
0	15
25	20
50	25

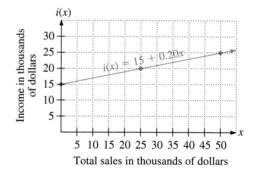

PRACTICE PROBLEM 2

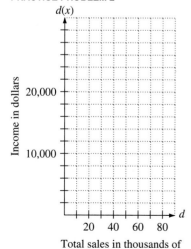

Practice Problem 2   A salesperson earns $10,000 a year plus a 15% commission on her total sales. Express her annual income in dollars as a function d of her total sales x. Graph her salary function. Use the values $d(0)$, $d(40,000)$, and $d(80,000)$. ■

Thus far most of our work has been with linear functions. Let's look at a variety of other functions and their graphs. We will begin with the absolute value function. What does the graph of the absolute value function look like? Recall that there are two solutions to the equation $|x| = 4$. That is, $|x| = 4$ when $x = -4$ and when $x = 4$. The solutions are -4 and 4. The ordered pairs are $(-4, 4)$ and $(4, 4)$. These two points will be four units above the x-axis on each side of the y-axis. That is, they will be *reflected* about the y-axis. To get a better idea of the shape of the graph, we need to plot more points.

EXAMPLE 3   Graph $g(x) = |x|$.

We will want to find the function values for five values of the independent variable x. Because of the nature of the function, we will choose both negative and positive values for x. The table of values and the resulting graph are as follows:

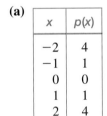

x	$g(x)$
-2	2
-1	1
0	0
1	1
2	2

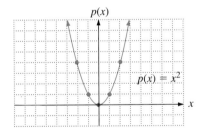

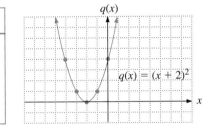

Practice Problem 3   Graph the function $h(x) = |x + 2|$. Use values for x from -4 to 0. ■

To Think About How are the graphs of $g(x) = |x|$ and $h(x) = |x + 2|$ the same? How are they different? What would the graph of $k(x) = |x - 2|$ look like?

Now let's take a look at a similar function, $p(x) = x^2$. What happens when $x = 2$? When $x = -2$? Notice that when x is 2 or -2 the value of the function is 4. The ordered pairs are $(-2, 4)$ and $(2, 4)$. This also looks like a graph that is reflected about the y-axis. We need to plot more points.

EXAMPLE 4   Graph. **(a)** $p(x) = x^2$ **(b)** $q(x) = (x + 2)^2$

For each function we will choose both negative and positive values of x. Since these are *not* linear functions, we use a curved line to connect the points. The tables of values and the resulting graphs are as follows:

(a)

x	$p(x)$
-2	4
-1	1
0	0
1	1
2	4

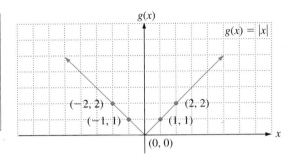

(b)

x	$q(x)$
-4	4
-3	1
-2	0
-1	1
0	4

Practice Problem 4   Graph. **(a)** $r(x) = (x - 2)^2$ **(b)** $s(x) = x^2 + 2$

(a)

x	$r(x)$

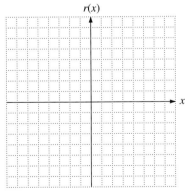

(b)

x	$s(x)$

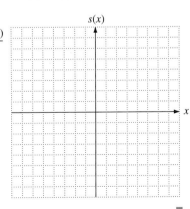

■

To Think About How are the graphs of each of the previous functions the same? How are they different? Describe how changing $p(x) = x^2$ to $q(x) = (x + 5)^2$ and to $s(x) = x^2 + 5$ affects the graph of each function.

Another interesting graph is the graph of the function $g(x) = x^3$. A quick look at $g(x)$ when x is 2 or -2 reveals that $g(x)$ is *not* reflected about the y-axis. That is, there are different function values for 2 and -2. Let's see.

$$g(x) = x^3$$

$$g(2) = 2^3 \qquad\qquad g(-2) = (-2)^3$$
$$= 8 \qquad\qquad\qquad = -8$$

EXAMPLE 5 Graph. **(a)** $g(x) = x^3$ **(b)** $h(x) = x^3 + 1$

We will pick five values for x, find the corresponding function values, and plot the five points to assist us in sketching the graph.

(a)

x	$g(x)$
-2	-8
-1	-1
0	0
1	1
2	8

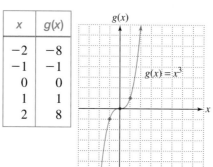

(b)

x	$h(x)$
-2	-7
-1	0
0	1
1	2
2	9

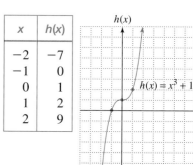

Practice Problem 5 Look at the function $f(x) = x^3 - 2$. What do you think the graph of $f(x) = x^3 - 2$ will look like? Make a table of values for the function and draw the graph.

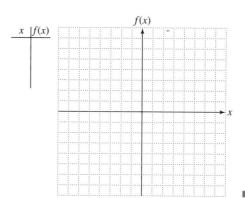

To Think About Describe how the graph of $f(x) = x^3 - 2$ is related to the graph of $g(x) = x^3$.

In Examples 1 through 5 each function has had a domain of all real numbers. We were free to choose any real number for x. However, if any specific function contains a variable in the denominator, the domain of that function will not include any value for which the denominator becomes zero.

EXAMPLE 6   Graph. $p(x) = \dfrac{4}{x}$

First we observe that we cannot choose x to be 0 because $\frac{4}{0}$ is not defined. (We can never divide by zero.) Therefore, the domain of this function is all real numbers except zero. To make a table of values, we will choose five values for x that are greater than 0 and five values for x that are less than 0.

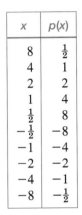
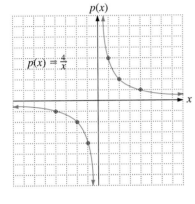

x	$p(x)$
8	$\frac{1}{2}$
4	1
2	2
1	4
$\frac{1}{2}$	8
$-\frac{1}{2}$	-8
-1	-4
-2	-2
-4	-1
-8	$-\frac{1}{2}$

The study of this type of function, called a *rational function*, will be continued in a more advanced course, such as college algebra or precalculus.

Practice Problem 6   Graph $q(x) = -\dfrac{4}{x}$.

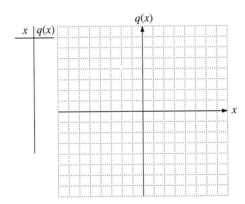

To Think About Describe how the graph of $q(x) = -\frac{4}{x}$ is related to the graph of $p(x) = \frac{4}{x}$.

2 Graphing Functions from a Table

Sometimes, when we study an event in daily life, we record data to better understand the event. In many cases we will not have an equation to work with, but rather a table of values. The graph of the values may help us to understand the function.

In the following example of a function, the number of items sold is the domain and the profit obtained by the sale is the range.

EXAMPLE 7   The marketing manager of a shoe company compiled the data below.

(a) Graph the function.

(b) From the graph, determine the profit from selling 4000 pairs of shoes in one month.

(c) What kind of profit would you expect to make from selling 0 pairs of shoes in a month? What value do you obtain on the graph for $x = 0$? What does this mean?

x Pairs of shoes sold in a month (in thousands of pairs)	$p(x)$ Monthly profit from the sales of shoes (in thousands of dollars)
3	5
5	9
7	13

(a) We plot the three ordered pairs and connect the points by a straight line. This means the function is a linear function.

(b) We find that the value $x = 4$ on the graph corresponds to the value $y = 7$. Thus we would expect that if we sold 4000 pairs of shoes we would have a profit of $7000 for the month.

(c) In any business we would not expect to make a profit at all if we sell no items. Looking at our graph, when $x = 0$, $y = -1$. Thus we would predict no profit, but rather a loss of $1000 in a month of zero sales of shoes. The loss may be due to fixed expenses such as rent and supplies.

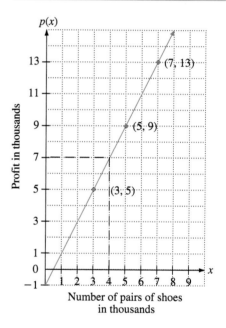

Number of pairs of shoes in thousands

Note: In a function such as this, the domain is all real numbers greater than or equal to zero. The range is all real numbers.

Practice Problem 7   An accountant reviewed a doctor's office for profitability by comparing the number of patients the doctor saw with the profit. The data are in the table on the right.

(a) Graph the function.

(b) Describe the function.

(c) What would the profit be if the doctor saw three patients per hour?

(d) What would the weekly loss be if the doctor saw zero patients per hour?

(e) Can we expect the function to increase indefinitely? Why or why not? ■

x Average number of patients per hour	$g(x)$ Weekly profit
6	$2000
9	$4000
12	$6000

PRACTICE PROBLEM 7

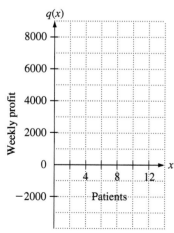

Given a set of data, we do not always know what the graph of the function will look like. Not all functions are linear. When points do not appear to lie on a line, we connect the points with a smooth curve.

Let's look at wind chill, the effect wind has on making the body colder. A combination of cold temperature and wind makes a body much colder than if the air were calm. For example, if it is 10 degrees Fahrenheit (°F) outside and the wind is blowing at 20 miles per hour (mph), the effect is the same as if the temperature were 24 degrees below zero! A wind chill table that describes this relationship is used in the next example.

EXAMPLE 8 The wind chill at 10 degrees Fahrenheit is given in the table at the right.

Wind Speed (mph)	Wind chill (°F)
0	10
5	7
10	−9
15	−18
20	−24
25	−29
30	−33
35	−35

(a) Graph the function.

(b) Estimate the wind chill when the wind is blowing at 23 mph.

(c) Estimate the wind chill when the wind is blowing at 40 mph.

(d) If the wind chill is −15°F, at what speed is the wind blowing?

(e) Based on your analysis of the graph, does an increase in wind speed result in a greater change in the wind chill at lower wind speeds or at higher wind speeds?

(a) First we need to determine which value is the independent variable and which value is the dependent variable. Since wind chill depends on wind speed, wind chill must be the dependent variable. These are the function values. Label the vertical axis Wind Chill and the horizontal axis Wind Speed. Plot the points. Use a smooth curve to connect the points.

(b) Find 23 along the horizontal axis. The function value for 23 on the curve is about −27°F.

(c) Since the last wind chill number in the table is −35, we need to extend the curve. This is called *extrapolation* and it is only an approximate value. At 40 mph the wind chill is about −37°F.

(d) Find −15 along the vertical scale. Move to the right until you intersect the curve. Then move up until you intersect the horizontal axis. This is the wind speed when the wind chill is −15. This is about 13 mph.

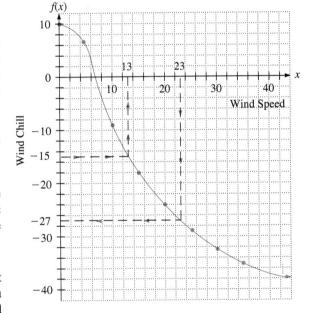

(e) The curve goes downward more quickly at lower wind speeds. Thus lower wind speeds have a greater effect on wind chill. For example, when the wind speed goes from 0 to 15 mph, the wind chill goes from 10 to −18°. This is a change of 28°. A change in wind speed from 20 to 35 mph only produces a change in wind chill of 11°.

f(x)

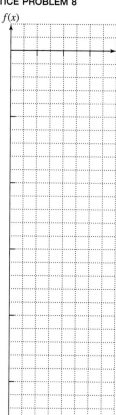

Practice Problem 8   The table for the wind chill factor at 0°F is given below.

Wind speed	0	5	10	15	20	25	30	35
Wind chill	0	−5	−22	−31	−39	−44	−48	−52

(a) Graph the function.

(b) Estimate the wind chill when the wind is blowing at 32 mph.

(c) Estimate the wind chill when the wind is blowing at 40 mph.

(d) If the wind chill is −24°F, at what speed is the wind blowing?

(e) What is the domain of the function? What is the range? ■

EXAMPLE 9   Graph the function $f(x)$ whose table of values is given below. The domain is $x \geq 2$ and the range is $f(x) \geq 0$.

(a) Determine an approximate value for $f(x)$ when $x = 8$.

(b) Determine an approximate value for x when $f(x) = 1.5$.

x	2	3	6	11
f(x)	0	1	2	3

(c) What do you notice about the graph as the values of x get larger?

    Assign the independent variable x to the horizontal axis. Assign the function values, the dependent variable, to the vertical axis. Plot the points and connect them with a smooth curve.

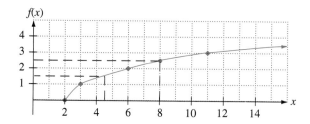

(a) Find $x = 8$ along the horizontal axis. Move up until you intersect the graph. Then move to the left until you intersect the vertical axis. This is $f(x)$ when x is 8. Read the scale. $f(x)$ is about 2.5.

(b) Find $f(x) = 1.5$ along the vertical axis. Move to the right until you intersect the graph. Then move down until you intersect the horizontal axis. This is x when $f(x)$ is 1.5. Read the scale. x is about 4.5.

(c) As the values of x get larger, the function values are increasing more slowly. In other words the *rate of change* decreases for larger values of x.

g(x)

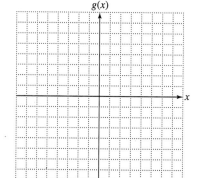

Practice Problem 9   Graph the function $g(x)$ whose table of values is given below. The domain is $x \leq 4$ and the range is $g(x) \geq 0$.

(a) Determine an approximate value for $g(x)$ when $x = -3$.

(b) Determine an approximate value for x when $g(x) = 1.5$.

x	4	3	0	−5
g(x)	0	1	2	3

(c) As x goes from 3 to 0 to −5, what property do you observe about the curve? ■

3.6 Exercises

Graph each function.

1. $f(x) = 2x + 1$

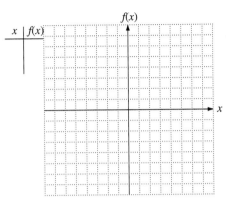

2. $f(x) = 2x - 1$

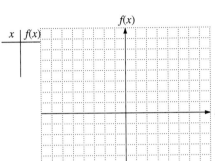

3. $f(x) = \dfrac{2}{3}x - 4$

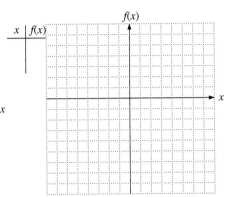

4. $f(x) = \dfrac{3}{4}x + 2$

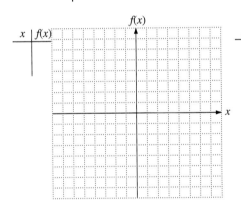

5. $f(x) = -\dfrac{1}{3}x + 6$

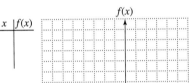

6. $f(x) = -\dfrac{1}{2}x - 1$

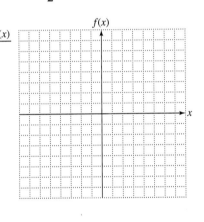

7. $f(x) = -\dfrac{1}{2}x + \dfrac{1}{2}$

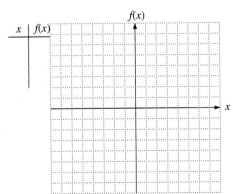

8. $f(x) = -\dfrac{2}{3}x - \dfrac{1}{2}$

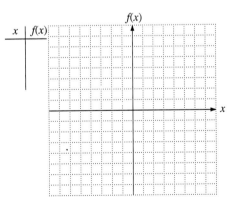

9. The population of Sommerville is 20,000 people, plus a growth of 400 people per year for every year since 1990. Express the population as a function $p(x)$, where x is the number of years since 1990. Obtain values for the function when $x = 0$, $x = 8$, $x = 12$, and $x = 16$. Graph the population function.

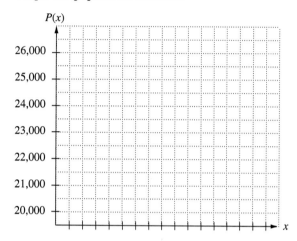

10. The cost of renting a new Chrysler Concorde at the airport is $35 for a day plus a cost of $0.10 per mile. Express the cost of renting a car for the day $c(x)$, where x is the number of miles. Obtain values for the function when $x = 0$, $x = 100$, $x = 200$, $x = 250$ miles. Graph the cost function.

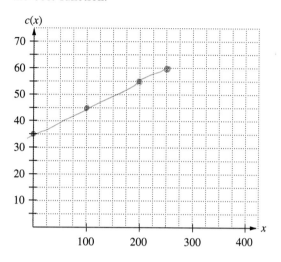

11. The R value of insulation in a house is a measure of its ability to resist the loss of heat from the house. The R value of fiberglass insulation is a linear function of its thickness in inches. One type of fiberglass insulation that is 6 inches thick has an R value of 19. The R value in general of this type of insulation is obtained by multiplying 3.2 by the thickness x measured in inches and then adding the result to -0.2. Express the R value as a function $R(x)$, where x is the thickness in inches. Obtain values of the function when $x = 0$, $x = 1$, $x = 3.5$, and $x = 6$. Graph the R function.

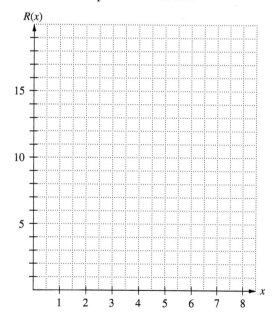

12. A wildlife biologist has determined that a certain stream in New Hampshire can support 45,000 fish if it is free of pollution. She has estimated that for every ton of pollutants in the stream 1500 fewer fish can be supported. Express the fish population p as a function $p(x)$, where x is the number of tons of pollutants found in the stream. Obtain values of the function when $x = 0$, $x = 10$, and $x = 30$. Graph the population function. What is the significance of the graph when $x = 30$?

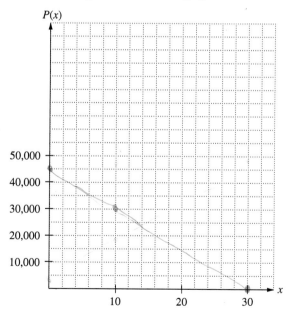

Graph each function.

13. $f(x) = |x - 1|$

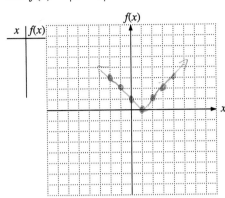

14. $g(x) = |x - 3|$

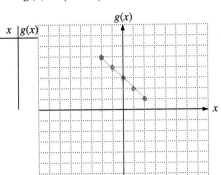

15. $f(x) = |x| + 2$

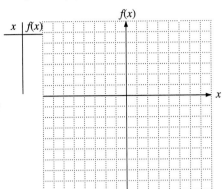

16. $g(x) = |x| - 5$

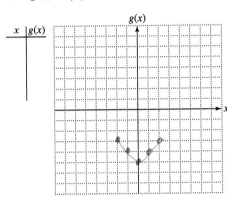

17. $f(x) = x^2 + 1$

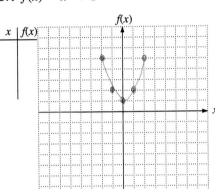

18. $g(x) = x^2 - 4$

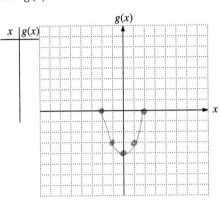

19. $f(x) = (x - 3)^2$

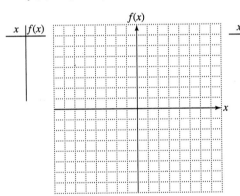

20. $g(x) = (x + 1)^2$

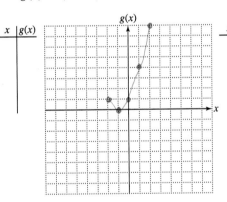

21. $f(x) = x^3 + 2$

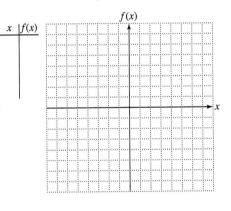

22. $g(x) = x^3 - 3$

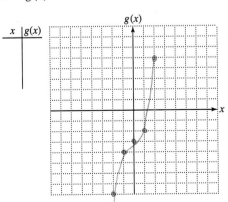

23. $p(x) = (x + 2)^3$

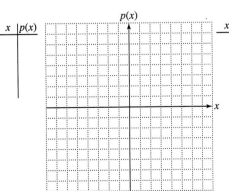

24. $s(x) = (x - 3)^3$

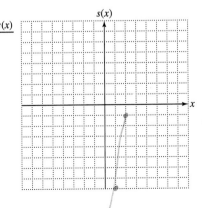

25. $f(x) = \dfrac{2}{x}$

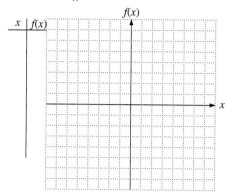

26. $g(x) = -\dfrac{3}{x}$

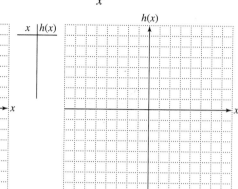

27. $h(x) = -\dfrac{6}{x}$

28. $t(x) = \dfrac{8}{x}$

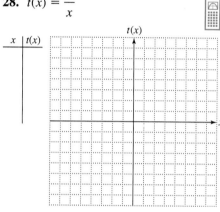

Optional Graphing Calculator Problem

29. On your graphing calculator, graph on the same set of axes: $y_1 = x^2$, $y_2 = 0.4x^2$, $y_3 = 2.6x^2$. What effect does the coefficient have on the graph?

In each of the following problems a table of values is given. Graph the function based on the table of values. Assume that both domain and range are all real numbers. After you have finished your graph in each case, estimate the value of $f(x)$ when $x = 2$.

30.

x	f(x)
−1	2
1	−2
3	−6

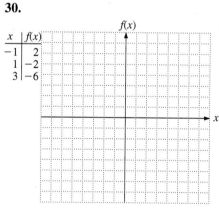

31.

x	f(x)
−5	−2
3	2
5	3

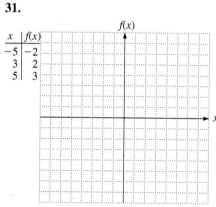

32.

x	f(x)
−3	−1
−2	2
−1	3
0	2
1	−1

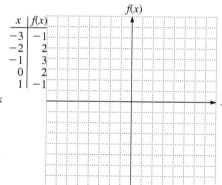

33.

x	f(x)
−2	0.25
−1	0.5
0	1
2.5	5.7
3	8

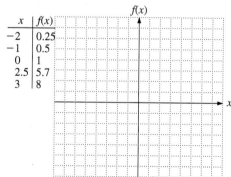

34.

x	f(x)
0	3
1	2.5
3	2.2
−1	4
−2	6

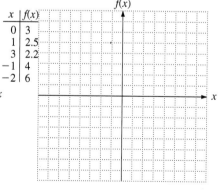

35.

x	f(x)
0	3
1	4
3	6
−1	4
−2	5

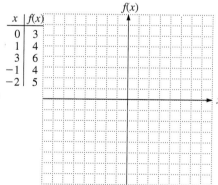

Applications

36. The following table shows the relationship between the number of computers manufactured each day and the company's profit for that day.

Number of computers made	20	30	40	50	60	70	80
Profit for the company $	0	6000	8000	9000	8000	6000	0

(a) Graph the function.
(b) How many computers should be made each day to achieve the highest profit?
(c) If the company always wants to earn a profit of $8000 or more, how many
computers should they always manufacture each day?
(d) What will the profit picture be if the company manufactures 82 computers per
day?
(e) Estimate the profit if the company manufactures 45 computers per day.

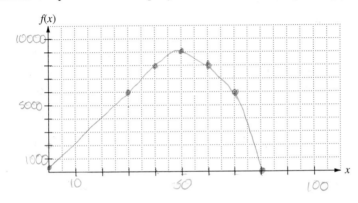

37. The following table is the result of studies of the buildup of carbon dioxide in the
atmosphere in eastern Canada.

Year	1960	1965	1970	1975	1980	1985	1990
Carbon dioxide in parts per million	320	323	327	333	340	348	358

(a) Graph the function.
(b) Estimate $c(x)$ when x is 1972.

(c) What would you estimate $c(x)$ will be in the
year 1995?
(d) How would you compare the rate of increase between 1960 and 1975 to the rate
of increase between 1975 and 1990?

(e) During what year did the concentration of
carbon dioxide reach 330 parts per million?

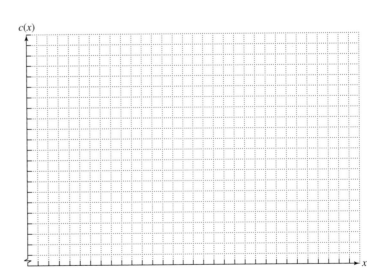

38. In hot weather, meteorologists measure a heat index to determine the relative comfort and safety of people subjected to higher temperatures and humidity. The table below gives the heat index at 80°F for a relative humidity of from 0% to 100%.

Relative humidity	0%	20%	40%	60%	80%	100%
Heat index at 80°F	73	77	79	82	86	91

(a) Graph the heat index at 80°F. Is the function linear? Why or why not?

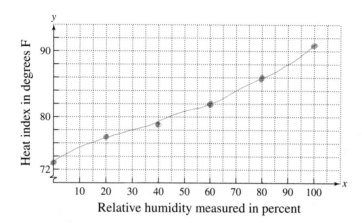

(b) Estimate the value of the function when the relative humidity is 50%
(c) For what levels of humidity is the increase in the heat index the fastest?
(d) For what level of humidity does 80° have a heat index of 80?
(e) What is the significance of the heat index being only 77 when the relative humidity is 20%

Cumulative Review Problems

Solve for x.

39. $2(3ax - 4y) = 5(ax + 3)$

40. $\dfrac{1}{2}(3x + 5y) = 2(x - 3y)$

41. $\dfrac{1}{3}x + \dfrac{1}{5}y = \dfrac{1}{10} + \dfrac{1}{2}x$

42. $0.12(x - 4) = 1.16x - 8.02$

43. Approximately 3.9 million square miles of the earth's land surface, (about 10% of the entire Earth), is under a permanent ice cover. The ice cover occurs 80% in Antarctica, 12% in Greenland, and 8% distributed among various polar islands and mountain peaks. How many square feet would be covered with ice in Antarctica?

44. A woodchuck breathes only ten times per hour while hibernating. How many times will it breathe during December, January, and February (during a leap year)?

Topic	Procedure	Examples
Graphing straight lines, p. 133.	An equation of the form $$Ax + By = C$$ has a graph that is a straight line. To graph such an equation, plot any three points—two give the line and the third checks it. (Where possible, use the x- and y-intercepts.)	$$2x + 3y = 12$$ Form is $Ax + By = C$, so it is a straight line. Let $x = 0$; then $y = 4$. Let $y = 0$; then $x = 6$. Let $x = 3$; then $y = 2$.
Intercepts, p. 134.	The number a is the x-intercept of a line if the line crosses the x-axis at $(a, 0)$. The number b is the y-intercept of a line if the line crosses the y-axis at $(0, b)$.	Find the intercepts of $7x - 2y = -14$. If $x = 0$, $-2y = -14$ and $y = 7$. The y-intercept is 7. If $y = 0$, $7x = -14$ and $x = -2$. The x-intercept is -2.
Slope, p. 141.	The *slope m* of any straight line that contains the points (x_1, y_1) and (x_2, y_2) is defined by $$m = \frac{y_2 - y_1}{x_2 - x_1}, \quad \text{where } x_2 \neq x_1$$	Find the slope of a line passing through $(6, -1)$ and $(-4, -5)$. $$m = \frac{-5 - (-1)}{-4 - 6} = \frac{-5 + 1}{-4 - 6} = \frac{-4}{-10} = \frac{2}{5}$$
Zero slope, p. 143.	All horizontal lines have *zero slope*. They can be written in the form $y = b$, where b is a real number.	What is the slope of $2y = 8$? $y = 4$. It is a horizontal line; the slope is zero.
No slope, p. 143.	All vertical lines have *no slope*. They can be written in the form $x = a$, where a is a real number.	What is the slope of $5x = -15$? $x = -3$. It is a vertical line. This line has *no slope*.
Parallel and perpendicular lines, pp. 144 and 145.	Two distinct lines with nonzero slopes m_1 and m_2, respectively, are 1. *Parallel* if $m_1 = m_2$. 2. *Perpendicular* if $m_1 = -\dfrac{1}{m_2}$.	Find the slope of a line parallel to $y = -\dfrac{3}{2}x + 6$. $$m = -\frac{3}{2}$$ Find the slope of a line perpendicular to $y = -4x + 7$. $m = \dfrac{1}{4}$
Standard form, p. 149.	The equation of a line is in *standard form* when it is written as $Ax + By = C$, where A, B, C are real numbers.	Place this equation in standard form: $y = -5(x + 6)$. $$y = -5x - 30$$ $$5x + y = -30$$
Slope–intercept form, p. 149.	The *slope–intercept form* of the equation of a line is $y = mx + b$, where the slope is m and the y-intercept is b.	Find the slope and y-intercept of $y = -\dfrac{7}{3}x + \dfrac{1}{4}$. The slope is $-\dfrac{7}{3}$; the y-intercept is $\dfrac{1}{4}$.
Point–slope form, p. 151.	The *point–slope form* of the equation of a line is $y - y_1 = m(x - x_1)$, where m is the slope and (x_1, y_1) are the coordinates of a point on the line.	Find an equation of the line passing through the points $(6, 0)$ and $(3, 4)$. $$m = \frac{4 - 0}{3 - 6} = -\frac{4}{3}$$ Then use the point–slope form. $$y - 0 = -\frac{4}{3}(x - 6)$$ $$y = -\frac{4}{3}x + 8$$

Topic	Procedure	Examples
Graphing the solution of linear inequalities in two variables, p. 159.	1. Replace the inequality symbol by an equals sign. This equation will be a boundary for the desired region. 2. Graph the line obtained in step 1, using a solid line if the original inequality contains a $\leq$ or $\geq$ symbol and a dashed line if the original inequality contains a $<$ or $>$ symbol. 3. Pick any point that does not lie on the boundary line. Substitute the coordinates into the original inequality. If you obtain a true inequality, shade the region on the side of the line containing the point. If you obtain a false inequality, shade the region on the side of the line opposite from the test point.	Graph the solution of $3y - 4x \geq 6$. 1. The boundary line is $3y - 4x = 6$. 2. The line passes through $(0, 2)$ and $(-1.5, 0)$. The inequality includes the boundary; draw a solid line. 3. Pick $(0, 0)$ as a test point. $$3y - 4x \geq 6$$ $$3(0) - 4(0) \overset{?}{\geq} 6$$ $$0 - 0 \overset{?}{\geq} 6$$ $$0 \ngeq 6$$ Our test point fails. We shade on the side that does not contain $(0, 0)$.
Finding the domain and the range of a relation, p. 165.	The set of all the first items of each ordered pair in a relation is called the *domain*. The set of all the second items of each ordered pair in a relation is called *range*.	Find the domain and range of the relation $$A = \{(5, 6), (1, 6), (3, 4), (2, 3)\}$$ Domain $= \{1, 2, 3, 5\}$ Range $= \{3, 4, 6\}$
Determining if a relation is a function, p. 167.	A *function* is a relation in which no two different ordered pairs have the same first coordinate.	Determine if each of the following are functions: $$B = \{(6, 7), (3, 0), (6, 4)\}$$ $$C = \{(-9, 3), (16, 4), (9, 3)\}$$ *B* is not a function. Two different pairs have the same first coordinate. They are $(6, 7)$ and $(6, 4)$. *C* is a function. There are no different pairs with the same first coordinate.
Determining if a graph represents the graph of a function, p. 168.	The graph of a function will have no two different ordered pairs with the same first coordinate.	This is a function. This is not a function. There are several different ordered pairs with the same first coordinate.
Function notation, p. 169.	Replace the *x* by the quantity within the parentheses and then simplify.	$$f(x) = 2x^2 - 3x + 4$$ Find $f(-2)$. $$f(-2) = 2(-2)^2 - 3(-2) + 4$$ $$= 2(4) - 3(-2) + 4$$ $$= 8 + 6 + 4 = 18$$
Graphing functions, p. 173.	Prepare a table of ordered pairs (if one is not provided) that satisfy the function equation. Graph these ordered pairs. Connect the ordered pairs by a line.	Graph $f(x) = |x| - 2$. First make a table. Then graph the ordered pairs and draw the graph. | x	$f(x)$
---	---
-2	0
-1	-1
0	-2
1	-1
2	0	

1. Find the value of a if $(a, -6)$ is a solution to the equation $7x - 2y - 5 = 0$.

2. Find the value of b if $(3, b)$ is a solution to the equation $-5x = 3(y + 4)$.

Graph the equation of the straight line determined by each of the following equations.

3. $y = -\dfrac{3}{2}x + 5$

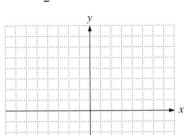

4. $y = -\dfrac{1}{4}x - 1$

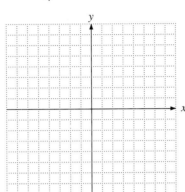

5. $2y - 6x + 4 = 0$

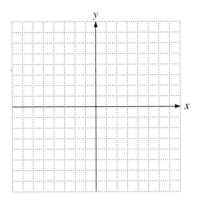

6. $5(x - y) = 10 - 3y$

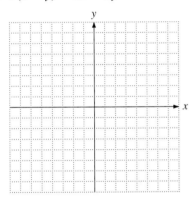

7. $5x - 6 = -2x + 8$

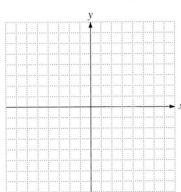

8. $-3y + 2 = -8y - 13$

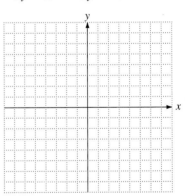

9. The profit in dollars for a microcomputer company is given by the equation $P = 140x - 2000$, where x is the number of microcomputers sold each day. How many microcomputers must be sold each day for the company to make a profit?

10. Find the slope of the line connecting $(-8, -4)$ and $(2, -3)$

11. A line passes through the points $(a, 6)$ and $(-2, 3)$. Find the value of a if $m = 2$.

12. A line is perpendicular to a line of zero slope. What is its slope?

13. Find the slope of a line perpendicular to the line passing through $\left(\frac{2}{3}, \frac{1}{3}\right)$ and $(4, 2)$.

14. Do the points $(3, -5)$, $(2, 1)$, and $(-2, 10)$ all lie on the same straight line? Explain.

15. Write the equation $3x + 2y = 5x - 6$ in slope–intercept form. Find the slope and y-intercept.

16. A line has a slope of $\frac{2}{3}$ and a y-intercept of -4. Write its equation in standard form.

17. Find the equation of the line in standard form that passes through $\left(\frac{1}{2}, -2\right)$ and has slope -4.

18. Find the equation of the line in standard form that passes through $(-3, 1)$ and has slope 0.

In Exercises 19 through 22, find the equation of the line satisfying the conditions given. Write your answer in standard form.

19. A line passing through $(5, 6)$ and $\left(-1, -\frac{1}{2}\right)$

20. A line that has no slope passing through $(-6, 5)$

21. A line parallel to $3x - 2y = 8$ and passing through $(5, 1)$.

22. A line perpendicular to $7x + 8y - 12 = 0$ passing through $(-2, 5)$.

Graph the region described by the inequality.

23. $y < 3x + 1$

24. $y < 2x + 4$

25. $y > -\frac{1}{2}x + 3$

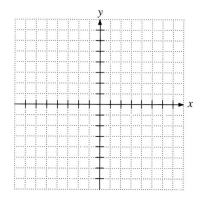

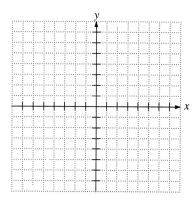

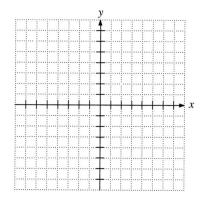

26. $y > -\frac{2}{3}x + 1$

27. $3x + 4y \le -12$

28. $5x + 3y \le -15$

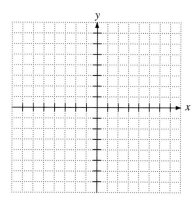

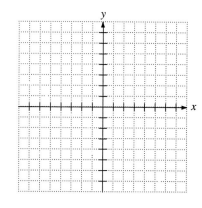

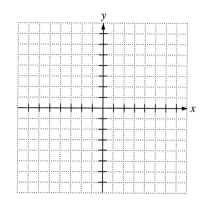

29. $x \leq 3y$

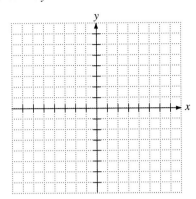

30. $3x - 5 < 7$

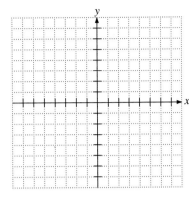

31. $5y - 2 > 3y - 10$

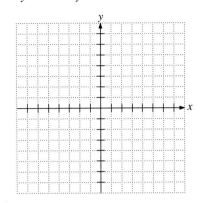

What is the domain and range of each relation. Is the relation a function?

32. $A = \{(0, 0), (1, 1), (2, 4), (3, 9), (1, 16)\}$

33. $B = \{(-20, 18), (-18, 16), (-16, 14), (-12, 18)\}$

Determine if the graph represents a function.

34.

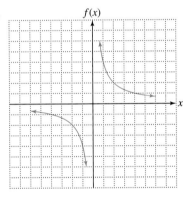

35.

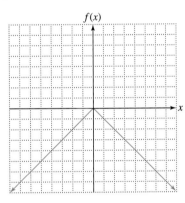

36.

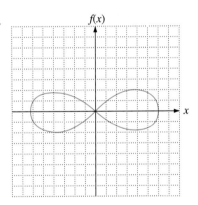

37. If $f(x) = 3x - 8$, find $f(-2)$.

38. If $g(x) = 2x^2 - 3x - 5$, find $g(-3)$.

39. If $h(x) = x^3 + 2x^2 - 5x + 8$, find $h(-1)$.

40. If $p(x) = |-6x - 3|$, find $p(3)$.

Graph each function.

41. $f(x) = 2|x - 1|$

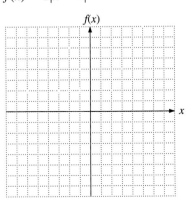

42. $g(x) = x^2 - 3$

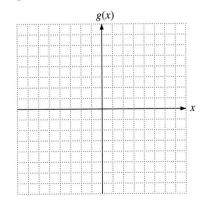

43. $h(x) = x^3 + 2$

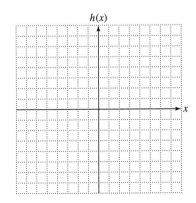

Graph the function based on the table of values. Estimate the value of f(x) when x = −2.

44.

x	f(x)
−1	5
−3	−3
−4	−4
−5	−3
−6	0
−7	5

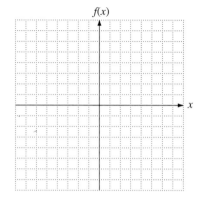

45.

x	f(x)
−3	4
−1	2
0	1
1	0
2	−1
3	0
4	1

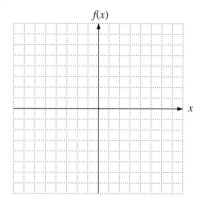

Complete the table of values for each function.

46. $f(x) = -\dfrac{4}{5}x + 3$

x	f(x)
−5	
0	
10	

47. $f(x) = 2x^2 - 3x + 4$

x	f(x)
−3	
0	
4	

48. $f(x) = 3x^3 - 4$

x	f(x)
−1	
0	
2	

49. $f(x) = \dfrac{7}{2x + 3}$

x	f(x)
−2	
0	
2	

Developing Your Study Skills

EXAM TIME: GETTING ORGANIZED

Studying adequately for an exam requires careful preparation. Begin early so that you will be able to spread your review over several days. Even though you may still be learning new material at this time, you can be reviewing concepts previously learned in the chapter. Giving yourself plenty of time for review will take the pressure off. You need this time to process what you have learned and to tie concepts together.

Adequate preparation enables you to feel confident and to think clearly with less tension and anxiety.

1. Graph the line with equation $2x + 3y = -10$. Plot at least three points.

x	y
-2	-2
1	-4
4	-6

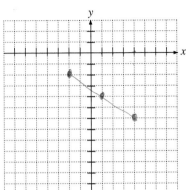

2. Find the slope of the line passing through $(2, -3)$ and $\left(\dfrac{1}{2}, -6\right)$.

3. Find the slope of the line $3x = -8y + 12$.

4. Write the equation in standard form of a line passing through $(5, -2)$ and $(-3, -1)$.

5. Write the equation in standard form of a line perpendicular to $6x - 7y - 1 = 0$ that passes through $(0, -2)$.

6. Write the equation of a horizontal line passing through $\left(-\dfrac{1}{3}, 2\right)$.

Graph the regions.

7. $y \geq -4x$

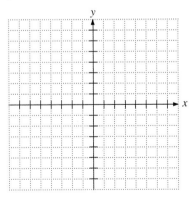

8. $4x - 2y < -6$

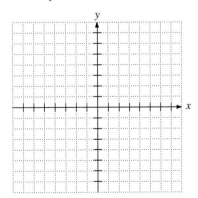

9. What are the domain and range of the following relation?

$$A = \{(0, 0), (1, 1), (1, -1), (2, 4), (2, -4)\}$$

1. _____

2. _____

3. _____

4. _____

5. _____

6. _____

7. _____

8. _____

9. _____

10. If $f(x) = 2x - 3$, find $f\left(\dfrac{3}{4}\right)$.

11. If $g(x) = \dfrac{1}{2}x^2 + 3$, find $g(-4)$.

12. If $h(x) = \left| -\dfrac{2}{3}x + 4 \right|$, find $h(-9)$.

13. If $p(x) = -2x^3 + 3x^2 + x - 4$, find $p(-2)$.

Graph each function.

14. $g(x) = 5 - x^2$

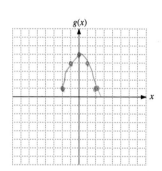

15. $h(x) = x^3 - 2$

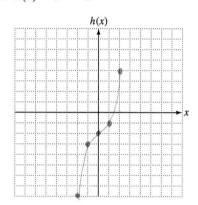

16. Graph the function defined by the following table which describes the distance y you can see across the ocean on a clear day if you are x feet above the water.

Height in feet, x	0	3	9	15
Distance in miles, $f(x)$	0	6	54	150

Based on your graph, how many miles can you see if you are 4 feet above the water?

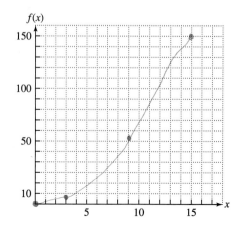

10. _____

11. _____

12. _____

13. _____

14. _____

15. _____

16. _____

Approximately one-half of this test covers the content of Chapters 1 and 2. The remainder covers the content of Chapter 3.

1. Name the property that justifies $(-6) + 6 = 0$.

2. Evaluate the expression.
$$3(4 - 6)^2 + \sqrt{16} + 12 \div (-3)$$

3. Simplify. $(3x^2y^{-3})^{-4}$

4. Simplify. $5x(2x - 3y) - 3(x^2 + 4)$

5. Solve for x. $|3x - 2| = 8$

6. Solve and graph your solution.
$$3(x - 2) > 6 \quad \text{or} \quad 5 - 3(x + 1) > 8$$

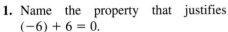

7. Solve for x. $3b = \frac{1}{2}(3x + 2y)$

8. A plastic insulator is made in a rectangular shape with a perimeter of 92 centimeters. The length is 1 centimeter longer than double the width. Find the dimensions of the insulator.

9. Sharim invested $3000 for one year. Part was invested at 5% and part at 8% interest. At the end of one year, he earned $189 in interest. How much did he invest at each rate?

10. Find the area of the following region, which is in the shape of a semicircle. Round your answer to the nearest hundredth. (Use $\pi = 3.14$)

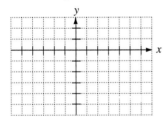

$r = 3$ inches

11. Graph the line $4x - 6y = 10$. Plot at least three points.

12. Find the slope of the line passing through $(6, 5)$ and $(-2, 1)$.

13. Write the equation of the line in standard form that passes through $(5, 1)$ and $(4, 3)$.

14. Write the equation of the line in standard form that passes through $(-2, -3)$ and is perpendicular to $y = \frac{2}{3}x - 4$.

15. What are the domain and range of the relation
$$\{(3, 7), (5, 8), (\tfrac{1}{2}, -1), (2, 2)\}$$
Is the relation a function?

16. $f(x) = -2x^2 - 4x + 1$. Find $f(-3)$.

1.	
2.	
3.	
4.	
5.	
6.	
7.	
8.	
9.	
10.	
11.	
12.	
13.	
14.	
15.	
16.	

Graph the following functions.

17. $p(x) = -\frac{1}{3}x + 2$

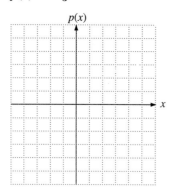

18. $h(x) = |x - 2|$

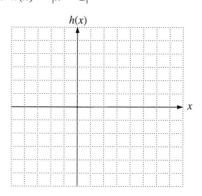

19. $r(x) = \dfrac{3}{x}$

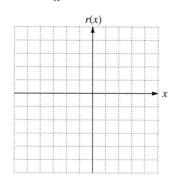

20. $f(x) = x^2 - 3$

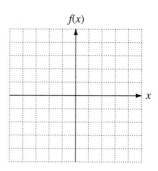

21. Graph the region. $y \le -\frac{3}{2}x + 3$

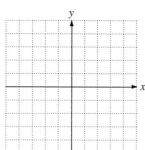

22. Complete the following table of values for $f(x)$.

$$f(x) = -2x^3 + 4$$

x	$f(x)$
-2	
0	
3	

17. _____

18. _____

19. _____

20. _____

21. _____

22. _____

Systems of Linear Equations and Inequalities

The streets and walkways of some of the famous historic cities of the world are paved with bricks and stones of various shapes and sizes. There is both art and science to laying bricks and stones in exactly the correct way to achieve an aesthetically pleasing pattern and a durable street. Could you use your mathematics skills to determine the correct number of bricks needed to achieve certain patterns? Turn to the Putting Your Skills to Work on page 213.

1. _____

2. _____

3. _____

4. _____

5. _____

6. _____

7. _____

If you are familiar with the topics in this chapter, take this test now. Check your answers with those in the back of the book. If an answer was wrong or you couldn't do a problem, study the appropriate section of the chapter.

If you are not familiar with the topics in this chapter, don't take this test now. Instead, study the examples, work the practice problems, and then take the test.

This test will help you identify those concepts that you have mastered and those that need more study.

Section 4.1

Find the solution to each system of equations. If there is no single solution to a system, state the reason.

1. $3x - 4y = 7$
$2x + y = 12$

2. $7x + 3y = 15$
$\dfrac{1}{3}x - \dfrac{1}{2}y = 2$

3. $2x = 3 + y$
$3y = 6x - 9$

Section 4.2

Find the solution to each system of equations.

4. $5x - 2y + z = -1$
$3x + y - 2z = 6$
$-2x + 3y - 5z = 7$

5. $x + y + 2z = 9$
$3x + 2y + 4z = 16$
$2y + z = 10$

Section 4.3

Use a system of linear equations to solve each problem.

6. A coach purchased two shirts and three pairs of pants for his team and paid $75.00. His assistant purchased three shirts and five pairs of pants at the same store for $121.00. What is the cost for a shirt? What is the cost for a pair of pants?

7. A biologist needs to use 21 milligrams of iron, 22 milligrams of vitamin B_{12}, and 26 milligrams of niacin for an experiment. She has available packets A, B, and C to meet these requirements. Packet A contains 3 milligrams of iron, 2 milligrams of vitamin B_{12}, and 4 milligrams of niacin. Packet B contains 2 milligrams of iron, 4 milligrams of vitamin B_{12}, and 5 milligrams of niacin. Packet C contains 2 milligrams of iron, 2 milligrams of vitamin B_{12}, and 1 milligram of niacin. How many of each packet should she use?

Section 4.4

Evaluate the determinant.

8. $\begin{vmatrix} 2 & -4 \\ 3 & -5 \end{vmatrix}$ **9.** $\begin{vmatrix} -4 & 3 \\ 7 & -5 \end{vmatrix}$ **10.** $\begin{vmatrix} 3 & 0 \\ 4 & 0 \end{vmatrix}$ **11.** $\begin{vmatrix} 5 & -1 & 0 \\ 3 & 2 & 1 \\ -1 & 4 & 2 \end{vmatrix}$

Section 4.5

Use Cramer's rule to solve each system of equations.

12. $2x + 7y = -10$
$5x + 6y = -2$

13. Solve for x **only:** $2x + y + 3z = 3$
$x - 2y + 2z = -2$
$3x - 4y = -20$

Section 4.6

Solve the following systems of inequalities by graphing.

14. $2x + 2y \geq -4$
$-3x + y \leq 2$

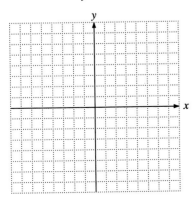

15. $x - 3y < -6$
$x < 3$

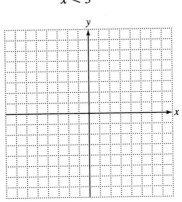

8. _____

9. _____

10. _____

11. _____

12. _____

13. _____

14. _____

15. _____

4.1 Systems of Equations in Two Variables

MathPro

Video 9

SSM

After studying this section, you will be able to:

1 *Determine whether an ordered pair is a solution to a system of two equations.*
2 *Solve a system of two linear equations by the graphing method.*
3 *Solve a system of two linear equations by the substitution method.*
4 *Solve a system of two linear equations by the addition (elimination) method.*
5 *Identify systems of equations that do not have a unique solution.*

1 *Determining Whether an Ordered Pair Is a Solution to a System of Two Equations*

In Chapter 3 we found that a linear equation containing two variables, such as $4x + 3y = 12$, has an unlimited number of ordered pairs (x, y) that satisfy it. For example, $(3, 0)$, $(0, 4)$, and $(-3, 8)$ all satisfy the equation $4x + 3y = 12$. We call two linear equations in *two* unknowns a **system of two equations in two variables.** Many such systems have exactly one solution. A **solution to a system** of two linear equations in two variables is an *ordered pair* that is a solution to *each* equation.

EXAMPLE 1   Determine whether $(3, -2)$ is a solution to the following system.

$$x + 3y = -3$$

$$4x + 3y = 6$$

We will begin by substituting $(3, -2)$ into the first equation to see if the ordered pair is a solution to the first equation.

$$3 + 3(-2) \stackrel{?}{=} -3$$

$$3 - 6 \stackrel{?}{=} -3$$

$$-3 = -3 \ \checkmark$$

Likewise, we will determine if $(3, -2)$ is a solution to the second equation.

$$4(3) + 3(-2) \stackrel{?}{=} 6$$

$$12 - 6 \stackrel{?}{=} 6$$

$$6 = 6 \ \checkmark$$

Since $(3, -2)$ is a solution to each equation in the system, it is a solution to the system itself.

It is important to remember that we cannot confirm that a particular ordered pair is in fact the solution to a system of two equations unless we have checked to see if the solution satisfies both equations. Merely checking one equation is not sufficient. Determining if an ordered pair is a solution to a system of equations requires that we must verify that the solution satisfies *both* equations.

Practice Problem 1   Determine if $(-3, 4)$ is a solution to the system

$$2x + 3y = 6$$

$$3x - 4y = 7 \ \blacksquare$$

2 Solving a System of Two Linear Equations by the Graphing Method

We could verify the solution of Example 1 by graphing each equation. If the lines intersect, the system has a unique solution. The point of intersection lies on both lines. Thus it is a solution to each equation and the solution to the system. We will illustrate this by graphing the equations in Example 1. Notice that the coordinates of the point of intersection are $(3, -2)$. The solution to the system is $(3, -2)$.

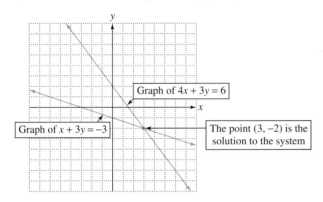

Therefore, one method that can be used to find the solution of a system of linear equations is to graph each line and to determine the solution from the point of intersection.

EXAMPLE 2 Solve this system of equations by graphing.

$$2x + 3y = 12$$
$$x - y = 1$$

Using the methods that we developed in Chapter 3, we graph each line and determine from the graph the point at which the two lines intersect.

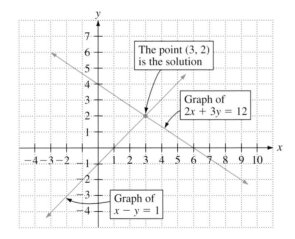

Finding the solution by the graphing method allows for some margin of error. Our graph of one or more lines could be off slightly. We verify that this answer is correct by the substitution of $x = 3$, $y = 2$ into the system of equations.

$x - y = 1$	$2x + 3y = 12$
$3 - 2 \overset{?}{=} 1$	$2(3) + 3(2) \overset{?}{=} 12$
$1 = 1$ ✓	$12 = 12$ ✓

Thus we have verified that the solution to the system is $(3, 2)$.

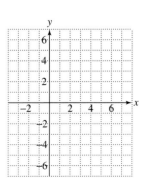

Practice Problem 2 Solve this system of equations by graphing. Check your solution.

$$3x + 2y = 10$$
$$x - y = 5 \blacksquare$$

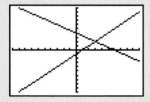
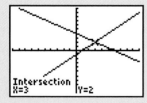
In most cases where the graphing method is employed, the two straight lines will intersect at one point. However, it is possible for a given system to have as its graph two parallel lines. In such a case there will be no solution. Such a system of equations is called *inconsistent* if it has no solution. Another possibility is that when we graph each equation in the system we obtain one line. In such a case there will be an infinite number of solutions. Any point that satisfies the first equation will also satisfy the second equation. A system of equations in two variables is *dependent* if it has infinitely many solutions.

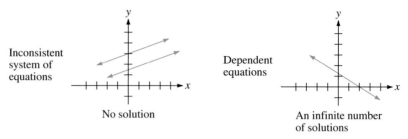

Inconsistent system of equations

No solution

Dependent equations

An infinite number of solutions

Since these situations are rarely encountered when the graphing method is used, we will discuss them in more detail after we have developed algebraic methods for solving a system of equations.

3 Solving a System of Two Linear Equations by the Substitution Method

One method of solving a system of linear equations in two variables is the **substitution method.** To use this method, we choose one equation and solve for one variable. It is usually best to solve for a variable that has a coefficient of $+1$ or -1. This will avoid unnecessary fractions. When we solve for one variable, we obtain an expression that contains the other variable. We will *substitute* this expression into the second equation. Thus we have one equation with one unknown, which we can easily solve.

EXAMPLE 3 Find the solution to the following system of equations. Use the substitution method.

$$x + 3y = -7 \quad \boxed{1}$$

$$4x + 3y = -1 \quad \boxed{2}$$

We can work with equation $\boxed{1}$ or equation $\boxed{2}$. Let's choose equation $\boxed{1}$ and solve for x. This gives us equation $\boxed{3}$.

$$x = -7 - 3y \quad \boxed{3}$$

Now we substitute this expression for x into equation $\boxed{2}$ and solve the equation for y.

$$4x + 3y = -1 \quad \boxed{2}$$
$$4(-7 - 3y) + 3y = -1$$
$$-28 - 12y + 3y = -1$$
$$-28 - 9y = -1$$
$$-9y = -1 + 28$$
$$-9y = +27$$
$$y = -3$$

Now we substitute $y = -3$ into equation 1, 2, or 3 to find x. Let's use 3:

$$x = -7 - 3(-3)$$
$$= -7 + 9$$
$$= 2$$

Therefore, our solution is $x = 2$ and $y = -3$. This is the ordered pair $(2, -3)$.

    Check: We must verify the solution in both of the *original* equations.

$$x + 3y = -7 \qquad\qquad 4x + 3y = -1$$
$$2 + 3(-3) \stackrel{?}{=} -7 \qquad\qquad 4(2) + 3(-3) \stackrel{?}{=} -1$$
$$2 - 9 \stackrel{?}{=} -7 \qquad\qquad 8 - 9 \stackrel{?}{=} -1$$
$$-7 = -7 \ \checkmark \qquad\qquad -1 = -1 \ \checkmark$$

Practice Problem 3   Use the substitution method to solve this system.

$$2x - y = 7$$
$$3x + 4y = -6 \ \blacksquare$$

We summarize the method here.

How to Solve a System of Two Equations by the Substitution Method

1. Choose one of the two equations and solve for one variable in terms of the other variable.
2. Substitute this expression from step 1 into the *other* equation.
3. You now have one equation with one variable. Solve this equation for that variable.
4. Substitute this value for the variable into an equation with two variables to obtain a value for the second variable.
5. Check each solution in each original equation.

Optional Graphing Calculator Note:

When solving the system of Example 3 by using a graphing calculator, you will first need to solve each equation for y. Equation 1 can be written as $y_1 = -\dfrac{1}{3}x - \dfrac{7}{3}$ or as $y_1 = \dfrac{-x - 7}{3}$. Likewise, equation 2 can be written as $y_2 = -\dfrac{4}{3}x - \dfrac{1}{3}$ or as $y_2 = \dfrac{-4x - 1}{3}$.

EXAMPLE 4 Solve the following system of equations.

$$\tfrac{1}{2}x - \tfrac{1}{4}y = -\tfrac{3}{4} \quad \boxed{1}$$

$$3x - 2y = -6 \quad \boxed{2}$$

First clear equation $\boxed{1}$ of fractions by multiplying each term by 4.

$$4\left(\frac{1}{2}x\right) - 4\left(\frac{1}{4}y\right) = 4\left(-\frac{3}{4}\right)$$

$$2x - \qquad y = \qquad -3 \quad \boxed{3}$$

The new system is now

$$2x - y = -3 \quad \boxed{3}$$

$$3x - 2y = -6 \quad \boxed{2}$$

Step 1 Let's solve equation $\boxed{3}$ for y.

$$-y = -3 - 2x$$

$$y = 3 + 2x$$

Step 2 Substitute this expression for y into equation $\boxed{2}$.

$$3x - 2(3 + 2x) = -6$$

Step 3 Solve this equation for x.

$$3x - 6 - 4x = -6$$

$$-6 - x = -6$$

$$-x = -6 + 6$$

$$-x = 0$$

$$x = 0$$

Step 4 Substitute $x = 0$ into equation $\boxed{2}$. (We choose to use one of the original equations we have not changed.)

$$3(0) - 2y = -6$$

$$-2y = -6$$

$$y = 3$$

So our solution is $x = 0$ and $y = 3$. We can also write this as $(0, 3)$.

Step 5 We must verify the solution in both original equations ($\boxed{1}$ and $\boxed{2}$).

$$\frac{1}{2}x - \frac{1}{4}y = -\frac{3}{4} \qquad\qquad 3x - 2y = -6$$

$$\frac{0}{2} - \frac{3}{4} \overset{?}{=} -\frac{3}{4} \qquad\qquad 3(0) - 2(3) \overset{?}{=} -6$$

$$\qquad\qquad\qquad\qquad -6 = -6 \ \checkmark$$

$$-\frac{3}{4} = -\frac{3}{4} \ \checkmark$$

Practice Problem 4 Use the substitution method to solve this system.

$$\frac{1}{2}x + \frac{2}{3}y = \quad 1$$

$$\frac{1}{3}x + y \quad = -1 \quad \blacksquare$$

4 Solving a System of Equations by the Addition Method

Another way to solve a system of two linear equations in two variables is to add the two equations so that a variable is eliminated. This technique is called the **addition method** or the **elimination method.** We usually have to multiply one or both of the equations by suitable factors so that we obtain the same coefficient of one variable (either x or y) in each equation.

EXAMPLE 5   Solve the following system by the addition method.

$$5x + 8y = -1 \quad \boxed{1}$$
$$3x + y = 7 \quad \boxed{2}$$

We can eliminate either the x or the y variable. Let's choose y. We multiply equation $\boxed{2}$ by -8. This gives

$$-8(3x) + (-8)(y) = -8(7)$$
$$-24x - 8y = -56 \quad \boxed{3}$$

We now add equations $\boxed{1}$ and $\boxed{3}$.

$$
\begin{array}{rl}
5x + 8y = & -1 \quad \boxed{1} \\
-24x - 8y = & -56 \quad \boxed{3} \\
\hline
-19x \quad = & -57
\end{array}
$$

Solve for x.

$$x = \frac{-57}{-19} = 3$$

Now substitute $x = 3$ into equation $\boxed{2}$ (or equation $\boxed{1}$):

$$3(3) + y = 7$$
$$9 + y = 7$$
$$y = -2$$

Our solution is $x = 3$ and $y = -2$, or $(3, -2)$.

$$
\begin{aligned}
\textit{Check:} \quad 5(3) + 8(-2) &\overset{?}{=} -1 \\
15 + (-16) &\overset{?}{=} -1 \\
-1 &= -1 \quad \checkmark \\
3(3) + (-2) &\overset{?}{=} 7 \\
9 + (-2) &\overset{?}{=} 7 \\
7 &= 7 \quad \checkmark
\end{aligned}
$$

Practice Problem 5   Use the addition method to solve this system.

$$-3x + y = 5$$
$$2x + 3y = 4 \quad \blacksquare$$

For convenience, we summarize the procedure here.

How to Solve a System of Two Linear Equations by the Addition (Elimination) Method

1. Arrange each equation in the form $ax + by = c$. (Remember, a, b, and c can be any real numbers.)
2. Multiply one or both equations by appropriate numbers so that either x or y is eliminated after you add the equations.
3. Add the two equations from step 2.
4. Solve this equation for the one variable.
5. Substitute this value into one of the *original* equations and solve it.
6. Check the solution in each original equation.

EXAMPLE 6 Solve the following system by the addition method.

$$3x + 2y = -8 \quad \boxed{1}$$
$$2x + 5y = 2 \quad \boxed{2}$$

To eliminate the variable x, we multiply equation $\boxed{1}$ by 2 and equation $\boxed{2}$ by -3. We now have the equivalent system

$$6x + 4y = -16$$
$$\underline{-6x - 15y = -6}$$
$$-11y = -22 \qquad \text{\textit{Add the equations.}}$$
$$y = 2 \qquad \text{\textit{Solve for y.}}$$

Substitute $y = 2$ into equation $\boxed{1}$.

$$3x + 2(2) = -8$$
$$3x + 4 = -8$$
$$3x = -12$$
$$x = -4$$

The solution to the system is $x = -4$ and $y = 2$, or $(-4, 2)$.

Check: Verify that this solution is correct.

Note: We could have easily eliminated the variable y in Example 6 by multiplying equation $\boxed{1}$ by 5 and equation $\boxed{2}$ by -2. Try it. Is the solution the same? Why?

Practice Problem 6 Use the addition (elimination) method to solve this system.

$$5x + 4y = 23$$
$$7x - 3y = 15 \quad \blacksquare$$

EXAMPLE 7 Solve the system

$$5x - 2y = 14 \quad \boxed{1}$$

$$3x + 4y = 11 \quad \boxed{2}$$

It is easier to eliminate y than x. If we eliminate x, we must multiply equation $\boxed{1}$ by 3 and equation $\boxed{2}$ by -5. To eliminate y, we need only multiply equation $\boxed{1}$ by 2.

$$2(5x) - 2(2y) = 2(14)$$

$$10x - 4y = 28 \quad \boxed{3}$$

So our system is

$$10x - 4y = 28$$

$$\underline{3x + 4y = 11}$$

$$13x \quad\quad = 39$$

$$x = 3$$

Now substitute this value into equation $\boxed{1}$.

$$5(3) - 2y = 14$$

$$-2y = -1$$

$$y = \frac{1}{2}$$

The solution is $x = 3$ and $y = \dfrac{1}{2}$, or $\left(3, \dfrac{1}{2}\right)$.

It is important to be able to verify a solution to a system of equations when the solution contains fractions. Make sure you can verify the following check on your own.

Check: $\quad 5(3) - 2\left(\dfrac{1}{2}\right) \stackrel{?}{=} 14$

$$15 - 1 \stackrel{?}{=} 14$$

$$14 = 14 \quad \checkmark$$

$$3(3) + 4\left(\dfrac{1}{2}\right) \stackrel{?}{=} 11$$

$$9 + 2 \stackrel{?}{=} 11$$

$$11 = 11 \quad \checkmark$$

Practice Problem 7 Use the addition method to solve this system.

$$3x + 12y = 25$$

$$2x - 6y = 12 \quad \blacksquare$$

So far we have examined only systems that have one solution. But other systems must also be considered. These systems can best be illustrated by several graphs. In general, the system of equations

$$ax + by = c$$

$$dx + ey = f$$

may have one solution, no solution, or an infinite number of solutions.

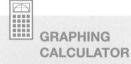

GRAPHING CALCULATOR

Identifying Systems of Equations

If there is a concern as to whether or not a system of equations has no solution, how can this be determined quickly using a graphing calculator? What is the most important thing to look for on the graph?

If the equations are dependent, how can we be sure of this by looking at the display of a graphing calculator? Why? Determine if there is one solution, no solution, or infinite number of solutions for each of the following systems.

1. $y_1 = -2x + 1$
   $3y_2 = -6x + 3$
2. $y_1 = 3x + 6$
   $y_2 = 3x + 2$
3. $y_1 = x - 3$
   $y_2 = -2x + 12$

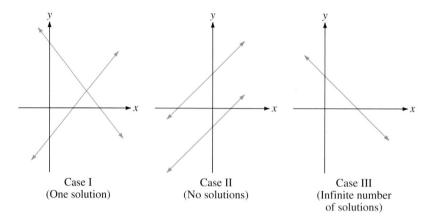

Case I
(One solution)

Case II
(No solutions)

Case III
(Infinite number
of solutions)

Case I: *One solution.* The two graphs intersect at one point, which is the solution. We say that the equations are *independent* and *consistent.* There is a point (an ordered pair) *consistent* with both equations.

Case II: *No solution.* The two graphs are parallel and so do not intersect. We say that the system of equations is *inconsistent* because there is no point consistent with both equations.

Case III: *An infinite number of solutions.* The graphs of each equation yield the same line. Every ordered pair on this line is a solution to both of the equations. We say that the equations are *dependent* and *consistent.*

EXAMPLE 8   If possible, solve the system.

$$2x + 8y = 16 \quad \boxed{1}$$

$$4x + 16y = -8 \quad \boxed{2}$$

To eliminate the variable y, we'll multiply $\boxed{1}$ by -2.

$$-2(2x) + (-2)(8y) = (-2)(16)$$

$$-4x - 16y = -32 \quad \boxed{3}$$

We now have the equivalent system

$$-4x - 16y = -32 \quad \boxed{3}$$

$$4x + 16y = -8 \quad \boxed{2}$$

When we add equations $\boxed{3}$ and $\boxed{2}$, we get

$$0 = -40$$

which, of course, is false. Thus we conclude that this system of equations is inconsistent, so **there is no solution**. Therefore, equations $\boxed{1}$ and $\boxed{2}$ do not intersect, as we can see on the graph.

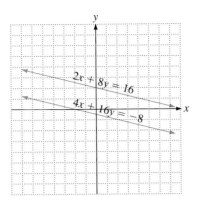

If we had used the substitution method to solve this system, we would still obtain a false statement. When you try to solve an inconsistent system of linear equations by any method, you will always obtain a mathematical equation that is not true.

Practice Problem 8   If possible, solve the system.

$$4x - 2y = 6$$
$$-6x + 3y = 9 \quad \blacksquare$$

EXAMPLE 9   If possible solve the system.

$$0.5x - 0.2y = \quad 1.3 \quad \boxed{1}$$
$$-1.0x + 0.4y = -2.6 \quad \boxed{2}$$

Although we could work directly with the decimals, it is easier to multiply each equation by the appropriate value (10, 100, and so on) so that the coefficients of the new system are integers. Therefore, we will multiply equations $\boxed{1}$ and $\boxed{2}$ by 10 to obtain the equivalent system

$$5x - 2y = \quad 13 \quad \boxed{3}$$
$$-10x + 4y = -26 \quad \boxed{4}$$

We can eliminate the variable y by multiplying each term of equation $\boxed{3}$ by 2.

$$\begin{array}{ll} 10x - 4y = \quad 26 & \boxed{5} \\ \underline{-10x + 4y = -26} & \boxed{4} \\ \qquad\quad 0 = \quad 0 & \textit{Add the equations.} \end{array}$$

This statement is always true. Hence the two equations are dependent, and there are an infinite number of solutions. Any solution satisfying equation $\boxed{1}$ will also satisfy equation $\boxed{2}$. For example $(3, 1)$ is a solution to equation $\boxed{3}$ (prove this). Hence it must also be a solution to equation $\boxed{4}$ (prove it). Thus the equations actually give the same line, as you can see on the graph.

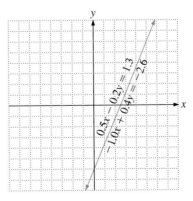

Practice Problem 9   If possible, solve the system.

$$0.3x - 0.9y = \quad 1.8$$
$$-0.4x + 1.2y = -2.4 \quad \blacksquare$$

4.1 Exercises

Determine whether the given ordered pair is a solution to the system of equations.

1. $(-2, -3)$ $\quad 2x - 4y = 8$
$\qquad\qquad\quad -2x + \;y = 1$

2. $(-5, -6)$ $\quad 3x - 5y = 15$
$\qquad\qquad\qquad -2x + \;y = \;\;4$

3. $(13, -10)$ $\quad 2x = 4 - 3y$
$\qquad\qquad\qquad\;\; y = 3 - x$

4. $(8, 6)$ $\quad 3x = 2y + 12$
$\qquad\qquad\quad 8y = -3x + 62$

5. $(\frac{1}{3}, -2)$ $\quad 6x + 7y = -12$
$\qquad\qquad\quad -3x + 5y = -11$

6. $(-2, \frac{4}{3})$ $\quad 4x + 3y = -4$
$\qquad\qquad\qquad x - 6y = -10$

Solve the system of equations by graphing. Check your solution.

7. $3x + y = 5$
$\quad\; 2x - y = 5$

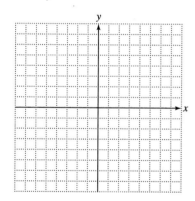

8. $3x + y = 2$
$\quad\; 2x - y = 3$

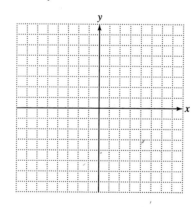

9. $2x + 3y = -6$
$\quad\;\; x - 3y = \;\;\; 6$

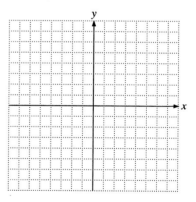

10. $2x + 3y = \;\;\; 6$
$\qquad 2x + \;\; y = -2$

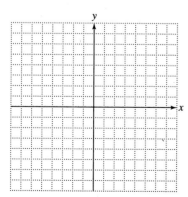

11. $y = -x + 3$
$\qquad 3x + 3y = -2$

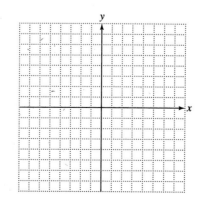

12. $y = \frac{1}{3}x - 2$
$\qquad -x + 3y = 9$

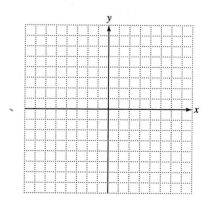

Find the solution to each system by the substitution method. Check your answers for Exercises 13 through 16.

13. $7x - 2y = 3$
$y = 4 - 2x$

14. $2x + 3y = 9$
$x = 2 + y$

15. $3x + 2y = -17$
$2x + y = 3$

16. $5x - 2y = 8$
$3x - y = 7$

17. $-x + 3y = -8$
$2x - y = 6$

18. $5x - 2y = 8$
$3x + y = 7$

★ **19.** $3a + 5b = -19$
$-a + 3b = -3$

★ **20.** $-3a + 4b = -18$
$2a + b = 1$

21. $\frac{1}{5}x - \frac{1}{2}y = 1$
$\frac{1}{5}x - 3y = -9$

22. $\frac{1}{5}x - \frac{1}{2}y = -1$
$\frac{3}{5}x - 3y = -9$

23. $3(x - 1) + 2(y + 2) = 6$
$4(x - 2) + y = 2$

24. $2(x - 2) + 3(y + 1) = -2$
$x + 2(y - 1) = -4$

Find the solution to each system by the addition (elimination) method. Check your answers for Exercises 25 through 28.

25. $9x + 2y = 2$
$3x + 5y = 5$

26. $x + 3y = 2$
$4x + 5y = 1$

27. $a + 2b = -1$
$2a - b = 3$

28. $2a + b = 3$
$a - 2b = -1$

29. $6s - 3t = 1$
$5s + 6t = 15$

30. $2s + 3t = 5$
$3s - 6t = 18$

31. $\frac{7}{2}x + \frac{5}{2}y = -4$
$\phantom{\frac{7}{2}}3x + \frac{2}{3}y = 1$

32. $\frac{4}{3}x - y = 4$
$\frac{3}{4}x - y = \frac{1}{2}$

33. $0.4x + 0.3y = 2.3$
$0.2x - 0.5y = 0.5$

34. $0.9x + 0.4y = 0$
$0.5x - 0.8y = 2.3$

★ **35.** $4(x + 1) + 2(y - 1) = 4$
$x + 3(y - 2) = 2$

★ **36.** $3(x + 1) + y = -7$
$2(x + 4) + 3(y + 1) = -5$

Mixed Practice

If possible, solve each system of equations. Use any method. If there is not a unique solution to a system, say why.

37. $2x + y = 4$
$\frac{2}{3}x + \frac{1}{4}y = 2$

38. $2x + 3y = 16$
$5x - \frac{3}{4}y = 7$

39. $0.2x = 0.1y - 1.2$
$2x - y = 6$

40. $0.1x - 0.6 = 0.3y$
$0.3x + 0.1y + 2.2 = 0$

41. $\begin{aligned} 5x - 7y &= 12 \\ -10x + 14y &= -24 \end{aligned}$

42. $\begin{aligned} 3x - 11y &= 9 \\ -9x + 33y &= 18 \end{aligned}$

43. $\begin{aligned} 8x + 9y &= 13 \\ 6x - 5y &= 45 \end{aligned}$

44. $\begin{aligned} -9x + 6y - 10 &= 0 \\ 6x &= 4y - 8 \end{aligned}$

45. $\begin{aligned} \frac{4}{5}b &= \frac{1}{5} + a \\ 15a - 12b &= 4 \end{aligned}$

46. $\begin{aligned} 3a - 2b &= \frac{3}{2} \\ \frac{3a}{2} &= \frac{3}{4} + b \end{aligned}$

47. $\begin{aligned} \frac{3}{4}x - y &= 4 \\ -\frac{1}{6}x + \frac{1}{3}y &= -1 \end{aligned}$

48. $\begin{aligned} \frac{2}{5}x + \frac{3}{5}y &= 1 \\ x - \frac{2}{3}y &= \frac{1}{3} \end{aligned}$

Find the solution to each system. Round your answer to four decimal places.

49. $\begin{aligned} 9.836x + \qquad y &= 12.824 \\ -8.073x - 5.982y &= 8.913 \end{aligned}$

50. $\begin{aligned} 0.0052x - 0.0093y &= 0.1256 \\ -0.0104x + 0.0521y &= 0.9315 \end{aligned}$

Optional Graphing Calculator Problems

On your graphing calculator, graph each system of equations on the same set of axes. Find the point of intersection to the nearest hundredth.

51. $y_1 = -1.7x + 3.8$ and $y_2 = 0.7x - 2.1$

52. $y_1 = -0.81x + 2.3$ and $y_2 = 1.6x + 0.8$

53. $\begin{aligned} 0.5x + 1.1y &= 5.5 \\ -3.1x + 0.9y &= 13.1 \end{aligned}$

54. $\begin{aligned} 5.86x + 6.22y &= -8.89 \\ -2.33x + 4.72y &= -10.61 \end{aligned}$

Cumulative Review Problems

55. 9 million tons of salt each year are applied to American highways for road de-icing. The cost of buying and applying the salt totals $200 million. How much is this per pound? Round your answer to the nearest cent.

56. In Boston, four-fifths of the automobiles that enter the city during rush hour will have to park in private or municipal parking lots. If there are 273,511 private or municipal lot spaces filled by cars entering the city during rush hour every morning, how many cars actually enter the city during rush hour? Round your answer to the nearest car.

Putting Your Skills to Work

THE MATHEMATICS OF BRICK LAYING

Suppose you wanted to help a friend put down a brick patio in his back yard using decorative red bricks that measure 8 inches by 4 inches. Let A = the area that you need to brick in, measured in square inches. The minimum number N of red bricks that are needed, if we allow for 10% of the area to be filled in by the cement between the bricks, is given by the equation $N = \dfrac{A - 0.10A}{32}$.

PROBLEMS FOR INDIVIDUAL INVESTIGATION

1. How many bricks will be needed to brick in a rectangular walkway that measures 48 inches wide and 120 inches long?

2. Write the equation in a slightly different form so that the area A is entered in square feet rather than square inches. Use the formula to find out how many bricks will be needed to brick in a square patio that measures 9 feet by 9 feet.

PROBLEMS FOR GROUP INVESTIGATION AND COOPERATIVE LEARNING

Together with some other members of your class see if you can determine the following.

3. To restore a brick street in Philadelphia a new shape of special brick is required that is x inches wide and y inches long. The bricks are laid in an alternating pattern so that the width of 20 bricks and the length of 45 bricks measures 630 inches. However, the width of 50 of the bricks and the length of 70 of the bricks measures 1116 inches. Set up a system of equations that describes this situation. Solve the system to determine the width and the length of this special brick.

4. In some very old cities of Germany, a large walkway brick is made that is 11 inches wide and z inches long. The width of 23 of these bricks and the length of 17 of these bricks totals a certain distance. However, the width of 45 of these bricks and the length of 36 of these bricks totals a length that is 508 inches greater than that original distance. Set up a system of equations that describes this situation. Solve the system to determine the length of the brick in inches. What is the length of that original distance?

 INTERNET CONNECTION: Go to `http://www.prenhall.com/tobey` to be connected.

Site: The Real McCoy Sizes and Shapes

This site gives the dimensions and coverage information for various brick sizes produced by a certain manufacturer. When calculating areas, be sure to choose the correct dimensions to multiply. (*Hint:* When you look at a brick wall, do you see "holes" in each brick?)

5. For Ontario brick, find the area of one brick in square meters and multiply by the given number of bricks per square meter. If Ontario brick is used for a walkway, what percentage of the surface will be filled in with cement instead of brick?

6. Repeat Problem 5 for Metric Jumbo brick. Why do you think your results for the two brick sizes differ?

7. For Ontario brick, write an equation similar to the one at the top of this page, where A is to be measured in square meters. Then write an equation for Metric Jumbo brick.

8. A brick wall is to measure 3 meters high by 7 meters long. How many Ontario bricks would be needed to build this wall? How many Metric Jumbo bricks?

4.2 Systems of Equations in Three Variables

MathPro   Video 10   SSM

After studying this section, you will be able to:

1 *Determine whether an ordered triple is the solution to a system of three equations in three variables.*

2 *Find the solution to a system of three linear equations in three variables if all coefficients are nonzero.*

3 *Find the solution to a system of three linear equations in three variables if some of the coefficients are zero.*

1 *Determining If an Ordered Triple Is the Solution to a System of Three Equations in Three Variables*

We are now going to study **systems of three linear equations with three variables** (unknowns). Many of these systems have exactly one solution. A **solution** to a system of three linear equations in three unknowns is an ordered triple of real numbers (x, y, z) that satisfies each equation in the system.

EXAMPLE 1   Determine whether $(2, -5, 1)$ is the solution to the system

$$3x + y + 2z = 3$$
$$4x + 2y - z = -3$$
$$x + y + 5z = 2$$

How can you prove that $(2, -5, 1)$ really is the solution to this system? We will substitute $x = 2$, $y = -5$, and $z = 1$ into each equation. If a true statement occurs each time, $(2, -5, 1)$ is a solution to each equation and hence the solution to the system. For the first equation,

$$3(2) + (-5) + 2(1) \overset{?}{=} 3$$
$$6 - 5 + 2 \overset{?}{=} 3$$
$$3 = 3 \quad \text{True statement.}$$

For the second,

$$4(2) + 2(-5) - 1 \overset{?}{=} -3$$
$$8 - 10 - 1 \overset{?}{=} -3$$
$$-3 = -3 \quad \text{True statement.}$$

Finally,

$$2 + (-5) + 5(1) \overset{?}{=} 2$$
$$2 - 5 + 5 \overset{?}{=} 2$$
$$2 = 2 \quad \text{True statement.}$$

Since we obtained three true statements, the ordered triple $(2, -5, 1)$ is a solution to the system.

Practice Problem 1   Determine whether $(3, -2, 2)$ is the solution to this system.

$$2x + 4y + z = 0$$
$$x - 2y + 5z = 17$$
$$3x - 4y + z = 19 \quad \blacksquare$$

To Think About Can we graph an equation in three variables? How? What would the graph look like? What would the graph of the system in Example 1 look like? Describe the graph of the solution.

2 Finding the Solution to a System of Three Linear Equations in Three Variables if All Coefficients Are Nonzero

One way to solve a system of three equations with three variables is to obtain from it a system of two equations in two variables; in other words, we eliminate one variable from both equations. We can then use the methods of Section 4.1 to solve the system. You can find the third variable (the one that was eliminated) by substituting the two variables that you have found into one of the original equations.

EXAMPLE 2   Find the solution to (that is, solve) the following system of equations.

$$-2x + 5y + z = 8 \quad \boxed{1}$$
$$-x + 2y + 3z = 13 \quad \boxed{2}$$
$$x + 3y - z = 5 \quad \boxed{3}$$

Let's eliminate z because it can be easily done by adding equations $\boxed{1}$ and $\boxed{3}$.

$$-2x + 5y + z = 8 \quad \boxed{1}$$
$$\underline{x + 3y - z = 5} \quad \boxed{3}$$
$$-x + 8y \quad\;\; = 13 \quad \boxed{4}$$

Now we need to choose a *different pair* from the original equations and once again eliminate the same variable. In other words, we have to use equations $\boxed{1}$ and $\boxed{2}$ or equations $\boxed{2}$ and $\boxed{3}$ and eliminate z. Let's multiply each term of equation $\boxed{3}$ by 3 (and call it equation $\boxed{6}$) and add the result to equation $\boxed{2}$.

$$-x + 2y + 3z = 13 \quad \boxed{2}$$
$$\underline{3x + 9y - 3z = 15} \quad \boxed{6}$$
$$2x + 11y \quad\;\; = 28 \quad \boxed{5}$$

We now can solve the system of two linear equations.

$$-x + 8y = 13 \quad \boxed{4}$$
$$2x + 11y = 28 \quad \boxed{5}$$

Multiply each term of equation $\boxed{4}$ by 2.

$$-2x + 16y = 26$$
$$\underline{2x + 11y = 28}$$
$$27y = 54 \qquad \textit{Add the equations.}$$
$$y = 2 \qquad \textit{Solve for y.}$$

Substituting $y = 2$ into $\boxed{4}$, we have

$$-x + 8(2) = 13$$
$$-x = -3$$
$$x = 3$$

Now substitute $x = 3$ and $y = 2$ into one of the original equations (any one will do) to solve for z. Let's use $\boxed{1}$.

$$-2x + 5y + z = 8$$
$$-2(3) + 5(2) + z = 8$$
$$-6 + 10 + z = 8$$
$$z = 4$$

The solution to the system is therefore $x = 3$, $y = 2$, $z = 4$, or $(3, 2, 4)$.

Check: Verify that $(3, 2, 4)$ satisfies *each* of the three *original* equations.

Practice Problem 2 Solve this system.

$$x + 2y + 3z = 4$$
$$2x + y - 2z = 3$$
$$3x + 3y + 4z = 10 \ \blacksquare$$

Here's a summary of the procedure that we just used.

How to Solve a System of Three Linear Equations in Three Unknowns

1. Use the addition method to eliminate any variable from any pair of equations. (The choice of variable is arbitrary.)
2. Use appropriate steps to eliminate the *same variable* from a *different pair* of equations. (If you don't eliminate the same variable, you will still have three unknowns.)
3. Solve the resulting system of two equations in two variables.
4. Substitute the values obtained in step 3 into one of the three original equations. Solve for the remaining variable.

It is helpful to write all equations in the form $Ax + By + Cz = D$ before using this four-step method.

EXAMPLE 3 Solve the system.

$$-4x + 2y - 3z \quad = 0$$
$$3x - 5y + 2z - 4 = 0$$
$$2x - 3y + 4z + 3 = 0$$

First we will transform each equation to an equivalent equation in the proper form.

$$-4x + 2y - 3z = \quad 0 \quad \boxed{1}$$
$$3x - 5y + 2z = \quad 4 \quad \boxed{2}$$
$$2x - 3y + 4z = -3 \quad \boxed{3}$$

Let's eliminate the variable x. Multiply each term of equation $\boxed{3}$ by 2 and call it equation $\boxed{4}$. Add the result to equation $\boxed{1}$.

$$-4x + 2y - 3z = \quad 0 \quad \boxed{1}$$
$$\underline{4x - 6y + 8z = -6 \quad \boxed{4}}$$
$$-4y + 5z = -6 \quad \boxed{5}$$

Now let's eliminate x by working with a different pair of equations. We will use equations $\boxed{2}$ and $\boxed{3}$. Multiply equation $\boxed{2}$ by 2 and call it equation $\boxed{6}$. Multiply equation $\boxed{3}$ by -3 and call it equation $\boxed{7}$. Now add these resulting equations.

$$6x - 10y + 4z = \quad 8 \quad \boxed{6}$$
$$\underline{-6x + 9y - 12z = \quad 9 \quad \boxed{7}}$$
$$-y - 8z = 17 \quad \boxed{8}$$

We will now solve the system

$$-4y + 5z = -6 \quad \boxed{5}$$
$$-y - 8z = 17 \quad \boxed{8}$$

We multiply each term of equation 8 by -4.

$$-4y + 5z = -6$$
$$4y + 32z = -68$$
$$\overline{37z = -74}$$
$$z = -2$$

We substitute this result into equation 8 to obtain the value for y.

$$-y - 8(-2) = 17$$
$$-y + 16 = 17$$
$$-y = 1$$
$$y = -1$$

Now we can substitute $y = -1, z = -2$ into any of the original equations that contain the three variables. We will use equation 3.

$$2x - 3y + 4z = -3$$
$$2x - 3(-1) + 4(-2) = -3$$
$$2x + 3 - 8 = -3$$
$$2x = 2$$
$$x = 1$$

The solution set is $x = 1, y = -1, z = -2$, or $(1, -1, -2)$.

Check: $-4(1) + 2(-1) - 3(-2) \overset{?}{=} 0$
$$-4 - 2 + 6 \overset{?}{=} 0$$
$$0 = 0 \checkmark$$

$3(1) - 5(-1) + 2(-2) - 4 \overset{?}{=} 0 \qquad\qquad 2(1) - 3(-1) + 4(-2) + 3 \overset{?}{=} 0$
$$3 + 5 - 4 - 4 \overset{?}{=} 0 \qquad\qquad\qquad 2 + 3 - 8 + 3 \overset{?}{=} 0$$
$$0 = 0 \checkmark \qquad\qquad\qquad\qquad\qquad 0 = 0 \checkmark$$

Practice Problem 3 Solve the system.

$$3x - y - 3z - 1 = 0$$
$$x - 5y + 5z - 3 = 0$$
$$2x - 3y + z - 2 = 0 \; \blacksquare$$

3 Finding the Solution to a System of Three Linear Equations in Three Variables if Some of the Coefficients Are Zero

If a system of three equations contains one or more equations of the form $Ax + By + Cz = 0$, where one of the values of A, B, or C is zero, then we will modify our approach slightly. We will select one equation that contains only two variables. Then we will take the remaining system of two equations and eliminate the variable that was missing in the equation that we selected.

EXAMPLE 4 Solve the system.

$$4x + 3y + 3z = 4 \quad \boxed{1}$$
$$3x \quad\quad + 2z = 2 \quad \boxed{2}$$
$$2x - 5y \quad\quad = -4 \quad \boxed{3}$$

Note that equation $\boxed{2}$ has no y term, and equation $\boxed{3}$ has no z term. Obviously, that makes our work easier. Let's work with equations $\boxed{2}$ and $\boxed{1}$ to obtain an equation that contains only x and y.

Step 1 Multiply equation $\boxed{1}$ by 2 and equation $\boxed{2}$ by -3 to obtain the system

$$8x + 6y + 6z = 8 \quad \boxed{4}$$
$$\underline{-9x \quad\quad - 6z = -6} \quad \boxed{5}$$
$$-x + 6y \quad\quad = 2 \quad \boxed{6}$$

Step 2 This step is already done, since equation $\boxed{3}$ has no z term.

Step 3 Now we can solve the system formed by equations $\boxed{3}$ and $\boxed{6}$.

$$2x - 5y = -4 \quad \boxed{3}$$
$$-x + 6y = 2 \quad \boxed{6}$$

If we multiply each term of equation $\boxed{6}$ by 2, we obtain the system

$$2x - 5y = -4$$
$$\underline{-2x + 12y = 4}$$
$$7y = 0 \quad \textit{Add.}$$
$$y = 0 \quad \textit{Solve for y.}$$

Substituting $y = 0$ in equation $\boxed{6}$, we find that

$$-x + 6(0) = 2$$
$$-x = 2$$
$$x = -2$$

Step 4 To find z, we substitute $x = -2$ and $y = 0$ into one of the original equations containing z. Since equation $\boxed{2}$ has only two variables, let's use it.

$$3x + 2z = 2$$
$$3(-2) + 2z = 2$$
$$2z = 8$$
$$z = 4$$

The solution to the system is $x = -2$, $y = 0$, and $z = 4$, or $(-2, 0, 4)$.

Check: Verify this solution by substituting these values into equations $\boxed{1}$, $\boxed{2}$, and $\boxed{3}$.

Practice Problem 4 Solve the system.

$$2x + y + z = 11$$
$$4y + 3z = -8$$
$$x - 5y \quad\quad = 2 \quad \blacksquare$$

A linear equation in three variables is a plane in three-dimensional space. A system of linear equations in three variables is three planes. The solution to a system is the point where the three planes intersect. This point is an ordered triple. See figure (a) in the margin. Notice that the point lies in each plane. If the system has no solution, then the planes do not intersect [see figure (b)] or all three planes do not intersect at any point(s) in common [see figures (c) and (d)]. Recall that the solution to the system must be a point that is in each plane. If the system has an infinite number of solutions, the three planes intersect in one line. See figure (e).

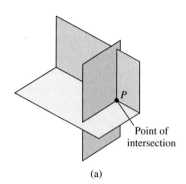

P
Point of intersection

(a)

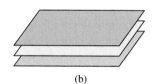

(b)

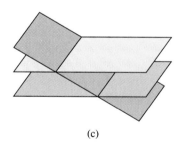

(c)

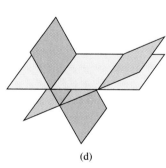

(d)

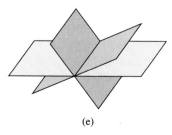

(e)

4.2 Exercises

1. Determine whether $(2, 1, -4)$ is a solution to the system.
$$2x - 3y + 2z = -7$$
$$x + 4y - z = 10$$
$$3x + 2y + z = 4$$

2. Determine whether $(5, -3, -2)$ is a solution to the system.
$$3x + 4y + 2z = -1$$
$$-x - 2y + 3z = -5$$
$$x + y + z = 0$$

Solve each system.

3. $$x + y + 2z = 0$$
$$2x - y - z = 1$$
$$x + 2y + 3z = 1$$

4. $$2x + y + 3z = 2$$
$$x - y + 2z = -4$$
$$x + 3y - z = 1$$

5. $$x + 2y - 3z = -11$$
$$-2x + y - z = -11$$
$$x + y + z = 6$$

6. $$-5x + 3y + 2z = 1$$
$$x + y + z = 7$$
$$2x - y + z = 7$$

7. $$3x - 2y + z = 4$$
$$x - 4y - z = 6$$
$$2x - y + 3z = 0$$

8. $$3x - y + 4z = 4$$
$$x - 2y + 2z = 4$$
$$2x + y - 3z = 5$$

9. $$x + 4y - z = -5$$
$$-2x - 3y + 2z = 5$$
$$x - \frac{2}{3}y + z = \frac{11}{3}$$

10. $$x - 4y + 4z = -1$$
$$-x + \frac{y}{2} - \frac{5}{2}z = -3$$
$$-x + 3y - z = 5$$

11. $$2x - 3y + 2z = -7$$
$$\frac{3}{2}x + y + \frac{1}{2}z = 2$$
$$x + 4y - z = 10$$

12. $$-2x + y - 2z = 0$$
$$4x + 2y + 2z = -6$$
$$3x - 2y + 2z = -3$$

13. $$a = 8 + 3b - 2c$$
$$4a + 2b - 3c = 10$$
$$c = 10 + b - 2a$$

14. $$a = c - b$$
$$3a - 2b + 6c = 1$$
$$c = 4 - 3b - 7a$$

15. $$3a + 3b + 2c = -3$$
$$-2a + 5b + 3c = 3$$
$$10a - 4b - 8c = 6$$

16. $$4a + 2b + 3c = 9$$
$$3a + 5b + 4c = 19$$
$$9a + 3b + 2c = 3$$

17. $$-x + 3y - 2z = 11$$
$$2x - 4y + 3z + 15 = 0$$
$$3x - 5y - 4z = 5$$

18. $$-3x - 2y + 3z = 2$$
$$2x - 5y + 2z + 2 = 0$$
$$4x - 3y + 4z = 10$$

Find the solution for each system of equations. Round your answers to five decimal places.

19.
$$x - 4y + 4z = -3.72186$$
$$-x + 3y - z = 5.98115$$
$$2x - y + 5z = 7.93645$$

20.
$$4x + 2y + 3z = 9$$
$$9x + 3y + 2z = 3$$
$$2.987x + 5.027y + 3.867z = 18.642$$

Solve each system.

21.
$$x + y = 1$$
$$y - z = -3$$
$$2x + 3y + z = 1$$

22.
$$y - 2z = 5$$
$$2x + z = -1$$
$$3x + y - z = 4$$

23.
$$-y + 2z = 1$$
$$x + y + z = 2$$
$$-x + 3z = 2$$

24.
$$-2x + y - 3z = 0$$
$$-2y - z = -1$$
$$x + 2y - z = 5$$

25.
$$x - 2y + z = 0$$
$$-3x - y = -6$$
$$y - 2z = -7$$

26.
$$x + 2z = 0$$
$$3x + 3y + z = 6$$
$$6y + 5z = -3$$

27.
$$3a - b + c = 6$$
$$4a + 2b - c = 12$$
$$a + 3c = 14$$

28.
$$a + 2b + c = 5$$
$$4a + 2c = 10$$
$$-b - 5c = -6$$

29.
$$-2x - y = -6$$
$$3y - 2z = -8$$
$$x + z = 5$$

30.
$$-3x - 4y = -15$$
$$2x - 5z = -3$$
$$-4y + 3z = -9$$

31.
$$2x = 2 - 3y$$
$$4x + 3y + z = 2$$
$$z = 2x - 6y$$

32.
$$3x = 11 + 2y - 4z$$
$$6x = 1 - 6y$$
$$z = 1 - 2y$$

Try to solve the system of equations. Explain your result in each case.

33.
$$2x + y = -3$$
$$2y + 16z = -18$$
$$-7x - 3y + 4z = 6$$

34.
$$6x - 2y + 2z = 2$$
$$4x + 8y - 2z = 5$$
$$-2x - 4y + z = -2$$

35.
$$3x + 3y - 3z = -1$$
$$4x + y - 2z = 1$$
$$-2x + 4y - 2z = -8$$

36.
$$-3x + 4y - z = -4$$
$$x + 2y + z = 4$$
$$-12x + 16y - 4z = -16$$

Cumulative Review Problems

37. Find the equation of a line in standard form passing through $(1, 4)$ and $(-2, 3)$.

38. Find the equation of a line in standard form that is perpendicular to $y = -\frac{2}{3}x + 4$ and passes through $(-4, 2)$.

39. A rancher in Australia has 346 horses, 545 sheep, and 601 cattle. He wants to purchase more animals so that he has 80% more cattle than horses, and 74% more sheep than horses. Round all answers to nearest whole number. **(a)** How many animals of each type will he have to buy? **(b)** How many animals of each type will he have after his purchases?

40. The current in a river moves at a speed of 3.5 miles per hour. A boat travels with the current 48 miles downstream in a total of 3 hours. What would be the speed of the boat if there were no current?

4.3 Applications of Systems of Linear Equations

After studying this section, you will be able to:

1. Solve an applied problem requiring the use of a system of two linear equations with two unknowns.
2. Solve an applied problem requiring the use of a system of three linear equations with three unknowns.

1 Solving an Applied Problem Requiring the Use of a System of Two Equations in Two Unknowns

We will now examine how a system of linear equations can assist us in solving applied problems.

EXAMPLE 1 For the visiting archaeologist lecture on campus, advance tickets cost $5 and tickets at the door cost $6. The ticket sales this year came to $4540. The department chairman wants to raise prices next year to $7 for advance tickets and $9 for tickets at the door. He said that if exactly the same number of people attended next year at these new price levels, the ticket sales would total $6560. If he is correct, how many tickets are sold in advance? How many tickets are sold at the door?

1. *Understand the problem.*
   Since we are looking for the number of tickets sold, we let

   $$x = \text{number of tickets bought in advance}$$

   $$y = \text{number of tickets bought at the door}$$

2. *Write a system of two equations in two unknowns.*
   If advance tickets cost $5, then the total sales will be $5x$; similarly, total sales of door tickets will be $6y$. Since the total sales of both types of tickets is $4540, we have

   $$5x + 6y = 4540$$

   By the same reasoning, we have

   $$7x + 9y = 6560$$

   Our system is thus

   $$5x + 6y = 4540 \quad \boxed{1}$$
   $$7x + 9y = 6560 \quad \boxed{2}$$

3. *Solve the system of equations and state the answer.*
   We will multiply each term of equation $\boxed{1}$ by -3 and also multiply each term of equation $\boxed{2}$ by 2 to obtain the following equivalent system.

   $$\begin{aligned} -15x - 18y &= -13,620 \quad \boxed{3} \\ \underline{14x + 18y} &= \underline{13,120} \quad \boxed{4} \\ -x &= -500 \end{aligned}$$

   Therefore, $x = 500$. Substituting $x = 500$ into equation $\boxed{1}$, we have

   $$5(500) + 6y = 4540$$
   $$6y = 2040$$
   $$y = 340$$

   Thus 500 advance tickets and 340 door tickets were sold.

4. *Check.*

We need to check our answers. Do they seem reasonable?

Would 500 advance tickets at $5 and 340 door tickets at $6 yield $4540?

$$5(500) + 6(340) \stackrel{?}{=} 4540$$

$$2500 + 2040 \stackrel{?}{=} 4540$$

$$4540 = 4540 \quad \checkmark$$

Would 500 advance tickets at $7 and 340 door tickets at $9 yield $6560?

$$7(500) + 9(340) \stackrel{?}{=} 6560$$

$$3500 + 3060 \stackrel{?}{=} 6560$$

$$6560 = 6560 \quad \checkmark$$

Therefore, our solution is correct.

Practice Problem 1 Coach Perez purchased baseballs at $6 each and bats at $21 each last week for the college baseball team. The total cost of the purchase was $318. This week he noticed that the same items are on sale. Baseballs are now $5 each and bats are $17. He found that if he had made the same purchase this week, it would have cost only $259. How many baseballs and how many bats did he buy last week? ■

EXAMPLE 2 An electronics firm makes two types of switching devices. Type A takes 4 minutes to make and requires $3 worth of materials. Type B takes 5 minutes to make and requires $5 worth of materials. When the production manager reviewed the latest batch, he found that it took 35 hours to make these switches with a materials cost of $1900. How many switches of each type were produced?

1. *Understand the problem.*

We are given a lot of information, but the major concern is to find out how many of the type A devices and the type B devices were produced. This becomes our starting point to define the variables we will use.

> Let A = the number of type A devices produced
>
> B = the number of type B devices produced

2. *Write a system of two equations.*

How should we construct the equations? What relationships exist between our variables (or unknowns)? According to the problem, the devices are related by time and by cost. So we set up one equation in terms of time (minutes in this case) and one in terms of cost (dollars). Each type A took 4 minutes to make, each type B took 5 minutes to make, and the total time was 2100 minutes. Each type A used $3 worth of materials, each type B used $5 worth of materials, and the total material cost was $1900. We can gather this information in a table. Making a table will help in forming the equations.

	Type A Devices	Type B Devices	Total
Number of minutes	4A	5B	2100
Cost of materials	3A	5B	1900

$$4A + 5B = 2100$$
$$3A + 5B = 1900$$

Therefore, we have the system

$$4A + 5B = 2100 \quad \boxed{1}$$

$$3A + 5B = 1900 \quad \boxed{2}$$

3. *Solve the system of equations and state the answers.*

Multiplying equation 2 by -1 and adding the equations, we find

$$A = 200$$

Substituting $A = 200$ into equation 1, we have

$$800 + 5B = 2100$$

$$5B = 1300$$

$$B = 260$$

Thus 200 type A devices and 260 type B devices were produced.

4. *Check.*

If each type A uses 4 minutes and each type B uses 5 minutes, does this amount to a total time of 2100 minutes?

$$4A + 5B = 2100$$

$$4(200) + 5(260) \stackrel{?}{=} 2100$$

$$800 + 1300 \stackrel{?}{=} 2100$$

$$2100 = 2100 \quad ✓$$

If each type A costs \$3 and each type B costs \$5, does this amount to a total cost of \$1900?

$$3A + 5B = 1900$$

$$3(200) + 5(260) \stackrel{?}{=} 1900$$

$$600 + 1300 \stackrel{?}{=} 1900$$

$$1900 = 1900 \quad ✓$$

Practice Problem 2   A furniture company makes both small and large chairs. It takes 30 minutes of machine time and 1 hour and 15 minutes of labor to build the small chair. The large chair needs 40 minutes of machine time and 1 hour and 20 minutes of labor. The company has 57 hours of labor time and 26 hours of machine time available each day. If all available time is used, how many chairs of each type can the company make? ∎

When we encounter motion problems involving rate, time, or distance, it is useful to recall the formula $D = RT$ or distance = (rate)(time).

EXAMPLE 3   An airplane travels between two cities that are 1500 miles apart. The trip against the wind takes 3 hours. The return trip with the wind takes $2\frac{1}{2}$ hours. What is the speed of the plane in still air (in other words, how fast could the plane go if there were no wind)? What is the speed of the wind?

1. *Understand the problem.*

Our unknowns are the speed of the plane in still air and the speed of the wind. Let

$$x = \text{the speed of the plane in still air}$$

$$y = \text{the speed of the wind}$$

Let's make a sketch to help us see how these speeds are related to one another. When we travel against the wind, the wind is slowing us down. Since the wind speed opposes the plane's speed in still air, we must subtract: $x - y$.

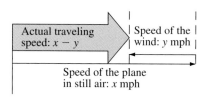

When we travel with the wind, the wind is helping the plane to travel forward. Thus the wind speed is added to the plane's speed in still air and we add $x + y$.

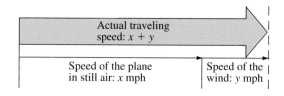

2. *Write a system of two equations.*

To help us write our equations, we organize the information in a chart. The chart will be based on the formula $RT = D$, which is (rate)(time) = distance.

	R $\cdot$	T	$= D$
Flying against the wind	$x - y$	3	1500
Flying with the wind	$x + y$	2.5	1500

Using each row of the chart, we obtain a system of equations.

$$(x - y)(3) = 1500$$

$$(x + y)(2.5) = 1500$$

If we remove the parentheses, we will obtain the system

$$3x - 3y = 1500 \quad \boxed{1}$$

$$2.5x + 2.5y = 1500 \quad \boxed{2}$$

3. *Solve the system of equations and state the answer.*
It will be helpful to clear equation 2 of decimal coefficients. This can be done by multiplying each term of equation 2 by 2.

$$3x - 3y = 1500 \quad \boxed{1}$$

$$5x + 5y = 3000 \quad \boxed{3}$$

If we multiply equation 1 by 5 and equation 3 by 3, we will obtain the system

$$15x - 15y = 7500$$

$$\underline{15x + 15y = 9000}$$

$$30x \quad\quad = 16,500$$

$$x = 550$$

Substituting this result in equation 1, we obtain

$$3(550) - 3y = 1500$$

$$1650 - 3y = 1500$$

$$-3y = -150$$

$$y = 50$$

Thus the speed of the plane in still air is 550 miles per hour and the speed of the wind is 50 miles per hour.

4. *Check.*
Left to the student.

Practice Problem 3 An airplane travels west from city *A* to city *B* against the wind. It takes 3 hours to travel 1950 kilometers. On the return trip the plane travels east from city *B* to city *C* a distance of 1600 kilometers in a time of 2 hours. On the return trip the plane travels with the wind. What is the speed of the plane in still air? What is the speed of the wind? ■

GRAPHING CALCULATOR

Exploration

This visual interpretation of two equations in two unknowns is sometimes most helpful. Study Example 3. Graph the two equations.

$3x - 3y = 1500$ and $2.5x + 2.5y = 1500$

What is the significance of the point of intersection? If you were an air traffic controller how would you interpret the linear equation $3x - 3y = 1500$? Why would this be useful? How would you interpret $2.5x + 2.5y = 1500$? How would this be useful?

EXAMPLE 4   A trucking firm has three sizes of trucks. The biggest trucks hold 10 tons of gravel; the next size holds 6 tons, and the smallest holds 4 tons. The firm's manager has 15 trucks available to haul 104 tons of gravel. However, to reduce fuel cost she wants to use two more of the fuel-efficient 10-ton trucks than she does the 6-ton trucks. Her assistant tells her that she has two more 10-ton trucks available than 6-ton trucks. How many trucks of each type should she use?

1. *Understand the problem.*
   Since we need to find three things (the number of 10-ton trucks, 6-ton trucks, and 4-ton trucks), it would be helpful to have three variables. Let

   $$x = \text{the number of 10-ton trucks used}$$

   $$y = \text{the number of 6-ton trucks used}$$

   $$z = \text{the number of 4-ton trucks used}$$

2. *Write a system of three equations.*
   We know that 15 trucks will be used; hence

   $$x + y + z = 15 \quad \boxed{1}$$

   How can we get our second equation? Well, we also know the *capacity* of each truck type, and we know the total tonnage to be hauled. The first type of truck hauls 10 tons, the second type 6 tons, and the third type 4 tons, and the total tonnage is 104 tons. Hence we can write

   $$10x + 6y + 4z = 104 \quad \boxed{2}$$

   We still need one more equation. What other given information can we use? The problem states that the manager wants to use two more 10-ton trucks than the number of 6-ton trucks. Thus

   $$x = 2 + y \quad \boxed{3}$$

   (We could have also written $x - y = 2$.) Hence our system of equations is

   $$\begin{aligned} x + y + z &= 15 \quad \boxed{1} \\ 10x + 6y + 4z &= 104 \quad \boxed{2} \\ x - y &= 2 \quad \boxed{3} \end{aligned}$$

3. *Solve the system of equations and state the answers.*
   Equation $\boxed{3}$ doesn't contain the variable z. Let's work with equations $\boxed{1}$ and $\boxed{2}$ to eliminate z. First we multiply equation $\boxed{1}$ by -4 and add it to equation $\boxed{2}$.

   $$\begin{array}{rcll} -4x - 4y - 4z &=& -60 & \boxed{4} \\ 10x + 6y + 4z &=& 104 & \boxed{2} \\ \hline 6x + 2y &=& 44 & \boxed{5} \end{array}$$

   Make sure you understand how we got equation $\boxed{5}$. Dividing each term of equation $\boxed{5}$ by 2 and adding to equation $\boxed{3}$ gives

   $$\begin{array}{rcll} 3x + y &=& 22 & \boxed{6} \\ x - y &=& 2 & \boxed{3} \\ \hline 4x &=& 24 \\ x &=& 6 \end{array}$$

For $x = 6$, equation **3** yields

$$6 - y = 2$$
$$4 = y$$

Now, using equation **1**, we get

$$6 + 4 + z = 15$$
$$z = 5$$

Thus the manager needs six 10-ton trucks, four 6-ton trucks, and five 4-ton trucks.

4. *Check.*
   Left to the student.

Practice Problem 4 A factory uses three machines to wrap boxes for shipment. Machines A, B, and C can wrap 260 boxes in 1 hour. If machine A runs 3 hours and machine B runs 2 hours, they can wrap 390 boxes. If machine B runs 3 hours and machine C runs 4 hours, 655 boxes can be wrapped. How many boxes per hour can each machine wrap? ∎

EXAMPLE 5 Electricians plan to lay three sizes of conduits (pipes) together under the floor of a building so that they can pull their wires from one end of the building to the other. The center of each pipe is labeled A, B, and C, as shown in our sketch. The distance from A to B is 15 centimeters. The distance from B to C is 13 centimeters. The distance from C to A is 16 centimeters. Find the radius of each pipe.

1. *Understand the problem.*
   We need to find the radius of each pipe. Draw and label a diagram to show the radius of each pipe.

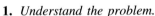
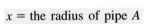

x = the radius of pipe A

y = the radius of pipe B

z = the radius of pipe C

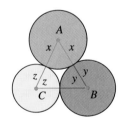

2. *Write a system of three equations.*
   Look at both diagrams. The distance from A to B is 15 centimeters. The distance from A to B can also be represented by $x + y$. Thus

$$x + y = 15 \quad \boxed{1}$$

The distance from B to C is 13 centimeters. Thus

$$y + z = 13 \quad \boxed{2}$$

The distance from C to A is 16 centimeters. Thus

$$z + x = 16 \quad \boxed{3}$$

We therefore have the system

$$x + y \quad\; = 15 \quad \boxed{1}$$
$$y + z = 13 \quad \boxed{2}$$
$$x \quad\; + z = 16 \quad \boxed{3}$$

3. *Solve the system of equations and state the answers.*

If we multiply equation 3 by −1 and add the result to equation 2, we obtain

$$-x \qquad - z = -16 \quad \boxed{4}$$
$$\underline{\qquad\quad y + z = \quad 13 \quad \boxed{2}}$$
$$-x + y \qquad = -3 \quad \boxed{5}$$

We then solve the system

$$-x + y = -3 \quad \boxed{5}$$
$$x + y = \ 15 \quad \boxed{1}$$

$y = 6$ and $x = 9$. Substituting these values into equation 3, we have $z = 7$. Thus, the radius of pipe A is 9 centimeters, the radius of pipe B is 6 centimeters, and the radius of pipe C is 7 centimeters.

4. *Check.*

Left to the student.

Practice Problem 5   An electrician places three wires next to each other inside a radio transmitter. The cross section of each wire is circular. The centers of the wires are labeled A, B, and C in this sketch. The distance between the center of A and the center of B is 13 millimeters. The distance between the center of B and the center of C is 22 millimeters. The distance from the outside edge of the wire at A to the outside edge of the wire at C is 52 millimeters, as shown in the figure. Find the radius of each wire.

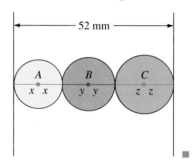

4.3 Exercises

Applications *Use a system of linear equations to solve each problem.*

1. An employment agency specializing in temporary construction help pays heavy equipment operators $40 per day and general laborers $25 per day. If 35 people were hired and the payroll was $1115, how many heavy equipment operators were employed? How many laborers?

2. A Broadway musical had a paid attendance of 530 people. Mezzanine tickets cost $9, and orchestra tickets cost $14. Ticket sales receipts totaled $5870. How many tickets of each type were sold?

3. Ninety-eight passengers rode in an Amtrak train from Boston to Denver. Tickets for regular coach seats cost $120. Tickets for sleeper car seats cost $290. The receipts for the trip totaled $19,750. How many passengers purchased each type of ticket?

4. The Tupper Farm has 450 acres of land allotted for raising corn and wheat. The cost to cultivate corn is $42 per acre. The cost to cultivate wheat is $35 per acre. The Tuppers have $16,520 available to cultivate these crops. How many acres of each crop should the Tuppers plant?

5. At Western State, full-time students pay a registration fee of $20, and part-time students pay $14. The registrar determined that registration fees of $9080 had to be billed to students. However, due to a programming error, the computer billed each full-time student $15 and each part-time student $12, for a total of $7140. How many full-time students have registered? How many part-time students?

6. For its sales staff, the Star Sales Company rents cars each month under a package deal from a local car dealer. Last month the company rented 25 compact cars and 10 intermediate cars for $13,750. This month, in a cost-saving move, it rented 32 compact cars and three intermediate cars for $12,700. What is the rental cost per month for each type of car?

7. An insurance company determined that it takes 20 hours to train a new salesperson on company procedure and 12 hours to train the salesperson on sales techniques. An experienced salesperson learns company procedure in 14 hours and sales techniques in 5 hours. Sales managers can devote 240 hours for training in sales techniques. The company extension school has 468 hours available to teach company procedure. How many experienced salespersons can be trained? How many new salespersons?

8. During a time study, a company that makes auto radar detectors found that its basic model required 3 hours to manufacture the inside components and 2 hours to manufacture the housing and controls. The advanced model needed 5 hours to manufacture the inside components and 3 hours to manufacture the housing and controls. The production division has 1050 hours available this week to manufacture inside components and 660 hours available to manufacture housing and controls. How many detectors of each type can be made?

9. To provide vitamin additives for patients, a staff hospital dietician has two prepackaged mixtures available. Mixture 1 contains 5 grams of vitamin C and 3 grams of niacin; mixture 2 contains 6 grams of vitamin C and 5 grams of niacin. On an average day she needs 87 grams of niacin and 117 grams of vitamin C. How many packets of each mixture will she need?

10. A farmer has several packages of fertilizer for his new grain crop. The old packages contain 50 pounds of long-term-growth supplement and 60 pounds of weed killer. The new packages contain 65 pounds of long-term-growth supplement and 45 pounds of weed killer. Using past experience, the farmer estimates that he needs 3125 pounds of long-term-growth supplement and 2925 pounds of weed killer for the fields. How many old packages of fertilizer and how many new packages of fertilizer should he use?

11. On Monday, Harold picked up 3 doughnuts and 4 large coffees for the office staff. He paid $4.91. On Tuesday, Melinda picked up 5 doughnuts and 6 large coffees for the office staff. She paid $7.59. What is the cost of one doughnut? What is the cost of one large coffee?

12. A local department store is preparing 4-color sales brochures to insert into the *Salem Evening News*. The printer has told them that he has a fixed charge to set up the printing of the brochure. Then he will charge them a specific per copy amount for each brochure printed. He quoted a price of $1350 for printing 5000 brochures and a price of $1750 for printing 7000 brochures. What is the fixed charge to set up the printing? What is the per copy cost for printing a brochure?

13. Against the wind a small plane flew 210 miles in 1 hour and 10 minutes. The return trip took only 50 minutes. What was the speed of the wind? What was the speed of the plane in still air?

14. With the current a motorboat can travel 84 miles in 2 hours. Against the current the same trip takes 3 hours. How fast can the boat travel in still water? What is the speed of the current?

15. It took Linda and Alice 4 hours to travel 24 miles downstream by canoe on Indian River. The next day they traveled for 6 hours upstream for 18 miles. What was the rate of the current? What was their average speed in still water?

16. Against the wind a commercial airline in South America flew 630 miles in 3 hours and 30 minutes. With a tailwind the return trip took 3 hours. What was the speed of the wind? What was the speed of the plane in still air?

17. Larry Bird scored 38 points in one of his final NBA games. During this game he did not have any free throws. He made several regular shots from the floor, which were worth 2 points. He made a few 3-point shots as well. He scored a total of 16 times in the game. How many 2-point baskets did he make? How many 3-point baskets did he make?

18. Charles Barclay scored 32 points in a recent basketball game without scoring any 3-point shots. He scored a total of 21 times. He made several free throws that were worth 1 point and several regular shots from the floor that were worth 2 points. How many free throws did he make? How many regular shots from the floor did he make?

19. Brenda has found her new Ford Escort gets 35 miles per gallon on the highway, but only 21 miles per gallon in city driving. She recently drove 420 miles on 14 gallons of gasoline. How many miles did she drive on the highway? How many miles of city driving did she do?

20. Carlos has found his new Neon gets 32 miles per gallon on the highway and 24 miles per gallon in city driving. He recently drove 432 miles on 16 gallons of gasoline. How many miles did he drive on the highway? How many miles of city driving did he do?

21. The White Mountain Ski Lodge has three types of vehicles. The large vans hold 15 passengers. The Dodge Mini-vans hold 7 passengers. The Ford Explorers each hold 5 passengers. The lodge has a total of 14 vehicles, and all together they can carry 98 passengers. The total number of large vans and Dodge Mini-vans is 6 vehicles. How many of each type of vehicle does the White Mountain Ski Lodge have?

22. Doug Camp's trucking firm has three types of trucks. One holds 10 tons of gravel, one holds 5 tons of gravel, and one holds 3 tons of gravel. Today this firm has 12 trucks on the road. They are available to haul 78 tons of gravel today. The total number of 5-ton trucks and 3-ton trucks on the road today is eight. How many of each type of truck is on the road?

23. A total of 300 people attended the high school play. The admission prices were $5 for adults, $3 for high school students, and $2 for any children not yet in high school. The ticket sales totaled $1010. The school principal suggested that next year they raise prices to $7 for adults, $4 for high school students, and $3 for children not yet in high school. He said that if exactly the same number of people attended next year at the higher prices ticket sales would total $1390. How many adults, high school students, and children not yet in high school attended this year?

24. The college conducted a CPR training class for students, faculty, and staff. Faculty were charged $10, staff were charged $8, and students were charged $2 to attend the class. A total of 400 people came. The receipts for all who attended totaled $2130. The college president remarked that if he charged faculty $15 and staff $10 and let students come free, the receipts this year would have been $2425. How many students, faculty, and staff came to the CPR training class?

25. A total of 12,000 passengers normally ride in the green line of the MBTA during the morning rush hour. When the token prices for a ride are $0.25 for children under 12, $1.00 for adults, and $0.50 for senior citizens, the revenue from these riders is $10,700. If the token prices were raised to $0.35 for children under 12, and $1.50 for adults, and the senior citizen price is unchanged, the expected revenue from these riders would be $15,820. How many riders in each category normally ride in the green line during the morning rush hour?

26. The owner of Danvers Ford found that he sold a total of 520 cars, Windstars, and Explorers last year. He paid the sales staff a commission of $100 for every car, $200 for every Windstar, and $300 for every Explorer sold. The total cost of these commissions last year was $87,000. In the coming year he is contemplating an increase so that the commission would be $150 for every car, $250 for every Windstar, but no change in the commission for every Explorer. If the sales were the same next year as this year that would incur a total cost of $106,500 in commissions. How many vehicles in each category were sold last year?

27. The Falmouth pumping station normally uses three pumps to provide drinking water for the town. When pumps A, B, and C are used for 2 hours, they can pump 74,000 gallons. The supervisor wanted to run pumps A and B for 4 hours, but pump B overheated and was shut down after 2 hours. During that time (that is, the 2 hours that pumps A and B ran, and the 2 additional hours that only pump A ran), 64,000 gallons of water were pumped. After cooling down, pump B was used for 5 hours and pump C for 4 hours to pump 120,000 gallons. How many gallons per hour can each pump handle?

28. A mathematician needs to find the radii of three circles. Each circle is drawn so that the circles are tangent (touching at only one point) (see the figure). The circles have centers at A, B, and C, respectively. The mathematician knows that $AB = 9$ centimeters, $AC = 7$ centimeters, and $CB = 8$ centimeters. Find the radius of each circle. *Hint:* Let $z = $ radius of circle C, let $y = $ radius of circle B, and let $x = $ radius of circle A.

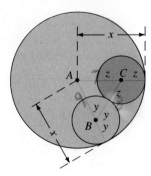

29. Sunshine Fruit Company packs three types of gift boxes of oranges, pink grapefruit, and white grapefruit. Box *A* contains 10 oranges, 3 pink grapefruit, and 3 white grapefruit. Box *B* contains 5 oranges, 2 pink grapefruit, and 3 white grapefruit. Box *C* contains 4 oranges, 1 pink grapefruit, and 2 white grapefruit. The shipping manager has available 51 oranges, 16 pink grapefruit, and 23 white grapefruit. How many gift boxes of each type can she prepare?

30. A company packs three types of packages in a large shipping box. Package type *A* weighs 0.5 pound, package type *B* weighs 0.25 pound, and package type *C* weighs 1.0 pound. Seventy-five packages weighing a total of 42 pounds were placed in the box. The number of type *A* packages was three less than the total number of types *C* and *B* combined. How many packages of each type were placed in the box?

31. This year the state highway department in Montana purchased 256 identical cars and 183 identical trucks for official use. The purchase price was $5,791,948. Due to a budget shortfall, next year the department plans to purchase only 64 cars and 107 trucks. It will be charged the same price for each car and for each truck. Next year it plans to spend $2,507,612. How much does the department pay for each car and for each truck?

32. A recent concert at Gordon College had a paid audience of 987 people. Advance tickets were $9.95 and tickets at the door were $12.95. A total of $10,738.65 was collected in ticket sales. How many of each type of ticket were sold?

To Think About

Use a system of four linear equations in four unknowns to solve the following problem.

33. A scientist at the University of Chicago is performing an experiment to determine how to increase the life span of mice through a controlled diet. The mice need 134 grams of carbohydrates, 150 grams of protein, 178 grams of fat, and 405 grams of moisture during the length of the experiment. The food is available in four packets, as shown in the table. How many packets of each type should the scientist use?

	Packet			
Contents	*A*	*B*	*C*	*D*
Carbohydrates	42	20	0	10
Protein	20	10	20	0
Fat	34	0	10	20
Moisture	50	35	30	40

Cumulative Review Problems

Solve for the variable indicated.

34. $\dfrac{1}{3}(4 - 2x) = \dfrac{1}{2}x - 3$

35. $0.06x + 0.15(0.5 - x) = 0.04$

36. $2(y - 3) - (2y + 4) = -6y$

4.4 Determinants

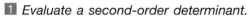

After studying this section, you will be able to:

MathPro   Video 11   SSM

1 *Evaluate a second-order determinant.*
2 *Evaluate a third-order determinant.*

1 *Evaluating a Second-order Determinant*

Mathematicians have developed techniques to solve systems of linear equations by focusing on the coefficients of the variables and the constants in the equations. The computational techniques can be easily carried out by computers or calculators. We will learn to do them by hand so that you will have a better understanding of what is involved.

To begin, we need to define a matrix and a determinant. A **matrix** is any rectangular array of numbers that is arranged in rows and columns. We use the symbol [ ] to indicate a matrix.

$$\begin{bmatrix} 3 & 2 & 4 \\ -1 & 4 & 0 \end{bmatrix}, \quad \begin{bmatrix} 4 & -3 \\ 2 & \frac{1}{2} \\ 1 & 5 \end{bmatrix}, \quad [-4 \quad 1 \quad 6], \quad \begin{bmatrix} \frac{1}{4} \\ 3 \\ -2 \end{bmatrix}$$

are matrices. If you have a graphing calculator, then you can enter the elements of a matrix into your calculator and store it for future use. Let's examine two systems of equations.

$$3x + 2y = 16 \qquad -6x \qquad = 18$$
$$\text{and}$$
$$x + 4y = 22 \qquad x + 3y = 9$$

We could write the coefficients of the variables in each of the above systems as a matrix.

$$\begin{matrix} 3x + 2y \\ x + 4y \end{matrix} \Rightarrow \begin{bmatrix} 3 & 2 \\ 1 & 4 \end{bmatrix}, \qquad \begin{matrix} -6x \\ x + 3y \end{matrix} \Rightarrow \begin{bmatrix} -6 & 0 \\ 1 & 3 \end{bmatrix}$$

Now we define a determinant. A **determinant** is a *square* arrangement of numbers. We use the symbol | | to indicate a determinant.

$$\begin{vmatrix} 3 & 2 \\ 1 & 4 \end{vmatrix} \quad \text{and} \quad \begin{vmatrix} -6 & 0 \\ 1 & 3 \end{vmatrix}$$

are determinants. The value of a determinant is a *real number* and is defined as follows.

Definition

The value of the second-order determinant $\begin{vmatrix} a & c \\ b & d \end{vmatrix}$ is $ad - bc$.

EXAMPLE 1 Find the value of each determinant. **(a)** $\begin{vmatrix} 2 & 3 \\ 1 & 5 \end{vmatrix}$ **(b)** $\begin{vmatrix} -4 & 2 \\ 3 & 6 \end{vmatrix}$

(a) $\begin{vmatrix} 2 & 3 \\ 1 & 5 \end{vmatrix} = (2)(5) - (1)(3) = 10 - 3 = 7$

(b) $\begin{vmatrix} -4 & 2 \\ 3 & 6 \end{vmatrix} = (-4)(6) - (3)(2) = -24 - 6 = -30$

Practice Problem 1 Evaluate the determinants.

(a) $\begin{vmatrix} 2 & -5 \\ 3 & 1 \end{vmatrix}$ **(b)** $\begin{vmatrix} -6 & -4 \\ -5 & 2 \end{vmatrix}$ ∎

When the determinant contains several negative numbers, be careful about signs.

EXAMPLE 2   Find the value of each determinant.

(a) $\begin{vmatrix} -6 & 2 \\ -1 & 4 \end{vmatrix}$ **(b)** $\begin{vmatrix} 0 & -3 \\ -2 & 6 \end{vmatrix}$

(a) $\begin{vmatrix} -6 & 2 \\ -1 & 4 \end{vmatrix} = (-6)(4) - (-1)(2) = -24 - (-2) = -24 + 2 = -22$

(b) $\begin{vmatrix} 0 & -3 \\ -2 & 6 \end{vmatrix} = (0)(6) - (-2)(-3) = 0 - (+6) = -6$

Practice Problem 2   Find the value of the determinant.

(a) $\begin{vmatrix} -7 & 3 \\ -4 & -2 \end{vmatrix}$ **(b)** $\begin{vmatrix} 5 & 6 \\ 0 & -5 \end{vmatrix}$ ■

The determinants may contain fractions or decimals.

EXAMPLE 3   Evaluate.

(a) $\begin{vmatrix} 0.5 & 0.8 \\ -9.3 & 1.5 \end{vmatrix}$ **(b)** $\begin{vmatrix} \frac{3}{5} & \frac{2}{5} \\ \frac{3}{4} & \frac{1}{10} \end{vmatrix}$

(a) $(0.5)(1.5) - (-9.3)(0.8) = 0.75 + 7.44 = 8.19$

(b) $(\frac{3}{5})(\frac{1}{10}) - (\frac{3}{4})(\frac{2}{5}) = \frac{3}{50} - \frac{3}{10} = \frac{3}{50} - \frac{15}{50} = -\frac{12}{50} = -\frac{6}{25}$

Practice Problem 3   Evaluate.

(a) $\begin{vmatrix} -0.6 & 1.3 \\ 0.5 & 5.1 \end{vmatrix}$ **(b)** $\begin{vmatrix} \frac{1}{6} & \frac{7}{12} \\ -\frac{3}{4} & \frac{1}{4} \end{vmatrix}$ ■

In physics and other areas of applied mathematics, a determinant is sometimes used to define an equation.

EXAMPLE 4   Solve for x in the equation.

$$\begin{vmatrix} 5 & -3 \\ 2x & 3x \end{vmatrix} = 42$$

By the definition of the value of a determinant,

$$(5)(3x) - (2x)(-3) = 42 \quad \textit{Solve for x.}$$

$$15x + 6x = 42$$

$$21x = 42$$

$$x = 2$$

Practice Problem 4   Solve for x in the equation.

$$\begin{vmatrix} x + 2 & 5 \\ x & -4 \end{vmatrix} = 46 \quad ■$$

Evaluating a Third-order Determinant

Third-order determinants have three rows and three columns. Again, each determinant has exactly one value.

Definition

The value of the third-order determinant

$$\begin{vmatrix} a_1 & b_1 & c_1 \\ a_2 & b_2 & c_2 \\ a_3 & b_3 & c_3 \end{vmatrix}$$

is

$$a_1 b_2 c_3 + b_1 c_2 a_3 + c_1 a_2 b_3 - a_3 b_2 c_1 - b_3 c_2 a_1 - c_3 a_2 b_1$$

Because this definition is difficult to memorize and cumbersome to use, we evaluate third-order determinants by a simpler method called **expansion by minors.** The **minor** of an element (number or letter) of a third-order determinant is the second-order determinant that remains after you delete the row and column in which the element appears.

EXAMPLE 5   Find **(a)** the minor of 6 and **(b)** the minor of -3 in the determinant.

$$\begin{vmatrix} 6 & 1 & 2 \\ -3 & 4 & 5 \\ -2 & 7 & 8 \end{vmatrix}$$

(a) Since the element 6 appears in the first row and the first column, we delete them.

$$\begin{vmatrix} 6 & 1 & 2 \\ -3 & 4 & 5 \\ -2 & 7 & 8 \end{vmatrix}$$

Therefore, the minor of 6 is

$$\begin{vmatrix} 4 & 5 \\ 7 & 8 \end{vmatrix}$$

(b) Since -3 appears in the first column and second row, we delete them.

$$\begin{vmatrix} 6 & 1 & 2 \\ -3 & 4 & 5 \\ -2 & 7 & 8 \end{vmatrix}$$

The minor of -3 is

$$\begin{vmatrix} 1 & 2 \\ 7 & 8 \end{vmatrix}$$

Practice Problem 5   Find **(a)** the minor of 3 and **(b)** the minor of -6 in the determinant.

$$\begin{vmatrix} 1 & 2 & 7 \\ -4 & -5 & -6 \\ 3 & 4 & -9 \end{vmatrix}$$ ∎

To evaluate a third-order determinant, we will use expansion by minors of elements in the first column; for example, we have

$$\begin{vmatrix} a_1 & b_1 & c_1 \\ a_2 & b_2 & c_2 \\ a_3 & b_3 & c_3 \end{vmatrix} = a_1 \begin{vmatrix} b_2 & c_2 \\ b_3 & c_3 \end{vmatrix} - a_2 \begin{vmatrix} b_1 & c_1 \\ b_3 & c_3 \end{vmatrix} + a_3 \begin{vmatrix} b_1 & c_1 \\ b_2 & c_2 \end{vmatrix}$$

Note that the signs alternate. We then evaluate the second-order determinant according to our definition.

EXAMPLE 6   Evaluate the determinant $\begin{vmatrix} 2 & 3 & 6 \\ 4 & -2 & 0 \\ 1 & -5 & -3 \end{vmatrix}$ by expanding it by minors of elements in the first column.

$$\begin{vmatrix} 2 & 3 & 6 \\ 4 & -2 & 0 \\ 1 & -5 & -3 \end{vmatrix} = 2\begin{vmatrix} -2 & 0 \\ -5 & -3 \end{vmatrix} - 4\begin{vmatrix} 3 & 6 \\ -5 & -3 \end{vmatrix} + 1\begin{vmatrix} 3 & 6 \\ -2 & 0 \end{vmatrix}$$

$$= 2[(-2)(-3) - (-5)(0)] - 4[(3)(-3) - (-5)(6)] + 1[(3)(0) - (-2)(6)]$$

$$= 2[6 - 0] - 4[-9 - (-30)] + 1[0 - (-12)]$$

$$= 2(6) - 4(21) + 1(12)$$

$$= 12 - 84 + 12$$

$$= -60$$

Practice Problem 6   Evaluate the determinant. $\begin{vmatrix} 1 & 2 & -3 \\ 2 & -1 & 2 \\ 3 & 1 & 4 \end{vmatrix}$ ∎

We may expand by minors about *any* row or column. The sign that precedes any term of the three-term expansion is given by the array

$$\begin{array}{ccc} + & - & + \\ - & + & - \\ + & - & + \end{array}$$

Note that the signs *alternate* for each row and column.

EXAMPLE 7   Use minors about the second row to evaluate.

$$\begin{vmatrix} 3 & -4 & 0 \\ -5 & 2 & 6 \\ -6 & 0 & -2 \end{vmatrix}$$

The products of the three elements in the second row with their minors are

$$5\begin{vmatrix} -4 & 0 \\ 0 & -2 \end{vmatrix} \qquad 2\begin{vmatrix} 3 & 0 \\ -6 & -2 \end{vmatrix} \qquad 6\begin{vmatrix} 3 & -4 \\ -6 & 0 \end{vmatrix}$$

Connecting these products with the signs from the second row of the sign array, we have

$$-(-5)\begin{vmatrix} -4 & 0 \\ 0 & -2 \end{vmatrix} + 2\begin{vmatrix} 3 & 0 \\ -6 & -2 \end{vmatrix} - 6\begin{vmatrix} 3 & -4 \\ -6 & 0 \end{vmatrix}$$

Now we do the computations:

$$+5[(-4)(-2) - (0)(0)] + 2[(3)(-2) - (-6)(0)] - 6[(3)(0) - (-6)(-4)]$$

$$= +5(+8) + 2(-6) - 6(-24) = +40 - 12 + 144$$

$$= 172$$

Practice Problem 7   Evaluate the determinant by minors about the second row.

$$\begin{vmatrix} 5 & 2 & 1 \\ 3 & 0 & -2 \\ -4 & -1 & 2 \end{vmatrix} \quad \blacksquare$$

Often a third-order determinant can be evaluated more quickly if we use a row or column that has at least one zero.

EXAMPLE 8   Evaluate. $\begin{vmatrix} -7 & 8 & 2 \\ 3 & -4 & 0 \\ -2 & 9 & 0 \end{vmatrix}$

If we evaluate by the third column, we will have two of the determinants multiplied by zero. The sign array of the third column is

$$\begin{matrix} + \\ - \\ + \end{matrix}$$

Expanding by the third column gives

$$+2\begin{vmatrix} 3 & -4 \\ -2 & 9 \end{vmatrix} - 0\begin{vmatrix} -7 & 8 \\ -2 & 9 \end{vmatrix} + 0\begin{vmatrix} -7 & 8 \\ 3 & -4 \end{vmatrix}$$

Thus

$$2\begin{vmatrix} 3 & -4 \\ -2 & 9 \end{vmatrix} = 2[(3)(9) - (-2)(-4)] = 2[27 - (+8)] = 2(19) = 38$$

Practice Problem 8   Evaluate the determinant. $\begin{vmatrix} 3 & 6 & -2 \\ -4 & 0 & -1 \\ 5 & 0 & 8 \end{vmatrix} \quad \blacksquare$

Developing Your Study Skills

EXAM TIME: THE NIGHT BEFORE

With adequate preparation, you can spend the night before the exam pulling together the final details.

1. Look over each section to be covered in the exam. Review the steps needed to solve each type of problem.

2. Review your list of terms, rules, and formulas that you are expected to know for the exam.

3. Take the Practice Test at the end of the chapter just as though you were taking the actual exam. Do not look in your text or get help in any way. Time yourself so that you know how long it takes you to complete the test.

4. Check the Practice Test. Redo the problems you missed.

5. Be sure you have ready the necessary supplies for taking your exam.

4.4 Exercises

Evaluate each determinant.

1. $\begin{vmatrix} 5 & 6 \\ 2 & 1 \end{vmatrix}$

2. $\begin{vmatrix} 3 & 4 \\ 1 & 8 \end{vmatrix}$

3. $\begin{vmatrix} 2 & -1 \\ 3 & 6 \end{vmatrix}$

4. $\begin{vmatrix} -4 & 2 \\ 1 & 5 \end{vmatrix}$

5. $\begin{vmatrix} -\frac{1}{2} & -\frac{2}{3} \\ 9 & 8 \end{vmatrix}$

6. $\begin{vmatrix} 10 & 4 \\ -\frac{3}{2} & -\frac{2}{5} \end{vmatrix}$

7. $\begin{vmatrix} -5 & 3 \\ -4 & -7 \end{vmatrix}$

8. $\begin{vmatrix} 2 & -3 \\ -4 & -6 \end{vmatrix}$

9. $\begin{vmatrix} 0 & -6 \\ 3 & -4 \end{vmatrix}$

10. $\begin{vmatrix} -5 & 0 \\ 2 & -7 \end{vmatrix}$

11. $\begin{vmatrix} 2 & -5 \\ -4 & 10 \end{vmatrix}$

12. $\begin{vmatrix} -3 & 6 \\ 7 & -14 \end{vmatrix}$

13. $\begin{vmatrix} 0 & 0 \\ -2 & 6 \end{vmatrix}$

14. $\begin{vmatrix} -4 & 0 \\ -3 & 0 \end{vmatrix}$

15. $\begin{vmatrix} 0.3 & 0.6 \\ 1.2 & 0.4 \end{vmatrix}$

16. $\begin{vmatrix} 0.1 & 0.7 \\ 0.5 & 0.8 \end{vmatrix}$

17. $\begin{vmatrix} 7 & 4 \\ b & -a \end{vmatrix}$

18. $\begin{vmatrix} \frac{1}{4} & \frac{3}{5} \\ \frac{2}{3} & \frac{1}{5} \end{vmatrix}$

19. $\begin{vmatrix} \frac{3}{7} & -\frac{1}{3} \\ -\frac{1}{4} & \frac{1}{2} \end{vmatrix}$

20. $\begin{vmatrix} -3 & y \\ -2 & x \end{vmatrix}$

Solve for x in each equation, where x represents a real number.

21. $\begin{vmatrix} x & 3 \\ 2 & -4 \end{vmatrix} = 5$

22. $\begin{vmatrix} -2 & x \\ 1 & 3 \end{vmatrix} = 2$

23. $\begin{vmatrix} 2x & 3x \\ 1 & 2 \end{vmatrix} = -7$

24. $\begin{vmatrix} -3 & 1 \\ 4x & x \end{vmatrix} = -21$

25. $\begin{vmatrix} x + 2 & 3x - 1 \\ -2 & 4 \end{vmatrix} = 2$

26. $\begin{vmatrix} -3 & 4 \\ x - 1 & 2x + 1 \end{vmatrix} = -3$

In the following determinant $\begin{vmatrix} 3 & -4 & 7 \\ -2 & 6 & 10 \\ 1 & -5 & 9 \end{vmatrix}$,

27. Find the minor of 3.

28. Find the minor of -2.

29. Find the minor of 10.

30. Find the minor of 9.

Evaluate each of the following determinants.

31. $\begin{vmatrix} 4 & 1 & 2 \\ 3 & -1 & 0 \\ 1 & 2 & 3 \end{vmatrix}$

32. $\begin{vmatrix} 2 & 3 & 1 \\ -3 & 1 & 0 \\ 2 & 1 & 4 \end{vmatrix}$

33. $\begin{vmatrix} -4 & 0 & -1 \\ 2 & 1 & -1 \\ 0 & 3 & 2 \end{vmatrix}$

34. $\begin{vmatrix} 3 & -4 & -1 \\ -2 & 1 & 3 \\ 0 & 1 & 4 \end{vmatrix}$

35. $\begin{vmatrix} \frac{1}{2} & 1 & -1 \\ \frac{3}{2} & 1 & 2 \\ 3 & 0 & -2 \end{vmatrix}$

36. $\begin{vmatrix} 1 & 2 & 3 \\ 4 & -2 & -1 \\ 5 & -3 & 2 \end{vmatrix}$

37. $\begin{vmatrix} 4 & 1 & 2 \\ -1 & -2 & -3 \\ 4 & -1 & 3 \end{vmatrix}$

38. $\begin{vmatrix} -\frac{1}{2} & 2 & 3 \\ \frac{5}{2} & -2 & -1 \\ \frac{3}{4} & -3 & 2 \end{vmatrix}$

39. $\begin{vmatrix} 2 & 0 & -2 \\ -1 & 0 & 2 \\ 3 & 4 & 3 \end{vmatrix}$

40. $\begin{vmatrix} 7 & 0 & 2 \\ 1 & 0 & -5 \\ 3 & 0 & 6 \end{vmatrix}$

41. $\begin{vmatrix} 6 & -4 & 3 \\ 1 & 2 & 4 \\ 0 & 0 & 0 \end{vmatrix}$

42. $\begin{vmatrix} 7 & 0 & 3 \\ 1 & 2 & 4 \\ 3 & 0 & -7 \end{vmatrix}$

▦ **Optional Graphing Calculator Problems**

If you have a graphing calculator, use the determinant function to evaluate the following.

43. $\begin{vmatrix} 1.3 & 1.8 & 2.5 \\ 7.9 & 5.3 & 6.0 \\ 1.7 & 1.8 & 2.8 \end{vmatrix}$

44. $\begin{vmatrix} 0.7 & 5.3 & 0.4 \\ 1.6 & 0.3 & 3.7 \\ 0.8 & 6.7 & 4.2 \end{vmatrix}$

45. $\begin{vmatrix} -55 & 17 & 19 \\ -62 & 23 & 31 \\ 81 & 51 & 74 \end{vmatrix}$

46. $\begin{vmatrix} 82 & -20 & 56 \\ 93 & -18 & 39 \\ 65 & -27 & 72 \end{vmatrix}$

To Think About

The area A of a triangle whose vertices are in counterclockwise order (x_1, y_1), (x_2, y_2), and (x_3, y_3) can be found from the formula

$$A = \frac{1}{2} \begin{vmatrix} x_1 & y_1 & 1 \\ x_2 & y_2 & 1 \\ x_3 & y_3 & 1 \end{vmatrix}$$

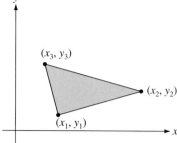

Find the area of each triangle, with the vertices as given (in counterclockwise order).

47. $(0, 1)$, $(3, 2)$, and $(1, 8)$

48. $(-2, 1)$, $(0, 3)$, and $(-1, 6)$

49. $(-1, -1)$, $(3, -6)$, and $(5, 2)$

50. $(-4, -4)$, $(2, -1)$, and $(-2, 3)$

The sign array for a 4 × 4 determinant is

$$\begin{array}{cccc} + & - & + & - \\ - & + & - & + \\ + & - & + & - \\ - & + & - & + \end{array}$$

We can evaluate all determinants of order 4 by the method of expansion by minors, using this sign array. The expansion has four numbers (the elements of a row or column), each multiplied by a third-order determinant. The appropriate sign must be in front of the four numbers. Expand by a row or column with at least one zero when possible. Evaluate the determinant.

51. $\begin{vmatrix} -3 & -2 & 1 & 0 \\ 2 & 1 & -1 & 0 \\ 3 & 4 & 2 & 1 \\ 2 & -1 & 1 & 2 \end{vmatrix}$

52. $\begin{vmatrix} 1 & 3 & 2 & 4 \\ -2 & 1 & 0 & 3 \\ 3 & 0 & 1 & 6 \\ 5 & 2 & 0 & 0 \end{vmatrix}$

If you have a graphing calculator, use the determinant function to evaluate the following.

53.
$$\begin{vmatrix} 123 & 407 & 166 & 358 \\ 728 & 625 & 99 & 818 \\ 236 & 688 & 86 & 909 \\ 158 & 754 & 76 & 661 \end{vmatrix}$$

54.
$$\begin{vmatrix} 0.31 & 0.76 & 1.55 & 0.82 \\ 5.31 & 8.05 & 3.72 & 8.91 \\ 4.98 & 5.77 & 4.03 & 9.33 \\ 0.56 & 0.93 & 6.37 & 7.11 \end{vmatrix}$$

Cumulative Review Problems

Simplify.

55. $5x^2(2x - 3) - 5x(7x^2 - x)$

56. $\dfrac{1}{2}a - 3y + 5z - \dfrac{1}{3}a + 5y - 2z$

57. $(3x^2y^3)(-2x^{-1}y^3)(4x^{-2}yz^2)$

58. $\dfrac{(4x^{-3}y^2)^3}{(3x^2y)^4}$

59. Louise Elton is a lawyer. Last month she collected $5320 in fees from three major clients: Eurovision, Traveland, and Unitech. She used 10% of her fees from Eurovision, 20% of her fees from Traveland, and 25% of her fees from Unitech to pay her monthly office rent of $881. Her fees from Eurovision were equal to the total fees from Traveland and Unitech combined. What was her income from each client last month?

60. Lora Connelly rented a Honda Accord from the local Honda dealer. She was charged a daily rate for the car, a daily rate for collision insurance, and a per mile charge. When she rented the car for 3 days with collision insurance and drove the car 400 miles she was charged $189. When she rented the car for 4 days with collision insurance and drove the car 360 miles she was charged $226. When she rented the car for 5 days without collision insurance and drove 520 miles she was charged $253. What was the dealer's daily rate, daily insurance cost, and the cost per mile to rent the car?

Developing Your Study Skills

PREVIEWING NEW MATERIAL

Part of your study time each day should consist of looking over the sections in your text that are to be covered the following day. You do not necessarily need to study and learn the material on your own, but a survey of the concepts, terminology, diagrams, and examples will help the new ideas seem more familiar as the instructor presents them. You can look for concepts that appear confusing or difficult and be ready to listen carefully for your instructor's explanations. You can be prepared to ask the questions that will increase your understanding. Previewing new material enables you to see what is coming and prepares you to be ready to absorb it.

4.5 Cramer's Rule (Optional)

MathPro

Video 11

SSM

After studying this section, you will be able to:

1 *Solve a system of two linear equations with two unknowns using Cramer's rule.*
2 *Solve a system of three linear equations with three unknowns using Cramer's rule.*

1 *Solving a System of Two Linear Equations with Two Unknowns Using Cramer's Rule*

We can solve a linear system of two equations with two unknowns by Cramer's rule. The rule is named for Gabriel Cramer, a Swiss mathematician who lived from 1704 to 1752. Cramer's rule expresses the solution for each variable of a linear system as the quotient of two determinants. Computer programs are available to solve a system of equations by Cramer's rule.

Cramer's Rule

The solution to

$$a_1 x + b_1 y = c_1$$
$$a_2 x + b_2 y = c_2$$

is

$$x = \frac{D_x}{D} \quad \text{and} \quad y = \frac{D_y}{D}, \qquad D \neq 0$$

where

$$D_x = \begin{vmatrix} c_1 & b_1 \\ c_2 & b_2 \end{vmatrix}, \qquad D_y = \begin{vmatrix} a_1 & c_1 \\ a_2 & c_2 \end{vmatrix}, \qquad D = \begin{vmatrix} a_1 & b_1 \\ a_2 & b_2 \end{vmatrix}$$

Cramer's rule may look complicated, but it soon becomes easy. Notice that D is just the determinant of the coefficients of the variables.

EXAMPLE 1 Solve by Cramer's rule.

$$-x + 2y = 9$$
$$5x + 3y = 7$$

The determinant of coefficients is

$$D = \begin{vmatrix} -1 & 2 \\ 5 & 3 \end{vmatrix} \qquad \begin{aligned} -x + 2y &= 9 \\ 5x + 3y &= 7 \end{aligned}$$

$$= (-1)(3) - (5)(2)$$

$$= -3 - 10$$

$$= -13$$

For the determinant of the x variable, just replace the x coefficients with the values on the right side of the equals sign, and use the coefficients of y.

$$D_x = \begin{vmatrix} 9 & 2 \\ 7 & 3 \end{vmatrix} \qquad \begin{aligned} -x + 2y &= 9 \\ 5x + 3y &= 7 \end{aligned}$$

$$= (9)(3) - (7)(2)$$

$$= 27 - 14$$

$$= 13$$

Likewise, for D_y, replace the y coefficients with the right-side values.

$$D_y = \begin{vmatrix} -1 & 9 \\ 5 & 7 \end{vmatrix} \qquad \begin{aligned} -x + 2y &= 9 \\ 5x + 3y &= 7 \end{aligned}$$

$$= (-1)(7) - (5)(9)$$

$$= -7 - 45$$

$$= -52$$

According to Cramer's rule,

$$x = \frac{D_x}{D} = \frac{13}{-13} = -1$$

$$y = \frac{D_y}{D} = \frac{-52}{-13} = 4$$

Thus the solution is $x = -1$ and $y = 4$.

Check: It is necessary to check the values of x and y by substitution into both equations

$$-x + 2y = 9$$

$$-(-1) + 2(4) \overset{?}{=} 9$$

$$1 + 8 \overset{?}{=} 9$$

$$9 = 9 \ \checkmark$$

$$5x + 3y = 7$$

$$5(-1) + 3(4) \overset{?}{=} 7$$

$$-5 + 12 \overset{?}{=} 7$$

$$7 = 7 \ \checkmark$$

Practice Problem 1 Find the solution to the system by Cramer's rule.

$$-2x + \ y = \ \ 8$$

$$3x + 4y = -1 \ \blacksquare$$

EXAMPLE 2 Solve by Cramer's rule.

$$-3x + y = 7$$
$$-4x - 3y = 5$$

$$D = \begin{vmatrix} -3 & 1 \\ -4 & -3 \end{vmatrix} \qquad D_x = \begin{vmatrix} 7 & 1 \\ 5 & -3 \end{vmatrix} \qquad D_y = \begin{vmatrix} -3 & 7 \\ -4 & 5 \end{vmatrix}$$

$$= (-3)(-3) - (-4)(1) \qquad = (7)(-3) - (5)(1) \qquad = (-3)(5) - (-4)(7)$$

$$= 9 - (-4) \qquad\qquad = -21 - 5 \qquad\qquad = -15 - (-28)$$

$$= 9 + 4 \qquad\qquad = -26 \qquad\qquad = -15 + 28$$

$$= 13 \qquad\qquad\qquad\qquad\qquad\qquad\qquad = 13$$

Hence

$$x = \frac{D_x}{D} = \frac{-26}{13} = -2$$

$$y = \frac{D_y}{D} = \frac{13}{13} = 1$$

The solution to the system is $x = -2$ and $y = 1$. Verify this.

Practice Problem 2 Solve by Cramer's rule.

$$5x + 3y = 17$$
$$2x - 5y = 13 \quad \blacksquare$$

2 *Solving a System of Three Linear Equations with Three Unknowns Using Cramer's Rule*

It is quite easy to extend Cramer's rule to three linear equations.

Cramer's Rule

The solution to the system

$$a_1x + b_1y + c_1z = d_1$$
$$a_2x + b_2y + c_2z = d_2$$
$$a_3x + b_3y + c_3z = d_3$$

is

$$x = \frac{D_x}{D}, \qquad y = \frac{D_y}{D}, \qquad z = \frac{D_z}{D}, \qquad D \neq 0$$

where

$$D = \begin{vmatrix} a_1 & b_1 & c_1 \\ a_2 & b_2 & c_2 \\ a_3 & b_3 & c_3 \end{vmatrix}, \qquad D_x = \begin{vmatrix} d_1 & b_1 & c_1 \\ d_2 & b_2 & c_2 \\ d_3 & b_3 & c_3 \end{vmatrix}$$

$$D_y = \begin{vmatrix} a_1 & d_1 & c_1 \\ a_2 & d_2 & c_2 \\ a_3 & d_3 & c_3 \end{vmatrix}, \qquad D_z = \begin{vmatrix} a_1 & b_1 & d_1 \\ a_2 & b_2 & d_2 \\ a_3 & b_3 & d_3 \end{vmatrix}$$

EXAMPLE 3 Use Cramer's rule to solve the system.

$$2x - y + z = 6$$
$$3x + 2y - z = 5$$
$$2x + 3y - 2z = 1$$

For consistency, we expand all determinants by the third row. (We could use any row or column.)

$$D = \begin{vmatrix} 2 & -1 & 1 \\ 3 & 2 & -1 \\ 2 & 3 & -2 \end{vmatrix} = 2 \begin{vmatrix} -1 & 1 \\ 2 & -1 \end{vmatrix} - 3 \begin{vmatrix} 2 & 1 \\ 3 & -1 \end{vmatrix} - 2 \begin{vmatrix} 2 & -1 \\ 3 & 2 \end{vmatrix}$$

$$= 2[+1 - (2)] - 3[-2 - (3)] - 2[4 - (-3)]$$
$$= 2[-1] - 3[-5] - 2[7] = -2 + 15 - 14 = -1$$

$$D_x = \begin{vmatrix} 6 & -1 & 1 \\ 5 & 2 & -1 \\ 1 & 3 & -2 \end{vmatrix} = 1 \begin{vmatrix} -1 & 1 \\ 2 & -1 \end{vmatrix} - 3 \begin{vmatrix} 6 & 1 \\ 5 & -1 \end{vmatrix} - 2 \begin{vmatrix} 6 & -1 \\ 5 & 2 \end{vmatrix}$$

$$= 1[+1 - (2)] - 3[-6 - (5)] - 2[12 - (-5)]$$
$$= 1[-1] - 3[-11] - 2[17] = -1 + 33 - 34 = -2$$

$$D_y = \begin{vmatrix} 2 & 6 & 1 \\ 3 & 5 & -1 \\ 2 & 1 & -2 \end{vmatrix} = 2 \begin{vmatrix} 6 & 1 \\ 5 & -1 \end{vmatrix} - 1 \begin{vmatrix} 2 & 1 \\ 3 & -1 \end{vmatrix} - 2 \begin{vmatrix} 2 & 6 \\ 3 & 5 \end{vmatrix}$$

$$= 2[-6 - (5)] - 1[-2 - (3)] - 2[10 - (18)]$$
$$= 2[-11] - 1[-5] - 2[-8] = -22 + 5 + 16 = -1$$

$$D_z = \begin{vmatrix} 2 & -1 & 6 \\ 3 & 2 & 5 \\ 2 & 3 & 1 \end{vmatrix} = 2 \begin{vmatrix} -1 & 6 \\ 2 & 5 \end{vmatrix} - 3 \begin{vmatrix} 2 & 6 \\ 3 & 5 \end{vmatrix} + 1 \begin{vmatrix} 2 & -1 \\ 3 & 2 \end{vmatrix}$$

$$= 2[-5 - (12)] - 3[10 - (18)] + 1[4 - (-3)]$$
$$= 2[-17] - 3[-8] + 1[7] = -34 + 24 + 7 = -3$$

$$x = \frac{D_x}{D} = \frac{-2}{-1} = 2 \qquad y = \frac{D_y}{D} = \frac{-1}{-1} = 1 \qquad z = \frac{D_z}{D} = \frac{-3}{-1} = 3$$

Check: Verify this solution.

Practice Problem 3 Find the solution to the system by Cramer's rule.

$$2x + 3y - z = -1$$
$$3x + 5y - 2z = -3$$
$$x + 2y + 3z = 2 \quad \blacksquare$$

GRAPHING CALCULATOR

Copying Matrices

If you are using a graphing calculator to evaluate the four determinants in Example 3 or similar problems, you should first enter matrix D into the calculator. Then copy the matrix using the copy function to three additional locations. Usually we store matrix D as matrix A. Then store a copy of it as matrix B, C, and D. Finally you use the edit function and modify one column of each of the matrices B, C, and D so that they become D_x, D_y, and D_z. This allows you to evaluate all four determinants in a minimum amount of time.

Cramer's rule is quite effective when we want the value of only one variable.

EXAMPLE 4 Use Cramer's rule to find the value of y in the system.

$$3x - y + 2z = 1$$
$$x \qquad + 4z = -1$$
$$3x - 2y \qquad = -1$$

First we write the system in the complete form, showing coefficients of 0.

$$3x - y + 2z = 2$$
$$x + 0y + 4z = -1$$
$$3x - 2y + 0z = -1$$

We will expand each determinant by the third row.

$$D = \begin{vmatrix} 3 & -1 & 2 \\ 1 & 0 & 4 \\ 3 & -2 & 0 \end{vmatrix} = 3\begin{vmatrix} -1 & 2 \\ 0 & 4 \end{vmatrix} + (+2)\begin{vmatrix} 3 & 2 \\ 1 & 4 \end{vmatrix} + 0\begin{vmatrix} 3 & -1 \\ 1 & 0 \end{vmatrix}$$

$$= 3[-4 - 0] + 2[12 - 2] + 0$$
$$= -12 + 20 = 8$$

$$D_y = \begin{vmatrix} 3 & 2 & 2 \\ 1 & -1 & 4 \\ 3 & -1 & 0 \end{vmatrix} = 3\begin{vmatrix} 2 & 2 \\ -1 & 4 \end{vmatrix} + (+1)\begin{vmatrix} 3 & 2 \\ 1 & 4 \end{vmatrix} + 0\begin{vmatrix} 3 & 2 \\ 1 & -1 \end{vmatrix}$$

$$= 3[8 - (-2)] + 1[12 - (2)] + 0 = 30 + 10 = 40$$

Thus

$$y = \frac{D_y}{D} = \frac{40}{8} = 5$$

Practice Problem 4 Use Cramer's rule to find the value of x in the system.

$$2x - 3y + 2z = -1$$
$$x + 2y \qquad = 14$$
$$-x \qquad + 3z = 5 \quad \blacksquare$$

Cramer's rule cannot be used for every system of linear equations. If the equations are dependent or if the system of equations is inconsistent, then the determinant of coefficients will be zero. Division by zero is not defined. In such a situation the system will not have a unique answer.

If $D = 0$, then the following are true:

1. If $D_x = 0$ and $D_y = 0$ (and $D_z = 0$, if there are three equations), then the equations are *dependent*. Such a system will have an infinite number of solutions.

2. If at least one of D_x or D_y is nonzero (and D_z if there are three equations), then the system of equations is *inconsistent*. Such a system will have no solution.

Math Exploration Topic

There is an additional method for solving systems of equations using matrix operations. This method is very efficient when computers or graphing calculators are used. It is located in Appendix B.

4.5 Exercises

Solve each system by Cramer's rule.

1. $x + 2y = 8$
   $2x + y = 7$

2. $x + 3y = 6$
   $2x + y = 7$

3. $5x + 4y = 10$
   $-x + 2y = 12$

4. $3x + 5y = 11$
   $2x + y = -2$

5. $x - 5y = 0$
   $x + 6y = 22$

6. $x - 3y = 4$
   $-3x + 4y = -12$

7. $4x + y = 0$
   $-2x - 5y = 9$

8. $9x - 7y = 6$
   $3x - 2y = 2$

9. $3x - 5y = 2$
   $6x - 11y = 5$

10. $-x - 3y = -14$
    $5x - 2y = 2$

11. $0.3x + 0.5y = 0.2$
    $0.1x + 0.2y = 0.0$

12. $0.5x + 0.3y = -0.7$
    $0.4x + 0.5y = -0.3$

Solve by Cramer's rule. Round your answers to four decimal places.

13. $52.9634x - 27.3715y = 86.1239$
    $31.9872x + 61.4598y = 44.9812$

14. $0.0076x + 0.0092y = 0.01237$
    $-0.5628x - 0.2374y = -0.7635$

In the following problems, clear parentheses. Then write each system so that both equations are in the form Ax + By = C. Then solve by Cramer's rule.

15. $1 + 2(x + 3) + y = 13$
$x - y + 3(y + 1) = -6$

16. $3(x + 1) = 4 - y$
$5 + 2(y - 1) = 3 + x$

17. $x = -\dfrac{5}{3}y - \dfrac{2}{3}$
$\dfrac{1}{5}y - \dfrac{2}{5}x = 2$

18. $x - \dfrac{5}{4}y = \dfrac{17}{4}$
$\dfrac{3}{10}x - \dfrac{1}{10}y = 1$

19. $\dfrac{1}{2}x + \dfrac{1}{3}y = 3$
$\dfrac{1}{8}y + \dfrac{1}{2} = \dfrac{3(x + 1)}{8} + \dfrac{1}{8}$

20. $\dfrac{1}{2}x - \dfrac{3}{8}y + \dfrac{1}{8} = 1$
$2(x + 1) - \dfrac{2}{3}y = x + 4$

Solve each system by Cramer's rule.

21. $2x + y + z = 4$
$x - y - 2z = -2$
$x + y - z = 1$

22. $x + 2y - z = -4$
$x + 4y - 2z = -6$
$2x + 3y + z = 3$

23. $2x + 2y + 3z = 6$
$x - y + z = 1$
$3x + y + z = 1$

24. $4x + y + 2z = 6$
$x + y + z = 1$
$-x + 3y - z = -5$

25. $x + 2y + z = 1$
$3x \quad - 4z = 8$
$3y + 5z = -1$

26. $3x + y + z = 2$
$2y + 3z = -6$
$2x - y \quad = -1$

27. $2x - y + z = -2$
$4x \quad + 5z = -2$
$3y + 4z = 2$

28. $2x + 5y + 3z = 1$
$x + 3y \quad = 8$
$4y - z = 5$

🖩 Optional Graphing Calculator Problems

Round your answers to the nearest thousandth.

29. $10x + 20y + 10z = -2$
$-24x - 31y - 11z = -12$
$61x + 39y + 28z = -45$

30. $121x + 134y + 101z = 146$
$315x - 112y - 108z = 426$
$148x + 503y + 516z = -127$

Solve only for the variable specified, using Cramer's rule.

31. Solve for z.

$$2x + y + 2z = -1$$
$$-x - 2y + 4z = -5$$
$$x + 8y + 8z = -1$$

32. Solve for z.

$$6x - 5y - 3z = -5$$
$$-2x + y + z = 1$$
$$3x - 2y + 3z = 1$$

33. Solve for y.

$$2x + 3y + 5z = -1$$
$$2x + y + z = 5$$
$$2y + 5z = -4$$

34. Solve for x.

$$x + 3y + 2z = 3$$
$$2x - 5y + z = 7$$
$$y + z = -3$$

Using the approach specified by Cramer's rule, find the value of each determinant. How many solutions are there for such a system of equations? What kind of a system of equations do we have in this case?

35.
$$2x - 7y = 7$$
$$-4x + 14y = 3$$

36.
$$-16x - 14y = 8$$
$$8x + 7y = -4$$

37.
$$2x - y - 3z = 1$$
$$x + 2y + 4z = 3$$
$$4x - 2y - 6z = 2$$

38.
$$3x + 2y + z = 7$$
$$x + y - z = 2$$
$$3x + 2y + z = 5$$

To Think About

Cramer's rule can be expanded to a system of four linear equations in four unknowns. The unique solution can be found if D ≠ 0. As with three equations, we need to find

$$w = \frac{D_w}{D}, \qquad x = \frac{D_x}{D}, \qquad y = \frac{D_y}{D}, \qquad z = \frac{D_z}{D}$$

by first finding the value of each of the fourth-order determinants: D, D_w, D_x, D_y, D_z

39. In the following system, use Cramer's rule to find z.

$$w + x + y + z = -1$$
$$2w + x + 2y + 3z = -3$$
$$4w + 3x + 4y + 2z = 4$$
$$w - x - y + 2z = -2$$

40. Solve for w, x, y, and z. *Round your answers to the nearest thousandth.*

$$28w + 35x - 18y + 40z = 60$$
$$60w + 32x + 28y = 400$$
$$30w + 15x + 18y + 66z = 720$$
$$26w - 18x - 15y + 75z = 125$$

Cumulative Review Problems

41. A rectangular parking lot has a perimeter of 100 meters. The length is 4 meters longer than twice the width. Find the dimensions of the parking lot.

42. Melinda rented a station wagon for five days from Rent a Wreck. She was charged $31 per day and 16¢ per mile. She paid $185.40 to the rental agency. How many miles did she drive?

43. Rita invested $4000 at Bank Boston. She invested part of the money at 6% and part at 8%. In one year she earned $292.00 in interest. How much did she invest at each rate?

44. Two points that lie on the line $y = ax + b$ are $(10, 22)$ and $(-5, -23)$. Find the values of a and b.

4.6 Systems of Linear Inequalities

MathPro   Video 11   SSM

After studying this section, you will be able to:

1 *Graph a system of linear inequalities.*

1 Graphing a System of Linear Inequalities

We have previously learned in Section 3.4 how to graph a linear inequality. We now consider how to graph a system of linear inequalities. The solution to a system of inequalities is the intersection of the solution sets of individual inequalities.

EXAMPLE 1   Graph the solution of

$$y \leq -3x + 2$$

$$-2x + y \geq -1$$

In this example, we will first graph each inequality separately. The graph of $y \leq -3x + 2$ is the region below the line $y = -3x + 2$ and the line itself.

The graph of $-2x + y \geq -1$ consists of the graph of the region above the line $-2x + y = -1$ and the line itself.

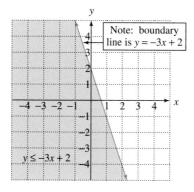

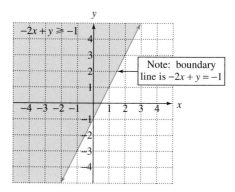

We will now place these graphs on one rectangular axis. The darker shaded region is the intersection of the two graphs. Thus the solution to the system of two inequalities is the darker shaded area and also the boundary lines that satisfy both inequalities.

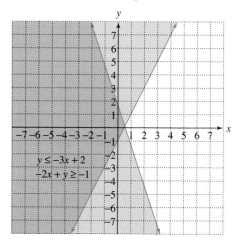

Practice Problem 1   Graph the solution of:

$$-2x + y \le -3$$
$$x + 2y \ge 4 \quad \blacksquare$$

PRACTICE PROBLEM 1

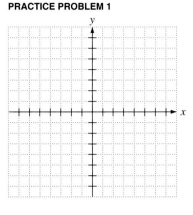

In general we do not use totally separate graphs, but rather do each sketch on one set of axes. We will illustrate that concept with the following example.

EXAMPLE 2   Graph the solution of:

$$y < 4$$
$$y > \tfrac{3}{2}x - 2$$

The region $y < 4$ is the area below the line $y = 4$. It does not include the line since we have the $<$ symbol. We thus use a dashed line to indicate that the boundary line is not part of the answer. The region $y > \tfrac{3}{2}x - 2$ is above the line $y = \tfrac{3}{2}x - 2$. Again we use the dashed line to indicate that the boundary line is not part of the answer. The final solution is the darkly shaded area. The solution does *not* include the dashed boundary lines.

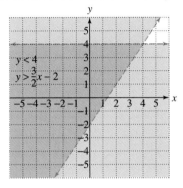

Practice Problem 2   Graph the solution of: $y > -1$

$$y < -\tfrac{3}{4}x + 2 \quad \blacksquare$$

PRACTICE PROBLEM 2

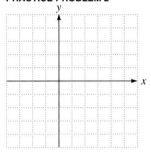

There are times when we require the exact location of the point where two boundary lines intersect. In these cases the boundary points are labeled on the final sketch of the solution.

EXAMPLE 3   Graph the solution to the following system of inequalities. Find the coordinates of any point where boundary lines intersect.

$$x + y \le 5$$
$$x + 2y \le 8$$
$$x \ge 0$$
$$y \ge 0$$

The region $x + y \le 5$ is the region under the line $x + y = 5$ and the line itself. The region $x + 2y \le 8$ is the region under the line $x + 2y = 8$ and the line itself. We can solve the system containing the equations $x + y = 5$ and $x + 2y = 8$ and find that their point of intersection is $(2, 3)$. The region $x \ge 0$ is the y-axis and all the region to the right of the y-axis. The region $y \ge 0$ is the x-axis and all the region above the x-axis. Thus the solution to the system is the shaded region, including the boundary lines that border the shaded region. The coordinates of the points where the boundary lines of the final solution intersect are called the **vertices.** Thus the vertices of the solution are $(0, 0)$, $(0, 4)$, $(2, 3)$, and $(5, 0)$.

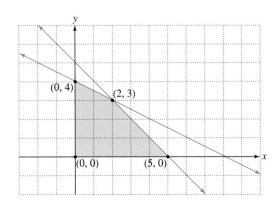

Practice Problem 3 Graph the solution to the following system of inequalities. Find the coordinates of any point where boundary lines intersect.

$$x + y \leq 6$$
$$3x + y \leq 12$$
$$x \geq 0$$
$$y \geq 0$$

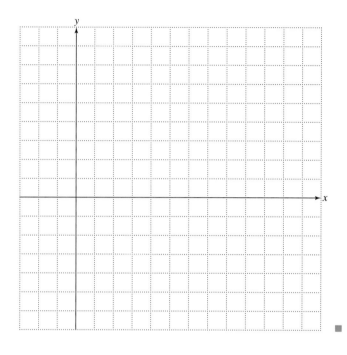

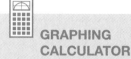
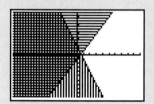
Developing Your Study Skills

EXAM TIME: HOW TO REVIEW

Reviewing adequately for an exam enables you to bring together the concepts you have learned over several sections. For your review, you will need to:

1. Reread your textbook. Make a list of any terms, rules, or formulas you need to know for the exam. Be sure you understand them all.

2. Reread your notes. Go over returned homework and quizzes. Redo the problems you missed.

3. Practice some of each type of problem covered in the chapter(s) you are to be tested on.

4. Use the end-of-chapter materials provided in your textbook. Read carefully through the Chapter Organizer. Do the Extra Practice sections. Take the Chapter Test. When you are finished, check your answers. Redo any problems you missed.

5. Get help if any concepts give you difficulty.

4.6 Exercises

Graph the solution for each of the following systems.

1. $y \geq 2x - 1$
$\quad x + y \leq 6$

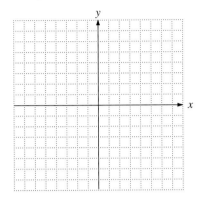

2. $y \leq 3x - 3$
$\quad 5x + 2y \geq 10$

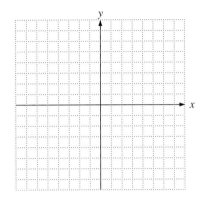

3. $y \geq -4x$
$\quad y \geq 3x - 2$

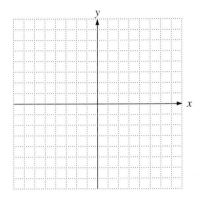

4. $y \geq x$
$\quad y \geq -x$

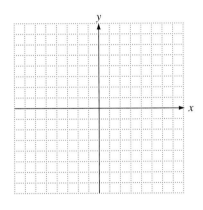

5. $y \geq 2x - 3$
$\quad y \leq \dfrac{2}{3}x$

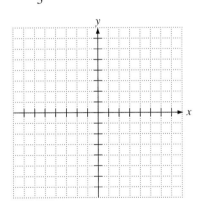

6. $y \leq 3x - 2$
$\quad y \geq \dfrac{1}{2}x$

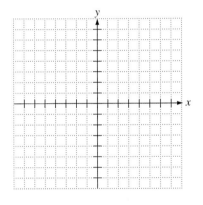

7. $\quad x - y \geq -1$
$\quad -3x - y \leq \quad 4$

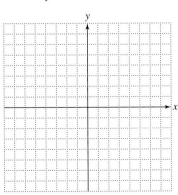

8. $2x + \quad y \leq 3$
$\quad x - 2y \leq 4$

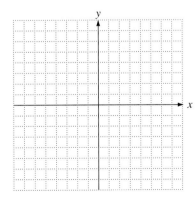

9. $x + 2y < 6$
$\qquad y < 3$

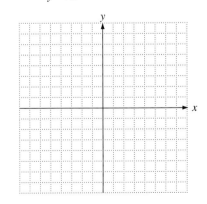

10. $2x + y < 8$
$y < 4$

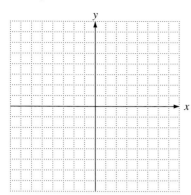

11. $3x + y > 6$
$x > 0$

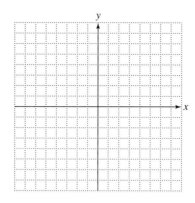

12. $2x - y > 4$
$x < 1$

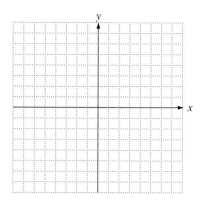

13. $y < \quad 4$
$x > -2$

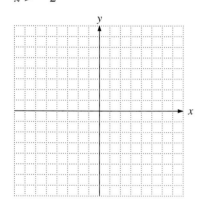

14. $y > -3$
$x < \quad 2$

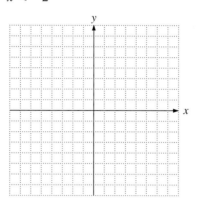

15. $3x + 2y < \quad 6$
$3x + 2y > -6$

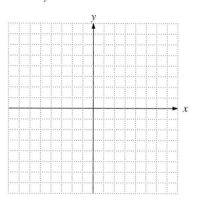

16. $2x - y < \quad 2$
$2x - y > -2$

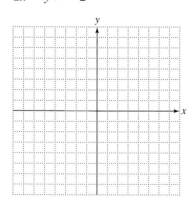

17. $\quad 2x - 4y \leq 8$
$-2x + \quad y \leq 2$

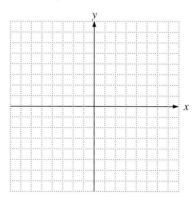

18. $5x - 2y \leq \quad 10$
$x - \quad y \geq -1$

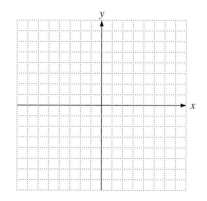

Graph the solution to the following systems of inequalities. Find the coordinates of any point where boundary lines intersect.
Hint: In Exercises 25 and 26, if the coordinates of a boundary point contain fractions, it is wise to obtain the point of intersection algebraically rather than depending solely on the graph.

19. $x + y \leq 4$
$3x + y \leq 6$

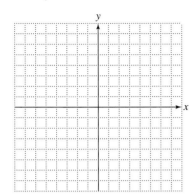

20. $x + y \leq 3$
$4x + y \leq 6$

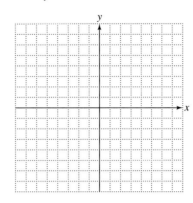

21. $x + 3y \leq 12$
$y < x$

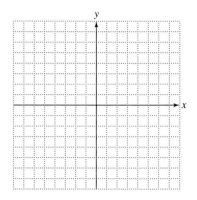

22. $x + 2y \leq 4$
$y < -x$

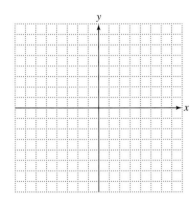

23. $x + y \geq 1$
$x - y \geq 1$
$x \geq 3$

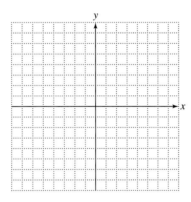

24. $x - y \leq \ \ 2$
$x + y \leq \ \ 2$
$x \geq -2$

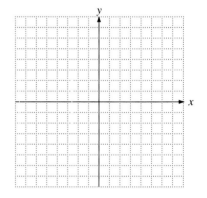

★ **25.** $2x - \ \ y \leq 2$
$2x + 3y \geq 6$
$x \geq 0$
$y \geq 0$

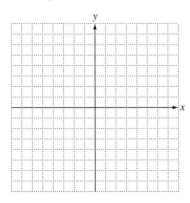

★ **26.** $3x + 2y \geq 4$
$x - \ \ y \leq 3$
$x \geq 0$
$y \geq 0$

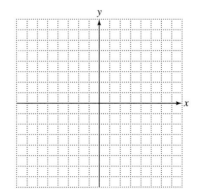

To Think About

Graph the region determined by each of the following systems

27. $y \le 3x + 6$
  $4y + 3x \le 3$
  $x \ge -2$
  $y \ge -3$

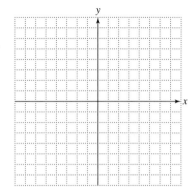

28. $x + y \le 100$
  $x + 3y \le 150$
  $x \ge 0$
  $y \ge 20$

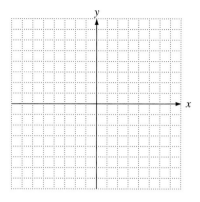

Cumulative Review

29. Find the slope of the line passing through $(3, -4)$ and $(-1, -2)$.

30. What is the slope and the y-intercept of the line defined by the equation $3x + 4y = -8$?

31. Find the equation of a line passing through $(2, 6)$ and $(-2, 1)$. Write the equation in standard form.

32. Find the equation of a line passing through $(5, 0)$ that is parallel to the line $y + 2x = 8$.

33. The Cape Cod Cinema took in $23,400 when showing films on 2 rainy days and 5 sunny days. The next week the cinema took in $25,800 on 4 rainy days and 3 sunny days. What is the average amount of money taken in when a film is shown at the Cape Cod Cinema on a rainy day? On a sunny day?

34. Illinois is establishing a number of new bicycle trails throughout the state. When a group of volunteers worked 3 days and a group of experienced professionals worked 4 days they were able to establish 389 feet of bicycle trails. When the same group of volunteers worked 5 days and the same group of experienced professionals worked 7 days they were able to establish 670 feet of bicycle trails. How many feet per day are established by the group of volunteers? By the group of experienced professionals?

35. Hector sells televisions at a local department store. He earns $200 per week plus a commission of 5% on his total sales. His brother Fernando sells automobile tires at a tire store. His salary is $100 per week plus a commission of 8% on his total sales. After they had both worked several weeks they discovered that they had each sold the same dollar value in sales. However, their total earnings were quite different. Hector had earned $7400 in the same amount of time that Fernando had earned $9200. How many weeks had they each worked? What was the dollar value of the amount of sales that each had made?

36. Two weeks ago Larry went for his office to Nick's Roast Beef and bought 3 roast beef sandwiches, 2 orders of french fries, and 3 sodas for $13.85. One week ago it was Alice's turn to take the trip to Nick's. She bought 4 roast beef sandwiches, 3 orders of french fries, and 5 sodas for $20.00. This week Roberta made the trip. She purchased 3 roast beef sandwiches, 3 french fries, and 4 sodas for $16.55. What is the cost of one roast beef sandwich? Of one order of french fries? Of one soda?

Topic	Procedure	Examples
Finding a solution to a system of equations by the graphing method, p. 201.	1. Determine the graph of the first equation. 2. Determine the graph of the second equation. 3. Approximate from your graph the point where the two lines intersect.	Solve the system by graphing. $$x + y = 6$$ $$2x - y = 6$$ We graph each line and determine from our sketch that the lines intersect at (4, 2). The solution is (4, 2) or $x = 4$, $y = 2$.
Solving a system of two linear equations by the substitution method, p. 202.	The substitution method is most appropriate when *at least one variable has a coefficient of 1 or −1.* 1. Solve for one variable in one of the equations. 2. In the other equation, replace that variable with the expression you obtained in step 1. 3. Solve the resulting equation. 4. Substitute the numerical value you obtain for a variable into the equation you found in step 1. 5. Solve this equation to find the other variable.	Solve. $2x + y = 11$ ① $x + 3y = 18$ ② $y = 11 - 2x$ from equation ①. Substitute this into equation ②. $$x + 3(11 - 2x) = 18$$ $$x + 33 - 6x = 18$$ $$-5x = -15$$ $$x = 3$$ Substitute $x = 3$ into $y = 11 - 2x$. $$y = 11 - 2(3) = 11 - 6 = 5$$ The solution is $x = 3$, $y = 5$.
Solving a system of two linear equations by the addition method, p. 205.	The addition method is most appropriate when the variables *all have coefficients other than 1 or −1.* 1. Multiply one or both equations by appropriate numerical values so that when the two resulting equations are added one variable is eliminated. 2. Solve the resulting equation. 3. Substitute the numerical value you obtain for the variable in one of the original equations. 4. Solve this equation to find the other variable.	Solve. $2x + 3y = 5$ ① $-3x - 4y = -2$ ② Multiply equation 1 by 3 and equation 2 by 2. $$6x + 9y = 15$$ $$\underline{-6x - 8y = -4}$$ $$y = 11$$ Substitute $y = 11$ into equation ①. $$2x + 3(11) = 5$$ $$2x + 33 = 5$$ $$2x = -28$$ $$x = -14$$ The solution is $x = -14$, $y = 11$.

Solving a system of three linear equations by algebraic methods, p. 214.	If there is one solution to a system of three linear equations in three unknowns, it may be obtained in the following manner.
	1. Choose two equations from the system.
	2. Multiply one or both of the equations by the appropriate constants so that by adding the two equations together one variable can be eliminated.
	3. Choose a *different* pair of the three original equations and eliminate the *same* variable using the procedure of step 2.
	4. Solve the system formed by the two equations resulting from steps 2 and 3 for both variables.
	5. Substitute the two values obtained in step 4 into one of the original three equations to find the third variable.
		Solve. $2x - y - 2z = -1$ ①
		       $x - 2y - z = 1$ ②
		       $x + y + z = 4$ ③
		Add equations 2 and 3 together to eliminate z.
		$$2x - y = 5 \quad ④$$
		Multiply equation 3 by 2 and add to equation 1.
		$$4x + y = 7 \quad ⑤$$
		Add equations 4 and 5.
		    $2x - y = 5$ ④
		    $\underline{4x + y = 7}$ ⑤
		      $6x = 12$
		       $x = 2$

Topic	Procedure	Examples
Solving a system of three linear equations by algebraic methods *(continued)*		Substitute $x = 2$ into $\boxed{5}$. $$4(2) + y = 7$$ $$y = -1$$ Substitute $x = 2$, $y = -1$ into $\boxed{3}$. $$2 + (-1) + z = 4$$ $$z = 3$$ The solution is $x = 2$, $y = -1$, $z = 3$.
Inconsistent system of equations, p. 208.	If there is *no solution* to a system of linear equations, the system of equations is inconsistent. When you try to solve an inconsistent system, you obtain an equation that is not true, like $0 = 5$.	Attempt to solve the system. $$4x + 3y = 10 \quad \boxed{1}$$ $$-8x - 6y = 5 \quad \boxed{2}$$ Multiply equation $\boxed{1}$ by 2 and add to equation $\boxed{2}$. $$8x + 6y = 20$$ $$\underline{-8x - 6y = 5}$$ $$0 = 25$$ But $0 \neq 25$; there is no solution. The system of equations is inconsistent.
Dependent equations, p. 208.	If there are an *infinite number of solutions* to a system of linear equations, at least one pair of equations is dependent. When you try to solve a system that contains dependent equations, you will obtain a math equation that is always true (such as $0 = 0$ or $3 = 3$). These equations are called identities.	Attempt to solve the system. $$x - 2y = -5 \quad \boxed{1}$$ $$-3x + 6y = 15 \quad \boxed{2}$$ Multiply equation $\boxed{1}$ by 3 and add to equation $\boxed{2}$. $$3x - 6y = -15$$ $$\underline{-3x + 6y = 15}$$ $$0 = 0$$ There are an infinite number of solutions. The equations are dependent.
Cramer's rule for a linear system containing two equations with two unknowns, p. 242.	The solution to $$ax + by = c$$ $$dx + ey = f$$ is $$x = \frac{D_x}{D} \quad \text{and} \quad y = \frac{D_y}{D}, \qquad \text{if } D \neq 0$$ where $$D_x = \begin{vmatrix} c & b \\ f & e \end{vmatrix}, \qquad D_y = \begin{vmatrix} a & c \\ d & f \end{vmatrix}, \qquad D = \begin{vmatrix} a & b \\ d & e \end{vmatrix}$$	Solve by Cramer's rule. $\quad 2x - y = 7$ $\qquad\qquad\qquad\qquad\quad 3x + 4y = 5$ $$D = \begin{vmatrix} 2 & -1 \\ 3 & 4 \end{vmatrix} = (2)(4) - (3)(-1) = 11$$ $$D_x = \begin{vmatrix} 7 & -1 \\ 5 & 4 \end{vmatrix} = (7)(4) - (5)(-1) = 33$$ $$D_y = \begin{vmatrix} 2 & 7 \\ 3 & 5 \end{vmatrix} = (2)(5) - (3)(7) = -11$$ $$x = \frac{33}{11} = 3, \qquad y = \frac{-11}{11} = -1$$
Cramer's rule for a linear system containing three equations with three unknowns, p. 244.	The solution to the system $$a_1x + b_1y + c_1z = d_1$$ $$a_2x + b_2y + c_2z = d_2$$ $$a_3x + b_3y + c_3z = d_3$$ is given by $$x = \frac{D_x}{D}, \qquad y = \frac{D_y}{D}, \qquad z = \frac{D_z}{D}, \qquad D \neq 0$$ $$D = \begin{vmatrix} a_1 & b_1 & c_1 \\ a_2 & b_2 & c_2 \\ a_3 & b_3 & c_3 \end{vmatrix}, \quad D_x = \begin{vmatrix} d_1 & b_1 & c_1 \\ d_2 & b_2 & c_2 \\ d_3 & b_3 & c_3 \end{vmatrix}$$ $$D_y = \begin{vmatrix} a_1 & d_1 & c_1 \\ a_2 & d_2 & c_2 \\ a_3 & d_3 & c_3 \end{vmatrix}, \quad D_z = \begin{vmatrix} a_1 & b_1 & d_1 \\ a_2 & b_2 & d_2 \\ a_3 & b_3 & d_3 \end{vmatrix}$$	Solve by Cramer's rule. $\quad 3x - y - 2z = 1$ $\qquad\qquad\qquad\qquad\quad 2x - y + 2z = 8$ $\qquad\qquad\qquad\qquad\quad\; x + 2y + 3z = 3$ We will expand the determinants by the first column. $$D = \begin{vmatrix} 3 & -1 & -2 \\ 2 & -1 & 2 \\ 1 & 2 & 3 \end{vmatrix}$$ $$= (3)\begin{vmatrix} -1 & 2 \\ 2 & 3 \end{vmatrix} + (-2)\begin{vmatrix} -1 & -2 \\ 2 & 3 \end{vmatrix} + (1)\begin{vmatrix} -1 & -2 \\ -1 & 2 \end{vmatrix}$$ $$= (3)[-3 - 4] + (-2)[-3 + 4] + (1)[-2 - 2]$$ $$= -21 + (-2) + (-4) = -27$$

Topic	Procedure	Examples
Cramer's rule for three equations *(continued)*		$D_x = \begin{vmatrix} 1 & -1 & -2 \\ 8 & -1 & 2 \\ 3 & 2 & 3 \end{vmatrix}$ $= (1)\begin{vmatrix} -1 & 2 \\ 2 & 3 \end{vmatrix} + (-8)\begin{vmatrix} -1 & -2 \\ 2 & 3 \end{vmatrix} + (3)\begin{vmatrix} -1 & -2 \\ -1 & 2 \end{vmatrix}$ $= (1)[-3-4] + (-8)[-3+4] + (3)[-2-2]$ $= -7 + (-8) + (-12) = -27$ In similar fashion, $D_y = \begin{vmatrix} 3 & 1 & -2 \\ 2 & 8 & 2 \\ 1 & 3 & 3 \end{vmatrix} = 54, \quad D_z = \begin{vmatrix} 3 & -1 & 1 \\ 2 & -1 & 8 \\ 1 & 2 & 3 \end{vmatrix} = -54$ $x = \dfrac{D_x}{D} = 1 \quad y = \dfrac{D_y}{D} = -2 \quad z = \dfrac{D_z}{D} = +2$
Cramer's rule with inconsistent systems, p. 246.	If $D = 0$ and some of D_x, D_y, D_z are nonzero, the system of equations is inconsistent. There is no solution.	Attempt to solve. $\quad 3x + 4y = 6$ $\qquad\qquad\qquad\quad -6x - 8y = 2$ $D = \begin{vmatrix} 3 & 4 \\ -6 & -8 \end{vmatrix} = -24 + 24 = 0$ $D_x = \begin{vmatrix} 6 & 4 \\ 2 & -8 \end{vmatrix} = -48 - 8 = -56$ $D = 0, \qquad D_x \neq 0$ The system is *inconsistent;* there is no solution.
Cramer's rule with dependent equations, p. 246.	If $D = 0$ and $D_x = D_y = D_z = 0$, the system of equations contains dependent equations. There are an infinite number of solutions.	Attempt to solve. $\quad x + 5y = -7$ $\qquad\qquad\qquad\quad -2x - 10y = 14$ $D = \begin{vmatrix} 1 & 5 \\ -2 & -10 \end{vmatrix} = -10 + 10 = 0$ $D_x = \begin{vmatrix} -7 & 5 \\ 14 & -10 \end{vmatrix} = 70 - 70 = 0$ $D_y = \begin{vmatrix} 1 & -7 \\ -2 & 14 \end{vmatrix} = 14 - 14 = 0$ Since $D = D_x = D_y = 0$, we know that these equations are dependent. There are an infinite number of solutions.
Graphing the solution for a system of inequalities in two variables, p. 452.	1. Replace the inequality symbol by an equals sign. 2. Sketch each boundary line. 3. Determine the region that satisfies each inequality individually. 4. Shade the common region that satisfies all the inequalities.	Graph. $\quad 3x + 2y \leq 10$ $\qquad\qquad -1x + 2y \geq 2$ 1–2. $3x + 2y \leq 10$ can be graphed more easily as $y \leq -\dfrac{3}{2}x + 5$. We draw a solid line. $-1x + 2y \geq 2$ can be graphed more easily as $y \geq \dfrac{1}{2}x + 1$. We draw a solid line. 3. We shade above $y \geq \dfrac{1}{2}x + 1$ and below $y \leq -\dfrac{3}{2}x + 5$ and include the boundary lines. 4. The common region is shaded.

Solve the following systems by graphing.

1. $x + 2y = 8$
$x - y = 2$

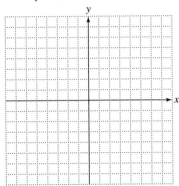

2. $x + y = 2$
$3x - y = 6$

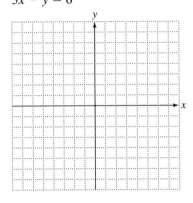

3. $2x + y = 6$
$3x + 4y = 4$

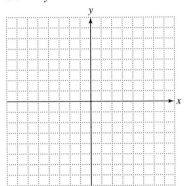

Solve the following systems by substitution.

4. $3x - 2y = -9$
$2x + y = 1$

5. $-6x - y = 1$
$3x - 4y = 31$

6. $4x + 3y = 10$
$5x - y = 3$

7. $-7x + y = -4$
$5x + 2y = 11$

Solve the following systems by addition.

8. $-2x + 5y = -12$
$3x + y = 1$

9. $-4x + 2y = -16$
$3x + 5y = -1$

10. $7x - 4y = 2$
$6x - 5y = -3$

11. $3x + 4y = -6$
$-2x + 7y = -25$

Solve by any appropriate method. If there is no unique solution, state why.

12. $2x + 4y = 9$
$3x + 6y = 8$

13. $x + 5y = 10$
$y = 2 - \dfrac{1}{5}x$

14. $7x + 6y = -10$
$2x + y = 0$

15. $3x + 4y = 1$
$9x - 2y = -4$

16. $x + \dfrac{1}{3}y = 1$
$\dfrac{1}{4}x - \dfrac{3}{4}y = -\dfrac{9}{4}$

17. $\dfrac{2}{3}x + y = \dfrac{14}{3}$
$\dfrac{2}{3}x - y = -\dfrac{22}{3}$

18. $9a + 10b = 7$
$6a - 4b = 10$

19. $3a + 5b = 8$
$2a + 4b = 3$

20. $x + 3 = 3y + 1$
$1 - 2(x - 2) = 6y + 1$

21. $10(x + 1) - 13 = -8y$
$4(2 - y) = 5(x + 1)$

22. $0.3x - 0.2y = 0.7$
$-0.6x + 0.4y = 0.3$

23. $0.2x - 0.1y = 0.8$
$0.1x + 0.3y = 1.1$

Solve by an appropriate method.

24. $3x - 2y - z = 3$
$2x + y + z = 1$
$-x - y + z = -4$

25. $-2x + y - z = -7$
$x - 2y - z = 2$
$6x + 4y + 2z = 4$

26. $2x + 5y + z = 3$
$x + y + 5z = 42$
$2x + y = 7$

27. $x + 2y + z = 5$
$3x - 8y = 17$
$2y + z = -2$

28.
$$2x - 4y + 3z = 0$$
$$x - 2y - 5z = 13$$
$$5x + 3y - 2z = 19$$

29.
$$5x + 2y + 3z = 10$$
$$6x - 3y + 4z = 24$$
$$-2x + y + 2z = 2$$

30.
$$3x + 2y \quad = 7$$
$$2x \quad + 7z = -26$$
$$5y + z = 6$$

31.
$$x - y \quad = 2$$
$$5x + 7y - 5z = 2$$
$$3x - 5y + 2z = -2$$

Use a system of linear equations to solve each of the following problems.

32. A plane flies 720 miles against the wind in 3 hours. The return trip with the wind takes only $2\frac{1}{2}$ hours. Find the speed of the wind. Find the speed of the plane in still air.

33. A local company has computerized its production line. Company officials have found that it takes 25 hours to train new employees to use the computerized equipment and 8 hours to review mathematics skills. Previously laid-off employees can be trained in 10 hours to use the computerized equipment and require only 3 hours to review mathematics skills. This month the management team has 275 hours available to train people to use computerized equipment and 86 hours available to review necessary mathematics skills. How many new employees can be trained? How many previously laid-off employees can be trained this month?

34. When the circus came to town, they hired general laborers at $70 per day and mechanics at $90 per day. Last year they paid $1950 for this temporary help for one day. This year they hired exactly the same number of people of each type. This year they are paying $80 for general laborers and $100 for mechanics for one day. This year they paid $2200 for temporary help. How many general laborers did they hire? How many mechanics did they hire?

35. A total of 590 tickets were sold for the circus matinee performance. Children's admission tickets were $6 and adult tickets were $11. The ticket receipts for the matinee performance were $4790. How many children's tickets were sold? How many adult tickets were sold?

36. A baseball coach bought two hats, five shirts, and four pairs of pants for $129. His assistant purchased one hat, one shirt, and two pairs of pants for $42. Next week the coach returned to buy two hats, three shirts, and one pair of pants for $63. What is the cost for each item?

37. For an experiment a scientist needs three food packets. Packet *A* has 2 grams of carbohydrates, 4 grams of protein, and 3 grams of fat. Packet *B* has 3 grams of carbohydrates, 1 gram of protein, and 1 gram of fat. Packet *C* has 4 grams of carbohydrates, 3 grams of protein, and 2 grams of fat. The experiment requires 29 grams of carbohydrates, 23 grams of protein, and 17 grams of fat. How many packets should she use?

38. Four jars of jelly, three jars of peanut butter, and five jars of honey cost $9.80. Two jars of jelly, two jars of peanut butter, and one jar of honey cost $4.20. Three jars of jelly, four jars of peanut butter, and two jars of honey cost $7.70. Find the cost for one jar of jelly, one jar of peanut butter, and one jar of honey.

39. The church youth group is planning a trip to Mount Washington. A total of 127 people need rides. The church has available buses that hold 40 passengers, and several parents have volunteered station wagons that hold 8 passengers or sedans that hold 5 passengers. The youth leader is planning to use nine vehicles to transport the people. One parent said that if they tripled the number of station wagons and doubled the number of sedans they would be able to transport 126 people. How many buses, station wagons, and sedans are they planning to use if they use nine vehicles?

Compute each determinant.

40. $\begin{vmatrix} 5 & -7 \\ 2 & -3 \end{vmatrix}$

41. $\begin{vmatrix} 0 & -\dfrac{1}{4} \\ 2 & -\dfrac{3}{2} \end{vmatrix}$

42. $\begin{vmatrix} -6 & -8 \\ -2 & 3 \end{vmatrix}$

43. $\begin{vmatrix} 12 & 3 \\ -4 & 7 \end{vmatrix}$

44. $\begin{vmatrix} 0.8 & -5.3 \\ -0.2 & 1.2 \end{vmatrix}$

45. $\begin{vmatrix} -\frac{3}{4} & -\frac{1}{5} \\ \frac{1}{4} & -\frac{5}{6} \end{vmatrix}$

46. $\begin{vmatrix} 0 & -3 & 2 \\ 1 & 5 & 3 \\ -2 & 1 & 4 \end{vmatrix}$

47. $\begin{vmatrix} 2 & 1 & 6 \\ 3 & 4 & -2 \\ 1 & 2 & 3 \end{vmatrix}$

48. $\begin{vmatrix} 6 & 1 & 0 \\ 2 & 3 & 0 \\ 6 & -2 & 1 \end{vmatrix}$

Use Cramer's rule to solve each system.

49. $-x - 5z = -5$
$13x + 2z = 2$

50. $x + y = 10$
$6x + 9y = 70$

51. $2x + 5y = 4$
$5x - 7y = -29$

52. $2x - 3y + 2z = 0$
$x + 2y - z = 2$
$2x + y + 3z = -1$

53. $x - 4y + 4z = -1$
$2x - y + 5z = -3$
$x - 3y + z = 4$

54. Solve for y.
$x - 2y + z = -5$
$2x + z = -10$
$y - z = 15$

55. Solve for x.
$\begin{vmatrix} -2 & 3 \\ x & 5x - 4 \end{vmatrix} = -2$

Solve by graphing each of the following systems of linear inequalities.

56. $x - y \le 3$
$y \le -\frac{1}{4}x + 2$

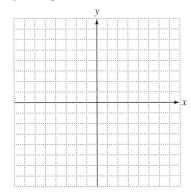

57. $-2x + 3y < 6$
$y > -2$

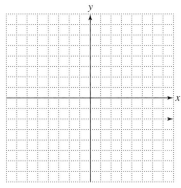

58. $x + y > 1$
$2x - y < 5$

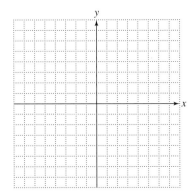

59. $x + y \ge 4$
$y \le x$
$x \le 6$

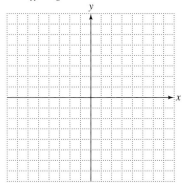

Solve each system of equations. If there is no solution to the system, give a reason.

1. $3x - 2y = -8$
$\quad x + 6y = \quad 4$

2. $6x - 2y = -2$
$\quad 3x + 4y = \quad 14$

3. $\dfrac{1}{4}a - \dfrac{3}{4}b = -1$
$\quad \dfrac{1}{3}a + \quad b = \quad \dfrac{5}{3}$

4. $7x - 1 = 3(1 + y)$
$\quad 1 - 6y = -7(2x + 1)$

5. $3x + 5y - 2z = -5$
$\quad 2x + 3y - \quad z = -2$
$\quad 2x + 4y + 6z = \quad 18$

6. $3x + 2y \quad\quad = 0$
$\quad 2x - \quad y + 3z = 8$
$\quad 5x + 3y + \quad z = 4$

Use a system of linear equations to solve the following problems.

7. A plane flew 1000 miles with a tailwind in 2 hours. The return trip against the wind took $2\dfrac{1}{2}$ hours. Find the speed of the wind and the speed of the plane in still air.

8. On an automobile assembly line, station wagons require 5 minutes for rustproofing, 4 minutes for painting, and 3 minutes for heat drying. Four-door sedans require 4 minutes for rustproofing, 3 minutes for painting, and 2 minutes for drying. Two-door sedans require 3 minutes for rustproofing, 3 minutes for painting, and 2 minutes for drying. The assembly-line supervisor wants to find an assembly plan to use 62 minutes of rustproofing time, 52 minutes of painting time, and 36 minutes of heat drying time. How many vehicles of each type should be sent down the assembly line?

1. _____

2. _____

3. _____

4. _____

5. _____

6. _____

7. _____

8. _____

Solve by Cramer's rule.

9. $5x - 3y = 3$
$7x + y = 25$

10. $\dfrac{1}{3}x + \dfrac{5}{6}y = 2$

$\dfrac{3}{5}x - y = -\dfrac{7}{5}$

11. Solve for z **only** by Cramer's rule.
$$x + 5y + 4z = -3$$
$$x - y - 2z = -3$$
$$x + 2y + 3z = -5$$

Solve by graphing the following system of linear inequalities.

12. $x + 2y \le 6$
$-2x + y \ge -2$

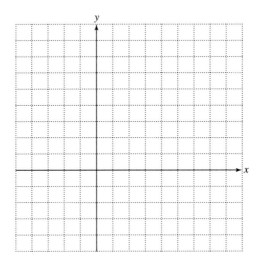

9. _____

10. _____

11. _____

12. _____

Approximately one-half of this test covers the content of Chapters 1–3. The remainder covers the content of Chapter 4.

1. State what property is illustrated.

$7 + 0 = 7$

2. Evaluate.

$\sqrt{25} + (2 - 3)^3 + 20 \div (-10)$.

3. Simplify. $(5x^{-2})(3x^{-4}y^2)$

4. Simplify. $2x - 4[x - 3(2x + 1)]$

5. Solve for P. $A = P(3 + 4rt)$

6. Solve for x. $\dfrac{1}{4}x + 5 = \dfrac{1}{3}(x - 2)$

7. Graph the line $4x - 8y = 10$. Plot at least three points.

8. Find the slope of a line passing through $(6, -1)$ and $(-4, -2)$.

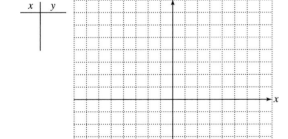

Solve the following linear inequalities and graph your solution on a number line.

9. $4x + 3 - 13x - 7 < 2(3 - 4x)$

10. $\dfrac{2x - 1}{3} \le 7$ and $2(x + 1) \ge 12$

1.
2.
3.
4.
5.
6.
7.
8.
9.
10.

11. _____

12. _____

13. _____

14. _____

15. _____

16. _____

17. _____

18. _____

19. _____

20. _____

11. Find the equation in standard form of a line passing through $(2, -3)$ and perpendicular to $5x + 6y = -2$.

12. A triangle has a perimeter of 69 meters. The second side is 7 meters longer than the first side. The third side is 6 meters shorter than double the length of the first side. Find the length of each side.

13. Victor invests $6000 in a bank. Part is invested at 7% interest and part at 9% interest. In one year Victor earns $510 in interest. How much did he invest at each amount?

14. Solve for (x, y).
$$5x + 2y = 2$$
$$4x + 3y = -4$$

15. Solve for (x, y, z).
$$2x + y - z = 4$$
$$x + 2y - 2z = 2$$
$$x - 3y + z = 4$$

16. Patricia bought five shirts and eight pairs of slacks for $345 at Super Discount Center. Joanna bought seven shirts and three pairs of the same slacks at the same store and her total was $237. How much did a shirt cost? How much did a pair of slacks cost?

17. Solve for (x, y) by Cramer's rule.
$$7x - 6y = 17$$
$$3x + y = 18$$

18. Solve for z **only.** Use Cramer's rule.
$$x + 3y + z = 5$$
$$2x - 3y - 2z = 0$$
$$x - 2y + 3z = -9$$

19. What happens when you attempt to solve $\quad -5x + 6y = 2$
$$10x - 12y = -4?$$ Why is this?

20. Solve the following system of inequalities by graphing.
$$x - y \geq -4$$
$$x + 2y \geq 2$$

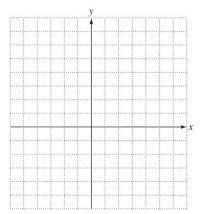

CHAPTER 5
Polynomials

R ow upon row of cannonballs appear at many historic military sites. The piles were placed in special geometric shapes that allowed military commanding officers to make a quick count of their supplies. Do you think you could quickly determine how many cannonballs are in a pile? Turn to page 324 to the Putting Your Skills to Work to test your ability.

If you are familiar with the topics in this chapter, take this test now. Check your answers with those in the back of the book. If an answer was wrong or you couldn't do a problem, study the appropriate section of the chapter.

If you are not familiar with the topics in this chapter, don't take this test now. Instead, study the examples, work the practice problems, and then take the test.

This test will help you identify those concepts that you have mastered and those that need more study.

Follow the directions. Simplify all answers.

Section 5.1

1. $(5x^2 - 3x + 2) + (-3x^2 - 5x - 8) - (x^2 + 3x - 10)$

2. $(x^2 - 3x - 4)(2x - 3)$ **3.** $(5a - 8)(a - 7)$

4. $(2y - 3)(2y + 3)$ **5.** $(3x^2 + 4)^2$

Sections 5.2 and 5.3

6. $(25x^3y^2 - 30x^2y^3 - 50x^2y^2) \div 5x^2y^2$ **7.** $(3y^3 - 5y^2 + 2y - 1) \div (y - 2)$

8. $(2x^4 + 9x^3 + 8x^2 - 9x - 10) \div (2x + 5)$

Section 5.4

Factor completely.

9. $24a^3b^2 + 36a^4b^2 - 60a^3b^3$ **10.** $3x(4x - 3y) - 2(4x - 3y)$

11. $10wx + 6xz - 15yz - 25wy$

Section 5.5

Factor.

12. $x^2 - 7xy + 10y^2$ **13.** $4y^2 - 4y - 15$

14. $28x^2 - 19xy + 3y^2$

1.

2.

3.

4.

5.

6.

7.

8.

9.

10.

11.

12.

13.

14.

Section 5.6

Factor.

15. $36x^2 - 60xy + 25y^2$ **16.** $121x^2 - 1$

17. $8x^3 - y^3$ **18.** $64x^3 + 27$

Section 5.7

*Factor if possible. Indicate the **prime** expressions.*

19. $x^3y^3 - 27y^6$ **20.** $2x^3 - 2x^2 - 24x$

21. $2x^2 + 8x - 3$ **22.** $81a^3 + 126a^2y + 49ay^2$

Section 5.8

*Solve by **factoring**.*

23. $12x^2 + x - 6 = 0$ **24.** $3x^2 + 5x = 7x^2 - 2x$

25. $(x + 5)(x - 3) = 2x + 1$

26. The area of a rectangle is 52 square meters. The length of the rectangle is 1 meter longer than three times its width. Find the length and width of the rectangle.

15.	_____
16.	_____
17.	_____
18.	_____
19.	_____
20.	_____
21.	_____
22.	_____
23.	_____
24.	_____
25.	_____
26.	_____

5.1 Addition, Subtraction, and Multiplication of Polynomials

MathPro

Video 12

SSM

After studying this section, you will be able to:

1. Identify types and degrees of polynomials.
2. Add and subtract polynomials.
3. Multiply polynomials.
4. Multiply two binomials by FOIL.
5. Multiply two binomials by $(a + b)(a - b) = a^2 - b^2$.
6. Multiply two binomials by $(a + b)^2 = a^2 + 2ab + b^2$.

1 Identifying Types and Degrees of Polynomials

A **polynomial** is an algebraic expression of one or more terms. A **term** is a number, a variable, or a product of numbers and variables. All the exponents of the variables must be nonnegative integers and there must be no division by a variable. Three types of polynomials that you will see often are **monomials, binomials,** and **trinomials.**

> 1. A **monomial** has *one* term.
> 2. A **binomial** has *two* terms.
> 3. A **trinomial** has *three* terms.

Here are some examples of polynomials.

	Monomial	Binomial	Trinomial	Polynomial
One variable	$8x^3$	$2y^2 + 3y$	$5x^2 + 2x - 6$	$x^4 + 2x^3 - x^2 + 9$
Two variables	$6x^2y$	$3x^2 - 5y^3$	$8x^2 + 5xy - 3y^2$	$x^3y + 5xy^2 + 3xy - 7y^5$
Three variables	$12uvw^3$	$11a^2b + 5c^2$	$4a^2b^4 + 7c^4 - 2a^5$	$3c^2 + 4c - 8d + 2e - e^2$

The following are *not* polynomials:

$$2x^{-3} + 5x^2 - 3 = 0$$

$$4ab^{1/2} = y$$

$$\frac{2}{x} + \frac{3}{y} = z$$

To Think About Give a reason why each expression is not a polynomial.

Polynomials are also classified by degree. The **degree of a term** is the sum of the exponents of its variables. The **degree of a polynomial** is the degree of the highest-degree term in the polynomial. If the polynomial has no variable, then it has degree zero.

EXAMPLE 1   Name the type of polynomial and give its degree.

(a) $p = 5x^6 + 3x^2 + 2$ **(b)** $g = 7x + 6$

(c) $h = 5x^2y + 3xy^3 + 6xy$ **(d)** $f = 7x^4y^5$

(a) p is a trinomial of degree 6.

(b) g is a binomial of degree 1.   *Remember, if no number appears as an exponent of a variable, the exponent is* 1.

(c) h is a trinomial of degree 4.

(d) f is a monomial of degree 9.

Practice Problem 1   State the type of polynomial and give its degree.

(a) $p = 3x^5 - 6x^4 + x^2$ **(b)** $g = 5x^2 + 2$

(c) $h = 3ab + 5a^2b^2 - 6a^4b$ **(d)** $f = 16x^4y^6$ ■

Many polynomials contain only one variable. A **polynomial in x** is an expression of the form

$$ax^n + bx^{n-1} + cx^{n-2} + \cdots + p$$

where n is a nonnegative integer and constants $a, b, c, \ldots, p$ are real numbers. We usually write polynomials in **descending order** of the variable. For example, the polynomial $4x^5 - 2x^3 + 6x^2 + 5x - 8$ is written in descending order.

2 Adding and Subtracting Polynomials

We can add and subtract polynomials by combining like terms and using the methods of Section 1.5.

EXAMPLE 2   Add the following polynomials.

$(5x^2 - 3x - 8) + (-3x^2 - 7x + 9)$

$\quad 5x^2 - 3x - 8 - 3x^2 - 7x + 9$ *We remove the parentheses and combine like terms.*

$= 2x^2 - 10x + 1$

Practice Problem 2   Add the following polynomials.

$(-7x^2 + 5x - 9) + (2x^2 - 3x + 5)$ ■

To subtract real numbers, we add the opposite of the second number to the first. Thus, for real numbers a and b we have $a - (+b) = a + (-b)$. Similarly for polynomials, to subtract polynomials, add the opposite of the second polynomial to the first.

EXAMPLE 3   Subtract. $(-5x^2 - 19x + 15) - (3x^2 - 4x + 13)$

$\quad (-5x^2 - 19x + 15) + (-3x^2 + 4x - 13)$ *We add the opposite of the second polynomial to the first polynomial.*

$= -8x^2 - 15x + 2$

Practice Problem 3   Subtract. $(2x^2 - 14x + 9) - (-3x^2 + 10x + 7)$ ■

EXAMPLE 4

(a) Add. $(4x^3 - 7x^2 + 2) + (7x^3 + 6x^2 - 8x - 5)$

(b) Subtract. $(4x^3 - 7x^2 + 2) - (7x^3 + 6x^2 - 8x - 5)$

(a) $4x^3 - 7x^2 + 2 + 7x^3 + 6x^2 - 8x - 5 = 11x^3 - x^2 - 8x - 3$

(b) $(4x^3 - 7x^2 + 2) + (-7x^3 - 6x^2 + 8x + 5) = -3x^3 - 13x^2 + 8x + 7$

Practice Problem 4

(a) Add. $(2x^3 - 6x^2 + 5x - 2) + (-7x^3 - 5x^2 + 3x + 4)$

(b) Subtract. $(2x^3 - 6x^2 + 5x - 2) - (-7x^3 - 5x^2 + 3x + 4)$ ∎

3 Multiplying Polynomials

The distributive property is the basis for multiplying polynomials. Recall that

$$a(b + c) = ab + ac$$

We can use this property to multiply a polynomial by a monomial.

$$3xy(5x^3 + 2x^2 - 4x + 1) = 3xy(5x^3) + 3xy(2x^2) - 3xy(4x) + 3xy(1)$$

$$= 15x^4y + 6x^3y - 12x^2y + 3xy$$

A similar procedure can be used to multiply two binomials.

$$(3x + 5)(6x + 7) = (3x + 5)6x + (3x + 5)7 \qquad \textit{We use the distributive property again.}$$

$$= (3x)(6x) + (5)(6x) + (3x)(7) + (5)(7)$$

$$= 18x^2 + 30x + 21x + 35$$

$$= 18x^2 + 51x + 35$$

A popular method for multiplying two binomials, called FOIL, will be introduced in Example 6.

The multiplication of a binomial and a trinomial is more involved. One way to multiply two polynomials is to write them vertically, as we do when multiplying two- and three-digit numbers. We then multiply them in the usual way.

EXAMPLE 5 Multiply. $(4x^2 - 2x + 3)(-3x + 4)$

$$
\begin{array}{r}
4x^2 - 2x + 3 \\
- 3x + 4 \\
\hline
16x^2 - 8x + 12 \\
-12x^3 + 6x^2 - 9x \\
\hline
-12x^3 + 22x^2 - 17x + 12
\end{array}
$$

Multiply $(4x^2 - 2x + 3)(+4)$.
Multiply $(4x^2 - 2x + 3)(-3x)$.
Add the two products.

Practice Problem 5 Multiply. $(2x^2 - 3x + 1)(x^2 - 5x)$ ∎

Another way to multiply polynomials is to multiply horizontally. Let's redo Example 5.

EXAMPLE 5 (REVISITED)   Multiply. $(4x^2 - 2x + 3)(-3x + 4)$

By the distributive law, we have
$$(4x^2 - 2x + 3)(-3x) + (4x^2 - 2x + 3)(4) = -12x^3 + 6x^2 - 9x + 16x^2 - 8x + 12$$
$$= -12x^3 + 22x^2 - 17x + 12$$

In actual practice you will find that you can do some of these steps mentally.

Practice Problem 5 (Revisited)   Multiply in a horizontal fashion.
$$(2x^2 - 3x + 1)(x^2 - 5x) \quad \blacksquare$$

---

4 Multiplying Two Binomials by FOIL

The FOIL method for multiplying two binomials has been developed to help you to keep track of the order of the terms to be multiplied. The acronym FOIL means

F	First
O	Outer
I	Inner
L	Last

That is, we multiply the first terms, then the outer terms, then the inner terms, and, finally, the last terms.

EXAMPLE 6   Multiply. $(5x + 2)(7x - 3)$

$$\begin{array}{c}\text{First}\qquad\text{Last}\quad\text{First + Outer + Inner + Last}\\ (5x + 2)(7x - 3) = 35x^2 - 15x + 14x - 6\\ \text{Inner}\qquad = 35x^2 - x - 6\\ \text{Outer}\end{array}$$

Practice Problem 6   Multiply. $(7x + 3)(2x - 5) \quad \blacksquare$

EXAMPLE 7   Multiply. $(7x^2 - 8)(2x - 3)$

$$\begin{array}{c}\text{First}\qquad\text{Last}\\ (7x^2 - 8)(2x - 3) = 14x^3 - 21x^2 - 16x + 24\\ \text{Inner}\\ \text{Outer}\end{array}$$

Note that in this case we were not able to combine the inner and outer product.

Practice Problem 7   Multiply. $(5a - 2b)(3c - 4d) \quad \blacksquare$

5 Multiplying $(a + b)(a - b)$

Products of the form $(a + b)(a - b)$ occur often and deserve special attention.

$$(a + b)(a - b) = a^2 - ab + ab - b^2 = a^2 - b^2$$

Notice that the middle terms, $-ab + ab$, when combined equal zero. The product is the difference of two squares, $a^2 - b^2$. This is always true when you multiply binomials of the form $(a + b)(a - b)$. You should memorize the formula.

$$(a + b)(a - b) = a^2 - b^2 \tag{3.1}$$

EXAMPLE 8   Multiply.   **(a)** $(2a - 9b)(2a + 9b)$ **(b)** $(5x^2 + 7y^3)(5x^2 - 7y^3)$

(a) $(2a - 9b)(2a + 9b) = (2a)^2 - (9b)^2$
$$= 4a^2 - 81b^2$$

(b) $(5x^2 + 7y^3)(5x^2 - 7y^3) = (5x^2)^2 - (7y^3)^2$
$$= 25x^4 - 49y^6$$

Of course, we could have used the FOIL method, but recognizing the special product allowed us to save time.

Practice Problem 8   Multiply.

(a) $(7x - 2y)(7x + 2y)$ **(b)** $(5a^2 + 8b^2)(5a^2 - 8b^2)$ ∎

6 Multiplying $(a + b)^2$

Another special product is the square of a binomial.

$$(a - b)^2 = (a - b)(a - b) = a^2 - ab - ab + b^2 = a^2 - 2ab + b^2$$

Once you understand the pattern, you should memorize these two formulas.

$$(a - b)^2 = a^2 - 2ab + b^2 \tag{3.2}$$

$$(a + b)^2 = a^2 + 2ab + b^2 \tag{3.3}$$

This procedure is also called *expanding a binomial*. **Note:** $(a - b)^2 \neq a^2 - b^2$ and $(a + b)^2 \neq a^2 + b^2$.

EXAMPLE 9   Multiply (that is, expand the binomials).

(a) $(5a - 8b)^2$ **(b)** $(3u + 11v^2)^2$ **(c)** $(10x^2 - 3y^2)^2$ **(d)** $(12x^3 + 1)^2$

(a) $(5a - 8b)^2 = (5a)^2 - 2(5a)(8b) + (8b)^2$
$$= 25a^2 - 80ab + 64b^2$$

(b) Here $a = 3u$ and $b = 11v^2$.

$$(3u + 11v^2)^2 = (3u)^2 + 2(3u)(11v^2) + (11v^2)^2$$

$$= 9u^2 + 66uv^2 + 121v^4$$

(c) Here $a = 10x^2$ and $b = 3y^2$.

$$(10x^2 - 3y^2)^2 = (10x^2)^2 - 2(10x^2)(3y^2) + (3y^2)^2$$

$$= 100x^4 - 60x^2y^2 + 9y^4$$

(d) Here $a = 12x^3$ and $b = 1$.

$$(12x^3 + 1)^2 = (12x^3)^2 + 2(12x^3)(1) + (1)^2$$

$$= 144x^6 + 24x^3 + 1$$

Practice Problem 9   Multiply.

(a) $(3x - 7y)^2$ **(b)** $(4u + 5v)^2$ **(c)** $(9a + 2b)^2$ **(d)** $(7x^2 - 3y^2)^2$ ∎

5.1 Exercises

Name the polynomial and give its degree.

1. $26x^3y - 35$
mult.

2. $12a + 15ab - 16$

3. $17x^3y^5z$
add.

4. $\dfrac{1}{2}xy^4 + \dfrac{1}{3}x^2y^5$

5. $\dfrac{3}{5}m^3n - \dfrac{2}{5}mn + \dfrac{1}{5}n^8$

6. $-27ab^3cd$

Perform addition or subtraction for the following polynomials.

7. $(x^2 + 5x - 2) + (-3x^2 - 7x + 10)$

8. $(x^2 + 2x - 12) + (7x^2 - 5x - 14)$

9. $(x^2 + 3x - 2) + (-2x^2 - 5x + 1) + (x^2 - x - 5)$

10. $(2x^2 - 5x - 1) + (3x^2 - 7x + 3) + (-5x^2 + x + 1)$

11. $(4x^3 - 6x^2 - 3x + 5.5) - (2x^3 + 3x^2 - 5x - 8.3)$

12. $(3x^3 + 2x^2 - 8x - 9.2) - (-5x^3 + x^2 - x - 12.7)$

13. $(5a^3 - 2a^2 - 6a + 8) + (5a + 6) - (-a^2 - a + 2)$

14. $(7a^2 - 2a + 6) + (-12a^3 - 6a + 5) - (3a^2 - a - 2)$

15. $\left(\dfrac{1}{2}x^2 - 7x\right) + \left(\dfrac{1}{3}x^2 + \dfrac{1}{4}x\right)$

16. $\left(\dfrac{1}{5}x^2 + 9x\right) + \left(\dfrac{4}{5}x^2 - \dfrac{1}{6}x\right)$

17. $(2.3x^3 - 5.6x^2 - 2) - (5.5x^3 - 7.4x^2 + 2)$

18. $(5.9x^3 + 3.4x^2 - 7) - (2.9x^3 - 9.6x^2 + 3)$

Multiply.

19. $2x(3x^2 - 5x + 1)$

20. $-5x(x^2 - 6x - 2)$

21. $-\dfrac{1}{3}xy(2x - 6y + 15)$

22. $4xy^2(x - y + 3)$

23. $(x + 5)(x - 6)$

24. $(x + 12)(x + 2)$

25. $(2x + 1)(3x + 2)$

26. $(2x - 1)(x + 5)$

27. $(7x - 6)(4x - 3)$

28. $(2x - 4)(9x - 5)$

29. $(5w + 2d)(3a - 4b)$

30. $(7a + 8b)(5d - 8w)$

31. $(3x - 2y)(-4x + y)$

32. $(-8x - 3y)(x + 2y)$

33. $(2r + 2s^2)(5r - 9s^2)$

34. $(-3r - 2s^2)(5r - 6s^2)$

35. $(2x - 3)(x^2 - x + 1)$

36. $(4x + 1)(2x^2 + x + 1)$

37. $(3x^2 - 2xy - 6y^2)(2x - y)$

38. $(5x^2 + 3xy - 7y^2)(3x - 2y)$

39. $(x^2 - 6x + 1)(2x^2 - 5x + 2)$

40. $(3x^2 - 2x - 4)(x^2 + 2x + 3)$

41. $(5a^3 - 3a^2 + 2a - 4)(a - 3)$

42. $(2b^3 - 5b^2 - 4b + 1)(2b - 1)$

43. $(r^2 + 3rs - 2s^2)(3r^2 - 4rs - 2s^2)$

44. $(m^2 - 6mp + 2p^2)(2m^2 - 4mp + 3p^2)$

Multiply mentally.

45. $(5x - 8y)(5x + 8y)$

46. $(2a - 7b)(2a + 7b)$

47. $(5a - 2b)^2$

48. $(6a + 5b)^2$

49. $(7m - 1)^2$

50. $(5r + 3)^2$

51. $(2x^2 + 1)(2x^2 - 1)$

52. $(1 - 7x^3)(1 + 7x^3)$

53. $(2a^2b^2 - 3)^2$

54. $(3x^2 - 5y^2)^2$

55. $(7x - 3y^2)(7x + 3y^2)$

56. $(5x + 8y^2)(5x - 8y^2)$

First multiply any two binomials in the problem; then multiply the result by the third binomial.

★ **57.** $(x + 2)(x - 3)(2x - 5)$ ★ **58.** $(x - 6)(x + 2)(3x + 2)$ ★ **59.** $(a + 3)(2 - a)(4 - 3a)$ ★ **60.** $(6 - 5a)(a + 1)(2 - 3a)$

Simplify.

 **61.** $(3.928x^2 - 5.617x) + (8.346x^2 - 9.098x) + (-1)(1.542x^2 - 3.986x)$

 **62.** $(52.613x + 49.408y)(34.078x - 28.231y)$

Applications

63. The area of the base of a rectangular box measures $2x^2 + 5x + 8$ cm^2. The height of the box measures $3x + 5$ cm. Find the volume of the box.

64. A rectangular garden has $3n^2 + 4n + 7$ flowers planted in each row. The garden has $2n + 5$ rows. Find the number of flowers in the garden.

Cumulative Review Problems

65. Collect like terms. $5x - 6y - 8x + 2y - 7x - 8y$

66. Simplify. $2\{3 - x[4 + x(2 - x)]\}$

In order to determine the strength for a certain anti-malaria medication, the concentration, in parts per million after time t, in hours, is given by the polynomial $-0.03t^2 + 78$.

67. Find the concentration after 3 hours.

68. Find the concentration after 30 hours.

69. Find the concentration after 50 hours.

70. Find the concentration after 50.9 hours.

5.2 Division of Polynomials

After studying this section, you will be able to:

1. *Divide a polynomial by a monomial.*
2. *Divide a polynomial by a polynomial.*

MathPro   Video 12   SSM

1 Dividing a Polynomial by a Monomial

The easiest type of polynomial division occurs when the divisor is a monomial. We do this type of division just as if we were dividing numbers. First we write the indicated division as the sum of separate fractions, and then we reduce (if possible) each fraction.

EXAMPLE 1   Divide. $(24x^2 - 18x + 2) \div 4$

First write the division in fractional form:

$$(24x^2 - 18x + 2) \div 4 = \frac{24x^2 - 18x + 2}{4}$$

Now divide each term of the polynomial in the numerator by the monomial in the denominator.

$$\frac{24x^2 - 18x + 2}{4} = \frac{24x^2}{4} - \frac{18x}{4} + \frac{2}{4}$$

$$= 6x^2 - \frac{9}{2}x + \frac{1}{2}$$

The answer may also be written in the form

$$6x^2 - \frac{9x}{2} + \frac{1}{2}$$

Both answers are considered correct.

Practice Problem 1   Divide. $(16x^3 - 8x^2 + 3x) \div 2x$ ■

EXAMPLE 2   Divide. $(15x^3 - 10x^2 + 40x) \div 5x$

$$\frac{15x^3 - 10x^2 + 40x}{5x} = \frac{15x^3}{5x} - \frac{10x^2}{5x} + \frac{40x}{5x}$$

$$= 3x^2 - 2x + 8$$

Practice Problem 2   Divide. $(-16x^4 + 16x^3 + 8x^2 + 64x) \div 8x$ ■

2 Division of a Polynomial by a Polynomial

When we divide polynomials by binomials or trinomials, we perform long division. This is much like dividing numbers. The polynomials must be in descending order.

First we write the problem in the form of long division.

$$2x + 3 \overline{)6x^2 + 17x + 12}$$

The divisor is $2x + 3$; the dividend is $6x^2 + 17x + 12$. Now we divide the first term of the dividend ($6x^2$) by the first term of the divisor ($2x$).

$$\boxed{3x} \qquad \boxed{6x^2 \div 2x = 3x}$$
$$2x + 3 \overline{)6x^2 + 17x + 12}$$

Now we multiply $3x$ (the first term of the quotient) by the divisor $2x + 3$.

$$
\begin{array}{r}
3x \\
2x + 3 \overline{)6x^2 + 17x + 12} \\
6x^2 + 9x \longleftarrow \boxed{\text{The product of } 3x(2x + 3).}
\end{array}
$$

Next, just as in arithmetic long division, we subtract this term from the dividend and bring down the next monomial.

$$
\begin{array}{r}
3x \\
2x + 3 \overline{)6x^2 + 17x + 12} \qquad \boxed{\text{Subtract } 6x^2 + 9x \text{ from } 6x^2 + 17x.}\\
\underline{6x^2 + 9x} \longleftarrow \\
8x + 12 \longleftarrow \boxed{\text{Bring down the next monomial.}}
\end{array}
$$

Now we divide the first term of this binomial ($8x$) by the first term of the divisor ($2x$).

$$
\begin{array}{r}
3x + \boxed{4} \qquad \boxed{8x \div 2x = 4} \\
2x + 3 \overline{)6x^2 + 17x + 12} \\
\underline{6x^2 + 9x} \\
8x + 12 \\
\underline{8x + 12} \longleftarrow \boxed{\text{The product } 4(2x + 3)} \\
0
\end{array}
$$

Note that we then multiplied $(2x + 3)(4)$ and subtracted, just as we did before. We continued this process until the remainder was zero. Thus we find that $\dfrac{6x^2 + 17x + 12}{2x + 3} = 3x + 4$.

GRAPHING CALCULATOR

Verifying Answers When Dividing Polynomials

One way to verify that the division was performed correctly is to graph $y_1 = 3x + 4$ and $y_2 = \dfrac{6x^2 + 17x + 12}{2x + 3}$

If the graphs appear to coincide then we have an independent verification that

$$\frac{6x^2 + 17x + 12}{2x + 3} = 3x + 4$$

Dividing a Polynomial by a Binomial or Trinomial

1. Write the division as in arithmetic. Write both polynomials in descending order; write missing terms with a coefficient of zero.

2. Divide the *first* term of the divisor into the first term of the dividend. The result is the first term of the quotient.

3. Multiply the first term of the quotient by *every* term in the divisor.

4. Write the product under the dividend (align like terms) and subtract.

5. Treat this difference as a new dividend. Repeat steps 2 through 4. Continue until the remainder is zero or is a polynomial of lower degree than the *first term* of the divisor.

6. If there is a remainder, write it as the numerator of a fraction with the divisor as the denominator. Add this fraction to the quotient.

EXAMPLE 3   Divide. $(6x^3 + 7x^2 + 3) \div (3x - 1)$

There is no x term in the dividend, so we write $0x$.

$$
\begin{array}{r}
2x^2 + 3x + 1 \\
3x - 1 \overline{)6x^3 + 7x^2 + 0x + 3} \\
\underline{6x^3 - 2x^2} \\
9x^2 + 0x \\
\underline{9x^2 - 3x} \\
3x + 3 \\
\underline{3x - 1} \\
4
\end{array}
$$

Note that we subtract $7x^2 - (-2x^2)$ to obtain $9x^2$.

Note that we subtract $0x - (-3x)$ to obtain $3x$.

The quotient is $2x^2 + 3x + 1$ with a remainder of 4. We may write this as

$$2x^2 + 3x + 1 + \frac{4}{3x - 1}$$

Check: $(3x - 1)(2x^2 + 3x + 1) + 4 \stackrel{?}{=} 6x^3 + 7x^2 + 3$

$$6x^3 + 7x^2 - 0x - 1 + 4 \stackrel{?}{=} 6x^3 + 7x^2 + 3$$

$$6x^3 + 7x^2 + 3 = 6x^3 + 7x^2 + 3 \quad \checkmark$$

Practice Problem 3   Divide. $(14x + 8x^2 - 14) \div (-3 + 4x)$ ∎

EXAMPLE 4   Divide. $\dfrac{64x^3 - 125}{4x - 5}$

This fraction is another way of writing the problem $(64x^3 - 125) \div (4x - 5)$.

Note that two terms are missing in the dividend. We write them with zero coefficients.

$$
\begin{array}{r}
16x^2 + 20x + 25 \\
4x - 5 \overline{)64x^3 + 0x^2 + 0x - 125} \\
\underline{64x^3 - 80x^2} \\
80x^2 + 0x \\
\underline{80x^2 - 100x} \\
100x - 125 \\
\underline{100x - 125} \\
0
\end{array}
$$

Note that $0x^2 - (-80x^2) = 80x^2$.

Note that $0x - (-100x) = 100x$.

The quotient is $16x^2 + 20x + 25$.
   Check: Verify that $(4x - 5)(16x^2 + 20x + 25) = 64x^3 - 125$.

Practice Problem 4   Divide. $(8x^3 + 27) \div (2x + 3)$ ∎

EXAMPLE 5 Divide. $(7x^3 - 10x - 7x^2 + 2x^4 + 8) \div (2x^2 + x - 2)$

Arrange the dividend in descending order before dividing.

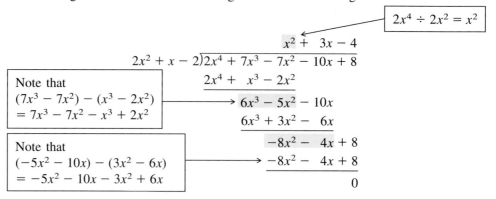

$2x^4 \div 2x^2 = x^2$

$$
\begin{array}{r}
x^2 + 3x - 4 \\
2x^2 + x - 2 \overline{) 2x^4 + 7x^3 - 7x^2 - 10x + 8} \\
2x^4 + x^3 - 2x^2 \\
6x^3 - 5x^2 - 10x \\
6x^3 + 3x^2 - 6x \\
-8x^2 - 4x + 8 \\
-8x^2 - 4x + 8 \\
0
\end{array}
$$

Note that
$(7x^3 - 7x^2) - (x^3 - 2x^2)$
$= 7x^3 - 7x^2 - x^3 + 2x^2$

Note that
$(-5x^2 - 10x) - (3x^2 - 6x)$
$= -5x^2 - 10x - 3x^2 + 6x$

The quotient is $x^2 + 3x - 4$.

Check: Verify that $(2x^2 + x - 2)(x^2 + 3x - 4) = 2x^4 + 7x^3 - 7x^2 - 10x + 8$.

Practice Problem 5 Divide. $(x^4 - 3x^3 + 3x + 4) \div (x^2 - 1)$ ■

Developing Your Study Skills

TAKING NOTES IN CLASS

An important part of mathematics studying is taking notes. In order to take meaningful notes, you must be an active listener. Keep your mind on what the instructor is saying, and be ready with questions whenever you do not understand something.

If you have previewed the lesson material, you will be prepared to take good notes. The important concepts will seem somewhat familiar. You will have a better idea of what needs to be written down. If you frantically try to write all that the instructor says or copy all the examples done in class, you may find your notes to be nearly worthless when you are home alone. You may find that you are unable to make sense of what you have written.

Write down *important* ideas and examples as the instructor lectures, making sure that you are listening and following the logic. Include any helpful hints or suggestions that your instructor gives you or refers to in your text. You will be amazed at how easily these are forgotten if they are not written down.

Successful note taking requires active listening and processing. Stay alert in class. You will realize the advantages of taking your own notes over copying those of someone else.

5.2 Exercises

Divide.

1. $(15x^2 + 20x - 30) \div 5$

2. $(14x^2 - 28x - 35) \div 7$

3. $(27x^4 - 9x^3 + 63x^2) \div 9x$

4. $(22x^4 + 33x^3 - 121x^2) \div 11x$

5. $\dfrac{4x^3 - 2x^2 + 5x}{2x}$

6. $\dfrac{4w^3 + 8w^2 - w}{4w}$

7. $\dfrac{18a^3b^2 + 12a^2b^2 - 4ab^2}{2ab^2}$

8. $\dfrac{25m^5n - 10m^4n + 15m^3n}{5m^3n}$

Divide. Check your answers for Problems 9 through 14.

9. $(15x^2 + 23x + 4) \div (5x + 1)$

10. $(12x^2 + 11x + 2) \div (4x + 1)$

11. $(28x^2 - 29x + 6) \div (4x - 3)$

12. $(30x^2 - 17x + 2) \div (5x - 2)$

13. $(6a^2 - 11a - 30) \div (3a + 5)$

14. $(18a^2 + 9a - 17) \div (3a + 4)$

15. $(x^3 - x^2 + 11x - 1) \div (x + 1)$

16. $(x^3 + 2x^2 - 3x + 2) \div (x + 1)$

17. $(2x^3 - x^2 - 7) \div (x - 2)$

18. $(4x^3 - 6x - 11) \div (2x - 4)$

19. $\dfrac{2x^3 + 13x^2 + 9x - 6}{2x + 3}$

20. $\dfrac{x^3 - 7x^2 + 13x - 15}{x - 5}$

21. $\dfrac{x^3 + 27}{x + 3}$

22. $\dfrac{x^3 - 64}{x - 4}$

23. $\dfrac{2x^4 - x^3 + 16x^2 - 4}{2x - 1}$

24. $\dfrac{4x^4 - 17x^2 + 14x - 3}{2x - 3}$

25. $\dfrac{15a^4 + 3a^3 + 4a^2 + 4}{3a^2 - 1}$

26. $\dfrac{2a^4 + 3a^3 + 4a^2 + 9a - 6}{a^2 + 3}$

27. $\dfrac{6t^4 - 5t^3 - 8t^2 + 16t - 8}{3t^2 + 2t - 4}$

28. $\dfrac{2t^4 + 5t^3 - 11t^2 - 20t + 12}{t^2 + t - 6}$

Divide.

  29. $(742.14x^2 - 124.362x + 3252.06) \div 12.6$

30. $(13.1x^2 + 9.32x - 9.58) \div (2.62x + 3.50)$

Optional Graphing Calculator Problems

If you have a graphing calculator verify that:

31. $\dfrac{2x^2 - x - 10}{2x - 5} = x + 2$

32. $\dfrac{4x^3 + 12x^2 + 7x - 3}{2x + 3} = 2x^2 + 3x - 1$

Cumulative Review Problems

Solve for x.

33. $3x + 4(3x - 5) = -x + 12$

34. $9x - 2x + 8 = 4x + 38$

35. $2(x - 3) - 13 = 17 - 3(x + 2)$

36. $5 - \dfrac{x}{4} = x - 1$

5.3 Synthetic Division

After studying this section, you will be able to:

1 *Use synthetic division to divide polynomials.*

MathPro   Video 12   SSM

1 *Using Synthetic Division to Divide Polynomials*

When dividing a polynomial by a binomial of the form $x + b$ or $x - b$, you may find a procedure known as **synthetic division** quite efficient. Notice the following division problems. The right-hand problem is the same as the left, but without the variables.

$$
\begin{array}{r}
3x^2 - 2x + 2 \\
x + 3\overline{)3x^3 + 7x^2 - 4x + 3} \\
\underline{3x^3 + 9x^2} \\
-2x^2 - 4x \\
\underline{-2x^2 - 6x} \\
2x + 3 \\
\underline{2x + 6} \\
-3
\end{array}
\qquad
\begin{array}{r}
3 \quad -2 \quad 2 \\
1 + 3\overline{)3 \quad 7 \quad -4 \quad 3} \\
\underline{3 \quad 9} \\
-2 \quad -4 \\
\underline{-2 \quad -6} \\
2 \quad 3 \\
\underline{2 \quad 6} \\
-3
\end{array}
$$

Eliminating the variables makes synthetic division efficient, and we can make the procedure simpler yet. Note that the colored numbers (3, −2, and 2) appear twice in the example above, once in the quotient and again in the subtraction. Synthetic division makes it possible to write each number only once. Also, in synthetic division we change the subtraction that division otherwise requires to addition. We do this by dropping the 1, which is the coefficient of x, and taking the opposite of the second number in the divisor. In our first example, this means dropping the 1 and changing 3 to −3. The following steps detail synthetic division.

Step 1

| Divisor, without the 1 and opposite sign | $-3\,\rfloor$ | 3   7   −4   3 | Dividend, without variables |

$$\underline{}$$
$$3$$

Step 2

$$
\begin{array}{r|rrrr}
-3 & 3 & 7 & -4 & 3 \\
  &   & -9 &   &   \\
\hline
  & 3 & -2 &   &
\end{array}
\qquad
\begin{array}{l}
\textit{Multiply } (-3)(3) = -9 \\
\textit{and add } 7 + (-9) = -2.
\end{array}
$$

Step 3

$$
\begin{array}{r|rrrr}
-3 & 3 & 7 & -4 & 3 \\
  &   & -9 & 6 &   \\
\hline
  & 3 & -2 & 2 &
\end{array}
\qquad
\begin{array}{l}
\textit{Multiply } (-3)(-2) = 6 \\
\textit{and add } -4 + 6 = 2.
\end{array}
$$

Step 4

$$
\begin{array}{r|rrrr}
-3 & 3 & 7 & -4 & 3 \\
  &   & -9 & 6 & -6 \\
\hline
  & 3 & -2 & 2 & -3
\end{array}
\qquad
\begin{array}{l}
\textit{Multiply } (-3)(2) = -6 \\
\textit{and add } 3 + (-6) = -3.
\end{array}
$$

$$3x^2 - 2x + 2 + \text{remainder of } -3 \qquad
\begin{array}{l}
\textit{Replace the variables,} \\
\textit{making sure their} \\
\textit{powers are correct.}
\end{array}$$

The result is read from the bottom row. Our answer is $3x^2 - 2x + 2 + \dfrac{-3}{x + 3}$.

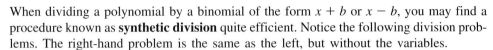

EXAMPLE 1   Divide by synthetic division. $(3x^3 - x^2 + 4x + 8) \div (x + 2)$

$$
\begin{array}{r|rrrr}
-2 & 3 & -1 & 4 & 8 \\
   &   & -6 & +14 & -36 \\
\hline
   & 3 & -7 & 18 & -28
\end{array}
$$

The quotient is $3x^2 - 7x + 18 + \dfrac{-28}{x + 2}$.

Practice Problem 1   Divide by synthetic division. $(x^3 - 3x^2 + 4x - 5) \div (x + 3)$ ■

When a term is missing in the sequence of descending powers of x, we use a zero to indicate the coefficient of that term.

EXAMPLE 2   Divide by synthetic division. $(3x^4 - 21x^3 + 31x^2 - 25) \div (x - 5)$

$$
\begin{array}{r|rrrrr}
5 & 3 & -21 & 31 & 0 & -25 \\
  &   & 15 & -30 & 5 & 25 \\
\hline
  & 3 & -6 & 1 & 5 & 0
\end{array}
$$
  Note that the remainder is zero.

The quotient is $3x^3 - 6x^2 + x + 5$.

Practice Problem 2   Divide by synthetic division.
$(2x^4 - x^2 + 5x - 12) \div (x - 3)$ ■

EXAMPLE 3   Divide by synthetic division. $(3x^4 - 4x^3 + 8x^2 - 5x - 5) \div (x - 2)$

$$
\begin{array}{r|rrrrr}
2 & 3 & -4 & 8 & -5 & -5 \\
  &   & 6 & 4 & 24 & 38 \\
\hline
  & 3 & 2 & 12 & 19 & 33
\end{array}
$$

The quotient is $3x^3 + 2x^2 + 12x + 19 + \dfrac{33}{x - 2}$.

Practice Problem 3   Divide by synthetic division.
$(2x^4 - 9x^3 + 5x^2 + 13x - 3) \div (x - 3)$ ■

5.3 Exercises

Divide by synthetic division.

1. $(x^2 + 2x - 63) \div (x - 7)$

2. $(x^2 + 2x - 80) \div (x - 8)$

3. $(2x^2 - 11x - 8) \div (x - 6)$

4. $(2x^2 - 15x - 23) \div (x - 9)$

5. $(3x^3 + x^2 - x + 4) \div (x + 1)$

6. $(3x^3 + 10x^2 + 6x - 4) \div (x + 2)$

7. $(x^3 + 7x^2 + 17x + 15) \div (x + 3)$

8. $(3x^3 - x^2 + 4x + 8) \div (x + 2)$

9. $(x^3 + 4x^2 - x + 5) \div (x - 2)$

10. $(4x^3 + x^2 - 3x - 1) \div (x - 1)$

11. $(x^3 - 2x^2 + 8) \div (x + 2)$

12. $(2x^3 + 7x^2 - 5) \div (x + 3)$

13. $(6x^4 + 15x^3 - 28x - 6) \div (x + 2)$

14. $(3x^4 - 25x^2 - 18) \div (x - 3)$

15. $(x^4 - 6x^3 + x^2 - 9) \div (x + 1)$

16. $(x^4 - 3x^3 - 11x^2 + 3x + 10) \div (x - 5)$

17. $(3x^5 + x - 1) \div (x + 1)$

18. $(2x^4 - x + 3) \div (x - 2)$

19. $(2x^5 + 5x^4 - 2x^3 + 2x^2 - 2x + 3) \div (x - 3)$

20. $(2x^5 - 3x^4 + x^3 - x^2 + 2x - 1) \div (x + 2)$

21. $(x^6 - 4) \div (x + 1)$

22. $(x^6 + 2x^4 - 5x + 11) \div (x - 2)$

23. $(x^3 + 2.5x^2 - 3.6x + 5.4) \div (x - 1.2)$

24. $(x^3 - 4.2x^2 - 8.8x + 3.7) \div (x + 1.8)$

★ **25.** $(x^4 + 3x^3 - 2x^2 + bx + 5) \div (x + 3)$ divides without any remainder. What is the value of b?

★ **26.** $(2x^4 + 12x^3 + ax^2 - 5x + 75) \div (x + 5)$ divides without any remainder. What is the value of a?

To Think About

How do we use synthetic division when the divisor is in the form $ax + b$? We divide the divisor by a to get $x + \dfrac{b}{a}$. After performing the division, we divide each term of the quotient by a. To divide $(2x^3 + 7x^2 - 5x - 4) \div (2x + 1)$ we would use $-\dfrac{1}{2}\bigg|\ 2\quad 7\quad -5\quad -4$ and then divide each term of the quotient by 2.

In Problems 27 and 28, divide by synthetic division.

27. $(2x^3 - 3x^2 + 6x + 4) \div (2x + 1)$

28. $(4x^3 - 6x^2 + 6) \div (2x + 3)$

29. When the divisor is of the form $ax + b$, why does this method discussed above work? What are we really doing when we divide the divisor and the quotient by the value a?

30. Why do we not have to divide the remainder by a when using this method?

Cumulative Review Problems

A total of 21 people were killed and 150 people injured in the Great Boston Molasses Flood in January 1919. A molasses storage tank burst and spilled 2 million gallons of molasses through the streets of Boston.

31. How many cubic feet of molasses were contained in the 2-million-gallon molasses tank? (Use 1 gallon ≈ 0.134 cubic feet.)

32. At one point the moving flood of molasses appeared as a huge cylindrically shaped object with a radius of 200 feet. At that point how deep was the molasses? (Round to the nearest tenth.)

5.4 Removing Common Factors; Factoring by Grouping

After studying this section, you will be able to:

1 *Remove the greatest common factor from a polynomial.*
2 *Factor a polynomial by the grouping method.*

MathPro   Video 13   SSM

We learned to multiply polynomials in Section 5.1. When two or more algebraic expressions (monomials, binomials, and so on) are multiplied, each expression is called a **factor.**

In the rest of this chapter, we will learn how to find the factors of a polynomial. Factoring is the opposite of multiplication and is an extremely important mathematical technique.

1 Removing the Greatest Common Factor

To remove a common factor, we make use of the distributive property.

$$ab + ac = a(b + c)$$

The greatest common factor simply means the largest factor in every term of the expression. It must contain

1. The largest possible numerical coefficient, and

2. The largest possible exponent for each variable

EXAMPLE 1   Remove the greatest common factor.

(a) $7x^2 - 14x$ **(b)** $40a^3 - 20a^2$

(a) $7x^2 - 14x = 7x(x - 2)$
  Be careful. The greatest common factor is $7x$, not 7.

(b) $40a^3 - 20a^2 = 20a^2(2a - 1)$
  The greatest common factor is $20a^2$.

Suppose we had written $10a(4a^2 - 2a)$ or $10a(2a)(2a - 1)$. Although we have factored the expression, we have not found the *greatest* common factor.

Practice Problem 1   Remove the greatest common factor.

(a) $19x^3 - 38x^2$ **(b)** $100a^4 - 50a^2$ ■

EXAMPLE 2   Remove the greatest common factor.

(a) $9x^2 - 18xy - 15y^2$ **(b)** $4a^3 - 12a^2b^2 - 8ab^3 + 6ab$

(a) $9x^2 - 18xy - 15y^2 = 3(3x^2 - 6xy - 5y^2)$
  The greatest common factor is 3.

(b) $4a^3 - 12a^2b^2 - 8ab^3 + 6ab = 2a(2a^2 - 6ab^2 - 4b^3 + 3b)$
  The greatest common factor is $2a$.

Practice Problem 2   Remove the greatest common factor.

(a) $21x^3 - 18x^2y + 24xy^2$

(b) $12xy^2 - 14x^2y + 20x^2y^2 + 36x^3y$ ■

EXAMPLE 3   Remove the greatest common factor.

(a) $55x^3y^5z^6 - 121x^4y^3z^5 + 33x^2y^5z^7$ **(b)** $140a^3b^2 - 210a^2b^3 + 70a^2b^2$

(a) $55x^3y^5z^6 - 121x^4y^3z^5 + 33x^2y^5z^7$
$= 11x^2y^3z^5(5xy^2z - 11x^2 + 3y^2z^2)$

> Be careful. Do you see why $11x^2y^3z^5$ is the greatest common factor?

(b) $140a^3b^2 - 210a^2b^3 + 70a^2b^2$

$= 70a^2b^2(2a - 3b + 1)$ Don't forget the 1 here.

> Do you see why we select $70a^2b^2$ as the greatest common factor?

Practice Problem 3   Remove the greatest common factor.

(a) $16a^5b^4 - 32a^3b^4 - 24a^6b^4$ **(b)** $50xyz^2 - 75x^2y^2z^2 + 100x^3y^2z^2$ ■

How do you know if you have factored correctly? You can do two things to verify your answer.

1. Examine the polynomial in the parentheses. Its terms should not have any remaining common factors.
2. Multiply the polynomial by the common factor. You should obtain the original expression.

In the remaining examples you will be asked to **factor** several polynomials. This merely means to find the factors that when multiplied give the polynomial as a product. Our method when asked to factor is to remove the greatest common factor.

EXAMPLE 4   Factor. $6x^3 - 9x^2y - 6x^2y^2$ Check your answer.

$$6x^3 - 9x^2y - 6x^2y^2 = 3x^2(2x - 3y - 2y^2)$$

Check:

1. $(2x - 3y - 2y^2)$ has no common factors. If it did, we would know that we had not removed the *greatest* common factor.
2. Multiply

$$3x^2(2x - 3y - 2y^2) = 6x^3 - 9x^2y - 6x^2y^2$$

> Observe that we do obtain the original polynomial.

Practice Problem 4   Factor. $9a^3 - 12a^2b^2 - 15a^4$ Check your answer. ■

The greatest common factor need not always be a monomial. It may be a binomial, or even a trinomial. For example,

$$5a(x + 3) + 2(x + 3) = (x + 3)(5a + 2)$$
$$5a(x + 4y) + 2(x + 4y) = (x + 4y)(5a + 2)$$

The common factors are binomials.

EXAMPLE 5 Factor.

(a) $2x(x + 5) - 3(x + 5)$ **(b)** $5a(a + b) - 2b(a + b) - 1(a + b)$

(a) $2x(x + 5) - 3(x + 5) = (x + 5)(2x - 3)$ | The common factor is $x + 5$. |
(b) $5a(a + b) - 2b(a + b) - 1(a + b) = (a + b)(5a - 2b - 1)$

| The common factor is $a + b$. |

Practice Problem 5 Factor. $7x(x + 2y) - 8y(x + 2y) - (x + 2y)$ ■

2 Factoring by Grouping

The common factors in our examples so far were already grouped inside parentheses, so it was easy to pick them out. However, this rarely happens, so we have to learn how to manipulate expressions to find the greatest common factor.

Polynomials of four terms can often be factored by the method of Example 5a. However, the parentheses are not always present in the original problem. When they are not present, we first group like terms. We then remove a common factor from the first two terms and a common factor from the second two terms. It should then be easy to find the greatest common factor.

EXAMPLE 6 Factor. $ax + 2ay + 2bx + 4by$

| Remove the greatest common factor (a) from the first two terms. |

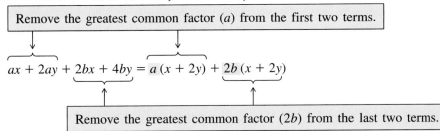

$$ax + 2ay + 2bx + 4by = a(x + 2y) + 2b(x + 2y)$$

| Remove the greatest common factor ($2b$) from the last two terms. |

Now we can see that $(x + 2y)$ is a common factor.

$$a(x + 2y) + 2b(x + 2y) = (x + 2y)(a + 2b)$$

Practice Problem 6 Factor. $bx + 5by + 2wx + 10wy$ ■

EXAMPLE 7 Factor. $2x^2 - 18y - 12x + 3xy$

First write the polynomial in this order. $2x^2 - 12x + 3xy - 18y$

| Remove the greatest common factor ($2x$) from the first two terms. |

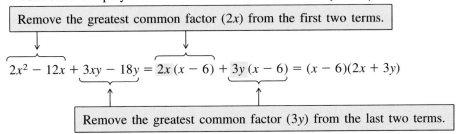

$$2x^2 - 12x + 3xy - 18y = 2x(x - 6) + 3y(x - 6) = (x - 6)(2x + 3y)$$

| Remove the greatest common factor ($3y$) from the last two terms. |

Practice Problem 7 Factor. $5x^2 - 12y + 4xy - 15x$ ■

If a problem can be factored by this method, we must rearrange the order of the four terms whenever necessary so that the first two terms do have a common factor.

EXAMPLE 8 Factor. $xy - 6 + 3x - 2y$

$xy + 3x - 2y - 6$	*Rearrange the terms so that the first two terms have a common factor.*
$= x(y + 3) - 2(y + 3)$	*Remove a common factor of x from the first two terms and −2 from the second two terms.*
$= (y + 3)(x - 2)$	*Remove the common factor y + 3.*

Practice Problem 8 Factor. $xy - 12 - 4x + 3y$ ∎

EXAMPLE 9 Factor. $2x^3 + 21 - 7x^2 - 6x$ Check your answer by multiplication.

$2x^3 - 7x^2 - 6x + 21$	*Rearrange the terms.*
$= x^2(2x - 7) - 3(2x - 7)$	*Remove a common factor from each group of two terms.*
$= (2x - 7)(x^2 - 3)$	*Remove the common factor 2x − 7.*

Check:

$$(2x - 7)(x^2 - 3) = 2x^3 - 6x - 7x^2 + 21 \quad \textit{Multiply the two binomials.}$$
$$= 2x^3 + 21 - 7x^2 - 6x \quad \textit{Rearrange the terms.}$$

The product is identical to the original expression.

Practice Problem 9 Factor. $2x^3 - 15 - 10x + 3x^2$ ∎

5.4 Exercises

Factor. (Be sure to remove the greatest common factor.)

1. $30 - 15y$

2. $16x - 16$

3. $xy - 3x^2y$

4. $7a^2 - 14a$

5. $b^2x^2 + bx + b$

6. $a^3b^2 + a^2b^3 + a^2b^2$

7. $2x^3 - 8x^2 + 12x$

8. $3x^4 - 6x^3 + 9x^2$

9. $9a^2b^2 - 36ab + 45ab^2$

10. $14x^2y - 35xy - 63x$

11. $-12ab^4c^3 - 8a^3b^2c^2 + 4ab^2c^2$

12. $20x^2y^2z^2 - 30x^3y^3z^2 + 25x^2yz^2$

13. $5x^3 - 10x^2 + 3$

14. $8x^3y - 4xy^2 + 5$

15. $12xy^3 - 24x^3y^2 + 36x^2y^4 - 60x^4y^3$

16. $15a^3b^3 + 6a^4b^3 - 9a^2b^3 + 30a^5b^3$

17. $3x(x + y) - 2(x + y)$

18. $5a(a + 3b) + 4(a + 3b)$

19. $5b(a - 3b) + 8(-3b + a)$

20. $4y(x - 5y) - 3(-5y + x)$

21. $3x(a + 5b) + (a + 5b)$

22. $2w(s - 3t) - (s - 3t)$

23. $2a^2(3x - y) - 5b^3(3x - y)$

24. $7a^3(5a + 4) - 2(5a + 4)$

25. $3x(5x + y) - 8y(5x + y) - (5x + y)$

26. $4w(y - 8x) + 5z(y - 8x) + (y - 8x)$

27. $x^3 + 5x^2 + 3x + 15$

28. $x^3 + 8x^2 + 2x + 16$

29. $4x + 4 - 3wx - 3w$

30. $ax + a - 7bx - 7b$

31. $a^2 - ay - 3a + 3y$

32. $6xy - 3x - 2py + p$

33. $5ax - 15ay - 2bx + 6by$

34. $4ax - 10ay - 2bx + 5by$

35. $t^2y - 25 - 5y + 5t^2$

36. $5bc - a^3 - b + 5a^3c$

37. $28x^2 + 6y^2w + 8xy^2 + 21xw$

38. $18ax - 6bx + 9ay^2 - 3by^2$

39. $12a^3 + c^3 - 3a^2c - 4ac^2$

40. $6x^3 + 35wy - 14x^2y - 15xw$

41. Remove the common factor 7.37 from each term.
$14.74x - 22.11y + 58.96$

42. Remove the common factor 9.81 from each term.
$19.62x^2 - 29.43w + 147.15z^2$

Cumulative Review Problems

Graph the equation in 43 and 44.

43. $6x - 2y = -12$

44. $y = \dfrac{2}{3}x - 2$

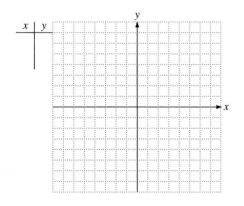

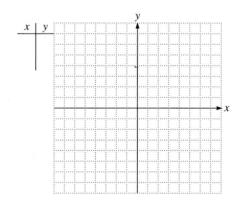

45. Find the slope of a line passing through $(6, -1)$ and $(2, 3)$.

46. Find the slope and y-intercept of $2y + 6x = -3$.

47. Reynaldo swims in a lap pool that is 50 meters long. He wants to swim at least one mile each day. How many laps (up and back the distance of the pool) will he have to swim, to reach his goal of one *mile* per day? (Use 1 mile = 1.61 kilometers.)

48. A student scored 82 on a test that had only 5-point fill-in questions and 4-point multiple choice questions. The student had 4 fill-in questions wrong. The maximum score on the test was 102 points. If the entire test had 22 questions, how many multiple choice questions did the student answer correctly?

5.5 Factoring Trinomials

After studying this section, you will be able to:

1. Factor trinomials of the form $x^2 + bx + c$.
2. Factor trinomials of the form $ax^2 + bx + c$.

1 Factoring Trinomials of the Form $x^2 + bx + c$

If we multiply $(x + 4)(x + 5)$, we obtain $x^2 + 9x + 20$. But suppose that we already have the polynomial $x^2 + 9x + 20$ and need to factor it. In other words, we need to find the expressions that when multiplied give us the polynomial. Let's use this example to find a general procedure.

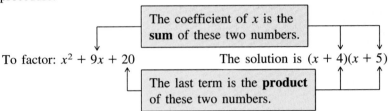

To factor: $x^2 + 9x + 20$ The solution is $(x + 4)(x + 5)$

The coefficient of x is the **sum** of these two numbers.

The last term is the **product** of these two numbers.

Factoring Trinomials of the Form $x^2 + bx + c$

1. The answer has the form $(x + m)(x + n)$, where m, n are real numbers.

2. The numbers m and n are chosen so that
   (a) $m \cdot n = c$
   (b) $m + n = b$

EXAMPLE 1   Factor. $x^2 + 8x + 15$

Because the trinomial has the form $x^2 + bx + c$, the answer has the form $(x + m)(x + n)$. Here $b = 8$ and $c = 15$. So we want to find the two numbers whose product is 15 but whose sum is 8. That is, $m \cdot n = 15$ and $m + n = 8$.

Factors of 15	Sum of the Factors
(15)(1)	$15 + 1 = 16$
(3)(5)	$3 + 5 = 8$ ✓

The factors whose sum is 8 are 3 and 5. Therefore, $m = 3$ and $n = 5$, and

$$x^2 + 8x + 15 = (x + 3)(x + 5)$$

Practice Problem 1   Factor. $x^2 + 14x + 48$ ∎

If the last term of the trinomial is positive and the middle term is negative, the two factors we seek will be negative numbers.

EXAMPLE 2 Factor. $x^2 - 14x + 24$

We want to find two numbers whose product is 24 but whose sum is -14. They will both be negative numbers.

Factors of 24	Sum of the Factors
$(-24)(-1)$	$-24 - 1 = -25$
$(-12)(-2)$	$-12 - 2 = -14$ ✓
$(-6)(-4)$	$-6 - 4 = -10$
$(-8)(-3)$	$-8 - 3 = -11$

The factors whose sum is -14 are -12 and -2. Thus,

$$x^2 - 14x + 24 = (x - 12)(x - 2)$$

Practice Problem 2 Factor. $x^2 - 10x + 21$ ∎

If the last term of the trinomial is negative, the two factors will be opposite in sign.

EXAMPLE 3 Factor. $x^2 + 11x - 26$

We want to find two numbers whose product is -26 but whose sum is $+11$. One number will be positive, the other negative.

Factors of -26	Sum of the Factors
$(-26)(1)$	$-26 + 1 = -25$
$(26)(-1)$	$26 - 1 = 25$
$(-13)(2)$	$-13 + 2 = -11$
$(13)(-2)$	$13 - 2 = 11$ ✓

The factors whose sum is $+11$ are -2 and 13. Thus,

$$x^2 + 11x - 26 = (x + 13)(x - 2)$$

Practice Problem 3 Factor. $x^2 - 13x - 48$ ∎

EXAMPLE 4 Factor. $x^4 - 2x^2 - 24$

You need to recognize that we can write this as $(x^2)^2 - 2(x^2) - 24$ or $y^2 + (-2)y + (-24)$, where $y = x^2$. So the factors will be $(y + m)(y + n)$. The two numbers whose product is -24 and whose sum is -2 are -6 and 4. Therefore, we have $(y - 6)(y + 4)$; but $y = x^2$, so our answer is

$$x^4 - 2x^2 - 24 = (x^2 - 6)(x^2 + 4)$$

Practice Problem 4 Factor. $x^4 + 9x^2 + 8$ ∎

Facts about Signs

To factor $x^2 + bx + c = (x + m)(x + n)$, we know certain facts about m and n.

1. m and n have the same sign if c is positive. (*Note:* We did *not* say they will have the same sign as c.)
   (a) They are positive if b is positive.
   (b) They are negative if b is negative.

2. m and n have opposite signs if c is negative. The larger number is positive if b is positive and negative if b is negative.

If you understand these sign facts, do Example 5. If not, review Examples 1 through 4.

EXAMPLE 5 Factor.

(a) $x^2 + 17x + 30$ **(b)** $x^2 - 11x + 28$
(c) $y^2 + 5y - 36$ **(d)** $x^4 - 4x^2 - 12$

(a) $x^2 + 17x + 30 = (x + 15)(x + 2)$ *Fact* 1(a).
(b) $x^2 - 11x + 28 = (x - 4)(x - 7)$ *Fact* 1(b).
(c) $y^2 + 5y - 36 = (y + 9)(y - 4)$ *The larger number (9) is positive because b is positive.*
(d) $x^4 - 4x^2 - 12 = (x^2 - 6)(x^2 + 2)$ *The larger number (6) is negative because b is negative.*

Practice Problem 5 Factor.

(a) $x^2 + 15x + 50$ **(b)** $x^2 - 12x + 35$
(c) $a^2 + 2a - 48$ **(d)** $x^4 + 2x^2 - 15$ ∎

Does the order in which we write the factors make any difference? In other words, is it true that $x^2 + bx + c = (x + n)(x + m)$? Since multiplication is commutative,

$$x^2 + bx + c = (x + n)(x + m) = (x + m)(x + n)$$

The order of the parentheses is not important.
    We can also factor trinomials that have more than one variable.

EXAMPLE 6 Factor.

(a) $x^2 - 21xy + 20y^2$ **(b)** $x^2 + 4xy - 21y^2$

(a) $x^2 - 21xy + 20y^2 = (x - 20y)(x - y)$

> The last terms in each factor contain the variable y.

(b) $x^2 + 4xy - 21y^2 = (x + 7y)(x - 3y)$

Practice Problem 6 Factor.

(a) $x^2 - 16xy + 15y^2$ **(b)** $x^2 + xy - 42y^2$ ∎

If each term of a trinomial has a common factor, you should remove the greatest common factor from each term first. Then you will be able to follow the factoring procedure we have used in the previous examples.

EXAMPLE 7 Factor $3x^2 - 30x + 48$

The factor 3 is common to all three terms of the polynomial. Factoring out the 3 gives us the following:

$$3x^2 - 30x + 48 = 3(x^2 - 10x + 16)$$

Now we continue to factor the trinomial in the usual fashion.

$$3(x^2 - 10x + 16) = 3(x - 8)(x - 2)$$

Practice Problem 7 Factor. $4x^2 - 44x + 72$ ■

2 Factoring Trinomials of the Form $ax^2 + bx + c$

Grouping Number Method

One way to factor a trinomial $ax^2 + bx + c$ is to write it as four terms and factor it by grouping as we did in Section 5.4. For example, the trinomial $2x^2 + 11x + 12$ can be written as $2x^2 + 3x + 8x + 12$.

$$2x^2 + 3x + 8x + 12 = x(2x + 3) + 4(2x + 3)$$
$$= (2x + 3)(x + 4)$$

We can factor all factorable trinomials of the form $ax^2 + bx + c$ in this way. Use the following procedure.

Grouping Number Method for Factoring Trinomials of the Form $ax^2 + bx + c = 0$

1. Obtain the grouping number ac.
2. Find the factors of the grouping number whose sum is b.
3. Use those two factors to write bx as the sum of the terms.
4. Factor by grouping.

EXAMPLE 8 Factor. $2x^2 + 19x + 24$

1. The grouping number is $(2)(24) = 48$.
2. The factors of 48 are

$$48 \cdot 1 \qquad 12 \cdot 4$$
$$24 \cdot 2 \qquad 8 \cdot 6$$
$$16 \cdot 3$$

We want the two factors of 48 whose sum is 19. Therefore, we select the factors 16 and 3.

3. We use the numbers 16 and 3 to write $19x$ as the sum of $16x$ and $3x$.

$$2x^2 + 19x + 24 = 2x^2 + 16x + 3x + 24$$

4. Factor by grouping.

$$2x^2 + 16x + 3x + 24 = 2x(x + 8) + 3(x + 8)$$
$$= (x + 8)(2x + 3)$$

Practice Problem 8 Factor. $3x^2 + 2x - 8$ ▪

EXAMPLE 9 Factor. $6x^2 + 7x - 5$

1. The grouping number is -30.
2. We want the factors of -30 whose sum is $+7$

$$-30 = (-30)(+1) \qquad -30 = (+5)(-6)$$
$$= (+30)(-1) \qquad\qquad = (-5)(+6)$$
$$= (+15)(-2) \qquad\qquad = (+3)(-10)$$
$$= (-15)(+2) \qquad\qquad = (-3)(+10)$$

3. Use -3 and 10 to write $6x^2 + 7x - 5$ with four terms.

$$6x^2 + 7x - 5 = 6x^2 - 3x + 10x - 5$$

4. Factor by grouping.

$$6x^2 - 3x + 10x - 5 = 3x(2x - 1) + 5(2x - 1)$$
$$= (2x - 1)(3x + 5)$$

Practice Problem 9 Factor. $10x^2 - 9x + 2$ ▪

If each of the three terms has a common factor, then prior to using the four-step procedure, we first remove the greatest common factor from each term of the trinomial.

EXAMPLE 10 Factor. $6x^3 - 26x^2 + 24x$

First we remove the common factor $2x$ from each term.

$$6x^3 - 26x^2 + 24x = 2x(3x^2 - 13x + 12)$$

Next we follow the four steps to factor $3x^2 - 13x + 12$.

1. The grouping number is 36.
2. We want the factors of 36 whose sum is -13. The two factors are -4 and -9.
3. We use -4 and -9 to write $3x^2 - 13x + 12$ with four terms.

$$3x^2 - 13x + 12 = 3x^2 - 4x - 9x + 12$$

4. Factor by grouping. Remember that we first removed the factor $2x$. This factor will remain and be part of the answer.

$$2x[3x^2 - 4x - 9x + 12] = 2x[x(3x - 4) - 3(3x - 4)]$$
$$= 2x(3x - 4)(x - 3)$$

Practice Problem 10 Factor. $9x^3 - 15x^2 - 6x$ ▪

Another way to factor trinomials $ax^2 + bx + c$ is by trial and error. This method has some advantages if the grouping number is large and we would have to list many factors. In the trial-and-error method we try different values and see which can be multiplied out to obtain the original expression.

EXAMPLE 11 Factor by trial and error. $2x^2 + 11x + 14$

We know we will use the factors of $2x^2$, which are $(2x)(x)$. (If each factor is to contain a variable, there is no other way to factor $2x^2$.) Our answer will thus be in the form

$$(2x + ?)(x + ?)$$

The two unknown numbers must be factors of 14, but we have several choices. We list them all and multiply out the inner and outer products to obtain the middle term of the product.

Possible Factors	Middle Term of Product
$(2x + 14)(x + 1)$	$+16x$
$(2x + 1)(x + 14)$	$+29x$
$(2x + 2)(x + 7)$	$+16x$
$(2x + 7)(x + 2)$	$+11x$

Thus,

$$2x^2 + 11x + 14 = (2x + 7)(x + 2)$$

Practice Problem 11 Factor by trial and error. $3x^2 + 31x + 10$ ■

If the last term is negative, there can be many more sign possibilities.

EXAMPLE 12 Factor by trial and error. $10x^2 - 49x - 5$

The first terms could have factors of $(10x)$ and (x) or $(5x)$ and $(2x)$. The second terms could have factors of $(+1)$ and (-5) or (-1) and $(+5)$. We list all the possibilities and look for one that will yield a middle term of $-49x$.

Possible Factors	Middle Term of Product
$(2x - 1)(5x + 5)$	$+5x$
$(2x + 1)(5x - 5)$	$-5x$
$(2x + 5)(5x - 1)$	$+23x$
$(2x - 5)(5x + 1)$	$-23x$
$(10x - 5)(x + 1)$	$+5x$
$(10x + 5)(x - 1)$	$-5x$
$(10x - 1)(x + 5)$	$+49x$
$(10x + 1)(x - 5)$	$-49x$

Thus,

$$10x^2 - 49x - 5 = (10x + 1)(x - 5)$$

As a check, it is always a good idea to multiply out the two binomials and see if you obtain the original expression.

We check by multiplying: $(10x + 1)(x - 5) = 10x^2 - 50x + 1x - 5$
$$= 10x^2 - 49x - 5$$

Practice Problem 12 Factor by trial and error. $8x^2 - 6x - 5$ ∎

EXAMPLE 13 Factor by trial and error. $6x^4 + x^2 - 12$

The first term of each factor must contain an x^2. Suppose that we try

Possible Factors	*Middle Term of Product*
$(2x^2 - 3)(3x^2 + 4)$	$-x^2$

We want a middle term that is opposite in sign since the original middle term was $+x^2$. In this case, we just need to reverse the signs of our possible factors. Do you see why? Therefore,

$$6x^4 + x^2 - 12 = (2x^2 + 3)(3x^2 - 4)$$

Practice Problem 13 Factor by trial and error. $6x^4 + 13x^2 - 5$ ∎

Developing Your Study Skills

KEEP TRYING

You may be one of those students who has had much difficulty with mathematics in the past and who is sure that you cannot do well in this course. Perhaps you are thinking, "I have never been any good at mathematics," or "I have always hated mathematics," or "Math always scares me," or "I have not had any math for so long that I have forgotten it all." You may even have picked up on the label "math anxiety" and attached it to yourself. That is most unfortunate, and it is time for you to reprogram your thinking. Replace those negative thoughts with more positive ones. You need to say things like, "I will give this math class my best shot," or "I can learn mathematics if I work at it," or "I will try to do better than I have done in previous math classes." You will be pleasantly surprised at the difference this more positive attitude makes!

We live in a highly technical world, and you cannot afford to give up on the study of mathematics. Dropping mathematics may prevent you from entering certain career fields that you may find interesting. You may not have to take math courses at as high a level as calculus, but such courses as finite math, college algebra, and trigonometry may be necessary. Learning mathematics can open new doors for you.

Learning mathematics is a process that takes time and effort. You will find that regular study and daily practice are necessary to strengthen your skills and to help you grow academically. This process will lead you toward success in mathematics. Then, as you become more successful, your confidence in your ability to do mathematics will grow.

5.5 Exercises

Factor each polynomial. In each case the coefficient of the first variable is 1.

1. $x^2 + 9x + 8$

2. $x^2 + 8x + 7$

3. $x^2 + 7x - 18$

4. $x^2 + 8x - 20$

5. $x^2 + x - 30$

6. $x^2 - x - 6$

7. $x^2 + 8x + 12$

8. $x^2 + 12x + 35$

9. $a^2 - 13a + 30$

10. $a^2 - 6a - 16$

11. $a^2 + 4a - 45$

12. $a^2 + 17a + 60$

13. $x^2 - 9xy + 20y^2$

14. $x^2 - 6xy - 27y^2$

15. $x^2 + 5xy - 14y^2$

16. $x^2 + 7xy + 10y^2$

17. $x^4 - 3x^2 - 40$

18. $x^4 + 6x^2 + 5$

19. $x^4 + 16x^2y^2 + 63y^4$

20. $x^4 - 6x^2 - 55$

Remove the greatest common factor from each term of the trinomial. Then factor the remaining trinomial.

21. $2x^2 + 26x + 44$

22. $2x^2 + 30x + 52$

23. $x^3 + 9x^2 - 36x$

24. $x^3 + 11x^2 - 42x$

Factor each polynomial. In each case the coefficient of the first variable is not *1. You may use either method we have studied.*

25. $2x^2 - 7x + 3$

26. $3x^2 + 22x + 7$

27. $30x^2 - x - 1$

28. $6x^2 + x - 1$

29. $6x^2 - 7x - 5$

30. $5x^2 - 13x - 28$

31. $3a^2 - 8a + 5$

32. $6a^2 + 11a + 3$

33. $8a^2 + 14a - 9$

34. $3a^2 - 20a + 12$

35. $6x^2 - 13x + 6$

36. $4x^2 + 4x - 15$

37. $2x^2 + 13x + 15$

38. $5x^2 - 8x - 4$

39. $3x^4 - 8x^2 - 3$

40. $6x^4 + 7x^2 - 5$

41. $6x^2 + 35xy + 11y^2$

42. $5x^2 + 12xy + 7y^2$

43. $2x^2 - 13xy + 15y^2$

44. $9x^2 - 13xy + 4y^2$

Remove the greatest common factor from each term of the trinomial. Then factor the remaining trinomial.

45. $4x^3 + 4x^2 - 15x$

46. $8x^3 + 6x^2 - 9x$

47. $18x^3 + 21x^2 + 6x$

48. $9x^3 + 30x^2 + 9x$

Mixed Practice

Factor each polynomial.

49. $x^2 - 2x - 63$

50. $x^2 + 6x - 40$

51. $6x^2 + x - 2$

52. $5x^2 + 17x + 6$

53. $x^2 - 20x + 51$

54. $x^2 - 20x + 99$

55. $15x^2 + x - 2$

56. $12x^2 - 5x - 3$

57. $2x^2 + 4x - 96$

58. $3x^2 + 9x - 84$

59. $18x^2 + 21x + 6$

60. $24x^2 + 26x + 6$

61. $4x^3 - 16x^2 - 48x$

62. $2x^3 - 6x^2 - 20x$

63. $6x^3 + 26x^2 - 20x$

64. $12x^3 - 14x^2 + 4x$

65. $3x^4 - 2x^2 - 5$

66. $6x^4 - 13x^2 - 5$

67. $7x^2 - 22xy + 3y^2$

68. $10x^2 - 17xy + 6y^2$

69. $x^6 - 10x^3 - 39$

70. $x^6 - 3x^3 - 70$

71. $6x^3 + 5x^2 - 4x$

72. $4x^3 + 4x^2 - 3x$

Cumulative Review Problems

73. Find the area of a circle of radius 3 inches.

74. Solve for b. $A = \dfrac{1}{2}(2a + 5b)$

75. Find the equation in standard form of a line parallel to $3y + x = 10$ that passes through $(6, -4)$.

76. Graph. $6x + 4y = -12$

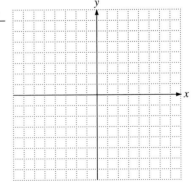

77. John and Carolyn Ciukaj have opened a new bicycle shop in Beverly. They want to have at least 120 bike racks and bicycle helmets in stock. Their wholesale cost for bike racks averages around $60 and for bicycle helmets around $70. They have available $7950 in capital to pay for stocking bike racks and bicycle helmets. How many of each should they stock?

78. Cheryl Finkelstein thinks the perfect cup of coffee is achieved when she adds 0.2 ounces of milk to her 8.0 ounces of coffee. However, then the cup is too full and often spills. So she always pours out exactly 0.2 ounces of coffee first and then adds the milk. Exactly how much milk should she add to obtain the perfect cup of coffee?

5.6 Special Cases of Factoring

MathPro

Video 14

SSM

After studying this section, you will be able to:

1 *Factor a binomial that is the difference of two squares.*
2 *Factor a perfect square trinomial.*
3 *Factor a binomial that is the sum or difference of two cubes.*

1 *Difference of Two Squares*

Recall the special product formula (3.1): $(a + b)(a - b) = a^2 - b^2$. We can use it now as a factoring formula.

Factoring the Difference of Two Squares
$$a^2 - b^2 = (a + b)(a - b)$$

EXAMPLE 1   Factor. $x^2 - 16$

In this case $a = x$ and $b = 4$ in the formula.

$$a^2 \; - \; b^2 \; = (a + b)(a - b)$$
$$(x)^2 - (4)^2 = (x + 4)(x - 4)$$

Practice Problem 1   Factor. $x^2 - 9$ ∎

EXAMPLE 2   Factor.   **(a)** $25x^2 - 36$ **(b)** $16x^2 - 81y^2$

In each case we will use the formula $a^2 - b^2 = (a + b)(a - b)$.
(a) $25x^2 - 36 = (5x)^2 - (6)^2 = (5x + 6)(5x - 6)$
(b) $16x^2 - 81y^2 = (4x)^2 - (9y)^2 = (4x + 9y)(4x - 9y)$

Practice Problem 2   Factor. $64x^2 - 121y^2$ ∎

EXAMPLE 3   Factor.   **(a)** $121x^6 - 1$ **(b)** $100w^4 - 9z^4$
(a) $121x^6 - 1 = (11x^3)^2 - (1)^2 = (11x^3 + 1)(11x^3 - 1)$
(b) $100w^4 - 9z^4 = (10w^2)^2 - (3z^2)^2 = (10w^2 + 3z^2)(10w^2 - 3z^2)$

Practice Problem 3   Factor.

(a) $49x^2 - 25y^4$ **(b)** $100x^6 - 1$ ∎

Whenever possible, common factors should be removed in the first step. Then the formula can be applied.

EXAMPLE 4 Factor. $75x^2 - 3$

We remove a common factor of 3 from each term.

$$75x^2 - 3 = 3(25x^2 - 1)$$
$$= 3(5x + 1)(5x - 1)$$

Practice Problem 4 Factor. $7x^2 - 28$ ∎

---

2 *Perfect Square Trinomials*

Recall our formulas for squaring a binomial:

$$(a - b)^2 = a^2 - 2ab + b^2 \qquad (3.2)$$
$$(a + b)^2 = a^2 + 2ab + b^2 \qquad (3.3)$$

We can use these formulas to factor a perfect square trinomial.

Perfect Square Factoring Formulas
$$a^2 - 2ab + b^2 = (a - b)^2$$
$$a^2 + 2ab + b^2 = (a + b)^2$$

Recognizing these special cases will save you a lot of time when factoring. How can we recognize a perfect square trinomial?

1. The first and last terms are perfect squares. (The numerical values are 1, 4, 9, 16, 25, 36, . . . , and the variables have an exponent that is an even whole number.)

2. The middle term is twice the product of the values that when squared give the first and last terms.

EXAMPLE 5 Factor. $25x^2 - 20x + 4$

Is this trinomial a perfect square? Yes.

1. The first and last terms are perfect squares.

$$25x^2 - 20x + 4 = (5x)^2 - 20x + (2)^2$$

2. The middle term is twice the product of the value $5x$ and the value 2. In other words, $2(5x)(2) = 20x$.

$$(5x)^2 - 2(5x)(2) + (2)^2 = (5x - 2)^2$$

Therefore, we can use the formula $a^2 - 2ab + b^2 = (a - b)^2$. Thus,

$$25x^2 - 20x + 4 = (5x - 2)^2$$

Practice Problem 5 Factor. $9x^2 - 30x + 25$ ∎

EXAMPLE 6 Factor. $16x^2 - 24x + 9$

1. The first and last terms are perfect squares because $(4x)^2 = 16x^2$ and $(3)^2 = 9$.
2. The middle term is twice the product of $(4x)(3)$. Therefore,

$$a^2 - 2ab + b^2 = (a - b)^2$$
$$16x^2 - 24x + 9 = (4x)^2 - 2(4x)(3) + (3)^2$$
$$16x^2 - 24x + 9 = (4x - 3)^2$$

Practice Problem 6 Factor. $25x^2 - 70x + 49$ ∎

EXAMPLE 7 Factor. $200x^2 + 360x + 162$

First we remove the common factor of 2.

$$200x^2 + 360x + 162 = 2(100x^2 + 180x + 81)$$
$$a^2 + 2ab + b^2 = (a + b)^2$$
$$2[100x^2 + 180x + 81] = 2[(10x)^2 + (2)(10x)(9) + (9)^2]$$
$$= 2(10x + 9)^2$$

Practice Problem 7 Factor. $242x^2 + 88x + 8$ ∎

EXAMPLE 8 Factor.

(a) $x^4 + 14x^2 + 49$ **(b)** $9x^4 + 30x^2y^2 + 25y^4$

(a) $x^4 + 14x^2 + 49 = (x^2)^2 + 2(x^2)(7) + (7)^2$
$$= (x^2 + 7)^2$$
(b) $9x^4 + 30x^2y^2 + 25y^4 = (3x^2)^2 + 2(3x^2)(5y^2) + (5y^2)^2$
$$= (3x^2 + 5y^2)^2$$

Practice Problem 8 Factor.   **(a)** $49x^4 + 28x^2 + 4$
(b) $36x^4 + 84x^2y^2 + 49y^4$ ∎

3 Sum or Difference of Two Cubes

There are also special formulas for factoring cubic binomials. Review your solutions to Problems 21 and 22 of Exercises 5.2. We see that the factors of $x^3 + 27$ are $(x + 3)(x^2 - 3x + 9)$ and that the factors of $x^3 - 64$ are $(x - 4)(x^2 + 4x + 16)$. Therefore, we can generalize this pattern and derive the following factoring formulas.

Sum and Difference of Cubes Factoring Formulas
$$a^3 + b^3 = (a + b)(a^2 - ab + b^2)$$
$$a^3 - b^3 = (a - b)(a^2 + ab + b^2)$$

EXAMPLE 9 Factor. $x^3 + 8$

Here $a = x$ and $b = 2$.

$$a^3 + b^3 = (a + b)(a^2 - ab + b^2)$$

$$x^3 + 8 = (x)^3 + (2)^3 = (x + 2)(x^2 - 2x + 4)$$

Practice Problem 9 Factor. $27x^3 + 64$ ∎

EXAMPLE 10 Factor. $125x^3 + y^3$

Here $a = 5x$ and $b = y$.

$$a^3 + b^3 = (a + b)(a^2 - ab + b^2)$$

$$125x^3 + y^3 = (5x)^3 + (y)^3 = (5x + y)(25x^2 - 5xy + y^2)$$

Practice Problem 10 Factor. $8x^3 + 125y^3$ ∎

EXAMPLE 11 Factor. $64x^3 - 27$

Here $a = 4x$ and $b = 3$.

$$a^3 - b^3 = (a - b)(a^2 + ab + b^2)$$

$$64x^3 - 27 = (4x)^3 - (3)^3 = (4x - 3)(16x^2 + 12x + 9)$$

Practice Problem 11 Factor. $64x^3 - 125y^3$ ∎

EXAMPLE 12 Factor. $125w^3 - 8z^6$

Here $a = 5w$ and $b = 2z^2$.

$$a^3 - b^3 = (a - b)(a^2 + ab + b^2)$$

$$125w^3 - 8z^6 = (5w)^3 - (2z^2)^3 = (5w - 2z^2)(25w^2 + 10wz^2 + 4z^4)$$

Practice Problem 12 Factor. $27w^3 - 125z^6$ ∎

EXAMPLE 13   Factor. $250x^3 - 2$

First we must remove the common factor of 2.

$$250x^3 - 2 = 2(125x^3 - 1)$$
$$= 2(5x - 1)(25x^2 + 5x + 1)$$
$$\uparrow$$

Note that this trinomial cannot be factored.

Practice Problem 13   Factor. $54x^3 - 16$ ∎

Sometimes, polynomials are combinations of two forms. For example, what should you do if a problem is the difference of two cubes *and* the difference of two squares? Usually, it's better to use the difference of two squares formula first. Then apply the difference of two cubes formula.

EXAMPLE 14   Factor. $x^6 - y^6$

We can write this binomial as $(x^2)^3 - (y^2)^3$ or as $(x^3)^2 - (y^3)^2$. Therefore, we can use either the difference of two cubes formula or the difference of two squares formula. It's usually better to use the difference of two squares formula first, so we'll do that.

$$x^6 - y^6 = (x^3)^2 - (y^3)^2$$

Here $a = x^3$ and $b = y^3$. Therefore,

$$(x^3)^2 - (y^3)^2 = (x^3 + y^3)(x^3 - y^3)$$

Now we use the sum of two cubes formula for the first factor and the difference of two cubes formula for the second factor.

$$x^3 + y^3 = (x + y)(x^2 - xy + y^2)$$
$$x^3 - y^3 = (x - y)(x^2 + xy + y^2)$$

Hence,

$$x^6 - y^6 = (x + y)(x^2 - xy + y^2)(x - y)(x^2 + xy + y^2)$$

Practice Problem 14   Factor. $64a^6 - 1$ ∎

You'll see these special cases of factoring often. You should memorize the following formulas.

Special Cases of Factoring

Difference of Two Squares

$$a^2 - b^2 = (a + b)(a - b)$$

Perfect Square Trinomial

$$a^2 - 2ab + b^2 = (a - b)^2$$
$$a^2 + 2ab + b^2 = (a + b)^2$$

Sum and Difference of Cubes

$$a^3 + b^3 = (a + b)(a^2 - ab + b^2)$$
$$a^3 - b^3 = (a - b)(a^2 + ab + b^2)$$

5.6 Exercises

Use the difference of two squares formula to factor. Be sure to remove any common factors.

1. $x^2 - 25$

2. $x^2 - 36$

3. $49x^2 - 4$

4. $100x^2 - 9$

5. $64x^2 - 1$

6. $81x^2 - 1$

7. $t^4 - 1$

8. $w^4 - z^4$

9. $81x^4 - 1$

10. $16x^4 - 1$

11. $49m^2 - 9n^2$

12. $36x^2 - 25y^2$

13. $1 - 81x^2y^2$

14. $1 - 49x^2y^2$

15. $81x^2 - 121$

16. $144x^2 - 9$

17. $32x^2 - 18$

18. $50x^2 - 8$

19. $5x - 20x^3$

20. $49x^3 - 36x$

Use the perfect square trinomial formula to factor. Be sure to remove any common factors.

21. $49x^2 - 14x + 1$

22. $9x^2 - 6x + 1$

23. $w^2 - 6w + 9$

24. $w^2 - 12w + 36$

25. $36x^2 - 12x + 1$

26. $4x^2 - 4x + 1$

27. $z^2 + 16z + 64$

28. $z^2 + 18x + 81$

29. $81w^2 + 36wt + 4t^2$

30. $25w^2 + 20wt + 4t^2$

31. $25x^2 - 40xy + 16y^2$

32. $49x^2 - 70xy + 25y^2$

33. $8x^2 + 24x + 18$

34. $128x^2 + 32x + 2$

35. $3x^3 - 24x^2 + 48x$

36. $50x^3 - 20x^2 + 2x$

Use the sum and difference of cubes factoring formulas to factor. Be sure to remove any common factors.

37. $8x^3 + 27$

38. $64x^3 + 27$

39. $x^3 + 125$

40. $x^3 + 64$

41. $64x^3 - 1$

42. $125x^3 - 1$

43. $125x^3 - 8$

44. $27x^3 - 64$

45. $1 - 27x^3$

46. $1 - 8x^3$

47. $64x^3 + 125$

48. $27x^3 + 216$

49. $64s^6 + t^6$

50. $125s^6 + t^6$

51. $6y^3 - 6$

52. $80y^3 - 10$

53. $64x^4 + 27x$

54. $64x^4 + 125x$

55. $x^5 - 8x^2y^3$

56. $x^5 - 27x^2y^3$

Mixed Practice   *Factor by the methods of this section.*

57. $25w^6 - 1$

58. $x^8 - 1$

59. $8w^6 + 8w^3 + 2$

60. $9w^4 + 12w^2 + 4$

61. $8a^3 - 27b^3$

62. $27w^3 + 125$

63. $125m^3 + 8n^3$

64. $64z^3 - 27w^3$

65. $9x^2 - 100y^2$

66. $49 - 64a^2b^2$

67. $4w^2 - 20wz + 25z^2$

68. $9x^2y^2 + 24xy + 16$

69. $36a^2 - 81b^2$

70. $121x^4 - 4y^2$

71. $64x^5 + x^2y^3z^3$

72. $w^4z^3 - 8wy^3$

73. $81x^4 - 36x^2 + 4$

74. $121 + 66y^2 + 9y^4$

75. $16x^4 - 81y^4$

76. $256x^4 - 1$

Try to factor the following four problems by using the formulas for the perfect square trinomial. Why can't the formulas be used? Then factor each problem correctly using an appropriate method.

77. $25x^2 + 25x + 4$

78. $16x^2 + 40x + 9$

79. $4x^2 - 15x + 9$

80. $36x^2 - 65x + 25$

To Think About   *Factor completely.*

81. $81x^{16} - 256$

82. $8x^{15} + 343y^{21}$

83. $121x^{16} - 110x^8y^{10} + 25y^{20}$

84. $m^6 - 64n^6$

Cumulative Review Problems

85. Peggy has an apartment in Atlanta with her husband Scott. Peggy ran the hot water tap for 5 minutes, realized it was too hot, and then ran the cold water tap for 4 minutes. She used 140 gallons of water. The water temperature was just right, and she enjoyed a relaxing bath. Later, while she was doing her math homework, her husband Scott got home from work. He ran the hot water tap for two minutes and then the cold water tap for three minutes. He is a much larger person, so he only used 84 gallons of water. He jumped in the tub and also enjoyed a relaxing bath. Scott asked Peggy if she knew how many gallons of water per minute came out of the cold water tap and the hot water tap. Peggy knew the answer, do you?

86. Belinda invested $4000 in mutual funds. In one year she earned $482. Part was invested at 14% and the remainder at 11%. How much did she invest at each rate?

87. A triangular circuit board has a perimeter of 66 centimeters. The first side is two-thirds as long as the second side. The third side is 14 centimeters shorter than the second side. Find the length of each side.

88. Three friends each bought a portable compact disc player. The total for the three purchases was $858. Melinda paid $110 more than Hector. Alice paid $86 less than Hector. How much did each person pay?

5.7 Completely Factoring a Polynomial

After studying this section, you will be able to:

1 *Recognize the type of factoring and accomplish the steps of factoring for any factorable polynomial.*

2 *Recognize polynomials that are prime.*

1 *Recognizing the Type of Factoring for Factorable Polynomials*

Not all polynomials have the convenient form of one of the special formulas. Most do not. The following procedure will help you to handle these common cases. You must practice this procedure until you can *recognize the various forms* and *determine which one to use.*

Completely Factoring a Polynomial

1. *First,* check for a common factor. Remove the greatest common factor (if there is one) before doing anything else.

2. **(a)** If the remaining polynomial has two terms, try to factor as one of the following:
   (1) The difference of two squares: $a^2 - b^2 = (a + b)(a - b)$ or
   (2) The difference of two cubes: $a^3 - b^3 = (a - b)(a^2 + ab + b^2)$ or
   (3) The sum of two cubes: $a^3 + b^3 = (a + b)(a^2 - ab + b^2)$
   (b) If the polynomial has three terms, try to factor it as one of the following:
   (1) A perfect square trinomial:
   $$a^2 + 2ab + b^2 = (a + b)^2 \quad \text{or} \quad a^2 - 2ab + b^2 = (a - b)^2 \quad \text{or}$$
   (2) A general trinomial of the form $x^2 + bx + c$ or the form $ax^2 + bx + c$
   (c) If the polynomial has four terms, try to factor by grouping.

EXAMPLE 1 Factor completely.

(a) $2x^2 - 18$ **(b)** $27x^4 - 8x$ **(c)** $27x^2 + 36xy + 12y^2$

(d) $2x^2 - 100x + 98$ **(e)** $6x^3 + 11x^2 - 10x$ **(f)** $5ax + 5ay - 20x - 20y$

(a) $2x^2 - 18 = 2(x^2 - 9)$ *Remove the common factor.*
$= 2(x + 3)(x - 3)$ *Use $a^2 - b^2 = (a + b)(a - b)$.*

(b) $27x^4 - 8x = x(27x^3 - 8)$ *Remove the common factor.*
$= x(3x - 2)(9x^2 + 6x + 4)$ *Use $a^3 - b^3 = (a - b)(a^2 + ab + b^2)$.*

(c) $27x^2 + 36xy + 12y^2 = 3(9x^2 + 12xy + 4y^2)$ *Remove the common factor.*
$= 3(3x + 2y)^2$ *Use $(a + b)^2 = a^2 + 2ab + b^2$.*

(d) $2x^2 - 100x + 98 = 2(x^2 - 50x + 49)$ *Remove the common factor.*
$= 2(x - 49)(x - 1)$ *The trinomial has the form $x^2 + bx + c$.*

(e) $6x^3 + 11x^2 - 10x = x(6x^2 + 11x - 10)$ *Remove the common factor.*
$= x(3x - 2)(2x + 5)$ *The trinomial has the form $ax^2 + bx + c$.*

(f) $5ax + 5ay - 20x - 20y = 5(ax + ay - 4x - 4y)$ *Remove the common factor.*
$= 5[a(x + y) - 4(x + y)]$ *Factor by grouping.*
$= 5(x + y)(a - 4)$ *Remove the common factor.*

Practice Problem 1 Factor completely.

(a) $7x^5 + 56x^2$ **(b)** $125x^2 + 50xy + 5y^2$ **(c)** $12x^2 - 75$

(d) $3x^2 - 39x + 126$ **(e)** $6ax + 6ay + 18bx + 18by$ **(f)** $6x^3 - x^2 - 12x$ ■

Can all polynomials be factored? No. Most polynomials cannot be factored. If a polynomial cannot be factored, it is said to be **prime.**

EXAMPLE 2   If possible, factor. $x^2 + 7x - 12$

This trinomial has the form $x^2 + bx + c$. To factor it, we must write it as $(x + ?)(x + ?)$ and try to find two numbers whose product is -12 and whose sum is $+7$. The chart shows there are none.

Factors of -12	*Sum of the Factors*
$(4)(-3)$	$+1$
$(-4)(3)$	-1
$(6)(-2)$	$+4$
$(2)(-6)$	-4
$(1)(-12)$	-11
$(-1)(12)$	$+11$

Therefore, the polynomial is *prime—it cannot be factored.*

Practice Problem 2   If possible, factor. $x^2 - 3x - 20$ ∎

EXAMPLE 3   If possible, factor. $6x^2 + 10x + 3$

The trinomial has the form $ax^2 + bx + c$. The grouping number is 18. If it can be factored, we must find two numbers whose product is 18 and whose sum is 10.

Factors of 18	*Sum of the Factors*
$(18)(1)$	19
$(6)(3)$	9
$(9)(2)$	11

There are no factors meeting the necessary conditions. Thus, the polynomial is *prime.* (If you use the trial-and-error method, try all the possible factors and show that none of them has a middle term of $10x$.)

Practice Problem 3   If possible, factor. $3x^2 - 10x + 4$ ∎

EXAMPLE 4   Factor if possible. $25x^2 + 49$

Unless there is a common factor that can be removed, binomials of the form $a^2 + b^2$ cannot be factored.
Therefore, $25x^2 + 49$ is *prime.*

Practice Problem 4   Factor if possible. $16x^2 + 81$ ∎

5.7 Exercises

Verbal and Writing Skills

1. In any factoring problem the first step is to _____.

2. If $x^2 + bx + c = (x + e)(x + f)$ and c is positive and b is negative, what can you say about the sign of e and of f?

Mixed Practice

Factor, if possible. Be sure to factor completely.

3. $3xy - 6yz$

4. $x^2 - x - 56$

5. $2x^2 + 5x - 3$

6. $ax - 2xy + 3aw - 6wy$

7. $8x^3 - 125y^3$

8. $27x^3 + 64y^3$

9. $x^2 + 2xy - xz$

10. $x^2 + 16$

11. $2ab^2 - 50a$

12. $x^3 - 11x^2 + 30x$

13. $3x^3 + 3x^2 - 6x$

14. $12y^3 - 36y^2 + 27y$

15. $6x^2 - 23x - 4$

16. $25x^2 - 40x + 16$

17. $x^2 - x + y - xy$

18. $ac + bc - a - b$

19. $16x^4 - 2x$

20. $1 - 16x^4$

21. $2x^5 - 16x^3 - 18x$

22. $8a^3b - 50ab^3$

23. $3x^4 + 27x^2$

24. $12x^3 + 3xy^2$

25. $6x^3 - 9x^2 - 15x$

26. $3x^2 + 2xy - 7y^2$

27. $4x^2 - 8x - 6$

28. $8x^2 + 10x - 12$

29. $y^2 - 5y + 7$

30. $2a^6 + 20a^5b + 50a^4b^2$

31. $2a^2 - 24a + 70$

32. $7a^2 + 14a - 168$

33. $-3ax + 2a^2y - 6ay + a^2x$

34. $8w^3 + awx + 4aw^2 + 2w^2x$

35. $64 + 49y^2$

36. $x^4 + 13x^3 + 36x^2$

37. $2x^4 - 3x^2 - 5$

38. $50x^2y^2 - 32y^2$

39. $12x^2 + 11x - 2$

40. $2x^5 - 3x^4 + x^3$

41. $54x^6y^6 - 2y^6$

42. $3x^6 + 7x^3 + 2$

43. $4x^4 + 20x^2y^4 + 25y^8$

44. $81x^4z^6 - 25y^8$

To Think About

45. $2xy + 3y + 6 + 4x = (2x + 3)(y + b)$. What is the value of b?

46. $25x^2 + ax + 169 = (5x - 13)^2$. What is the value of a?

Cumulative Review Problems

Solve the following inequalities.

47. $3x - 2 \leq -5 + 2(x - 3)$

48. $|2 + 5x - 3| < 2$

49. $\left| \dfrac{1}{3}(5 - 4x) \right| > 4$

50. $x - 4 \geq 7$ or $4x + 1 \leq 17$

Use the bar graph at the right to answer the following.

51. Examining this entire time period, what was the average value of the net receipts for a two-year period for the Republican party?

52. Examining this entire time period, what was the average percent of increase from one two year period to the next for the Democrat party? (*Hint:* First find the percentage of increase from 1985–86 and 1989–90, and then find the percentage of increase from 1989–90 to 1993–94. Then average the two values.)

53. Assuming the percentage of increase obtained in Problem 52, what would be the expected net receipts for the Democratic Party in the 1997–98 period?

54. If the percentage of increase from the period 1989–90 to the period 1993–94 continued at the same rate what would be the expected net receipts for the Democrat party in the 1997–98 period?

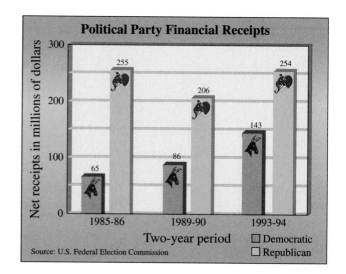

Political Party Financial Receipts

Source: U.S. Federal Election Commission

5.8 Solving Equations and Problems Using Polynomials

After studying this section, you will be able to:

MathPro   Video 14   SSM

1 *Factor to find the roots of a quadratic equation.*

2 *Solve applied problems that involve a factorable quadratic equation.*

1 Factoring to Find the Roots of a Quadratic Equation

Up to now, we have solved only first-degree equations. In this section we will solve quadratic, or second-degree, equations.

> **Definition**
>
> A second-degree equation of the form $ax^2 + bx + c = 0$, where a, b, c are real numbers and $a \neq 0$, is a **quadratic equation.** $ax^2 + bx + c = 0$ is the **standard form** of a quadratic equation.

Before solving a quadratic equation, we will first write the equation in standard form. Although it is not necessary that a, b, and c be integers, the equation is usually written this way. So we'll write out quadratic equations so that a, b, and c are integers.

The key to solving quadratic equations by factoring is the zero factor property. If we multiply two real numbers, the resulting product will be zero if one or both of the factors is zero. Thus, if the product of two real numbers is zero, at least one of the factors must be zero. We will state this property of real numbers formally.

> **Zero Factor Property**
>
> For all real numbers a, b:
>
> $$\text{If } a \cdot b = 0, \text{ then } a = 0 \text{ or } b = 0 \text{ or both} = 0.$$

EXAMPLE 1   Solve the equation $x^2 + 15x = 100$.

When we say "solve the equation" or "find the roots," we mean find the values of x that satisfy the equation.

$$x^2 + 15x - 100 = 0 \qquad \text{\textit{Get 0 on one side.}}$$
$$(x + 20)(x - 5) = 0 \qquad \text{\textit{Factor the trinomial.}}$$
$$x + 20 = 0 \quad \textit{or} \quad x - 5 = 0 \qquad \text{\textit{Set each factor equal to 0.}}$$
$$x = -20 \qquad\qquad x = 5 \qquad \text{\textit{Solve each equation.}}$$

Check: $x^2 + 15x = 100$ Use the *original* equation.

$$x = -20: \quad (-20)^2 + 15(-20) \overset{?}{=} 100$$
$$400 - 300 \overset{?}{=} 100$$
$$100 = 100 \quad \checkmark \quad \text{It checks.}$$

$$x = 5: \quad (5)^2 + 15(5) \overset{?}{=} 100$$
$$25 + 75 \overset{?}{=} 100$$
$$100 = 100 \quad \checkmark \quad \text{It checks.}$$

Thus, $x = 5$ and $x = -20$ are both roots of the quadratic equation $x^2 + 15x = 100$.

Practice Problem 1   Find the roots of $x^2 + x = 56$. ∎

For convenience, we list the steps we have employed to solve the quadratic equation.

1. Put the quadratic equation in standard form and, if possible, *factor* the quadratic expression.
2. Set each factor equal to zero.
3. Solve the resulting equations to find each root. (A quadratic equation has two roots.)
4. Check your solutions.

It is extremely important to remember that, when you are placing the quadratic equation in standard form, one side of the equation must be zero. Several algebraic operations may be necessary to obtain that desired result before we can factor the polynomial.

EXAMPLE 2   Find the roots of $6x^2 + 4 = 7(x + 1)$.

$$6x^2 + 4 = 7x + 7 \qquad \textit{Apply the distributive property.}$$

$$6x^2 - 7x - 3 = 0 \qquad \textit{Put the equation in standard form.}$$

$$(2x - 3)(3x + 1) = 0 \qquad \textit{Factor the trinomial.}$$

$$2x - 3 = 0 \quad or \quad 3x + 1 = 0 \qquad \textit{Set each factor equal to 0.}$$

$$2x = 3 \qquad\qquad 3x = -1 \qquad \textit{Solve the equations.}$$

$$x = \frac{3}{2} \qquad\qquad x = -\frac{1}{3}$$

Check: $6x^2 + 4 = 7(x + 1)$ Use the *original* equation.

$$x = \frac{3}{2}: \quad 6\left(\frac{3}{2}\right)^2 + 4 \stackrel{?}{=} 7\left(\frac{3}{2} + 1\right)$$

$$6\left(\frac{9}{4}\right) + 4 \stackrel{?}{=} 7\left(\frac{5}{2}\right)$$

$$\frac{27}{2} + 4 \stackrel{?}{=} \frac{35}{2}$$

$$\frac{27}{2} + \frac{8}{2} \stackrel{?}{=} \frac{35}{2}$$

$$\frac{35}{2} = \frac{35}{2} \quad ✓$$

It checks, so $x = \frac{3}{2}$ is a root. Verify that $x = -\frac{1}{3}$ is also a root.

If you are using a calculator to check your roots, you can complete the check more rapidly using $x = 1.5$ and $x = -0.33333333$. The latter value is approximate, so some round off error is expected.

Practice Problem 2   Find the roots of $12x^2 - 11x + 2 = 0$. ■

GRAPHING CALCULATOR

Finding Roots

Not all quadratic equations are as easy to factor as Example 2. Suppose you are asked to find the roots of

$10x(x + 1) = 83x + 12{,}012$

This can be written in the form:

$10x^2 - 73x - 12{,}012 = 0$

This can be factored to obtain:

$(5x + 156)(2x - 77) = 0$

$x = -\dfrac{156}{5}$ and $x = \dfrac{77}{2}$

In decimal form the answers are:

$x = -31.2$ and $x = 38.5$

However you can use the graphing calculator to graph

$y = 10x^2 - 73x - 12{,}012$

By setting an appropriate viewing window and then using the Zoom and Trace features of a graphing calculator, you can find the two places where $y = 0$.

Some calculators have a command that will find the zeros (roots) of a graph. They are $x = -31.2$ and $x = 38.5$.

In similar fashion graph each equation using the form $y = ax^2 + bx + c$ to find the roots.

$10x^2 - 189x - 12{,}834 = 0$

$10x(x + 2) = 11{,}011 - 193x$

EXAMPLE 3 Find the roots of $3x^2 - 5x = 0$.

$$3x^2 - 5x = 0 \qquad \text{\textit{The equation is already in standard form.}}$$

$$x(3x - 5) = 0 \qquad \text{\textit{Factor.}}$$

$$x = 0 \quad or \quad 3x - 5 = 0 \qquad \text{\textit{Set each factor equal to 0.}}$$
$$3x = 5$$
$$x = \frac{5}{3}$$

Check: Verify that $x = 0$ and $x = \frac{5}{3}$ are roots of $3x^2 - 5x = 0$.

Practice Problem 3 Find the roots of $7x^2 - 14x = 0$. ∎

EXAMPLE 4 Solve. $7x^2 + 2x = 3x(x - 4)$

$$7x^2 + 2x = 3x^2 - 12x \qquad \text{\textit{Distributive property.}}$$

$$7x^2 - 3x^2 + 2x + 12x = 0 \qquad \text{\textit{Get 0 on one side.}}$$

$$4x^2 + 14x = 0 \qquad \text{\textit{Combine like terms.}}$$

$$2x(2x + 7) = 0 \qquad \text{\textit{Factor.}}$$

$$2x = 0 \quad or \quad 2x + 7 = 0 \qquad \text{\textit{Set each factor equal to 0.}}$$

$$\frac{2x}{2} = \frac{0}{2} \qquad\qquad 2x = -7 \qquad \text{\textit{Solve each equation.}}$$
$$x = -\frac{7}{2}$$
$$x = 0$$

Check: Verify that $x = 0$ and $x = -\frac{7}{2}$ are roots of $7x^2 + 2x = 3x(x - 4)$.

Practice Problem 4 Solve. $5x^2 - 7x = 2x(x - 3)$ ∎

EXAMPLE 5 Solve. $9x(x - 1) = 3x - 4$

$$9x^2 - 9x = 3x - 4 \qquad \text{\textit{Remove parentheses.}}$$

$$9x^2 - 9x - 3x + 4 = 0 \qquad \text{\textit{Get 0 on one side.}}$$

$$9x^2 - 12x + 4 = 0 \qquad \text{\textit{Combine like terms.}}$$

$$(3x - 2)^2 = 0 \qquad \text{\textit{Factor.}}$$

$$3x - 2 = 0 \quad or \quad 3x - 2 = 0$$
$$3x = 2 \qquad\qquad 3x = 2$$
$$x = \frac{2}{3} \qquad\qquad x = \frac{2}{3}$$

We obtain the same answer twice. This value is called a **double root.**

Practice Problem 5 Solve. $16x(x - 2) = 8x - 25$ ∎

The zero factor property can be extended to a polynomial equation of degree greater than 2. In the following example, we will find the three roots of a third-degree polynomial equation.

EXAMPLE 6 Solve. $2x^3 = 24x - 8x^2$

$2x^3 + 8x^2 - 24x = 0$	*Get 0 on one side of the equation.*
$2x(x^2 + 4x - 12) = 0$	*Remove the common factor 2x.*
$2x(x + 6)(x - 2) = 0$	*Factor the trinomial.*
$2x = 0 \qquad x + 6 = 0 \qquad x - 2 = 0$	*Zero factor property.*
$x = 0 \qquad\quad x = -6 \qquad\quad x = 2$	*Solve for x.*

The solutions are $x = 0$, $x = -6$, and $x = 2$.

Practice Problem 6 Solve. $3x^3 + 6x^2 = 45x$ ■

2 *Solving Applied Problems That Involve a Factorable Quadratic Equation*

Some applied situations lead to a factorable quadratic equation. Using the methods developed in this section, we can solve these types of problems.

EXAMPLE 7 A racing sailboat has a triangular sail. Find the base and altitude of the triangle that has an area of 35 square meters and a base that is 3 meters shorter than the altitude.

1. *Understand the problem.*
   We draw a sketch and recall the formula for the area of a triangle.

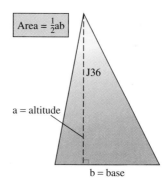

Let a = the length in meters of the altitude.

Let $a - 3$ = the length in meters of the base.

2. *Write an equation.*

$$A = \frac{1}{2}ab$$

$$35 = \frac{1}{2}a(a - 3)$$ *Replace A (area) by 35.*
Replace b (base) by a − 3.

3. *Solve the equation and state the answer.*

$70 = a(a - 3)$ *Multiply each side by 2.*

$70 = a^2 - 3a$ *Remove parentheses.*

$0 = a^2 - 3a - 70$ *Subtract 70 from each side.*

$0 = (a - 10)(a + 7)$

$a = 10 \quad or \quad a = -7$

The altitude of a triangle must be a positive number, so we disregard -7. Thus,

$$\text{altitude} = a = 10 \text{ meters}$$

$$\text{base} = a - 3 = 7 \text{ meters}$$

The altitude of the triangle measures 10 meters in length and the base of the triangle measures 7 meters in length. The sail is 10 meters tall and has a base of 7 meters.

4. *Check.* Is the base 3 meters shorter than the altitude?

$$10 - 3 = 7 \quad \checkmark \quad \text{Yes}$$

Is the area of the triangle 35 square meters?

$$A = \frac{1}{2}ab$$

$$A = \frac{1}{2}(10)(7) = 5(7) = 35 \quad \checkmark \quad \text{Yes}$$

Practice Problem 7   A racing sailboat has a triangular sail. Find the base and the altitude of the triangle if the area of the triangle is 52 square feet and the altitude is 5 feet longer than the base. ∎

Developing Your Study Skills

GETTING HELP

Getting the right kind of help at the right time can be a key ingredient in being successful in mathematics. When you have gone to class on a regular basis, taken careful notes, methodically read your textbook, and diligently done your home-work—all of which means making every effort possible to learn the mathematics—you may find that you are still having difficulty. If this is the case, then you need to seek help. Make an appointment with your instructor to find out what help is available to you. The instructor, tutoring services, a mathematics lab, video tapes, and computer software may be among resources you can draw on.

Once you discover the resources available in your school, you need to take advantage of them. Do not put it off, or you will find yourself getting behind. You cannot afford that. When studying mathematics, you must keep up with your work.

EXAMPLE 8 A square panel that holds fuses was used by a car manufacturer last year. The square panel that is used this year has an area that is greater by 72 square centimeters. The length of each side of the new panel was 3 centimeters more than double the length of the panel for last year's car. Find the dimensions of each panel.

1. *Understand the problem.*
   We draw a sketch of each square panel and recall that the area of each panel is obtained by squaring its side.

   Let x = the length in centimeters of the panel in last year's car

   Let $2x + 3$ = the length in centimeters of the panel in this year's car

   The area of the larger square is 72 square centimeters greater than the area of the smaller square.

2. *Write an equation.*

   Larger square is 72 square centimeters greater in area than the smaller square.

   $$(2x + 3)^2 = 72 + x^2$$

3. *Solve the equation and state the answer.*

$4x^2 + 12x + 9 = 72 + x^2$	*Remove parentheses.*
$4x^2 + 12x + 9 - x^2 - 72 = 0$	*Get 0 on one side of the equation.*
$3x^2 + 12x - 63 = 0$	*Simplify.*
$3(x^2 + 4x - 21) = 0$	*Remove the common factor of 3.*
$3(x + 7)(x - 3) = 0$	*Factor the trinomial.*
$x + 7 = 0 \qquad x - 3 = 0$	*Use the zero factor property.*
$x = -7 \qquad x = 3$	*Solve for x.*

   A fuse box cannot measure -7 centimeters, so we reject the negative answer. We use $x = 3$; then $2x + 3 = 2(3) + 3 = 6 + 3 = 9$.
      Thus the old fuse box measured 3 centimeters on a side. The new fuse box measured 9 centimeters on a side.

4. *Check.* Verify that each answer is correct.

Practice Problem 8 Last year Grandpa Jones had a small square garden. This year he has a square garden that is 112 square feet larger in area. The side of the new garden is 2 feet longer than triple the side of the old garden. Find the dimensions of each garden. ■

5.8 Exercises

Find all the roots and check your answer.

1. $x^2 + 11x + 10 = 0$

2. $x^2 - 8x + 15 = 0$

3. $x^2 - 13x = -36$

4. $x^2 - x = 2$

5. $5x^2 - 6x = 0$

6. $3x^2 + 5x = 0$

7. $2x^2 - 7x + 5 = 0$

8. $2x^2 + 3x - 5 = 0$

9. $3x^2 - 2x - 8 = 0$

10. $4x^2 - 13x + 3 = 0$

11. $6x^2 + 11x = 10$

12. $9x^2 + 9x = -2$

13. $8x^2 = 11x - 3$

14. $5x^2 = 11x - 2$

15. $x(x - 6) = 27$

16. $x(x - 21) = 22$

17. $4(x - 5) = (1 + x)(1 - x)$

18. $8(1 - x) = -1(4 + x^2)$

19. $x^2 + \dfrac{5}{3}x = \dfrac{2}{3}x$

20. $x^2 - \dfrac{5}{2}x = \dfrac{x}{2}$

21. $2x^2 + 3x = -14x^2 + 5x$

22. $5x^2 - 2x = x^2 - 8x$

23. $x^3 + 5x^2 + 6x = 0$

24. $x^3 + 11x^2 + 18x = 0$

25. $x^3 = x^2 + 20x$

26. $x^3 = 2x^2 + 24x$

27. $2x^3 - 18x = 0$

28. $2x^3 - 32x = 0$

29. $3x^3 + 15x^2 = 42x$

30. $2x^3 + 6x^2 = 20x$

31. $4x^2 + 7x = \dfrac{15}{2}$

32. $\dfrac{5}{2}x^2 - 8x = -\dfrac{3}{2}$

33. $2(x + 3) = -3x + 2(x^2 - 3)$

34. $2(x^2 - 4) - 3x = 4x - 11$

35. $18x^2 = 3x$

36. $20x^2 = 5x$

37. $6x^2 + 3 = 5x^2 + 4x + 3$

38. $4x^2 + 5x - 2 = 3x^2 - 2$

39. $4x(3x + 7) = 2(x - 6)$

40. $2x(x - 1) = 2(3 + x)$

To Think About

41. The equation $x^2 + bx - 12 = 0$ has a solution of $x = -4$. What is the value of b? What is the other solution of the equation?

42. The equation $2x^2 - 3x + c = 0$ has a solution of $x = -\frac{1}{2}$. What is the value of c? What is the other solution of the equation?

Applications

Solve the following applied problems.

43. The local child-care center has big, colorful geometric-designs in its carpet. A bright orange triangle has an area of 242 square centimeters. The altitude of the orange triangle is 22 centimeters longer than twice its base. Find the length of the altitude and the length of the base.

44. The area of the triangular logo on a laptop computer is 87 square millimeters. The altitude of the logo is 5 millimeters longer than 4 times the base. Find the measurements of the altitude and the base.

45. During half-time at the Superbowl, one of the performers will sing on a triangular platform that measures 119 square yards. The base of the triangular stage is 6 yards longer than four times the length of the altitude. What are the dimensions of the triangular stage?

46. The area of a triangular neon billboard advertising the local mall is 104 square feet. The base of the triangle is 2 feet longer than triple the length of the altitude.
(a) What are the dimensions of the triangular billboard in feet?

(b) What are the dimensions of the triangular billboard in yards?

47. The area of a rectangular mouse pad is 480 square centimeters. Its length is 16 centimeters shorter than twice its width.
(a) What are the length and width of the mouse pad in centimeters?
(b) What are the length and width of the mouse pad in millimeters?

48. The area of a rectangular desk telephone is 896 square centimeters. The length of the rectangular telephone is 4 centimeters longer than its width.
(a) What are the length and width, in centimeters, of the desk telephone?
(b) What are the length and width, in millimeters, of the desk telephone?

49. A rare square Turkish rug, belonging to the family of President John F. Kennedy, is being auctioned off. The area of the Turkish rug is 96 feet larger than its perimeter. Find the length in feet of the side.

50. The backstage dressing room of the most famous circus in the world is in the shape of a square. The area of the square dressing room is 165 more than its perimeter. Find the length of the side.

51. The volume of a rectangular solid can be written as $V = LWH$, where L is the length, W is the width, and H is the height. If a rectangular solid has a volume of 260 cubic feet, a height of 5 feet, and the length is 1 foot longer than triple the width, find the length and width of the solid.

52. The volume of a rectangular solid can be written as $V = LWH$, where L is the length, W is the width, and H is the height. If a rectangular solid has a volume of 390 cubic meters, a height of 6 meters, and the length is 3 meters longer than double the width, find the length and width of the solid.

53. In the northern Atlantic Ocean a certain fishing area is in the shape of a rectangle. The length of the rectangle is 7 miles longer than double its width. The area of this rectangle is 85 square miles. What are the dimensions of this rectangle?

54. In planning for the first trip to the moon, NASA surveyed a rectangular area of 54 square miles. The length of the rectangle was 3 miles less than double the width. What were the dimensions of this potential landing area?

★ 55. Media One has a square target that receives signal transmissions from a television tower. They recently enlarged the square target so that it has 176 square centimeters more area than the old target. The new target has a side that is 1 cm longer than double the width of the old side. What are the dimensions of the old and the new targets?

★ 56. The Parad family has a square vegetable garden. They increased the size of the garden and created a new larger garden that has an area that is 84 square feet larger than the old garden. The new garden has a side that is 2 feet longer than double the length of the old side. What are the dimensions of both the old and the new gardens?

Use the following formula for Problems 53 through 56. A manufacturer finds that the profit in dollars for manufacturing n units is $P = 2n^2 - 19n - 10$. (Assume that n is a positive integer.)

57. How many units are produced when the profit is $410?

58. How many units are produced when the profit is $0?

59. How many units are produced when there is a loss of $52 ($P = -52$)?

60. How many units are produced when there is a loss of $34 ($P = -34$)?

Cumulative Review Problems

Simplify. Do not leave negative exponents in your answer.

61. $(2x^3y^2)^3(5xy^2)^2$

62. $\dfrac{(2a^3b^2)^3}{16a^5b^8}$

63. $(-3x^{-2}y^4z)^{-2}$

64. $\left(\dfrac{5xy^{-2}}{2x^{-3}y}\right)^3$

Putting Your Skills to Work

USING POLYNOMIALS TO COUNT CANNONBALLS

Hundreds of years ago, soldiers built forts with cannon emplacements. It was desirable to have a pile of cannon balls stacked next to each cannon. These stacks often had a triangular shape. The base was a triangle and each succeeding row was a triangle. Each row of the triangular pile on the right would look like this:

Top row   Middle row   Bottom row

The number of cannon balls that would be found in a three-layer triangular pile would be 10 cannon balls. Military historians sometimes use the cannon ball polynomial to find the number of cannon balls that are in a triangular pile. The cannon ball polynomial can be written as $\frac{1}{6}x^3 + \frac{1}{2}x^2 + \frac{1}{3}x$, where x is the number of layers. For example, if $x = 3$ we would obtain

$$\frac{1}{6}(3)^3 + \frac{1}{2}(3)^2 + \frac{1}{3}(3) = \frac{1}{6}(27) + \frac{1}{2}(9) + \frac{1}{3}(3) = \frac{9}{2} + \frac{9}{2} + 1 = 10$$

This is correct since we have already determined that with three layers there would be 10 cannon balls.

PROBLEMS FOR INDIVIDUAL INVESTIGATION

1. Use the cannon ball polynomial to find the number of cannon balls in a triangular pile with 4 rows, then with 5 rows.

2. Factor the cannon ball polynomial. (*Hint:* first remove a common factor of x.)

PROBLEMS FOR GROUP INVESTIGATION AND COOPERATIVE GROUP ACTIVITY

Together with members of your class, investigate the following.

Consider the case where cannon balls were stacked in a *square pile*. Each succeeding row consists of cannon balls in a square array.

3. How many cannon balls would be in a square pile of 3 rows? How many cannon balls would be in a square pile of 4 rows?

4. Can you find a pattern to determine how many cannon balls would be in a square pile of x rows?

INTERNET CONNECTION: Go to http://www.prenhall.com/tobey to be connected.

Site: Simple Cubic and Related Structures

This site shows the basis for a crystal lattice structure. By repeating the structure shown in all directions, you can imagine the arrangement of atoms within a crystal.

5. How does the structure shown relate to the two methods of stacking cannonballs shown on this page? Explain.

6. Explore the different types of crystal structures discussed in this Web site. What other structures could theoretically be used to stack cannonballs?

Topic	Procedure	Examples
Adding and subtracting polynomials, p. 273.	Combine like terms following the rules of signs.	$(5x^2 - 6x - 8) + (-2x^2 - 5x + 3) = 3x^2 - 11x - 5$ $(3a^2 - 2ab - 5b^2) - (-7a^2 + 6ab - b^2)$ $\quad = (3a^2 - 2ab - 5b^2) + (7a^2 - 6ab + b^2)$ $\quad = 10a^2 - 8ab - 4b^2$
Multiplying polynomials, p. 274.	1. Multiply each term of the first polynomial by each term of the second polynomial. 2. Combine like terms.	$2x^2(3x^2 - 5x - 6) = 6x^4 - 10x^3 - 12x^2$ $(3x + 4)(2x - 7) = 6x^2 - 21x + 8x - 28$ $\qquad\qquad\qquad = 6x^2 - 13x - 28$ $(x - 3)(x^2 + 5x + 8)$ $\quad = x^3 + 5x^2 + 8x - 3x^2 - 15x - 24$ $\quad = x^3 + 2x^2 - 7x - 24$
Division of a polynomial by a monomial, p. 279.	1. Write the division as the sum of separate fractions. 2. If possible, reduce the separate fractions.	$(16x^3 - 24x^2 + 56x) \div (-8x)$ $\quad = \dfrac{16x^3}{-8x} + \dfrac{-24x^2}{-8x} + \dfrac{56x}{-8x}$ $\quad = -2x^2 + 3x - 7$
Dividing a polynomial by a binomial or a trinomial, p. 280.	1. Write the division as in arithmetic. Write both polynomials in descending order; write any missing terms with a coefficient of zero. 2. Divide the *first* term of the divisor into the first term of the dividend. The result is the first term of the quotient. 3. Multiply the first term of the quotient by *every* term in the divisor. 4. Write this product under the dividend (align like terms) and subtract. 5. Treat this difference as a new dividend. Repeat steps 2 to 4. Continue until the remainder is zero or is a polynomial of lower degree than the *first term* of the divisor.	Divide. $(6x^3 + 5x^2 - 2x + 1) \div (3x + 1)$ $\begin{array}{r} 2x^2 + \ x - 1 \\ 3x+1{\overline{)6x^3 + 5x^2 - 2x + 1}} \\ \underline{6x^3 + 2x^2} \\ 3x^2 - 2x \\ \underline{3x^2 + \ x} \\ -3x + 1 \\ \underline{-3x - 1} \\ 2 \end{array}$ The quotient is $2x^2 + x - 1 + \dfrac{2}{3x + 1}$.
Synthetic division (optional topic), p. 285.	Synthetic division can be used if the divisor is in the form $(x - b)$ or $(x + b)$. 1. Write the coefficients of the terms in descending order of the dividend. Write any missing terms with a coefficient of zero. 2. The division will be of the form $(x - b)$ or $(x + b)$. Write down the opposite of b to the left. 3. Bring down the first coefficient to the bottom row. 4. Multiply each coefficient on the bottom by the opposite of b and add it to the upper coefficient. 5. Continue until the bottom row is filled.	Divide. $(3x^5 - 2x^3 + x^2 - x + 7) \div (x + 2)$ $\begin{array}{r\|rrrrrr} -2 & 3 & 0 & -2 & 1 & -1 & 7 \\ & & -6 & 12 & -20 & 38 & -74 \\ \hline & 3 & -6 & 10 & -19 & 37 & \boxed{-67} \end{array}$ The quotient is $3x^4 - 6x^3 + 10x^2 - 19x + 37 + \dfrac{-67}{x + 2}$.
Removing a common factor, p. 289.	Remove the greatest common factor from each term. Many factoring problems are two steps, of which this is the first.	$5x^3 - 25x^2 - 10x = 5x(x^2 - 5x - 2)$ $20a^3b^2 - 40a^4b^3 + 30a^3b^3 = 10a^3b^2(2 - 4ab + 3b)$
Factoring the difference of two squares, p. 304.	$a^2 - b^2 = (a + b)(a - b)$	$9x^2 - 1 = (3x + 1)(3x - 1)$ $8x^2 - 50 = 2(4x^2 - 25) = 2(2x + 5)(2x - 5)$
Factoring the perfect square trinomial, p. 305.	$a^2 + 2ab + b^2 = (a + b)^2$ $a^2 - 2ab + b^2 = (a - b)^2$	$16x^2 + 40x + 25 = (4x + 5)^2$ $18x^2 + 120xy + 200y^2 = 2(9x^2 + 60xy + 100y^2)$ $\qquad\qquad\qquad\qquad\qquad = 2(3x + 10y)^2$ $4x^2 - 36x + 81 = (2x - 9)^2$ $25a^3 - 10a^2b + ab^2 = a(25a^2 - 10ab + b^2)$ $\qquad\qquad\qquad\qquad = a(5a - b)^2$

Topic	Procedure	Examples
Factoring the sum and difference of two cubes, p. 306.	$a^3 + b^3 = (a + b)(a^2 - ab + b^2)$ $a^3 - b^3 = (a - b)(a^2 + ab + b^2)$	$8x^3 + 27 = (2x + 3)(4x^2 - 6x + 9)$ $250x^3 + 2y^3 = 2(125x^3 + y^3)$ $\qquad = 2(5x + y)(25x^2 - 5xy + y^2)$ $27x^3 - 64 = (3x - 4)(9x^2 + 12x + 16)$ $125y^4 - 8y = y(125y^3 - 8)$ $\qquad = y(5y - 2)(25y^2 + 10y + 4)$
Factoring the trinomials of the form $x^2 + bx + c$, p. 295.	The factors will be of the form $(x + m)(x + n)$, where $m \cdot n = c$ and $m + n = b$.	$x^2 - 7x + 12 = (x - 4)(x - 3)$ $3x^2 - 36x + 60 = 3(x^2 - 12x + 20) =$ $\qquad\qquad\qquad 3(x - 2)(x - 10)$ $x^2 + 2x - 15 = (x + 5)(x - 3)$ $2x^2 - 44x - 96 = 2(x^2 - 22x - 48)$ $\qquad = 2(x - 24)(x + 2)$
Factoring the trinomials of the form $ax^2 + bx + c$, p. 298.	Use the trial-and-error method or the grouping number method.	$2x^2 + 7x + 3 = (2x + 1)(x + 3)$ $8x^2 - 26x + 6 = 2(4x^2 - 13x + 3) =$ $\qquad\qquad\qquad 2(4x - 1)(x - 3)$ $7x^2 + 20x - 3 = (7x - 1)(x + 3)$ $5x^3 - 18x^2 - 8x = x(5x^2 - 18x - 8) =$ $\qquad\qquad\qquad x(5x + 2)(x - 4)$
Factoring by grouping, p. 291.	1. Make sure that the first two terms have a common factor; otherwise, rearrange the order. 2. Divide problem into two parts. Remove the common factor from each part. 3. Remove the common binomial in parentheses. 4. Place the remaining terms in other parentheses.	$6xy - 8y + 3xw - 4w$ $2y(3x - 4) + w(3x - 4) = (3x - 4)(2y + w)$
Solving a quadratic equation by factoring, p. 315.	1. Put the equation in standard form. 2. Factor, if possible. 3. Set each factor equal to 0. 4. Solve each of the resulting equations.	Solve. $(x + 3)(x - 2) = 5(x + 3)$ $x^2 - 2x + 3x - 6 = 5x + 15$ $x^2 + x - 6 = 5x + 15$ $x^2 + x - 5x - 6 - 15 = 0$ $x^2 - 4x - 21 = 0$ $(x - 7)(x + 3) = 0$ $x = 7 \ \text{ or } \ x = -3.$

Chapter 5 Review Problems

Perform the indicated operations.

1. $(x^2 - 3x + 5) + (-2x^2 - 7x + 8)$

2. $(-4x^2y - 7xy + y) + (5x^2y + 2xy - 9y)$

3. $(-6x^2 + 7xy - 3y^2) - (5x^2 - 3xy - 9y^2)$

4. $(-13x^2 + 9x - 14) - (-2x^2 - 6x + 1)$

5. $(7x - 2) + (5 - 3x) + (2 - 2x)$

6. $(5x - 2x^2 - x^3) - (2x - 3 + 5x^2)$

Multiply.

7. $3xy(x^2 - xy + y^2)$

8. $(3x^2 + 1)(2x - 1)$

9. $(5x^2 + 3)^2$

10. $(x - 3)(2x - 5)(x + 2)$

11. $(x^2 - 3x + 1)(-2x^2 + x - 2)$

12. $(3x - 5)(3x^2 + 2x - 4)$

Divide.

13. $(25x^3y - 15x^2y - 100xy) \div (-5xy)$

14. $(12x^2 - 16x - 4) \div (3x + 2)$

15. $(2x^3 - 7x^2 + 2x + 8) \div (x - 2)$

16. $(3y^3 - 2y + 5) \div (y - 3)$

17. $(15a^4 - 3a^3 + 4a^2 + 4) \div (3a^2 - 1)$

18. $(x^4 - x^3 - 7x^2 - 7x - 2) \div (x^2 - 3x - 2)$

19. $(2x^4 - 13x^3 + 16x^2 - 9x + 20) \div (x - 5)$

20. $(3x^4 + 5x^3 - x^2 + x - 2) \div (x + 2)$

21. $(4x^4 + 12x^3 - x^2 - x + 2) \div (x + 3)$

22. $(2x^4 - 9x^3 + 5x^2 + 13x - 3) \div (x - 3)$

Factor, if possible. Be sure to factor completely.

23. $x^2 + 15x + 36$

24. $5x^2 - 11x + 2$

25. $9x^2 - 121$

26. $36x^2 + 25$

27. $x^2 - 8wy + 4xw - 2xy$

28. $x^3 + 8x^2 + 12x$

29. $2x^2 - 7x - 3$

30. $x^2 + 6xy - 27y^2$

31. $27x^4 - x$

32. $21a^2 + 20ab + 4b^2$

33. $-3a^3b^3 + 2a^2b^4 - a^2b^3$

34. $a^4b^4 + a^3b^4 - 6a^2b^4$

35. $3x^4 - 5x^2 - 2$

36. $2x^4 + 20x^2 - 48$

37. $9a^2b + 15ab - 14b$

38. $2x^4 + 7x^2 - 6$

39. $12x^2 + 12x + 3$

40. $4y^4 - 13y^3 + 9y^2$

41. $y^4 + 2y^3 - 35y^2$

42. $4x^2y^2 - 12x^2y - 8x^2$

43. $3x^4 - 7x^2 - 6$

44. $a^2 + 5ab^3 + 4b^6$

45. $3x^2 - 12 - 8x + 2x^3$

46. $2x^4 - 12x^2 - 54$

47. $8a + 8b - 4bx - 4ax$

48. $8x^4 + 34x^2y^2 + 21y^4$

49. $4x^3 + 10x^2 - 6x$

50. $2a^2x - 15ax + 7x$

51. $16x^4y^2 - 56x^2y + 49$

52. $128x^3y - 2xy$

53. $26x^3y - 13xy^3 + 52x^2y^4$

54. $5xb - 28y + 4by - 35x$

55. $27abc^2 - 12ab$

56. $5a^6 + 40a^3b^3$

57. $50x^4 - 100x^3 + 64x^2$

58. $60x^2 - 100xy + 15y^2$

Solve the following equations.

59. $5x^2 - 9x - 2 = 0$

60. $2x^2 - 11x + 12 = 0$

61. $(2x - 1)(3x - 5) = 20$

62. $7x^2 = 21x$

63. $x^3 + 7x^2 = -12x$

64. $3x^2 + 14x + 3 = -1 + 4(x + 1)$

Use a quadratic equation to solve each of the following problems.

65. The area of a triangle is 77 square meters. The altitude of the triangle is 3 meters longer than the base of the triangle. Find the base and the altitude of the triangle.

66. A park has an area of 40 square miles. The park is in the shape of a rectangle. The length of the rectangle is 2 miles less than triple the width. Find the dimensions of the park.

67. A square is constructed for sound insulation in a restaurant. It does not provide enough insulation, so a larger square is constructed. The larger square has 24 square yards more insulation. The side of the larger square is 3 yards longer than double the side of the smaller square. Find the dimensions of each square.

68. The profit in dollars of a scientific calculator manufacturing plant to operate for one hour is given by the equation $P = 3x^2 - 7x - 10$, where x is the number of calculators assembled in one hour. Find the number of calculators that should be made in one hour if the hourly profit is to be $30.00.

Developing Your Study Skills

EXAM TIME: TAKING THE EXAM

Allow yourself plenty of time to get to your exam. You may even find it helpful to arrive a little early, in order to collect your thoughts and ready yourself. This will help you feel more relaxed.

After you get your exam, you will find it helpful to do the following:

1. Take two or three moderately deep breaths. Inhale, then exhale slowly. You will feel your entire body begin to relax.

2. Write down on the back of the exam any formulas or ideas that you need to remember.

3. Look over the entire test quickly in order to pace yourself and use your time wisely. Notice how many points each problem is worth. Spend more time on items of greater worth.

4. Read directions carefully, and be sure to answer all questions clearly. Keep your work neat and easy to read.

5. Ask your instructor about anything that is not clear to you.

6. Work the problems and answer the questions that are easiest for you first. Then come back to the more difficult ones.

7. Do not get bogged down on one problem too long, because it may jeopardize your chances of finishing other problems. Leave the tough problem and come back to it when you have time later.

8. Check your work. This will help you to catch minor errors.

9. Stay calm if others leave before you do. You are entitled to use the full amount of allotted time.

Combine.

1. $(3x^2y - 2xy^2 - 6) + (5 + 2xy^2 - 7x^2y)$

2. $(5a^2 - 3) - (2 + 5a) - (4a - 3)$

Multiply.

3. $2x^2(x - 3y)$

4. $(2x - 3y^2)^2$

5. $(x^2 + 6x - 2)(x^2 - 3x - 4)$

Divide.

6. $(-15x^3 - 12x^2 + 21x) \div (-3x)$

7. $(2x^4 - 7x^3 + 7x^2 - 9x + 10) \div (2x - 5)$

8. $(x^3 - x^2 - 5x + 2) \div (x + 2)$

9. $(x^4 + x^3 - x - 3) \div (x + 1)$

10. $(2x^5 - 7x^4 - 15x^2 - x + 5) \div (x - 4)$

Factor, if possible.

11. $121x^2 - 25y^2$

12. $9x^2 + 30xy + 25y^2$

13. $x^3 - 26x^2 + 48x$

14. $24x^2 + 10x - 4$

15. $4x^3y + 8x^2y^2 + 4x^2y$

16. $x^2 - 6wy + 3xy - 2wx$

1. _____

2. _____

3. _____

4. _____

5. _____

6. _____

7. _____

8. _____

9. _____

10. _____

11. _____

12. _____

13. _____

14. _____

15. _____

16. _____

17. $2x^2 - 3x + 2$

18. $3x^4 + 36x^3 + 60x^2$

19. $18x^2 + 3x - 15$

20. $25x^2y^4 - 16y^4$

21. $54a^4 - 16a$

22. $9x^5 - 6x^3y + xy^2$

23. $3x^4 + 17x^2 + 10$

24. $x^2 - 8xy + 12y^2$

25. $3x - 10ay + 6y - 5ax$

26. $16x^4 - 1$

Solve the following equations.

27. $x^2 = 5x + 14$

28. $3x^2 - 11x - 4 = 0$

29. $7x^2 + 6x = 8x$

Use a quadratic equation to solve the following problem.

30. The area of a triangular road sign is 70 square inches. The altitude of the triangle is 4 inches less than the base of the triangle. Find the altitude and the base of the triangle.

17. _____

18. _____

19. _____

20. _____

21. _____

22. _____

23. _____

24. _____

25. _____

26. _____

27. _____

28. _____

29. _____

30. _____

Approximately one-half of this test covers the content of Chapters 1 through 4. The remainder covers the content of Chapter 5.

1. What property is illustrated by the equation $3(5 \cdot 2) = (3 \cdot 5)2$?

2. Evaluate. $\dfrac{2 + 6(-2)}{(2 - 4)^3 + 3}$

3. Evaluate. $2\sqrt{16} + 3\sqrt{49}$

4. Solve for x. $5x + 7y = 2$

5. Solve for x.
$2(3x - 1) - 4 = 2x - (6 - x)$

6. Find the slope of the line passing through $(-2, -3)$ and $(1, 5)$.

7. Graph. $y = -\dfrac{2}{3}x + 4$

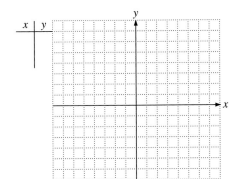

8. Graph the region $3x - 4y \geq -12$.

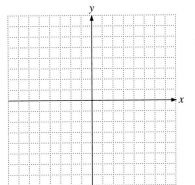

9. Solve the inequality.
$-3(x + 2) < 5x - 2(4 + x)$

10. What are the dimensions of a rectangle with a perimeter of 46 meters if the length is 5 meters longer than twice the width?

11. Combine. $(2a^2 - 3ab + 4b^2) - (-3a^2 + 6ab - 8b^2)$

Multiply and simplify your answer.

12. $-3xy^2(2x + 3y - 5xy)$

13. $(5x - 2)(2x^2 - 3x - 4)$

1. _____

2. _____

3. _____

4. _____

5. _____

6. _____

7. _____

8. _____

9. _____

10. _____

11. _____

12. _____

13. _____

Divide.

14. $(-21x^3 + 14x^2 - 28x) \div (7x)$

15. $(2x^3 - 3x^2 + 3x - 4) \div (x - 2)$

Factor, if possible.

16. $2x^3 - 10x^2$

17. $64x^2 - 49$

18. $9x^3 - 24x^2 + 16x$

19. $25x^2 + 60x + 36$

20. $3x^2 - 15x - 42$

21. $2x^2 + 24x + 40$

22. $16x^2 + 9$

23. $6x^3 + 11x^2 + 3x$

24. $27x^4 + 64x$

25. $14mn + 7n + 10mp + 5p$

Solve for x.

26. $3x^2 - 4x - 4 = 0$

27. $x^2 + 11x = 26$

28. A hospital has paved a triangular parking lot for emergency helicopter landings. The area of the triangle is 68 square meters. The altitude of the triangle is 1 meter longer than double the base of the triangle. Find the altitude and the base of this triangular region.

14. _____

15. _____

16. _____

17. _____

18. _____

19. _____

20. _____

21. _____

22. _____

23. _____

24. _____

25. _____

26. _____

27. _____

28. _____

Rational Algebraic Expressions and Equations

The number of cell phones in use in the world today is increasing significantly every day. This modern form of communication is not available everywhere, since it is dependent on the signal strength from transmitting towers. The strength of this signal is altered drastically as the distance from the cell phone to the transmitting tower is increased. Do you think you could measure the strength of the signal? Turn to the Putting Your Skills to Work on page 372.

If you are familiar with the topics in this chapter, take this test now. Check your answers with those in the back of the book. If an answer was wrong or you couldn't do a problem, study the appropriate section of the chapter.

If you are not familiar with the topics in this chapter, don't take this test now. Instead, study the examples, work the practice problems, and then take the test.

This test will help you to identify those concepts that you have mastered and those that need more study.

Section 6.1

Simplify.

1. $\dfrac{49x^2 - 9y^2}{7x^2 + 4xy - 3y^2}$ **2.** $\dfrac{2x^3 + 3x^2 - x}{x - 5x^2 - 6x^3}$

3. $\dfrac{2a^2 + 5a + 3}{a^2 + a + 1} \cdot \dfrac{a^3 - 1}{2a^2 + a - 3} \cdot \dfrac{6a - 30}{3a + 3}$

4. $\dfrac{5x^3y^2}{x^2y + 10xy^2 + 25y^3} \div \dfrac{2x^4y^5}{3x^3 - 75xy^2}$

Section 6.2

Add or subtract these fractions. Simplify your answer.

5. $\dfrac{x}{3x - 6} - \dfrac{4}{3x}$ **6.** $\dfrac{2}{x + 5} + \dfrac{3}{x - 5} + \dfrac{7x}{x^2 - 25}$

7. $\dfrac{y + 1}{y^2 + y - 12} - \dfrac{y - 3}{y^2 + 7y + 12}$

Section 6.3

Simplify each complex fraction.

8. $\dfrac{\dfrac{1}{12x} + \dfrac{5}{3x}}{\dfrac{2}{3x^2}}$ **9.** $\dfrac{\dfrac{x}{4x^2 - 1}}{3 - \dfrac{2}{2x + 1}}$

Section 6.4

Solve for the variable and check your solution. If there is no solution, say so.

10. $\dfrac{3}{y + 5} - \dfrac{1}{y - 5} = \dfrac{5}{y^2 - 25}$ **11.** $\dfrac{1}{6y} - \dfrac{4}{9} = \dfrac{4}{9y} - \dfrac{1}{2}$

Section 6.5

Solve for the variable indicated.

12. $\dfrac{d_1}{d_2} = \dfrac{w_1}{w_2}$ for d_2 **13.** $I = \dfrac{nE}{nr + R}$ for n

Set up a proportion and use it to find the desired quantity.

14. A house 49 feet tall casts a shadow 14 feet long. At the same time a nearby flagpole casts a 9-foot shadow. How tall is the flagpole?

15. A drawing on a 3- by 5-inch card is projected on a wall. A $\frac{3}{4}$-inch line on the card is 2 feet long on the wall. What are the dimensions of the outline of the 3 by 5 card on the wall?

1. _____

2. _____

3. _____

4. _____

5. _____

6. _____

7. _____

8. _____

9. _____

10. _____

11. _____

12. _____

13. _____

14. _____

15. _____

6.1 Simplifying, Multiplying, and Dividing Rational Expressions

After studying this section, you will be able to:

 Simplify a rational expression.
 Simplify the product of two or more rational expressions.
3 Simplify the quotient of two or more rational expressions.

1 Simplifying a Rational Expression

You may recall from Chapter 1 that a *rational number* is an exact quotient $\frac{a}{b}$ of two numbers a and b, $b \neq 0$. A **rational expression** is an expression of the form $\frac{P}{Q}$, where P and Q are polynomials, and Q is not zero. For example,

$$\frac{7}{x + 2} \quad \text{and} \quad \frac{x + 5}{x - 3}$$

are rational expressions. Since the denominator cannot equal zero, the first expression is undefined if $x + 2 = 0$. This occurs when $x = -2$. We say that x can be any real number except -2. Look at the second expression. When is this expression undefined? Why? For what values of x is the expression defined? Note that the numerator can be zero, because any fraction $\frac{0}{a}$ ($a \neq 0$) is just 0.

The domain of a rational expression is the set of values that can be used to replace the variable. Thus the domain of $\frac{7}{x + 2}$ is all real numbers except -2. The domain of $\frac{x + 5}{x - 3}$ is all real numbers except 3.

EXAMPLE 1 Find the domain. $\dfrac{x - 7}{x^2 + 8x - 20}$

The domain will be all real numbers except those that make the denominator equal to zero. What value(s) will make the denominator equal to zero? We determine this by solving the equation $x^2 + 8x - 20 = 0$

$$(x + 10)(x - 2) = 0 \quad \textit{Factor.}$$

$$x + 10 = 0 \quad \text{or} \quad x - 2 = 0 \quad \textit{Use the zero factor property.}$$

$$x = -10 \quad\quad\quad x = 2 \quad \textit{Solve for x.}$$

The domain is all real numbers except -10 and 2.

Practice Problem 1 Find the domain. $\dfrac{4x + 3}{x^2 - 9x - 22}$ ∎

We have learned that fractions can be simplified (or reduced to lowest terms) by factoring the numerator and denominator into prime factors and dividing by the common factor. For example,

$$\frac{15}{25} = \frac{3 \cdot \cancel{5}}{5 \cdot \cancel{5}} = \frac{3}{5}$$

We can do this by using the *basic rule of fractions*.

GRAPHING CALCULATOR

Finding Domains

To find the domain in Example 1, graph

$$y = \frac{x - 7}{x^2 + 8x - 20} \text{ on a}$$

graphing calculator. Since graphing calculators try to connect the points in a graph, it may not be easy to see where the expression is not defined.

Display:

Since the domain will be all values except those that make the denominator equal to zero, we can graph $y = x^2 + 8x - 20$ to identify the values that are not in the domain.

Display:

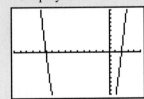

Use the Zoom and Trace feature or the zero command to find where $y = 0$.

Try to find the domain of the following expression. Round to the nearest tenth.

$$\frac{x + 9}{2.1x^2 + 5.2x - 3.1}$$

Basic Rule of Fractions

For any polynomials a, b, c,

$$\frac{ac}{bc} = \frac{a}{b}, \qquad \text{where } b, c \neq 0$$

What we are doing is factoring out a common factor of 1 ($\frac{c}{c} = 1$). We have

$$\frac{ac}{bc} = \frac{a}{b} \cdot \frac{c}{c} = \frac{a}{b} \cdot 1 = \frac{a}{b}$$

Note that c must be a factor of the numerator *and* the denominator. Thus, the basic rule of fractions simply says that we may remove a *common* factor out of the numerator and denominator. This is the same as dividing the numerator and denominator of the fraction by the same nonzero value. The nonzero value can be a number or an algebraic expression.

EXAMPLE 2 Simplify. $\dfrac{6x^2 - 8x}{2x^2 + 14x}$

$$\frac{6x^2 - 8x}{2x^2 + 14x} = \frac{2x(3x - 4)}{2x(x + 7)} \qquad \text{\textit{Factor the numerator and denominator.}}$$

$$= \frac{2x}{2x} \cdot \frac{3x - 4}{x + 7} \qquad \text{\textit{Use the definition of multiplication of fractions.}}$$

$$= 1 \cdot \frac{3x - 4}{x + 7} \qquad \text{\textit{Identify } $\frac{2x}{2x} = 1$.}$$

$$= \frac{3x - 4}{x + 7} \qquad \text{\textit{Use the basic rule of fractions.}}$$

Practice Problem 2 Simplify. $\dfrac{15x^2 - 18x}{6x^2 + 3x}$ ∎

EXAMPLE 3 Simplify. $\dfrac{2a^2 - ab - b^2}{a^2 - b^2}$

$$\frac{(2a + b)(a - b)}{(a + b)(a - b)} = \frac{2a + b}{a + b} \cdot 1 = \frac{2a + b}{a + b}$$

As you become more familiar with this basic rule, you won't have to write out every step. We did so here to show the application of the rule. We cannot simplify this fraction any further.

Practice Problem 3 Simplify. $\dfrac{x^2 - 36y^2}{x^2 - 3xy - 18y^2}$ ∎

EXAMPLE 4 Simplify. $\dfrac{2x - 3y}{2x^2 - 7xy + 6y^2}$

$$\frac{(2x - 3y)\,1}{(2x - 3y)(x - 2y)} = \frac{1}{x - 2y} \qquad \text{\textit{Note: Do you see why it is necessary to have a 1 in the numerator of the answer?}}$$

Practice Problem 4 Simplify. $\dfrac{9x^2y}{3xy^2 + 6x^2y}$ ∎

EXAMPLE 5 Simplify. $\dfrac{2x^2 + 2x - 12}{x^3 + 7x^2 + 12x}$

$$\frac{2x^2 + 2x - 12}{x^3 + 7x^2 + 12x} = \frac{2(x^2 + x - 6)}{x(x^2 + 7x + 12)} = \frac{2\,(x + 3)\,(x - 2)}{x\,(x + 3)\,(x + 4)} = \frac{2(x - 2)}{x(x + 4)}$$

We usually leave the answer in factored form.

Practice Problem 5 Simplify. $\dfrac{2x^2 - 8x - 10}{2x^2 - 20x + 50}$ ∎

Be alert for situations where each term of one factor is opposite in sign from each term of another factor. In such cases you should factor -1 or another negative quantity from one polynomial.

EXAMPLE 6 Simplify. **(a)** $\dfrac{-2x + 14y}{x^2 - 5xy - 14y^2}$ **(b)** $\dfrac{49 - x^2}{x^2 - 15x + 56}$

(a) $\dfrac{-2x + 14y}{x^2 - 5xy - 14y^2} = \dfrac{-2(x - 7y)}{(x + 2y)(x - 7y)}$ *Remove -2 as a common factor from each term of the numerator and factor the denominator.*

$= \dfrac{-2}{x + 2y}$ *Use the basic rule of fractions.*

(b) $\dfrac{49 - x^2}{x^2 - 15x + 56} = \dfrac{(7 + x)(7 - x)}{(x - 8)(x - 7)}$

$= \dfrac{(7 + x)(-1)(-7 + x)}{(x - 8)(x - 7)}$ *Remove common factor of -1 from $(7 - x)$.*

$= \dfrac{(7 + x)(-1)(x - 7)}{(x - 8)(x - 7)}$ *Use the commutative property of addition: $-7 + x = x - 7$.*

$= \dfrac{-1(x + 7)}{x - 8}$ *Use the basic rule of fractions and the commutative property of multiplication and addition. Remove parentheses.*

$= \dfrac{-x - 7}{x - 8}$

Practice Problem 6 Simplify. **(a)** $\dfrac{-3x + 6y}{x^2 - 7xy + 10y}$ **(b)** $\dfrac{36 - x^2}{x^2 - 10x + 24}$ ∎

EXAMPLE 7 Simplify. $\dfrac{25y^2 - 16x^2}{8x^2 - 14xy + 5y^2}$

$$\frac{(5y + 4x)(5y - 4x)}{(4x - 5y)(2x - y)} = \frac{(5y + 4x)\,(5y - 4x)}{-1\,(-4x + 5y)\,(2x - y)} = \frac{5y + 4x}{-1(2x - y)} = \frac{5y + 4x}{y - 2x}$$

Observe that $4x - 5y = -1(-4x + 5y)$.

Practice Problem 7 Simplify. $\dfrac{7a^2 - 23ab + 6b^2}{4b^2 - 49a^2}$ ∎

Simplifying the Product of Two or More Rational Expressions

Multiplication of rational expressions follows the same rule as multiplication of integer fractions. However, it is particularly helpful to use the basic rule of fractions to simplify wherever possible.

Multiplying Rational Expressions

For any polynomials a, b, c, d,

$$\frac{a}{b} \cdot \frac{c}{d} = \frac{ac}{bd}, \qquad \text{where } b, d \neq 0$$

EXAMPLE 8   Multiply. $\dfrac{2x^2 - 4x}{x^2 - 5x + 6} \cdot \dfrac{x^2 - 9}{2x^4 + 14x^3 + 24x^2}$

We first use the basic rule of fractions; that is, we factor (if possible) the numerator and denominator and remove common factors.

$$\frac{2x(x-2)}{(x-2)(x-3)} \cdot \frac{(x+3)(x-3)}{2x^2(x^2+7x+12)} = \frac{2x(x-2)(x+3)(x-3)}{(2x)x(x-2)(x-3)(x+3)(x+4)}$$

$$= \frac{2x}{2x} \cdot \frac{1}{x} \cdot \frac{x-2}{x-2} \cdot \frac{x+3}{x+3} \cdot \frac{x-3}{x-3} \cdot \frac{1}{x+4}$$

$$= 1 \cdot \frac{1}{x} \cdot 1 \cdot 1 \cdot 1 \cdot \frac{1}{x+4}$$

$$= \frac{1}{x(x+4)} \quad \text{or} \quad \frac{1}{x^2+4x}$$

Although either form of the answer is correct, we usually use the factored form.

Practice Problem 8   Multiply. $\dfrac{2x^2 + 5xy + 2y^2}{4x^2 - y^2} \cdot \dfrac{2x^2 + xy - y^2}{x^2 + xy - 2y^2}$ ∎

EXAMPLE 9   Multiply. $\dfrac{7x + 7y}{4ax + 4ay} \cdot \dfrac{8a^2x^2 - 8b^2x^2}{35ax^3 - 35bx^3}$

$$\frac{7(x+y)}{4a(x+y)} \cdot \frac{8x^2(a^2 - b^2)}{35x^3(a-b)} = \frac{7(x+y)}{4a(x+y)} \cdot \frac{8x^2(a+b)(a-b)}{35x^3(a-b)}$$

$$= \frac{\overset{1}{\cancel{7}(x+y)}}{4a\cancel{(x+y)}} \cdot \frac{\overset{2}{\cancel{8}}\overset{}{x^2}(a+b)\cancel{(a-b)}}{\underset{5x}{\cancel{35}\cancel{x^3}}\cancel{(a-b)}}$$

$$= \frac{2(a+b)}{5ax} \quad \text{or} \quad \frac{2a+2b}{5ax}$$

Note that we shortened our steps by not writing out every factor of 1 as we did in Example 8. Either way is correct.

Practice Problem 9   Multiply. $\dfrac{2x^3 - 3x^2}{3x^2 + 3x} \cdot \dfrac{9x + 36}{10x^2 - 15x}$ ∎

3 Simplifying the Quotient of Two Rational Expressions

When we divide fractions, we take the **reciprocal** of the second fraction and then multiply the fractions. (Remember that the reciprocal of a fraction $\frac{m}{n}$ is $\frac{n}{m}$. Thus the reciprocal of $\frac{2}{3}$ is $\frac{3}{2}$, and the reciprocal of $\frac{3x}{11y^2}$ is $\frac{11y^2}{3x}$.) We divide algebraic fractions in the same way. They are usually called rational expressions.

Dividing Rational Expressions

For any polynomials a, b, c, d,

$$\frac{a}{b} \div \frac{c}{d} = \frac{a}{b} \cdot \frac{d}{c}, \qquad \text{where } b, c, d \neq 0$$

EXAMPLE 10   Divide. $\dfrac{4x^2 - y^2}{x^2 + 4xy + 4y^2} \div \dfrac{4x - 2y}{3x + 6y}$

We take the reciprocal of the second fraction and multiply the fractions.

$$\frac{4x^2 - y^2}{x^2 + 4xy + 4y^2} \cdot \frac{3x + 6y}{4x - 2y} = \frac{(2x + y)(2x - y)}{(x + 2y)(x + 2y)} \cdot \frac{3(x + 2y)}{2(2x - y)}$$

$$= \frac{3(2x + y)}{2(x + 2y)} \quad \text{or} \quad \frac{6x + 3y}{2x + 4y}$$

Practice Problem 10   Divide. $\dfrac{8x^3 + 27}{64x^3 - 1} \div \dfrac{4x^2 - 9}{16x^2 + 4x + 1}$ ∎

EXAMPLE 11   Divide. $\dfrac{24 + 10x - 4x^2}{2x^2 + 13x + 15} \div (2x - 8)$

We take the reciprocal of the second fraction and multiply the fractions. The reciprocal of $(2x - 8)$ is $\frac{1}{2x - 8}$.

$$\frac{-4x^2 + 10x + 24}{2x^2 + 13x + 15} \cdot \frac{1}{2x - 8} = \frac{-2(2x^2 - 5x - 12)}{(2x + 3)(x + 5)} \cdot \frac{1}{2(x - 4)}$$

$$= \frac{\overset{-1}{\cancel{-2}}(x - 4)(2x + 3)}{(2x + 3)(x + 5)} \cdot \frac{1}{\underset{1}{\cancel{2}(x - 4)}} = \frac{-1}{x + 5} \quad \text{or} \quad -\frac{1}{x + 5}$$

Practice Problem 11   Divide. $\dfrac{4x^2 - 9}{2x^2 + 11x + 12} \div (-6x + 9)$ ∎

6.1 Exercises

Find the domain of each of the following rational expressions.

1. $\dfrac{5x + 6}{2x - 6}$

2. $\dfrac{3x - 8}{4x + 20}$

3. $\dfrac{-7x + 2}{x^2 - 5x - 36}$

4. $\dfrac{-8x + 9}{x^2 + 10x - 24}$

Simplify completely.

5. $\dfrac{21x^3y}{14x^2y^2}$

6. $\dfrac{5ab^3}{25a^2b^2}$

7. $\dfrac{3x^2 - 24x}{3x^2 + 12x}$

8. $\dfrac{10x^2 + 15x}{35x^2 - 5x}$

9. $\dfrac{4x^2}{6x^2 - 8x}$

10. $\dfrac{12x^2}{24x^2 - 18x}$

11. $\dfrac{21y^2 - 35y}{-7y}$

12. $\dfrac{25x^3 + 20x^2}{-5x^2}$

13. $\dfrac{3a^3b + 6a^2b}{3a^3b + 12a^2b}$

14. $\dfrac{x^2y^3 + 4xy^3}{x^2y^3 - 3xy^2}$

15. $\dfrac{2x + 10}{2x^2 - 50}$

16. $\dfrac{x^2 - 16}{2x - 8}$

17. $\dfrac{7a + 14b}{5a + 10b}$

18. $\dfrac{9x - 9y}{11x - 11y}$

19. $\dfrac{2x - 6}{3x^2 - x^3}$

20. $\dfrac{4y - 2y^2}{5y - 10}$

21. $\dfrac{2y^2 - 8}{2y + 4}$

22. $\dfrac{x + 2}{7x^2 - 28}$

23. $\dfrac{30x - x^2 - x^3}{x^3 - x^2 - 20x}$

24. $\dfrac{2x^2 - x^3 - x^4}{x^4 - x^3}$

25. $\dfrac{2y^2 + 2y - 12}{y^2 + 2y - 8}$

26. $\dfrac{y^2 + 6y + 9}{2y^2 + y - 15}$

27. $\dfrac{2y^2 - y - 15}{9 - y^2}$

28. $\dfrac{25 - a^2}{3a^2 - 13a - 10}$

Simplify and multiply.

29. $\dfrac{12x^3y}{6xy^5} \cdot \dfrac{5xy^2}{25x^4y^4}$

30. $\dfrac{25ab^4}{125ab^3} \cdot \dfrac{10a^5b^4}{14a^3b}$

31. $\dfrac{3a^2}{a + 2} \cdot \dfrac{a^2 - 4}{3a}$

32. $\dfrac{5x^2}{x^2 - 4} \cdot \dfrac{x^2 + 4x + 4}{10x^3}$

33. $\dfrac{x^2 + 3x + 8}{4x - 4} \cdot \dfrac{2x - 2}{x^2 + 3x + 8}$

34. $\dfrac{x - 5}{10x - 2} \cdot \dfrac{25x^2 - 1}{x^2 - 10x + 25}$

35. $\dfrac{x^2 - 5xy - 24y^2}{x - y} \cdot \dfrac{x^2 + 6xy - 7y^2}{x + 3y}$

36. $\dfrac{x - 3y}{x^2 + 3xy - 18y^2} \cdot \dfrac{x^2 + xy - 30y^2}{x - 5y}$

37. $\dfrac{2y^2 - 5y - 12}{4y^2 + 8y + 3} \cdot \dfrac{2y^2 + 7y + 3}{y^2 - 16}$

38. $\dfrac{6y^2 + y - 1}{6y^2 + 5y + 1} \cdot \dfrac{3y^2 + 4y + 1}{3y^2 + 2y - 1}$

39. $\dfrac{x^3 - 125}{x^5 y} \cdot \dfrac{x^3 y^2}{x^2 + 5x + 25}$

40. $\dfrac{3a^3 b^2}{8a^3 - b^3} \cdot \dfrac{4a^2 + 2ab + b^2}{12ab^4}$

Simplify and divide.

41. $\dfrac{y^2 + 2y}{6y} \div \dfrac{y^2 - 4}{3y^2}$

42. $\dfrac{3y + 12}{8y^3} \div \dfrac{9y + 36}{16y^3}$

43. $\dfrac{(4a + 5)^2}{3a^2 - 10a - 8} \div \dfrac{4a + 5}{(a - 4)^2}$

44. $\dfrac{2a^2 - 7a - 15}{(a + 4)^2} \div \dfrac{(a - 5)^2}{a + 4}$

45. $\dfrac{x^2 - xy - 6y^2}{x^2 + 2} \div (x^2 + 2xy)$

46. $\dfrac{x^2 - 5x + 4}{2x - 8} \div (3x^2 - 3x)$

47. $\dfrac{9 - y^2}{y^2 + 5y + 6} \div \dfrac{2y - 6}{5y + 10}$

48. $\dfrac{4 - 2y}{2y + y^2} \div \dfrac{3y^2 - 12}{2y^2 + 8y + 8}$

49. $\dfrac{3a^2 - 27b^2}{2a^2 - 5ab - 3b^2} \div \dfrac{6a^2 - 21ab + 9b^2}{4a^2 - b^2}$

50. $\dfrac{4a^2 - 9b^2}{6a^2 - 5ab - 6b^2} \div \dfrac{4a^2 + 12ab + 9b^2}{2a^2 + ab - 3b^2}$

Mixed Practice

Perform the indicated operation. When no operation is indicated, simplify the rational expression completely.

51. $\dfrac{7x}{y^2} \div 21x^3$

52. $\dfrac{2y^4}{10x^2} \cdot \dfrac{5x^3}{4y^3}$

53. $\dfrac{3x^2 - 2x}{6x - 4}$

54. $\dfrac{-28x^5 y}{35x^6 y^2}$

55. $\dfrac{x^2y - 49y}{x^2y^3} \cdot \dfrac{3x^2y - 21xy}{x^2 - 14x + 49}$

56. $\dfrac{x^2 + 6x + 9}{2x^2y - 18y} \div \dfrac{6xy + 18y}{3x^2y - 27y}$

57. $\dfrac{-8 + 6x - x^2}{2x^3 - 8x}$

58. $\dfrac{x^2 - 7x + 10}{xy^6} \div \dfrac{x^2 - 11x + 30}{x^2y^5}$

59. $\dfrac{a^2 - a - 12}{2a^2 + 5a - 12}$

60. $\dfrac{5y^3 - 45y}{6 - y - y^2}$

61. $\dfrac{x^3 - 25x}{x^2 - 6x + 5} \cdot \dfrac{2x^2 - 2}{4x^2} \div \dfrac{x^2 + 5x}{7x + 7}$

62. $\dfrac{3x^2 + 3x - 60}{2x - 8} \cdot \dfrac{25x^3}{x^3 + 3x^2 - 10x} \cdot \dfrac{x^2 - 7x + 10}{30x^2}$

Optional Graphing Calculator Problems

Try to find the domain of the following expressions. Round your value to the nearest tenth.

63. $\dfrac{2x + 5}{3.6x^2 + 1.8x - 4.3}$

64. $\dfrac{5x - 4}{1.6x^2 - 1.3x - 5.9}$

Cumulative Review Problems

Graph the straight line. Plot at least three points.

65. $y = -\dfrac{3}{2}x + 4$

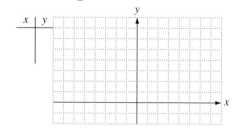

66. $6x - 3y = -12$

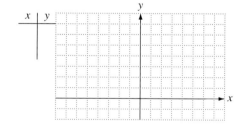

67. Find the equation of a line in standard form that passes through $(0, 5)$ and $(-2, -3)$.

68. Find the equation of a line in standard form that is perpendicular to $3x + 4y = 12$ and passes through $(-4, 4)$.

69. A directory assistance operator averages 2 inquiries per minute. If she works a 38-hour week, and takes one hour each day for lunch, (away from the phone), how many inquiries will she have answered in one week?

70. A hair stylist is able to service 21 customers in a 7-hour shift if he only *cuts* hair. The same hair stylist is able to service 14 customers in a 7-hour shift if he *cuts and colors* hair. He is able to service 18 customers in a 7-hour shift if he only *colors* hair. Can he feasibly service 12 cuts, 13 cuts and color, and 5 colors during one shift?

6.2 Addition and Subtraction of Rational Expressions

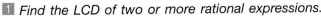

After studying this section, you will be able to:

1 *Find the LCD of two or more rational expressions.*
2 *Add or subtract two or more rational expressions.*

MathPro Video 15 SSM

1 *Finding the LCD of Two or More Rational Expressions*

Recall that to add or subtract fractions the denominators must be the same. If the denominators are not the same, we use the basic rule of fractions to change one or both fractions to equivalent fractions with the same or common denominators. For example,

$$\frac{3}{7} + \frac{2}{7} = \frac{5}{7}, \qquad \frac{2}{3} + \frac{3}{4} = \frac{2}{3} \cdot \frac{4}{4} + \frac{3}{4} \cdot \frac{3}{3} = \frac{8}{12} + \frac{9}{12} = \frac{17}{12}$$

How did we know that 12 was the least common denominator (LCD)? The least common denominator of two or more fractions is the product of the different prime factors in each denominator. If factors are repeated, we use the highest power of each factor.

$$3 = 3$$
$$4 = 2 \cdot 2$$
$$\text{LCD} = 3 \cdot 2 \cdot 2 = 12$$

This same technique is used to add or subtract rational expressions written as algebraic fractions.

How to Find the LCD

1. Factor each denominator completely into prime factors.
2. The LCD is the product of all the **different prime factors.**
3. If a factor occurs more than once in the denominator, you must use the highest power of that factor. (In other words, take the factor that occurs the greatest number of times in the denominator.)

Now we'll do some sample problems.

EXAMPLE 1   Find the LCD of these fractions

$$\frac{7}{x^2 - 4}, \qquad \frac{2}{x - 2}$$

Step 1 We factor each denominator completely (into prime factors).

$$(x^2 - 4) = (x + 2)(x - 2)$$
$$x - 2 \text{ cannot be factored.}$$

Step 2 The LCD is the product of all the *different* factors. (Note that it is *not* the product of *all* the factors.) The different factors are $x + 2$ and $x - 2$. So the LCD is $(x + 2)(x - 2)$.

Step 3 No factor occurs more than once in any denominator.

Practice Problem 1   Find the LCD. $\dfrac{8}{x^2 - x - 12}, \dfrac{3}{x - 4}$ ∎

EXAMPLE 2 Find the LCD of these fractions. $\dfrac{7}{12xy^2}, \dfrac{4}{15x^3y}$

Step 1 We factor each denominator.

$$12xy^2 = 2 \cdot 2 \cdot 3 \cdot x \cdot y \cdot y$$
$$15x^3y = \qquad\quad 3 \cdot 5 \cdot x \cdot x \cdot x \cdot y$$

Step 2 Our LCD will require each of the different factors that appear in either denominator. They are 2, 3, 5, x, and y.

Step 3 The factor 2 and the factor y each occur twice in $12xy^2$. The factor x occurs three times in $15x^3y$.
Thus the LCD $= 2 \cdot 2 \cdot 3 \cdot 5 \cdot x \cdot x \cdot x \cdot y \cdot y$
$= 60x^3y^2$

Practice Problem 2 Find the LCD of these fractions. $\dfrac{2}{15x^3y^2}, \dfrac{13}{25xy^3}$ ■

2 Adding or Subtracting Two or More Rational Expressions

We can add and subtract fractional algebraic expressions with the same denominator just as we do in arithmetic: We simply add or subtract the numerators.

Addition and Subtraction of Rational Expressions

For any polynomials a, b, c,

$$\frac{a}{b} + \frac{c}{b} = \frac{a+c}{b}, \qquad b \neq 0$$

$$\frac{a}{b} - \frac{c}{b} = \frac{a-c}{b}, \qquad b \neq 0$$

EXAMPLE 3 Add. $\dfrac{3}{2x^2y} + \dfrac{8}{2x^2y}$

$$\frac{3}{2x^2y} + \frac{8}{2x^2y} = \frac{11}{2x^2y}$$

Practice Problem 3 Add. $\dfrac{7}{3a^2b} + \dfrac{-2+a}{3a^2b}$ ■

EXAMPLE 4 Subtract. $\dfrac{5x+2}{(x+3)(x-4)} - \dfrac{6x}{(x+3)(x-4)}$

$$\frac{5x+2}{(x+3)(x-4)} - \frac{6x}{(x+3)(x-4)} = \frac{-x+2}{(x+3)(x-4)}$$

Practice Problem 4 Subtract. $\dfrac{4x}{(x+6)(2x-1)} - \dfrac{3x+1}{(x+6)(2x-1)}$ ■

If the two rational expressions have different denominators, then we will first need to find the LCD. Then we must rewrite each fraction as an equivalent fraction that has the LCD for a denominator.

EXAMPLE 5 Add the rational expressions. $\dfrac{7}{(x + 2)(x - 2)} + \dfrac{2}{x - 2}$

The LCD $= (x + 2)(x - 2)$.
Now before we can add the fractions, we must rewrite our fractions as fractions with the LCD.

$$\frac{7}{(x + 2)(x - 2)} + \frac{2}{x - 2} \cdot \frac{x + 2}{x + 2}$$

Since $\frac{x + 2}{x + 2} = 1$, we have not changed the *value* of the fraction. We are simply writing it in another equivalent form. Thus, we now have

$$\frac{7}{(x + 2)(x - 2)} + \frac{2(x + 2)}{(x + 2)(x - 2)} = \frac{7 + 2(x + 2)}{(x + 2)(x - 2)}$$

$$= \frac{7 + 2x + 4}{(x + 2)(x - 2)}$$

$$= \frac{2x + 11}{(x + 2)(x - 2)}$$

Practice Problem 5 Add the rational expressions. $\dfrac{8}{(x - 4)(x + 3)} + \dfrac{3}{x - 4}$ ∎

EXAMPLE 6 Add. $\dfrac{4}{x + 6} + \dfrac{5}{6x}$

The LCD $= 6x(x + 6)$. We need to multiply each fraction by 1 to obtain two equivalent fractions that have the LCD for the denominator. In the first fraction we will multiply by $1 = \frac{6x}{6x}$. In the second fraction we will multiply by $1 = \frac{x + 6}{x + 6}$.

$$\frac{4(6x)}{(x + 6)(6x)} + \frac{5(x + 6)}{6x(x + 6)}$$

$$\frac{24x}{6x(x + 6)} + \frac{5x + 30}{6x(x + 6)} = \frac{29x + 30}{6x(x + 6)}$$

Practice Problem 6 Add. $\dfrac{5}{x + 4} + \dfrac{3}{4x}$ ∎

EXAMPLE 7 Add. $\dfrac{7}{2x^2y} + \dfrac{3}{xy^2}$

You should be able to see that the LCD of these fractions is $2x^2y^2$. Hence

$$\frac{7}{2x^2y} \cdot \frac{y}{y} + \frac{3}{xy^2} \cdot \frac{2x}{2x} = \frac{7y}{2x^2y^2} + \frac{6x}{2x^2y^2} = \frac{6x + 7y}{2x^2y^2}$$ ∎

Practice Problem 7 Add. $\dfrac{7}{4ab^3} + \dfrac{1}{3a^3b^2}$ ∎

 GRAPHING CALCULATOR

Exploration

You can verify the answer to Example 6 with a graphing calculator.

Graph $y_1 = \dfrac{4}{x + 6} + \dfrac{5}{6x}$

and $y_2 = \dfrac{29x + 30}{6x(x + 6)}$ on the same set of axes. Since y_1 is equivalent to y_2 you will obtain exactly the same curve.

Verify on your graphing calculator that y_1 *is* or *is not* equivalent to y_2 for each of the following equations.

1. $y_1 = \dfrac{x}{x + 3} - \dfrac{3 - x}{x^2 - 9}$

   and $y_2 = \dfrac{x + 1}{x + 3}$

2. $y_1 = \dfrac{3x}{2x - 3} +$

   $\dfrac{3x + 6}{2x^2 + x - 6}$

   and $y_2 = \dfrac{3x + 3}{2x + 3}$

EXAMPLE 8   Add. $\dfrac{2}{x-5} + \dfrac{3x}{x^2 - 2x - 15} + \dfrac{3}{x+3}$

After we factor the denominator of the second fraction, we see that the LCD is $(x-5)(x+3)$.

$$\frac{2}{x-5} \cdot \boxed{\frac{x+3}{x+3}} + \frac{3x}{(x-5)(x+3)} + \frac{3}{x+3} \cdot \boxed{\frac{x-5}{x-5}}$$

$$= \frac{2x+6}{(x-5)(x+3)} + \frac{3x}{(x-5)(x+3)} + \frac{3x-15}{(x-5)(x+3)}$$

$$= \frac{8x-9}{(x-5)(x+3)}$$

Practice Problem 8   Add. $\dfrac{7}{x+4} + \dfrac{-5x}{x^2 - x - 20} + \dfrac{2}{x-5}$ ■

When two rational expressions are subtracted, we must be very careful of the signs in the numerator of the second fraction.

EXAMPLE 9   Subtract. $\dfrac{2}{x^2 + 3x + 2} - \dfrac{4}{x^2 + 4x + 3}$

$$x^2 + 3x + 2 = (x+1)(x+2)$$

$$x^2 + 4x + 3 = (x+1)(x+3)$$

Therefore, the LCD is $(x+1)(x+2)(x+3)$. We now have

$$\frac{2}{(x+1)(x+2)} \cdot \boxed{\frac{x+3}{x+3}} - \frac{4}{(x+1)(x+3)} \cdot \boxed{\frac{x+2}{x+2}}$$

$$= \frac{2x+6}{(x+1)(x+2)(x+3)} - \frac{4x+8}{(x+1)(x+2)(x+3)} = \frac{2x+6-4x-8}{(x+1)(x+2)(x+3)}$$

$$= \frac{-2x-2}{(x+1)(x+2)(x+3)} = \frac{-2\cancel{(x+1)}}{\cancel{(x+1)}(x+2)(x+3)}$$

$$= \frac{-2}{(x+2)(x+3)}$$

Study this problem carefully. Be sure you understand the reason for each step. You'll see this type of problem often.

Practice Problem 9   Subtract. $\dfrac{4x+2}{x^2 + x - 12} - \dfrac{3x+8}{x^2 + 6x + 8}$ ■

The following example involves repeated factors in the denominator. Read it through carefully to be sure you understand each step.

EXAMPLE 10 Subtract. $\dfrac{2x + 1}{25x^2 + 10x + 1} - \dfrac{6x}{25x + 5}$

Step 1 Factor each denominator into prime factors.

$$25x^2 + 10x + 1 = (5x + 1)^2$$

$$25x + 5 = 5(5x + 1)$$

Step 2 The different factors are 5 and $5x + 1$. However, $5x + 1$ appears to the first power *and* to the second power. So we need step 3.

Step 3 We must use the *highest* power of the common factor. In this example the highest power is 2, so we use $(5x + 1)^2$.

The LCD is the product of the *different* prime factors, and each factor has the highest power of all the factors. So our LCD is $5(5x + 1)^2$. First we write our problem in factored form.

$$\frac{2x + 1}{(5x + 1)^2} - \frac{6x}{5(5x + 1)}$$

Next, we must multiply our fractions by the appropriate factor to change them to equivalent fractions with the LCD. We write

$$\frac{2x + 1}{(5x + 1)^2} \cdot \frac{5}{5} - \frac{6x}{5(5x + 1)} \cdot \frac{5x + 1}{5x + 1} = \frac{5(2x + 1) - 6x(5x + 1)}{5(5x + 1)^2}$$

$$= \frac{10x + 5 - 30x^2 - 6x}{5(5x + 1)^2}$$

$$= \frac{-30x^2 + 4x + 5}{5(5x + 1)^2}$$

Practice Problem 10 Subtract. $\dfrac{7x - 3}{4x^2 + 20x + 25} - \dfrac{3x}{4x + 10}$ ∎

Caution: Adding and subtracting rational expressions is somewhat difficult. You should take great care in selecting the LCD. Students often find they have made a careless error in picking the LCD. Likewise, great care should also be taken to copy correctly all + and − signs. It is very easy to make a sign error when combining the equivalent fractions. Try to work very neatly and very carefully. A little extra diligence will show significant results in terms of more problems worked correctly.

6.2 Exercises

Find the LCD.

1. $\dfrac{2}{x^2y}, \dfrac{5}{7y}$

2. $\dfrac{3}{x^3y}, \dfrac{5}{2x}$

3. $\dfrac{7}{x+2}, \dfrac{9}{2x}$

4. $\dfrac{-4}{3y}, \dfrac{5}{y-3}$

5. $\dfrac{1}{x^2+7xy+12y^2}, \dfrac{3}{x+4y}$

6. $\dfrac{2}{4x^2-8x+3}, \dfrac{1}{2x-1}$

7. $\dfrac{11x}{(2x+5)^3}, \dfrac{3x-1}{(x+6)(2x+5)^2}$

8. $\dfrac{12xy}{(x+5)(x-2)}, \dfrac{2xy}{(x+5)^4(x-2)}$

9. $\dfrac{3x}{5x^2-7x}, \dfrac{5x}{5x^2-12x+7}$

10. $\dfrac{5x}{3x^2+2x}, \dfrac{2x}{6x^2-5x-6}$

11. $\dfrac{5y}{9y^2-49}, \dfrac{2y+1}{9y^2-42y+49}$

12. $\dfrac{2y+1}{16y^2-25}, \dfrac{3y-1}{16y^2+40y+25}$

13. $\dfrac{3}{8x^2}, \dfrac{5x}{6x^2+18x}, \dfrac{x+2}{x^2-5x-24}$

14. $\dfrac{5}{2x^2-50}, \dfrac{2x}{x^2-2x-35}, \dfrac{11x}{4x^2}$

Add or subtract the fractions and simplify your answer.

15. $\dfrac{2}{x+2} - \dfrac{3}{2x}$

16. $\dfrac{5}{3x+y} + \dfrac{2}{3xy}$

17. $\dfrac{12}{5x^2} + \dfrac{2}{5xy}$

18. $\dfrac{7}{4ab} + \dfrac{3}{4b^2}$

19. $\dfrac{3}{x^2-7x+12} + \dfrac{5}{x^2-4x}$

20. $\dfrac{2}{x^2-9} + \dfrac{3}{x-3}$

21. $\dfrac{7x}{49x^2-25y^2} + \dfrac{3}{7x-5y}$

22. $\dfrac{3}{x^2-3xy+2y^2} + \dfrac{5x}{x-2y}$

23. $\dfrac{3y}{y^2+y-2} + \dfrac{2}{y^2-4y+3}$

24. $\dfrac{2y}{y^2 - 7y + 10} + \dfrac{3y}{y^2 - 8y + 15}$

25. $\dfrac{3a}{3a^2 - 12} + \dfrac{5}{2a^2 + 4a}$

26. $\dfrac{2b}{3b^2 - 48} + \dfrac{2}{4b + b^2}$

27. $\dfrac{y + 2}{2y - 3} - \dfrac{4}{y + 3}$

28. $\dfrac{y - 5}{y + 2} - \dfrac{2y}{4y - 1}$

29. $\dfrac{2x - 1}{x - 6} - 1$

30. $\dfrac{8x + 3}{2x - 1} - 3$

31. $\dfrac{3x}{x^2 + 3x - 10} - \dfrac{2x}{x^2 + x - 6}$

32. $\dfrac{1}{x^2 - x - 2} - \dfrac{3}{x^2 + 2x + 1}$

33. $\dfrac{2x + 1}{x^2 + 5x + 6} - \dfrac{3}{x^2 - 4}$

34. $\dfrac{5x - 2}{x^2 - 9} - \dfrac{2}{x^2 - 2x - 3}$

35. $\dfrac{8y^2}{y^3 - 16y} - \dfrac{4y}{y^2 - 4y}$

36. $\dfrac{2y - 46}{2y^2 - 2y - 40} + \dfrac{4}{2y - 10}$

37. $a + 3 + \dfrac{2}{3a - 5}$

38. $a - 2 + \dfrac{3}{2a + 1}$

39. $\dfrac{2}{x^2 + 3xy + 2y^2} + \dfrac{4}{x^2 - xy - 2y^2} - \dfrac{3}{x^2 - 4y^2}$

40. $\dfrac{2}{x^2 - 5xy + 6y^2} + \dfrac{1}{x^2 - 7xy + 12y^2} - \dfrac{3}{x^2 - 6xy + 8y^2}$

To Think About

Perform the operations indicated.

41. $3x - 2 + \dfrac{5x}{3x - 2} + \dfrac{2x^2}{(3x - 2)^2}$

42. $\left[x + 1 + \dfrac{1}{x - 1}\right] \div \left[\dfrac{1}{x} + \dfrac{1}{x - 1}\right]$

Cumulative Review Problems

43. In the 1920s and 1930s, Charlie Chaplin was one of the most celebrated men of his era. During a visit to London, England, he received 73,000 letters in just two days. If he had 5 people reading his letters, and each person was able to read 10 letters per hour, how long would it have taken his assistants to help him to read these letters?

44. There were 985 new prescription drugs introduced into the United States between 1940 and 1976. Of these the United States originated 630.
  (a) What percentage of these prescription drugs were formulated in the United States?
  (b) If 7% of the new prescription drugs came from Switzerland, how many actual medications would have been formulated there?

45. Alreda, Tony, and Melissa each purchased a car. The total cost of the cars was $26,500. Alreda purchased a used car that cost $1500 more than one purchased by Tony. Melissa purchased a car that cost $1000 more than double the cost of Tony's car. How much did each car cost?

46. A chemist at Argonne Laboratories must combine a mixture that is 15% acid with a mixture that is 30% acid to obtain 60 liters of a mixture that is 20% acid. How much of each kind should he use?

47. A Four Winns speedboat traveled up the Hudson River against the current a distance of 75 kilometers in 5 hours. The return trip with the current took only 3 hours. Find the speed of the current and the speed of the speedboat in still water.

48. Hop Tyep invested $3000 in high-quality bonds for one year. He earned $236 in interest. Part of the money was invested at 7% and part at 9%. How much was invested at each rate?

6.3 Complex Rational Expressions

After studying this section, you will be able to:

1 *Simplify complex rational expressions.*

1 Simplifying Complex Rational Expressions

A complex rational expression is a large fraction that has at least one fraction in the numerator or in the denominator or in both the numerator and the denominator. The following are three examples of complex rational expressions.

$$\frac{7 + \frac{1}{x}}{x + 2}, \qquad \frac{2}{\frac{x}{y} + 3}, \qquad \frac{\frac{a+b}{7}}{\frac{1}{x} + \frac{1}{x+a}}$$

There are two ways to simplify complex rational expressions. You can use whichever method you like.

EXAMPLE 1   Simplify. $\dfrac{x + \dfrac{1}{x}}{\dfrac{1}{x} + \dfrac{3}{x^2}}$

METHOD 1

1. Simplify numerator and denominator.

$$x + \frac{1}{x} = \frac{x^2 + 1}{x}$$

$$\frac{1}{x} + \frac{3}{x^2} = \frac{x+3}{x^2}$$

2. Divide the numerator by the denominator.

$$\frac{\dfrac{x^2+1}{x}}{\dfrac{x+3}{x^2}} = \frac{x^2+1}{x} \div \frac{x+3}{x^2}$$

$$= \frac{x^2+1}{x} \cdot \frac{x^2}{x+3}$$

$$= \frac{x^2+1}{\cancel{x}} \cdot \frac{\cancel{x^2}^{x}}{x+3}$$

$$= \frac{x(x^2+1)}{x+3}$$

3. The result is already simplified.

METHOD 2

1. Find the LCD. The LCD is x^2.
2. Multiply the numerator and denominator by the LCD. Use the distributive property.

$$\frac{x + \dfrac{1}{x}}{\dfrac{1}{x} + \dfrac{3}{x^2}} \cdot \frac{x^2}{x^2} = \frac{x^3 + x}{x + 3}$$

3. Simplify.

$$\frac{x^3+x}{x+3} = \frac{x(x^2+1)}{x+3}$$

Practice Problem 1   Simplify. $\dfrac{y + \dfrac{3}{y}}{\dfrac{2}{y^2} + \dfrac{5}{y}}$ ∎

MathPro   Video 15   SSM

GRAPHING CALCULATOR

Exploration

You can verify the answer for Example 1 on a graphing calculator. Graph y_1 and y_2 on the same axis

$$y_1 = \frac{x + \dfrac{1}{x}}{\dfrac{1}{x} + \dfrac{3}{x^2}}$$

$$y_2 = \frac{x(x^2+1)}{x+3}$$

The domain of y_1 is more restrictive. If we use the domain of y_1 then the graphs of y_1 and y_2 should be identical.

Verify on your graphing calculator that y_1 is or is not equivalent to y_2 in each of the following equations.

1. $y_1 = \dfrac{1 + \dfrac{3}{x+2}}{1 + \dfrac{6}{x-1}}$

   $y_2 = \dfrac{x-1}{x+2}$

2. $y_1 = \dfrac{\dfrac{1}{x+1} - \dfrac{1}{x}}{\dfrac{1}{x}}$

   $y_2 = \dfrac{1}{x+1}$

EXAMPLE 2 Simplify. $\dfrac{\dfrac{1}{2x+6}+\dfrac{3}{2}}{\dfrac{3}{x^2-9}+\dfrac{x}{x-3}}$

Method 1: Combining Fractions in Both Numerator and Denominator

1. Simplify the numerator and denominator, if possible, by combining quantities to obtain one fraction in the numerator and one fraction in the denominator.

2. Divide the numerator by the denominator (that is, multiply the numerator by the reciprocal of the denominator).

3. Simplify the expression.

METHOD 1

1. Simplify the numerator.

$$\frac{1}{2x+6}+\frac{3}{2}=\frac{1}{2(x+3)}+\frac{3}{2}$$

$$=\frac{1}{2(x+3)}+\frac{3(x+3)}{2(x+3)}$$

$$=\frac{1+3x+9}{2(x+3)}$$

$$=\frac{3x+10}{2(x+3)}$$

Simplify the denominator.

$$\frac{3}{x^2-9}+\frac{x}{x-3}=\frac{3}{(x+3)(x-3)}+\frac{x}{x-3}$$

$$=\frac{3}{(x+3)(x-3)}+\frac{x(x+3)}{(x+3)(x-3)}$$

$$=\frac{x^2+3x+3}{(x+3)(x-3)}$$

2. Divide the numerator by the denominator.

$$\frac{3x+10}{2(x+3)}\div\frac{x^2+3x+3}{(x+3)(x-3)}=\frac{3x+10}{2\cancel{(x+3)}}\cdot\frac{\cancel{(x+3)}(x-3)}{x^2+3x+3}$$

$$=\frac{(3x+10)(x-3)}{2(x^2+3x+3)}$$

3. Simplify. The answer is already simplified.

Method 2: Multiplying Each Term by the LCD of All Individual
Fractions

1. Find the LCD of all the algebraic fractions in the numerator and denominator.
2. Multiply the numerator and denominator of the complex fraction by the LCD.
3. Simplify the result.

METHOD 2

1. To find the LCD, we factor:

$$\frac{\dfrac{1}{2x+6}+\dfrac{3}{2}}{\dfrac{3}{x^2-9}+\dfrac{x}{x-3}}=\frac{\dfrac{1}{2(x+3)}+\dfrac{3}{2}}{\dfrac{3}{(x+3)(x-3)}+\dfrac{x}{x-3}}$$

The LCD of the two fractions in the numerator and the two in the denominator is

$$2(x+3)(x-3)$$

2. Multiply the numerator and denominator by the LCD.

$$\frac{\dfrac{1}{2(x+3)}+\dfrac{3}{2}}{\dfrac{3}{(x+3)(x-3)}+\dfrac{x}{x-3}}\cdot\frac{2(x+3)(x-3)}{2(x+3)(x-3)}$$

$$=\frac{\dfrac{1}{2(x+3)}\cdot 2(x+3)(x-3)+\dfrac{3}{2}\cdot 2(x+3)(x-3)}{\dfrac{3}{(x+3)(x-3)}\cdot 2(x+3)(x-3)+\dfrac{x}{x-3}\cdot 2(x+3)(x-3)}$$

$$=\frac{x-3+3(x+3)(x-3)}{6+2x(x+3)}$$

$$=\frac{3x^2+x-30}{2x^2+6x+6}=\frac{(3x+10)(x-3)}{2(x^2+3x+3)}$$

3. Simplify. The answer is already simplified.

By either Method 1 or 2 we can leave the answer in factored form, or we can multiply it out to obtain

$$\frac{3x^2+x-30}{2x^2+6x+6}$$

Practice Problem 2 Simplify. $\dfrac{\dfrac{4}{16x^2-1}+\dfrac{3}{4x+1}}{\dfrac{x}{4x-1}+\dfrac{5}{4x+1}}$ ∎

EXAMPLE 3 Simplify by Method 1. $\dfrac{x+3}{\dfrac{9}{x}-x}$

$$\frac{x+3}{\dfrac{9}{x}-x} = \frac{x+3}{\dfrac{9}{x}-\dfrac{x}{1}\cdot\dfrac{x}{x}} = \frac{\dfrac{x+3}{1}}{\dfrac{9}{x}-\dfrac{x^2}{x}} = \frac{x+3}{\dfrac{9-x^2}{x}} = \frac{x+3}{1}\div\frac{9-x^2}{x} = \boxed{\frac{x+3}{1}\cdot\frac{x}{9-x^2}}$$

$$= \frac{\cancel{x+3}}{1}\cdot\frac{x}{\cancel{(3+x)}(3-x)} = \frac{x}{3-x}$$

Practice Problem 3 Simplify by Method 1. $\dfrac{4+\dfrac{1}{x+3}}{\dfrac{2}{x^2+4x+3}}$ ■

EXAMPLE 4 Simplify by Method 2. $\dfrac{\dfrac{3}{x+2}+\dfrac{1}{x}}{\dfrac{3}{y}-\dfrac{2}{x}}$

The LCD of the numerator is $x(x+2)$. The LCD of the denominator is xy. Thus the LCD of the complex fraction is $xy(x+2)$. Thus

$$\frac{\dfrac{3}{x+2}+\dfrac{1}{x}}{\dfrac{3}{y}-\dfrac{2}{x}}\cdot\frac{\boxed{xy(x+2)}}{\boxed{xy(x+2)}} = \frac{3xy+xy+2y}{3x(x+2)-2y(x+2)}$$

$$= \frac{4xy+2y}{(x+2)(3x-2y)}$$

$$= \frac{2y(2x+1)}{(x+2)(3x-2y)}$$

Practice Problem 4 Simplify by Method 2. $\dfrac{\dfrac{7}{y+3}-\dfrac{3}{y}}{\dfrac{2}{y}+\dfrac{5}{y+3}}$ ■

6.3 Exercises

Simplify the complex fractions by any method.

1. $\dfrac{\dfrac{3}{x} + \dfrac{3}{y}}{\dfrac{6}{x^2 y^2}}$

2. $\dfrac{\dfrac{8}{x} + \dfrac{8}{y}}{\dfrac{2}{x^2}}$

3. $\dfrac{\dfrac{3x + 2}{x - 2y}}{\dfrac{5x - 6}{x - 2y}}$

4. $\dfrac{\dfrac{x^2 - 4}{x + 2}}{\dfrac{3x + 6}{x + 2}}$

5. $\dfrac{1 - \dfrac{25}{y^2}}{\dfrac{5}{y} + 1}$

6. $\dfrac{1 - \dfrac{9}{y^2}}{1 + \dfrac{3}{y}}$

7. $\dfrac{\dfrac{y}{6} - \dfrac{1}{2y}}{\dfrac{3}{2y} - \dfrac{1}{y}}$

8. $\dfrac{\dfrac{1}{3y} + \dfrac{1}{6y}}{\dfrac{1}{2y} + \dfrac{3}{4y}}$

9. $\dfrac{\dfrac{2}{y^2 - 9}}{\dfrac{3}{y + 3} + 1}$

10. $\dfrac{\dfrac{2}{y + 4}}{\dfrac{3}{y - 4} - \dfrac{1}{y^2 - 16}}$

11. $\dfrac{\dfrac{7}{x} - 2}{\dfrac{x^2 + 1}{4}}$

12. $\dfrac{\dfrac{3}{x} + x}{\dfrac{x^2 + 2}{3}}$

13. $\dfrac{6}{2x - \dfrac{10}{x - 4}}$

14. $\dfrac{-8}{\dfrac{6x}{x - 1} - 4}$

15. $\dfrac{\dfrac{1}{2x + 3} + \dfrac{2}{4x^2 + 12x + 9}}{\dfrac{5}{2x^2 + 3x}}$

16. $\dfrac{\dfrac{3}{5x - 2} - \dfrac{2}{25x^2 - 4}}{\dfrac{7x}{5x^2 - 2x}}$

17. $\dfrac{\dfrac{2}{a + b} + \dfrac{3}{a - b}}{\dfrac{1}{2a} + \dfrac{3}{2b}}$

18. $\dfrac{\dfrac{7}{a + 3b} + \dfrac{2}{a - 3b}}{\dfrac{b}{3a} + \dfrac{a}{2b}}$

19. $\dfrac{\dfrac{1}{x+a} - \dfrac{1}{x}}{a}$

20. $\dfrac{\dfrac{1}{x-a} - \dfrac{1}{x}}{a}$

21. $\dfrac{\dfrac{6}{y^2 - 3y - 4} + 1}{1 - \dfrac{1}{y+1}}$

22. $\dfrac{1 + \dfrac{1}{y-2}}{\dfrac{6}{y^2 + 3y - 10} - \dfrac{1}{y-2}}$

To Think About

23. $1 - \dfrac{1}{1 - \dfrac{1}{y-2}}$

24. $\dfrac{x}{1 + \dfrac{1}{x}} + \dfrac{2x}{2 + \dfrac{2}{x}}$

Cumulative Review Problems

Solve for x.

25. $|2 - 3x| = 4$

26. $\left| \dfrac{1}{2}(5 - x) \right| = 5$

27. $|7x - 3 - 2x| < 6$

28. $|0.6x + 0.3| \geq 1.2$

29. The price of building an interstate highway project in the late 1970s near New York City cost $4000 per inch. **(a)** How much money was spent per mile? **(b)** Now a new interstate highway connector will be built in the same area beginning in 2001 for $660 million per mile. If the land acquisition cost was $570 million how many miles can be built if the total budget limit is $4,860,000,000?

30. Jan Robbins and her family are being relocated by her corporation from Dallas to Stockholm, Sweden. The moving company charges $2.50 per kilogram for air freight and $1.30 per kilogram for belongings shipped by ocean freighter. Jan needs to ship 5600 kilograms of her belongings to Stockholm. Her corporation will only pay $9380 for the shipments. How many kilograms should she ship by airfreight and how many kilograms by ocean freighter.

6.4 Rational Equations

After studying this section, you will be able to:

1 *Solve a rational equation that has a solution and be able to check the solutions.*
2 *Identify those rational equations that have no solution.*

MathPro   Video 16   SSM

1 *Solving a Rational Equation*

A rational equation is an equation that has one or more rational expressions as terms. To solve rational equations, we find the LCD of all fractions in the equation and multiply each side of the equation by the LCD. We then solve the resulting linear equation.

EXAMPLE 1   Solve. $\dfrac{9}{4} - \dfrac{1}{2x} = \dfrac{4}{x}$ Check your solution.

First we multiply each side of the equation by the LCD, which is $4x$.

$$4x \left(\frac{9}{4} - \frac{1}{2x} \right) = 4x \left(\frac{4}{x} \right)$$

$$\cancel{4}x \left(\frac{9}{\cancel{4}} \right) - \overset{2}{\cancel{4x}} \left(\frac{1}{\cancel{2x}} \right) = 4\cancel{x} \left(\frac{4}{\cancel{x}} \right) \qquad \textit{Use the distributive property.}$$

$$9x - 2 = 16 \qquad\qquad \textit{Simplify.}$$

$$9x = 18 \qquad\qquad \textit{Collect like terms.}$$

$$x = 2 \qquad\qquad \textit{Divide each side by the coefficient of x.}$$

Check: $\dfrac{9}{4} - \dfrac{1}{2(2)} \overset{?}{=} \dfrac{4}{2}$

$\dfrac{9}{4} - \dfrac{1}{4} \overset{?}{=} 2$

$\dfrac{8}{4} \overset{?}{=} 2$

$2 = 2$ ✓ It checks.

Practice Problem 1   Solve and check. $\dfrac{4}{3x} + \dfrac{x+1}{x} = \dfrac{1}{2}$ ∎

Usually, we combine the first two steps of the problem and only show the step of multiplying each term of the equation by the LCD. We will follow this approach in the remaining examples in this section.

This is one more illustration of the need to understand a mathematical principle rather than merely copying down a step without understanding. Because we understand the distributive property we can move directly to simplifying a rational equation by multiplying each term of the equation by the LCD.

GRAPHING CALCULATOR

Solving Rational Equations

To solve Example 2 on your graphing calculator find the point of intersection of

$$y_1 = \frac{2}{3x + 6}$$

and $y_2 = \dfrac{1}{6} - \dfrac{1}{2x + 4}$.

Use the Zoom and Trace features or the intersection command to find that the solution is $x = 5.00$ (to the nearest hundredth). What two difficulties do you observe in the graph that make this exploration more challenging? How can these be overcome?

Use this method to find the solution to the following equations on your graphing calculator. (Round your answer to the nearest hundredth).

1. $\dfrac{3}{x + 3} + \dfrac{5}{x + 4}$

$$= \dfrac{12x + 19}{x^2 + 7x + 12}$$

2. $\dfrac{x - 2.84}{x + 1.12} = \dfrac{x - 5.93}{x + 5.06}$

EXAMPLE 2 Solve. $\dfrac{2}{3x + 6} = \dfrac{1}{6} - \dfrac{1}{2x + 4}$

$$\frac{2}{3(x + 2)} = \frac{1}{6} - \frac{1}{2(x + 2)} \qquad \text{\textit{Factor each denominator.}}$$

$$\overset{2}{\cancel{6(x + 2)}}\left[\frac{2}{\cancel{3(x + 2)}}\right] = \cancel{6}(x + 2)\left[\frac{1}{\cancel{6}}\right] - \overset{3}{\cancel{6(x + 2)}}\left[\frac{1}{\cancel{2(x + 2)}}\right] \qquad \begin{array}{l}\textit{Multiply each} \\ \textit{term by the LCD} \\ 6(x + 2).\end{array}$$

$$4 = x + 2 - 3 \qquad \textit{Simplify.}$$

$$4 = x - 1 \qquad \textit{Collect like terms.}$$

$$5 = x \qquad \textit{Solve for x.}$$

Check: Verify that $x = 5$ is the solution.

Practice Problem 2 Solve and check. $\dfrac{1}{3x - 9} = \dfrac{1}{2x - 6} - \dfrac{5}{6}$ ∎

EXAMPLE 3 Solve. $\dfrac{y^2 - 10}{y^2 - y - 20} = 1 + \dfrac{7}{y - 5}$

$$\frac{y^2 - 10}{(y - 5)(y + 4)} = 1 + \frac{7}{y - 5} \qquad \begin{array}{l}\textit{Factor each denominator.} \\ \textit{Multiply each term by the LCD } (y - 5)(y + 4)\end{array}$$

$$\cancel{(y - 5)(y + 4)}\left[\frac{y^2 - 10}{\cancel{(y - 5)(y + 4)}}\right] = (y - 5)(y + 4)[1] + (y - 5)\cancel{(y + 4)}\left[\frac{7}{y \cancel{- 5}}\right]$$

$$y^2 - 10 = (y - 5)(y + 4)(1) + 7(y + 4) \qquad \textit{Remove common factors.}$$

$$y^2 - 10 = y^2 - y - 20 + 7y + 28 \qquad \textit{Simplify.}$$

$$y^2 - 10 = y^2 + 6y + 8 \qquad \textit{Collect like terms.}$$

$$-10 = 6y + 8 \qquad \textit{Subtract } y^2 \textit{ from each side.}$$

$$-18 = 6y \qquad \textit{Add } -8 \textit{ to each side.}$$

$$-3 = y \qquad \textit{Divide each side by coefficient of y.}$$

Check: $\dfrac{(-3)^2 - 10}{(-3)^2 - (-3) - 20} \overset{?}{=} 1 + \dfrac{7}{-3 - 5}$

$$\frac{9 - 10}{9 + 3 - 20} \overset{?}{=} 1 + \frac{7}{-8}$$

$$\frac{-1}{-8} \overset{?}{=} 1 - \frac{7}{8}$$

$$\frac{1}{8} = \frac{1}{8} \checkmark$$

Practice Problem 3 Solve. $\dfrac{y^2 + 4y - 2}{y^2 - 2y - 8} = 1 + \dfrac{4}{y - 4}$ ∎

---

2 Identifying Equations with No Solution

Some equations have no solution. This can happen in two distinct ways. In the first case, when you attempt to solve the equation, you obtain a contradiction, such as $0 = 1$. This occurs because the variable "drops out" of the equation. No solution can be obtained. In the second case, we may solve an equation to get an *apparent* solution, but this "solution" may not satisfy the original equation. We call the apparent solution an **extraneous solution.** An equation that yields an extraneous solution has no solution.

Case 1: The Variable Drops Out

In this case, when you attempt to solve the equation, the coefficient of the variable term becomes zero. Thus you are left with a statement such as $0 = 1$, which you are certain is false. In such a case you know that there is no value for the variable that could make $0 = 1$, hence there cannot be any solution.

EXAMPLE 4 Solve. $\dfrac{z + 1}{z^2 - 3z + 2} + \dfrac{3}{z - 1} = \dfrac{4}{z - 2}$

$\dfrac{z + 1}{(z - 2)(z - 1)} + \dfrac{3}{z - 1} = \dfrac{4}{z-2}$ *Factor to find the LCD $(z - 2)(z - 1)$. Then multiply each term by LCD.*

$\cancel{(z-2)(z-1)}\left[\dfrac{z + 1}{\cancel{(z-2)(z-1)}}\right] + (z - 2)\cancel{(z-1)}\left[\dfrac{3}{\cancel{z-1}}\right] = \cancel{(z-2)}(z - 1)\left[\dfrac{4}{\cancel{z-2}}\right]$

$z + 1 + 3(z - 2) = 4(z - 1)$ *Divide out common factors.*

$z + 1 + 3z - 6 = 4z - 4$ *Simplify.*

$4z - 5 = 4z - 4$ *Collect like terms.*

$4z - 4z = -4 + 5$ *Obtain variable terms on one side and constant values on the other.*

$0z = 1$

$0 = 1$

Of course, $0 \neq 1$. Therefore, no value of z makes the original equation true. Hence the equation has **no solution.**

Practice Problem 4 Solve. $\dfrac{2x - 1}{x^2 - 7x + 10} + \dfrac{3}{x - 5} = \dfrac{5}{x - 2}$ ∎

Fortunately, these types of equations occur infrequently. Although you need to know how to identify them, you will not have to deal with them often.

EXAMPLE 5 Solve. $\dfrac{4y}{y+3} - \dfrac{12}{y-3} = \dfrac{4y^2 + 36}{y^2 - 9}$

$\dfrac{4y}{y+3} - \dfrac{12}{y-3} = \dfrac{4y^2 + 36}{(y+3)(y-3)}$ *Factor each denominator to find the LCD* $(y+3)(y-3)$. *Multiply each term by LCD.*

$(y+3)(y-3)\left[\dfrac{4y}{y+3}\right] - (y+3)(y-3)\left[\dfrac{12}{y-3}\right] = (y+3)(y-3)\left[\dfrac{4y^2 + 36}{(y+3)(y-3)}\right]$

$4y(y-3) - 12(y+3) = 4y^2 + 36$ *Divide out common factors.*

$4y^2 - 12y - 12y - 36 = 4y^2 + 36$ *Remove parentheses.*

$4y^2 - 24y - 36 = 4y^2 + 36$ *Collect like terms.*

$-24y - 36 = 36$ *Subtract* $4y^2$ *from each side.*

$-24y = 72$ *Add* 36 *to each side.*

$y = \dfrac{72}{-24}$ *Divide each side by* -24

$y = -3$

Check: $\dfrac{4(-3)}{-3+3} - \dfrac{12}{-3-3} \overset{?}{=} \dfrac{4(-3)^2 + 36}{(-3)^2 - 9}$

$\dfrac{-12}{0} - \dfrac{12}{-6} \overset{?}{=} \dfrac{36 + 36}{0}$

You cannot divide by zero. Division by zero is not defined. A value of a variable that makes a denominator in the original equation zero is not a solution to the equation. Thus this equation has **no solution.**

Sometimes you may find that you do not have sufficient time for a complete check, but you still wish to make sure that you do not have a "no solution" situation. In those instances you can do a quick analysis to be sure that your obtained value for the variable does not make a denominator zero. If you were solving the equation

$\dfrac{4x - 1}{x^2 + 5x - 14} = \dfrac{1}{x - 2} - \dfrac{2}{x + 7}$ you would know immediately that you could not have

$x = 2$ or $x = -7$ as a solution. Do you see why?

Practice Problem 5   Solve and check. $\dfrac{y}{y-2} - 3 = 1 + \dfrac{2}{y-2}$ ∎

6.4 Exercises

Solve the equations and check your solution. If there is no solution, say so.

1. $\dfrac{2}{x} + \dfrac{3}{2x} = \dfrac{7}{6}$

2. $\dfrac{1}{x} + \dfrac{2}{3x} = \dfrac{1}{3}$

3. $3 - \dfrac{2}{x} = \dfrac{1}{4x}$

4. $\dfrac{5}{3x} + 2 = \dfrac{1}{x}$

5. $\dfrac{1}{x} - 3 = \dfrac{4}{x}$

6. $\dfrac{1}{y} + 2 = \dfrac{3}{y}$

7. $\dfrac{3}{x} - \dfrac{5}{4} = 1$

8. $\dfrac{2}{x} - \dfrac{3}{5} = 1$

9. $\dfrac{3}{y} = \dfrac{9}{2y - 1}$

10. $\dfrac{5}{y} = \dfrac{9}{2y + 1}$

11. $\dfrac{y + 6}{y + 3} - 2 = \dfrac{3}{y + 3}$

12. $4 - \dfrac{8x}{x + 1} = \dfrac{8}{x + 1}$

13. $\dfrac{3}{x} + \dfrac{4}{2x} = \dfrac{4}{x - 1}$

14. $\dfrac{1}{2x} + \dfrac{5}{x} = \dfrac{3}{x - 1}$

15. $\dfrac{4}{x - 2} = \dfrac{-1}{x + 3}$

16. $\dfrac{2}{x + 4} = \dfrac{-7}{x - 5}$

17. $\dfrac{x - 3}{x + 1} = \dfrac{x - 2}{x + 6}$

18. $\dfrac{x - 2}{x + 4} = \dfrac{x + 1}{x + 10}$

19. $\dfrac{3y}{y + 1} + \dfrac{4}{y - 2} = 3$

20. $\dfrac{4y}{y + 2} + \dfrac{2}{y - 1} = 4$

21. $1 + \dfrac{x + 1}{x^2 + 2x - 3} = \dfrac{x + 2}{x + 3}$

22. $\dfrac{3}{2x - 1} + \dfrac{3}{2x + 1} = \dfrac{8x}{4x^2 - 1}$

23. $\dfrac{6}{x} - \dfrac{3}{x^2 - x} = \dfrac{7}{x - 1}$

24. $\dfrac{5 - x}{x^2 - 1} + \dfrac{7}{x + 1} = \dfrac{6}{x}$

25. $\dfrac{3}{y + 1} - \dfrac{7}{5} = \dfrac{1}{5y + 5}$

26. $\dfrac{2}{3} + \dfrac{5}{y - 4} = \dfrac{y + 6}{3y - 12}$

27. $1 - \dfrac{10}{z - 3} = \dfrac{-5}{3z - 9}$

28. $\dfrac{3}{2} + \dfrac{2}{2z - 8} = \dfrac{1}{z - 4}$

29. $\dfrac{4}{y + 5} - \dfrac{32}{y^2 - 25} = \dfrac{-2}{y - 5}$

30. $\dfrac{-12}{y^2 - 9} - \dfrac{1}{y + 3} = \dfrac{1}{y - 3}$

31. $\dfrac{4}{z^2 - 9} = \dfrac{2}{z^2 - 3z}$

32. $\dfrac{z^2 + 16}{z^2 - 16} = \dfrac{z}{z + 4} - \dfrac{4}{z - 4}$

33. $\dfrac{1}{2x - 1} + \dfrac{2}{x - 5} = \dfrac{-22}{2x^2 - 11x + 5}$

34. $\dfrac{5}{2x + 3} + \dfrac{-4}{3x - 4} = \dfrac{3x}{6x^2 + x - 12}$

35. $\dfrac{6}{x + 1} = \dfrac{6}{x - 1} + \dfrac{-9}{x^2 - 1}$

36. $\dfrac{2}{2x + 3} + \dfrac{12}{4x^2 - 4x - 15} = \dfrac{2}{2x - 5}$

Verbal and Writing Skills

37. In what situations will a rational equation have no solution?

38. What does "extraneous solution" mean? What must we do to determine if a solution is an extraneous solution?

Optional Graphing Calculator Problems

Solve each equation. Round your answer to the nearest tenth.

39. $\dfrac{5}{x + 3.6} - \dfrac{4.2}{x - 7.6} = \dfrac{3.3}{x^2 - 4x - 27.36}$

40. $\dfrac{153.8}{x^2 + 4.9x - 39.56} = \dfrac{75.3}{x + 9.2} + \dfrac{84.2}{x - 4.3}$

Cumulative Review Problems

Factor completely.

41. $7x^2 - 63$

42. $2x^2 + 20x + 50$

43. $64x^3 - 27y^3$

44. $3x^2 - 13x + 14$

6.5 Applications: Formulas and Ratio Problems

After studying this section, you will be able to:

MathPro   Video 16   SSM

1 *Solve a formula for a particular variable.*
2 *Solve ratio problems.*

1 Solving a Formula for a Particular Variable

In science, economics, business, and mathematics, we use formulas that contain rational expressions. We often have to solve these formulas for a specific variable in terms of the other variables.

EXAMPLE 1 Solve for r. $S = \dfrac{a}{1 - r}$

This formula states that the limit S of an infinite geometric series is given by the first term a divided by the expression $1 - r$, where r is the common ratio and $|r| < 1$.

$$S(1 - r) = \left[\frac{a}{1 - r}\right](1 - r) \qquad \textit{Multiply both sides by the LCD } (1 - r).$$

$$S - Sr = a \qquad\qquad \textit{Simplify.}$$

$$-Sr = a - S \qquad\qquad \textit{Isolate the term containing } r.$$

$$r = \frac{a - S}{-S} \qquad\qquad \textit{Divide each side by } -S.$$

To avoid a negative sign in the denominator, the answer may be expressed as

$$r = -\frac{a - S}{S} \quad \text{or} \quad r = \frac{-a + S}{S}$$

Practice Problem 1 Solve for r. $P = \dfrac{A}{1 + r}$

This formula is used to determine the principal P that must be invested for 1 year to obtain an amount A if the simple interest rate expressed in decimal form is r. ■

EXAMPLE 2 Solve for a. $\dfrac{1}{f} = \dfrac{1}{a} + \dfrac{1}{b}$

This formula is used in optics in the study of light passing through a lens. It relates the focal length f of the lens to the distance a of an object from the lens and the distance b of the image from the lens.

$$abf\left[\frac{1}{f}\right] = abf\left[\frac{1}{a}\right] + abf\left[\frac{1}{b}\right] \qquad \textit{Multiply each term by the LCD abf.}$$

$$ab = bf + af \qquad\qquad \textit{Simplify.}$$

$$ab - af = bf \qquad\qquad \textit{Obtain all the terms containing } a \textit{ on one side of the equation.}$$

$$a(b - f) = bf \qquad\qquad \textit{Factor.}$$

$$a = \frac{bf}{b - f} \qquad\qquad \textit{Divide each side by } b - f.$$

Practice Problem 2 Solve for t. $\dfrac{1}{t} = \dfrac{1}{c} + \dfrac{1}{d}$

This formula relates the total amount of time t in hours that is required for two workers to complete a job working together if one worker can complete it alone in c hours and the other worker in d hours. ∎

EXAMPLE 3 The gravitational force F between two masses m_1 and m_2 a distance d apart is

$$F = \frac{Gm_1m_2}{d^2}$$

Solve for m_2.

The subscripts on the variable m mean that m_1 and m_2 are *different*. (The m stands for "mass.")

$d^2[F] = \cancel{d^2}\left[\dfrac{Gm_1m_2}{\cancel{d^2}}\right]$ *Multiply each side by the LCD d^2.*

$d^2F = Gm_1m_2$ *Simplify.*

$\dfrac{d^2F}{Gm_1} = \dfrac{Gm_1m_2}{Gm_1}$ *Divide each side by the coefficient of m_2, which is Gm_1.*

$\dfrac{d^2F}{Gm_1} = m_2$

Practice Problem 3 The number of telephone calls C between two cities of populations p_1 and p_2 that are a distance d apart may be represented by the formula

$$C = \frac{Bp_1p_2}{d^2}$$

Solve this equation for p_1. ∎

2 Solving Problems Involving Ratio and Proportion

The **ratio** of two values is the first value divided by the second value.

> The **ratio** of a to b ($b \neq 0$) is written as $\dfrac{a}{b}$ or $a \div b$ or $a : b$.

EXAMPLE 4 Smithville College has 88 faculty, 24 administrators, and 2056 students. What is the student-to-faculty ratio?

There are 2056 students to 88 faculty. So the ratio of students to faculty is $\frac{2056}{88} = \frac{257}{11}$ or $257 : 11$.

Practice Problem 4 Last month the Smithville plant produced 414 perfect parts and 36 defective parts. What is the ratio of perfect parts to defective parts? ∎

A **proportion** is an equation that says that two ratios are equal. If a has the same ratio to b as c has to d we can write the proportion equation $\frac{a}{b} = \frac{c}{d}$. We will use proportions to solve Examples 5 and 6.

EXAMPLE 5   On a long trip, Susan's car needed 7 gallons of gas to go 180 miles. If her car continues to consume gas at this rate, how many miles can she travel on 11 gallons of gas? Round your answer to the nearest whole mile.

Let x = number of miles she can travel.

Our first ratio is $\frac{7}{180}$ because it took 7 gallons of gas to drive 180 miles.

Our second ratio is $\frac{11}{x}$ because we want to know how far the car will travel on 11 gallons of gas.

Our proportion is

number of gallons of gas	$\rightarrow$	$\dfrac{7}{180}$	$=$	$\dfrac{11}{x}$	$\leftarrow$	number of gallons of gas
number of miles driven	$\nearrow$				$\nwarrow$	number of miles driven

$$\frac{7}{180} = \frac{11}{x} \qquad \textit{Now we solve for x.}$$

$$\cancel{180}x\left[\frac{7}{\cancel{180}}\right] = 180\cancel{x}\left[\frac{11}{\cancel{x}}\right] \qquad \textit{Multiply each side of the equation by the LCD 180x.}$$

$$7x = 1980$$

$$x = \frac{1980}{7}$$

$$x \approx 282.86 \approx 283 \text{ miles}$$

She can travel approximately 283 miles more.

Practice Problem 5   If there are 13 milliliters of acid in 37 milliliters of solution, how much acid will be contained in 296 milliliters of solution? ∎

EXAMPLE 6   A company plans to employ 910 people with a ratio of two managers for every 11 workers. How many managers should be hired? How many workers?

If we let x = the number of managers, then $910 - x$ = the number of workers. We are given the ratio of managers to workers, so let's set up our proportion in that way.

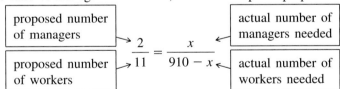

The LCD is $11(910 - x)$. Multiplying by the LCD, we get

$$\cancel{11}(910 - x)\left[\frac{2}{\cancel{11}}\right] = 11(\cancel{910 - x})\left[\frac{x}{\cancel{910 - x}}\right]$$

$$2(910 - x) = 11x$$

$$1820 - 2x = 11x$$

$$1820 = 13x$$

$$140 = x$$

$$910 - x = 910 - 140 = 770$$

The number of managers needed is 140. The number of workers needed is 770.

Practice Problem 6 Western University has 168 faculty. The university always maintains a student-to-faculty ratio of 21:2. How many students should they enroll to maintain that ratio? ∎

The next example concerns *similar triangles*. These triangles have corresponding angles equal and corresponding sides that are *proportional* (not equal). Similar triangles are frequently used to determine distances that cannot be conveniently measured. For example, in this sketch,

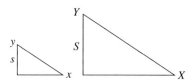

x and X are corresponding angles, y and Y are corresponding angles, and s and S are corresponding sides. Hence angle x = angle X, angle y = angle Y, and side s is proportional to side S (again, note that we did not say that side s is equal to side S). So, really, one triangle is just a magnification of the other triangle.

EXAMPLE 7 A helicopter is hovering to an unknown distance above an 850-foot building. A man watching the helicopter is 500 feet from the base of the building and 11 feet from a flagpole that is 29 feet tall. The man's line of sight to the helicopter is directly above the flagpole, as you can see on this sketch. How far above the building is the helicopter? Round your answer to the nearest foot.

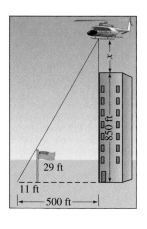

1. *Understand the problem.*
   Can you see the two triangles in the diagram in the margin? For convenience, we separate them out in the sketch on the right. We want to find the distance x. Are the triangles similar? The angles at the bases of the triangles are equal. (Why?) It follows, then, the top angles must also be equal. (Remember, the angles of any triangle add up to 180°.) Since the angles are equal, the triangles are similar and the sides are proportional.

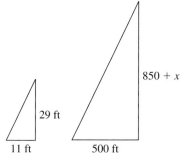

2. *Write an equation.*
   We can set up our proportion like this:

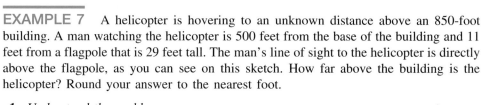

3. *Solve the equation and state the answer.*
   The LCD is $29(850 + x)$.

$$29(850 + x)\left[\frac{11}{29}\right] = 29(850 + x)\left[\frac{500}{850 + x}\right]$$

$$11(850 + x) = 29(500)$$

$$9350 + 11x = 14{,}500$$

$$11x = 5150$$

$$x = \frac{5150}{11} = 468.\overline{18}$$

So the helicopter is about 468 feet above the building.

Practice Problem 7 Solve the problem in the same way as Example 7 for a man watching 450 feet from the base of a 900-foot building as shown in the figure in the margin. This flagpole is 35 feet tall and the man is 10 feet from the flagpole. ∎

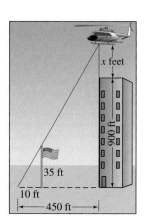

We will encounter some challenging problems involving similar triangles in Exercises 65 and 66.

We will sometimes encounter problems where two or more people or machines are working together to complete a certain task. These type of problems are sometimes called *work problems*. In general, these types of problems can be analyzed by using the concept

$$\boxed{\begin{array}{c}\text{Part of task done} \\ \text{by first person}\end{array}} + \boxed{\begin{array}{c}\text{part of task done} \\ \text{by second person}\end{array}} = \boxed{\begin{array}{c}\text{1 (one complete} \\ \text{task finished)}\end{array}}$$

We will also use a general idea about the rate at which something is done. If Robert can do a task in 3 hours, then he can do $\frac{1}{3}$ of the task in 1 hour. If Susan can do the same task in 2 hours, then she can do $\frac{1}{2}$ of the task in 1 hour. In general, if a person can do a task in t hours, then that person can do $\frac{1}{t}$ of the task in 1 hour.

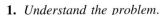

EXAMPLE 8 Robert can paint the kitchen in 3 hours. Susan can paint the kitchen in 2 hours. How long will it take them to paint the kitchen if Robert and Susan work together?

1. *Understand the problem.*
   Robert can paint $\frac{1}{3}$ of the kitchen in 1 hour.
   Susan can paint $\frac{1}{2}$ of the kitchen in 1 hour.
   We do not know how long it will take them working together, so we let x = the number of hours it takes them to paint the kitchen working together. To assist us, we will construct a table that relates the data. We will use the concept that (rate)(time) = amount done.

	(rate) ·	(time) =	amount
	Rate of Work per Hour	Time Worked in Hours	Amount of Task Done
Robert	$\frac{1}{3}$	x	$\frac{x}{3}$
Susan	$\frac{1}{2}$	x	$\frac{x}{2}$

2. *Write an equation.*

$$\boxed{\begin{array}{c}\text{Amount of work done} \\ \text{by Robert in } x \text{ hours}\end{array}} + \boxed{\begin{array}{c}\text{Amount of work done} \\ \text{by Susan in } x \text{ hours}\end{array}} = \boxed{\begin{array}{c}\text{1 completed} \\ \text{job}\end{array}}$$

$$\frac{x}{3} + \frac{x}{2} = 1$$

3. *Solve the equation and state the answer.*
   Multiply each side of the equation by the LCD and use the distributive property.

$$6\left(\frac{x}{3}\right) + 6\left(\frac{x}{2}\right) = 6(1)$$

$$2x + 3x = 6$$

$$5x = 6$$

$$x = \frac{6}{5} = 1\frac{1}{5} = 1.2 \text{ hours}$$

Robert and Susan can paint the kitchen in 1.2 hours working together.

Practice Problem 8 Alfred can mow the huge lawn at his Vermont farm with his new lawn mower in 4 hours. His young son can mow the lawn with the old lawn mower in 5 hours. If they work together, how long will it take them to mow the lawn? ∎

6.5 Exercises

Solve for the variable indicated.

1. $t = \dfrac{d}{r}$; for r **2.** $I = \dfrac{E}{R}$; for R **3.** $P = \dfrac{A}{1 + rt}$; for t **4.** $P = \dfrac{A}{1 + rt}$; for r

5. $\dfrac{1}{f} = \dfrac{1}{a} + \dfrac{1}{b}$; for b **6.** $\dfrac{1}{f} = \dfrac{1}{a} + \dfrac{1}{b}$; for f **7.** $R = \dfrac{ab}{3x}$; for x **8.** $A^2 = \dfrac{Q}{5M}$; for M

9. $F = \dfrac{xy + xz}{2}$; for x **10.** $A = \dfrac{ha + hb}{2}$; for h **11.** $I = \dfrac{nE}{R + nr}$; for n **12.** $T = \dfrac{24I}{B + Bn}$; for B

13. $\dfrac{A}{2\pi r} = b + h$; for r **14.** $\dfrac{3V}{\pi h} = r^2$; for h **15.** $\dfrac{E}{e} = \dfrac{R + r}{r}$; for e **16.** $\dfrac{E}{e} = \dfrac{Rer}{r}$; for R

17. $\dfrac{P_1 V_1}{T_1} = \dfrac{P_2 V_2}{T_2}$; for T_1 **18.** $\dfrac{P_1 V_1}{T_1} = \dfrac{P_2 V_2}{T_2}$; for T_2 **19.** $F = \dfrac{Gm_1 m_2}{d^2}$; for d^2 **20.** $F = \dfrac{Gm_1 m_2}{d^2}$; for G

21. $E = T_1 - \dfrac{T_1}{T_2}$; for T_1 **22.** $E = T_1 - \dfrac{T_1}{T_2}$; for T_2 **23.** $m = \dfrac{y_2 - y_1}{x_2 - x_1}$; for x_1 **24.** $m = \dfrac{y_2 - y_1}{x_2 - x_1}$; for x_2

25. $S = \dfrac{V_1 t + V_2 t}{2}$; for t **26.** $S = \dfrac{V_1 t + V_2 t}{2}$; for V_1 **27.** $Q = \dfrac{kA(t_1 - t_2)}{L}$; for t_2 **28.** $Q = \dfrac{kA(t_1 - t_2)}{L}$; for A

29. $\dfrac{T_2 W}{T_2 - T_1} = q$; for T_2 **30.** $d = \dfrac{LR_2}{R_2 + R_1}$; for R_1 **31.** $V = \dfrac{mv}{m + M}$; for m **32.** $V = \dfrac{mv}{m + M}$; for M

Round your answer to four decimal places.

33. Solve for T. $\dfrac{1.98\,V}{1.96\,V_0} = 0.983 + 5.936(T - T_0)$

34. Solve for r_1. $\dfrac{1}{R} = \dfrac{1}{r_1} + \dfrac{1}{0.368} + \dfrac{1}{0.736}$.

Applications

In Exercises 35 through 64, round your answer to the nearest hundredth, unless otherwise directed.

35. A map of Afghanistan has a scale that states that 4 centimeters = 3 kilometers. The distance on the map between two settlements is 7.5 centimeters. How many kilometers apart are the two settlements?

36. An indoor/outdoor sports center at the Manchester Athletic Club is expanding its facilities. The architects have drawn up a blueprint of the new clubhouse with a scale of 5:100. The drawing of a racket ball observation area is 7 inches by 11 inches long on the drawing. What are the length and width of the racket ball observation area?

37. A speed of 60 miles per hour (mph) is equivalent to a speed of 88 kilometers per hour (km/h). Paul is driving in Quebec where the speed limit is 100 km/h. Convert this speed limit to mph.

38. Hakim discovered that 40 centiliters of water evaporated completely in 2.5 minutes from a pan boiling at 100° Celsius. How long would it take 140 centiliters of water to evaporate completely?

39. Major League Baseball is very picky when it comes to inspecting the actual balls used in the game. The supervisors checked a box containing 48 balls. Of the 48, 7 were found to be unacceptable. If there are 8 boxes of balls, each containing 48 balls, *yet to be inspected,* and the supervisors assume that the ratio is the same,
(a) How many balls will pass inspection?
(b) How many balls will be defective?

40. At the annual Golden Retriever Convention last year, 450 pounds of dry dog food were provided to feed 156 dogs. This year, after word got out on the Internet, 320 dogs showed up with their owners. How much dry dog food did the convention need to feed the dogs? (Round this answer to the nearest pound.)

41. 35 grizzly bears were captured and radio-tagged by wildlife personnel in the Yukon. The bears were released back into the wild. Later that same year, 50 were captured, and 22 had tags. Estimate the number of grizzly bears in that part of the Yukon. (Round this answer to the nearest whole number.)

42. Thirty alligators were captured in the Louisiana bayou, (swamp), and tagged by National Park officials. The alligators were put back in the bayou. One month later, 50 alligators were captured from the same part of the bayou and 18 were tagged. Estimate to the nearest whole number how many alligators are in that part of Louisiana.

43. A car can travel 106.25 kilometers on 21 liters of gas. How far can the same car go on 37.5 liters of gas?

44. A fence post at a Brazilian ranch is 3.75 meters long and weighs 148 kilograms. How much would a similar fence post weigh if it were 7 meters long?

45. Pierre La Bonte, an artist, is practicing his skill by trying to paint a picture from a photograph. The photograph measures three inches wide by five inches long. In order to keep all dimensions the same, what should be the width of the new painting if the length is 36 inches?

46. A farmer, Everett Hatfield, needed a larger pasture, so he cut down trees around his old pasture, which measured 500 yards wide by 800 yards long. The larger pasture has the *same ratio* of width to length. The new width is 1200 yards. What is the new length?

47. A radio receiver weighing 40 kilograms on earth weighs only 6.4 kilograms on the moon. How much does a receiver weigh on earth if it weighs 20 kilograms on the moon?

48. At a depth of 120 feet, the water pressure on a submarine is 52 pounds per square inch. What is the pressure at 300 feet below the surface?

49. A city auto dealer took a poll to find out how many people preferred station wagons. The ratio of those preferring station wagons to those preferring sedans was 3:11. If the dealer wanted to order 224 sedans and station wagons, how many of each should he order?

50. The ratio of detectives to patrol officers at Center City is 2:9. The police force has 187 detectives and patrol officers. How many are detectives? How many are patrol officers?

51. When Jino D'Alessandro retired, his doctor told him to exercise in such a way that he ran 2 miles for every 7 miles he walked. If he plans to cover 63 miles in total each week, how many miles should he walk? How many miles should he run?

52. The harbormaster said that last year the ratio of powerboats to sailboats moored in the harbor was 4:9. If a total of 78 boats were moored in the harbor, how many boats were powerboats? How many boats were sailboats?

53. A twelve-foot marble statue in Italy casts a shadow that is 15 feet long. At the same time of day, a wall casts a shadow that is 8 feet long. How high is the wall?

54. A three-foot-tall child casts a shadow that is 4.8 feet long. At the same time of day, a building casts a shadow that extends 177 feet. How tall is the building?

55. A nutritionist told Maggie that for health reasons, she should eat fish 7 times for every 4 times she eats red meat. If over the next five months she eats red meat 112 times, how many times should she eat fish?

56. In a rural area of Oklahoma, the ratio of cars to trucks is 5:11. If there are a total of 2128 vehicles in this region, how many cars and how many trucks are in this rural area?

57. Matt can plow out all the driveways on his street in Duluth, Minnesota with his new 4-wheel-drive truck in 6 hours. His neighbor using a snow blade on a lawn tractor can plow out the same number of driveways in 9 hours. How long would it take them to do the work together?

58. The new mechanic at Speedy Lube can perform 30 oil changes on cars coming through the Lube Center in 4 hours. His assistant can perform the same number of oil changes in 6 hours. How long would it take them to do the work together?

59. Tori can clean the house in 6 hours. Brenda can do the same job in 5 hours. How long would it take Tori and Brenda to clean the house together?

60. The flight attendant at the gate can check in 200 passengers in 8 minutes. Her assistant can check in the same number of passengers in 11 minutes. How long would it take them to check people in together?

61. An experienced judge at a dog show is able to judge 180 dogs in 25 minutes. A new judge is able to do the same task in 45 minutes. How long would it take if the judges work together?

62. Suzanna is able to plant flower bulbs in her garden in 4 hours. Her neighbor, Barbara, who has the same size garden, is able to plant her flower bulbs in 7 hours. How long would it take them if they consolidate their efforts and work together?

63. A lumberjack working with cousin Fred can split a cord of seasoned oak firewood in 3 hours. The lumberjack can split the wood alone without any help from Fred in 4 hours. How long would it take Fred if he works alone without the lumberjack?

64. Pipe A and pipe B can together fill a swimming pool in 2 hours. If only pipe A is used, the pool can be filled in 3 hours. How long would it take for only pipe B to fill the pool?

To Think About

To find the width of a river, a hiking club laid out a triangular pattern of measurements. See the figure. Use your knowledge of similar triangles to solve Problems 65 and 66.

65. If the observer stands at the point shown in the figure, then $a = 2$ feet, $b = 5$ feet, and $c = 116$ feet. What is the width of the river?

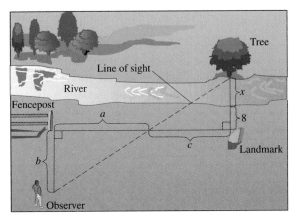

66. What is the width of the river if $a = 3$ feet, $b = 8$ feet, and $c = 297$ feet?

Cumulative Review Problems

Factor completely.

67. $8x^2 - 6x - 5$

68. $x^2 - 2x - 120$

69. $25x^2 - 90xy + 81y^2$

70. $5yx^2 - 20y$

Developing Your Study Skills

WHY IS REVIEW NECESSARY?

You master a course in mathematics by learning the concepts one step at a time. There are basic concepts like addition, subtraction, multiplication, and division of whole numbers that are considered the foundation upon which all of mathematics is built. These must be mastered first. Then the study of mathematics is built step by step upon this foundation, each step supporting the next. The process is a carefully designed procedure, and so no steps can be skipped. A student of mathematics needs to realize the importance of this building process to succeed.

Because learning new concepts depends on those previously learned, students often need to take time to review. The reviewing process will strengthen the understanding and application of concepts that are weak due to lack of mastery or passage of time. Review at the right time on the right concepts can strengthen previously learned skills and make progress possible.

Timely, periodic review of previously learned mathematical concepts is absolutely necessary in order to master new concepts. You may have forgotten a concept or grown a bit rusty in applying it. Reviewing is the answer. Make use of any review sections in your textbook, whether they are assigned or not. Look back to previous chapters whenever you have forgotten how to do something. Study the examples and practice some exercises to refresh your understanding.

Be sure that you understand and can perform the computations of each new concept. This will enable you to move successfully on to the next ones.

Remember, mathematics is a step-by-step building process. Learn one concept at a time, skipping none, and reinforce and strengthen with review whenever necessary.

Putting Your Skills to Work

SIGNAL STRENGTH OF CELL PHONES

The strength of the signal that is sent by a cell phone transmitting tower to an individual phone can be described by a rational equation. Suppose a particular transmitting tower sends out a signal that has an intensity of 12 watts per square yard at a distance of 3 miles from a cell phone. The intensity of the signal can be approximated by the equation

$$I = \frac{108}{d^2}$$

where I is the intensity of the signal measured in watts per square yard and d is the distance measured in miles from the cell phone to the transmitting tower.

PROBLEMS FOR INDIVIDUAL INVESTIGATION

1. What would be the intensity of the signal if the transmitting tower were 6 miles from the cell phone? If the distance is doubled what happens to the intensity of the signal?

2. Graph the equation $I = \frac{108}{d^2}$ using the values $d = 2, 3,$ 4, 6, 8, and 10. Based on your graph what would be the value of d for the intensity to be 4 watts per square yard?

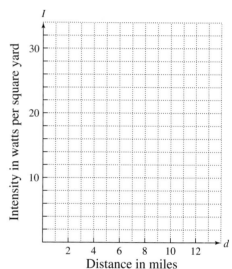

PROBLEMS FOR GROUP INVESTIGATION AND COOPERATIVE GROUP LEARNING

Together with other members of your class see if you can answer the following.

3. How could the equation be modified if distance was measured in kilometers instead of miles? (Use the variable k.)

4. How could the equation be modified if the distance was measured in kilometers instead of miles and the Intensity was measured in watts per square meter? (Use the variable k.)

INTERNET CONNECTION: Go to `http://www.prenhall.com/tobey` to be connected.

Site: Pagers OnLine Coverage Maps

Pagers use transmitting towers like the ones used by cell phones. This site shows the coverage area for a certain pager service provider. Choose several states or regions and view the detailed coverage maps.

5. If there were only one transmitter in a particular state, what shape would you expect the coverage area to have? What other factors might affect the shape of the coverage area? Estimate the number of transmitters for one of the states you chose to view closely.

6. Use the equation at the top of this page to estimate the signal strength required to operate a pager. This will give only a very rough estimate, since the correct equation varies depending on the signal strength of the transmitter and other factors. (You may need to look at another map in order to determine the scale of the coverage map; one possible source for another map is http://www.mapsonus.com/.)

Topic	Procedure	Examples
Simplifying fractions where numerator and denominator are polynomials, p. 336.	1. Factor the numerator and denominator, if possible. 2. Any *factor* that is common to both numerator and denominator can be divided out. This is an application of the basic rule of fractions. *Basic rule of fractions:* For any polynomials a, b, c (where $b \neq 0$ and $c \neq 0$). $$\frac{ac}{bc} = \frac{a}{b}$$	Simplify. $\dfrac{6x^2 - 14x + 4}{6x^2 - 20x + 6}$ $\dfrac{2(3x^2 - 7x + 2)}{2(3x^2 - 10x + 3)} = \dfrac{2(3x-1)(x-2)}{2(3x-1)(x-3)} = \dfrac{x-2}{x-3}$
Multiplying rational algebraic expressions, p. 338.	1. Factor all numerators and denominators, if possible. 2. Any *factor* that is common to a numerator of one factor and the denominator of the same fraction or any fraction that is multiplied by it can be divided out. 3. Write the indicated product of the remaining factors in the numerator. Write the indicated product of the remaining factors in the denominator. $$\frac{a}{b} \cdot \frac{c}{d} = \frac{ac}{bd}$$	Multiply. $\dfrac{x^2 - 4x}{6x - 12} \cdot \dfrac{3x^2 - 6x}{x^3 + 3x^2}$ $\dfrac{x(x-4)}{6(x-2)} \cdot \dfrac{3x(x-2)}{x^2(x+3)} = \dfrac{x-4}{2(x+3)}$ or $\dfrac{x-4}{2x+6}$
Dividing rational algebraic expressions, p. 339.	1. Invert the second fraction and multiply it by the first fraction. $$\frac{a}{b} \div \frac{c}{d} = \frac{a}{b} \cdot \frac{d}{c}$$ 2. Apply the steps for multiplying rational algebraic expressions.	Divide. $\dfrac{6x^2 - 5x - 6}{24x^2 + 13x - 2} \div \dfrac{4x^2 + x - 3}{8x^2 + 7x - 1}$ $\dfrac{6x^2 - 5x - 6}{24x^2 + 13x - 2} \cdot \dfrac{8x^2 + 7x - 1}{4x^2 + x - 3}$ $= \dfrac{(3x+2)(2x-3)}{(3x+2)(8x-1)} \cdot \dfrac{(8x-1)(x+1)}{(x+1)(4x-3)} = \dfrac{2x-3}{4x-3}$
Adding rational algebraic expressions, p. 344.	1. If all fractions have a common denominator, add the numerators and place the result over the common denominator. $$\frac{a}{c} + \frac{b}{c} = \frac{a+b}{c}$$ 2. If the fractions do not have a common denominator, factor the denominators (if necessary) and determine the least common denominator (LCD). 3. Multiply each fraction by the necessary value so that each fraction becomes an equivalent fraction with the LCD as the denominator. 4. Add the numerators and place the result over the common denominator. 5. Simplify, if possible.	Add. $\dfrac{7x}{x^2 - 9} + \dfrac{x+2}{x+3}$ $\dfrac{7x}{(x+3)(x-3)} + \dfrac{x+2}{x+3}$ The LCD $= (x+3)(x-3)$. $\dfrac{7x}{(x+3)(x-3)} + \dfrac{x+2}{x+3} \cdot \dfrac{x-3}{x-3}$ $= \dfrac{7x}{(x+3)(x-3)} + \dfrac{x^2 - x - 6}{(x+3)(x-3)}$ $= \dfrac{x^2 + 6x - 6}{(x+3)(x-3)}$
Subtracting rational algebraic expressions, p. 344.	Follow the procedures of adding rational algebraic expressions except that you subtract the second numerator from the first after each fraction has the LCD as the denominator. $$\frac{a}{c} - \frac{b}{c} = \frac{a-b}{c}$$	Subtract. $\dfrac{4x}{3x - 2} - \dfrac{5x}{x + 4}$ The LCD $= (3x-2)(x+4)$. $\dfrac{4x}{3x-2} \cdot \dfrac{x+4}{x+4} - \dfrac{5x}{x+4} \cdot \dfrac{3x-2}{3x-2}$ $= \dfrac{4x^2 + 16x}{(3x-2)(x+4)} - \dfrac{15x^2 - 10x}{(x+4)(3x-2)}$ $= \dfrac{-11x^2 + 26x}{(3x-2)(x+4)}$

Topic	Procedure	Examples
Simplifying a complex fraction by Method 1, p. 351.	1. Simplify the numerator and denominator, if possible, by combining quantities to obtain one fraction in the numerator and one fraction in the denominator. 2. Divide the numerator by the denominator. (That is, multiply the numerator by the reciprocal of the denominator.) 3. Simplify the result.	Simplify by Method 1. $\dfrac{4 - \dfrac{1}{x^2}}{\dfrac{2}{x} + \dfrac{1}{x^2}}$ $Step\ 1:\ \dfrac{\dfrac{4x^2}{x^2} - \dfrac{1}{x^2}}{\dfrac{2x}{x^2} + \dfrac{1}{x^2}} = \dfrac{\dfrac{4x^2 - 1}{x^2}}{\dfrac{2x + 1}{x^2}}$ $Step\ 2:\ \dfrac{4x^2 - 1}{x^2} \cdot \dfrac{x^2}{2x + 1}$ $Step\ 3:\ \dfrac{(2x + 1)(2x - 1)}{\cancel{x^2}} \cdot \dfrac{\cancel{x^2}}{2x + 1} = 2x - 1$
Simplifying a complex fraction by Method 2, p. 351.	1. Find the LCD of the rational expressions in the numerator and the denominator. 2. Multiply the numerator and the denominator of the complex fraction by the LCD. 3. Simplify the results.	Simplify by Method 2. $\dfrac{4 - \dfrac{1}{x^2}}{\dfrac{2}{x} - \dfrac{1}{x^2}}$ $Step\ 1:$ The LCD of the fractions is x^2. $Step\ 2:\ \dfrac{\left[4 - \dfrac{1}{x^2}\right]x^2}{\left[\dfrac{2}{x} - \dfrac{1}{x^2}\right]x^2} = \dfrac{4(x^2) - \left(\dfrac{1}{x^2}\right)(x^2)}{\left(\dfrac{2}{x}\right)(x^2) - \left(\dfrac{1}{x^2}\right)(x^2)}$ $= \dfrac{4x^2 - 1}{2x - 1}$ $Step\ 3:\ \dfrac{(2x + 1)(2x - 1)}{(2x - 1)} = 2x + 1$
Solving rational equations, p. 357.	1. Determine the LCD of all denominators in the equation. 2. Multiply each term in the equation by the LCD. 3. Simplify and remove parentheses. 4. Collect any like terms. 5. Solve for the variable. 6. Check your answer. Be sure that the value you obtained does not make any fraction in the original equation have a value of 0 in the denominator. If so, there is no solution.	Solve. $\dfrac{4}{y - 1} + \dfrac{-y + 5}{3y^2 - 4y + 1} = \dfrac{9}{3y - 1}$ The LCD = $(y - 1)(3y - 1)$. $(y - 1)(3y - 1)\left[\dfrac{4}{y - 1}\right]$ $+ (y - 1)(3y - 1)\left[\dfrac{-y + 5}{(y - 1)(3y - 1)}\right]$ $= (y - 1)(3y - 1)\left[\dfrac{9}{3y - 1}\right]$ $4(3y - 1) + (-y) + 5 = 9(y - 1)$ $12y - 4 - y + 5 = 9y - 9$ $11y + 1 = 9y - 9$ $11y - 9y = -9 - 1$ $2y = -10$ $y = -5$ *Check:* $\dfrac{4}{-5 - 1} + \dfrac{-(-5) + 5}{3(-5)^2 - 4(-5) + 1} \overset{?}{=} \dfrac{9}{3(-5) - 1}$ $\dfrac{4}{-6} + \dfrac{10}{96} \overset{?}{=} \dfrac{9}{-16}$ $-\dfrac{9}{16} = -\dfrac{9}{16}$ ✓

Topic	Procedure	Examples
Solving formulas containing fractions for a specified variable, p. 363.	1. Remove any parentheses. 2. Multiply each term of the equation by the LCD. 3. Add or subtract a quantity to each side of the equation so that only terms containing the desired variable are on one side of the equation while all other terms are on the other side. 4. If there are two or more unlike terms containing the desired variable, remove that variable as a common factor. 5. Divide each side of the equation by the coefficient of the desired variable. 6. Simplify, if possible.	Solve for n. $v = c\left(1 - \dfrac{t}{n}\right)$ $v = c - \dfrac{ct}{n}$ $n(v) = n(c) - n\left(\dfrac{ct}{n}\right)$ $nv = nc - ct$ $nv - nc = -ct$ $n(v - c) = -ct$ $n = \dfrac{-ct}{v - c}$ or $\dfrac{ct}{-v + c}$
Solving proportions in applied problems, p. 364.	1. Determine a given ratio in the problem for which both values are known. 2. Determine a similar ratio where only one value is known. Describe the other value by a variable. 3. Determine how the ratios may be made into one equation. 4. Solve the resulting equation.	The student to faculty ratio at Central University is 25:2. If there are 3700 students at the university, how many faculty members are there? The ratio of students to faculty: $\dfrac{25}{2}$ Let f = the number of faculty. students $\longrightarrow$ $\dfrac{25}{2} = \dfrac{3700}{f}$ $\longleftarrow$ students faculty $\longrightarrow$ $\quad\quad$ $\longleftarrow$ faculty $\quad\quad 25f = 7400$ $\quad\quad\quad f = 296$ There are 296 faculty members.

Chapter 6 Review Problems

Simplify.

1. $\dfrac{6x^3 - 9x^2}{12x^2 - 18x}$

2. $\dfrac{12x^4}{3x^5 - 15x^2}$

3. $\dfrac{28a^3b^3}{35a^6b^2}$

4. $\dfrac{a^2 - a - 20}{a^2 - 2a - 15}$

5. $\dfrac{14 - 19y - 3y^2}{3y^2 - 23y + 14}$

6. $\dfrac{ax + 2a - bx - 2b}{3x^2 - 12}$

7. $\dfrac{a^4 - 1}{a^4 + 3a^2 + 2}$

8. $\dfrac{6x^2y + 6xy - 36y}{3x^2y - 15xy + 18y}$

9. $\dfrac{4x^2 - 1}{x^2 - 4} \cdot \dfrac{2x^2 + 4x}{4x + 2}$

10. $\dfrac{3y}{4xy - 6y^2} \cdot \dfrac{2x - 3y}{12xy}$

11. $\dfrac{y^2 + 8y - 20}{y^2 + 6y - 16} \cdot \dfrac{y^2 + 3y - 40}{y^2 + 6y - 40}$

12. $\dfrac{3x^3y}{x^2 + 7x + 12} \cdot \dfrac{x^2 + 8x + 15}{6xy^2}$

13. $\dfrac{2x + 12}{3x - 15} \div \dfrac{2x^2 - 6x - 20}{x^2 - 10x + 25}$

14. $\dfrac{6x^2 - 6a^2}{3x^2 + 3} \div \dfrac{x^4 - a^4}{a^2x^2 + a^2}$

15. $\dfrac{y^4 - 1}{1 + y^2} \cdot \dfrac{y^2 + 8y + 15}{y^2 - 2y + 1} \cdot \dfrac{1 - y^2}{y^2 + 10y + 25}$

16. $\dfrac{y^2 + y - 20}{y^2 - 4y + 4} \cdot \dfrac{y^2 + y - 6}{12 + y - y^2} \cdot \dfrac{10 - 5y}{2y + 10}$

17. $\dfrac{9y^2 - 3y - 2}{6y^2 - 13y - 5} \div \dfrac{3y^2 + 10y - 8}{2y^2 + 13y + 20}$

18. $\dfrac{4a^2 + 12a + 5}{2a^2 - 7a - 13} \div (4a^2 + 2a)$

Add or subtract the fractions and simplify your answers.

19. $\dfrac{4}{xy^2} + \dfrac{3}{x^2y} - \dfrac{2}{x^2y^2}$

20. $\dfrac{5}{x - 3} + \dfrac{2}{3x + 1}$

21. $\dfrac{5}{4x} + \dfrac{-3}{x + 4}$

22. $\dfrac{x - 5}{2x + 1} - \dfrac{x + 1}{x - 2}$

23. $\dfrac{4}{y + 5} + \dfrac{3y + 2}{y^2 - 25}$

24. $\dfrac{2y - 1}{12y} - \dfrac{3y + 2}{9y}$

25. $\dfrac{y^2 - 4y - 19}{y^2 + 8y + 15} - \dfrac{2y - 3}{y + 5}$

26. $\dfrac{4y}{y^2 + 2y + 1} + \dfrac{3}{y^2 - 1}$

27. $\dfrac{a}{5 - a} - \dfrac{2}{a + 3} + \dfrac{2a^2 - 2a}{a^2 - 2a - 15}$

28. $\dfrac{5}{a^2 + 3a + 2} + \dfrac{6}{a^2 + 4a + 3} - \dfrac{7}{a^2 + 5a + 6}$

29. $4a + 3 - \dfrac{2a + 1}{a + 4}$

30. $\dfrac{1}{a} + \dfrac{1}{3a} + 3a + 2$

Simplify the complex fractions.

31. $\dfrac{\dfrac{2}{x} + \dfrac{3}{y}}{\dfrac{7}{xy}}$

32. $\dfrac{\dfrac{1}{x} + \dfrac{3}{2y}}{\dfrac{1}{4y} + \dfrac{7}{2y}}$

33. $\dfrac{\dfrac{5}{x} + 1}{1 - \dfrac{25}{x^2}}$

34. $\dfrac{\dfrac{4}{x + 3}}{\dfrac{2}{x - 2} - \dfrac{1}{x^2 + x - 6}}$

35. $\dfrac{\dfrac{y}{y+1}+\dfrac{1}{y}}{\dfrac{y}{y+1}-\dfrac{1}{y}}$

36. $\dfrac{\dfrac{10}{a+2}-5}{\dfrac{4}{a+2}-2}$

37. $\dfrac{\dfrac{2}{x+4}-\dfrac{1}{x^2+4x}}{\dfrac{3}{2x+8}}$

38. $\dfrac{\dfrac{y^2}{y^2-x^2}-1}{x+\dfrac{xy}{x-y}}$

39. $y-\dfrac{y}{1+\dfrac{1}{1-\dfrac{1}{y}}}$

40. $\dfrac{\dfrac{3}{x}-\dfrac{2}{x+1}}{\dfrac{5}{x^2+5x+4}-\dfrac{1}{x+4}}$

Solve for the variable and check your solution. If there is no solution, say so.

41. $\dfrac{3}{7}+\dfrac{4}{x+1}=1$

42. $\dfrac{3}{2}=1-\dfrac{1}{x-1}$

43. $\dfrac{1}{x+2}-\dfrac{1}{x}=\dfrac{-2}{x}$

44. $\dfrac{3}{x-2}+\dfrac{8}{x+3}=\dfrac{6}{x-2}$

45. $\dfrac{5}{2a}=\dfrac{2}{a}-\dfrac{1}{12}$

46. $\dfrac{1}{2a}=\dfrac{2}{a}-\dfrac{3}{8}$

47. $\dfrac{1}{y}+\dfrac{1}{2y}=2$

48. $\dfrac{5}{y^2}+\dfrac{7}{y}=\dfrac{6}{y^2}$

49. $\dfrac{a+2}{2a+6}=\dfrac{3}{2}-\dfrac{3}{a+3}$

50. $\dfrac{5}{a+5}+\dfrac{a+4}{2a+10}=\dfrac{3}{2}$

51. $\dfrac{1}{x+2}-\dfrac{5}{x-2}=\dfrac{-15}{x^2-4}$

52. $\dfrac{y+1}{y^2+2y-3}-\dfrac{1}{y+3}=\dfrac{1}{y-1}$

Solve for the variable indicated.

53. $\dfrac{N}{V}=\dfrac{m}{M+N}$; for M

54. $m=\dfrac{y-y_0}{x-x_0}$; for x

55. $\dfrac{P_1 V_1}{T_1}=\dfrac{P_2 V_2}{T_2}$; for T_1

56. $\dfrac{1}{f}=\dfrac{1}{a}+\dfrac{1}{b}$; for a

57. $S=\dfrac{V_1 t+V_2 t}{2}$; for t

58. $A=\dfrac{12I}{p+3pr}$; for p

59. $d=\dfrac{LR_2}{R_2+R_1}$; for R_2

60. $\dfrac{S-P}{Pr}=t$; for r

Solve the following problems. If necessary, round your answer to the nearest hundredth.

61. A company tested a random sample of 50 calculators and found 3 defective ones. In a batch of 950 calculators, how many were probably defective?

62. The ratio of kilograms to pounds is $5:11$. How much does a 143-pound man weigh in kilograms?

63. How long will it take a pump to empty a 4900-gallon pool if the same pump can empty a 3500-gallon pool in 4 hours?

64. In a sanctuary a sample of 100 wild rabbits is tagged and released by the wildlife management team. In a few weeks, after they have mixed with the general rabbit population, a sample of 40 rabbits is caught and 8 have a tag. Estimate the population of rabbits in the sanctuary.

65. The ratio of a picture's width to length is $5:7$. If the length of the picture is 21 centimeters, what is the width?

66. The ratio of officers to state troopers is $2:9$. If there are 154 men and women on the force, how many are officers?

67. The scale on a maritime sailing chart shows that 2 centimeters is equivalent to 7 nautical miles. A boat captain lays out a course on the chart for 3.5 cm. How many nautical miles will this be?

68. A 7-foot-tall tree casts a shadow that is 6 feet long. At the same time of day a building casts a shadow that is 156 feet long. How tall is the building?

69. Donna can trim all the hedges around her house with an electric hedge trimmer in 6 hours. Norma can do the same job in 9 hours using a manual trimmer. How long would it take them to do the hedge trimming around the house if they worked together?

70. Mr. Jensen has delivered mail for the university for the last 10 years. It takes him 3 hours to deliver all the mail to each campus office. When he was out sick, Mr. Sherf took over the task and found it took him 7 hours. How long would it take the two men working together?

71. If the hot water faucet at Mike's house is left on, it takes 15 minutes to fill the jacuzzi. If the cold water faucet is left on, it takes 10 minutes to fill the jacuzzi. How many minutes will it take if both faucets are left on?

72. In the summer it takes Dominic 12 hours to paint the barn. It takes his young son 18 hours to paint the barn. How many hours would it take if they both worked together?

Simplify.

1. $\dfrac{x^3 + 3x^2 + 2x}{x^3 - 2x^2 - 3x}$

2. $\dfrac{y^2 - 4}{y^3 + 8}$

3. $\dfrac{2y^2 + 7y - 4}{y^2 + 2y - 8} \cdot \dfrac{2y^2 - 8}{3y^2 + 11y + 10}$

4. $\dfrac{4 - 2x}{3x^2 - 2x - 8} \div \dfrac{2x^2 + x - 1}{9x + 12}$

5. $\dfrac{3x + 8}{x^2 - 25} - \dfrac{5}{x - 5}$

6. $\dfrac{2}{x^2 + 5x + 6} + \dfrac{3x}{x^2 + 6x + 9}$

7. $\dfrac{\dfrac{4}{y + 2} - 2}{5 - \dfrac{10}{y + 2}}$

8. $\dfrac{\dfrac{1}{x} - \dfrac{3}{x + 2}}{\dfrac{2}{x^2 + 2x}}$

1. _____

2. _____

3. _____

4. _____

5. _____

6. _____

7. _____

8. _____

Solve for the variable and check your answer. If no solution exists, so state.

9. $\dfrac{5}{3} = \dfrac{x+2}{x}$

10. $2 + \dfrac{x}{x+4} = \dfrac{3x}{x-4}$

11. $\dfrac{1}{2y+4} - \dfrac{1}{6} = \dfrac{-2}{3y+6}$

12. $\dfrac{3}{2x+3} - \dfrac{1}{2x-3} = \dfrac{2}{4x^2-9}$

13. Solve for W. $h = \dfrac{S - 2WL}{2W + 2L}$

14. Solve for b. $\dfrac{4}{a} = \dfrac{3}{b} + \dfrac{2}{c}$

15. A successful business has 1400 employees. A high-speed printer requires 30 minutes to print the payroll. If the company expands to 2450 employees, how long will it take to print the payroll?

16. A speed of 88 feet per second is equivalent to 60 miles per hour. The police investigated a recent accident in which the car was determined to be traveling at 110 feet per second. How many miles per hour was the car traveling?

Approximately one-half of this test covers the content of Chapters 1–5. The remainder covers the content of Chapter 6.

1. Simplify. $\left(\dfrac{3x^{-2}y^3}{z^4}\right)^{-2}$

2. Solve for x. $\dfrac{2}{3}(3x - 1) = \dfrac{2}{5}x + 3$

3. Graph the straight line.
$-6x + 2y = -12$.

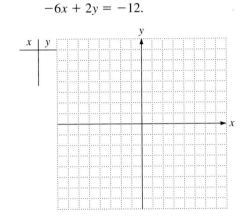

4. Find the equation in standard form of a line parallel to $5x - 6y = 8$ that passes through $(-1, -3)$.

5. Brenda invested $7000 in two accounts at the bank. One account earns 5% interest. The other earns 8% interest. She earned $539 interest in one year. How much was invested at each rate?

6. Solve for x and graph.
$3(2 - 6x) > 4(x + 1) + 24$

7. Evaluate. $2x^2 - 3x - 4y^2$ when $x = -2$ and $y = 3$.

8. Solve for x. $|3x - 4| \leq 10$

Factor.

9. $27x^3 + 125y^3$

10. $81x^3 - 90x^2y + 25xy^2$

Solve for x.

11. $x^2 + 12x + 20 = 0$

12. $3x^2 - 11x - 4 = 0$

1. _____

2. _____

3. _____

4. _____

5. _____

6. _____

7. _____

8. _____

9. _____

10. _____

11. _____

12. _____

Simplify.

13. $\dfrac{7x^2 - 28}{x^2 + 6x + 8}$

14. $\dfrac{2x^2 + x - 1}{2x^2 - 9x + 4} \cdot \dfrac{3x^2 - 12x}{6x + 15}$

15. $\dfrac{x^3 + 27}{x^2 + 7x + 12} \div \dfrac{x^2 - 6x + 9}{2x^2 + 13x + 20}$

16. $\dfrac{5}{2x - 8} - \dfrac{3x}{x^2 - 9x + 20}$

17. $\dfrac{\dfrac{1}{2x + 1} + 1}{4 - \dfrac{3}{4x^2 - 1}}$

18. $\dfrac{3}{x - 6} + \dfrac{4}{x + 4}$

Solve for the variable and check your answer.

19. $\dfrac{1}{2x + 3} - \dfrac{4}{4x^2 - 9} = \dfrac{3}{2x - 3}$

20. $\dfrac{1}{4x} - \dfrac{3}{2x} = \dfrac{5}{8}$

21. Solve for b. $H = \dfrac{3b + 2x}{5 - 4b}$

22. At a certain time of day the Bay Bridge handles 1650 cars in 12 minutes. How many cars will it handle in 1 hour?

13. _____

14. _____

15. _____

16. _____

17. _____

18. _____

19. _____

20. _____

21. _____

22. _____

Rational Exponents and Radicals

Many Americans buy a car and keep it for many years, but some are not aware of how the car depreciates in value each year. Suppose you knew how much you paid for a car and knew what it was worth after 6 years. Could you find the rate of depreciation? Can you tell which cars have a greater rate of depreciation? Please turn to the Putting Your Skills to Work on page 435.

Pretest Chapter 7

If you are familiar with the topics in this chapter, take this test now. Check your answers with those in the back of the book. If an answer was wrong or you couldn't do a problem, study the appropriate section of the chapter.

   If you are not familiar with the topics in this chapter, don't take this test now. Instead, study the examples, work the practice problems, and then take the test.

   This test will help you to identify those concepts that you have mastered and those that need more study.

Section 7.1

1. Multiply and simplify your answer. $(-3x^{1/4}y^{1/2})(-2x^{-1/2}y^{1/3})$

Simplify.

2. $\left(\dfrac{27x^2y^{-5}}{x^{-4}y^4}\right)^{2/3}$

3. $\dfrac{-18x^{-2}y^2}{-3x^{-5}y^{1/3}}$

4. $(-4x^{-1/4}y^{1/3})^3$

Section 7.2

Evaluate.

5. $81^{-3/4}$

6. $\sqrt{169} + \sqrt[3]{-64}$

7. $\sqrt[3]{27a^{12}b^6c^{15}}$

Section 7.3

8. Simplify. $\sqrt[4]{32x^8y^{15}}$

9. Combine like terms where possible. $3\sqrt{48y^3} - 2\sqrt[3]{16} + 3\sqrt[3]{54} - 5y\sqrt{12y}$

Section 7.4

10. Multiply and simplify. $(3\sqrt{3} - 5\sqrt{6})(\sqrt{12} - 3\sqrt{6})$

11. Rationalize the denominator; simplify your answer. $\dfrac{6}{\sqrt[3]{9x}}$

12. Rationalize the denominator; simplify your answer. $\dfrac{\sqrt{2} + \sqrt{3}}{\sqrt{2} - \sqrt{3}}$

Section 7.5

Solve and check your solution(s).

13. $\sqrt{3x - 5} + 1 = x$

14. $\sqrt{2x + 3} - \sqrt{x - 2} = 2$

Section 7.6

Perform the operations indicated.

15. $(3 - 2i) - (-1 + 3i)$

16. $i^{15} + \sqrt{-25}$

17. $(3 + 5i)^2$

18. $\dfrac{3 + 2i}{2 + 3i}$

Section 7.7

19. If y varies directly with x^2, and $y = 18$ when $x = 3$, find the value of y when $x = 5$.

20. If y varies inversely with x, and $y = 12$ when $x = 6$, find the value of y when $x = 10$.

1. _____

2. _____

3. _____

4. _____

5. _____

6. _____

7. _____

8. _____

9. _____

10. _____

11. _____

12. _____

13. _____

14. _____

15. _____

16. _____

17. _____

18. _____

19. _____

20. _____

7.1 Rational Exponents

After studying this section, you will be able to apply the laws of exponents to:

1 *Simplify expressions with rational exponents.*
2 *Adding rational expressions.*
3 *Factor an expression with rational exponents.*

MathPro   Video 16   SSM

1 *Simplifying Expressions with Rational Exponents*

Before studying this section, you may need to review Section 1.4. For convenience, we list the rules of exponents that we learned there.

$$x^m x^n = x^{m+n} \qquad\qquad x^0 = 1$$

$$\frac{x^m}{x^n} = x^{m-n} \qquad\qquad (x^m)^n = x^{mn}$$

$$x^{-n} = \frac{1}{x^n} \qquad\qquad (xy)^n = x^n y^n$$

$$\frac{x^{-n}}{y^{-m}} = \frac{y^m}{x^n} \qquad\qquad \left(\frac{x}{y}\right)^n = \frac{x^n}{y^n}$$

To ensure that you understand these rules, study carefully Example 1 and work Practice Problem 1.

EXAMPLE 1 Simplify. $\left(\dfrac{5xy^{-3}}{2x^{-4}y}\right)^{-2}$

$$\left(\frac{5xy^{-3}}{2x^{-4}y}\right)^{-2} = \frac{(5xy^{-3})^{-2}}{(2x^{-4}y)^{-2}} \qquad \left(\frac{x}{y}\right)^n = \frac{x^n}{y^n}.$$

$$= \frac{5^{-2}x^{-2}(y^{-3})^{-2}}{2^{-2}(x^{-4})^{-2}y^{-2}} \qquad (xy)^n = x^n y^n.$$

$$= \frac{5^{-2}x^{-2}y^6}{2^{-2}x^8 y^{-2}} \qquad (x^m)^n = x^{mn}.$$

$$= \frac{5^{-2}}{2^{-2}} \cdot \frac{x^{-2}}{x^8} \cdot \frac{y^6}{y^{-2}}$$

$$= \frac{2^2}{5^2} \cdot x^{-2-8} \cdot y^{6+2} \qquad \frac{x^{-n}}{y^{-m}} = \frac{y^m}{x^n}; \ \frac{x^m}{x^n} = x^{m-n}.$$

$$= \frac{4}{25}x^{-10}y^8$$

The answer can also be written as $\frac{4y^8}{25x^{10}}$. Explain why.

Practice Problem 1 Simplify. $\left(\dfrac{3x^{-2}y^4}{2x^{-5}y^2}\right)^{-3}$ ∎

SIDELIGHT When to use the rule $\frac{x^{-n}}{y^{-m}} = \frac{y^m}{x^n}$ is entirely up to you. In Example 1, we could have begun by writing

$$\left(\frac{5xy^{-3}}{2x^{-4}y}\right)^{-2} = \left(\frac{5x\,x^4}{2y\,y^3}\right)^{-2} = \left(\frac{5x^5}{2y^4}\right)^{-2}$$

Complete the steps to simplify this expression.

Likewise, in the fourth step in Example 1, we could have written

$$\frac{5^{-2}x^{-2}y^6}{2^{-2}x^8y^{-2}} = \frac{2^2y^6y^2}{5^2x^8x^2}$$

Complete the steps to simplify this expression. Are the two answers above the same as the answer in Example 1? Why or why not?

We generally begin to simplify a rational expression with exponents by raising a power to a power because sometimes negative powers become positive. The order in which you use the rules of exponents is up to you. Work carefully. Keep track of your exponents and where you are as you simplify the rational expression.

These rules for exponents can also be extended to include rational exponents—that is, exponents that are fractions. As you recall, rational numbers are of the form $\frac{a}{b}$, where a and b are integers and b does not equal zero. We will write fractional exponents using diagonal lines. Thus, we will write $\frac{5}{6}$ as 5/6 and $\frac{a}{b}$ as a/b throughout this chapter when writing fractional exponents. For now we restrict the base to *positive* real numbers. Later we will talk about negative bases.

EXAMPLE 2   Simplify.

(a) $(x^{2/3})^4$ **(b)** $\dfrac{x^{5/6}}{x^{1/6}}$ **(c)** $x^{2/3} \cdot x^{-1/3}$ **(d)** $5^{3/7} \cdot 5^{2/7}$

We will not write out every step or every rule of exponents that we use. You should be able to follow the solutions.

(a) $(x^{2/3})^4 = x^{(2/3)(4/1)} = x^{8/3}$ **(b)** $\dfrac{x^{5/6}}{x^{1/6}} = x^{5/6-1/6} = x^{4/6} = x^{2/3}$

(c) $x^{2/3} \cdot x^{-1/3} = x^{2/3-1/3} = x^{1/3}$ **(d)** $5^{3/7} \cdot 5^{2/7} = 5^{3/7+2/7} = 5^{5/7}$

Practice Problem 2   Simplify.

(a) $(x^4)^{3/8}$ **(b)** $\dfrac{x^{3/7}}{x^{2/7}}$ **(c)** $x^{-7/5} \cdot x^{4/5}$ ∎

Sometimes the fractional exponents will not have the same denominator. Remember you need to change the fractions to equivalent fractions with the same denominator when the rule of exponents requires you to add or to subtract the fractions.

EXAMPLE 3   Simplify. Express your answer with positive exponents only.

(a) $\dfrac{3x^{1/4}}{x^{1/3}}$ **(b)** $(2x^{1/2})(3x^{1/3})$ **(c)** $\dfrac{18x^{1/4}y^{-1/3}}{-6x^{-1/2}y^{1/6}}$

(a) $\dfrac{3x^{1/4}}{x^{1/3}} = 3x^{1/4-1/3} = 3x^{3/12-4/12} = 3x^{-1/12} = \dfrac{3}{x^{1/12}}$

(b) $(2x^{1/2})(3x^{1/3}) = 6x^{1/2+1/3} = 6x^{3/6+2/6} = 6x^{5/6}$

(c) $-3x^{1/4-(-1/2)}y^{-1/3-1/6} = -3x^{1/4+2/4}y^{-2/6-1/6}$

$$= -3x^{3/4}y^{-3/6}$$

$$= -3x^{3/4}y^{-1/2} \quad \textit{Notice that we simplified the exponent}$$

$$= \frac{-3x^{3/4}}{y^{1/2}} \quad -\frac{3}{6}. \textit{ The factor } y^{-3/6} \textit{ became } y^{-1/2}.$$

Practice Problem 3   Simplify. Express your answer with positive exponents only.

(a) $\dfrac{x^4}{4x^{1/2}}$ **(b)** $(-3x^{1/4})(2x^{1/2})$ **(c)** $\dfrac{13x^{1/12}y^{-1/4}}{26x^{-1/3}y^{1/2}}$ ∎

EXAMPLE 4   Multiply and simplify. $-2x^{5/6}(3x^{1/2} - 4x^{-1/3})$

$$-2x^{5/6}(3x^{1/2} - 4x^{-1/3}) = -6x^{5/6+1/2} + 8x^{5/6-1/3}$$

$$= -6x^{5/6+3/6} + 8x^{5/6-2/6}$$

$$= -6x^{8/6} + 8x^{3/6}$$

$$= -6x^{4/3} + 8x^{1/2}$$

Practice Problem 4   Multiply and simplify. $-3x^{1/2}(2x^{1/4} + 3x^{-1/2})$ ∎

Numerical values raised to a rational power can sometimes be simplified if they are placed in exponent form.

EXAMPLE 5   Evaluate.   **(a)** $(25)^{3/2}$ **(b)** $(27)^{2/3}$

(a) $(25)^{3/2} = (5^2)^{3/2} = 5^{2/1 \cdot 3/2} = 5^3 = 125$

(b) $(27)^{2/3} = (3^3)^{2/3} = 3^{3/1 \cdot 2/3} = 3^2 = 9$

Practice Problem 5   Evaluate. **(a)** $(4)^{5/2}$ **(b)** $(27)^{4/3}$ ∎

2 Adding Rational Expressions

Adding rational expressions sometimes requires several steps. Sometimes this involves removing negative exponents. Suppose you write $2x^{-1/2}$ as $\frac{2}{x^{1/2}}$. This is a rational expression. Recall that to add rational expressions you need to have a common denominator. Take time to look at the steps needed to write $2x^{-1/2} + x^{1/2}$ as one term.

EXAMPLE 6   Write as one fraction with positive exponents. $2x^{-1/2} + x^{1/2}$

$$2x^{-1/2} + x^{1/2} = \frac{2}{x^{1/2}} + \frac{x^{1/2} \cdot x^{1/2}}{x^{1/2}} = \frac{2}{x^{1/2}} + \frac{x^1}{x^{1/2}} = \frac{2+x}{x^{1/2}}$$

Practice Problem 6   Write as one fraction with only positive exponents.
$3x^{1/3} + x^{-1/3}$ ∎

3 Factoring Expressions with Rational Exponents

To factor expressions, we need to be able to recognize common factors. If the terms of the expression contain exponents, we look for the same exponential factor in each term. For example, in the expression $6x^5 + 4x^3 - 8x^2$, the common factor of each term is $2x^2$. We can factor $6x^5 + 4x^3 - 8x^2$ by removing the common factor $2x^2$ from each term. This then becomes $2x^2(3x^3 + 2x - 4)$.

We do exactly the same thing when we factor expressions with rational exponents. The key is to identify the exponent of the common factor. In the expression $6x^{3/4} + 4x^{1/2} - 8x^{1/4}$, the exponent of the common factor is $\frac{1}{4}$ because $\frac{1}{4}$ can be subtracted from the exponent of each term. The common factor is $2x^{1/4}$. Thus we factor the expression $6x^{3/4} + 4x^{1/2} - 8x^{1/4}$ as $2x^{1/4}(3x^{1/2} + 2x^{1/4} - 4)$. We do not always need to factor out the greatest common factor. In the following examples we simply factor out a common factor.

EXAMPLE 7 Remove a common factor of $2x$ from $2x^{3/2} + 4x^{5/2}$.

We rewrite the exponents of each term so that we can see that each term contains the factor $2x$ or $2x^{2/2}$.

$$2x^{3/2} + 4x^{5/2} = 2x^{2/2+1/2} + 4x^{2/2+3/2}$$

$$= 2(x^{2/2})(x^{1/2}) + 4(x^{2/2})(x^{3/2})$$

$$= 2x(x^{1/2} + 2x^{3/2})$$

Practice Problem 7 Remove a common factor of $4y$ from $4y^{3/2} - 8y^{5/2}$. ∎

EXAMPLE 8 Remove a common factor of $x^{-1/2}$ from $x^{3/2} - 5x^{-1/2}$.

We rewrite the exponent of each term as a sum that includes $-\frac{1}{2}$.

$$x^{3/2} - 5x^{-1/2} = x^{-1/2+4/2} - 5x^{-1/2} \qquad \textit{Think } -\tfrac{1}{2} + \tfrac{4}{2} = \tfrac{3}{2}.$$

$$= (x^{-1/2})(x^{4/2}) - 5(x^{-1/2})$$

$$= x^{-1/2}(x^{4/2} - 5)$$

$$= x^{-1/2}(x^2 - 5)$$

Practice Problem 8 Remove a common factor of $x^{-3/2}$ from $x^{5/2} - 2x^{-3/2}$. ∎

For convenience we list here the properties of exponents we have discussed in this section.

When x, y are **positive real numbers** and a, b are **rational numbers**

$$x^a x^b = x^{a+b} \qquad \frac{x^a}{x^b} = x^{a-b} \qquad x^0 = 1$$

$$x^{-a} = \frac{1}{x^a} \qquad \frac{x^{-a}}{y^{-b}} = \frac{y^b}{x^a}$$

$$(x^a)^b = x^{ab} \qquad (xy)^a = x^a y^a \qquad \left(\frac{x}{y}\right)^a = \frac{x^a}{y^a}$$

7.1 Exercises

Simplify.

1. $\left(\dfrac{4x^2y^{-3}}{x}\right)^2$

2. $\left(\dfrac{3xy^{-2}}{x^3}\right)^2$

3. $\left(\dfrac{a^{-1}b}{-2b^2}\right)^3$

4. $\left(\dfrac{a^{-3}b^2}{3b}\right)^3$

5. $(x^{3/4})^2$

6. $(x^{4/3})^6$

7. $(y^6)^{1/3}$

8. $(y^8)^{1/2}$

9. $\dfrac{x^{7/12}}{x^{1/12}}$

10. $\dfrac{x^{7/8}}{x^{3/8}}$

11. $\dfrac{x^3}{x^{1/2}}$

12. $\dfrac{x^2}{x^{1/3}}$

13. $x^{1/7} \cdot x^{3/7}$

14. $x^{3/5} \cdot x^{1/5}$

15. $y^{3/5} \cdot y^{-1/10}$

16. $y^{7/10} \cdot y^{-1/5}$

17. $3^{1/6} \cdot 3^{5/6}$

18. $4^{3/4} \cdot 4^{1/4}$

19. $5^{-3/10} \cdot 5$

20. $6^{-3/8} \cdot 6^{1/2}$

Write each expression with positive exponents.

21. $x^{-3/4}$

22. $x^{-5/6}$

23. $a^{-5/6}b^{1/3}$

24. $a^{-5/8}b^{1/4}$

25. $6^{-1/2}$

26. $4^{-1/3}$

27. $2a^{-1/4}$

28. $3^{-2/5} \cdot 2^{1/3}$

Mixed Practice

Simplify and express your answer with positive exponents. Evaluate the numerical expressions.

29. $(x^{1/2}y^{1/3})(x^{1/3}y^{2/3})$

30. $(x^{-1/3}y^{2/3})(x^{1/3}y^{1/4})$

31. $(7x^{1/3}y^{1/4})(-2x^{1/4}y^{-1/6})$

32. $(8x^{-1/5}y^{1/3})(-3x^{-1/4}y^{1/6})$

33. $5^{-2/3} \cdot 5^{1/6}$

34. $7^{3/4} \cdot 7^{-1/4}$

35. $\dfrac{2x^{1/5}}{x^{-1/2}}$

36. $\dfrac{3y^{2/3}}{y^{-1/4}}$

37. $\dfrac{-20x^2y^{-1/5}}{5x^{-1/2}y}$

38. $\dfrac{12x^{-2/3}y}{-6xy^{-3/4}}$

39. $\left(\dfrac{8a^2b^6}{a^{-1}b^3}\right)^{1/3}$

40. $\left(\dfrac{16a^5b^{-2}}{a^{-1}b^{-6}}\right)^{1/2}$

41. $\left(\dfrac{a^5b^{-5}}{81ab^3}\right)^{3/4}$

42. $\left(\dfrac{a^7b^3}{32a^2b^{-7}}\right)^{2/5}$

43. $(7x^{2/3}y^{1/4}z^{3/2})^2$

44. $(5x^{-1/2}y^{1/3}z^{4/5})^3$

45. $(5^4x^4y^2z^6)^{1/2}$

46. $(3^3x^6y^3z^{12})^{1/3}$

47. $x^{2/3}(x^{4/3} - x^{1/5})$

48. $x^{-1/4}(x^{2/3} + x^{3/4})$

49. $a^{5/6}(a^{-1/2} + 3a^{1/5})$

50. $a^{2/3}(a^{1/2} + 4a^{-1/4})$

51. $(2a^{1/5}b^{2/3})^{1/4}$

52. $(5a^{2/5}b^{1/8})^{4/5}$

53. $\dfrac{(x^{-5/6}x^4)^{3/4}}{x^{2/3}}$

54. $\dfrac{(x^{1/3}x^3)^{3/2}}{x^{-1/4}}$

55. $(25)^{1/2}$

56. $(27)^{2/3}$ **57.** $(16)^{3/4}$ **58.** $(4)^{3/2}$

59. $(27)^{2/3} + (16)^{3/2}$ **60.** $(81)^{3/4} + (25)^{1/2}$

61. $(5.8276x^{1/4}y^{-2/3})(-3.0761x^{5/4}y^{-4/3})$ **62.** $\dfrac{-82.32206x^8y^{1/2}}{-17.89610x^{-6}y^{1/4}}$

Write each expression as one fraction with positive exponents.

63. $3y^{1/2} + y^{-1/2}$ **64.** $2y^{1/3} + y^{-2/3}$ **65.** $y^{-2/3} + 4^{1/3}$ **66.** $3^{-1/2} + y^{-1/2}$

Remove a common factor of $3x^{1/2}$.

67. $-12x^{3/2} + 6x^{5/2}$ **68.** $27x^{5/2} - 3x^{3/2}$

Remove a common factor of $2a$.

69. $10a^{5/4} - 4a^{8/5}$ **70.** $6a^{4/3} - 8a^{3/2}$

To Think About

71. What is the value of a if $x^a \cdot x^{1/4} = x^{-1/8}$?   **72.** What is the value of b if $x^b \div x^{1/3} = x^{-1/12}$?

Applications

The radius needed to create a sphere with a given volume V can be approximated by using the equation $r = 0.62(V)^{1/3}$. Find the radius of the spheres with the following volume.

73. 27 cubic meters   **74.** 64 cubic meters

Cumulative Review Problems

Solve for x.   *Solve for b.*

75. $-4(x + 1) = \dfrac{1}{3}(3 - 2x)$ **76.** $A = \dfrac{h}{2}(a + b)$

Giving a young patient the wrong amount of medication can have serious and even fatal consequences. One method used by doctors, nurses, and pharmacists to verify the correct dosage of a prescription drug for a child is the use of the formula

$$y = \frac{ax}{a + 12}$$

where y = the child's dosage, x = adult dosage, and a = the age of the child in years.

77. If the adult dosage is 400 milligrams, how much should a 7-year-old child receive? Round your answer to the nearest milligram.

78. The adult dosage is 250 milligrams, and a certain child was assigned a dosage level of 75 milligrams, how old should the child be for the dosage be correct? Round your answer to the nearest year.

7.2 Radical Expressions

After studying this section, you will be able to:

1. Evaluate radical expressions.
2. Change radical expressions to expressions with rational exponents.
3. Change expressions with rational exponents to radical expressions.
4. Evaluate higher-order radicals containing a variable radicand that represents a negative number.

MathPro   Video 17   SSM

1 Evaluating Radical Expressions

In Section 1.3 we studied simple radical expressions called square roots. The **square root** of a number is *one* of that number's two equal factors. That is, since $3 \cdot 3 = 9$, 3 is a square root of 9. But $(-3) \cdot (-3) = 9$, so -3 is also a square root. We call the positive square root the **principal** square root.

The symbol $\sqrt{}$ is called a **radical sign.** We use it to denote positive square roots (and higher-order roots also). A negative square root is written $-\sqrt{}$. Thus we write

$$\sqrt{9} = 3 \qquad -\sqrt{9} = -3$$
$$\sqrt{64} = 8 \qquad \text{(because } 8 \cdot 8 = 64\text{)}$$
$$\sqrt{121} = 11 \qquad \text{(because } 11 \cdot 11 = 121\text{)}$$

Now because $\sqrt{9} = \sqrt{3 \cdot 3} = \sqrt{3^2} = 3$, we can say that

---

Definition of Square Root

If x is a positive real number, then $\sqrt{x}$ is the *positive* (or principal) *square root* of x; in other words, $(\sqrt{x})^2 = x$.

Note that x must be *positive*. Why? Suppose we want to find $\sqrt{-36}$. We must find two equal factors that when multiplied together give -36. Are there any? No, because

$$6 \cdot 6 = 36$$
$$(-6)(-6) = 36$$

So there is no real number that we can square to get -36.

We call $\sqrt[n]{x}$ a **radical expression.** The $\sqrt{}$ symbol is the **radical sign,** the x is the **radicand,** and the n is the **index** of the radical. When no number for n appears in the radical expression, it is understood that 2 is the index, which means we are looking for the square root. For example, in the radical expression $\sqrt{25}$, with no number given for the index, n, we take the index to be 2. Thus $\sqrt{25}$ is the principal square root of 25.

We can extend the notion of square root to higher-order roots, such as cube roots, fourth roots, and so on. A cube root of a number is one of that number's three equal factors. The index of the radical, n, is 3, and the radical sign is $\sqrt[3]{}$. Similarly, a fourth root of a number is one of that number's four equal factors. The index of the radical, n, is 4, and the radical sign is $\sqrt[4]{}$. For example,

$$\sqrt[3]{27} = 3 \qquad \text{because } 3 \cdot 3 \cdot 3 = 3^3 = 27$$
$$\sqrt[3]{8} = 2 \qquad \text{because } 2 \cdot 2 \cdot 2 = 2^3 = 8$$
$$\sqrt[4]{81} = 3 \qquad \text{because } 3 \cdot 3 \cdot 3 \cdot 3 = 3^4 = 81$$
$$\sqrt[5]{32} = 2 \qquad \text{because } 2 \cdot 2 \cdot 2 \cdot 2 \cdot 2 = 2^5 = 32$$
$$\sqrt[3]{-64} = -4 \qquad \text{because } (-4)(-4)(-4) = (-4)^3 = -64$$

You should be able to see the pattern here.

$$\sqrt[3]{27} = \sqrt[3]{3^3} = 3$$
$$\sqrt[4]{81} = \sqrt[4]{3^4} = 3$$
$$\sqrt[5]{32} = \sqrt[5]{2^5} = 2$$
$$\sqrt[6]{729} = \sqrt[6]{3^6} = 3$$
$$\sqrt[3]{-64} = \sqrt[3]{(-4)^3} = -4$$

In these cases, we see that $\sqrt[n]{x^n} = x$. We now give the following definition.

Definition of Higher-order Roots

1. If x is a *nonnegative* real number, then $\sqrt[n]{x}$ is a positive nth root and has the property that

$$(\sqrt[n]{x})^n = x$$

2. If x is a *negative* real number, then
   (a) $(\sqrt[n]{x})^n = x$ when n is an *odd integer*,
   (b) $(\sqrt[n]{x})^n$ is *not* a real number   when n is an *even integer*.

EXAMPLE 1   If possible, find the root of each negative number. If there is no real number answer, say so.

(a) $\sqrt[3]{-216}$ (b) $\sqrt[5]{-32}$ (c) $\sqrt[4]{-16}$ (d) $\sqrt[6]{-64}$

(a) $\sqrt[3]{-216} = \sqrt[3]{(-6)^3} = -6$ (b) $\sqrt[5]{-32} = \sqrt[5]{(-2)^5} = -2$

(c) $\sqrt[4]{-16}$ is not a real number because n is even and x is negative.

(d) $\sqrt[6]{-64}$ is not a real number because n is even and x is negative.

Practice Problem 1   If possible, find the roots. If there is no real number answer, say so.

(a) $\sqrt[3]{216}$ (b) $\sqrt[5]{32}$ (c) $\sqrt[3]{-8}$ (d) $\sqrt[4]{-81}$ ∎

2 Changing Radical Expressions to Expressions with Rational Exponents

Now we want to extend our definition of roots to rational exponents. By the laws of exponents we know that

$$x^{1/2} \cdot x^{1/2} = x^{1/2 + 1/2} = x^1 = x$$

Since $x^{1/2}x^{1/2} = x$, $x^{1/2}$ is one of x's two equal factors. This tells us that $x^{1/2}$ must be a square root of x. That is, $x^{1/2} = \sqrt{x}$. Is this true? By the definition of square root, $(\sqrt{x})^2 = x$. Does $(x^{1/2})^2 = x$? Using the law of exponents

$$(x^{1/2})^2 = x^{(1/2)(2)} = x^1 = x$$

We conclude that indeed

$$x^{1/2} = \sqrt{x}$$

In the same way we can write

$$x^{1/3} \cdot x^{1/3} \cdot x^{1/3} = x \qquad x^{1/3} = \sqrt[3]{x}$$
$$x^{1/4} \cdot x^{1/4} \cdot x^{1/4} \cdot x^{1/4} = x \qquad x^{1/4} = \sqrt[4]{x}$$
$$\vdots \qquad\qquad\qquad \vdots$$
$$\underbrace{x^{1/n} \cdot x^{1/n} \cdots x^{1/n}}_{n \text{ factors}} = x \qquad x^{1/n} = \sqrt[n]{x}$$

Therefore, we are ready to define fractional exponents in general.

Definition

If n is a positive integer and x is a positive real number, then

$$x^{1/n} = \sqrt[n]{x}$$

EXAMPLE 2   Change to rational exponents and simplify.

(a) $\sqrt[4]{x^4}$ **(b)** $\sqrt[5]{(32)^5}$ **(c)** $\sqrt[6]{2^6}$ **(d)** $\sqrt[7]{x^{14}}$

(a) $\sqrt[4]{x^4} = (x^4)^{1/4} = x^{4/4} = x^1 = x$

(b) $\sqrt[5]{(32)^5} = (32^5)^{1/5} = 32^{5/5} = 32^1 = 32$

(c) $\sqrt[6]{2^6} = (2^6)^{1/6} = 2^{6/6} = 2^1 = 2$

(d) $\sqrt[7]{x^{14}} = (x^{14})^{1/7} = x^{14/7} = x^2$

Practice Problem 2   Change to rational exponents and simplify.

(a) $\sqrt[3]{x^3}$ **(b)** $\sqrt[4]{y^4}$ **(c)** $\sqrt[5]{(8)^5}$ **(d)** $\sqrt[6]{x^{12}}$ ∎

EXAMPLE 3   Replace all radicals with rational exponents.

(a) $\sqrt[3]{x^2}$ **(b)** $(\sqrt[5]{w})^7$ **(c)** $\sqrt[4]{(ab)^3}$ **(d)** $\sqrt[4]{\sqrt[3]{x}}$

(a) $\sqrt[3]{x^2} = x^{2/3}$ **(b)** $(\sqrt[5]{w})^7 = w^{7/5}$

(c) $\sqrt[4]{(ab)^3} = (ab)^{3/4}$ or $a^{3/4}b^{3/4}$ **(d)** $\sqrt[4]{\sqrt[3]{x}} = \sqrt[4]{x^{1/3}} = (x^{1/3})^{1/4} = x^{1/12}$

Practice Problem 3   Replace all radicals with rational exponents.

(a) $\sqrt[4]{x^3}$ **(b)** $\sqrt[5]{(xy)^7}$ **(c)** $(\sqrt[5]{z})^4$ ∎

EXAMPLE 4   Evaluate or simplify. Assume all variables are positive.

(a) $\sqrt[5]{32x^{10}}$ **(b)** $\sqrt[3]{-125x^9}$ **(c)** $(16x^4)^{3/4}$ **(d)** $25(x^8y^4)^{5/2}$

(a) $\sqrt[5]{32x^{10}} = (2^5x^{10})^{1/5} = 2x^2$ **(b)** $\sqrt[3]{-125x^9} = [(-5)^3x^9]^{1/3} = -5x^3$

(c) $(16x^4)^{3/4} = (2^4x^4)^{3/4} = 2^3x^3 = 8x^3$

(d) $25(x^8y^4)^{5/2} = 25x^{8/1 \cdot 5/2}y^{4/1 \cdot 5/2} = 25x^{20}y^{10}$

Practice Problem 4   Evaluate or simplify. Assume all variables are positive.

(a) $\sqrt[4]{81x^{12}}$ **(b)** $\sqrt[3]{-27x^6}$ **(c)** $(32x^5)^{3/5}$ **(d)** $16(x^6y^8)^{7/2}$ ∎

---

3 *Changing Expressions with Rational Exponents to Radical Expressions*

Sometimes we need to change an expression with rational exponents to radical expressions. This is especially helpful when we need to evaluate an expression with rational exponents, because the value of the radical form of the expression is more recognizable. For example, because of our experience with radicals, we know that $\sqrt{25} = 5$. It is not as easy to see that $25^{1/2} = 5$. We begin, therefore, by changing expressions with rational exponents to radical expressions. Recall that

$$x^{1/n} = \sqrt[n]{x}$$

Again, using the laws of exponents, we know that

$$x^{m/n} = (x^m)^{1/n} = (x^{1/n})^m$$

So we can make the following general definition.

EXAMPLE 5   Change to radical form.

(a) $(xy)^{5/7}$ **(b)** $w^{-2/3}$ **(c)** $3x^{3/4}$ **(d)** $(3x)^{3/4}$

(a) $(xy)^{5/7} = \sqrt[7]{(xy)^5} = \sqrt[7]{x^5 y^5}$ **(b)** $w^{-2/3} = \dfrac{1}{w^{2/3}} = \dfrac{1}{\sqrt[3]{w^2}}$

or $(xy)^{5/7} = (\sqrt[7]{xy})^5$

(c) $3x^{3/4} = 3\sqrt[4]{x^3}$ **(d)** $(3x)^{3/4} = \sqrt[4]{(3x)^3} = \sqrt[4]{27x^3}$

or $(3x)^{3/4} = (\sqrt[4]{3x})^3$

Practice Problem 5   Change to radical form.

(a) $x^{3/4}$ **(b)** $y^{-1/3}$ **(c)** $(2x)^{4/5}$ **(d)** $2x^{4/5}$ ■

EXAMPLE 6   Change to radical form and evaluate.

(a) $16^{3/2}$ **(b)** $125^{2/3}$ **(c)** $(-27)^{2/3}$

(d) $(-16)^{5/2}$ **(e)** $144^{-1/2}$ **(f)** $32^{-4/5}$

(a) $16^{3/2} = (\sqrt{16})^3 = (4)^3 = 64$

(b) $125^{2/3} = (\sqrt[3]{125})^2 = (5)^2 = 25$

(c) $(-27)^{2/3} = (\sqrt[3]{-27})^2 = (-3)^2 = 9$

(d) $(-16)^{5/2} = (\sqrt{-16})^5$; however, $\sqrt{-16}$ is not a real number.

(e) $144^{-1/2} = \dfrac{1}{144^{1/2}} = \dfrac{1}{\sqrt{144}} = \dfrac{1}{12}$

(f) $32^{-4/5} = \dfrac{1}{32^{4/5}} = \dfrac{1}{(\sqrt[5]{32})^4} = \dfrac{1}{(2)^4} = \dfrac{1}{16}$

Practice Problem 6   Change to radical form and evaluate.

(a) $8^{2/3}$ **(b)** $4^{3/2}$ **(c)** $(-8)^{4/3}$

(d) $(-25)^{3/2}$ **(e)** $100^{-3/2}$ **(f)** $32^{-3/5}$ ■

4 *Evaluating Higher-order Radicals Containing a Variable Radicand That Represents a Negative Number*

We now consider the case where the radicand is negative. Refer to the definition for higher-order roots on page 392.

Definition

When x is a **negative real number,** or if we cannot determine if x is positive or negative, then

$$\sqrt[n]{x^n} = |x|, \qquad \text{when } n \text{ is an } even \text{ positive integer}$$

$$\sqrt[n]{x^n} = x \qquad \text{when } n \text{ is an } odd \text{ positive integer}$$

EXAMPLE 7   Evaluate; x may be a positive or negative real number.

(a) $\sqrt[3]{(-2)^3}$ **(b)** $\sqrt[4]{(-2)^4}$ **(c)** $\sqrt[5]{x^5}$ **(d)** $\sqrt[6]{x^6}$

(a) $\sqrt[3]{(-2)^3} = -2$ because the index is odd .
(b) $\sqrt[4]{(-2)^4} = |-2| = 2$ because the index is even .
(c) $\sqrt[5]{x^5} = x$ because the index is odd .
(d) $\sqrt[6]{x^6} = |x|$ because the index is even .

Practice Problem 7   Evaluate; y and w may be positive or negative real numbers.

(a) $\sqrt[5]{(-3)^5}$ **(b)** $\sqrt[4]{(-5)^4}$ **(c)** $\sqrt[4]{w^4}$ **(d)** $\sqrt[7]{y^7}$ ∎

EXAMPLE 8   Simplify. Assume that x and y may be positive or negative real numbers.

(a) $\sqrt{49x^2}$ **(b)** $\sqrt[4]{81y^{16}}$ **(c)** $\sqrt[3]{27x^6y^9}$

In both (a) and (b) we observe that the index is an even positive number. We will need the absolute value in these types of situations.

(a) $\sqrt{49x^2} = 7|x|$
(b) $\sqrt[4]{81y^{16}} = 3|y^4|$

Note that since we know that this is positive (anything to the fourth power will be positive), we can write this as $3y^4$ without the absolute value symbol. Thus, $\sqrt[4]{81y^{16}} = 3y^4$.

In (c) the index is an odd integer. The absolute value is never needed in such a case.

(c) $\sqrt[3]{27x^6y^9} = \sqrt[3]{(3)^3(x^2)^3(y^3)^3} = 3x^2y^3$

Practice Problem 8   Simplify. Assume that x and y may be positive or negative real numbers.

(a) $\sqrt{36x^2}$ **(b)** $\sqrt[3]{125x^3y^6}$ **(c)** $\sqrt[4]{16y^8}$ ∎

7.2 Exercises

Verbal and Writing Skills

1. In a simple sentence, explain what a square root is.

2. In a simple sentence, explain what a cube root is.

3. Give an example to show why the cube root of a negative number is a negative number.

4. Give an example to show why it is not possible to find a real number that is the square root of a negative number.

Evaluate if possible.

5. $\sqrt{64}$

6. $\sqrt{100}$

7. $\sqrt{25} + \sqrt{49}$

8. $\sqrt{16} + \sqrt{81}$

9. $-\sqrt{\dfrac{1}{9}}$

10. $-\sqrt{\dfrac{4}{25}}$

11. $\sqrt[3]{27}$

12. $\sqrt[3]{64}$

13. $\sqrt[3]{-125}$

14. $\sqrt[3]{-8}$

15. $\sqrt[4]{625}$

16. $\sqrt[4]{81}$

17. $-\sqrt[6]{64}$

18. $\sqrt[5]{-32}$

19. $\sqrt[7]{-128}$

20. $-\sqrt[8]{1}$

21. $\sqrt[4]{-81}$

22. $\sqrt[6]{-64}$

23. $\sqrt[5]{(8)^5}$

24. $\sqrt[6]{(9)^6}$

25. $\sqrt[8]{(5)^8}$

26. $\sqrt[7]{(11)^7}$

27. $\sqrt[3]{-\dfrac{1}{64}}$

28. $\sqrt[3]{-\dfrac{8}{27}}$

For Problems 29 through 82, assume that variables represent positive real numbers.

Replace all radicals with rational exponents.

29. $\sqrt[3]{9}$

30. $\sqrt{5}$

31. $\sqrt[5]{2x}$

32. $\sqrt[4]{3y}$

33. $\sqrt[7]{(a+b)^3}$

34. $\sqrt[9]{(a-b)^5}$

35. $\sqrt{\sqrt[3]{x}}$

36. $\sqrt[5]{\sqrt{y}}$

37. $(\sqrt[6]{3x})^5$

38. $(\sqrt[5]{2x})^3$

Simplify.

39. $\sqrt[4]{x^8}$

40. $\sqrt[4]{x^{12}}$

41. $\sqrt[5]{(17)^5}$

42. $\sqrt[5]{(13)^5}$

43. $\sqrt[3]{x^3 y^6}$

44. $\sqrt[4]{a^8 b^4}$

45. $\sqrt[8]{y^{40}}$

46. $\sqrt[9]{y^{36}}$

47. $\sqrt{36x^8 y^4}$

48. $\sqrt{49x^2 y^8}$

49. $\sqrt[3]{8a^6 b^{15}}$

50. $\sqrt[3]{64a^9 b^6}$

51. $\sqrt[3]{-125x^{30}}$

52. $\sqrt[3]{-27x^{45}}$

Change to radical form.

53. $x^{3/4}$

54. $y^{2/3}$

55. $7^{-2/3}$

56. $5^{-3/5}$

57. $(2a+b)^{5/7}$

58. $(x+3y)^{4/7}$

59. $(-x)^{3/5}$

60. $(-y)^{5/7}$

61. $(2xy)^{3/5}$

62. $(3ab)^{2/7}$

Mixed Practice

Evaluate or simplify.

63. $4^{3/2}$

64. $27^{2/3}$

65. $(-8)^{1/3}$

66. $\left(\dfrac{1}{49}\right)^{1/2}$

67. $\left(\dfrac{16}{81}\right)^{3/4}$

68. $(-125)^{2/3}$

69. $(25x^4)^{-1/2}$

70. $(36y^8)^{-1/2}$

71. $\sqrt[4]{81a^8b^{12}}$

72. $\sqrt[4]{16a^{20}b^{16}}$

73. $\sqrt{121x^4}$

74. $\sqrt{49x^8}$

75. $\sqrt{100a^8b^{10}}$

76. $\sqrt{64a^{20}b^8}$

77. $\sqrt{36x^6y^8z^{10}}$

78. $\sqrt{100x^{10}y^{12}z^2}$

79. $\sqrt[3]{216a^3b^9c^{12}}$

80. $\sqrt[3]{-125a^6b^{15}c^{21}}$

81. $\sqrt[4]{16x^8y^{12}} + \sqrt[4]{81x^8y^{12}}$

82. $\sqrt[5]{32a^5b^{15}} + \sqrt[5]{a^5b^{15}}$

Evaluate.

83. $\sqrt[3]{12{,}167}$

84. 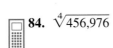 $\sqrt[4]{456{,}976}$

85. 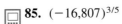 $(-16{,}807)^{3/5}$

86. 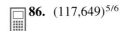 $(117{,}649)^{5/6}$

Simplify. Assume that the variables represent any positive or negative real number.

87. $\sqrt{25x^2}$

88. $\sqrt{100x^2}$

89. $\sqrt[3]{-8x^6}$

90. $\sqrt[3]{-27x^9}$

91. $\sqrt[4]{x^8y^{16}}$

92. $\sqrt[4]{x^{16}y^{40}}$

93. $\sqrt[4]{a^{12}b^4}$

94. $\sqrt[4]{a^4b^{20}}$

95. $\sqrt{4x^8y^4}$

96. $\sqrt{49a^{12}b^4}$

Applications

A company finds that the daily cost of producing appliances at one of its factories is $C = 120\sqrt[3]{n} + 375$, where n is the number of parts produced in a day and C is the cost in dollars.

97. Find the cost if 343 parts are produced per day.

98. Find the cost if 216 parts are produced per day.

Cumulative Review Problems

99. If in 1993 the world produced 3.45×10^{17} Btu of energy, how many Btu of energy were produced that year in the Middle East?

100. If the total world production of energy increases by 20% of its present level by the year 2015 but that of North America increases by only 5% of its present value, what percent of the World Energy Production in 2015 will be produced in North America?

World Energy Production 1993

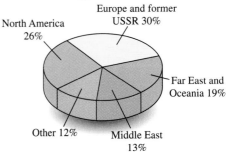

Source: U.S. Energy Information Administration

7.3 Simplifying, Adding, and Subtracting Radicals

MathPro   Video 17   SSM

After studying this section, you will be able to:

1. *Simplify a radical by using the product rule.*
2. *Add and subtract like radical terms.*
3. *Approximate square roots.*

1 Simplifying a Radical by Using the Product Rule

When we simplify a radical, we want to get an equivalent expression with the smallest possible quantity in the radicand. We can use the product rule for radicals to simplify radicals.

Product Rule for Radicals

For all positive real numbers a, b and positive integers n,

$$\sqrt[n]{a}\sqrt[n]{b} = \sqrt[n]{ab}$$

You should be able to derive the product rule from your knowledge of the laws of exponents. We have

$$\sqrt[n]{ab} = (ab)^{1/n} = a^{1/n}b^{1/n} = \sqrt[n]{a}\sqrt[n]{b}$$

Throughout the remainder of this chapter, assume that all variables in any radicand represent positive numbers, unless a specific statement is made to the contrary.

EXAMPLE 1   Simplify. $\sqrt{32}$

Solution 1: $\sqrt{32} = \sqrt{16 \cdot 2} = \sqrt{16}\sqrt{2} = 4\sqrt{2}$

Solution 2: $\sqrt{32} = \sqrt{4 \cdot 8} = \sqrt{4}\sqrt{8} = 2\sqrt{8} = 2\sqrt{4 \cdot 2} = 2\sqrt{4}\sqrt{2} = 4\sqrt{2}$

Although we obtained the same answer both times, the first solution is much shorter. You should try to use the largest factor that is a perfect square when you use the product rule.

Practice Problem 1   Simplify. $\sqrt{20}$ ∎

EXAMPLE 2   Simplify.   **(a)** $\sqrt{125}$ **(b)** $\sqrt{48}$

(a) $\sqrt{125} = \sqrt{25}\sqrt{5} = 5\sqrt{5}$ **(b)** $\sqrt{48} = \sqrt{16}\sqrt{3} = 4\sqrt{3}$

Practice Problem 2   Simplify.   **(a)** $\sqrt{75}$ **(b)** $\sqrt{27}$ ∎

EXAMPLE 3   Simplify.   **(a)** $\sqrt[3]{16}$ **(b)** $\sqrt[3]{-81}$

(a) $\sqrt[3]{16} = \sqrt[3]{8}\sqrt[3]{2} = 2\sqrt[3]{2}$ **(b)** $\sqrt[3]{-81} = \sqrt[3]{-27}\sqrt[3]{3} = -3\sqrt[3]{3}$

Practice Problem 3   Simplify.   **(a)** $\sqrt[3]{24}$ **(b)** $\sqrt[3]{-108}$ ∎

EXAMPLE 4 Simplify. $\sqrt[4]{48}$

$$\sqrt[4]{48} = \sqrt[4]{16}\sqrt[4]{3} = 2\sqrt[4]{3}$$

Practice Problem 4 Simplify. $\sqrt[4]{64}$ ■

EXAMPLE 5 Simplify.

(a) $\sqrt{27x^3y^4}$ **(b)** $\sqrt{100a^4b^7c^0}$ **(c)** $\sqrt[3]{16x^4y^3z^6}$ **(d)** $\sqrt[4]{10{,}000x^9y^8w^5}$

(a) $\sqrt{27x^3y^4} = \sqrt{9 \cdot 3 \cdot x^2 \cdot x \cdot y^4} = \sqrt{9x^2y^4}\sqrt{3x}$ *Factor out the perfect squares.*

$$= 3xy^2\sqrt{3x}$$

(b) $\sqrt{100a^4b^7c^0} = \sqrt{100 \cdot a^4 \cdot b^6 \cdot b \cdot 1} = \sqrt{100a^4b^6}\sqrt{b}$ *Recall that $c^0 = 1$.*

$$= 10a^2b^3\sqrt{b}$$

(c) $\sqrt[3]{16x^4y^3z^6} =$

$\sqrt[3]{8 \cdot 2 \cdot x^3 \cdot x \cdot y^3 \cdot z^6} = \sqrt[3]{8x^3y^3z^6}\sqrt[3]{2x}$ *Factor out the perfect cubes.*

$$= 2xyz^2\sqrt[3]{2x}$$ *Why is z^6 a perfect cube?*

(d) $\sqrt[4]{10{,}000x^9y^8w^5} = \sqrt[4]{10{,}000 \cdot x^8 \cdot x \cdot y^8 \cdot w^4 \cdot w} = \sqrt[4]{10{,}000x^8y^8w^4}\sqrt[4]{xw}$

$$= 10x^2y^2w\sqrt[4]{xw}$$

Practice Problem 5 Simplify.

(a) $\sqrt{45x^6y^7}$ **(b)** $\sqrt{81a^4b^0c^7}$ **(c)** $\sqrt{27a^7b^8c^9}$ **(d)** $\sqrt{256a^8b^{11}c^{17}}$ ■

---

2 *Adding and Subtracting Radicals*

Only like **radicals** can be added or subtracted. Two radicals are like if they have the same radicand and the same index. $2\sqrt{5}$ and $3\sqrt{5}$ are like radicals. $2\sqrt{5}$ and $2\sqrt{3}$ are not like radicals; $2\sqrt{5}$ and $2\sqrt[3]{5}$ are not like radicals. When we combine radicals, we combine like terms by using the distributive property.

EXAMPLE 6 Combine. **(a)** $2\sqrt{5} + 3\sqrt{5} - 4\sqrt{5}$ **(b)** $5\sqrt[3]{x} - 8\sqrt[3]{x} - 12\sqrt[3]{x}$

(a) $2\sqrt{5} + 3\sqrt{5} - 4\sqrt{5} = (2 + 3 - 4)\sqrt{5} = 1\sqrt{5} = \sqrt{5}$

(b) $5\sqrt[3]{x} - 8\sqrt[3]{x} - 12\sqrt[3]{x} = (5 - 8 - 12)\sqrt[3]{x} = -15\sqrt[3]{x}$

Practice Problem 6 Combine.

(a) $5\sqrt[3]{7} - 12\sqrt[3]{7} - 6\sqrt[3]{7}$ **(b)** $19\sqrt{xy} + 5\sqrt{xy} - 10\sqrt{xy}$ ■

Sometimes, when you simplify the radicand, you may find you have like radicals.

EXAMPLE 7 Combine. $5\sqrt{3} - \sqrt{27} + 2\sqrt{48}$

$$5\sqrt{3} - \sqrt{27} + 2\sqrt{48} = 5\sqrt{3} - \sqrt{9}\sqrt{3} + 2\sqrt{16}\sqrt{3}$$
$$= 5\sqrt{3} - 3\sqrt{3} + 2(4)\sqrt{3}$$
$$= 5\sqrt{3} - 3\sqrt{3} + 8\sqrt{3}$$
$$= 10\sqrt{3}$$

Practice Problem 7 Combine. $4\sqrt{2} - 5\sqrt{50} - 3\sqrt{98}$ ■

EXAMPLE 8 Combine. $6\sqrt{x} + 4\sqrt{12x} - \sqrt{75x} + 3\sqrt{x}$

$$6\sqrt{x} + 4\sqrt{12x} - \sqrt{75x} + 3\sqrt{x} = 6\sqrt{x} + 4\sqrt{4}\sqrt{3x} - \sqrt{25}\sqrt{3x} + 3\sqrt{x}$$
$$= 6\sqrt{x} + 8\sqrt{3x} - 5\sqrt{3x} + 3\sqrt{x}$$
$$= 6\sqrt{x} + 3\sqrt{x} + 8\sqrt{3x} - 5\sqrt{3x}$$
$$= 9\sqrt{x} + 3\sqrt{3x}$$

Practice Problem 8 Combine. $4\sqrt{2x} + \sqrt{18x} - 2\sqrt{125x} - 6\sqrt{20x}$ ■

EXAMPLE 9 Combine. $2\sqrt[3]{81x^3y^4} + 3xy\sqrt[3]{24y}$

$$2\sqrt[3]{81x^3y^4} + 3xy\sqrt[3]{24y} = 2\sqrt[3]{27x^3y^3}\sqrt[3]{3y} + 3xy\sqrt[3]{8}\sqrt[3]{3y}$$
$$= 2(3xy)\sqrt[3]{3y} + 3xy(2)\sqrt[3]{3y}$$
$$= 6xy\sqrt[3]{3y} + 6xy\sqrt[3]{3y}$$
$$= 12xy\sqrt[3]{3y}$$

Practice Problem 9 Combine. $3x\sqrt[3]{54x^4} - 3\sqrt[3]{16x^7}$ ■

3 Approximating Square Roots

In practical situations it is useful to find a rational number that approximates a square root that is irrational. This can be done on any calculator that has a square root key. Table A-1 in the back of the book may also be used. To find $\sqrt{193}$ on a calculator, enter 193 and then push the square root key. 193 $\boxed{\sqrt{}}$ 13.892444. If you are using a graphing calculator, push the $\boxed{\text{2nd}}$ function key; then push the square root key and 193 $\boxed{\text{ENTER}}$. $\boxed{\text{2nd}}$ $\boxed{\sqrt{}}$ 193 $\boxed{\text{ENTER}}$ 13.89244399. Depending on your calculator, more or fewer digits will be displayed. We could say that $\sqrt{193} \approx 13.892444$. If we use Table A-1 in the back of the book, we have $\sqrt{193} \approx 13.892$.

EXAMPLE 10 Approximate. $\sqrt{250}$

Using a calculator, we have

$$250 \boxed{\sqrt{}} \quad 15.811388$$

Thus we conclude that $\sqrt{250} \approx 15.811388$.

Using a graphing calculator, we have $\boxed{\text{2nd}}$ $\boxed{\sqrt{}}$ 250 $\boxed{\text{ENTER}}$ 15.8113883. Thus $\sqrt{250} \approx 15.8113883$.

If we use Table A-1, we will need to do an extra step since the table only goes to 200. We have

$$\sqrt{250} = \sqrt{25}\sqrt{10} = 5\sqrt{10} \approx 5(3.162) \approx 15.810$$

Because of round-off errors, there is a slight difference in our answers.

Practice Problem 10 Approximate. $\sqrt{300}$ ■

7.3 Exercises

Simplify. Assume that all variables are positive real numbers. If no real value is possible, so state.

1. $\sqrt{50}$

2. $\sqrt{20}$

3. $\sqrt{18}$

4. $\sqrt{75}$

5. $\sqrt{120}$

6. $\sqrt{80}$

7. $\sqrt{81}$

8. $\sqrt{108}$

9. $\sqrt{9x^3}$

10. $\sqrt{16x^5}$

11. $\sqrt{60a^4b^5}$

12. $\sqrt{45a^3b^8}$

13. $\sqrt{98x^5y^6z}$

14. $\sqrt{24xy^8z^3}$

15. $\sqrt[3]{8}$

16. $\sqrt[3]{27}$

17. $\sqrt[3]{108}$

18. $\sqrt[3]{128}$

19. $\sqrt[3]{56y}$

20. $\sqrt[3]{54x}$

21. $\sqrt[3]{8a^3b^8}$

22. $\sqrt[3]{125a^6b^2}$

23. $\sqrt[3]{56x^{10}y^{12}}$

24. $\sqrt[3]{72x^5y^{20}}$

25. $\sqrt[3]{-16a^6b^3c^{12}}$

26. $\sqrt[3]{-40a^3b^7c^{14}}$

27. $\sqrt[4]{48x^8y^{13}}$

28. $\sqrt[4]{32x^5y^{10}}$

29. $\sqrt[4]{81kp^{23}}$

30. $\sqrt[4]{16k^{12}p^{18}}$

31. $\sqrt[5]{-32x^5y^6}$

32. $\sqrt[5]{-243x^4y^{10}}$

33. $\sqrt[6]{8x^6y^{12}}$

34. $\sqrt[6]{64x^6y^8}$

To Think About

35. $\sqrt[4]{1792} = a\sqrt[4]{7}$. What is the value of a?

36. $\sqrt[3]{3072} = b\sqrt[3]{6}$. What is the value of b?

Combine. Assume that all variables represent positive real numbers.

37. $\sqrt{49} + \sqrt{100}$

38. $\sqrt{25} + \sqrt{81}$

39. $\sqrt{3} + 7\sqrt{3} - 2\sqrt{3}$

40. $\sqrt{11} - 5\sqrt{11} + 3\sqrt{11}$

41. $2\sqrt{12} - \sqrt{3}$

42. $3\sqrt{50} - \sqrt{2}$

43. $-2\sqrt{50} + \sqrt{32} - 3\sqrt{8}$

44. $-\sqrt{12} + 2\sqrt{48} - \sqrt{75}$

45. $-5\sqrt{45} + 6\sqrt{20} + 3\sqrt{5}$

46. $-7\sqrt{10} + 4\sqrt{40} - 8\sqrt{90}$

47. $\sqrt{44} - 3\sqrt{63x} + 4\sqrt{28x}$

48. $\sqrt{75x} + 2\sqrt{108x} - 6\sqrt{3x}$

49. $x\sqrt{20} - 3\sqrt{125x^2}$

50. $-3\sqrt{45a^6} + a\sqrt{80a^4}$

51. $\sqrt[3]{16} + 3\sqrt[3]{54}$

52. $\sqrt[3]{128} - 4\sqrt[3]{16}$

53. $-2\sqrt[3]{125x^3y^4} + 3y^2\sqrt[3]{8x^3}$

54. $2x\sqrt[3]{40xy} - 3\sqrt[3]{5x^4y}$

55. $5\sqrt[3]{16a^4b} + \sqrt{45a} - 3a\sqrt[3]{54ab} - 2\sqrt{80a}$

56. $-2\sqrt{72x^3y} + \sqrt[3]{3x^4y} - 3x\sqrt{50xy} + 4x\sqrt[3]{81xy}$

Use a calculator with a square root key or Table B to approximate the following square roots. Round your answer to three decimal places.

57. $\sqrt{5}$

58. $\sqrt{6}$

59. $\sqrt{156}$

60. $\sqrt{142}$

61. $\sqrt{500}$

62. $\sqrt{600}$

63. $\sqrt{344}$

64. $\sqrt{468}$

To Think About

65. Use a calculator to show that
$\sqrt{48} + \sqrt{27} + \sqrt{75} = 12\sqrt{3}$.

66. Use a calculator to show that
$\sqrt{98} + \sqrt{50} + \sqrt{128} = 20\sqrt{2}$.

Applications

Approximate the amount of current in amps I (amperes) drawn by an appliance in your home using the formula
$I = \sqrt{\dfrac{P}{R}}$ *where P is the power measured in watts and R is the resistance measured in ohms. Round your answer to three decimal places.*

67. What is the current I if $P = 500$ watts and $R = 10$ ohms?

68. What is the current I if $P = 480$ watts and $R = 8$ ohms?

Cumulative Review Problems

Factor completely.

69. $81x^2y - 25y$

70. $16x^3 - 56x^2y + 49xy^2$

The FDA has recently made the recommendation that an adult's minimum daily intake of the mineral phosphorus should be 1 gram. A small serving of scallops (6 average scallops) has 0.2 gram of phosphorus while one small serving of skim milk (one cup) has 0.25 gram of phosphorus.

71. If you eat only scallops how many servings would you need to obtain the minimum daily requirement of phosphorus? If you drink only skim milk how many servings would you need to obtain the daily requirement for phosphorus?

72. If the number of servings of scallops and the number of servings of skim milk totals 4.5 servings, how many of each would you need to meet the minimum daily requirement of phosphorus?

7.4 Multiplication and Division of Radicals

After studying this section, you will be able to:

1. *Multiply radical expressions.*
2. *Divide radical expressions.*
3. *Simplify radical expressions by rationalizing the denominator.*

MathPro · Video 17 · SSM

1 Multiplying Radical Expressions

We use the product rule for radicals to multiply radical expressions. Recall $\sqrt[n]{a}\sqrt[n]{b} = \sqrt[n]{ab}$.

EXAMPLE 1 Multiply. **(a)** $(\sqrt{3})(\sqrt{5})$ **(b)** $(3\sqrt{2})(5\sqrt{11x})$

(a) $(\sqrt{3})(\sqrt{5}) = \sqrt{3 \cdot 5} = \sqrt{15}$

(b) $(3\sqrt{2})(5\sqrt{11x}) = (3)(5)\sqrt{2 \cdot 11x} = 15\sqrt{22x}$

Practice Problem 1 Multiply. **(a)** $\sqrt{7}\sqrt{5}$ **(b)** $(-4\sqrt{2})(-3\sqrt{13x})$ ∎

EXAMPLE 2 Multiply. **(a)** $\sqrt{5}(\sqrt{2} + 5\sqrt{5})$ **(b)** $\sqrt{6x}(\sqrt{3} + \sqrt{2x} + \sqrt{5})$

(a) $\sqrt{5}(\sqrt{2} + 5\sqrt{5}) = \sqrt{5}\sqrt{2} + \sqrt{5}(5\sqrt{5})$
$$= \sqrt{10} + 5\sqrt{25} = \sqrt{10} + 5(5) = \sqrt{10} + 25$$

(b) $\sqrt{6x}(\sqrt{3} + \sqrt{2x} + \sqrt{5}) = (\sqrt{6x})(\sqrt{3}) + (\sqrt{6x})(\sqrt{2x}) + (\sqrt{6x})(\sqrt{5})$
$$= \sqrt{18x} + \sqrt{12x^2} + \sqrt{30x}$$
$$= \sqrt{9}\sqrt{2x} + \sqrt{4x^2}\sqrt{3} + \sqrt{30x}$$
$$= 3\sqrt{2x} + 2x\sqrt{3} + \sqrt{30x}$$

Practice Problem 2 Multiply.

(a) $\sqrt{3}(\sqrt{6} - 5\sqrt{7})$ **(b)** $\sqrt{2x}(\sqrt{5} + 2\sqrt{3x} + \sqrt{8})$ ∎

To multiply two binomials containing radicals, we can use the distributive property. Most students find that the FOIL approach is helpful in remembering how to quickly find the four products.

EXAMPLE 3 Multiply. $(\sqrt{2} + 3\sqrt{5})(2\sqrt{2} - \sqrt{5})$

By FOIL

$$(\sqrt{2} + 3\sqrt{5})(2\sqrt{2} - \sqrt{5}) = 2\sqrt{4} - \sqrt{10} + 6\sqrt{10} - 3\sqrt{25}$$
$$= 4 + 5\sqrt{10} - 15$$
$$= -11 + 5\sqrt{10}$$

By the distributive property

$$(\sqrt{2} + 3\sqrt{5})(2\sqrt{2} - \sqrt{5}) = (\sqrt{2} + 3\sqrt{5})(2\sqrt{2}) - (\sqrt{2} + 3\sqrt{5})\sqrt{5}$$
$$= (\sqrt{2})(2\sqrt{2}) + (3\sqrt{5})(2\sqrt{2}) - (\sqrt{2})(\sqrt{5}) - (3\sqrt{5})(\sqrt{5})$$
$$= 2\sqrt{4} + 6\sqrt{10} - \sqrt{10} - 3\sqrt{25}$$
$$= 4 + 5\sqrt{10} - 15$$
$$= -11 + 5\sqrt{10}$$

Practice Problem 3 Multiply. $(\sqrt{7} + 4\sqrt{2})(2\sqrt{7} - 3\sqrt{2})$ ∎

EXAMPLE 4   Multiply. $(7 - 3\sqrt{2})(4 - \sqrt{3})$

$$(7 - 3\sqrt{2})(4 - \sqrt{3}) = 28 - 7\sqrt{3} - 12\sqrt{2} + 3\sqrt{6}$$

Practice Problem 4   Multiply. $(2 - 5\sqrt{5})(3 - 2\sqrt{2})$ ■

EXAMPLE 5   Multiply. $(\sqrt{7} + \sqrt{3x})^2$

Solution 1: We can use the FOIL method or the distributive property.

$$\begin{aligned}(\sqrt{7} + \sqrt{3x})(\sqrt{7} + \sqrt{3x}) &= \sqrt{49} + \sqrt{21x} + \sqrt{21x} + \sqrt{9x^2} \\ &= 7 + \sqrt{21x} + \sqrt{21x} + 3x \\ &= 7 + 2\sqrt{21x} + 3x\end{aligned}$$

Solution 2: We could also use this formula from Chapter 4.

$$(a + b)^2 = a^2 + 2ab + b^2$$

where $a = \sqrt{7}$ and $b = \sqrt{3x}$. Then

$$\begin{aligned}(\sqrt{7} + \sqrt{3x})^2 &= (\sqrt{7})^2 + 2\sqrt{7}\sqrt{3x} + (\sqrt{3x})^2 \\ &= 7 + 2\sqrt{21x} + 3x\end{aligned}$$

Practice Problem 5   Multiply. $(\sqrt{5x} + \sqrt{10})^2$ Use the approach that seems easiest to you. ■

EXAMPLE 6   Multiply. $(\sqrt{3x + 1} + 2)^2$

Solution 1:

$$\begin{aligned}(\sqrt{3x + 1} + 2)(\sqrt{3x + 1} + 2) &= (\sqrt{3x + 1})^2 + 2\sqrt{3x + 1} + 2\sqrt{3x + 1} + 4 \\ &= 3x + 1 + 4\sqrt{3x + 1} + 4 \\ &= 3x + 5 + 4\sqrt{3x + 1}\end{aligned}$$

Solution 2: Use the formula $(a + b)^2 = a^2 + 2ab + b^2$, $a = \sqrt{3x + 1}$ and $b = 2$.

$$\begin{aligned}(\sqrt{3x + 1} + 2)^2 &= (\sqrt{3x + 1})^2 + 2(\sqrt{3x + 1})(2) + (2)^2 \\ &= 3x + 1 + 4\sqrt{3x + 1} + 4 \\ &= 3x + 5 + 4\sqrt{3x + 1}\end{aligned}$$

Practice Problem 6   Multiply. $(\sqrt{4x - 1} + 3)^2$ ■

EXAMPLE 7 Multiply. **(a)** $\sqrt[3]{3x}(\sqrt[3]{x^2} + 3\sqrt[3]{4y})$

(b) $(\sqrt[3]{2y} + \sqrt[3]{4})(2\sqrt[3]{4y^2} - 3\sqrt[3]{2})$

(a) $\sqrt[3]{3x}(\sqrt[3]{x^2} + 3\sqrt[3]{4y}) = (\sqrt[3]{3x})(\sqrt[3]{x^2}) + 3(\sqrt[3]{3x})(\sqrt[3]{4y})$

$$= \sqrt[3]{3x^3} + 3\sqrt[3]{12xy}$$
$$= x\sqrt[3]{3} + 3\sqrt[3]{12xy}$$

(b) $(\sqrt[3]{2y} + \sqrt[3]{4})(2\sqrt[3]{4y^2} - 3\sqrt[3]{2}) = 2\sqrt[3]{8y^3} - 3\sqrt[3]{4y} + 2\sqrt[3]{16y^2} - 3\sqrt[3]{8}$

$$= 2(2y) - 3\sqrt[3]{4y} + 2\sqrt[3]{8}\sqrt[3]{2y^2} - 3(2)$$
$$= 4y - 3\sqrt[3]{4y} + 4\sqrt[3]{2y^2} - 6$$

Practice Problem 7 Multiply.

(a) $\sqrt[3]{2x}(\sqrt[3]{4x^2} + 3\sqrt[3]{y})$ **(b)** $(\sqrt[3]{7} + \sqrt[3]{x^2})(2\sqrt[3]{49} - \sqrt[3]{x})$ ∎

2 Dividing Radical Expressions

We can use the laws of exponents to develop a rule for dividing two radicals.

$$\sqrt[n]{\frac{a}{b}} = \left(\frac{a}{b}\right)^{1/n} = \frac{a^{1/n}}{b^{1/n}} = \frac{\sqrt[n]{a}}{\sqrt[n]{b}}$$

This quotient rule is very useful. We will now state it more formally.

Quotient Rule for Radicals

For all positive real numbers a, b and positive integers n,

$$\frac{\sqrt[n]{a}}{\sqrt[n]{b}} = \sqrt[n]{\frac{a}{b}}$$

Sometimes it will be best to change $\sqrt[n]{\frac{a}{b}}$ to $\frac{\sqrt[n]{a}}{\sqrt[n]{b}}$. Sometimes it will be best to change $\frac{\sqrt[n]{a}}{\sqrt[n]{b}}$ to $\sqrt[n]{\frac{a}{b}}$. To use the quotient rule for radicals, you need to have a good number sense. You should know your squares up to 15^2 and your cubes up to 5^3.

EXAMPLE 8 Divide.

(a) $\sqrt{\dfrac{9}{25}}$ **(b)** $\dfrac{\sqrt{48}}{\sqrt{3}}$ **(c)** $\sqrt[3]{\dfrac{125}{8}}$ **(d)** $\dfrac{\sqrt{28x^5y^3}}{\sqrt{7x}}$

(a) $\sqrt{\dfrac{9}{25}} = \dfrac{\sqrt{9}}{\sqrt{25}} = \dfrac{3}{5}$

(b) $\dfrac{\sqrt{48}}{\sqrt{3}} = \sqrt{\dfrac{48}{3}} = \sqrt{16} = 4$

(c) $\sqrt[3]{\dfrac{125}{8}} = \dfrac{\sqrt[3]{125}}{\sqrt[3]{8}} = \dfrac{5}{2}$

(d) $\dfrac{\sqrt{28x^5y^3}}{\sqrt{7x}} = \sqrt{\dfrac{28x^5y^3}{7x}} = \sqrt{4x^4y^3} = 2x^2y\sqrt{y}$

Practice Problem 8 Divide.

(a) $\sqrt{\dfrac{144}{16}}$ **(b)** $\dfrac{\sqrt{75}}{\sqrt{3}}$ **(c)** $\sqrt[3]{\dfrac{27}{64}}$ **(d)** $\dfrac{\sqrt{54a^3b^7}}{\sqrt{6b^5}}$ ∎

3 *Simplifying Radical Expressions by Rationalizing the Denominator*

Recall that to simplify a radical we want to get the smallest possible quantity in the radicand. Whenever possible, we find the square root of a perfect square. Thus to simplify $\sqrt{\frac{7}{16}}$ we have

$$\sqrt{\frac{7}{16}} = \frac{\sqrt{7}}{\sqrt{16}} = \frac{\sqrt{7}}{4}$$

Notice that the denominator does not contain a square root. The expression $\frac{\sqrt{7}}{4}$ is in simplest form.

Let's look at $\sqrt{\frac{16}{7}}$. We have

$$\sqrt{\frac{16}{7}} = \frac{\sqrt{16}}{\sqrt{7}} = \frac{4}{\sqrt{7}}$$

Notice that the denominator contains a square root. If an expression contains a square root in the denominator, it is not considered to be simplified. How can we rewrite $\frac{4}{\sqrt{7}}$ as an equivalent expression that does not contain the $\sqrt{7}$ in the denominator? Since $\sqrt{7}\sqrt{7} = 7$, we can multiply the numerator and the denominator by the radical in the denominator.

$$\frac{4}{\sqrt{7}} \cdot \frac{\sqrt{7}}{\sqrt{7}} = \frac{4\sqrt{7}}{7}$$

This expression is considered to be in simplest form. We call this process rationalizing the denominator.

Rationalizing the denominator is the process of transforming a fraction with one or more radicals in the denominator into an equivalent fraction without a radical in the denominator.

EXAMPLE 9   Simplify by rationalizing the denominator. $\dfrac{3}{\sqrt{2}}$

$$\frac{3}{\sqrt{2}} = \frac{3}{\sqrt{2}} \cdot \boxed{\frac{\sqrt{2}}{\sqrt{2}}} \qquad \textit{Since } \frac{\sqrt{2}}{\sqrt{2}} = 1.$$

$$= \frac{3\sqrt{2}}{\sqrt{4}} \qquad\qquad \textit{Product rule of radicals.}$$

$$= \frac{3\sqrt{2}}{2}$$

Practice Problem 9   Simplify by rationalizing the denominator. $\dfrac{7}{\sqrt{3}}$ ■

We can rationalize the denominator either before or after we multiply the numerator and the denominator by a radical.

EXAMPLE 10   Simplify. $\dfrac{3}{\sqrt{12x}}$

Solution 1: First we simplify the radical in the denominator and then we multiply.

$$\frac{3}{\sqrt{12x}} = \frac{3}{\sqrt{4}\sqrt{3x}} = \frac{3}{2\sqrt{3x}} \cdot \boxed{\frac{\sqrt{3x}}{\sqrt{3x}}} = \frac{\cancel{3}\sqrt{3x}}{2(\cancel{3}x)} = \frac{\sqrt{3x}}{2x}$$

Solution 2: We can multiply numerator and denominator by a value that will make the denominator a perfect square.

$$\frac{3}{\sqrt{12x}} = \frac{3}{\sqrt{12x}} \cdot \frac{\sqrt{3x}}{\sqrt{3x}}$$

$$= \frac{3\sqrt{3x}}{\sqrt{36x^2}} \qquad \text{Since } \sqrt{12x}\sqrt{3x} = \sqrt{36x^2}.$$

$$= \frac{3\sqrt{3x}}{6x} = \frac{\sqrt{3x}}{2x}$$

Practice Problem 10 Simplify. $\dfrac{8}{\sqrt{20x}}$ ∎

If the radical expression has a fraction, it is not considered to be simplified. We can use the quotient rule of radicals and then rationalize the denominator to simplify the radical. We have already rationalized denominators when they contain square roots. Now we will rationalize the denominator when the denominator contains radical expressions that are cube roots or higher-order roots.

EXAMPLE 11 Simplify. $\sqrt[3]{\dfrac{2}{3x^2}}$

Solution 1: $\sqrt[3]{\dfrac{2}{3x^2}} = \dfrac{\sqrt[3]{2}}{\sqrt[3]{3x^2}}$ *Quotient rule of radicals.*

$$= \frac{\sqrt[3]{2}}{\sqrt[3]{3x^2}} \cdot \frac{\sqrt[3]{9x}}{\sqrt[3]{9x}}$$
Multiply the denominator by an appropriate value so that the denominator will be a perfect cube.

$$= \frac{\sqrt[3]{18x}}{\sqrt[3]{27x^3}}$$
Observe that we can evaluate the cube root in the denominator.

$$= \frac{\sqrt[3]{18x}}{3x}$$

Solution 2: $\sqrt[3]{\dfrac{2}{3x^2}} = \sqrt[3]{\dfrac{2}{3x^2} \cdot \dfrac{9x}{9x}}$

$$= \sqrt[3]{\frac{18x}{27x^3}}$$

$$= \frac{\sqrt[3]{18x}}{\sqrt[3]{27x^3}}$$

$$= \frac{\sqrt[3]{18x}}{3x}$$

Practice Problem 11 Simplify. $\sqrt[3]{\dfrac{6}{5x}}$ ∎

If the radical expression contains a sum or difference with radicals, we multiply the numerator and denominator by the **conjugate** of the denominator. For example, the conjugate of $x + \sqrt{y}$ is $x - \sqrt{y}$; similarly, the conjugate of $x - \sqrt{y}$ is $x + \sqrt{y}$. What is the conjugate of $3 + \sqrt{2}$? It is $3 - \sqrt{2}$. How about $\sqrt{11} + \sqrt{xyz}$? It is $\sqrt{11} - \sqrt{xyz}$.

> **Conjugates**
>
> The expressions $\sqrt{x} + \sqrt{y}$ and $\sqrt{x} - \sqrt{y}$ are called *conjugates*. Each expression is the conjugate of the other expression.

Multiplying by conjugates is simply an application of the formula

$$(a + b)(a - b) = a^2 - b^2$$

For example,

$$(\sqrt{x} + \sqrt{y})(\sqrt{x} - \sqrt{y}) = (\sqrt{x})^2 - (\sqrt{y})^2 = x - y$$

EXAMPLE 12   Simplify. $\dfrac{4}{3 + \sqrt{2}}$

$$\frac{5}{3 + \sqrt{2}} = \frac{5}{3 + \sqrt{2}} \cdot \boxed{\frac{3 - \sqrt{2}}{3 - \sqrt{2}}} \qquad \textit{Multiply numerator and denominator by the conjugate of } 3 + \sqrt{2}.$$

$$= \frac{15 - 5\sqrt{2}}{(3)^2 - (\sqrt{2})^2}$$

$$= \frac{15 - 5\sqrt{2}}{9 - 2} = \frac{15 - 5\sqrt{2}}{7}$$

Practice Problem 12   Simplify. $\dfrac{4}{2 + \sqrt{5}}$ ∎

EXAMPLE 13   Simplify. $\dfrac{\sqrt{7} + \sqrt{3}}{\sqrt{7} - \sqrt{3}}$

The conjugate of $\sqrt{7} - \sqrt{3}$ is $\sqrt{7} + \sqrt{3}$.

$$\frac{\sqrt{7} + \sqrt{3}}{\sqrt{7} - \sqrt{3}} \cdot \frac{\sqrt{7} + \sqrt{3}}{\sqrt{7} + \sqrt{3}} = \frac{\sqrt{49} + 2\sqrt{21} + \sqrt{9}}{(\sqrt{7})^2 - (\sqrt{3})^2}$$

$$= \frac{7 + 2\sqrt{21} + 3}{7 - 3}$$

$$= \frac{10 + 2\sqrt{21}}{4}$$

$$= \frac{2(5 + \sqrt{21})}{2 \cdot 2}$$

$$= \frac{\cancel{2}(5 + \sqrt{21})}{\cancel{2} \cdot 2}$$

$$= \frac{5 + \sqrt{21}}{2}$$

Practice Problem 13   Simplify. $\dfrac{\sqrt{11} + \sqrt{2}}{\sqrt{11} - \sqrt{2}}$ ∎

7.4 Exercises

Multiply and simplify. Assume that all variables represent positive numbers.

1. $(\sqrt{3})(\sqrt{5})$

2. $(\sqrt{6})(\sqrt{11})$

3. $(2\sqrt{6})(-3\sqrt{2})$

4. $(-4\sqrt{5})(2\sqrt{10})$

5. $(x\sqrt{2})(4\sqrt{5})$

6. $(y\sqrt{3})(5\sqrt{7})$

7. $(2\sqrt{xy})(\sqrt{2x})$

8. $(3\sqrt{5x})(\sqrt{xy})$

9. $(3\sqrt{2xy^2})(5\sqrt{6x^3y^2})$

10. $(8\sqrt{5x^4y})(2\sqrt{3xy^3})$

11. $\sqrt{10}(\sqrt{5} + 3\sqrt{10})$

12. $\sqrt{6}(2\sqrt{6} - 5\sqrt{2})$

13. $2\sqrt{x}(3\sqrt{x} - 4\sqrt{5})$

14. $5\sqrt{y}(\sqrt{2y} + 3\sqrt{5})$

15. $(2\sqrt{3} + \sqrt{6})(\sqrt{3} - \sqrt{6})$

16. $(5\sqrt{2} + \sqrt{10})(\sqrt{2} - \sqrt{10})$

17. $(2\sqrt{3} + \sqrt{2})(2\sqrt{3} - 4\sqrt{2})$

18. $(3\sqrt{3} + \sqrt{5})(\sqrt{3} - 2\sqrt{5})$

19. $(\sqrt{7} + 4\sqrt{5x})(2\sqrt{7} + 3\sqrt{5x})$

20. $(\sqrt{6} + 3\sqrt{3y})(5\sqrt{6} + 2\sqrt{3y})$

21. $(\sqrt{x} - 4)(\sqrt{x} - 2)$

22. $(3 - \sqrt{x})(2 - \sqrt{x})$

23. $(\sqrt{3} + 2\sqrt{2})(\sqrt{5} + \sqrt{3})$

24. $(3\sqrt{5} + \sqrt{3})(\sqrt{2} + 2\sqrt{5})$

25. $(\sqrt{x} - 2\sqrt{3x})(\sqrt{x} + 2\sqrt{3x})$

26. $(2\sqrt{x} + \sqrt{5x})(2\sqrt{x} - \sqrt{5x})$

27. $(\sqrt{5} - 2\sqrt{6})^2$

28. $(\sqrt{3} + 4\sqrt{7})^2$

29. $(\sqrt{3x + 4} + 3)^2$

30. $(\sqrt{2x + 1} - 2)^2$

31. $(2\sqrt{a} - 5\sqrt{b})^2$

32. $(5\sqrt{a} + 4\sqrt{b})^2$

33. $(\sqrt[4]{6x})(3\sqrt[4]{8x^7})$

34. $(\sqrt[4]{9y^3})(2\sqrt[4]{27y^3})$

35. $(\sqrt[3]{x^2})(3\sqrt[3]{4x} - 4\sqrt[3]{x^5})$

36. $(2\sqrt[3]{x})(\sqrt[3]{4x^2} - \sqrt[3]{14x})$

37. $(\sqrt[3]{x} + \sqrt[3]{2x^2})(3\sqrt[3]{x} - \sqrt[3]{4x})$

38. $(4\sqrt[3]{4x} - \sqrt[3]{2x^2})(\sqrt[3]{x} + 2\sqrt[3]{2x})$

To Think About

39. What answer do you obtain when you multiply $(\sqrt[3]{6} + \sqrt[3]{5})(\sqrt[3]{36} - \sqrt[3]{30} + \sqrt[3]{25})$? Why?

40. What answer do you obtain when you multiply $(\sqrt[3]{9} - \sqrt[3]{4})(3\sqrt[3]{3} + \sqrt[3]{36} + 2\sqrt[3]{2})$? Why?

Divide and simplify. Assume that all variables represent positive numbers.

41. $\sqrt{\dfrac{49}{25}}$

42. $\sqrt{\dfrac{16}{36}}$

43. $\sqrt[3]{\dfrac{64}{125}}$

44. $\sqrt[3]{\dfrac{216}{27}}$

45. $\sqrt{\dfrac{4x^3}{81y^4}}$

46. $\sqrt{\dfrac{9a^5}{64}}$

47. $\sqrt[3]{\dfrac{8x^5y^6}{27}}$

48. $\sqrt[3]{\dfrac{125a^3b^4}{64}}$

49. $\dfrac{\sqrt{250x^3}}{\sqrt{10x}}$

50. $\dfrac{\sqrt{72a^5}}{\sqrt{2a}}$

51. $\dfrac{\sqrt[3]{24x^3y^5}}{\sqrt[3]{3y^2}}$

52. $\dfrac{\sqrt[3]{250a^4b^6}}{\sqrt[3]{2a}}$

Simplify by rationalizing the denominator.

53. $\dfrac{3}{\sqrt{2}}$

54. $\dfrac{5}{\sqrt{7}}$

55. $\sqrt{\dfrac{x}{8}}$

56. $\sqrt{\dfrac{y}{12}}$

57. $\sqrt{\dfrac{16}{5}}$

58. $\sqrt{\dfrac{9}{7}}$

59. $\dfrac{\sqrt{2a}}{\sqrt{b}}$

60. $\dfrac{\sqrt{3a}}{\sqrt{b}}$

61. $\dfrac{1}{\sqrt{5y}}$

62. $\dfrac{1}{\sqrt{3x}}$

63. $\dfrac{3\sqrt{3}}{\sqrt{8x}}$

64. $\dfrac{5\sqrt{2}}{\sqrt{12y}}$

65. $\sqrt{\dfrac{5y}{7x^2}}$

66. $\sqrt{\dfrac{8x}{3y^2}}$

67. $\dfrac{2}{\sqrt{6} + \sqrt{3}}$

68. $\dfrac{5}{\sqrt{2} + \sqrt{7}}$

69. $\dfrac{\sqrt{3}}{\sqrt{5} - 2}$

70. $\dfrac{\sqrt{7}}{\sqrt{7} - 1}$

71. $\dfrac{\sqrt{x}}{\sqrt{3x} + \sqrt{2}}$

72. $\dfrac{\sqrt{x}}{\sqrt{5} + \sqrt{2x}}$

73. $\dfrac{\sqrt{5} + \sqrt{3}}{\sqrt{5} - \sqrt{3}}$

74. $\dfrac{\sqrt{11} - \sqrt{5}}{\sqrt{11} + \sqrt{5}}$

75. $\dfrac{\sqrt{3x} - \sqrt{2y}}{\sqrt{3x} + \sqrt{y}}$

76. $\dfrac{\sqrt{x} + \sqrt{y}}{\sqrt{x} - 2\sqrt{y}}$

77. $\dfrac{5\sqrt{3} - 3\sqrt{2}}{3\sqrt{2} - 2\sqrt{3}}$

78. $\dfrac{2\sqrt{6} + \sqrt{5}}{3\sqrt{6} - \sqrt{5}}$

79. $\dfrac{x\sqrt{5} + 1}{\sqrt{5} + 2}$

80. $\dfrac{y\sqrt{2} - 1}{2\sqrt{2} + 1}$

81. $\dfrac{5}{8 - \sqrt{6}}$

82. $\dfrac{3}{7 - \sqrt{2}}$

83. $\dfrac{3}{\sqrt[4]{2x^3}}$

84. $\dfrac{6}{\sqrt[4]{8x}}$

85. $\dfrac{\sqrt[3]{x^2}}{\sqrt[3]{7x^2}}$

86. $\dfrac{\sqrt[3]{6y^4}}{\sqrt[3]{4x^5}}$

87. $\dfrac{2x^2y}{\sqrt[5]{4x^2y^4}}$

88. $\dfrac{3xy^2}{\sqrt[5]{8xy^3}}$

89. A student rationalized the denominator of $\dfrac{\sqrt{6}}{2\sqrt{3} - \sqrt{2}}$ and
obtained $\dfrac{\sqrt{3} + 3\sqrt{2}}{5}$. Find a decimal approximation of each expression. Are the decimals equal? Did the student do the work correctly?

90. A student rationalized the denominator of $\dfrac{\sqrt{5}}{\sqrt{5} + \sqrt{3}}$
and obtained $\dfrac{5 - \sqrt{15}}{2}$. Find a decimal approximation of each expression. Are the decimals equal? Did the student do the work correctly?

*In calculus, students are sometimes required to rationalize the **numerator** of an expression. In this case the numerator will not have a radical in the answer. Rationalize the numerator in each of the following.*

91. $\dfrac{\sqrt{5}}{7}$

92. $\dfrac{\sqrt{6}}{10}$

93. $\dfrac{\sqrt{3} + 2\sqrt{7}}{8}$

94. $\dfrac{\sqrt{5} - 4\sqrt{3}}{6}$

Applications

A triangular piece of lawn is fertilized. The cost is $0.18 per square foot. Find the cost to fertilize the lawn if:

95. The base of the triangle is $\sqrt{21}$ feet and the altitude is $\sqrt{50}$ feet. Round answer to nearest cent.

96. The base of the triangle is $\sqrt{17}$ feet and the altitude is $\sqrt{40}$ feet. Round answer to nearest cent.

Cumulative Review Problems

Solve for x and y.

97. $2x + 3y = 13$
$5x - 2y = 4$

Solve for x, y, and z.

98. $3x - y - z = 5$
$2x + 3y - z = -16$
$x + 2y + 2z = -3$

99. A strong cup of coffee contains about 200 milligrams of caffeine. A cup of strong tea contains 80 milligrams of caffeine. Juanita drinks one cup of each every day. However, she resolved on January 1 to reduce her intake of caffeine to less than 18 milligrams per day. On January 2 she cuts her consumption of both coffee and tea in half. Three days later she again cuts her consumption in half. She continues this pattern of reduction. On what day will she reach her goal?

100. Juanita's husband Carlos has several cups of coffee and tea each day. On January 1, he had 11 cups in total. He consumed a total of 1480 milligrams of caffeine on January 1. Using the information in problem 99 find out how many cups of coffee and how many cups of tea he consumed. If he cuts his consumption of coffee and tea in half on January 2, and continues to cut his consumption of coffee and tea in half every four days, how long will it take him to reach his goal of less than 24 milligrams per day?

7.5 Radical Equations

After studying this section, you will be able to:

1. Solve a radical equation that requires squaring each side once.
2. Solve a radical equation that requires squaring each side twice.

1 Solving a Radical Equation by Squaring Each Side Once

A radical equation is an equation with a variable in one or more of the radicals. $3\sqrt{x} = 8$ and $\sqrt{3x - 1} = 5$ are radical equations. We solve radical equations by raising each side of the equation to the appropriate power. In other words, we square both sides if the radicals are square roots, we cube both sides if the radicals are cube roots, and so on. Once you have done this, solving for the unknown becomes routine.

EXAMPLE 1   Solve. $\sqrt{2x + 6} = 4$

$$(\sqrt{2x + 6})^2 = (4)^2 \qquad \textit{Square each side.}$$
$$2x + 6 = 16 \qquad \textit{Simplify.}$$
$$2x = 10 \qquad \textit{Add } -6 \textit{ to each side.}$$
$$x = 5 \qquad \textit{Divide each side by 2.}$$

Check: If $x = 5$, we substitute 5 for x in the original equation.

$$\sqrt{2(5) + 6} \overset{?}{=} 4$$
$$\sqrt{10 + 6} \overset{?}{=} 4$$
$$\sqrt{16} \overset{?}{=} 4$$
$$4 = 4 \quad \checkmark$$

It checks. Therefore, $x = 5$ is the solution to the radical equation.

Practice Problem 1   Solve and check your solution. $\sqrt{3x + 1} = 5$ ∎

Sometimes after squaring each side a quadratic equation is obtained. In this case we set one side equal to zero and use the zero factor method that we developed in section 5.8. After solving the equation, *always* check your answers to see if extraneous solutions have been introduced.

EXAMPLE 2   Solve. $\sqrt{2x + 9} = x + 3$

$$(\sqrt{2x + 9})^2 = (x + 3)^2 \qquad \textit{Square each side.}$$
$$2x + 9 = x^2 + 6x + 9 \qquad \textit{Simplify.}$$
$$0 = x^2 + 4x \qquad \textit{Collect all terms on one side.}$$
$$0 = x(x + 4) \qquad \textit{Factor.}$$
$$x = 0 \quad or \quad x = -4 \qquad \textit{Solve for x.}$$

Check: $\sqrt{2x + 9} = x + 3$

for $x = 0$: $\sqrt{2(0) + 9} \overset{?}{=} 0 + 3$ for $x = -4$: $\sqrt{2(-4) + 9} \overset{?}{=} -4 + 3$

$\sqrt{9} \overset{?}{=} 3$ $\sqrt{1} \overset{?}{=} -1$

$3 = 3$ ✓ $1 \neq -1$

Therefore, $x = 0$ is the only solution to this equation.

Practice Problem 2   Solve and check your solution(s). $\sqrt{3x - 8} = x - 2$ ∎

> ### GRAPHING CALCULATOR
>
> **Solving Radical Equations**
>
> On a graphing calculator Example 2 can be solved in two ways. Let $y_1 = \sqrt{2x + 9}$ and let $y_2 = x + 3$. Use your graphing calculator to determine where y_1 intersects y_2. What value of x do you obtain? Now let $y = \sqrt{2x + 9} - x - 3$ and find the value of x when $y = 0$. What value of x do you obtain? Which method seems more efficient?
>
> Use the method above that you have found most efficient to solve the following equations and round your answer to the nearest tenth.
>
> $\sqrt{x + 9.5} = x - 2.3$
>
> $\sqrt{6x + 1.3} = 2x - 1.5$

As you begin to solve more complicated radical equations, it is important to make sure that one radical expression is alone on one side of the equation. This is often referred to as *isolating the radical term.*

EXAMPLE 3   Solve. $\sqrt{10x + 5} - 1 = 2x$

$$\sqrt{10x + 5} = 2x + 1 \qquad\qquad \text{Isolate the radical term.}$$

$$(\sqrt{10x + 5})^2 = (2x + 1)^2 \qquad\qquad \text{Square each side.}$$

$$10x + 5 = 4x^2 + 4x + 1 \qquad\qquad \text{Simplify.}$$

$$0 = 4x^2 - 6x - 4 \qquad\qquad \text{Set the equation equal to zero.}$$

$$0 = 2(2x^2 - 3x - 2) \qquad\qquad \text{Remove the common factor.}$$

$$0 = 2(2x + 1)(x - 2)$$

$$2x + 1 = 0 \quad or \quad x - 2 = 0 \qquad\qquad \text{Solve for x.}$$

$$2x = -1 \qquad\qquad x = 2$$

$$x = -\frac{1}{2}$$

Check: $\sqrt{10x + 5} - 1 = 2x$

$x = -\dfrac{1}{2}:$ $\sqrt{10\left(-\dfrac{1}{2}\right) + 5} - 1 \overset{?}{=} 2\left(-\dfrac{1}{2}\right)$ $x = 2:$ $\sqrt{10(2) + 5} - 1 \overset{?}{=} 2(2)$

$\sqrt{-5 + 5} - 1 \overset{?}{=} -1$ $\sqrt{25} - 1 \overset{?}{=} 4$

$\sqrt{0} - 1 \overset{?}{=} -1$ $5 - 1 \overset{?}{=} 4$

$-1 = -1$ ✓ $4 = 4$ ✓

Both answers check, so $x = -\dfrac{1}{2}$ and $x = 2$ are roots of the equation.

Practice Problem 3   Solve and check your solutions. $\sqrt{x + 4} = x + 4$ ∎

2 Solving a Radical Equation by Squaring Each Side Twice

In some problems, we must square each side twice in order to remove all the radicals. It is important to isolate at least one radical before squaring each side.

EXAMPLE 4   Solve. $\sqrt{5x + 1} - \sqrt{3x} = 1$

$$\sqrt{5x + 1} = 1 + \sqrt{3x} \qquad\qquad \text{Isolate one of the radicals.}$$

$$(\sqrt{5x + 1})^2 = (1 + \sqrt{3x})^2 \qquad\qquad \text{Square each side.}$$

$$5x + 1 = (1 + \sqrt{3x})(1 + \sqrt{3x})$$

$$5x + 1 = 1 + 2\sqrt{3x} + 3x$$

$$2x = 2\sqrt{3x} \qquad\qquad \text{Isolate the remaining radical.}$$

$$x = \sqrt{3x} \qquad\qquad \text{Divide each side by 2.}$$

$$(x)^2 = (\sqrt{3x})^2 \qquad \textit{Square each side.}$$

$$x^2 = 3x$$

$$x^2 - 3x = 0 \qquad \textit{Set the equation equal to zero.}$$

$$x(x - 3) = 0$$

$$x = 0 \quad or \quad x - 3 = 0 \qquad \textit{Solve for x.}$$

$$x = 3$$

Check: $\sqrt{5x + 1} - \sqrt{3x} = 1$

$x = 0$: $\sqrt{5(0) + 1} - \sqrt{3(0)} \overset{?}{=} 1$

$\sqrt{1} - \sqrt{0} \overset{?}{=} 1$

$1 = 1$ ✓

$x = 3$: $\sqrt{5(3) + 1} - \sqrt{3(3)} \overset{?}{=} 1$

$\sqrt{16} - \sqrt{9} \overset{?}{=} 1$

$1 = 1$ ✓

Both answers check. The solutions are $x = 0$ and $x = 3$.

Practice Problem 4   Solve and check your solution(s). $\sqrt{2x + 5} - 2\sqrt{2x} = 1$ ∎

EXAMPLE 5   Solve. $\sqrt{2y + 5} - \sqrt{y - 1} = \sqrt{y + 2}$

$$(\sqrt{2y + 5} - \sqrt{y - 1})^2 = (\sqrt{y + 2})^2$$

$$(\sqrt{2y + 5} - \sqrt{y - 1})(\sqrt{2y + 5} - \sqrt{y - 1}) = y + 2$$

$$2y + 5 - 2\sqrt{(y - 1)(2y + 5)} + y - 1 = y + 2$$

$$-2\sqrt{(y - 1)(2y + 5)} = -2y - 2$$

$$\sqrt{(y - 1)(2y + 5)} = y + 1 \qquad \begin{array}{l}\textit{Divide each side by} \\ -2.\end{array}$$

$$(\sqrt{2y^2 + 3y - 5})^2 = (y + 1)^2 \qquad \textit{Square each side.}$$

$$2y^2 + 3y - 5 = y^2 + 2y + 1$$

$$y^2 + y - 6 = 0 \qquad \begin{array}{l}\textit{Set the equation equal} \\ \textit{to zero.}\end{array}$$

$$(y + 3)(y - 2) = 0$$

$$y = -3 \quad or \quad y = 2$$

Check: Verify that $y = 2$ is a valid solution, but $y = -3$ is not a valid solution.

Practice Problem 5   Solve and check your solutions.

$\sqrt{y - 1} + \sqrt{y - 4} = \sqrt{4y - 11}$ ∎

7.5 Exercises

Exercises

Verbal and Writing Skills

1. Before squaring each side of a radical equation what step should be taken first?

2. Why do we have to check the solutions when we solve radical equations?

Solve each radical equation. Check your solution(s).

3. $\sqrt{x-2} = 7$

4. $\sqrt{x-4} = 6$

5. $\sqrt{3x+7} = 4$

6. $\sqrt{2x+5} = 5$

7. $x = \sqrt{6x+7}$

8. $\sqrt{7x+8} = x$

9. $\sqrt{5x+1} = x+1$

10. $\sqrt{2x+1} = x-7$

11. $y - \sqrt{y-3} = 5$

12. $\sqrt{2y-4} + 2 = y$

13. $\sqrt{y+1} - 1 = y$

14. $5 + \sqrt{2y+5} = y$

15. $x - 2\sqrt{x-3} = 3$

16. $2\sqrt{4x+1} + 5 = x+9$

17. $\sqrt{x^2 - 8x} = 3$

18. $\sqrt{x^2 + 36} = 10$

19. $\sqrt[3]{2x+3} = 2$

20. $\sqrt[3]{3x-6} = 3$

21. $2\sqrt[3]{x-1} = \sqrt[3]{x^2 + 2x}$

22. $\sqrt[3]{x^2 + 17x} = 3\sqrt[3]{x}$

23. $\sqrt[3]{1-7x} - 4 = 0$

24. $\sqrt[3]{4x-3} - 5 = 0$

25. $\sqrt{x+4} = 1 + \sqrt{x-3}$

26. $\sqrt{5x+1} = 1 + \sqrt{3x}$

27. $\sqrt{x+6} = 1 + \sqrt{x+2}$

28. $\sqrt{x-1} = 4\sqrt{x+1}$

29. $\sqrt{x-5} = 2 + \sqrt{x+3}$

30. $\sqrt{3x+1} - \sqrt{x-4} = 3$

31. $\sqrt{2x-1} + 2 = \sqrt{3x+10}$

32. $\sqrt{3x+3} + \sqrt{x-1} = 4$

33. $\sqrt{2x+9} - \sqrt{x+1} = 2$

34. $\sqrt{4x+6} = \sqrt{x+1} - \sqrt{x+5}$ **35.** $\sqrt{3x+4} + \sqrt{x+5} = \sqrt{7-2x}$ **36.** $\sqrt{2x+6} = \sqrt{7-2x} + 1$

37. $2\sqrt{x} - \sqrt{x-5} = \sqrt{2x-2}$ **38.** $\sqrt{x+\sqrt{x+2}} = 2$ **39.** $\sqrt{3-2\sqrt{x}} = \sqrt{x}$

40. $\sqrt{3+\sqrt{y}} = 1 + \sqrt{y}$ **41.** $\sqrt{3y-5} = \sqrt{2\sqrt{y+1}}$ ★**42.** $\sqrt[4]{2x^2+3x-8} = -1$

Optional Graphing Calculator Problems

Solve for x. Round your answer to four decimal places.

43. $x = \sqrt{5.326x - 1.983}$

44. $\sqrt[3]{5.62x + 9.93} = 1.47$

Applications

45. In a sudden stop, a car traveling on wet pavement at a speed V will produce skid marks of length S feet according to the formula $V = 2\sqrt{3S}$.
  (a) Solve the equation for S.
  (b) Use your result from (a) to find the length of the skid mark S if the car is traveling at 18 miles per hour.

46. If a sailor is looking from an observation platform H feet above sea level, he can see a distance of D miles, according to the formula $D = \sqrt{\dfrac{3H}{2}}$.
  (a) Solve the equation for H.
  (b) Use the result from (a) to find the height H that the sailor must be above the water line to see a distance of 5 miles away.

To Think About

47. The solution to the equation $\sqrt{x^2 - 4x + c} = x - 1$ is $x = 4$. What is the value of c?

48. The solution to the equation $\sqrt{x+b} - \sqrt{x} = -2$ is $x = 16$. What is the value of b?

Cumulative Review Problems

Simplify.

49. $(4^3 x^6)^{2/3}$ **50.** $(2^{-3} x^{-6})^{1/3}$ **51.** $\sqrt[3]{-216 x^6 y^9}$ **52.** $\sqrt[4]{64 x^{12} y^{16}}$

53. The Mississippi Magic paddle boat can travel 12 mph in still water. After traveling for 3 hours downstream with the current, it takes 5 hours to return to its original starting point going upstream against the current. What is the speed of the current?

54. Louise Elton rides the ski lift for 1.75 miles to the top of Mount Gray. Immediately, she then skis directly down the mountain. The ski trail winding down the mountain is 2.5 miles long. If she skis five times as fast as the lift runs and the round trip takes 45 minutes, find the rate at which she skis.

7.6 Complex Numbers

MathPro   Video 18   SSM

After studying this section, you will be able to:

1. *Simplify expressions involving complex numbers.*
2. *Add and subtract complex numbers.*
3. *Multiply complex numbers.*
4. *Evaluate complex numbers of the form i^n.*
5. *Divide two complex numbers.*

1 Simplifying Expressions Involving Complex Numbers

Until now we have not been able to solve an equation such as $x^2 = -4$ because there is no *real* number that satisfies this equation. However, this equation *does* have a solution, but the solution is not a real number. It is an **imaginary** number.

We define a new number

$$i = \sqrt{-1} \quad \text{or} \quad i^2 = -1$$

Then

$$\sqrt{-4} = \sqrt{4(-1)} = \sqrt{4}\sqrt{-1} = \sqrt{4}i = 2i$$

Thus one solution to the equation $x^2 + 4 = 0$ is $x = 2i$. Let's check it.

$$x^2 + 4 = 0$$
$$(2i)^2 + 4 \stackrel{?}{=} 0$$
$$4i^2 + 4 \stackrel{?}{=} 0$$
$$4(-1) + 4 \stackrel{?}{=} 0$$
$$0 = 0 \quad \checkmark$$

The value $-2i$ is also a solution. You should verify this. A **complex number** is a combination of a real number and an imaginary number.

Now we formalize our definitions and give some examples of imaginary numbers.

Definition of Imaginary Number
The imaginary number i is defined as
$$i = \sqrt{-1} \quad \text{and} \quad i^2 = -1$$

Definition
For all positive real numbers a,
$$\sqrt{-a} = \sqrt{-1}\sqrt{a} = i\sqrt{a}$$

EXAMPLE 1 Simplify. **(a)** $\sqrt{-36}$ **(b)** $\sqrt{-17}$

(a) $\sqrt{-36} = \sqrt{-1}\sqrt{36} = (i)(6) = 6i$
(b) $\sqrt{-17} = \sqrt{-1}\sqrt{17} = i\sqrt{17}$

Practice Problem 1 Simplify. **(a)** $\sqrt{-49}$ **(b)** $\sqrt{-31}$ ∎

To avoid confusing $\sqrt{17}i$ with $\sqrt{17i}$, we write the i before the radical. That is, we write $i\sqrt{17}$.

EXAMPLE 2 Simplify. **(a)** $\sqrt{-12}$ **(b)** $\sqrt{-45}$

(a) $\sqrt{-12} = \sqrt{-1}\sqrt{12} = i\sqrt{12} = i\sqrt{4}\sqrt{3} = 2i\sqrt{3}$

(b) $\sqrt{-45} = \sqrt{-1}\sqrt{45} = i\sqrt{45} = i\sqrt{9}\sqrt{5} = 3i\sqrt{5}$

Practice Problem 2 Simplify. **(a)** $\sqrt{-98}$ **(b)** $\sqrt{-150}$ ∎

The rule $\sqrt{a}\sqrt{b} = \sqrt{ab}$ requires that $a \geq 0$ and $b \geq 0$. Therefore, we cannot use our product rule when the radicands are negative unless we first use the definition of $\sqrt{-1}$. Recall that

$$\sqrt{-1} \cdot \sqrt{-1} = i \cdot i = i^2 = -1$$

EXAMPLE 3 Multiply. $\sqrt{-16} \cdot \sqrt{-25}$

First we must use the definition $\sqrt{-1} = i$. Thus

$$(\sqrt{-16})(\sqrt{-25}) = (i\sqrt{16})(i\sqrt{25})$$
$$= i^2(4)(5)$$
$$= -1(20) \qquad i^2 = -1.$$
$$= -20$$

Practice Problem 3 Change to the i notation and multiply. $\sqrt{-8} \cdot \sqrt{-2}$ ∎

Now we formally define a complex number.

> **Definition**
>
> A number that can be placed in the form $a + bi$, where a and b are real numbers, is a **complex number.**

Under this definition every real number is also a complex number. For example, the real number 5 can be written as $5 + 0i$. Therefore, 5 is a complex number. In similar fashion, the imaginary number $2i$ can be written as $0 + 2i$. Thus $2i$ is a complex number. Thus the set of complex numbers includes the set of real numbers and the set of imaginary numbers.

> **Definition**
>
> Two complex numbers $a + bi$ and $c + di$ are equal if and only if $a = c$ and $b = d$.

This definition means that two complex numbers are equal only if their real parts are equal *and* their imaginary parts are equal.

EXAMPLE 4 Find real numbers x and y if $x + 3i\sqrt{7} = -2 + yi$.

By our definition, the real parts must be equal, so x must be -2; the imaginary parts must also be equal, so y must be $3\sqrt{7}$.

Practice Problem 4 Find real numbers x and y if $-7 + 2yi\sqrt{3} = x + 6i\sqrt{3}$. ∎

 GRAPHING CALCULATOR

Complex Numbers

Some graphing calculators, such as the TI-83, have a complex number mode. If your graphing calculator has this capability, you will be able to use it to do complex number operations. First you must use the Mode command to transfer selection from "Real" to "Complex" or "a + bi." To verify your status, try to find $\sqrt{-7}$ on your graphing calculator. If you obtain an approximate answer of "2.645751311 i" then your calculator is operating in the complex number mode. If you obtain "ERROR: NONREAL ANSWER" then your calculator is not operating in the complex number mode.

2 Adding or Subtracting Complex Numbers

> ### Adding and Subtracting Complex Numbers
> For all real numbers a, b, c, and d,
> $$(a + bi) + (c + di) = (a + c) + (b + d)i$$
> $$(a + bi) - (c + di) = (a - c) + (b - d)i$$

In other words, we add (or subtract) the real parts, and we add (or subtract) the imaginary parts.

EXAMPLE 5 (a) Add. $(7 - 3i) + (-4 - 2i)$ (b) Subtract. $(6 - 2i) - (3 - 5i)$

(a) $(7 - 3i) + (-4 - 2i) = (7 - 4) + (-3 - 2)i = 3 - 5i$

(b) $(6 - 2i) - (3 - 5i) = (6 - 2i) + (-3 + 5i) = (6 - 3) + (-2 + 5)i = 3 + 3i$

Practice Problem 5

(a) Add. $(-6 - 2i) + (5 - 8i)$ (b) Subtract. $(3 - 4i) - (-2 - 18i)$ ∎

3 Multiplying Complex Numbers

The procedure for multiplying complex numbers is similar to the procedure for multiplying polynomials. We might expect this to be true. We will see that the complex numbers obey the associative, commutative, and distributive properties.

EXAMPLE 6 Multiply. $(7 - 6i)(2 + 3i)$

Use FOIL.

$$(7 - 6i)(2 + 3i) = (7)(2) + (7)(3i) + (-6i)(2) + (-6i)(3i)$$
$$= 14 + 21i - 12i - 18i^2$$
$$= 14 + 21i - 12i - 18(-1)$$
$$= 14 + 21i - 12i + 18$$
$$= 32 + 9i$$

Practice Problem 6 Multiply. $(4 - 2i)(3 - 7i)$ ∎

EXAMPLE 7 Multiply. $3i(4 - 5i)$

Use the distributive property.

$$3i(4 - 5i) = (3)(4)i + (3)(-5)i^2$$
$$= 12i - 15i^2$$
$$= 12i - 15(-1)$$
$$= 15 + 12i$$

Practice Problem 7 Multiply. $-2i(5 + 6i)$ ∎

It is important to rewrite any square roots with negative radicands using the i notation for complex numbers before attempting to do any multiplication.

EXAMPLE 8 Multiply and simplify your answer.

(a) $\sqrt{-25} \cdot \sqrt{-36}$ **(b)** $2i(3 + 4i + \sqrt{-2})$

(a) Before any other steps are taken we must first rewrite the expression using i notation.
$$\sqrt{-25} \cdot \sqrt{-36} = (i\sqrt{25})(i\sqrt{36})$$
Then we can finish the calculations using the properties of complex numbers.
$$= (5i)(6i)$$
$$= 30i^2$$
$$= 30(-1)$$
$$= -30$$

(b) Before any other steps are taken we must first write $\sqrt{-2}$ as $i\sqrt{2}$. Thus we have
$$2i(3 + 4i + \sqrt{-2}) = 2i(3 + 4i + i\sqrt{2})$$
$$= 6i + 8i^2 + 2i^2\sqrt{2}$$
$$= 6i + 8(-1) + 2(-1)\sqrt{2}$$
$$= -8 - 2\sqrt{2} + 6i$$

Notice the order we use to write the final answer:
integer + irrational number + imaginary number.

Practice Problem 8 Multiply. Then simplify your answer.

(a) $\sqrt{-50} \cdot \sqrt{-4}$ **(b)** $-3i(2 - 3i + \sqrt{-4})$ ∎

4 Evaluating Complex Numbers of the Form i^n

How would you evaluate i^n, where n is any positive integer? Where n is any negative integer? We look for a pattern. We have defined
$$i^2 = -1$$
We could write
$$i^3 = i^2 \cdot i = (-1)i = -i$$
and, similarly,
$$i^4 = i^2 \cdot i^2 = (-1)(-1) = +1$$
$$i^5 = i^4 \cdot i = (+1)i = +i$$
We notice that $i^5 = i$. Let's look at i^6.
$$i^6 = i^4 \cdot i^2 = (+1)(-1) = -1$$
We begin to see a pattern that starts with i and repeats itself for i^5. Will $i^7 = -i$? Why or why not?

Values	$i = i$	$i^5 = i$	$i^9 = i$
of i^n	$i^2 = -1$	$i^6 = -1$	$i^{10} = -1$
	$i^3 = -i$	$i^7 = -i$	$i^{11} = -i$
	$i^4 = +1$	$i^8 = +1$	$i^{12} = +1$

We can use this pattern to evaluate powers of i.

EXAMPLE 9 Evaluate. **(a)** i^{36} **(b)** i^{27} **(c)** i^{38}

(a) $i^{36} = (i^4)^9 = (1)^9 = 1$

(b) $i^{27} = (i^{24+3}) = (i^{24})(i^3) = (i^4)^6(i^3) = 1^6(-i) = -i$

This suggests a quick method for evaluating powers of i. Divide the exponent by 4. i^4 raised to any power will be 1. Then use the first column of the values of i^n chart on page 421 to evaluate the remainder.

(c) $i^{38} = (i^{36+2}) = 1(i^2) = 1(-1) = -1$

Practice Problem 9 Evaluate. **(a)** i^{42} **(b)** i^{53} **(c)** i^{103} ∎

▨ **GRAPHING
CALCULATOR**

Complex Operations

When doing complex number operations on a graphing calculator the answer will usually be displayed as an approximate value in decimal form. Try Example 10 on your graphing calculator by entering $(7 + i) \div (3 - 2i)$. You should obtain an approximate answer of: $1.461538462 + 1.307692308\, i$.

⑤ Dividing Two Complex Numbers

The complex numbers $a + bi$ and $a - bi$ are called **conjugates.** The product of two complex conjugates is always a real number.

$$(a + bi)(a - bi) = a^2 - abi + abi + b^2i^2$$
$$= a^2 - b^2(-1)$$
$$= a^2 + b^2$$

To divide two complex numbers, we want to remove any expression involving i from the denominator. So when dividing two complex numbers, we multiply the numerator and denominator by the conjugate of the denominator. This is just what we did when we rationalized the denominator in a radical expression.

EXAMPLE 10 Divide. $\dfrac{7 + i}{3 - 2i}$

$$\frac{(7 + i)}{(3 - 2i)} \cdot \frac{(3 + 2i)}{(3 + 2i)} = \frac{21 + 14i + 3i + 2i^2}{9 - 4i^2} = \frac{21 + 17i + 2(-1)}{9 - 4(-1)} = \frac{21 + 17i - 2}{9 + 4}$$

$$= \frac{19 + 17i}{13} \quad \text{or} \quad \frac{19}{13} + \frac{17}{13}i$$

Practice Problem 10 Divide. $\dfrac{4 + 2i}{3 + 4i}$ ∎

EXAMPLE 11 Divide. $\dfrac{3 - 2i}{4i}$

The conjugate of $0 + 4i$ is $0 - 4i$ or simply $-4i$.

$$\frac{(3 - 2i)}{(4i)} \cdot \frac{(-4i)}{(-4i)} = \frac{-12i + 8i^2}{-16i^2} = \frac{-12i + 8(-1)}{-16(-1)}$$

$$= \frac{-8 - 12i}{16} = \frac{\cancel{4}(-2 - 3i)}{\cancel{4} \cdot 4}$$

$$= \frac{-2 - 3i}{4} \quad \text{or} \quad -\frac{1}{2} - \frac{3}{4}i$$

Practice Problem 11 Divide. $\dfrac{5 - 6i}{-2i}$ ∎

7.6 Exercises

Verbal and Writing Skills

1. Does $x^2 = -9$ have a real number solution? Why or why not?

2. Describe a complex number and give an example(s).

3. Are the complex numbers $2 + 3i$ and $3 + 2i$ equal? Why or why not?

4. Describe in your own words how to add or subtract complex numbers.

Simplify. Express in terms of i.

5. $\sqrt{-49}$ **6.** $\sqrt{-100}$ **7.** $\sqrt{-19}$ **8.** $\sqrt{-30}$

9. $\sqrt{-50}$ **10.** $\sqrt{-48}$ **11.** $\sqrt{-63}$ **12.** $\sqrt{-52}$

13. $-\sqrt{-81}$ **14.** $-\sqrt{-36}$ **15.** $2 + \sqrt{-3}$ **16.** $5 + \sqrt{-7}$

17. $-3 + \sqrt{-24}$ **18.** $-6 - \sqrt{-32}$ **19.** $\sqrt{-9} + \sqrt{-16}$ **20.** $\sqrt{-25} + \sqrt{-4}$

Find the real numbers x and y.

21. $x - 3i = 5 + yi$ **22.** $x - 6i = 7 + yi$ **23.** $-8 - 2yi = x - 4i$

24. $x - 5i\sqrt{2} = 3 + yi$ **25.** $23 + yi = 17 - x + 3i$ **26.** $2 + x - 11i = 19 + yi$

27. $5 + 3i - x + 2yi = 8 - 4i$ **28.** $7 - 4i + x + 4yi = -3 + 7i$

Perform the addition or subtraction.

29. $(7 - 8i) + (-12 - 4i)$ **30.** $(-3 + 5i) + (23 - 7i)$ **31.** $\left(-\dfrac{3}{2} + \dfrac{1}{2}i\right) + \left(\dfrac{5}{2} - \dfrac{3}{2}i\right)$

32. $\left(\dfrac{3}{4} - \dfrac{3}{4}i\right) + \left(\dfrac{9}{4} + \dfrac{5}{4}i\right)$ **33.** $(12 - 3i) - (5 + 3i)$ **34.** $(20 + 5i) - (6 - 3i)$

35. $(1.3 - 0.6i) - (0.7 - 0.3i)$ **36.** $(-2.1 + 1.3i) - (-1.8 + 3.5i)$

Multiply and simplify your answer. Place in i notation before doing any other operations.

37. $(2 + 3i)(2 - i)$ **38.** $(4 - 6i)(2 + i)$ **39.** $\left(-\dfrac{2}{3} + \dfrac{1}{2}i\right)\left(\dfrac{3}{2} - \dfrac{1}{2}i\right)$

40. $\left(3 + \dfrac{3}{2}i\right)\left(\dfrac{8}{3} - 4i\right)$ **41.** $4 - 3(-10 + 2i)$ **42.** $6 - 5(7 - 6i)$

43. $2i(5i - 6)$

44. $4i(7 - 2i)$

45. $(6 - 3i)^2$

46. $(5 - 4i)^2$

47. $(6 + 7i)^2$

48. $(8 + 2i)^2$

49. $(i\sqrt{3})(i\sqrt{7})$

50. $(i\sqrt{2})(i\sqrt{6})$

51. $(\sqrt{-3})(\sqrt{-2})$

52. $(\sqrt{-5})(\sqrt{-3})$

53. $(\sqrt{-36})(\sqrt{-4})$

54. $(\sqrt{-25})(\sqrt{-9})$

55. $3i(4 + \sqrt{-3})$

56. $4i(5 + \sqrt{-2})$

57. $-5i(5 - 6i + \sqrt{-7})$

★ **58.** $(3 + \sqrt{-2})(4 + \sqrt{-5})$

★ **59.** $(2 + \sqrt{-3})(6 + \sqrt{-2})$

★ **60.** $(7 - \sqrt{-3})(2 + \sqrt{-1})$

Evaluate.

61. i^{17}

62. i^{21}

63. i^{24}

64. i^{16}

65. i^{46}

66. i^{83}

67. $i^{30} + i^{28}$

68. $i^{26} + i^{24}$

Divide.

69. $\dfrac{2 + i}{3 - i}$

70. $\dfrac{4 + 2i}{2 - i}$

71. $\dfrac{3i}{4 + 2i}$

72. $\dfrac{-2i}{3 + 5i}$

73. $\dfrac{8 - 6i}{3i}$

74. $\dfrac{5 + 2i}{4i}$

75. $\dfrac{7}{5 - 6i}$

76. $\dfrac{3}{4 + 2i}$

Optional Graphing Calculator Problems

Perform each operation to obtain an approximate answer.

77. $(29.3 + 56.2i)^2$

78. $(0.34 - 0.72i)(1.93 - 2.52i)$

79. $\dfrac{196 - 34.8i}{24.9 + 56.4i}$

80. $\dfrac{361 + \sqrt{-256}}{422 - \sqrt{-315}}$

Applications

The impedance Z in an alternating current circuit (like the one used in your home and in your classroom) is given by the formula $Z = V/I$, where V is the voltage and I is the current.

81. Find the value of Z if $V = 3 + 2i$ and $I = 3i$.

82. Find the value of Z if $V = 4 + 2i$ and $I = -3i$.

7.7 Variation

After studying this section, you will be able to:

1. Solve problems requiring the use of direct variation.
2. Solve problems requiring the use of inverse variation.
3. Solve problems requiring the use of joint or combined variation.

MathPro   Video 18   SSM

1 Solving Problems Using Direct Variation

Many times in daily life we observe how a change in one quantity produces a change in another. If we order one large pepperoni pizza, we pay $8.95. If we order two of the same pizza, we pay $17.90. For three of the same pizza, it is $26.85. The change in the number of pizzas we order results in a corresponding change in the price we pay.

Notice that the price we pay for each pizza stays the same. That is, each pizza costs $8.95. The number of pizzas changes and the corresponding price for the order changes. From our experience with functions and with equations, we see that $y = \$8.95x$, where the price, y, depends on the number of pizzas, x. We see that the variable y is a constant multiple of x. The two variables are said to vary directly. That is, y varies directly with x. We can write a general equation that represents this idea. If y varies directly with x, then $y = kx$, where k represents a constant.

When we solve problems using direct variation, we are usually not given the value of the constant of variation, k. This is something that we must find. Usually all we are given is a point of reference. That is, we are given the value of y for a specific value of x. Using this information, we can find k.

EXAMPLE 1   If y varies directly with x and $y = 25$ when $x = 15$, find y when $x = 20$.

We first find the constant by substituting in the known values of $x = 15$ and $y = 25$.

$$y = kx$$

$$25 = k(15)$$

$$\frac{25}{15} = \frac{15k}{15} \qquad \textit{Divide each side by } 15.$$

$$\frac{5}{3} = k$$

Next we use the equation with the k value known.

$$y = \frac{5}{3}x$$

We want to find y when $x = 20$.

$$y = \frac{5}{3}(20)$$

$$y = \frac{100}{3}$$

Practice Problem 1   If y varies directly with x and $y = 9$ when $x = \frac{1}{2}$, find the value of y when $x = \frac{2}{3}$. ∎

EXAMPLE 2   The time of a pendulum's swing varies directly with the square root of its length. If the pendulum is 1 foot long when the time is 0.2 second, find the time if the length is 4 feet.

Let t = the time and L = the length.
We then have the equation

$$t = k\sqrt{L}$$

We can evaluate k by substituting $L = 1$ and $t = 0.2$ into the equation.

$$t = k\sqrt{L}$$
$$0.2 = k(\sqrt{1})$$
$$0.2 = k \qquad \textit{Because } \sqrt{1} = 1.$$

Now we know the value of k and can write the equation more completely.

$$t = 0.2\sqrt{L}$$

When $L = 4$, we have

$$t = 0.2\sqrt{4}$$
$$t = (0.2)(2)$$
$$t = 0.4 \text{ second}$$

Practice Problem 2   In some racing cars the maximum speed varies directly with the square root of the horsepower of the engine. If the maximum speed of a car with 256 horsepower is 128 miles per hour, what is the maximum speed of a car with 225 horse-power? ■

2 Solving Problems Involving Inverse Variation

In some cases when one variable increases, another variable decreases. For example, as the amount of money you earn each year increases, the percentage of your income that you get to keep after taxes decreases. If one variable is a constant multiple of the reciprocal of the other, the two variables are said to vary inversely. If y varies inversely with x, we can express this by the equation $y = \dfrac{k}{x}$, where k is the constant of variation.

EXAMPLE 3   If y varies inversely with x and $y = 12$ when $x = 5$, find y when $x = 14$.

If y varies inversely with x, we can write the equation $y = \dfrac{k}{x}$. We can find the value of k by substituting the values $y = 12$ and $x = 5$.

$$12 = \frac{k}{5}$$
$$60 = k$$

We can now write the equation

$$y = \frac{60}{x}$$

To find the value of y when $x = 14$, we substitute 14 for x in the equation.

$$y = \frac{60}{14}$$
$$y = \frac{30}{7}$$

Practice Problem 3   If y varies inversely with x and $y = 45$ when $x = 16$, find the value of y when $x = 36$. ■

EXAMPLE 4 The amount of light from a light source varies inversely with the square of the distance to the light source. If an object receives 6.25 lumens when the light source is 8 meters away, how much light will it receive if it is 4 meters away?

Let L = the amount of light and d = the distance to the light source.

Since the amount of light varies inversely with the *square of the distance* to the light source, we have

$$L = \frac{k}{d^2}$$

Substituting the known values of $L = 6.25$ when $d = 8$, we can find the value of k.

$$6.25 = \frac{k}{8^2}$$

$$6.25 = \frac{k}{64}$$

$$400 = k$$

We are now able to write a more specific equation,

$$L = \frac{400}{d^2}$$

We will use this to find L when $d = 4$ meters.

$$L = \frac{400}{4^2}$$

$$L = \frac{400}{16}$$

$$L = 25 \text{ lumens}$$

Check: Does this answer seem reasonable? Would we expect to have more light if we move closer to the light source? ✓

Practice Problem 4 If the amount of power in an electrical circuit is held constant, the resistance in the circuit varies inversely with the square of the amount of current. If the amount of current is 0.01 ampere, the resistance is 800 ohms. What is the resistance if the amount of current is 0.04 ampere? ∎

3 Solving Problems Involving Joint or Combined Variation

Sometimes a quantity depends on the variation of two or more variables.

EXAMPLE 5 y varies directly with x and z and inversely with d^2. When $x = 7$, $z = 3$, and $d = 4$, the value of y is 20. Find the value of y when $x = 5$, $z = 6$, and $d = 2$.

We can write the equation

$$y = \frac{kxz}{d^2}$$

To find the value of k, we substitute into the equation $y = 20$, $x = 7$, $z = 3$, and $d = 4$.

$$20 = \frac{k(7)(3)}{4^2}$$

$$20 = \frac{21k}{16}$$

$$320 = 21k$$

$$\frac{320}{21} = k$$

Now we substitute $\dfrac{320}{21}$ for k in our original equation.

$$y = \frac{\dfrac{320}{21}xz}{d^2} \quad \text{or} \quad y = \frac{320xz}{21d^2}$$

We use this equation to find y for the known values of x, z, and d. We want to find y when $x = 5$, $z = 6$, and $d = 2$.

$$y = \frac{320(5)(6)}{21(2)^2}$$

$$y = \frac{\overset{80}{\cancel{320}}(5)\overset{2}{\cancel{(6)}}}{\underset{7}{\cancel{21}}\underset{1}{(\cancel{4})}}$$

$$y = \frac{800}{7}$$

Practice Problem 5 y varies directly with z and w^2 and inversely with x. $y = 20$ when $z = 3$, $w = 5$, and $x = 4$. Find y when $z = 4$, $w = 6$, and $x = 2$. ∎

Many applied problems involve joint variation. For example, a cylindrical cement column has a safe load that varies directly with the diameter raised to the fourth power and inversely with the square of its length.

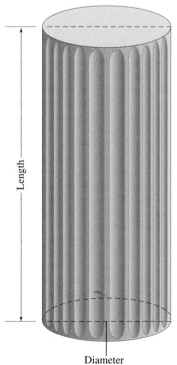

Length

Diameter

7.7 Exercises

Verbal and Writing Skills

1. Give an example in everyday life of direct variation and write an equation as a mathematical model.

2. The general equation $y = kx$ means that y varies _____ with x. k is called the _____ of variation.

3. If y varies inversely with x, we write the equation _____.

4. Write a mathematical model for the following situation: The strength of a rectangular beam varies jointly with its width and the square of its depth.

Round all answers to nearest tenth.

5. If y varies directly with x and $y = 8$ when $x = 20$, find y when $x = 4$.

6. If y varies directly with x and $y = 30$ when $x = 2$, find y when $x = 5$.

7. If y varies directly with x^2 and $y = 4.5$ when $x = 3$, find y when $x = 5$.

8. If y varies directly with x^3 and $y = 32$ when $x = 4$, find y when $x = 7$.

9. The marine biology submarine was searching the waters for blue whales at 50 feet below the surface, where it experiences a pressure of 21 pounds per square inch (psi). If the pressure of water on a submerged object varies directly with the distance beneath the surface, how much pressure would the submarine experience if it had to dive to 170 feet?

10. The distance to stop a car varies directly with the square of its speed. A car that is traveling 30 miles per hour can stop in 40 feet. What distance will it take to stop if it is traveling 60 miles per hour?

11. When an object is dropped, the distance it falls in feet varies directly with the square of the time of the drop in seconds. If a box falls 4 feet in 0.5 second, how far will it fall in 6 seconds?

12. A veterinarian specializing in marine biology is faced with operating on a dolphin. A cube-shaped aquarium has been built, which will allow the doctor to operate while keeping most of the dolphin's body submerged. The time it takes to fill the tank with water varies directly with the cube of the side of the container. If a cube 2 meters on each side can be filled in 7 minutes, how long will it take to fill the dolphin tank, which is 3.5 meters on each side?

13. If y varies inversely with x and $y = 20$ when $x = 1.5$, find y when $x = 25$.

14. If y varies inversely with x and $y = 20$ when $x = 25$, find y when $x = 16$.

15. If y varies inversely with the square of x and $y = 30$ when $x = 6$, find y when $x = 8$.

16. If y varies inversely with the cube of x and $y = 20$ when $x = 4$, find y when $x = 3$.

17. Engineers have decided that part of the structure housing certain elements of a satellite will use 9 inches of special insulation. You know that the heat loss through a certain type of insulation varies inversely as the thickness of the insulation. If the heat loss through 6 inches of the insulation is 2000 Btu/hour, how much heat will you lose with the 9 inches on the satellite?

18. The weight of an object on Earth's surface varies inversely with the square of its distance from the center of Earth. An object weighs 1000 pounds on Earth's surface. This is approximately 4000 miles from the center of Earth. How much would an object weigh 4500 miles from the center of Earth?

19. Police officers can detect speeding by using variation. The speed of a car varies inversely with the time to cover a certain fixed distance. Between two certain points in the highway, a car travels 45 miles per hour in 6 seconds. What is the speed of a car that travels the same distance in 9 seconds?

20. If the voltage in an electric circuit is kept at the same level, the current varies inversely with the resistance. The current measures 40 amperes when the resistance is 270 ohms. Find the current when the resistance is 100 ohms.

21. y varies jointly with w, z, and h. If $y = 50$ when $w = 3$, $z = 4$, and $h = 6$, find y when $w = 2$, $z = 3$, and $h = 5$. Do your calculations to the nearest tenth.

22. y varies jointly with w and z. If $y = 100$ when $w = \frac{1}{2}$ and $z = 38$, find y when $w = 4$ and $z = 6$. Do your calculations to the nearest tenth.

23. y varies directly with x and z and inversely with w^2. If $y = 18$ when $x = 2.5$, and $z = 6$, and $w = 2$, find y when $x = 6.5$, $z = 3$, and $w = 6$.

24. y varies directly with x and z and inversely with the square root of w. If $y = 10$ when $x = 6$, $z = 3$, and $w = 4$, find y when $x = 8$, $z = 10$, and $w = 9$.

25. Atmospheric drag tends to slow down moving objects. Atmospheric drag varies jointly with an object's surface area A and velocity v. If a Dodge Intrepid, traveling at a speed of 45 mph, with a surface area of 37.8 square feet, experiences a drag of 222 newtons, how fast must a Dodge Caravan, with a surface area of 55 square feet, travel in order to experience a drag force of 450 newtons?

26. The attraction F of two masses m_1 and m_2 varies directly with the product of m_1 and m_2 and inversely with the square of the distance between the two bodies. If a force of 10 pounds attracts two bodies weighing 80 tons and 100 tons that are 100 miles apart, how great will be the force if the two bodies weighed 8 tons and 15 tons and were 20 miles apart?

27. The field intensity of a magnetic field varies directly with the force acting on it and inversely with the strength of the pole. If the intensity of the magnetic field is 4 oersteds when the force is 700 dynes and the strength of the pole is 200, find the intensity of the field if the force is 500 dynes and the strength of the pole is 250.

28. The force on a blade of a wind generator varies jointly with the product of the blade's area and the square of the wind velocity. The force of the wind is 20 pounds when the area is 3 square feet and the velocity is 30 feet per second. Find the force when the area is increased to 5 square feet and the velocity is reduced to 25 feet per second.

Cumulative Review Problems

Solve each of the following equations or word problems.

29. $x^2 - 15x + 56 = 0$

30. $2x^2 + 5x - 3 = 0$

31. $3x^2 - 8x + 4 = 0$

32. $4x^2 = -28x + 32$

33. In Champaign, Illinois, the sales tax is 6.25%. Donny bought an amplifier for his stereo that cost $488.75 after tax. What was the original price of the amplifier?

34. It takes 7.5 gallons of white paint to properly paint lines on 3 tennis courts. How much paint is needed to paint 22 tennis courts?

Topic	Procedure	Examples
Multiplication of variables with rational exponents, p. 385.	$x^m x^n = x^{m+n}$	$(3x^{1/5})(-2x^{3/5}) = -6x^{4/5}$
Division of variables with rational exponents, p. 385.	$\dfrac{x^m}{x^n} = x^{m-n}, \qquad n \neq 0$	$\dfrac{-16x^{3/20}}{24x^{5/20}} = -\dfrac{2x^{-1/10}}{3}$
Removing negative exponents, p. 385.	$x^{-n} = \dfrac{1}{x^n}, \qquad m, n \neq 0$ $\dfrac{x^{-n}}{y^{-m}} = \dfrac{y^m}{x^n}$	Write with positive exponents: $3x^{-4} = \dfrac{3}{x^4}$ $\dfrac{2x^{-6}}{5y^{-8}} = \dfrac{2y^8}{5x^6}$ $4^{-2} = \dfrac{1}{4^2} = \dfrac{1}{16}$
Zero exponent, p. 385.	$x^0 = 1$ (if $x \neq 0$)	$(3x^{1/2})^0 = 1$
Raising a variable with an exponent to a power, p. 385.	$(x^m)^n = x^{mn}$ $(xy)^n = x^n y^n$ $\left(\dfrac{x}{y}\right)^n = \dfrac{x^n}{y^n}, y \neq 0$	$(x^{-1/2})^{-2/3} = x^{1/3}$ $(3x^{-2}y^{-1/2})^{2/3} = 3^{2/3}x^{-4/3}y^{-1/3}$ $\left(\dfrac{4x^{-2}}{3^{-1}y^{-1/2}}\right)^{1/4} = \dfrac{4^{1/4}x^{-1/2}}{3^{-1/4}y^{-1/8}}$
Multiplication of expressions with rational exponents, p. 387.	Add exponents whenever variables with same base are multiplied.	$x^{2/3}(x^{1/3} - x^{1/4}) = x^{3/3} - x^{2/3+1/4} = x - x^{11/12}$
Higher-order roots, p. 392.	If x is a nonnegative real number, $\sqrt[n]{x}$ is a positive nth root and has the property that $(\sqrt[n]{x})^n = x$ If x is a negative real number, $(\sqrt[n]{x})^n = x$ when n is an odd integer. If x is a negative real number, $(\sqrt[n]{x})^n$ is not a real number when n is an even integer.	$\sqrt[3]{27} = 3$ because $3^3 = 27$. $\sqrt[5]{-32} = -2$ because $(-2)^5 = -32$. $\sqrt[4]{-16}$ is *not* a real number.
Rational exponents and radicals, p. 393.	For positive integers m and n and x being any real number for which $x^{1/n}$ is defined, $x^{m/n} = (\sqrt[n]{x})^m = \sqrt[n]{x^m}$ $x^{1/n} = \sqrt[n]{x}$	Write as a radical. $x^{3/7} = \sqrt[7]{x^3}$, $3^{1/5} = \sqrt[5]{3}$ Write as an expression with a fractional exponent. $\sqrt[4]{w^3} = w^{3/4}$ Evaluate. $\quad 25^{3/2} = (\sqrt{25})^3 = (5)^3 = 125$
Higher-order roots and absolute value, p. 395.	$\sqrt[n]{x^n} = \lvert x \rvert$ when n is an even positive integer. $\sqrt[n]{x^n} = x$ when n is an odd positive integer.	$\sqrt[6]{x^6} = \lvert x \rvert$ $\sqrt[5]{x^5} = x$
Evaluation of higher-order roots, p. 395.	Use exponent notation.	$\sqrt[5]{-32x^{15}} = \sqrt[5]{(-2)^5 x^{15}}$ $= [(-2)^5 x^{15}]^{1/5} = (-2)^1 x^3 = -2x^3$

Topic	Procedure	Examples
Simplification of radicals, p. 398.	Use product rule: $$\sqrt[n]{a}\,\sqrt[n]{b} = \sqrt[n]{ab}$$	Simplify when $x > 0$, $y > 0$. $$\sqrt{75x^3} = \sqrt{25x^2}\sqrt{3x}$$ $$= 5x\sqrt{3x}$$ $$\sqrt[3]{16x^5y^6} = \sqrt[3]{8x^3y^6}\sqrt[3]{2x^2}$$ $$= 2xy^2\sqrt[3]{2x^2}$$
Combining radicals, p. 399.	Simplify radicals and combine them if they have the same index and the same radicand.	Combine. $$2\sqrt{50} - 3\sqrt{98} = 2\sqrt{25}\sqrt{2} - 3\sqrt{49}\sqrt{2}$$ $$= 2(5)\sqrt{2} - 3(7)\sqrt{2}$$ $$= 10\sqrt{2} - 21\sqrt{2} = -11\sqrt{2}$$
Multiplying radicals, p. 403.	1. Multiply coefficients outside radical and then multiply radicands. 2. Simplify answer.	$$(2\sqrt{3})(4\sqrt{5}) = 8\sqrt{15}$$ $$2\sqrt{6}(\sqrt{2} - 3\sqrt{12}) = 2\sqrt{12} - 6\sqrt{72}$$ $$= 2\sqrt{4}\sqrt{3} - 6\sqrt{36}\sqrt{2}$$ $$= 4\sqrt{3} - 36\sqrt{2}$$ $(\sqrt{2} + \sqrt{3})(2\sqrt{2} - \sqrt{3})$ Use FOIL $$= 2\sqrt{4} - \sqrt{6} + 2\sqrt{6} - \sqrt{9}$$ $$= 4 + \sqrt{6} - 3$$ $$= 1 + \sqrt{6}$$
Simplifying quotients of radicals, p. 405.	$$\sqrt[n]{\frac{a}{b}} = \frac{\sqrt[n]{a}}{\sqrt[n]{b}}$$	$$\sqrt[3]{\frac{5}{27}} = \frac{\sqrt[3]{5}}{\sqrt[3]{27}} = \frac{\sqrt[3]{5}}{3}$$
Rationalizing denominators, p. 406.	Multiply numerator and denominator by a value that eliminates the radical in the denominator.	$$\frac{2}{\sqrt{7}} \cdot \frac{\sqrt{7}}{\sqrt{7}} = \frac{2\sqrt{7}}{7}$$ $$\frac{3}{\sqrt{5} + \sqrt{2}} \cdot \frac{\sqrt{5} - \sqrt{2}}{\sqrt{5} - \sqrt{2}} = \frac{3\sqrt{5} - 3\sqrt{2}}{(\sqrt{5})^2 - (\sqrt{2})^2}$$ $$= \frac{3\sqrt{5} - 3\sqrt{2}}{5 - 2}$$ $$= \frac{3\sqrt{5} - 3\sqrt{2}}{3}$$ $$= \sqrt{5} - \sqrt{2}$$
Solving radical equations, p. 413.	1. Perform algebraic operations to obtain one radical by itself on one side of the equation. 2. If the equation contains square roots, square each side of the equation. Otherwise, raise each side to the appropriate power for third- and higher-order roots. 3. Simplify, if possible. 4. If the equation still contains a radical, repeat steps 1 to 3. 5. Solve the resulting equation. 6. Check all apparent solutions! Solutions to radical equations must be verified.	Solve. $$x = \sqrt{2x + 9} - 3$$ $$x + 3 = \sqrt{2x + 9}$$ $$(x + 3)^2 = (\sqrt{2x + 9})^2$$ $$x^2 + 6x + 9 = 2x + 9$$ $$x^2 + 6x - 2x + 9 - 9 = 0$$ $$x^2 + 4x = 0$$ $$x(x + 4) = 0$$ $x = 0$ or $x = -4$ Check: $x = 0$: $\quad 0 \overset{?}{=} \sqrt{2(0) + 9} - 3$ $\quad\quad\quad 0 \overset{?}{=} \sqrt{9} - 3$ $\quad\quad\quad 0 = 3 - 3 \;\checkmark$ $x = -4$: $\quad -4 \overset{?}{=} \sqrt{2(-4) + 9} - 3$ $\quad\quad\quad\quad -4 \overset{?}{=} \sqrt{1} - 3$ $\quad\quad\quad\quad -4 \neq -2$ $x = 0$ is the only solution.

Topic	Procedure	Examples
Simplifying imaginary numbers, p. 418.	Use $i = \sqrt{-1}$ and $i^2 = -1$.	$\sqrt{-16} = \sqrt{-1}\sqrt{16} = 4i$ $\sqrt{-18} = \sqrt{-1}\sqrt{18} = i\sqrt{9}\sqrt{2} = 3i\sqrt{2}$
Adding and subtracting complex numbers, p. 420.	Combine real coefficients and imaginary coefficients separately.	$(5 + 6i) + (2 - 4i) = 7 + 2i$ $(-8 + 3i) - (4 - 2i) = -8 + 3i - 4 + 2i$ $= -12 + 5i$
Multiplying complex numbers, p. 420.	Use the FOIL method and $i^2 = -1$.	$(5 - 6i)(2 - 4i) = 10 - 20i - 12i + 24i^2$ $= 10 - 32i + 24(-1)$ $= 10 - 32i - 24$ $= -14 - 32i$
Dividing complex numbers, p. 422.	Multiply numerator and denominator by conjugate of denominator.	$\dfrac{5 + 2i}{4 - i} = \dfrac{5 + 2i}{4 - i} \cdot \dfrac{4 + i}{4 + i} = \dfrac{20 + 5i + 8i + 2i^2}{16 - i^2}$ $= \dfrac{20 + 13i + 2(-1)}{16 - (-1)}$ $= \dfrac{20 + 13i - 2}{16 + 1}$ $= \dfrac{18 + 13i}{17}$ or $\dfrac{18}{17} + \dfrac{13}{17}i$
Raising i to a power, p. 421.	$i^1 = i$ $i^2 = -1$ $i^3 = -i$ $i^4 = 1$	Evaluate. $i^{27} = i^{24} \cdot i^3$ $= (i^4)^6 \cdot i^3$ $= (1)^6(-i)$ $= -i$
Direct variation, p. 425.	If y varies directly with x, there is a constant of variation, k, such that $y = kx$. Once k is determined, other values of y or x can easily be computed.	y varies directly with x. When $x = 2$, $y = 7$. $y = kx$ $7 = k(2)$ *Substitute.* $k = \dfrac{7}{2}$ *Solve.* $y = \dfrac{7}{2}x$ What is y when $x = 18$? $y = \dfrac{7}{2}x = \dfrac{7}{2} \cdot 18 = 63$
Inverse variation, p. 426.	If y varies inversely with x, the constant k is such that $$y = \dfrac{k}{x}$$	y varies inversely with x. When x is 5, y is 12. What is y when x is 30? $y = \dfrac{k}{x}$ $12 = \dfrac{k}{5}$ *Substitute.* $k = 60$ *Solve.* $y = \dfrac{60}{x}$ *Substitute.* When $x = 30$, $y = \dfrac{60}{30} = 2$.

Putting Your Skills to Work

DETERMINING THE RATE OF DEPRECIATION

The rate of annual depreciation for an automobile varies quite a bit depending on the brand name of the car. However, automobiles that are considered to be of very high quality tend to depreciate each year at approximately the same rate. To determine the annual depreciation rate for these types of cars the formula

$$R = 1 - \left(\frac{F}{P}\right)^{1/n}$$

where R is the annual depreciation rate
   F is the final price that the car can be sold for after n years
   P is the purchase price of the car when it is new
   and n is the number of years the car was owned where $n \geq 1$

PROBLEMS FOR INDIVIDUAL INVESTIGATION AND ANALYSIS

1. Nancy originally purchased her Honda Accord for $16,500. If Nancy kept the car for 6 years and sold it for $7500, what was the annual depreciation rate?

2. Phil originally purchased his Honda Civic for $12,000. If he kept the car for 7 years and sold it for $5500, what was the annual depreciation rate?

PROBLEMS FOR GROUP INVESTIGATION AND COOPERATIVE GROUP LEARNING

Together with some members of your class see if you can determine the following.

3. Dave and Elsie Bagley purchased an Oldsmobile Ninety-Eight for $20,000. They kept the car for 5 years and sold it for $12,000. **(a)** What was the annual depreciation rate? **(b)** If they had kept the car for 8 years and sold it for $9000 would the rate of depreciation be more or less?

4. Draw a bar graph that shows the value of a Dodge Caravan that was purchased for $18,000 and depreciates at the rate of 11% per year. Construct the bar graph to cover 5 years from the date of purchase.

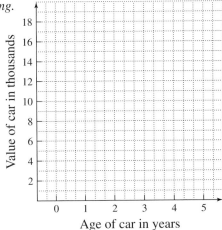

INTERNET CONNECTION: Go to `http://www.prenhall.com/tobey` to be connected.

Site: Used Car Pricing Report (Consumers Car Club)

This site provides pricing information for used cars. When you enter a year and model of car, you will learn the original list price of the car as well as current wholesale and retail prices.

5. For the car model of your choice, calculate the depreciation rates for a 5-year-old car, an 8-year-old car, and an 11-year-old car.

6. Choose several 6-year-old car models, and calculate the depreciation rate for each. Compare your results. See if you can explain any similarities or differences in the rates of depreciation you calculated for various cars.

In all problems assume that the variables represent positive real numbers, unless otherwise stated. Simplify using only positive exponents in your answer.

1. $(3xy^{1/2})(5x^2y^{-3})$

2. $(-7x^{1/3}y^{1/3})(4xy^{2/3})$

3. $\dfrac{8x^6}{-2x^{5/2}}$

4. $\dfrac{27x^{5/3}}{3x^{-1/3}}$

5. $(8x^2y^{-3})^{2/3}$

6. $(25a^3b^4)^{1/2}$

7. $5^{1/4} \cdot 5^{1/2}$

8. $(2a^{1/3}b^{1/4})(-3a^{1/2}b^{1/2})$

9. $\dfrac{6x^{2/3}y^{1/10}}{12x^{1/6}y^{-1/5}}$

10. $(2x^{-1/5}y^{1/10}z^{4/5})^{-5}$

11. $\left(\dfrac{49a^3b^6}{a^{-7}b^4}\right)^{1/2}$

12. $a^{1/5}(2a^{3/4} - 3a^{1/10})$

13. $\dfrac{(x^{3/4}y^{2/5})^{1/2}}{x^{-1/8}}$

14. $\left(\dfrac{27x^{5n}}{x^{2n-3}}\right)^{1/3}$

15. $(6^{5/3})^{3/7}$

16. Combine as one fraction containing only positive exponents. $2x^{1/3} + x^{-2/3}$

17. Remove a common factor of $3x$ from $6x^{3/2} - 9x^{1/2}$.

In Problems 18 through 29, assume that all variables are positive real numbers.

18. Write in exponential form. $\sqrt{\sqrt[5]{2x}}$

19. Write in radical form. $(2x + 3y)^{4/9}$

20. Evaluate. $\sqrt[3]{125} + \sqrt[4]{81}$

21. Explain the difference between $-\sqrt[6]{64}$ and $\sqrt[6]{-64}$.

Evaluate or simplify each expression.

22. $27^{-4/3}$

23. $\left(\dfrac{4}{9}\right)^{3/2}$

24. $\sqrt{99x^3y^6z^{10}}$

25. $\sqrt[3]{-56a^8b^{10}c^{12}}$

26. $\sqrt[4]{16x^8y^3z^{11}}$

27. $\sqrt[5]{x^7y^{10}z^{23}}$

28. $\sqrt{144x^{10}y^{12}z^0}$

29. $\sqrt[3]{125a^9b^6c^{300}}$

*In 30–35, assume that x and y can be either **positive** or **negative** real numbers. Simplify the following.*

30. $\sqrt[3]{y^3}$

31. $\sqrt{y^2}$

32. $\sqrt[4]{x^4y^4}$

33. $\sqrt[5]{x^{10}}$

34. $\sqrt[3]{x^{21}}$

35. $\sqrt{x^8}$

Combine where possible.

36. $\sqrt{50} + 2\sqrt{32} - \sqrt{8}$

37. $\sqrt{28} - 4\sqrt{7} + 5\sqrt{63}$

38. $\sqrt[3]{8} + 3\sqrt[3]{16} - 4\sqrt[3]{54}$

39. $\sqrt[3]{2y^4} + 3y\sqrt[3]{16y}$

40. $2\sqrt{32x} - 5x\sqrt{2} + \sqrt{18x} + 2\sqrt{8x^2}$

41. $2\sqrt[3]{5x^3y} + 6x\sqrt{3y} - 5x\sqrt[3]{40y} - \sqrt{75x^2y}$

Multiply and simplify.

42. $(5\sqrt{12})(3\sqrt{6})$

43. $3\sqrt{x}(2\sqrt{8x} - 3\sqrt{48})$

44. $(5\sqrt{2} + \sqrt{3})(\sqrt{2} - 2\sqrt{3})$

45. $(5\sqrt{6} - 2\sqrt{2})(\sqrt{6} - \sqrt{2})$

46. $(2\sqrt{5} - 3\sqrt{6})^2$

47. $(\sqrt[3]{2x} + \sqrt[3]{6})(\sqrt[3]{4x^2} - \sqrt[3]{y})$

Using a calculator or a square root table, approximate each of the following to the nearest thousandth.

48. $\sqrt{140}$

49. $\sqrt{270}$

50. $4 + \sqrt{31}$

Rationalize the denominator and simplify the expression.

51. $\sqrt{\dfrac{3x^2}{y}}$

52. $\dfrac{2}{\sqrt{3y}}$

53. $\dfrac{3\sqrt{7x}}{\sqrt{21x}}$

54. $\dfrac{2}{\sqrt{6} - \sqrt{5}}$

55. $\dfrac{\sqrt{x}}{3\sqrt{x} + \sqrt{y}}$

56. $\dfrac{\sqrt{5}}{\sqrt{7} - 3}$

57. $\dfrac{2\sqrt{3} + \sqrt{6}}{\sqrt{3} + 2\sqrt{6}}$

58. $\dfrac{5\sqrt{2} - \sqrt{3}}{\sqrt{6} - \sqrt{3}}$

59. $\dfrac{3\sqrt{x} + \sqrt{y}}{\sqrt{x} - \sqrt{y}}$

60. $\dfrac{2xy}{\sqrt[3]{16xy^5}}$

61. $\dfrac{5\sqrt{3}}{3\sqrt{2}}$

62. $\sqrt{\dfrac{7}{5}}$

63. Simplify. $\sqrt{-16} + \sqrt{-45}$

64. Find x and y. $2x - 3i + 5 = yi - 2 + \sqrt{6}$

Simplify by performing the operation indicated.

65. $(-12 - 6i) + (3 - 5i)$

66. $(2 - i) - (12 - 3i)$

67. $(7 + 3i)(2 - 5i)$

68. $(8 - 4i)^2$

69. $2i(3 + 4i)$

70. $3 - 4(2 + i)$

71. Evaluate. i^{34}

72. Evaluate. i^{57}

Divide.

73. $\dfrac{7 - 2i}{3 + 4i}$

74. $\dfrac{5 - 2i}{1 - 3i}$

75. $\dfrac{4 - 3i}{5i}$

76. $\dfrac{12}{3 - 5i}$

77. $\dfrac{10 - 4i}{2 + 5i}$

Solve and check your solution(s).

78. $\sqrt{3x + 4} = 5$

79. $\sqrt{9 - 4x} = 1$

80. $2\sqrt{6x + 1} = 10$

81. $\sqrt[3]{3x - 1} = \sqrt[3]{5x + 1}$

82. $\sqrt{2x + 1} = 2x - 5$

83. $1 + \sqrt{3x + 1} = x$

84. $\sqrt{3x + 1} - \sqrt{2x - 1} = 1$

85. $\sqrt{7x + 2} = \sqrt{x + 3} + \sqrt{2x - 1}$

86. If y varies directly with x and $y = 16$ when $x = 5$, find the value of y when $x = 3$.

87. If y varies directly with x and $y = 5$ when $x = 20$, find the value of y when $x = 50$.

88. The distance it takes to stop a car varies directly with the square of its speed. A car traveling on wet pavement can stop in 50 feet when traveling at 30 miles per hour. How long will it take the car to stop if it is traveling at 55 miles per hour?

89. The time it takes a falling object to drop a given distance varies directly with the square root of the distance traveled. A steel ball takes 2 seconds to drop a distance of 64 feet. How many seconds will it take to drop a distance of 196 feet?

90. If y varies inversely with x and $y = 8$ when $x = 3$, find the value of y when $x = 48$.

91. The volume of a gas varies inversely with the pressure of the gas on its container. If a pressure of 24 pounds per square inch corresponds to a volume of 70 cubic inches, what pressure is needed to correspond to a volume of 100 cubic inches?

92. Suppose that y varies directly with x and inversely with the square of z. When $x = 8$ and $z = 4$, then $y = 1$. Find y when $x = 6$ and $z = 3$.

93. The capacity of a cylinder varies directly with the height and the square of the radius. A cylinder with a radius of 3 centimeters and a height of 5 centimeters has a capacity of 50 cubic centimeters. What is the capacity of a cylinder with a height of 9 centimeters and a radius of 4 centimeters?

Simplify.

1. $(2x^{1/2}y^{1/3})(-3x^{1/3}y^{1/6})$

2. $\dfrac{7x^3}{4x^{3/4}}$

3. $(8x^{1/3})^{3/2}$

4. $6^{1/5} \cdot 6^{3/5}$

Evaluate.

5. $8^{-2/3}$

6. $64^{3/2}$

Simplify. Assume that all variables are positive.

7. $\sqrt[3]{250x^4y^6}$

8. $\sqrt{64x^6y^5}$

9. $\sqrt{75a^4b^9}$

Combine like terms where possible.

10. $3\sqrt{48} - \sqrt[3]{54x^5} + 2\sqrt{27} + 2x\sqrt[3]{16x^2}$

11. $\sqrt{32} - 3\sqrt{8} + 2\sqrt{72}$

Multiply and simplify.

12. $2\sqrt{3}(3\sqrt{6} - 5\sqrt{2})$

13. $(5\sqrt{3} - \sqrt{6})(2\sqrt{3} + 3\sqrt{6})$

Rationalize the denominator.

14. $\dfrac{8}{\sqrt{20x}}$

15. $\sqrt{\dfrac{5}{x}}$

16. $\dfrac{5 + 2\sqrt{3}}{4 - \sqrt{3}}$

1. _____

2. _____

3. _____

4. _____

5. _____

6. _____

7. _____

8. _____

9. _____

10. _____

11. _____

12. _____

13. _____

14. _____

15. _____

16. _____

Solve and check your solution(s).

17. $\sqrt{3x - 2} = x$

18. $5 + \sqrt{x + 15} = x$

19. $5 - \sqrt{x - 2} = \sqrt{x + 3}$

Simplify by using the properties of complex numbers.

20. $(8 + 2i) - 3(2 - 4i)$

21. $i^{18} + \sqrt{-16}$

22. $(3 - 2i)(4 + 3i)$

23. $\dfrac{2 + 5i}{1 - 3i}$

24. $(6 + 3i)^2$

25. i^{35}

26. If y varies inversely with x and $y = 9$ when $x = 2$, find the value of y when $x = 6$.

27. Suppose y varies directly with x and inversely with the square of z. When $x = 8$ and $z = 4$, then $y = 3$. Find y when $x = 5$ and $z = 6$.

28. The distance to stop a car varies directly with the square of its speed. A car traveling on pavement can stop in 30 feet when traveling at 30 miles per hour. How long will it take to stop the car if it is traveling at 50 miles per hour?

17. _____

18. _____

19. _____

20. _____

21. _____

22. _____

23. _____

24. _____

25. _____

26. _____

27. _____

28. _____

Approximately one-half of this test covers the content of Chapters 1–6. The remainder covers the content of Chapter 7.

1. Identify what property of real numbers is illustrated by the equation $7 + (2 + 3) = (7 + 2) + 3$.

2. Remove parentheses and collect like terms. $2a(3a^3 - 4) - 3a^2(a - 5)$

3. Perform in the proper order.
$7(12 - 14)^3 - 7 + 3 \div (-3)$

4. Solve for x. $y = \dfrac{2}{3}x - 8$

5. Graph. $3x - 5y = 15$

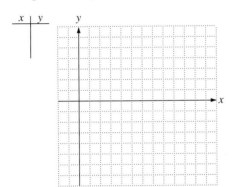

6. Factor completely. $16x^2 + 24x - 16$

7. Solve for x, y, and z.
$$x + 4y - z = 10$$
$$3x + 2y + z = 4$$
$$2x - 3y + 2z = -7$$

8. Combine. $\dfrac{7x}{x^2 - 2x - 15} - \dfrac{2}{x - 5}$

9. The length of a rectangle is 3 meters longer than twice its width. The perimeter of the rectangle is 48 meters. Find the dimensions of the rectangle.

10. Solve for b. $56x + 2 = 8b + 4x$

1. _____

2. _____

3. _____

4. _____

5. _____

6. _____

7. _____

8. _____

9. _____

10. _____

Simplify.

11. $\dfrac{2x^{-3}y^{-4}}{4x^{-5/2}y^{7/2}}$

12. $(3x^{-1/2}y^2)^{-1/3}$

13. Evaluate. $64^{-1/3}$

14. Simplify. $\sqrt[3]{40x^5y^9}$

15. Combine like terms.
$\sqrt{80x} + 2\sqrt{45x} - 3\sqrt{20x}$

16. Multiply and simplify.
$(2\sqrt{3} - 5\sqrt{2})(\sqrt{3} + 4\sqrt{2})$

17. Rationalize the denominator.
$\dfrac{\sqrt{3} + 2}{2\sqrt{3} - 5}$

18. Simplify. $i^{21} + \sqrt{-16} + \sqrt{-49}$

19. Simplify. $(2 + 3i)^2$

20. Simplify. $\dfrac{1 + 4i}{1 + 3i}$

Solve for x and check your solutions.

21. $x - 3 = \sqrt{3x + 1}$

22. $1 + \sqrt{x + 1} = \sqrt{x + 2}$

23. If y varies directly with the square of x and $y = 12$ when $x = 2$, find the value of y if $x = 5$.

24. The amount of light provided by a light bulb varies inversely with the square of the distance from the light bulb. A light bulb provides 120 lumens at a distance of 10 feet from the light. How many lumens are provided if the distance to the light is 15 feet?

11.	
12.	
13.	
14.	
15.	
16.	
17.	
18.	
19.	
20.	
21.	
22.	
23.	
24.	

Quadratic Equations and Inequalities

Have you ever wondered how engineers design highways to maximize traffic flow? Can you imagine some of the mathematics involved if you wanted to put the entire highway underground in the middle of a city like Boston? If this idea interests you please turn to the Putting Your Skills to Work on page 479.

If you are familiar with the topics in this chapter, take this test now. Check your answers with those in the back of the book. If an answer was wrong or you couldn't do a problem, study the appropriate section of the chapter.

If you are not familiar with the topics in this chapter, don't take this test now. Instead, study the examples, work the practice problems, and then take the test.

This test will help you to identify those concepts that you have mastered and those that need more study.

1. _____

2. _____

3. _____

4. _____

5. _____

6. _____

7. _____

8. _____

9. _____

10. _____

Section 8.1

1. Solve by the square root property. $3x^2 + 5 = 41$

2. Solve by completing the square. $2x^2 - 4x - 3 = 0$

Section 8.2

Solve by the quadratic formula.

3. $4x^2 - 4x - 7 = 0$

4. $5x(x + 1) = 1 + 6x$

5. Solve for the *nonreal* complex roots of the equation. $4x^2 = -12x - 17$

Sections 8.1 and 8.2

Solve by any method.

6. $12x^2 + x - 6 = 0$

7. $3x^2 + 5x = 7x^2 - 2x$

8. $\dfrac{18}{x} + \dfrac{12}{x + 1} = 9$

Section 8.3

Solve for any real roots and check your answers.

9. $x^6 - 7x^3 - 8 = 0$

10. $w^{4/3} - 6w^{2/3} + 8 = 0$

Section 8.4

11. Solve for x. Assume that w is a positive constant.

$$3x^2 + 2wx + 8w = 0$$

12. The area of a rectangle is 52 square meters. The length of the rectangle is 1 meter longer than three times its width. Find the length and width of the rectangle.

Section 8.5

13. Find the vertex and the intercepts of the quadratic function $f(x) = 3x^2 + 6x - 9$.

14. Draw a graph of the quadratic function $g(x) = -x^2 + 6x - 5$. Label vertex and intercepts.

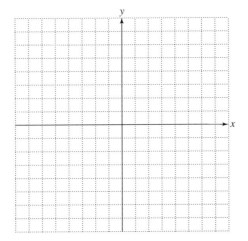

Section 8.6

Solve each quadratic inequality and graph your solution.

15. $x^2 - x - 6 > 0$

16. $2x^2 + 9x \leq -9$

17. Use a square root table or a calculator to approximate to the nearest tenth the solution for the quadratic inequality.

$$3x^2 + 2x - 6 < 0$$

11.

12.

13.

14.

15.

16.

17.

8.1 Quadratic Equations

After studying this section, you will be able to:

1 *Solve quadratic equations by the square root property.*
2 *Solve quadratic equations by completing the square.*

1 *Solving Quadratic Equations by the Square Root Property*

You recall that an equation that can be written in the form $ax^2 + bx + c = 0$, where a, b, and c are real numbers and $a \neq 0$, is called a **quadratic equation.** We have previously solved quadratic equations using the zero-factor property. This has allowed us to solve $x^2 - 7x + 12 = 0$ by factoring it as $(x - 3)(x - 4) = 0$ and solving to find $x = 3$ and $x = 4$. In this chapter we will develop new methods of solving quadratic equations.

The first method is often called the square root property.

> ### The Square Root Property
>
> If $x^2 = a$, then $x = \pm\sqrt{a}$ for all real numbers a.

The notation $\pm\sqrt{a}$ is a shorthand way of writing "$+\sqrt{a}$ or $-\sqrt{a}$." The symbol $\pm$ is read "plus or minus." We can justify this property by using the zero-factor property. If we write $x^2 = a$ in the form $x^2 - a = 0$, we can factor it to obtain $(x + \sqrt{a})(x - \sqrt{a}) = 0$ and thus $x = -\sqrt{a}$ or $x = +\sqrt{a}$. This can be written more compactly as $x = \pm\sqrt{a}$.

EXAMPLE 1   Solve and check. $x^2 = 15$

If $x^2 = 15$, then $x = \pm\sqrt{15}$. The roots are $\sqrt{15}$ and $-\sqrt{15}$.
Check: $(\sqrt{15})^2 \stackrel{?}{=} 15$ $(-\sqrt{15})^2 \stackrel{?}{=} 15$

$15 = 15$ ✓ $15 = 15$ ✓

Practice Problem 1   Solve and check. $x^2 = 29$ ∎

EXAMPLE 2   Solve and check. $x^2 - 36 = 0$

If we add 36 to each side, we have $x^2 = 36$.

$$x = \pm\sqrt{36}$$

$$x = \pm 6$$

Thus the two roots are 6 and -6.
    Check: $6^2 = 36$ and $(-6)^2 = 36$

Practice Problem 2   Solve and check. $x^2 - 121 = 0$ ∎

EXAMPLE 3   Solve. $x^2 = 48$

If $x^2 = 48$, then

$$x = \pm\sqrt{48} = \pm\sqrt{16 \cdot 3}$$

$$x = \pm 4\sqrt{3}$$

The roots are $4\sqrt{3}$ and $-4\sqrt{3}$.

Practice Problem 3   Solve. $x^2 = 18$ ∎

EXAMPLE 4 Solve and check. $3x^2 + 2 = 77$

$$3x^2 = 75$$
$$x^2 = 25$$
$$x = \pm\sqrt{25}$$
$$x = \pm 5$$

The roots are 5 and -5.

Check: $\qquad 3(5)^2 + 2 \overset{?}{=} 77 \qquad\qquad\qquad 3(-5)^2 + 2 \overset{?}{=} 77$

$\qquad\qquad\quad 3(25) + 2 \overset{?}{=} 77 \qquad\qquad\qquad 3(25) + 2 \overset{?}{=} 77$

$\qquad\qquad\quad\; 75 + 2 \overset{?}{=} 77 \qquad\qquad\qquad\quad 75 + 2 \overset{?}{=} 77$

$\qquad\qquad\qquad\;\; 77 = 77 \; \checkmark \qquad\qquad\qquad\qquad 77 = 77 \; \checkmark$

Practice Problem 4 Solve. $5x^2 + 1 = 46$ ∎

Sometimes we obtain roots that are complex numbers.

EXAMPLE 5 Solve and check. $4x^2 = -16$

$$x^2 = -4$$
$$x = \pm\sqrt{-4}$$
$$x = \pm 2i \qquad \textit{Simplify using } \sqrt{-1} = i.$$

The roots are $2i$ and $-2i$.

Check: $\qquad 4(2i)^2 \overset{?}{=} -16 \qquad\qquad\qquad 4(-2i)^2 \overset{?}{=} -16$

$\qquad\qquad\quad 4(4i^2) \overset{?}{=} -16 \qquad\qquad\qquad 4(4i^2) \overset{?}{=} -16$

$\qquad\qquad\quad 4(-4) \overset{?}{=} -16 \qquad\qquad\qquad 4(-4) \overset{?}{=} -16$

$\qquad\qquad\quad -16 = -16 \; \checkmark \qquad\qquad\quad -16 = -16 \; \checkmark$

Practice Problem 5 Solve and check. $3x^2 = -27$ ∎

EXAMPLE 6 Solve. $(4x - 1)^2 = 5$

$$4x - 1 = \pm\sqrt{5}$$
$$4x = 1 \pm \sqrt{5}$$
$$x = \frac{1 \pm \sqrt{5}}{4}$$

The roots are $x = \dfrac{1 + \sqrt{5}}{4}$ and $x = \dfrac{1 - \sqrt{5}}{4}$.

Practice Problem 6 Solve. $(2x + 3)^2 = 7$ ∎

2 *Solving Quadratic Equations by Completing the Square*

Often, a quadratic equation cannot be factored (or it may be difficult to factor). So we use another method of solving the equation, called **completing the square.** When we complete the square, we are changing the polynomial to a perfect square trinomial. The form of the equation then becomes $(x + d)^2 = e$.

Now we already know that

$$(x + d)^2 = x^2 + 2dx + d^2$$

Notice three things about the quadratic equation on the right-hand side.

1. The coefficient of the quadratic term (x^2) is 1.
2. The coefficient of the linear (x) terms is $2d$.
3. The constant term (d^2) is the square of *half* the coefficient of the linear term.

For example, in the trinomial $x^2 + 6x + 9$, the coefficient of the linear term is 6 and the constant term is $\left(\frac{6}{2}\right)^2 = (3)^2 = 9$.

For the trinomial $x^2 - 10x + 25$, the coefficient of the linear term is -10 and the constant term is $\left(\frac{-10}{2}\right)^2 = (-5)^2 = 25$.

What number n makes the trinomial $x^2 + 12x + n$ a perfect square?

$$n = \left(\frac{12}{2}\right)^2 = 6^2 = 36$$

Hence the trinomial $x^2 + 12x + 36$ is a perfect square trinomial and can be written as $(x + 6)^2$.

Now let's solve some equations.

EXAMPLE 7   Solve and check. $x^2 + 6x + 1 = 0$

Step 1   First we put the equation in the form $ax^2 + bx = c$ by adding -1 to each side of the equation. Thus we will obtain

$$x^2 + 6x = -1$$

Step 2   We want to complete the square of $x^2 + 6x$. That is, we want to add a constant term to $x^2 + 6x$ so that we get a perfect square trinomial. We do this by taking half the coefficient of x and squaring it.

$$\left(\frac{6}{2}\right)^2 = 3^2 = 9$$

Adding 9 to $x^2 + 6x$ gives the perfect square trinomial $x^2 + 6x + 9$, which we factor to $(x + 3)^2$. *But* we cannot just add 9 to the left side of our equation unless we also add 9 to the right side. (Why?) We now have

$$x^2 + 6x + 9 = -1 + 9$$

Step 3   Now we factor.

$$(x + 3)^2 = 8$$

Step 4   We now use the square root property.

$$(x + 3) = \pm\sqrt{8}$$
$$x + 3 = \pm 2\sqrt{2}$$

Step 5   Next we solve for x by adding -3 to each side of the equation.

$$x = -3 \pm 2\sqrt{2}$$

The roots are $x = -3 + 2\sqrt{2}$ and $-3 - 2\sqrt{2}$.

Step 6 We *must* check our solution in the *original* equation (not the perfect square trinomial we constructed).

$$x^2 + 6x + 1 = 0 \qquad\qquad\qquad x^2 + 6x + 1 = 0$$

$$(-3 + 2\sqrt{2})^2 + 6(-3 + 2\sqrt{2}) + 1 \stackrel{?}{=} 0 \qquad (-3 - 2\sqrt{2})^2 + 6(-3 - 2\sqrt{2}) + 1 \stackrel{?}{=} 0$$

$$9 - 12\sqrt{2} + 8 - 18 + 12\sqrt{2} + 1 \stackrel{?}{=} 0 \qquad 9 + 12\sqrt{2} + 8 - 18 - 12\sqrt{2} + 1 \stackrel{?}{=} 0$$

$$18 - 18 - 12\sqrt{2} + 12\sqrt{2} \stackrel{?}{=} 0 \qquad\qquad 18 - 18 + 12\sqrt{2} - 12\sqrt{2} \stackrel{?}{=} 0$$

$$0 = 0 \ \checkmark \qquad\qquad\qquad\qquad 0 = 0 \ \checkmark$$

Practice Problem 7 Solve by completing the square. $x^2 + 8x + 3 = 0$ ∎

Let us summarize for future reference the six steps we have performed in order to solve a quadratic equation by completing the square.

Completing the Square

1. Put the equation in the form $ax^2 + bx = c$. If $a \neq 1$, divide each term by a.
2. Square $\dfrac{b}{2}$ and add $\left(\dfrac{b}{2}\right)^2$ to both sides of the equation.
3. Factor the left side (a perfect square trinomial).
4. Use the square root property.
5. Solve the equations.
6. Check solutions in the *original* equation.

EXAMPLE 8 Complete the square to solve. $x^2 - 5x - 24 = 0$

First we add 24 to each side of the equation to obtain the form $ax^2 + bx = c$. Next we must determine what is the number $\left(\frac{b}{2}\right)^2$.

$$x^2 - 5x + \underline{\ \ ?\ \ } = 24 + \underline{\ \ ?\ \ }$$

One-half of −5 squared is
$$\left[\frac{1}{2}(-5)\right]^2 = \left[-\frac{5}{2}\right]^2 = \frac{25}{4}.$$

$$x^2 - 5x + \frac{25}{4} = 24 + \frac{25}{4}$$

Add $\dfrac{25}{4}$ to each side.

$$\left(x - \frac{5}{2}\right)^2 = \frac{96}{4} + \frac{25}{4}$$

Factor the left side.

$$\left(x - \frac{5}{2}\right)^2 = \frac{121}{4}$$

Simplify.

$$\left(x - \frac{5}{2}\right) = \pm\sqrt{\frac{121}{4}}$$

Take the square root of each side.

$$x - \frac{5}{2} = \pm\frac{11}{2}$$

Simplify.

$$x = \frac{5}{2} \pm \frac{11}{2} = \frac{5 \pm 11}{2}$$

So $x = \dfrac{5 + 11}{2} = 8$ and $x = \dfrac{5 - 11}{2} = -3$. Can you verify that these are roots of the given equation?

Practice Problem 8 Solve by completing the square. $x^2 + 3x - 18 = 0$ ∎

EXAMPLE 9   Solve. $3x^2 - 8x + 1 = 0$

$$3x^2 - 8x = -1 \qquad \textit{Add } -1 \textit{ to each side.}$$

$$\frac{3x^2}{3} - \frac{8x}{3} = -\frac{1}{3} \qquad \textit{Divide each term by 3. (Remember that the coefficient of the quadratic term must be 1.)}$$

$$x^2 - \frac{8}{3}x + \frac{16}{9} = -\frac{1}{3} + \frac{16}{9}$$

$$\left(x - \frac{4}{3}\right)^2 = \frac{13}{9}$$

$$x - \frac{4}{3} = \pm\sqrt{\frac{13}{9}}$$

$$x - \frac{4}{3} = \pm\frac{\sqrt{13}}{3}$$

$$x = \frac{4}{3} \pm \frac{\sqrt{13}}{3}$$

$$x = \frac{4 \pm \sqrt{13}}{3}$$

Check: For $x = \dfrac{4 + \sqrt{13}}{3}$

$$3\left(\frac{4 + \sqrt{13}}{3}\right)^2 - 8\left(\frac{4 + \sqrt{13}}{3}\right) + 1 \overset{?}{=} 0$$

$$\frac{16 + 8\sqrt{13} + 13}{3} - \frac{32 + 8\sqrt{13}}{3} + 1 \overset{?}{=} 0$$

$$\frac{16 + 8\sqrt{13} + 13 - 32 - 8\sqrt{13}}{3} + 1 \overset{?}{=} 0$$

$$\frac{29 - 32}{3} + 1 \overset{?}{=} 0$$

$$-\frac{3}{3} + 1 \overset{?}{=} 0$$

$$-1 + 1 = 0 \ \checkmark$$

See if you can check the solution $x = \dfrac{4 - \sqrt{13}}{3}$.

Practice Problem 9   Solve by completing the square. $2x^2 + 4x + 1 = 0$ ■

8.1 Exercises

Solve the equations by using the square root property. Express any complex numbers using i notation.

1. $x^2 = 144$ **2.** $x^2 = 81$ **3.** $x^2 - 20 = 0$ **4.** $x^2 - 50 = 0$ **5.** $2x^2 - 26 = 0$

6. $3x^2 + 1 = 28$ **7.** $2x^2 - 10 = 0$ **8.** $4x^2 + 3 = 43$ **9.** $\dfrac{x^2}{2} - 4 = 14$ **10.** $\dfrac{x^2}{3} + 1 = 4$

11. $x^2 = -25$ **12.** $x^2 = -36$ **13.** $6x^2 + 4 = 4x^2$ **14.** $2x^2 + 24 = 0$ **15.** $(x - 3)^2 = 12$

16. $(x + 2)^2 = 18$ **17.** $(2x + 1)^2 = 7$ **18.** $(3x + 2)^2 = 5$ **19.** $(4x - 3)^2 = 36$ **20.** $(5x - 2)^2 = 25$

21. $(3x - 4)^2 = 8$ **22.** $(2x + 7)^2 = 27$ **23.** $(x + 1/2)^2 = 7/4$ **24.** $(x - 1/3)^2 = 5/9$

Solve the equations by completing the square. Simplify your answer. Express any complex numbers using i notation.

25. $x^2 - 4x = 11$ **26.** $x^2 - 2x = 5$ **27.** $x^2 + 10x + 5 = 0$ **28.** $x^2 + 6x + 2 = 0$

29. $x^2 - 8x = 17$ **30.** $x^2 - 12x = 4$ **31.** $x^2 - 6x + 4 = 0$ **32.** $x^2 + 20x + 10 = 0$

33. $\dfrac{x^2}{2} + \dfrac{5}{2}x = 2$ **34.** $\dfrac{x^2}{3} - \dfrac{x}{3} = 3$ **35.** $2y^2 + 10y = -11$ **36.** $6y^2 - 6y = 3$

37. $4x^2 - 4x - 3 = 0$ **38.** $2x^2 - 7x + 4 = 0$ **39.** $2y^2 - y = 6$ **40.** $2y^2 - y = 15$

41. $x^2 + 1 = x$ **42.** $2x^2 + 2 = 3x$ **43.** $3x^2 + 8x + 3 = 2$ **44.** $4x^2 + 7x + 2 = 3$

45. $3x^2 - 8x + 7 = 0$ **46.** $7x^2 - 5x + 2 = 0$

To Think About

47. Check the solution $x = -1 + \sqrt{6}$ in the equation $x^2 + 2x - 5 = 0$.

48. Check the solution $x = 2 + \sqrt{3}$ in the equation $x^2 - 4x + 1 = 0$.

Applications

The time a basketball player spends in the air when shooting a basket is called the hang time. The vertical leap L measured in feet is related to the hang time t measured in seconds by the equation $L = 4t^2$.

49. If Larry Bird during his career as a Boston Celtics player often displayed a leap of 3.1 feet, find his hang time for that leap.

50. Shaquile O'Neal of the Los Angeles Lakers has often shown a vertical leap of 3.3 feet. Find his hang time for that leap.

The formula $D = 16t^2$ is used to approximate the distance in feet that an object falls in t seconds.

51. A parachutist jumps from an airplane, falls for 3600 feet, and then opens her parachute. For how many seconds was the parachutist falling before opening the parachute?

52. How long would it take an object to fall to the ground from a helicopter hovering at 1936 feet above the ground?

Cumulative Review Problems

Evaluate.

53. $\sqrt{b^2 - 4ac}$; $b = 4$, $a = 3$, $c = -4$

54. $\sqrt{b^2 - 4ac}$; $b = -5$, $a = 2$, $c = -3$

55. $5x^2 - 6x + 8$; $x = -2$

56. $2x^2 + 3x - 5$; $x = -3$

8.2 The Quadratic Formula and Solutions to Quadratic Equations

MathPro Video 19 SSM

After studying this section, you will be able to:

1 *Solve a quadratic equation by using the quadratic formula.*
2 *Use the discriminant to determine the nature of the roots of a quadratic equation.*
3 *Write a quadratic equation, given the solutions of the equation.*

1 *Solving a Quadratic Equation by Using the Quadratic Formula*

The last method we'll study for solving quadratic equations is the **quadratic formula.** This method works for *any* quadratic equation.

The quadratic formula is developed from completing the square. We begin with the standard form of the equation:

$$ax^2 + bx + c = 0$$

To complete the square, we want the equation to be in the *form $x^2 + dx = e$*. Thus we divide by a.

$$\frac{ax^2}{a} + \frac{b}{a}x + \frac{c}{a} = 0$$

$$x^2 + \frac{b}{a}x = -\frac{c}{a}$$

Now we complete the square by adding $\left(\frac{b}{2a}\right)^2$ to each side.

$$x^2 + \frac{b}{a}x + \left(\frac{b}{2a}\right)^2 = -\frac{c}{a} + \left(\frac{b}{2a}\right)^2$$

Factor the left side and write the right-hand side as one fraction.

$$\left(x + \frac{b}{2a}\right)^2 = \frac{b^2 - 4ac}{4a^2}$$

Now we use the square root property.

$$x + \frac{b}{2a} = \pm\sqrt{\frac{b^2 - 4ac}{4a^2}}$$

Now we solve for x and simplify.

$$x = -\frac{b}{2a} \pm \sqrt{\frac{b^2 - 4ac}{4a^2}}$$

$$= \frac{-b \pm \sqrt{b^2 - 4ac}}{2a}$$

This is the quadratic formula.

Quadratic Formula

For all equations $ax^2 + bx + c = 0$,

$$x = \frac{-b \pm \sqrt{b^2 - 4ac}}{2a} \qquad \text{where } a \neq 0$$

EXAMPLE 1 Solve by using the quadratic formula. $2x^2 + 3x - 7 = 0$

Substituting $a = 2$, $b = 3$, and $c = -7$ into the quadratic formula

$$x = \frac{-b \pm \sqrt{b^2 - 4ac}}{2a}$$

gives

$$x = \frac{-3 \pm \sqrt{(3)^2 - 4(2)(-7)}}{2(2)}$$

$$x = \frac{-3 \pm \sqrt{9 + 56}}{4}$$

$$x = \frac{-3 \pm \sqrt{65}}{4}$$

Practice Problem 1 Solve by the quadratic formula. $3x^2 + 4x - 5 = 0$ ∎

EXAMPLE 2 Solve by using the quadratic formula. $x^2 + 8x = -3$

The standard form is $x^2 + 8x + 3 = 0$, $a = 1$, $b = 8$, and $c = 3$. Then

$$x = \frac{-b \pm \sqrt{b^2 - 4ac}}{2a}$$

$$= \frac{-8 \pm \sqrt{8^2 - 4(1)(3)}}{2(1)}$$

$$= \frac{-8 \pm \sqrt{64 - 12}}{2} = \frac{-8 \pm \sqrt{52}}{2} = \frac{-8 \pm \sqrt{4}\sqrt{13}}{2}$$

$$= \frac{-8 \pm 2\sqrt{13}}{2} = \frac{\cancel{2}(-4 \pm \sqrt{13})}{\cancel{2}}$$

$$= -4 \pm \sqrt{13}$$

Practice Problem 2 Solve by the quadratic formula. $x^2 + 5x = -1 + 2x$ ∎

EXAMPLE 3 Solve by using the quadratic formula. $3x^2 - x - 2 = 0$

Here $a = 3$, $b = -1$, and $c = -2$.

$$x = \frac{-b \pm \sqrt{b^2 - 4ac}}{2a}$$

$$= \frac{-(-1) \pm \sqrt{(-1)^2 - 4(3)(-2)}}{2(3)}$$

$$= \frac{1 \pm \sqrt{1 + 24}}{6} = \frac{1 \pm \sqrt{25}}{6}$$

$$x = \frac{1 + 5}{6} = \frac{6}{6} \quad or \quad x = \frac{1 - 5}{6} = -\frac{4}{6}$$

$$x = 1 \qquad\qquad\qquad x = -\frac{2}{3}$$

Practice Problem 3 Solve by the quadratic formula. $2x^2 + 7x + 6 = 0$ ∎

EXAMPLE 4   Solve by using the quadratic formula. $2x^2 - 48 = 0$

This equation is equivalent to $2x^2 - 0x - 48 = 0$. Therefore, we know that $a = 2$, $b = 0$, and $c = -48$.

$$x = \frac{-b \pm \sqrt{b^2 - 4ac}}{2a}$$

$$= \frac{-0 \pm \sqrt{(0)^2 - 4(2)(-48)}}{2(2)}$$

$$= \frac{\pm\sqrt{384}}{4}$$

$$= \frac{\pm\sqrt{64}\sqrt{6}}{4} = \frac{\pm 8\sqrt{6}}{4}$$

$$= \pm 2\sqrt{6}$$

Practice Problem 4   Solve by the quadratic formula. $2x^2 - 26 = 0$ ■

Placing a Quadratic Equation in Standard Form.

When the quadratic equation contains fractions, eliminate them by multiplying each term by the LCD. Then put the equation into standard form.

EXAMPLE 5   Solve by using the quadratic formula. $\dfrac{2x}{x + 2} = 1 - \dfrac{3}{x + 4}$

You may think that this equation isn't quadratic because it doesn't look like one. That's why we clear fractions first. The LCD is $(x + 2)(x + 4)$.

$$\frac{2x}{x+2}(x+2)(x+4) = 1(x+2)(x+4) - \frac{3}{x+4}(x+2)(x+4)$$

$$2x(x + 4) = (x + 2)(x + 4) - 3(x + 2)$$

$$2x^2 + 8x = x^2 + 6x + 8 - 3x - 6 \qquad \textit{Now you can see that the}$$
$$\textit{equation really is quadratic.}$$

$$2x^2 + 8x = x^2 + 3x + 2$$

$$x^2 + 5x - 2 = 0$$

Now the equation is in standard form, so we use the quadratic formula to solve it, where $a = 1$, $b = 5$, and $c = -2$.

$$x = \frac{-5 \pm \sqrt{5^2 - 4(1)(-2)}}{2(1)} = \frac{-5 \pm \sqrt{25 + 8}}{2}$$

$$x = \frac{-5 \pm \sqrt{33}}{2}$$

Practice Problem 5   Solve by the quadratic formula. $\dfrac{1}{x} + \dfrac{1}{x - 1} = \dfrac{5}{6}$ ■

Some quadratic equations will have solutions that are not real numbers. You should use i notation to simplify the solutions of nonreal complex numbers.

EXAMPLE 6 Solve and simplify your answer. $8x^2 - 4x + 1 = 0$

$a = 8$, $b = -4$, and $c = 1$.

$$x = \frac{-(-4) \pm \sqrt{(-4)^2 - 4(8)(1)}}{2(8)}$$

$$= \frac{4 \pm \sqrt{16 - 32}}{16} = \frac{4 \pm \sqrt{-16}}{16}$$

$$= \frac{4 \pm 4i}{16} = \frac{4(1 \pm i)}{16} = \frac{1 \pm i}{4}$$

Practice Problem 6 Solve by the quadratic formula. $2x^2 - 4x + 5 = 0$ ∎

You may have noticed that complex roots come in pairs. In other words, if $a + bi$ is a solution of a quadratic equation, its conjugate $a - bi$ is also a solution.

2 Using the Discriminant to Determine the Nature of the Roots of a Quadratic Equation

In using the quadratic formula so far, we have solved quadratic equations that had two real roots (sometimes the roots were rational, and sometimes they were irrational). We have solved equations like Example 6 with nonreal complex numbers. This occurs when the expression $b^2 - 4ac$ is negative. Recall that $b^2 - 4ac$ is the radicand in the quadratic equation

$$x = \frac{-b \pm \sqrt{b^2 - 4ac}}{2a}$$

The expression $b^2 - 4ac$ is called the **discriminant.** Depending on whether the discriminant is positive, zero, or negative, the roots of the quadratic equation will be rational, irrational, or complex. We summarize the type of solutions in the following table.

If the discriminant $b^2 - 4ac$ is:	Then the quadratic equation $ax^2 + bx + c = 0$ where a, b, c are integers will have:
A positive number that is also a perfect square	Two different rational solutions (such an equation can always be factored)
A positive number that is not a perfect square	Two different irrational solutions
Zero	One rational solution
Negative	Two complex solutions containing i (they will be complex conjugates)

EXAMPLE 7 What type of solutions does the equation $2x^2 - 9x - 35 = 0$ have? Do not solve the equation.

$a = 2$, $b = -9$, and $c = -35$. Thus

$$b^2 - 4ac = (-9)^2 - 4(2)(-35) = 361$$

Since the discriminant is positive, the equation has two real (rational or irrational) roots.

Since $(19)^2 = 361$ we see that 361 is a perfect square. Thus the equation has two different rational solutions. This type of quadratic equation can always be factored.

Practice Problem 7 Use the discriminant to find what type of solutions the equation $9x^2 + 12x + 4 = 0$ has. Do not solve the equation. ■

EXAMPLE 8 Use the discriminant to determine the type of solutions for each of the following equations.

(a) $3x^2 - 4x + 2 = 0$ **(b)** $5x^2 - 3x - 5 = 0$

(a) Here $a = 3$, $b = -4$, and $c = 2$. Thus

$$b^2 - 4ac = (-4)^2 - 4(3)(2)$$

$$= 16 - 24 = -8$$

Since the discriminant is negative, the equation will have two complex solutions containing i.

(b) Here $a = 5$, $b = -3$, and $c = -5$. Thus

$$b^2 - 4ac = (-3)^2 - 4(5)(-5)$$

$$= 9 + 100 = 109$$

Since this positive number is not a perfect square, the equation will have two different irrational solutions.

Practice Problem 8 Use the discriminant to determine the type of solutions for each of the following equations.

(a) $x^2 - 4x + 13 = 0$ **(b)** $9x^2 + 6x + 7 = 0$ ■

3 Writing a Quadratic Equation Given the Solutions of the Equation

By using the zero-product rule in reverse, we can find a quadratic equation that contains two given solutions. To illustrate, if $x = 3$ and $x = 7$, then we could write the equation $(x - 3)(x - 7) = 0$ and therefore a quadratic equation that has these two solutions is $x^2 - 10x + 21 = 0$. This answer is not unique. Any constant multiple of $x^2 - 10x + 21 = 0$ would also have roots of $x = 3$ and $x = 7$. Thus $2x^2 - 20x + 42 = 0$ also has roots of $x = 3$ and $x = 7$.

EXAMPLE 9 Find a quadratic equation whose roots are 5 and -2.

$$x = 5 \qquad x = -2$$

$$x - 5 = 0 \qquad x + 2 = 0$$

$$(x - 5)(x + 2) = 0$$

$$x^2 - 3x - 10 = 0$$

Practice Problem 9 Find a quadratic equation whose roots are -10 and -6. ■

EXAMPLE 10 Find a quadratic equation whose solutions are $x = 3i$ and $x = -3i$.

First we write the two equations

$$x - 3i = 0 \quad \text{and} \quad x + 3i = 0$$

$$(x - 3i)(x + 3i) = 0$$

$$x^2 + 3ix - 3ix - 9i^2 = 0$$

$$x^2 - 9(-1) = 0 \qquad \text{Use } i^2 = -1.$$

$$x^2 + 9 = 0$$

Practice Problem 10 Find a quadratic equation whose solutions are $x = 2i\sqrt{3}$ and $x = -2i\sqrt{3}$. ■

EXAMPLE 11 Find a quadratic equation whose roots are $-\dfrac{3}{5}$ and $\dfrac{1}{4}$. It would be helpful to have an equation that does not contain fractions.

$$x = -\frac{3}{5}$$

$$5x = -3 \qquad \textit{Multiply each side by 5.}$$

$$5x + 3 = 0 \qquad \textit{Add 3 to each side.}$$

In a similar fashion we can transform the second equation.

$$x = \frac{1}{4}$$

$$4x = 1 \qquad \textit{Multiply each side by 4.}$$

$$4x - 1 = 0 \qquad \textit{Add } -1 \textit{ to each side.}$$

Therefore, we can write

$$(5x + 3)(4x - 1) = 0$$

$$20x^2 - 5x + 12x - 3 = 0$$

$$20x^2 + 7x - 3 = 0$$

Check: We can verify that $x = -\dfrac{3}{5}$ and $x = \dfrac{1}{4}$ are roots of the equation $20x^2 + 7x - 3 = 0$.

When $x = -\dfrac{3}{5}$ $\qquad 20\left(-\dfrac{3}{5}\right)^2 + 7\left(-\dfrac{3}{5}\right) - 3 \stackrel{?}{=} 0$

$$20\left(\frac{9}{25}\right) + 7\left(-\frac{3}{5}\right) - 3 \stackrel{?}{=} 0$$

$$\frac{36}{5} + \left(-\frac{21}{5}\right) - 3 \stackrel{?}{=} 0$$

$$\frac{15}{5} - 3 \stackrel{?}{=} 0$$

$$0 = 0 \quad \checkmark$$

When $x = \dfrac{1}{4}$ $\qquad 20\left(\dfrac{1}{4}\right)^2 + 7\left(\dfrac{1}{4}\right) - 3 \stackrel{?}{=} 0$

$$20\left(\frac{1}{16}\right) + 7\left(\frac{1}{4}\right) - 3 \stackrel{?}{=} 0$$

$$\frac{5}{4} + \frac{7}{4} - 3 \stackrel{?}{=} 0$$

$$\frac{12}{4} - 3 \stackrel{?}{=} 0$$

$$0 = 0 \quad \checkmark$$

Practice Problem 11 Find a quadratic equation whose roots are $\dfrac{4}{5}$ and $\dfrac{3}{7}$. ∎

8.2 Exercises

Verbal and Writing Skills

1. How is the quadratic formula used to solve a quadratic equation?

2. The discriminant is the expression _____.

3. If the discriminant is zero, then the quadratic equation will have _____ solution(s).

4. If the discriminant is a perfect square, then the quadratic equation will have _____ solution(s).

Solve by the quadratic formula. Simplify your answer. Use i notation for nonreal complex numbers.

5. $x^2 + 5x - 10 = 0$

6. $x^2 + 3x - 20 = 0$

7. $2x^2 + x - 4 = 0$

8. $5x^2 - x - 1 = 0$

9. $x^2 = \frac{2}{3}x$

10. $\frac{4}{5}x^2 = x$

11. $x^2 + 8x + 13 = 0$

12. $x^2 - 2x - 17 = 0$

13. $6x^2 - x - 1 = 0$

14. $4x^2 + 11x - 3 = 0$

15. $4x^2 + 3x - 2 = 0$

16. $6x^2 - 2x - 1 = 0$

17. $3x^2 + 1 = 8$

18. $5x^2 - 1 = 5$

19. $2x^2 - 7x + 4 = x - 1$

20. $2x^2 - 2x + 2 = 7$

21. $3x^2 + 5x + 1 = 5x + 4$

22. $2x^2 - 7x - 3 = 9 - 7x$

23. $(x - 3)(x + 2) = 1$

27. $\dfrac{1}{x+3} + \dfrac{1}{x} = \dfrac{1}{4}$

28. $\dfrac{1}{x+2} + \dfrac{1}{x} = \dfrac{1}{3}$

29. $\dfrac{1}{y} + \dfrac{1}{y-4} = \dfrac{5}{6}$

30. $\dfrac{1}{y} + \dfrac{2}{y+3} = \dfrac{1}{4}$

31. $\dfrac{1}{15} + \dfrac{3}{y} = \dfrac{4}{y+1}$

32. $\dfrac{1}{4} + \dfrac{6}{y+2} = \dfrac{6}{y}$

33. $x^2 - 4x + 8 = 0$

34. $x^2 - 2x + 4 = 0$

35. $2x^2 + 15 = 0$

36. $5x^2 = -3$

37. $3x^2 - 8x + 7 = 0$

38. $3x^2 - 4x + 6 = 0$

39. $\dfrac{1}{2}x + \dfrac{3}{4} = -\dfrac{1}{2}x^2$

40. $x^2 = \dfrac{1}{3}x - \dfrac{4}{3}$

Use the discriminant to find what type of solutions (two rational, two irrational, one rational, or two nonreal complex) each of the following equations has. Do not solve the equation.

41. $x^2 - 6x + 17 = 0$

42. $2x^2 + 4x + 3 = 0$

43. $3x^2 + 4x = 2$

44. $4x^2 - 20x + 25 = 0$

45. $2x^2 + 10x + 8 = 0$

46. $2x^2 - 7x - 4 = 0$

47. $9x^2 + 4 = 12x$

48. $5x^2 - 8x - 2 = 0$

Write a quadratic equation having the given solutions.

49. $5, -11$

50. $13, -2$

51. $-6, -10$

52. $-5, -12$

53. $6i, -6i$

54. $4i, -4i$

55. $2i\sqrt{5}, -2i\sqrt{5}$

56. $3i\sqrt{2}, -3i\sqrt{2}$

57. $3, -\dfrac{5}{2}$

58. $-2, \dfrac{5}{6}$

Solve for x by using the quadratic formula. Approximate your answers to four decimal places.

59. $3x^2 + 5x - 9 = 0$

60. $1.2x^2 - 12.3x - 4.2 = 0$

61. $20.6x^2 - 73.4x + 41.8 = 0$

62. $0.162x^2 + 0.094x - 0.485 = 0$

To Think About

63. The solutions to the equation $ax^2 + 3x - 6 = 0$ are $x = \dfrac{-3 \pm \sqrt{57}}{4}$. What is the value of a?

64. The solutions to the equation $2x^2 + 7x + c = 0$ are $x = \dfrac{-7 \pm \sqrt{65}}{4}$. What is the value of c?

65. Verify that $y = \dfrac{2 + \sqrt{3}}{2}$ is a solution of the equation $4y^2 = 8y - 1$.

66. Verify that $y = \dfrac{3 + \sqrt{2}}{2}$ is a solution of the equation $4y^2 = 12y - 7$.

Cumulative Review Problems

Simplify.

67. $9x^2 - 6x + 3 - 4x - 12x^2 + 8$

68. $3y(2 - y) + \dfrac{1}{5}(10y^2 - 15y)$

69. Last year, Cecile, a professional mountain bike racer, purchased 3 new padded riding suits to protect her from injury and compress her muscles while riding. In addition, she purchased 2 pairs of racing goggles. The cost for these items was $343. This year, suits cost $10 more and goggles cost $5 more than last year. This year she purchased 2 new suits and 3 pairs of goggles for $312. How much did each suit cost last year? How much did each pair of goggles cost last year?

70. Music Galaxy sells compact discs, cassettes, and everything you could possibly want from a music supply superstore. The management plans to expand its compact disc section. Presently, it takes 50 feet of an inner security fence to enclose the rectangular section. The expansion plans call for tripling the width and doubling the length. The new CD section will need 118 feet of inner security fencing. What is the length and width of the original compact disc section?

8.3 Equations That Can Be Transformed to Quadratic Form

MathPro

Video 20

SSM

After studying this section, you will be able to:

1 *Solve equations of higher degree that can be transformed to quadratic form.*
2 *Solve equations with fractional exponents that can be transformed to quadratic form.*

1 *Solving Equations of Higher Degree*

Some higher-order equations can be solved by writing them in the form of a quadratic equation. An equation is **quadratic in form** if we can substitute a linear term for the lowest-degree variable and get an equation of the form $ay^2 + by + c = 0$.

EXAMPLE 1   Solve. $x^4 - 13x^2 + 36 = 0$

Let $y = x^2$. Then $y^2 = x^4$. Thus our new equation is

$$y^2 - 13y + 36 = 0 \qquad \textit{Replace } x^2 \textit{ by } y \textit{ and } x^4 \textit{ by } y^2.$$

$$(y - 4)(y - 9) = 0 \qquad \textit{Factor.}$$

$$y - 4 = 0 \quad \textit{or} \quad y - 9 = 0 \qquad \textit{Solve for } y.$$

$$y = 4 \qquad\qquad y = 9 \qquad \textit{These are \textbf{not} the roots to the original}$$
$$x^2 = 4 \qquad\qquad x^2 = 9 \qquad \textit{equation. We must replace } y \textit{ by } x^2.$$

$$x = \pm\sqrt{4} \qquad\qquad x = \pm\sqrt{9}$$

$$x = \pm 2 \qquad\qquad x = \pm 3$$

Thus there are *four* solutions to the original equation: $x = +2, x = -2, x = +3, x = -3$. This is clearly possible, since the degree of the equation is 4.

Practice Problem 1   Solve. $x^4 - 5x^2 - 36 = 0$ ▪

EXAMPLE 2   Solve for all real roots. $2x^6 - x^3 - 6 = 0$

Let $y = x^3$. Then $y^2 = x^6$. Thus we have

$$2y^2 - y - 6 = 0 \qquad \textit{Replace } x^3 \textit{ by } y \textit{ and } x^6 \textit{ by } y^2.$$

$$(2y + 3)(y - 2) = 0 \qquad \textit{Factor.}$$

$$2y + 3 = 0 \quad \textit{or} \quad y - 2 = 0 \qquad \textit{Solve for } y.$$

$$y = -\frac{3}{2} \qquad\qquad y = 2$$

$$x^3 = -\frac{3}{2} \quad \textit{or} \quad x^3 = 2 \qquad \textit{Replace } y \textit{ by } x^3.$$

$$x = \sqrt[3]{-\frac{3}{2}} \qquad x = \sqrt[3]{2} \qquad \textit{Take the cube root of each side of the equation.}$$

$$x = \frac{\sqrt[3]{-12}}{2} \qquad\qquad \textit{Simplify } \sqrt[3]{-\frac{3}{2}} \textit{ by rationalizing.}$$

Check these solutions.

Practice Problem 2   Solve for all real roots. $x^6 - 5x^3 + 4 = 0$ ▪

EXAMPLE 3   Solve. $x^{2/3} - 3x^{1/3} + 2 = 0$ Check your solutions.

Let $y = x^{1/3}$ and $y^2 = x^{2/3}$.

$y^2 - 3y + 2 = 0$	*Replace $x^{1/3}$ by y, and $x^{2/3}$ by y^2.*
$(y - 2)(y - 1) = 0$	*Factor.*
$y - 2 = 0$ *or* $y - 1 = 0$	
$y = 2$ $y = 1$	*Solve for y.*
$x^{1/3} = 2$ *or* $x^{1/3} = 1$	*Replace y by $x^{1/3}$.*
$(x^{1/3})^3 = (2)^3$ $(x^{1/3})^3 = (1)^3$	*Cube each side of the equation.*
$x = 8$ $x = 1$	

Check: $x^{2/3} - 3x^{1/3} + 2 = 0$

$x = 8$: $(8)^{2/3} - 3(8)^{1/3} + 2 \overset{?}{=} 0$

$(\sqrt[3]{8})^2 - 3(\sqrt[3]{8}) + 2 \overset{?}{=} 0$

$(2)^2 - 3(2) + 2 \overset{?}{=} 0$

$4 - 6 + 2 \overset{?}{=} 0$

$0 = 0$ ✓

$x = 1$: $(1)^{2/3} - 3(1)^{1/3} + 2 \overset{?}{=} 0$

$(\sqrt[3]{1})^2 - 3\sqrt[3]{1} + 2 \overset{?}{=} 0$

$1 - 3 + 2 \overset{?}{=} 0$

$0 = 0$ ✓

The types of problems that we are solving in Section 8.3 are more difficult to solve. Part of the difficulty lies in the fact that the problems have different numbers of solutions. A fourth-degree equation such as Equation 1 has four different solutions. A sixth-degree equation such as Example 2 has only two solutions. However, some sixth-degree equations sometimes will have as many as six solutions. The equation that we examined in Example 3 has only two solutions. However, other problems of this type with fractional exponents may have one solution or even no solution at all. It is good to take some time to carefully examine the steps of the solution and to determine if it seems logical to have obtained the number of solutions that you have.

If you have access to a graphing program on a computer such as Derive or Maple, these can be very helpful in determining or verifying the solutions to these types of problems. Of course if you have a graphing calculator this can be most helpful, particularly in verifying the value of the solution and the number of solutions.

Optional Graphing Calculation Exploration: To illustrate the point if you have a graphing calculator, verify your work for Example 3 by graphing the equation

$$y = x^{(\frac{2}{3})} - 3x^{(\frac{1}{3})} + 2$$

Determine from your graph if the curve does in fact cross the x-axis (the places where $y = 0$) when $x = 1$ and $x = 8$. You will have to carefully select the window so that you can see the behavior of the curve more clearly. For this equation a useful window is $[-1, 12, -1, 2]$. Remember for most graphing calculators you will need to surround the exponents with the parentheses symbols.

Practice Problem 3   Solve and check. $3x^{4/3} - 5x^{2/3} + 2 = 0$ ∎

EXAMPLE 4   Solve. $2x^{1/2} = 5x^{1/4} + 12$ Check your solutions.

$$2x^{1/2} - 5x^{1/4} - 12 = 0 \qquad \textit{Place in standard form.}$$

$$2y^2 - 5y - 12 = 0 \qquad \textit{Replace } x^{1/4} \textit{ by } y \textit{ and } x^{1/2} \textit{ by } y^2.$$

$$(2y + 3)(y - 4) = 0 \qquad \textit{Factor.}$$

$$2y = -3 \qquad or \qquad y = 4$$

$$y = -\frac{3}{2} \qquad\qquad\qquad\qquad \textit{Solve for y.}$$

$$\boxed{x^{1/4} = -\frac{3}{2}} \qquad or \qquad \boxed{x^{1/4} = 4} \qquad \textit{Replace y by } x^{1/4}.$$

$$(x^{1/4})^4 = \left(-\frac{3}{2}\right)^4 \qquad (x^{1/4})^4 = (4)^4 \qquad \textit{Solve for x.}$$

$$x = \frac{81}{16} \qquad\qquad\qquad x = 256$$

Check: $2x^{1/2} - 5x^{1/4} - 12 = 0$

$x = \frac{81}{16}: \quad 2\left(\frac{81}{16}\right)^{1/2} - 5\left(\frac{81}{16}\right)^{1/4} - 12 \stackrel{?}{=} 0 \qquad x = 256: \quad 2(256)^{1/2} - 5(256)^{1/4} - 12 \stackrel{?}{=} 0$

$$2\left(\frac{9}{4}\right) - 5\left(\frac{3}{2}\right) - 12 \stackrel{?}{=} 0 \qquad\qquad 2(16) - 5(4) - 12 \stackrel{?}{=} 0$$

$$\frac{9}{2} - \frac{15}{2} - 12 \stackrel{?}{=} 0 \qquad\qquad\qquad 32 - 20 - 12 \stackrel{?}{=} 0$$

$$\boxed{-15 \neq 0} \qquad\qquad\qquad\qquad 0 = 0 \;\checkmark$$

$x = 256$ is the **only valid solution.**

$x = \dfrac{81}{16}$ is extraneous and not a valid solution.

Practice Problem 4   Solve. $3x^{1/2} = 8x^{1/4} - 4$ Check your solutions. ∎

Now that we have covered four basic examples, this same general substitution technique can be extended to other types of equations. In each case by using $y =$ the appropriate value, a quadratic equation can be obtained. The following table lists some substitutions that would be appropriate.

If you want to solve:	Then you would use the substitution:
$x^4 - 13x^2 + 36 = 0$	$y = x^2$
$2x^6 - x^3 - 6 = 0$	$y = x^3$
$x^{2/3} - 3x^{1/3} + 2 = 0$	$y = x^{1/3}$
$6(x - 1)^{-2} + (x - 1)^{-1} - 2 = 0$	$y = (x - 1)^{-1}$
$(2x^2 + x)^2 + 4(2x^2 + x) + 3 = 0$	$y = 2x^2 + x$
$\left(\dfrac{1}{x - 1}\right)^2 + \dfrac{1}{x - 1} - 6 = 0$	$y = \dfrac{1}{x - 1}$
$2x - 5x^{1/2} + 2 = 0$	$y = x^{1/2}$

A collection of problems of these general types is provided in the exercise set.

8.3 Exercises

Solve. Express any nonreal complex numbers with i notation.

1. $x^4 - 9x^2 + 20 = 0$

2. $x^4 - 11x^2 + 18 = 0$

3. $2x^4 + 5x^2 - 12 = 0$

4. $2x^4 - x^2 - 3 = 0$

5. $3x^4 = 10x^2 + 8$

6. $5x^4 = 4x^2 + 1$

In Problems 7 through 10, find all valid real roots for each equation.

7. $x^6 - 7x^3 - 8 = 0$

8. $x^6 - 3x^3 - 4 = 0$

9. $2x^6 - 7x^3 - 4 = 0$

10. $12x^6 + 5x^3 - 2 = 0$

Solve for real roots.

11. $x^8 = 3x^4 - 2$

12. $x^8 = 7x^4 - 12$

13. $3x^8 + 13x^4 = 10$

14. $3x^8 - 10x^4 = 8$

15. $x^{2/3} + 2x^{1/3} - 8 = 0$

16. $x^{2/3} + x^{1/3} - 12 = 0$

17. $x^{2/3} + 2x^{1/3} - 3 = 0$

18. $x^{2/3} + 9x^{1/3} = -18$

19. $2x^{1/2} - 5x^{1/4} - 3 = 0$

20. $3x^{1/2} - 14x^{1/4} - 5 = 0$

21. $2x^{1/2} - x^{1/4} - 6 = 0$

22. $2x^{1/2} - x^{1/4} - 1 = 0$

23. $x^{2/5} + x^{1/5} - 2 = 0$

24. $2x^{2/5} + 7x^{1/5} + 3 = 0$

In each problem make an appropriate substitution to make the equation quadratic in form. Find all real or complex values for x.

25. $(x^2 + x)^2 - 5(x^2 + x) = -6$

26. $(x^2 - 2x)^2 + 2(x^2 - 2x) = 3$

27. $x - 4x^{1/2} - 21 = 0$

28. $x - 6x^{1/2} + 8 = 0$

29. $10x^{-2} + 7x^{-1} + 1 = 0$

30. $20x^{-2} + 9x^{-1} + 1 = 0$

31. $\dfrac{2}{(x-1)^2} + \dfrac{3}{x-1} = 2$

32. $\dfrac{3}{(x-2)^2} - \dfrac{4}{(x-2)} + 1 = 0$

Find the real values of x to the nearest hundredth.

33. $1.23x^4 - 13.16x^2 - 45.08 = 0$

34. $0.97x^{1/2} - 5.02x^{1/4} + 5.96 = 0$

To Think About

Solve. Find all valid real roots for each equation.

35. $15 - \dfrac{2x}{x-1} = \dfrac{x^2}{x^2 - 2x + 1}$

36. $4 - \dfrac{x^3 + 1}{x^3 + 6} = \dfrac{x^3 - 3}{x^3 + 2}$

Cumulative Review Problems

Simplify.

37. $\sqrt{8x} + 3\sqrt{2x} - 4\sqrt{50x}$

38. $\sqrt{27x} + 5\sqrt{3x} - 2\sqrt{48x}$

Multiply and simplify.

39. $3\sqrt{2}(\sqrt{5} - 2\sqrt{6})$

40. $(\sqrt{2} + \sqrt{6})(3\sqrt{2} - 2\sqrt{5})$

41. How much greater is the average salary of a man than a woman in each of these occupations? Express your answer to the nearest tenth of a percent.

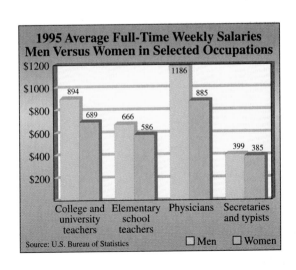

1995 Average Full-Time Weekly Salaries Men Versus Women in Selected Occupations

Source: U.S. Bureau of Statistics   ☐ Men   ☐ Women

42. Approximately how many extra years would a woman college teacher have to work to earn the same lifetime earnings from teaching as her male counterpart if he taught for 30 years?

8.4 Formulas and Applications

MathPro

Video 20

SSM

After studying this section, you will be able to:

1 *Solve a quadratic equation containing several variables.*
2 *Solve problems requiring the use of the Pythagorean theorem.*
3 *Solve applied problems requiring the use of a quadratic equation.*

1 *Solving a Quadratic Equation Containing Several Variables*

In mathematics, physics, and engineering we must often solve an equation for a variable in terms of other variables. You recall we solved linear equations in several variables in Section 2.2. We will now examine several cases where the variable that we are solving for is squared. If the variable we are solving for is squared, and there is no other term containing that variable, then the quadratic equation can be solved by using the square root property.

EXAMPLE 1 The surface area of a sphere is given by $A = 4\pi r^2$. Solve this equation for r. (You do not need to rationalize the denominator.)

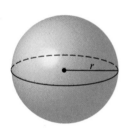

$$A = 4\pi r^2$$

$$\frac{A}{4\pi} = r^2$$

$$\pm \sqrt{\frac{A}{4\pi}} = r \qquad \text{Use the square root property.}$$

$$\pm \frac{1}{2}\sqrt{\frac{A}{\pi}} = r \qquad \text{Simplify.}$$

Since the radius of a sphere must be a positive value, we use only the principal root.

$$\frac{1}{2}\sqrt{\frac{A}{\pi}} = r$$

Practice Problem 1 The volume of a cylindrical cone is $V = \frac{1}{3}\pi r^2 h$. Solve this equation for r. (You do not need to rationalize the denominator.) ∎

EXAMPLE 2 Solve for x. $3x^2 + 5 = 2b^2$

$$3x^2 = 2b^2 - 5 \qquad \text{Add } -5 \text{ to each side.}$$

$$x^2 = \frac{2b^2 - 5}{3} \qquad \text{Divide each side by 3.}$$

$$x = \pm \sqrt{\frac{2b^2 - 5}{3}} \qquad \text{Use the square root property.}$$

Practice Problem 2 Solve for x. $5x^2 - 8 = 3x^2 + 2w$ ∎

Some quadratic equations containing many variables can be solved for one variable by factoring.

EXAMPLE 3   Solve for y. $y^2 - 2yz - 15z^2 = 0$

$$(y + 3z)(y - 5z) = 0 \qquad \textit{Factor.}$$

$$y + 3z = 0 \qquad y - 5z = 0 \qquad \textit{Set each factor equal to 0.}$$

$$y = -3z \qquad y = 5z \qquad \textit{Solve for y.}$$

Practice Problem 3   Solve for y. $2y^2 + 9wy + 7w^2 = 0$ ∎

Sometimes the quadratic formula is required in order to solve the equation.

EXAMPLE 4   Solve for x. $2x^2 + 3wx - 4z = 0$

We use the quadratic formula where the variable is considered to be x and the letters w and z are considered constants. Thus $a = 2$, $b = 3w$, and $c = -4z$.

$$x = \frac{-b \pm \sqrt{b^2 - 4ac}}{2a}$$

$$= \frac{-3w \pm \sqrt{(3w)^2 - 4(2)(-4z)}}{2(2)} = \frac{-3w \pm \sqrt{9w^2 + 32z}}{4}$$

Note that this answer cannot be simplified.

Practice Problem 4   Solve for y. $3y^2 + 2fy - 7g = 0$ ∎

EXAMPLE 5   The formula for the curved surface area S of a right circular cone of altitude h and base radius r is $S = \pi r \sqrt{r^2 + h^2}$. Solve for r^2.

$$S = \pi r \sqrt{r^2 + h^2}$$

$$\frac{S}{\pi r} = \sqrt{r^2 + h^2} \qquad \textit{Isolate the radical.}$$

$$\frac{S^2}{\pi^2 r^2} = r^2 + h^2 \qquad \textit{Square both sides.}$$

$$\frac{S^2}{\pi^2} = r^4 + h^2 r^2 \qquad \textit{Multiply each term by } r^2.$$

$$0 = r^4 + h^2 r^2 - \frac{S^2}{\pi^2} \qquad \textit{Subtract } S^2/\pi^2.$$

We can make this equation quadratic in form by letting $y = r^2$. Then

$$0 = y^2 + h^2 y - \frac{S^2}{\pi^2}$$

By the quadratic formula

$$y = \frac{-h^2 \pm \sqrt{(h^2)^2 - 4(1)\left(-\dfrac{S^2}{\pi^2}\right)}}{2}$$

$$= \frac{-h^2 \pm \sqrt{\dfrac{\pi^2 h^4}{\pi^2} + \dfrac{4S^2}{\pi^2}}}{2}$$

$$= \frac{-h^2 \pm \dfrac{1}{\pi}\sqrt{\pi^2 h^4 + 4S^2}}{2}$$

$$= \frac{-\pi h^2 \pm \sqrt{\pi^2 h^4 + 4S^2}}{2\pi}$$

Since $y = r^2$, we have

$$r^2 = \frac{-\pi h^2 \pm \sqrt{\pi^2 h^4 + 4S^2}}{2\pi}$$

Practice Problem 5   The formula for the number of diagonals d in a polygon of n sides is $d = \dfrac{n^2 + 2n}{2}$. Solve for n. ∎

2 *Solving Problems Requiring the Use of the Pythagorean Theorem*

A most useful formula is the Pythagorean theorem for right triangles.

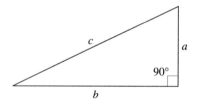

> **Pythagorean Theorem**
> If c is the longest side of a right triangle, then $a^2 + b^2 = c^2$.

The longest side of a right triangle is called the **hypotenuse.** The two other sides are called the **legs** of the triangle.

EXAMPLE 6

(a) In the Pythagorean formula $a^2 + b^2 = c^2$, solve for a.

(b) Find the value of a if $c = 13$ and $b = 5$.

(a) $a^2 = c^2 - b^2$ *Subtract b^2 from each side.*

  $a = \pm\sqrt{c^2 - b^2}$ *Use the square root property.*

Since a, b, and c are positive numbers representing the lengths of the sides of a triangle, we use only the positive root, $a = \sqrt{c^2 - b^2}$.

(b) $a = \sqrt{c^2 - b^2}$

  $= \sqrt{(13)^2 - (5)^2} = \sqrt{169 - 25} = \sqrt{144} = 12$
  Thus $a = 12$.

Practice Problem 6

(a) In the Pythagorean formula, solve for b.

(b) Find the value of b if $c = 26$ and $a = 24$. ∎

EXAMPLE 7 The perimeter of a triangular piece of land is 12 miles. One leg of the triangle is 1 mile longer than the other leg. Find the length of each boundary of the land if the triangle is a right triangle.

1. *Understand the problem.*
   Draw a picture.

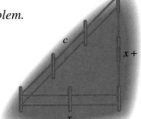

2. *Write an equation.*
   We can use the Pythagorean theorem. First, we want only one variable in our equation (right now, both c and x are not known).
   We are given that the perimeter is 12 miles, so

$$x + (x + 1) + c = 12$$

Thus

$$c = -2x + 11$$

By the Pythagorean theorem,

$$x^2 + (x + 1)^2 = (-2x + 11)^2$$

3. *Solve the equation and state the answer.*

$$x^2 + (x + 1)^2 = (-2x + 11)^2$$

$$x^2 + x^2 + 2x + 1 = 4x^2 - 44x + 121$$

$$0 = 2x^2 - 46x + 120$$

$$0 = x^2 - 23x + 60$$

By the quadratic formula,

$$x = \frac{-(-23) \pm \sqrt{(-23)^2 - 4(1)(60)}}{2(1)}$$

$$x = \frac{23 \pm \sqrt{289}}{2}$$

$$x = \frac{23 \pm 17}{2}$$

$$x = \frac{40}{2} = 20 \quad or \quad x = \frac{6}{2} = 3$$

The only answer that makes sense is $x = 3$. Thus the sides of the triangle are $x = 3$, $x + 1 = 3 + 1 = 4$, and $-2x + 11 = -2(3) + 11 = 5$. The answer $x = 20$ cannot be right because the perimeter (the sum of *all* the sides) was only 12. The triangle thus has sides of 3 miles, 4 miles, and 5 miles. The longest boundary of this triangular piece of land is 5 miles. The other two boundaries are 4 miles and 3 miles.

Notice that we could have factored the quadratic equation instead of using the quadratic formula. $x^2 - 23x + 60 = 0$ can be written as $(x - 20)(x - 3) = 0$.

4. *Check.*

Is the perimeter 12 miles?

$$5 + 4 + 3 = 12 \quad ✓$$

Is one leg 1 mile longer than the other?

$$4 = 3 + 1 \quad ✓$$

Practice Problem 7   The perimeter of a triangular piece of land is 30 miles. One leg of the triangle is 7 miles shorter than the other leg. Find the length of each boundary of the land if the triangle is a right triangle. ■

3 Solving Applied Problems Involving the Use of a Quadratic Equation

Many types of area problems require the use of a quadratic equation in order to obtain the solution. Our next two examples illustrate this outcome.

EXAMPLE 8   The radius of an old circular pipe under a roadbed is 10 inches. Designers wanted to replace it with a smaller pipe and decided they could use a pipe with an area 36π square inches smaller. What should the radius of the new pipe be?

First we need the formula for the area of a circle.

$$A = \pi r^2$$

where A is the area and r is the radius. The area of the old pipe is

$$A_{\text{old}} = \pi(10)^2$$

$$= 100\pi$$

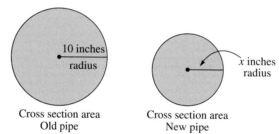

Cross section area   Cross section area
Old pipe   New pipe

Let x = the radius of the new pipe

$$(\text{area of old pipe}) - (\text{area of new pipe}) = 36\pi$$

$$100\pi \quad - \quad \pi x^2 \quad = 36\pi$$

$$64\pi = \pi x^2 \qquad \text{Add } \pi x^2 \text{ to each side and}$$
$$\text{subtract } 36\pi \text{ from each side.}$$

$$\frac{64\pi}{\pi} = \frac{\pi x^2}{\pi} \qquad \text{Divide each side by } \pi.$$

$$64 = x^2$$

$$\pm 8 = x \qquad \text{Use the square root property.}$$

Since the radius must be positive, we select $x = 8$. The radius of the new pipe is 8 inches.

Practice Problem 8   Redo Example 8 if the radius under the roadbed is 6 inches and the designers wanted to replace it with a pipe with an area that is 45π square inches larger. What should the radius of the new pipe be? ■

EXAMPLE 9 Find the base and altitude of a triangle that has an area of 35 square meters if the base is 3 meters shorter than the altitude.

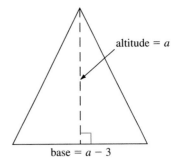
altitude = a

base = a − 3

The area of a triangle is given by

$$A = \frac{1}{2}ab$$

Let a = the length in meters of the altitude. Then $a - 3$ = the length in meters of the base.

$35 = \frac{1}{2}a(a - 3)$	*Replace A (area) by 35.*
	Replace b (base) by a − 3.
$70 = a(a - 3)$	*Multiply each side by 2.*
$70 = a^2 - 3a$	*Use the distributive property.*
$0 = a^2 - 3a - 70$	*Subtract 70 from each side.*
$0 = (a - 10)(a + 7)$	
$a = 10 \quad or \quad a = -7$	

The side of a triangle must be a positive number, so we disregard -7. Thus

$$\text{altitude} = a = 10 \text{ meters}$$

$$\text{base} = a - 3 = 7 \text{ meters}$$

Practice Problem 9 The length of a rectangle is 3 feet less than twice the width. The area of the rectangle is 54 square feet. Find the dimensions of the rectangle. ∎

We will now examine a few word problems that require the use of the concept that distance = (rate)(time) or $d = rt$.

EXAMPLE 10 When Barbara was training for a bicycle race, she rode a total of 135 miles on Monday and Tuesday. On Monday she rode for 75 miles in the rain. On Tuesday she rode 5 miles per hour faster because the weather was better. Her total cycling time for the two days was 8 hours. Find her speed for each day.

It would be helpful to organize a few basic facts.
We can find each distance. If she rode 75 miles on Monday and a total of 135 miles during two days, then she rode $135 - 75 = 60$ miles on Tuesday.
Let x = the cycling rate in miles per hour on Monday. Since she rode 5 miles an hour faster on Tuesday, we let $x + 5$ = the cycling rate in miles per hour on Tuesday.

Since distance divided by rate = time, $(\frac{d}{r} = t)$, we can determine that the time she cycled on Monday was $\frac{75}{x}$ and the time she cycled on Tuesday was $\frac{60}{x + 5}$.

We put these facts in a table.

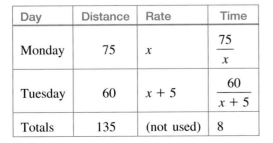

Day	Distance	Rate	Time
Monday	75	x	$\frac{75}{x}$
Tuesday	60	$x + 5$	$\frac{60}{x + 5}$
Totals	135	(not used)	8

Since the total cycling time was 8 hours, we have

time cycling Monday + time cycling Tuesday = 8 hours

$$\frac{75}{x} + \frac{60}{x+5} = 8$$

The LCD of this equation is $x(x+5)$. Multiply each term by the LCD.

$$\cancel{x}(x+5)\left(\frac{75}{\cancel{x}}\right) + x\cancel{(x+5)}\left(\frac{60}{\cancel{x+5}}\right) = x(x+5)(8)$$

$$75(x+5) + 60x = 8x(x+5)$$

$$75x + 375 + 60x = 8x^2 + 40x$$

$$0 = 8x^2 - 95x - 375$$

$$0 = (x-15)(8x+25)$$

$$x - 15 = 0 \quad or \quad 8x + 25 = 0$$

$$x = 15 \qquad\qquad x = \frac{-25}{8}$$

We disregard the negative answer. The cyclist does not have a negative rate of speed—unless he was pedaling backward! Thus $x = 15$, so Barbara's rate of speed on Monday was 15 mph; and $x + 5 = 20$, so Barbara's rate of speed on Tuesday was 20 mph.

Practice Problem 10   Carlos traveled in his car at a constant speed for 150 miles on a secondary road. Then he traveled 10 mph faster on a better road for 240 miles. If Carlos drove for 7 hours, find the car's speed for each part of the trip. ■

EXAMPLE 11   The McSwiggins were driving down a crowded highway to a new home 50 miles away from their present home. Ms. McSwiggin drove the family car, and Mr. McSwiggin drove the rental moving van. Ms. McSwiggin drove 5 mph faster than her husband and arrived at their new home $\frac{1}{2}$ hour sooner than he did. Find the average speed of each vehicle.

Let r = the speed in miles per hour traveled by Mr. McSwiggin.
Since his wife traveled 5 miles per hour faster, her speed is $r + 5$.

Once again we can determine the time by using the fact that $\dfrac{d}{r} = t$.

Let's set up the following table:

Driver	Distance	Rate	Time
Ms. McSwiggin	50	$r + 5$	$\dfrac{50}{r+5}$
Mr. McSwiggin	50	r	$\dfrac{50}{r}$

In this example we do *not* know the total time. However, we know that Ms. McSwiggin took $\frac{1}{2}$ hour less time than Mr. McSwiggin.

Ms. McSwiggin's time | is | $\frac{1}{2}$ hour less than

Mr. McSwiggin's time

$$\frac{50}{r+5} = \frac{50}{r} - \frac{1}{2}$$

Now we multiply each side of the equation by the LCD $= 2r(r+5)$.

$$2r(r+5)\frac{50}{r+5} = 2r(r+5)\frac{50}{r} - 2r(r+5)\frac{1}{2}$$

$$2r(50) = 2(50)(r+5) - r(r+5)$$

$$100r = 100r + 500 - r^2 - 5r$$

$$r^2 + 5r - 500 = 0 \qquad\qquad \textit{Simplify.}$$

In a problem like this, try to factor to obtain $(r + 25)(r - 20)$. If you are not able to obtain the factors in a reasonable time, then, using the quadratic formula,

$$r = \frac{-5 \pm \sqrt{5^2 - (4)(1)(-500)}}{2(1)} = \frac{-5 \pm \sqrt{2025}}{2} = \frac{-5 \pm 45}{2}$$

$$r = \frac{-5 + 45}{2} = 20 \quad or \quad r = \frac{-5 - 45}{2} = -25$$

Since the McSwiggins were not driving backward, we disregard the negative answer. So Mr. McSwiggin drove at the quite safe speed of 20 mph, and Ms. McSwiggin drove at 25 mph.

Practice Problem 11 Redo Example 11 if the new house is 100 miles away and the family car is driven 25 miles per hour faster than the rental moving van. Find the average speed of each vehicle if the car takes 2 hours less than the rental van. ∎

8.4 Exercises

Solve for the variable specified. Assume that all other variables are nonzero.

1. $S = 16t^2$, for t

2. $E = mc^2$, for c

3. $V = \pi r^2 h$, for r

4. $H = 0.4nd^2$, for d

5. $3H = \dfrac{1}{2}ax^2$, for x

6. $5B = \dfrac{2}{3}hx^2$, for x

7. $4y^2 + 5 = 2A$, for y

8. $9x^2 - 2 = 3B$, for x

9. $F = \dfrac{kbV^2}{r}$, for V

10. $P = \dfrac{gtw^2}{x}$, for w

11. $V = \pi(r^2 + R^2)h$, for r

12. $H = b(a^2 + w^2)$, for w

13. $L = \dfrac{kd^4}{h^2}$, for h

14. $B = \dfrac{3gy^4}{a^2}$, for y

15. $x^2 + 3bx - 10b^2 = 0$, for x

16. $y^2 - 4yw - 45w^2 = 0$, for y

17. $\dfrac{2}{3}a^2 - \dfrac{5}{3}ab = b^2$, for a

18. $\dfrac{w^2}{5} - 3y^2 = \dfrac{2wy}{5}$

19. $P = EI - RI^2$, for I

20. $A = P(1 + r)^2$, for r

21. $5w^2 + 2bw - 8 = 0$, for w

22. $7w^2 - 5bw + 3 = 0$, for w

23. $S = 2\pi rh + \pi r^2$, for r

24. $B = 3abx^2 - 5x$, for x

25. $(a + 1)x^2 + 5x + 2w = 0$, for x

26. $(b - 2)x^2 - 3x + 5y = 0$, for x

In Problems 27 through 38 use the Pythagorean theorem to find the missing side.

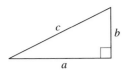

27. $b = 7$, $a = \sqrt{3}$; find c

28. $a = 2\sqrt{3}$, $b = 3$; find c

29. $c = 13$, $b = 5$; find a

30. $c = 17$, $a = 15$; find b

31. $c = \sqrt{21}$, $a = \sqrt{5}$

32. $c = \sqrt{34}$, $b = \sqrt{19}$

33. $c = 12$, $b = 2a$; find b and a

34. $c = 15$, $a = 2b$; find b and a

35. A racing sailboat is sailing on a course where the three straight line distances sailed form a right triangle. One leg of the triangle represents a distance of 12 miles. The other leg of the triangle is 6 miles shorter than the hypotenuse. What is the length of the hypotenuse of this triangle? What is the length of the other leg?

36. Tony Pitkin has a cornfield in South Dakota that is shaped like a right triangle. The hypotenuse of this triangle is a distance on the field of 10 miles. One leg of the triangular shaped field is 2 miles shorter than the other leg. Find the length of the other two sides of the field.

37. An airplane flew from London a distance of 11 kilometers due north. The plane banked to the right and flew the second leg of the journey. Finally, the plane then banked to the right and flew back to the starting point. The final leg of this journey is 3 miles longer than the second leg. The entire journey is shaped like a right triangle with the 11-kilometer distance serving as the hypotenuse. How many miles were the second and third legs of the journey? Round your answer to the nearest hundredth of a mile.

38. A Norwegian freighter heads 12 miles due east heading for the United States. The vessels turns to the left abruptly and travels for a certain distance seeking to avoid an underwater obstruction. Having discovered engine trouble, the captain turns abruptly again and heads for home port, which is 4 miles farther away than the certain distance. The entire trip is shaped like a right triangle, with the 12-mile distance serving as the hypotenuse. What were the lengths of each of the other two distances? Round your answer to the nearest hundredth of a mile.

39. The area of a rectangular wall of a barn is 126 square feet. Its length is 4 feet longer than twice its width. Find the length and width of the wall of the barn.

40. The area of a rectangular tennis court is 140 square meters. Its length is 6 meters less than twice its width. Find the length and width of the tennis court.

41. The area of a triangular flag is 72 square centimeters. Its altitude is 2 centimeters longer than twice its base. Find the height of the altitude and the length of the base.

42. The area of a triangle is 96 square centimeters. Its altitude is 2 centimeters longer than five times the base. Find the lengths of the altitude and the base.

43. Benita traveled at a constant speed for 160 miles on an old road. She then traveled 5 miles per hour faster on a newer road for 90 miles. If she drove for 6 hours, find the car's speed for each part of the trip.

44. Roberto drove at a constant speed in a rainstorm for 225 miles. He took a break and the rain stopped. He then drove 150 miles at a speed that was 5 miles per hour faster than his previous speed. If he drove for 8 hours, find the car's speed for each part of the trip.

45. A driver drove his heavily loaded truck from the company warehouse to a delivery point at 35 mph. He unloaded the truck and drove back to the warehouse at 45 mph. The total trip took 5 hours 20 minutes. How far is the delivery point from the warehouse?

46. Bob drove from home to work at 50 mph. After work the traffic was heavier, and he drove home at 45 mph. His driving time to and from work was 1 hour 16 minutes. How far does he live from his job?

To Think About

47. Solve for w. $w = \dfrac{12b^2}{\dfrac{5}{2}w + \dfrac{7}{2}b + \dfrac{21}{2}}$

48. The formula $A = P(1 + r)^2$ gives the amount A in dollars that will be obtained in 2 years at an annual compound interest rate of r if P dollars are invested. If you invest $P = \$1400$ and it grows to $\$1514.24$ in 2 years, what is the annual interest rate r?

Cumulative Review Problems

Rationalize the denominators.

49. $\dfrac{4}{\sqrt{3x}}$

50. $\dfrac{5\sqrt{6}}{2\sqrt{5}}$

51. $\dfrac{3}{\sqrt{x} + \sqrt{y}}$

52. $\dfrac{2\sqrt{3}}{\sqrt{3} - \sqrt{6}}$

53. $\dfrac{3ab}{\sqrt[3]{8ab^2}}$

Putting Your Skills to Work

A MATHEMATICAL LOOK AT TRAFFIC FLOW

Boston, Massachusetts, is in the midst of a huge highway construction project known as the "Big Dig." This ambitious project has already spent or contracted $5.2 billion and is likely to more than double that amount when finished. The basic idea is to replace a 6-lane elevated road called the Central Artery that was built to handle 75,000 vehicles per day but now is jammed with 190,000 vehicles per day. The replacement will be an 8- to 10-lane underground expressway that can easily handle the present traffic level. The following information is based on preliminary data from the Massachusetts Highway Department.

PROBLEMS FOR INDIVIDUAL INVESTIGATION AND ANALYSIS

Round answers to nearest whole number.

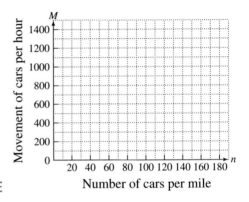

1. One model for traffic flow for the present Central Artery is given by the equation $M = -0.2 n^2 + 34.3 n$, where M is the movement of cars per hour and n is the number of cars per mile driving down the road. Find the traffic movement when $n = 0, 40, 80, 86, 120, 140, 160, 171$ cars per mile.

2. Graph the data you obtained in Problem 1. For what number of cars per mile is the greatest movement of cars on the road?

PROBLEMS FOR GROUP INVESTIGATION AND COOPERATIVE GROUP LEARNING

Together with other members of your class determine the following.

3. In one model prepared by the engineers, the movement of cars is expected to be given by the equation $M = -0.19n^2 + 38.6n$. Graph the curve and make an estimation of what number of cars per mile will create the greatest movement of cars on the highway using that equation.

4. Another possible model for traffic flow with the new underground expressway says that the movement of cars would be determined by the equation $M = an^2 + bn$ such that the equation would pass through the ordered pairs $(n, M) = (180, 468)$ and $(n, M) = (100, 1700)$. Use a system of equations to find the values of a and b, and write the equation.

INTERNET CONNECTION: Go to `http://www.prenhall.com/tobey` to be connected.

Site: The Big Dig: A Big Lesson (Beacon Hill Institute)

This page gives information and opinions regarding the Big Dig.

5. Sketch a graph showing the estimated cost of the Big Dig as a function of time. If you can find more recent cost estimates, include them in your graph.

6. Give several reasons why cost estimates have increased since the project was proposed.

8.5 Quadratic Functions

MathPro Video 21 SSM

After studying this section, you will be able to:

1 *Find the vertex and intercepts of a quadratic function.*
2 *Graph a quadratic function.*

1 Finding the Vertex and the Intercepts of a Quadratic Function

In Section 3.6 we graphed functions such as $p(x) = x^2$ and $g(x) = (x + 2)^2$. We will now study quadratic functions in more detail.

Definition of a Quadratic Function

$$f(x) = ax^2 + bx + c, \qquad a \neq 0$$

Graphs of quadratic functions of this form will always be a parabola opening upward if $a > 0$ or downward if $a < 0$. The **vertex** of a parabola is the lowest point on a parabola opening upward or the highest point on a parabola opening downward. The vertex will occur when $x = \dfrac{-b}{2a}$. To find the y value, or $f(x)$ when $x = \dfrac{-b}{2a}$, next we find $f\left(\dfrac{-b}{2a}\right)$. Therefore we can say that the quadratic function has a vertex at $\left(\dfrac{-b}{2a}, f\left(\dfrac{-b}{2a}\right)\right)$.

It is helpful to know the x-intercepts and the y-intercept when graphing the quadratic function.

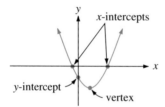 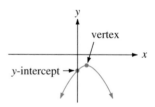

A quadratic function will always have one y-intercept. However, it may have zero, one, or two x-intercepts. Why?

When Graphing Quadratic Functions of the Form
$f(x) = ax^2 + bx + c, a \neq 0$

1. The coordinates of the vertex are $\left(\dfrac{-b}{2a}, f\left(\dfrac{-b}{2a}\right)\right)$.

2. The y-intercept is at $f(0)$.

3. The x-intercepts (if they exist) occur when $f(x) = 0$.

They can be found by the quadratic formula at all times and sometimes by factoring.

Since we may replace y by $f(x)$, the graph is equivalent to the graph of $y = ax^2 + bx + c$.

EXAMPLE 1   Find the coordinates of the vertex and the intercepts of the quadratic function $f(x) = x^2 - 8x + 15$.

For this function $a = 1$, $b = -8$, $c = 15$.

Step 1 The vertex occurs at $x = \dfrac{-b}{2a}$. Thus

$$x = \frac{-(-8)}{2(1)} = \frac{8}{2} = 4$$

The vertex has an x-coordinate of 4. To find the y-coordinate, we evaluate $f(4)$.

$$f(4) = 4^2 - 8(4) + 15 = 16 - 32 + 15 = -1$$

Thus, the vertex is $(4, -1)$.

Step 2 The y-intercept is at $f(0)$. We evaluate $f(0)$ to find the y-coordinate when x is 0.

$$f(0) = 0^2 - 8(0) + 15 = 15$$

The y-intercept is $(0, 15)$.

Step 3 If there are x-intercepts, they will occur when $f(x) = 0$. That is when $x^2 - 8x + 15 = 0$. We solve for x.

$$(x - 5)(x - 3) = 0$$

$$x - 5 = 0 \qquad x - 3 = 0$$

$$x = 5 \qquad\quad x = 3$$

Thus we conclude that the x-intercepts are $(5, 0)$ and $(3, 0)$. We list these four important points for the function in table form.

Name	x	$f(x)$
Vertex	4	-1
y-intercept	0	15
x-intercept	5	0
x-intercept	3	0

Practice Problem 1   Find the coordinates of the vertex and the intercepts of the quadratic function $f(x) = x^2 - 6x + 5$. ∎

2 *Graphing a Quadratic Function*

It is helpful to find the vertex and the intercepts of a quadratic function before graphing it.

EXAMPLE 2   Find the vertex and the intercepts. Then graph the function $f(x) = x^2 + 2x - 4$.

Here $a = 1$, $b = 2$, $c = -4$. Since $a > 0$, the parabola opens *upward*.

Step 1 We find the vertex.

$$x = \frac{-b}{2a} = \frac{-2}{2(1)} = \frac{-2}{2} = -1$$

$$f(-1) = (-1)^2 + 2(-1) - 4 = 1 + (-2) - 4 = -5$$

The vertex is $(-1, -5)$.

GRAPHING CALCULATOR

Finding the x-intercepts and the Vertex

You can use a graphing calculator to find the x-intercepts and vertex of a quadratic function. To find the intercepts of the quadratic function $f(x) = x^2 - 4x + 3$, graph $y = x^2 - 4x + 3$ on a graphing calculator using an appropriate window.
Display:

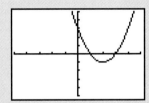

Next you can use the Trace and Zoom features or zero command of your calculator to find the x-intercepts.

You can also use the Zoom and Trace features to determine the vertex.

Some calculators have a feature that will calculate the maximum or minimum point on the graph. Use the feature that calculates the minimum point to find the vertex.
Display:

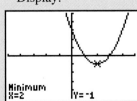

Thus the vertex is at $(2, -1)$.

Step 2 We find the y-intercept. The y-intercept is at $f(0)$.

$$f(0) = (0)^2 + 2(0) - 4 = -4$$

The y-intercept is $(0, -4)$.

Step 3 We find the x-intercepts. The x-intercepts occur when $f(x) = 0$.
We set $x^2 + 2x - 4 = 0$ and solve for x. The equation does not factor so we use the quadratic formula.

$$x = \frac{-b \pm \sqrt{b^2 - 4ac}}{2a} = \frac{-2 \pm \sqrt{2^2 - 4(1)(-4)}}{2(1)} = \frac{-2 \pm \sqrt{20}}{2} = -1 \pm \sqrt{5}$$

To aid our graphing, we will approximate the value of x to the nearest tenth by using a Square Root Table or by using a scientific calculator.

$$1 \boxed{+/-} \boxed{+} 5 \boxed{\sqrt{}} \boxed{=} \qquad 1.236068$$

$$x \approx 1.2$$

$$1 \boxed{+/-} \boxed{-} 5 \boxed{\sqrt{}} \boxed{=} \qquad -3.236068$$

$$x \approx -3.2$$

The x-intercepts are approximately $(-3.2, 0)$ and $(1.2, 0)$.

We have found that the vertex is $(-1, -5)$; the y-intercept is $(0, -4)$; and the x-intercepts are approximately $(-3.2, 0)$ and $(1.2, 0)$. We connect these points by a smooth curve to graph the parabola.

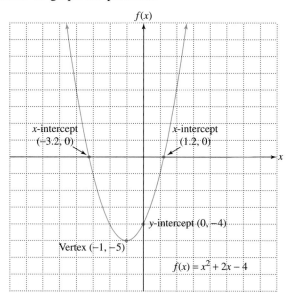

PRACTICE PROBLEM 2

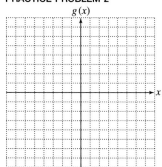

Practice Problem 2 Find the vertex and the intercepts. Then graph the function $g(x) = x^2 - 2x - 2$. ∎

EXAMPLE 3 Find the vertex and the intercepts. Then graph the function $f(x) = -2x^2 + 4x - 3$.

Here $a = -2$, $b = 4$, $c = -3$. Since $a < 0$, the parabola opens *downward*. The vertex occurs when $x = \dfrac{-b}{2a}$.

$$x = \frac{-4}{2(-2)} = \frac{-4}{-4} = 1$$

$$f(1) = -2(1)^2 + 4(1) - 3 = -2 + 4 - 3 = -1$$

The vertex is $(1, -1)$.

The y-intercept is at $f(0)$.

$$f(0) = -2(0)^2 + 4(0) - 3 = -3$$

The y-intercept is $(0, -3)$.

If there are any x-intercepts, they will occur when $f(x) = 0$. We set $-2x^2 + 4x - 3 = 0$ and use the quadratic formula to solve for x.

$$x = \frac{-4 \pm \sqrt{4^2 - 4(-2)(-3)}}{2(-2)} = \frac{-4 \pm \sqrt{-8}}{-4}$$

Because $\sqrt{-8}$ yields imaginary numbers, there are no real roots. There are no x-intercepts for the graph of the function. That is, the graph does not intersect the x-axis.

We know that the parabola opens *downward*. Thus the vertex is a maximum value at $(1, -1)$. Since this graph has no x-intercepts, we will look for three additional points to help us in drawing the graph. Try $f(2)$, $f(3)$, and $f(-1)$.

$$f(2) = -2(2)^2 + 4(2) - 3 = -8 + 8 - 3 = -3$$

$$f(3) = -2(3)^2 + 4(3) - 3 = -18 + 12 - 3 = -9$$

$$f(-1) = -2(-1)^2 + 4(-1) - 3 = -2 - 4 - 3 = -9$$

We plot the vertex, the y-intercept, and the points $(2, -3)$, $(3, -9)$ and $(-1, -9)$.

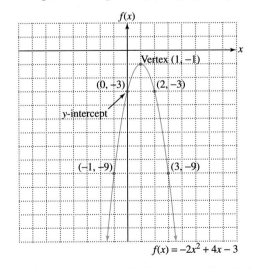

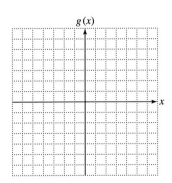

Practice Problem 3 Find the vertex and the intercepts. Graph the function $g(x) = -2x^2 - 8x - 6$. ∎

8.5 Exercises

Find the coordinates of the vertex and the intercepts of the following quadratic functions. When necessary approximate the x-intercepts to the nearest tenth.

1. $f(x) = x^2 - 8x - 20$

2. $f(x) = x^2 - 4x - 12$

3. $g(x) = -x^2 - 4x + 12$

4. $g(x) = x^2 + 10x - 24$

5. $p(x) = 3x^2 + 12x + 3$

6. $p(x) = 2x^2 + 4x + 1$

7. $r(x) = -3x^2 - 2x - 6$

8. $r(x) = -3x^2 - 4x - 3$

9. $s(x) = -2x^2 + 5x + 4$

10. $s(x) = -2x^2 + 6x + 5$

11. $f(x) = 2x^2 + x - 15$

12. $f(x) = 2x^2 + 3x - 9$

13. $h(x) = 5x^2 + 10x + 11$

14. $h(x) = 4x^2 + 8x + 38$

In the following problems, find the vertex, the y-intercept, the x-intercepts (if any exist) and then graph the function.

15. $f(x) = x^2 - 6x + 8$

16. $f(x) = x^2 + 6x + 8$

17. $g(x) = x^2 + 2x - 8$

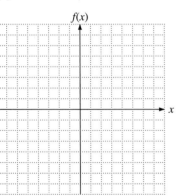

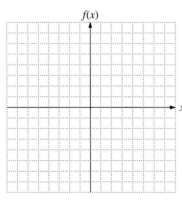

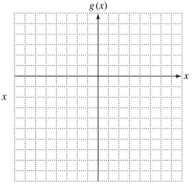

18. $g(x) = x^2 - 2x - 8$

19. $p(x) = -x^2 + 3x - 2$

20. $p(x) = -x^2 - 3x - 2$

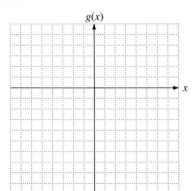

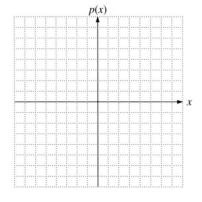

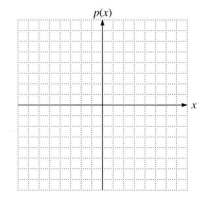

21. $r(x) = x^2 + 4x + 6$

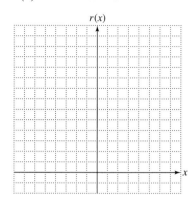

22. $r(x) = -x^2 + 4x - 5$

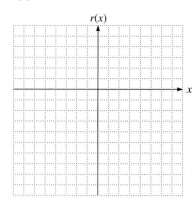

23. $f(x) = x^2 - 6x + 5$

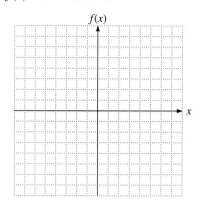

24. $f(x) = x^2 - 4x + 4$

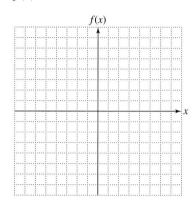

25. $g(x) = -x^2 + 6x - 9$

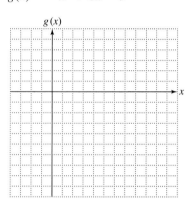

26. $g(x) = 2x^2 - 2x + 1$

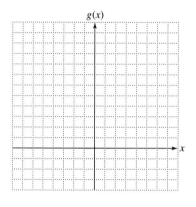

Applications

27. Susan throws a softball into the air at a speed of 32 feet per second upward from a 40-foot platform. The distance upward that the ball travels is given by the function $d(t) = -16t^2 + 32t + 40$. What is the maximum height of the softball? How many seconds does it take to reach the ground after first being thrown upward? (Round to nearest tenth.)

28. Henry is standing on a platform overlooking a baseball stadium. It is 160 feet above the playing field. When he throws a baseball upward at 64 feet per second, the distance d from the ground is given by the function $d(t) = -16t^2 + 64t + 160$. What is the maximum height of the baseball if he throws it upward? How many seconds does it take until the ball finally hits the ground? (Round to nearest tenth.)

Optional Graphing Calculator Problems

Find the vertex and intercepts for each of the following. If your answers are not exact, round your answers to the nearest tenth.

29. $y = x^2 - 4.4x + 7.59$

30. $y = x^2 + 7.8x + 13.8$

31. Graph $y = 2.3x^2 - 5.4x - 1.6$. Find the x-intercepts to the nearest tenth.

32. Graph $y = -4.6x^2 + 7.2x - 2.3$. Find the x-intercepts to the nearest tenth.

8.6 Quadratic Inequalities

MathPro   Video 21   SSM

After studying this section, you will be able to:

1 *Solve a factorable quadratic inequality.*
2 *Solve a nonfactorable quadratic inequality.*

1 *Solving a Factorable Quadratic Inequality*

We now solve quadratic inequalities such as $x^2 - 2x - 3 > 0$ and $2x^2 + x - 15 < 0$. A quadratic inequality has the form $ax^2 + bx + c < 0$ (or replace $<$ by $>$, $\leq$, $\geq$), where $a \neq 0$, and a, b, and c are real numbers. We use our knowledge of solving quadratic equations to solve quadratic inequalities.

Let's solve the inequality $x^2 - 2x - 3 > 0$. We want to find two points called the **critical points.** To do this we replace the inequality by an equals sign and solve the resulting equation.

$$x^2 - 2x - 3 = 0$$

$$(x + 1)(x - 3) = 0 \qquad \textit{Factor.}$$

$$\boxed{x + 1 = 0} \quad or \quad \boxed{x - 3 = 0} \quad \textit{Zero-product rule.}$$

$$x = -1 \qquad\qquad x = 3$$

These two solutions form critical points that divide the number line into three segments.

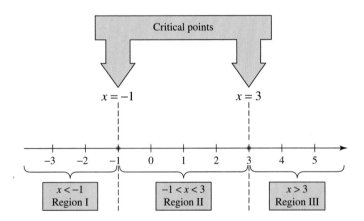

We will show as an exercise that all values of x in a given segment produce results that are greater than zero, or else all values of x in a given segment produce results that are less than zero.

To solve the quadratic inequality, we pick an arbitrary test point in each region and see if it is greater than zero, since we want to find where $x^2 - 2x - 3$ is > 0. If one point in a region satisfies the inequality, than *all* points in the region satisfy the inequality. We will test $x^2 - 2x - 3$ for three values of x.

$\boxed{x < -1,\ \textit{Region I:}}$ A sample point is $x = -2$

$$(-2)^2 - 2(-2) - 3 = 4 + 4 - 3 = 5 > 0$$

$\boxed{-1 < x < 3,\ \textit{Region II:}}$ A sample point is $x = 0$

$$(0)^2 - 2(0) - 3 = 0 + 0 - 3 = -3 < 0$$

$\boxed{x > 3,\ \textit{Region III:}}$ A sample point is $x = 4$

$$(4)^2 - 2(4) - 3 = 16 - 8 - 3 = 5 > 0$$

Thus we see $x^2 - 2x - 3 > 0$ when $x < -1$ *or* $x > 3$. No points in region II satisfy the inequality. The graph of the solution is

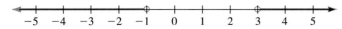

We summarize our method.

Solving a Quadratic Inequality

1. Replace the inequality symbol by an equals sign. Solve the resulting equation to find the critical points.
2. Separate the number line into three distinct regions on each side of these critical points.
3. Evaluate the quadratic expression at a test point in each region.
4. Determine which regions satisfy the original conditions of the quadratic inequality.

EXAMPLE 1   Solve and graph. $x^2 - 10x + 24 > 0$

1. Replace the inequality symbol by an equals sign. Solve the resulting equation to find the critical points $x = 4$ and $x = 6$.

$$x^2 - 10x + 24 = 0$$
$$(x - 4)(x - 6) = 0$$
$$x - 4 = 0 \quad or \quad x - 6 = 0$$
$$x = 4 \qquad \qquad x = 6$$

2. Separate the number line into distinct regions on each side of the critical points. Pick a test point in each region.

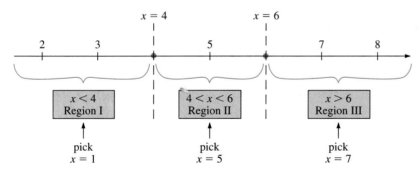

3. Evaluate the quadratic expression at the test point in each of the regions.

$$x^2 - 10x + 24 \quad \cdot$$

$\boxed{x < 4,\ \textit{Region I:}}$ We pick a sample point $x = 1$

$$(1)^2 - 10(1) + 24 = 1 - 10 + 24 = 15 > 0$$

$\boxed{4 < x < 6,\ \textit{Region II:}}$ We pick a sample point $x = 5$

$$(5)^2 - 10(5) + 24 = 25 - 50 + 24 = -1 < 0$$

$\boxed{x > 6,\ \textit{Region III:}}$ $x = 7$

$$(7)^2 - 10(7) + 24 = 49 - 70 + 24 = 3 > 0$$

Solving Quadratic Inequalities

To solve Example 2 on a graphing calculator, graph $y = 2x^2 + x - 6$, zoom in on the two x-intercepts, and use the trace feature to find that value of x where y changes from negative to positive or zero. You can use the Zero command if your calculator has it. Can you verify from your graph that the solution to Example 2 is $-2 \leq x \leq 1.5$? Use your graphing calculator and this method to solve the following problems. Round your answers to the nearest hundredth.

1. $3x^2 - 3x - 60 \geq 0$
2. $2.1x^2 + 4.3x - 29.7 > 0$
3. $15.3x^2 - 20.4x + 6.8 \geq 0$
4. $16.8x^2 > -16.8x - 35.7$

If you use a graphing calculator and you think there is no solution to a quadratic inequality, how can you *verify* this from the graph? Why is this possible?

PRACTICE PROBLEM 2

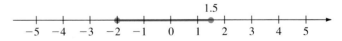

---

4. Determine which regions satisfy the original conditions of the quadratic inequality.

$$x^2 - 10x + 24 > 0 \text{ when } x < 4 \text{ or when } x > 6$$

The graph of the solution is

Practice Problem 1 Solve and graph $x^2 - 2x - 8 < 0$. ∎

EXAMPLE 2 Solve and graph. $2x^2 + x - 6 \leq 0$

$$2x^2 + x - 6 = 0$$
$$(2x - 3)(x + 2) = 0$$
$$2x - 3 = 0 \qquad or \quad x + 2 = 0$$
$$2x = 3 \qquad\qquad\qquad x = -2$$
$$x = \frac{3}{2} = 1.5$$

The critical points are $x = -2$ and $x = 1.5$. Now we arbitrarily pick a test point in each region.

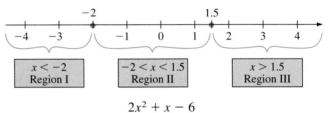

$$2x^2 + x - 6$$

Region I: We pick $x = -3$
$$2(-3)^2 + (-3) - 6 = 18 - 3 - 6 = 9 > 0$$

Region II: We pick $x = 0$
$$2(0)^2 + (0) - 6 = 0 + 0 - 6 = -6 < 0$$

Region III: We pick $x = 2$
$$2(2)^2 + (2) - 6 = 8 + 2 - 6 = 4 > 0$$

Since our inequality is $\leq$ and not just $<$, we need to include the critical points. Thus $2x^2 + x - 6 \leq 0$ when $-2 \leq x \leq 1.5$. The graph of our solution is

Practice Problem 2 Solve and graph. $3x^2 - x - 2 \geq 0$. ∎

2 Solving a Nonfactorable Quadratic Inequality

If the quadratic expression does not factor, then we will use the quadratic formula to obtain the critical points.

EXAMPLE 3 Solve and graph $x^2 + 4x > 6$. Approximate your answer to the nearest tenth.

First we write $x^2 + 4x - 6 > 0$. Now $x^2 + 4x - 6 = 0$ does not factor, so we use the quadratic formula to find the critical points.

$$x = \frac{-4 \pm \sqrt{4^2 - 4(1)(-6)}}{2(1)} = \frac{-4 \pm \sqrt{16 + 24}}{2}$$

$$= \frac{-4 \pm \sqrt{40}}{2} = \frac{-4 \pm 2\sqrt{10}}{2} = -2 \pm \sqrt{10}$$

Using a calculator or our table of square roots, we find that

$$-2 + \sqrt{10} \approx -2 + 3.162 \approx 1.162 \text{ or about } 1.2$$

$$-2 - \sqrt{10} \approx -2 - 3.162 \approx -5.162 \text{ or about } -5.2$$

We will see if $x^2 + 4x - 6 > 0$.

Region I: $x = -6$

$$(-6)^2 + 4(-6) - 6 = 36 - 24 - 6 = 6 > 0$$

Region II: $x = 0$

$$(0)^2 + 4(0) - 6 = 0 + 0 - 6 = -6 < 0$$

Region III: $x = 2$

$$(2)^2 + 4(2) - 6 = 4 + 8 - 6 = 6 > 0$$

Thus $x^2 + 4x > 6$ when $x^2 + 4x - 6 > 0$, and that occurs when $x < -2 - \sqrt{10}$ or $x > -2 + \sqrt{10}$. Rounding to the nearest tenth, our answer is

$$x < -5.2 \quad \text{or} \quad x > 1.2$$

Practice Problem 3 Solve and graph. $x^2 + 2x < 7$. Round your answer to the nearest tenth. ■

PRACTICE PROBLEM 3

8.6 Exercises

Verbal and Writing Skills

1. In solving a quadratic inequality, why is it necessary to find the critical points?

2. In solving a quadratic inequality, what is the difference between solving a problem like $ax^2 + bx + c > 0$ and a problem like $ax^2 + bx + c \geq 0$?

Solve and graph.

3. $x^2 + x - 12 < 0$

4. $x^2 + 2x - 15 < 0$

5. $2x^2 - 5x > 0$

6. $3x^2 + 11x < 0$

7. $2x^2 + x - 3 < 0$

8. $6x^2 - 5x + 1 < 0$

9. $x^2 \geq 4$

10. $x^2 + 8x + 12 \geq 0$

Solve.

11. $2x^2 \leq 11x + 6$

12. $2x^2 \leq -13x + 7$

13. $20 - x - x^2 > 0$

14. $28 - 3x - x^2 > 0$

15. $6x^2 - 5x > 6$

16. $3x^2 + 17x > -10$

17. $2x^2 + 7x - 4 \leq 0$

18. $5x^2 + 13x - 6 \leq 0$

19. $-x^2 - 5x + 36 \geq 0$

20. $-x^2 - 7x + 44 \geq 0$

21. $x^2 - 4x \leq -4$

22. $x^2 - 6x \leq -9$

Solve each of the following quadratic inequalities if possible. Round your answers to the nearest tenth.

23. $x^2 - 2x > 4$

24. $x^2 + 6x > 8$

25. $x^2 - 6x < -7$

26. $x^2 < 2x + 1$

27. $2x^2 \geq 6x - 3$

28. $4x^2 \geq 6x - 1$

29. $x^2 + 3x + 8 \leq 0$

30. $x^2 + 2x + 6 \leq 0$

Applications

In Problems 31 and 32 a projectile is fired vertically with an initial velocity of 640 feet per second. The distance s in feet above the ground after t seconds is given by the equation $s = -16t^2 + 640t$.

31. For what range of time (measured in seconds) will the height s be less than 4800 feet?

32. For what range of time (measured in seconds) will the height s be greater than 6000 feet?

In Problems 33 and 34, the profit of a manufacturing company is determined by the number x of units manufactured each day. (a) Find when the profit is greater than zero. (b) Find the daily profit when 50 units are manufactured. (c) Find the daily profit when 60 units are manufactured.

33. Profit $= -25(x^2 - 280x + 4000)$

34. Profit $= -20(x^2 - 220x + 2400)$

Optional Graphing Calculator Problems

Round your answer to the nearest tenth.

35. $12.3x^2 + 4.9x - 2.8 > 0$

36. $3.9x^2 - 6.8x - 4.6 \leq 0$

Cumulative Review Problems

37. The university's synchronized swimming team will not let Mona participate unless she passes biology with a C (70 or better) average. There are 6 tests in the semester, and she has failed (0) the first one. She decided to find a tutor. Since then, she received an 81, 92, and 80 on the next three tests. What must her minimum scores be on the last two tests to pass the course with a minimum grade of 70 and participate in synchronized swimming?

38. In a huge bowl at a college party, there are 360 ounces of mixed potato chips, peanuts, pretzels, and popcorn. There are 70 more ounces of peanuts than potato chips. There are twice as many ounces of pretzels as ounces of popcorn. There are 10 more ounces of popcorn than potato chips. How many ounces of each ingredient are in the snack mix?

Topic	Procedure	Examples
Solving a quadratic equation by using the square root property, p. 446.	If $x^2 = a$, then $x = \pm\sqrt{a}$.	Solve. $2x^2 - 50 = 0$ $\quad 2x^2 = 50$ $\quad\quad x^2 = 25$ $\quad\quad x = \pm\sqrt{25}$ $\quad\quad x = \boxed{\pm 5}$
Solving a quadratic equation by completing the square, p. 448.	1. Put the equation in the form $ax^2 + bx = -c$. 2. If $a \neq 1$, divide each term of the equation by a. 3. Square half of the numerical coefficient of the linear term. Add the result to both sides of the equation. 4. Factor the left side; then take the square root of both sides of the equation. 5. Solve each resulting equation for x.	Solve. $2x^2 - 4x - 1 = 0$ $2x^2 - 4x = 1$ $\dfrac{2x^2}{2} - \dfrac{4x}{2} = \dfrac{1}{2}$ $x^2 - 2x + \underline{\quad} = \dfrac{1}{2} + \underline{\quad}$ $x^2 - 2x + 1 = \dfrac{1}{2} + 1$ $(x - 1)^2 = \dfrac{3}{2}$ $x - 1 = \pm\sqrt{\dfrac{3}{2}}$ $\quad\quad = \dfrac{\pm\sqrt{6}}{2}$ $\boxed{x = 1 \pm \dfrac{1}{2}\sqrt{6}}$
Placing a quadratic equation in standard form, p. 455.	A quadratic equation in standard form is an equation of the form $ax^2 + bx + c = 0$, where a, b, and c are real numbers and $a \neq 0$. It is often necessary to remove parentheses and clear away fractions by multiplying each term of the equation by the LCD to obtain standard form.	Place in standard form: $\dfrac{2}{x - 3} + \dfrac{x}{x + 3} = \dfrac{5}{x^2 - 9}$ $(x + 3)(x - 3)\left[\dfrac{2}{x - 3}\right] + (x + 3)(x - 3)\left[\dfrac{x}{x + 3}\right]$ $= (x + 3)(x - 3)\left[\dfrac{5}{(x + 3)(x - 3)}\right]$ $2(x + 3) + x(x - 3) = 5$ $2x + 6 + x^2 - 3x = 5$ $\boxed{x^2 - 1x + 1 = 0}$
Solve a quadratic equation by using the quadratic formula, p. 453.	If $ax^2 + bx + c = 0$, where $a \neq 0$, $$x = \frac{-b \pm \sqrt{b^2 - 4ac}}{2a}$$ 1. Put the equation in standard form. 2. Determine the values of a, b, and c. 3. Substitute the values of a, b, and c into the formula. 4. Simplify the result to obtain the values of x. 5. Any imaginary solutions to the quadratic equation should be simplified by using the definition $\sqrt{-a} = i\sqrt{a}$, where $a > 0$.	Solve. $2x^2 = 3x - 2$ $2x^2 - 3x + 2 = 0$ $a = 2, b = -3, c = 2$ $x = \dfrac{-(-3) \pm \sqrt{(-3)^2 - 4(2)(2)}}{2(2)}$ $= \dfrac{3 \pm \sqrt{9 - 16}}{4}$ $= \dfrac{3 \pm \sqrt{-7}}{4}$ $= \boxed{\dfrac{3 \pm i\sqrt{7}}{4}}$
Equations that can be transferred to quadratic form, p. 462.	1. Replace the variable with the smallest exponent by y. 2. If possible, replace the variable with the largest exponent by y^2. (You will need to verify that this is correct.) 3. Solve the resulting equation for y. 4. Replace y by the substitution used in step 1. 5. Solve the resulting equation for x. 6. Check your solution, in the *original* equation.	Solve. $x^{2/3} - x^{1/3} - 2 = 0$ Let $y = x^{1/3}$ and $y^2 = x^{2/3}$. $y^2 - y - 2 = 0$ $(y - 2)(y + 1) = 0$ $y = 2 \quad\text{or}\quad y = -1$ $x^{1/3} = 2 \quad\text{or}\quad x^{1/3} = -1$ $(x^{1/3})^3 = 2^3 \quad\quad (x^{1/3})^3 = (-1)^3$ $x = 8 \quad\quad\quad x = -1$

Chapter Organizer

Topic	Procedure	Examples
Equations in quadratic form continued.		*Check.* $x = 8$: $\;(8)^{2/3} - (8)^{1/3} - 2 \stackrel{?}{=} 0$ $2^2 - 2 - 2 \stackrel{?}{=} 0$ $4 - 4 \stackrel{?}{=} 0$ ✓ $x = -1$: $\;(-1)^{2/3} - (-1)^{1/3} - 2 \stackrel{?}{=} 0$ $(-1)^2 - (-1) - 2 \stackrel{?}{=} 0$ $1 + 1 - 2 = 0$ ✓ Both $x = 8$ and $x = -1$ are solutions.
Solving quadratic equations containing two or more variables, p. 468.	Treat the letter to be solved for as a variable, but treat all other letters as constants. Solve the equation by factoring, by using the square root property, or by the quadratic formula.	Solve for x. **(a)** $6x^2 - 11xw + 4w^2 = 0$ **(b)** $4x^2 + 5b = 2w^2$ **(c)** $2x^2 + 3xz - 10z = 0$ **(a)** By factoring: $(3x - 4w)(2x - w) = 0$ $3x - 4w = 0 \quad$ or $\quad 2x - w = 0$ $x = \dfrac{4w}{3} \qquad\qquad x = \dfrac{w}{2}$ **(b)** Using the square root property: $4x^2 = 2w^2 - 5b$ $x^2 = \dfrac{2w^2 - 5b}{4}$ $x = \pm\sqrt{\dfrac{2w^2 - 5b}{4}} = \pm\dfrac{1}{2}\sqrt{2w^2 - 5b}$ **(c)** By the quadratic formula, $a = 2$, $b = 3z$, $c = -10z$: $x = \dfrac{-3z \pm \sqrt{9z^2 + 80z}}{4}$
The Pythagorean theorem, p. 470.	In any right triangle, if c is the length of the hypotenuse while a and b are the lengths of the two legs, then $$c^2 = a^2 + b^2$$	Find a if $c = 7$ and $b = 5$. $49 = a^2 + 25$ $49 - 25 = a^2$ $24 = a^2$ $\sqrt{24} = a$ $2\sqrt{6} = a$
Graphing quadratic functions, p. 481.	To graph quadratic functions of the form $f(x) = ax^2 + bx + c$, $a \neq 0$ **1.** Find the vertex at $\left(\dfrac{-b}{2a}, f\left(\dfrac{-b}{2a}\right)\right)$ **2.** Find the y-intercept at $f(0)$. **3.** Find the x-intercepts if they exist. Set $f(x) = 0$ and solve for x.	Graph $f(x) = x^2 + 6x + 8$. $x = \dfrac{-6}{2} = -3$ $f(-3) = (-3)^2 + 6(-3) + 8 = -1$ The vertex is $(-3, -1)$. $f(0) = (0)^2 + 6(0) + 8 = 8$ The y-intercept is $(0, 8)$. $x^2 + 6x + 8 = 0$ $(x + 2)(x + 4) = 0$ $x = -2, x = -4$ The x-intercepts are $(-2, 0)$ and $(-4, 0)$.

Topic	Procedure	Examples
Solving quadratic inequalities in one variable, p. 486.	**1.** Replace the inequality by an equals sign. Solve the resulting equation to find the critical points. **2.** Separate the number line into distinct regions on each side of these critical points. **3.** Evaluate the quadratic expression at a test point in each region. **4.** Determine which regions satisfy the original conditions of the quadratic inequality.	Solve and graph. $3x^2 + 5x - 2 > 0$ **1.** $3x^2 + 5x - 2 = 0$ $(3x - 1)(x + 2) = 0$ $3x - 1 = 0 \qquad x + 2 = 0$ $x = \dfrac{1}{3} \qquad\quad x = -2$ Critical points are -2 and $\dfrac{1}{3}$. **2.** $x < -2$ Region I $-2 < x < \frac{1}{3}$ Region II $x > \frac{1}{3}$ Region III pick $x = -3$ pick $x = 0$ pick $x = 3$ **3.** Pick $x = -3$ Pick $x = 0$ Pick $x = 3$ $3x^2 + 5x - 2$ *Region I:* $x = -3$ $3(-3)^2 + 5(-3) - 2 = 27 - 15 - 2 = 10 > 0$ *Region II:* $x = 0$ $3(0)^2 + 5(0) - 2 = 0 + 0 - 2 = -2 < 0$ *Region III:* $x = 3$ $3(3)^2 + 5(3) - 2 = 27 + 15 - 2 = 40 > 0$ **4.** We know that the expression is greater than zero (that is, $3x^2 + 5x - 2 > 0$) when $$x < -2 \text{ or } x > \frac{1}{3}.$$

Chapter 8 Review Problems

Solve each of the following problems by the specified method. Simplify all answers.

Solve by the square root property.

1. $3x^2 = 54$

2. $(x + 2)^2 = 25$

Solve by completing the square.

3. $x^2 + 8x + 13 = 0$

4. $4x^2 - 8x + 1 = 0$

Solve by the quadratic formula.

5. $3x^2 - 10x + 6 = 0$

6. $x^2 - 6x - 4 = 0$

Solve by any appropriate method and simplify your answer. Express any nonreal complex solutions using i notation.

7. $3x^2 - 8x + 6 = 0$

8. $4x^2 - 12x + 9 = 0$

9. $11x - 7x^2 = 2x$

10. $8x^2 - 26x + 15 = 0$

11. $12x^2 + 35 = 8x^2 + 15$

12. $9x^2 + 27 = 0$

13. $x^2 + 4x + 4 = 36$

14. $25x^2 - 10x + 1 = 1$

15. $2x^2 - 15 = -x$

16. $6x^2 + 12x - 24 = 0$

17. $4x^2 - 3x + 2 = 0$

18. $3x^2 + 5x + 1 = 0$

19. $3x(3x + 2) - 2 = 3x$

20. $10x(x - 2) + 10 = 2x$

21. $\dfrac{(x + 2)^2}{5} + 2x = -9$

22. $\dfrac{(x - 2)^2}{20} + x = -3$

23. $\dfrac{5}{6}x^2 - x + \dfrac{1}{3} = 0$

24. $\dfrac{4}{5}x^2 + x + \dfrac{1}{5} = 0$

25. $y + \dfrac{5}{3y} + \dfrac{17}{6} = 0$

26. $\dfrac{19}{y} - \dfrac{15}{y^2} + 10 = 0$

27. $\dfrac{15}{y^2} - \dfrac{2}{y} = 1$

28. $y - 18 + \dfrac{81}{y} = 0$

29. $(3y + 2)(y - 1) = 7(-y + 1)$

30. $y(y + 1) + (y + 2)^2 = 4$

31. $\dfrac{2x}{x + 3} + \dfrac{3x - 1}{x + 1} = 3$

32. $\dfrac{4x + 1}{2x + 5} + \dfrac{3x}{x + 4} = 2$

Determine the nature of each of the following quadratic equations. Do not solve the equation. Find the discriminant in each case and determine if the equation has **(a)** one rational solution, **(b)** two rational solutions, **(c)** two irrational solutions, or **(d)** two nonreal complex solutions.

33. $2x^2 + 5x - 3 = 0$

34. $3x^2 - 7x - 12 = 0$

35. $4x^2 - 6x + 5 = 0$

36. $25x^2 - 20x + 4 = 0$

Write a quadratic equation having the given numbers as solutions.

37. 5, −5

38. 3*i*, −3*i*

39. $4\sqrt{2}$, $-4\sqrt{2}$

40. −3/4, −1/2

Use a calculator or a square root table to approximate the solutions to the nearest tenth for each of the following quadratic equations.

41. $3x^2 + 9x + 4 = 0$

42. $5x^2 - 2x - 9 = 0$

Solve for any valid real roots.

43. $x^4 - 6x^2 + 8 = 0$

44. $2x^6 - 5x^3 - 3 = 0$

45. $3x^{1/2} - 11x^{1/4} = 4$

46. $x^{2/3} + 9x^{1/3} = -8$

47. $(2x - 5)^2 + 4(2x - 5) + 3 = 0$

48. $1 + 4x^{-8} = 5x^{-4}$

Solve for the variable specified. Assume that all radical expressions obtained have a positive radicand.

49. $A = \dfrac{2B^2 C}{3H}$, for *B*

50. $2H = 3g(a^2 + b^2)$, for *b*

51. $20d^2 - xd - x^2 = 0$, for *d*

52. $yx^2 - 3x - 7 = 0$, for *x*

53. $3y^2 - 4ay + 2a = 0$, for *y*

54. $PV = 5x^2 + 3y^2 + 2x$, for *x*

Use the Pythagorean theorem to find the missing side. Assume that c is the length of the hypotenuse of a right triangle and a and b are the lengths of the legs. Leave your answer as a radical in simplified form.

55. $a = 3\sqrt{2}$, $b = 2$; find *c*

56. $c = 16$, $b = 4$; find *a*

57. A plane is 6 miles away from an observer. The plane is exactly 5 miles above the ground. The plane is directly above a car. How far is the car from the observer? Round your answer to the nearest tenth of a mile.

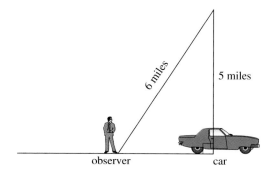

58. The area of a rectangle is 203 square meters. Its length is 1 meter longer than four times its width. Find the length and width of the rectangle.

59. The area of a triangle is 70 square centimeters. Its altitude is 6 meters longer than twice the length of the base. Find the dimensions of the altitude and base.

60. Jessica had driven at a constant speed for 200 miles. Then it started to rain. So for the next 90 miles she traveled 5 miles per hour slower. The entire trip took 6 hours of driving time. Find her speed for each part of the trip.

61. John rode in a motorboat for 60 miles at constant cruising speed to get to his fishing grounds. Then for 5 miles he trolled to catch fish. His trolling speed was 15 miles per hour slower than his cruising speed. The trip took 4 hours. Find his speed for each part of the trip.

62. Mr. and Mrs. Gomez are building a rectangular garden that is 10 feet by 6 feet. Around the outside of the garden, they will build a brick walkway. They have 100 square feet of brick. How wide should they make the brick walkway? Round your answer to the nearest tenth of a foot.

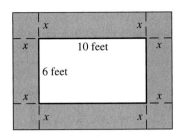

63. The local YMCA is building a rectangular swimming pool that is 40 feet by 30 feet. The builders want to make a walkway around the pool made of a nonslip cement surface. They want to use 296 square feet of nonslip cement surface. How wide should the walkway be?

Find the vertex and the intercepts of the following quadratic functions.

64. $f(x) = x^2 + 10x + 25$

65. $g(x) = -x^2 + 6x - 11$

In the following problems, find the vertex, the y-intercept, the x-intercepts (if any exist) and then graph the function.

66. $f(x) = x^2 + 4x + 3$

67. $f(x) = x^2 + 6x + 5$

68. $f(x) = -x^2 + 6x - 5$

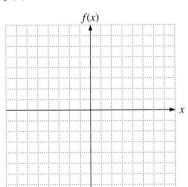

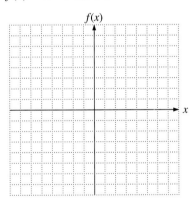

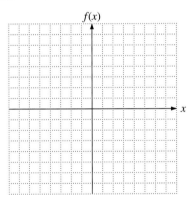

69. A salesman for an electronics store finds that in one month he can sell $(1200 - x)$ compact disc players that each sell for x dollars. Write a function for the revenue. What is the price x that will result in the maximum revenue for the store?

70. A model rocket is launched upward from a platform 40 feet above the ground. The height of the rocket h is given at any time t in seconds by the function $h(t) = -16t^2 + 400t + 40$. Find the maximum height of the rocket. What is the amount of time t that it will take to go through its complete flight and then hit the ground? (Assume the rocket does NOT have a parachute.) Round your answer to the nearest tenth.

Solve and graph your solution.

71. $x^2 + 7x - 18 < 0$

72. $x^2 + 4x - 21 < 0$

73. $x^2 - 9x + 20 > 0$

74. $x^2 - 11x + 28 > 0$

Solve, if possible, each of the following. Approximate, if necessary, any irrational solutions to the nearest tenth.

75. $3x^2 - 5x - 2 \le 0$ **76.** $2x^2 - 5x - 3 \le 0$ **77.** $9x^2 - 4 > 0$ **78.** $16x^2 - 25 > 0$

79. $x^2 - 9x > 4 - 7x$ **80.** $4x^2 - 8x \le 12 + 5x^2$ **81.** $x^2 + 13x > 16 + 7x$ **82.** $3x^2 - 12x > -11$

83. $4x^2 + 12x + 9 < 0$

84. $-2x^2 + 7x + 12 \le -3x^2 + x$

85. $(x + 4)(x - 2)(3 - x) > 0$

86. $(x + 1)(x + 4)(2 - x) < 0$

Solve the quadratic equations and simplify your answers. Use i notation for any imaginary numbers.

1. $5x^2 - 7x = 0$

2. $3x^2 + 5x = 2$

3. $\dfrac{3x}{2} - \dfrac{8}{3} = \dfrac{2}{3x}$

4. $x(x - 8) + 16 = 8(x - 6)$

5. $5x^2 - 2 = 48$

6. $\dfrac{2x}{2x + 1} - \dfrac{6}{4x^2 - 1} = \dfrac{x + 1}{2x - 1}$

7. $2x(x - 3) = -3$

8. $2x^2 - 6x + 5 = 0$

Solve for any valid real roots.

9. $x^4 - 9x^2 + 14 = 0$

10. $3x^{-2} - 11x^{-1} - 20 = 0$

11. $x^{2/3} - 2x^{1/3} - 12 = 0$

1. _____

2. _____

3. _____

4. _____

5. _____

6. _____

7. _____

8. _____

9. _____

10. _____

11. _____

Solve for the variable specified.

12. $B = \dfrac{xyw}{z^2}$, for z

13. $5y^2 + 2by + 6w = 0$, for y

14. Find the hypotenuse of a right triangle if the lengths of its legs are 6 and $2\sqrt{3}$.

15. The area of a rectangle is 80 square miles. Its length is 1 mile longer than three times its width. Find its length and width.

16. Shirley and Bill paddled a canoe at a constant speed for 6 miles. They rested, had lunch, and then paddled 1 mile per hour faster for an additional 3 miles. The entire trip took 4 hours. How fast did they paddle during each part of the trip?

17. Find the vertex and the intercepts. Then graph $f(x) = -x^2 - 6x - 5$.

Solve.

18. $-3x^2 + 10x + 8 \geq 0$ **19.** $2x^2 + 3x \geq 27$

20. Approximate to the nearest tenth using a calculator or a square root table a solution for $x^2 + 3x - 7 > 0$.

12. _____

13. _____

14. _____

15. _____

16. _____

17. _____

18. _____

19. _____

20. _____

Approximately one-half of this test is based on the content of Chapters 1–7. The remainder is based on the content of Chapter 8.

1. Simplify. $(-3x^{-2}y^3)^4$

2. Collect like terms.

$$\frac{1}{2}a^3 - 2a^2 + 3a - \frac{1}{4}a^3 - 6a + a^2$$

3. Solve for y. $a(2y + b) = 3ay - 4$

4. Graph. $6x - 3y = -12$

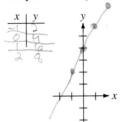

5. Write the equation of a line parallel to $2y + x = 8$ and passing through $(6, -1)$.

6. Find the volume of a sphere of radius 2 inches.

7. Factor. $125x^3 - 27y^3$

8. Simplify. $\sqrt{48x^4y^5}$

9. Multiply. $(3 + \sqrt{2})(\sqrt{6} + \sqrt{3})$

10. Rationalize the denominator.

$$\frac{5}{\sqrt{7}}$$

Solve and simplify your answer. Use the i notation for imaginary numbers.

11. $5x^2 + 12x = 15x$

12. $3x^2 + 14x = 5$

13. $44 = 3(2x - 3)^2 + 8$

14. $3 - \frac{4}{x} + \frac{5}{x^2} = 0$

1. _____

2. _____

3. _____

4. _____

5. _____

6. _____

7. _____

8. _____

9. _____

10. _____

11. _____

12. _____

13. _____

14. _____

15. _____

16. _____

17. _____

18. _____

19. _____

20. _____

21. _____

22. _____

23. _____

24. _____

Solve and check.

15. $\sqrt{x - 12} = \sqrt{x} - 2$

16. $x^{2/3} + 9x^{1/3} + 18 = 0$

Solve for y.

17. $2y^2 + 5wy - 7z = 0$

18. $3y^2 + 16z^2 = 5w$

19. The hypotenuse of a right triangle is $\sqrt{31}$. One leg of a triangle is 4. Find the length of the other leg.

20. A triangle has an area of 45 square meters. The altitude is 3 meters longer than three times the length of the base. Find each dimension.

Given the quadratic function $f(x) = -x^2 + 8x - 12$:

21. Find the vertex and the intercepts of $f(x)$.

22. Graph the function.

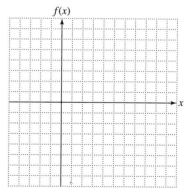

Solve each of the following quadratic inequalities.

23. $x^2 > -2x + 15$

24. $6x^2 - x \leq 2$

CHAPTER 9

The Conic Sections

Recently the Hale-Bopp comet passed within 122 million miles of Earth. The tail of the comet could be seen with the naked eye by millions of people all over the planet. At its fastest point in the elliptical orbit path, the comet was moving at 98,000 miles per hour. Do you think you could measure some of the distances involved in the long journey of this comet? Turn to the Putting Your Skills to Work on page 547.

If you are familiar with the topics in this chapter, take this test now. Check your answers with those in the back of the book. If an answer was wrong or you couldn't do a problem, study the appropriate section of the chapter.

If you are not familiar with the topics in this chapter, don't take this test now. Instead, study the examples, work the practice problems, and then take the test.

This test will help you identify those concepts that you have mastered and those that need more study.

1. _____

2. _____

3. _____

4. _____

5. _____

Section 9.1

1. Find the distance between $(-6, -2)$ and $(-3, 4)$.

2. Write the equation of a circle in standard form with a center at $(8, -2)$ and a radius of $\sqrt{7}$.

3. Put the equation of the circle $x^2 + y^2 - 2x - 4y + 1 = 0$ in standard form. Find its center and radius and sketch the graph.

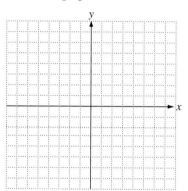

Section 9.2

Graph each parabola. Write the equation in standard form.

4. $x = (y + 1)^2 + 2$

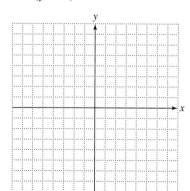

5. $x^2 = y - 4x - 1$

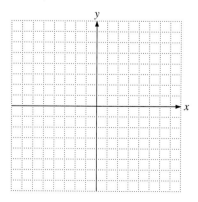

Section 9.3

Graph each ellipse. Write the equation in standard form.

6. $4x^2 + y^2 - 36 = 0$

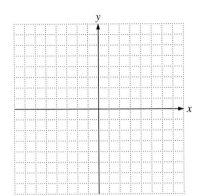

7. $\dfrac{(x+3)^2}{25} + \dfrac{(y-1)^2}{16} = 1$

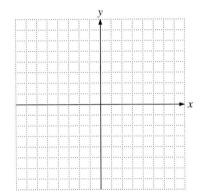

Section 9.4

Graph each hyperbola. Write the equation in standard form.

8. $25y^2 - 9x^2 = 225$

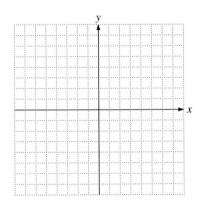

9. $\dfrac{(x-2)^2}{4} - \dfrac{(y+1)^2}{9} = 1$

Section 9.5

Solve each nonlinear system.

10. $y = x^2 + 1$
$4y^2 = 4 - x^2$

11. $x^2 + y^2 = 25$
$3x + 4y = 0$

6. _____

7. _____

8. _____

9. _____

10. _____

11. _____

9.1 The Distance Formula and the Circle

 MathPro   Video 22   SSM

After studying this section, you will be able to:

1 *Find the distance between two points.*
2 *Find the center and radius of a circle and graph the circle if the equation is in standard form.*
3 *Write the equation of a circle in standard form given its center and radius.*
4 *Place an equation of a circle in standard form.*

In this chapter we'll talk about four special geometric figures—the circle, the parabola, the ellipse, and the hyperbola. These shapes are called **conic sections** because they can be formed by slicing a cone with a plane. We'll discuss their equations and graphs. The equation of any conic is of degree 2.

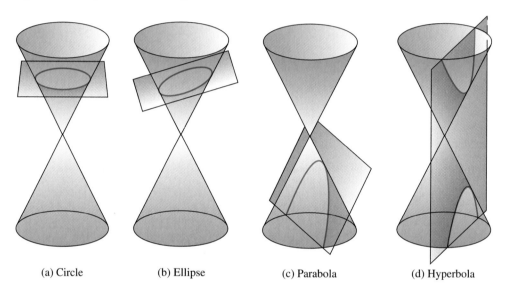

(a) Circle (b) Ellipse (c) Parabola (d) Hyperbola

Conic sections are an important and interesting subject. An entire branch of mathematics, called analytic geometry, is concerned with conic sections. Conic sections can be found in applications of physics and engineering. Satellite transmission "dishes" have parabolic shapes, the orbits of planets are ellipses and of comets are hyperbolas; the path of a ball, rocket, or bullet is a parabola (if we neglect air resistance).

1 *Finding the Distance Between Two Points*

Before we investigate the conic sections, we need to know how to find the distance between two points in the *x-y* plane. We will derive a *distance formula* and use it to find the equations for the conic sections.

Recall from Chapter 1 that to find the distance between two points we simply subtract the values of the points. For example, the distance from −3 to 5 on the *x*-axis is

$$|5 - (-3)| = |5 + 3| = 8$$

Remember that absolute value is another name for distance. We could have written

$$|-3 - (5)| = |-8| = 8$$

Similarly, the distance from −3 to 5 on the *y*-axis is

$$|5 - (-3)| = 8$$

We use this simple fact to find the distance between two points in the x-y plane. Let $A(x_1, y_1)$ and $B(x_2, y_2)$ be points on a graph. First we draw a horizontal line through A, and then we draw a vertical line through B. (We could have drawn a horizontal line through B and a vertical line through A.) The lines intersect at point $C(x_2, y_1)$. Why are the coordinates x_2, y_1? The distance from A to C is $x_2 - x_1$ and from B to C is $y_2 - y_1$.

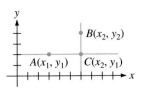

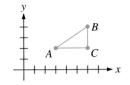

Now, if we draw a line from A to B, we have right triangle ABC. We can use the Pythagorean theorem to find the length (distance) of the line. By the Pythagorean theorem,

$$(AB)^2 = (AC)^2 + (BC)^2$$

Let's rename the distance AB as d. Then

$$d^2 = (x_2 - x_1)^2 + (y_2 - y_1)^2$$

and

$$d = \sqrt{(x_2 - x_1)^2 + (y_2 - y_1)^2}$$

This is the distance formula.

Distance Formula

The distance between two points (x_1, y_1) and (x_2, y_2) is

$$d = \sqrt{(x_2 - x_1)^2 + (y_2 - y_1)^2}$$

EXAMPLE 1 Find the distance between $(3, -4)$ and $(-2, -5)$.

To use the formula, we arbitrarily let $(x_1, y_1) = (3, -4)$ and $(x_2, y_2) = (-2, -5)$.

$$d = \sqrt{(x_2 - x_1)^2 + (y_2 - y_1)^2}$$
$$= \sqrt{[-2 - 3]^2 + [-5 - (-4)]^2}$$
$$= \sqrt{(-5)^2 + (-5 + 4)^2}$$
$$= \sqrt{(-5)^2 + (-1)^2}$$
$$= \sqrt{25 + 1} = \sqrt{26}$$

Practice Problem 1 Find the distance between $(-6, -2)$ and $(3, 1)$. ∎

The choice of which point is (x_1, y_1) and which point is (x_2, y_2) is up to you. We would obtain exactly the same answer in Example 1 if $(x_1, y_1) = (-2, -5)$ and if $(x_2, y_2) = (3, -4)$. Try it for yourself and see if you obtain the same result.

In the next example we are given the distance between two points. We are asked to find the value of x in the first coordinate.

EXAMPLE 2 Find x if the distance between $(x, 3)$ and $(2, 5)$ is $\sqrt{5}$.

$$\sqrt{5} = \sqrt{(2 - x)^2 + (5 - 3)^2} = \sqrt{(2 - x)^2 + (2)^2} = \sqrt{4 - 4x + x^2 + 4}$$
$$\sqrt{5} = \sqrt{x^2 - 4x + 8}$$

Squaring each side, we have
$$5 = x^2 - 4x + 8$$
$$0 = x^2 - 4x + 3$$
$$0 = (x - 3)(x - 1)$$
$$x = 3 \quad \text{and} \quad x = 1$$

The two solutions are $x = 3$ and $x = 1$. The points $(3, 3)$ and $(1, 3)$ are $\sqrt{5}$ units from $(2, 5)$.

Practice Problem 2 Find y if the distance between $(2, y)$ and $(1, 7)$ is $\sqrt{10}$ units. ∎

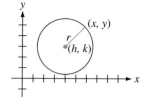

GRAPHING CALCULATOR

Graphing Circles

A graphing calculator is designed to graph *functions*. In order to graph a circle you need to separate it into two "halves," each of which is a function. Thus in order to graph the circle in Example 3, first solve for *y*.

$$(y - 3)^2 = 25 - (x - 2)^2$$

$$y - 3 = \pm\sqrt{25 - (x - 2)^2}$$

$$y = 3 \pm \sqrt{25 - (x - 2)^2}$$

Now graph the two functions

$$y_1 = 3 + \sqrt{25 - (x - 2)^2}$$
(the upper half of the circle)

$$y_2 = 3 - \sqrt{25 - (x - 2)^2}$$
(the lower half of the circle)

To get the circle to appear correctly, use a "square" window setting. Window settings will vary depending on the calculator.
Display:

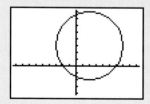

---

2 *Finding the Center and Radius of a Circle and Graphing the Circle*

A **circle** is defined as the set of all points in a plane that are at a fixed distance from a point in that plane. The fixed distance is called the **radius,** and the point is called the **center** of a circle.

We can use the distance formula to find the equation of a circle. Let a circle of radius *r* have a center at (h, k). For any point (x, y) on the circle, the distance formula tells us that

$$\sqrt{(x - h)^2 + (y - k)^2} = r$$

Squaring each side gives

$$(x - h)^2 + (y - k)^2 = r^2$$

This is the equation of a circle with center at (h, k) and radius *r*.

Equation of a Circle in Standard Form

The equation of a circle with center at (h, k) and radius *r* is

$$(x - h)^2 + (y - k)^2 = r^2$$

EXAMPLE 3   Find the center and radius of the circle $(x - 2)^2 + (y - 3)^2 = 25$. Then sketch its graph.

From the equation of a circle

$$(x - h)^2 + (y - k)^2 = r^2$$

we see that $(h, k) = (2, 3)$. Thus the center of the circle is at $(2, 3)$. Since $r^2 = 25$, the radius of the circle is $r = 5$. The graph of this circle is shown on the right.

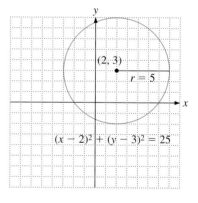

Practice Problem 3   Find the center and radius of the circle $(x + 1)^2 + (y + 2)^2 = 9$. Then sketch its graph.

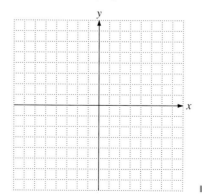

---

3 Writing the Equation of a Circle Given the Center and Radius

We can write the equation of a circle in standard form if we are given the center and the radius. We use the general equation of a circle.

EXAMPLE 4   Write the equation of a circle with center $(-1, 3)$ and radius $\sqrt{5}$. Put your answer in standard form.

We are given that $(h, k) = (-1, 3)$ and $r = \sqrt{5}$. Thus $(x - h)^2 + (y - k)^2 = r^2$ becomes

$$[x - (-1)]^2 + [y - 3]^2 = (\sqrt{5})^2$$
$$(x + 1)^2 + (y - 3)^2 = 5$$

Practice Problem 4   Write the equation of a circle with center at $(-5, 0)$ and radius $= \sqrt{3}$. Leave your answer in standard form. ■

4 Placing the Equation of a Circle in Standard Form

The standard form of the equation of a circle helps us to sketch the graph of the circle. Sometimes the equation of a circle is not given in standard form and we need to rewrite the equation.

EXAMPLE 5   Write the equation of the circle $x^2 + 2x + y^2 + 6y + 6 = 0$ in standard form. Find the radius and center of the circle and sketch its graph.

The general equation of a circle is

$$(x - h)^2 + (y - k)^2 = r^2$$

If we multiply out the terms in the equation, we get

$$(x^2 - 2hx + h^2) + (y^2 - 2ky + k^2) = r^2$$

Comparing this with the equation we were given,

$$(x^2 + 2x) + (y^2 + 6y) = -6$$

suggests that we complete the square to put the equation in standard form.

$$x^2 + 2x + \underline{\hspace{1cm}} + y^2 + 6y + \underline{\hspace{1cm}} = -6$$
$$x^2 + 2x + 1 \quad + y^2 + 6y + 9 \quad = -6 + 1 + 9$$
$$x^2 + 2x + 1 + y^2 + 6y + 9 = 4$$
$$(x + 1)^2 + (y + 3)^2 = 4$$

Thus, the center is at $(-1, -3)$ and the radius is 2. The sketch of the circle is shown.

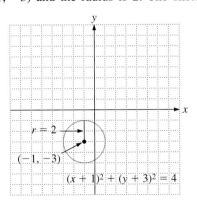

Practice Problem 5   Put the equation of the circle $x^2 + 4x + y^2 + 2y - 20 = 0$ in standard form. Find its radius and center and sketch its graph. ■

GRAPHING CALCULATOR

Exploration

One way to graph Example 5 is to write the equation as a quadratic in y and then employ the quadratic formula.

$$y^2 + 6y + (x^2 + 2x + 6) = 0$$

$$ay^2 + by + c = 0$$

$$a = 1, \; b = 6,$$
$$c = x^2 + 2x + 6$$

$$y = \frac{-6 \pm \sqrt{36 - 4(1)(x^2 + 2x + 6)}}{2(1)}$$

$$y = \frac{-6 \pm \sqrt{12 - 8x - 4x^2}}{2}$$

Thus we have the two "halves" of the curve.

$$y_1 = \frac{-6 + \sqrt{12 - 8x - 4x^2}}{2}$$

$$y_2 = \frac{-6 - \sqrt{12 - 8x - 4x^2}}{2}$$

Graph y_1 and y_2 on one axis to obtain the graph of the following circle:

$$x^2 + y^2 + 6x - 4y - 12 = 0$$

From your graph, estimate the value of the radius, and the coordinates of the center. Some graphing calculators have a feature to get a background grid for your graphs. If you have this, use it to help find the coordinates of the center.

PRACTICE PROBLEM 5

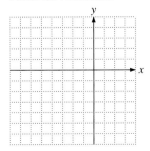

9.1 Exercises

Verbal and Writing Skills

1. Explain how you would find the distance from -2 to 4 on the y-axis.

2. Explain how you would find the distance between $(3, -1)$ and $(-4, 0)$ in the x-y plane.

3. $(x - 1)^2 + (y + 2)^2 = 9$ is the equation of a circle. What is the center and the radius of the circle?

4. $x^2 - 6x + y^2 - 2y = 6$ is the equation of a circle. Explain how you would change the equation to standard form.

Find the distance between each pair of points. Simplify your answer.

5. $(2, 3)$ and $(3, 5)$

6. $(4, 1)$ and $(6, 3)$

7. $(6, -2)$ and $(-1, 3)$

8. $(-4, 1)$ and $(2, -3)$

9. $(-2, -6)$ and $(3, -4)$

10. $(-7, -1)$ and $(3, -2)$

11. $(-4, 3)$ and $(-1, 7)$

12. $(-1, -1)$ and $(3, 2)$

13. $(3, 9)$ and $(-2, -3)$

14. $(8, 4)$ and $(-4, -1)$

15. $(0, -3)$ and $(4, 1)$

16. $(-5, -6)$ and $(2, 0)$

17. $\left(\frac{1}{3}, \frac{3}{5}\right)$ and $\left(\frac{7}{3}, \frac{1}{5}\right)$

18. $\left(-\frac{1}{4}, \frac{1}{7}\right)$ and $\left(\frac{3}{4}, \frac{6}{7}\right)$

19. $(1.3, 2.6)$ and $(-5.7, 1.6)$

20. $(8.2, 3.5)$ and $(6.2, -0.5)$

21. $(2\sqrt{3}, -1)$ and $(\sqrt{3}, -2)$

22. $(4, \sqrt{2})$ and $(-1, 3\sqrt{2})$

23. Find the distance between $(5.23, -1.67)$ and $(2.98, 3.05)$.

24. Find the distance between $(8.67, -5.33)$ and $(-2.58, 5.82)$.

Find the specified variable so that the distance between them is as given.

25. $(7, 2)$ and $(1, y)$ is 10

26. $(3, y)$ and $(3, -5)$ is 9

27. $(1.5, 2)$ and $(0, y)$ is 2.5

28. $\left(1, \frac{15}{2}\right)$ and $\left(x, -\frac{1}{2}\right)$ is 10

29. $(7, 3)$ and $(x, 6)$ is $\sqrt{10}$

30. $(4, 5)$ and $(2, y)$ is $\sqrt{5}$

Applications

Use the following information to solve Problems 31 and 32. An airport is located at point O. A short-range radar tower is located at point R. The maximum range at which the radar can detect a plane is 4 miles out from point R.

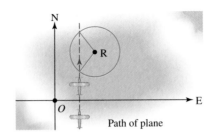

31. Assume that *R* is 6 miles east of *O* and 6 miles north of *O*. In other words, *R* is located at the point (6, 6). An airplane is flying parallel to and 4 miles east of the north axis. In other words, the plane is flying along the path $x = 4$. What is the *greatest distance* north of the airport at which the plane can still be detected by the radar at *R*? Round your answer to the nearest tenth of a mile.

32. Assume that *R* is 5 miles east of *O* and 7 miles north of *O*. In other words, *R* is located at the point (5, 7). An airplane is flying parallel to and 2 miles east of the north axis. In other words, the plane is flying along the path $x = 2$. What is the *shortest distance* north of the airport at which the plane can be detected by the radar at *R*? Round your answer to the nearest tenth of a mile.

Write in standard form the equation of the circle with the given center and radius.

33. Center (7, 12); $r = 15$

34. Center (2, 5); $r = 7$

35. Center (7, −4); $r = 2$

36. Center (5, −14); $r = 6$

37. Center (−1, −7); $r = \sqrt{5}$

38. Center (−3, −5); $r = \sqrt{2}$

39. Center (−3.5, 0); $r = 6$

40. Center (0, −5); $r = 11$

41. Center (0, 0); $r = 12$

42. Center (0, −4.5); $r = 4$

43. Center (26.8, 29.2); $r = 46.53$

44. Center (−35.82, 88.37); $r = 29.63$

Give the center and radius of each circle. Then graph it.

45. $x^2 + y^2 = 9$

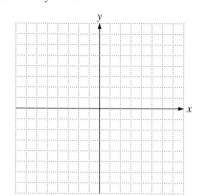

46. $x^2 + y^2 = 25$

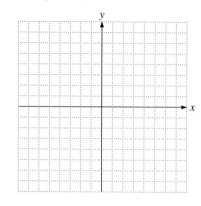

47. $(x - 3)^2 + (y - 2)^2 = 4$

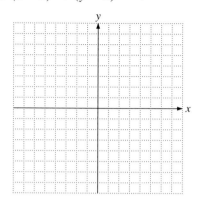

48. $(x - 5)^2 + (y - 3)^2 = 16$

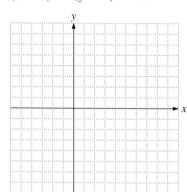

49. $(x + 2)^2 + (y - 3)^2 = 25$

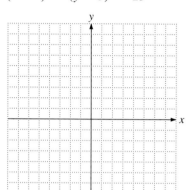

50. $\left(x - \dfrac{3}{2}\right)^2 + (y + 2)^2 = 9$

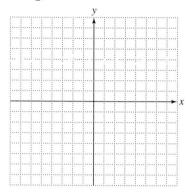

51. $(x + 3)^2 + y^2 = 11$

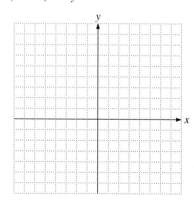

52. $x^2 + (y + 4)^2 = 7$

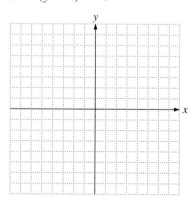

Put each equation in standard form, using the approach of Example 5. Find the center and radius of each circle.

53. $x^2 + y^2 + 6x - 4y - 3 = 0$

54. $x^2 + y^2 + 8x - 6y - 24 = 0$

55. $x^2 + y^2 - 12x + 2y - 12 = 0$

56. $x^2 + y^2 + 4x - 4y + 7 = 0$

57. $x^2 + y^2 + 12x + 10 = 0$

58. $x^2 + y^2 - 6x - 21 = 0$

59. A Ferris wheel has a radius of 25.1 feet. The height of the tower t is 29.7 feet. The distance from the origin to the base of the tower is 42.7 feet. Find the equation of the circle in standard form.

60. A Ferris wheel has a radius of 25.3 feet. The height of the tower t is 31.8 feet. The distance from the origin to the base of the tower is 44.8 feet. Find the equation of the circle in standard form.

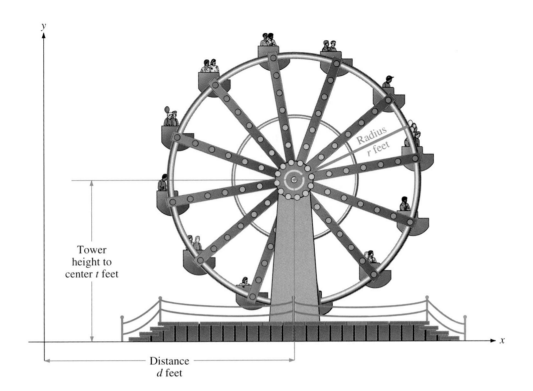

Radius r feet

Tower height to center t feet

Distance d feet

Optional Graphing Calculator Problems: *Graph each circle using your graphing calculator.*

61. $(x - 5.32)^2 + (y + 6.54)^2 = 47.28$

62. $x^2 + 9.56x + y^2 - 7.12y + 8.9995 = 0$

Cumulative Review Problems

Solve the following quadratic equations by factoring.

63. $3x^2 - 5x + 2 = 0$

64. $9 + \dfrac{3}{x} = \dfrac{2}{x^2}$

Solve the following quadratic equations by using the quadratic formula.

65. $4x^2 + 2x = 1$

66. $5x^2 - 6x - 7 = 0$

The Parabola

MathPro

Video 22

SSM

After studying this section, you will be able to:

1 *Graph a vertical parabola.*
2 *Graph a horizontal parabola.*
3 *Place the equation of a parabola in standard form.*

If we pass a plane through a cone parallel to, but not touching, the edge of the cone, we form a parabola. A **parabola** is defined as the set of points that is the same distance from some fixed line (called the **directrix**) and some fixed point (called the **focus**) that is *not* on the line.

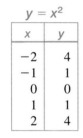

The shape of a parabola is a common one. For example, the cables that are used to support the weight of a bridge are in the shape of parabolas.

The simplest form for the equation is one variable = (another variable)². That is $y = x^2$ or $x = y^2$. We will make a table of values for each equation, plot the points, and draw a graph. For the first equation we choose values for x and find y. For the second equation we choose values for y and find x.

$y = x^2$

x	y
−2	4
−1	1
0	0
1	1
2	4

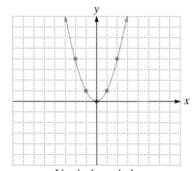

Vertical parabola

$x = y^2$

x	y
4	−2
1	−1
0	0
1	1
4	2

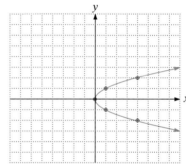

Horizontal parabola

Notice that the graph of $y = x^2$ is symmetric about the y-axis. That is, if you fold the graph along the y-axis, the two parts of the curve would coincide. For this parabola, the y-axis is the **axis of symmetry.** What is the axis of symmetry for the parabola $x = y^2$? Every parabola has an axis of symmetry; this axis can be *any* line. The point at which the parabola crosses the axis of symmetry is the vertex. What are the coordinates of the **vertex** for $y = x^2$? for $x = y^2$?

1 *Graphing a Vertical Parabola*

EXAMPLE 1   Graph the parabola $y = x^2 + 3$. Find the coordinates of the vertex and the axis of symmetry.

We take a moment to look at the equation and notice that this is the equation of a vertical parabola since the x term is squared. We make a table of values, plot points, and draw the graph.

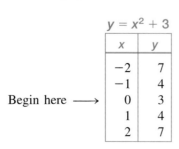

$y = x^2 + 3$

x	y
−2	7
−1	4
0	3
1	4
2	7

Begin here ⟶

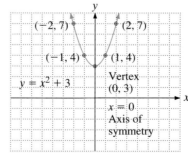

The vertex is (0, 3) and the axis of symmetry is the y-axis.

Practice Problem 1 Graph $y = 3 - x^2$. Identify the vertex and the axis of symmetry. ■

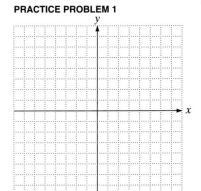

To Think About The equation $y = 3 - x^2$ can be written as $y = -x^2 + 3$. How does the term $+3$ affect the graph of the parabola? What effect does the negative sign in front of the squared term have on the graph? What would the graph of $y = -x^2 - 2$ look like? Verify by making a table of values and plotting points.

EXAMPLE 2 Graph $y = (x - 2)^2$. Identify the vertex and the axis of symmetry.

We make a table of values. We begin with $x = 2$ in the middle of the table of values because $(2 - 2)^2 = 0$. That is, when $x = 2$, $y = 0$. We then fill in the x- and y-values above and below $x = 2$. We plot the points and draw the graph.

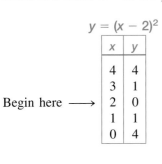

$y = (x - 2)^2$

x	y
4	4
3	1
2	0
1	1
0	4

Begin here ⟶ 2 0

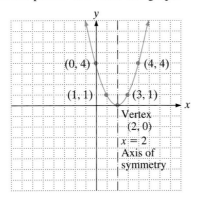

The vertex is $(2, 0)$ and the axis of symmetry is the line $x = 2$.

Practice Problem 2 Graph $y = -(x + 3)^2$. Identify the vertex and the axis of symmetry. ■

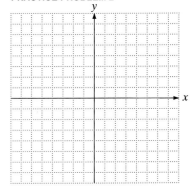

To Think About Compare the graph in Example 2 and Practice Problem 2 to the graph of $y = x^2$. These graphs have shifted along the x-axis. How can we determine this by looking at the equations?

Let us look again at the parabolas we have graphed. Notice that the graph of $y = x^2 + 3$ is similar to the graph of $y = x^2$ except that it has been shifted 3 units up. The graph of $y = (x - 2)^2$ is similar to the graph of $y = x^2$ except that it has been shifted 2 units to the right. These moves are combined in the following example.

EXAMPLE 3 Graph $y = (x - 2)^2 + 3$. Find the vertex, the axis of symmetry, and the y-intercept.

This graph looks just like the graph of $y = x^2$, except that it is shifted 2 units to the right and 3 units up. The vertex is $(2, 3)$. The axis of symmetry is $x = 2$. We can find the y-intercept by putting $x = 0$ into the equation. We get

$$y = (0 - 2)^2 + 3 = 4 + 3 = 7$$

Thus the y-intercept is $(0, 7)$.

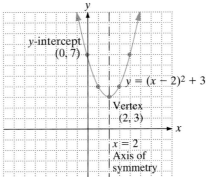

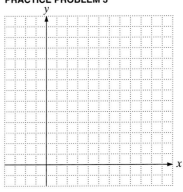

Practice Problem 3 Graph the parabola $y = (x - 6)^2 + 4$. ■

Section 9.2   *The Parabola*   **515**

Based on examples we have studied we will discuss the properties of the Standard Form of Parabolas.

> ### Standard Form of Vertical Parabolas
>
> **1.** The graph of $y = a(x - h)^2 + k$, $a \neq 0$, is a vertical parabola.
> **2.** The parabola opens upward $\smile$ if $a > 0$ and downward $\frown$ if $a < 0$.
> **3.** The vertex of the parabola is (h, k).
> **4.** The axis of symmetry is the line $x = h$.
> **5.** The y-intercept is the point where the parabola crosses the y-axis (where $x = 0$).

We can use these steps to graph a parabola. If we want greater accuracy, we should also plot a few other points.

EXAMPLE 4   Graph $y = -\dfrac{1}{2}(x + 3)^2 - 1$.

Step 1   The equation has the form $y = a(x - h)^2 + k$, where $a = -\frac{1}{2}$, $h = -3$, and $k = -1$, so it is a vertical parabola.

Step 2 $a < 0$; so the parabola opens downward.

Step 3   We have $h = -3$ and $k = -1$:

$$y = a(x - h)^2 + k$$

$$y = -\frac{1}{2}[x - (-3)]^2 + (-1)$$

Therefore, the vertex of the parabola is $(-3, -1)$.

Step 4   The axis of symmetry is the line $x = -3$.

   We plot a few points on either side of the axis of symmetry. We try $x = -1$, because $(-1 + 3)^2$ is 4 and $-\frac{1}{2}(4)$ is an integer. We avoid fractions. When $x = -1$, $y = -\frac{1}{2}(-1 + 3)^2 - 1 = -3$. Thus the point is $(-1, -3)$. The image of this point on the other side of the axis of symmetry is $(-5, -3)$. We now try $x = 1$. When $x = 1$, $y = -\frac{1}{2}(1 + 3)^2 - 1 = -9$. Thus the point is $(1, -9)$. The image of this point on the other side of the axis of symmetry is $(-7, -9)$.

Step 5   When $x = 0$,

$$y = -\frac{1}{2}(0 + 3)^2 - 1$$

$$= -\frac{1}{2}(9) - 1$$

$$= -4.5 - 1 = -5.5$$

Thus the y-intercept is $(0, -5.5)$.

The graph is shown on the right.

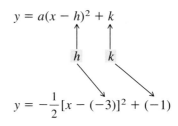

$$y = -\tfrac{1}{2}(x + 3)^2 - 1$$

PRACTICE PROBLEM 4

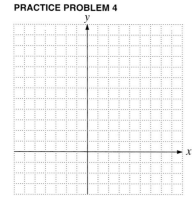

Practice Problem 4   Graph $y = \dfrac{1}{4}(x - 2)^2 + 3$. ■

2 Graphing Horizontal Parabolas

Recall that the equation of a horizontal parabola is $x = y^2$, where the squared term is the y-variable. Horizontal parabolas open to the left or right. They are symmetric to the x-axis or to a line parallel to the x-axis. We will now look at examples of horizontal parabolas.

EXAMPLE 5   Graph $x = -2y^2$.

Notice that the y-term is squared. This means that the parabola is horizontal. We make a table of values, plot points, and draw the graph. To make the table of values, we choose values for y and find x. Since there are no shifts, we begin with $y = 0$.

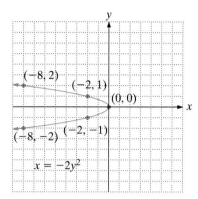

$x = -2y^2$

x	y
-8	-2
-2	-1
0	0
-2	1
-8	2

←—— Begin here.

The parabola $x = -2y^2$ has a vertex at $(0, 0)$. The axis of symmetry is the x-axis.

To Think About Compare this graph to the graph of $x = y^2$. How is it different? How is it the same? What does the coefficient -2 in the equation $x = -2y^2$ do to the graph of the equation $x = y^2$?

Practice Problem 5   Graph the parabola $x = -2y^2 + 4$. ∎

Now we can make the same type of observations for horizontal parabolas as we did for vertical ones.

PRACTICE PROBLEM 5

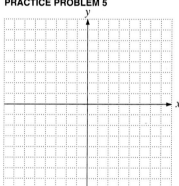

Standard Form of Horizontal Parabolas

1. The graph of $x = a(y - k)^2 + h$ is a horizontal parabola if $a \neq 0$.
2. The parabola opens to the right ⟨ if $a > 0$ and opens to the left ⟩ if $a < 0$.
3. The vertex of the parabola is (h, k).
4. The axis of symmetry is the line $y = k$.
5. The x-intercept is the point where the parabola crosses the x-axis (where $y = 0$).

EXAMPLE 6   Graph $x = (y - 3)^2 - 5$. Find the vertex, the axis of symmetry, and the x-intercept.

Step 1   The equation has the form $x = a(y - k)^2 + h$, where $a = 1$, $k = 3$, and $h = -5$, so it is a horizontal parabola.

Step 2 $a > 0$, so the parabola opens to the right.

Step 3   We have $k = 3$ and $h = -5$:

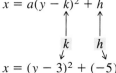

$$x = a(y - k)^2 + h$$

$$x = (y - 3)^2 + (-5)$$

Therefore, the vertex is $(-5, 3)$.

Step 4 The line $y = 3$ is the axis of symmetry.

We look for a few points on either side of the axis of symmetry. We will try y values close to the vertex $(-5, 3)$. We try $y = 4$ and $y = 2$. When $y = 4$, $x = (4 - 3)^2 - 5 = -4$. When $y = 2$, $x = (2 - 3)^2 - 5 = -4$. Thus the points are $(-4, 4)$ and $(-4, 2)$. Remember to list the x-values first in the coordinate pair. We try $y = 5$ and $y = 1$. When $y = 5$, $x = (5 - 3)^2 - 5 = -1$. When $y = 1$, $x = (1 - 3)^2 - 5 = -1$. Thus the points are $(-1, 5)$ and $(-1, 1)$. You may prefer to find one point, graph it, and find its image on the other side of the axis of symmetry, as was done in the previous example. We decided to look for both pairs of points using the equation. We plot the points and draw the graph.

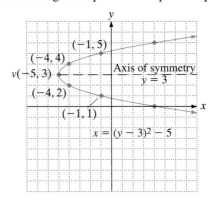

Step 5 When $y = 0$,

$$x = (0 - 3)^2 - 5 = 9 - 5 = 4$$

Thus, the x-intercept is $(4, 0)$.

Notice that the graph also crosses the y-axis. You can find the y-intercepts by setting $x = 0$ and solving the resulting quadratic equation. Try it.

Practice Problem 6 Graph the parabola $x = -(y + 1)^2 - 3$. Find the vertex, axis of symmetry, and the x-intercept. ∎

3 Placing the Equation of a Parabola in Standard Form

So far, all the equations we have graphed have been in standard form. This rarely happens in the real world. How do you suppose we put the quadratic equation $y = ax^2 + bx + c$ in the standard form of $y = a(x - h)^2 + k$? We do so by completing the square.

EXAMPLE 7 Place the equation $x = y^2 + 4y + 1$ in standard form. Then graph.

Since the y term is squared, we have a horizontal parabola, so the standard form is

$$x = a(y - k)^2 + h$$

Now

$$x = y^2 + 4y + \underline{\quad} - \underline{\quad} + 1$$

$$= y^2 + 4y + \left(\frac{4}{2}\right)^2 - \left(\frac{4}{2}\right)^2 + 1 \qquad \text{Complete the square.}$$

$$= (y^2 + 4y + 4) - 3 \qquad \text{Simplify.}$$

$$= (y + 2)^2 - 3 \qquad \text{Standard form.}$$

We see that $a = 1$, $k = -2$, and $h = -3$. Since a is positive, the parabola opens to the right. The vertex is $(-3, -2)$. The axis of symmetry is $y = -2$. If we let $y = 0$, we find that the x-intercept is $(1, 0)$. The graph is in the margin on the left.

PRACTICE PROBLEM 6

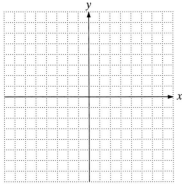

Practice Problem 7 Place in standard form and graph the equation $x = y^2 - 6y + 13$. ∎

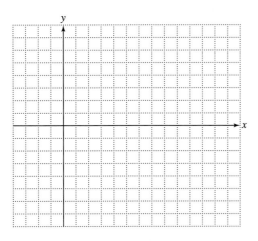

GRAPHING CALCULATOR

Graphing Parabolas

Graphing horizontal parabolas such as the one in Example 7 on a graphing calculator requires dividing the curve into two "halves." Vertical parabolas can be graphed immediately on a graphing calculator. Why is this? How can you tell if it is necessary to divide a curve into two halves? Graph the following on a graphing calculator. Use the quadratic formula when needed.

1. $y^2 + 8x - 4y = 28$

2. $4x^2 - 4x + 32y = 47$

EXAMPLE 8 Place the equation $y = 2x^2 - 4x - 1$ in standard form. Then graph.

This time the x term is squared, so we have a vertical parabola. The standard form is

$$y = a(x - h)^2 + k$$

We need to complete the square.

$$y = 2(x^2 - 2x + \underline{\quad}) - \underline{\quad} - 1$$
$$= 2[x^2 - 2x + (1)^2] - 2(1)^2 - 1$$
$$= 2(x - 1)^2 - 3$$

The parabola opens up ($a > 0$), the vertex is $(1, -3)$, the axis of symmetry is $x = 1$, and the y-intercept is $(0, -1)$.

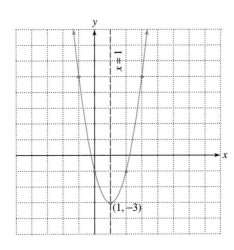

Practice Problem 8 Place in standard form and graph $y = 2x^2 + 8x + 9$. ∎

PRACTICE PROBLEM 8

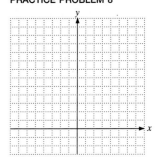

9.2 Exercises

Verbal and Writing Skills

1. The graph of $y = x^2$ is symmetric to the _____. The graph of $x = y^2$ is symmetric to the _____.

2. What is the axis of symmetry of the parabola $x = \frac{1}{2}(y + 5)^2 - 1$?

3. What is the vertex of the parabola $y = 2(x - 3)^2 + 4$?

4. How does the coefficient -6 affect the graph of the parabola $y = -6x^2$?

Graph each parabola and label the vertex. Find the y-intercept. Place x and y axes at a convenient place for your graph.

5. $y = \dfrac{1}{2}x^2$

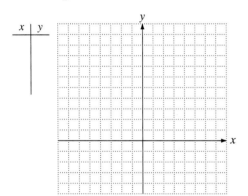

6. $y = \dfrac{1}{3}x^2$

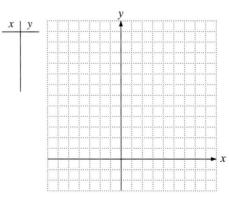

7. $y = -3x^2$

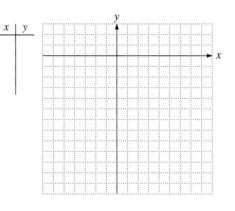

8. $y = -4x^2$

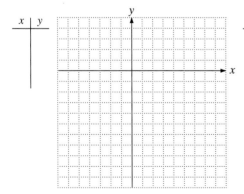

9. $y = x^2 - 6$

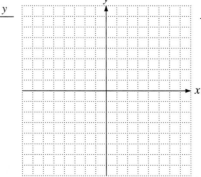

10. $y = x^2 + 2$

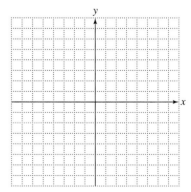

11. $y = -2x^2 + 4$

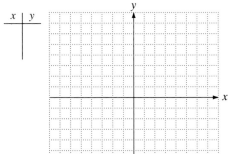

12. $y = -3x^2 + 1$

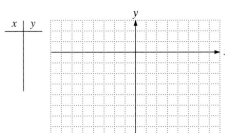

13. $y = (x - 5)^2 + 1$

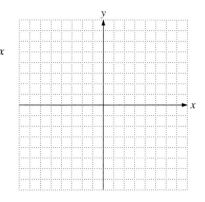

14. $y = (x - 4)^2 + 3$

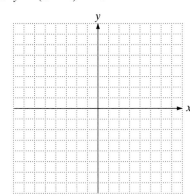

15. $y = -(x + 3)^2 + 1$

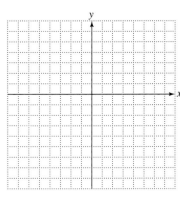

16. $y = -(x + 1)^2 + 4$

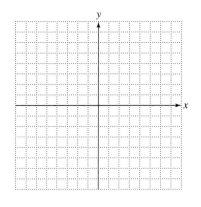

17. $y = 2(x - 1)^2 + 2$

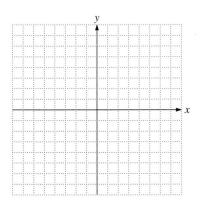

18. $y = 2(x - 3)^2 + 1$

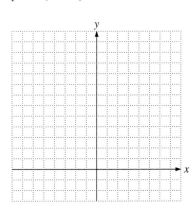

19. $y = \frac{1}{2}(x + 4)^2 - 2$

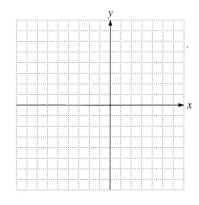

20. $y = \frac{1}{2}(x + 3)^2 - 4$

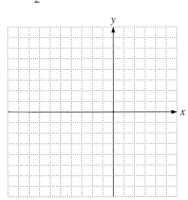

21. $y = -3(x + 2)^2 + 5$

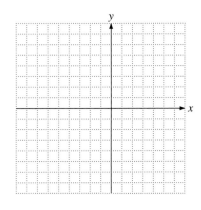

22. $y = -2(x + 3)^2 - 1$

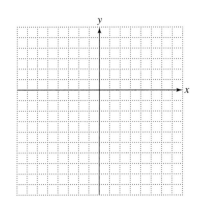

Graph each parabola and label the vertex. Find the x-intercept.

23. $x = \frac{1}{3}y^2 + 1$

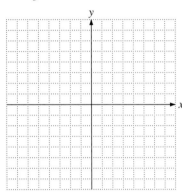

24. $x = \frac{1}{4}y^2 - 2$

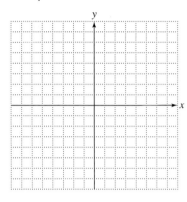

25. $x = -4y^2$

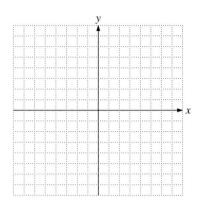

26. $x = -3y^2$

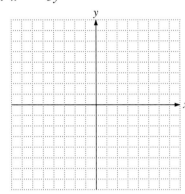

27. $x = (y - 2)^2 + 3$

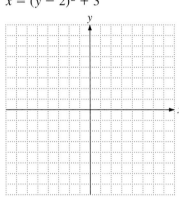

28. $x = (y - 4)^2 + 1$

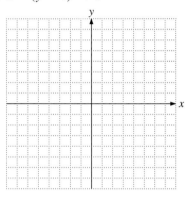

29. $x = -3(y + 1)^2 - 2$

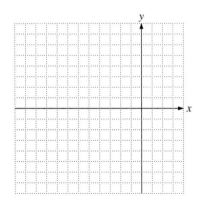

30. $x = -2(y + 3)^2 - 1$

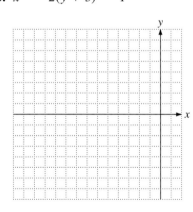

31. $x = \frac{1}{2}(y + 1)^2 + 2$

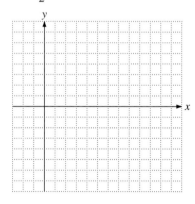

32. $x = \frac{1}{2}(y + 2)^2 + 1$

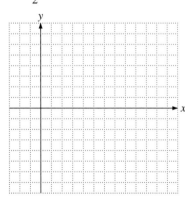

33. $x = -(y - 1)^2 + 3$

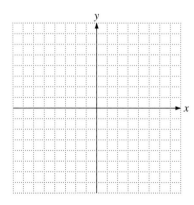

34. $x = -(y - 2)^2 + 1$

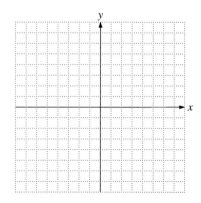

*Put each equation in standard form. Determine (**a**) if the parabola is horizontal or vertical, (**b**) the way it opens, and (**c**) the vertex.*

35. $y = x^2 + 6x + 10$

36. $y = x^2 - 4x - 1$

37. $y = -2x^2 + 4x - 3$

38. $y = -2x^2 + 4x + 5$

39. $x = y^2 + 10y + 23$

40. $x = y^2 + 8y + 9$

Applications

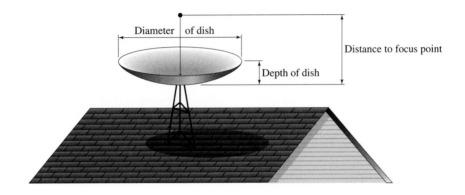

41. Find the equation of a satellite dish of the form $y = ax^2$ where the bottom of the dish passes through $(0, 0)$, the diameter of the dish is 24 inches, and the depth of the dish is 4 inches.

42. Find the equation of a satellite dish of the form $y = ax^2$ where the bottom of the dish passes through $(0, 0)$, the diameter of the dish is 32 inches, and the depth of the dish is 8 inches.

43. The distance p to the focus point of the satellite dish is given by the equation $a = 1/4p$. Find the distance p for the equation obtained in Problem 41.

44. The distance p to the focus point of the satellite dish is given by the equation $a = 1/4p$. Find the distance p for the equation obtained in Problem 42.

Optional Graphing Calculator Problems

Find the vertex and y-intercept of each parabola. Find the two x-intercepts.

45. $y = 2x^2 + 6.48x - 0.1312$

46. $y = -3x^2 + 33.66x - 73.5063$

Applications

By writing a quadratic equation in the form $y = a(x - h)^2 + k$, we can find the maximum or minimum value of the equation and the value of x for which it occurs. Remember, the equation $y = a(x - h)^2 + k$ is a vertical parabola. For $a > 0$, the parabola opens up. Thus the y-coordinate of the vertex is the smallest (or minimum) value of x. Similarly, when $a < 0$, the parabola opens downward, so the y-coordinate of the vertex is the maximum value of the equation. Since the vertex occurs at (h, k), the maximum value of the equation occurs when $x = h$. Then

$$y = -a(x - h)^2 + k = a(0) + k = k$$

For example, suppose the weekly profit of a manufacturing company in dollars is $P = -2(x - 45)^2 + 2300$ for x units manufactured. The maximum profit per week is $2300 and is attained when 45 units are manufactured. Use this approach for problems 47 through 50.

47. Find the maximum monthly profit for a company that manufactures x items. The profit equation is $P = -3x^2 + 240x + 31{,}200$. How many items must be produced each month to attain maximum profit?

48. Find the maximum monthly profit for a company that manufactures x items. The profit equation is $P = -2x^2 + 200x + 47{,}000$. How many items must be produced each month to attain maximum profit?

49. The effective yield from a grove of orange trees is described by the equation $E = x(900 - x)$, where x is the number of orange trees per acre. What is the maximum effective yield? How many orange trees per acre should be planted to achieve maximum yield?

50. A research pharmacologist has determined that sensitivity (S) to a drug depends on the dosage d, given by the equation $S = 650d - 2d^2$ where d is in milligrams. What is the maximum sensitivity that will occur? What dosage will produce that maximum sensitivity?

Cumulative Review Problems

Simplify.

51. $\sqrt{50x^3}$

52. $\sqrt[3]{40x^3y^4}$

Add.

53. $\sqrt{98x} + x\sqrt{8} - 3\sqrt{50x}$

54. $\sqrt[3]{16x^4} + 4x\sqrt[3]{2} - 8x\sqrt[3]{54}$

55. Matthew drove from work to his home at 40 mph. The next morning, an accident on the road delayed him for 15 minutes. The driving time plus the delay was 56 minutes. How far does Matthew live from his job?

56. A driver delivering eggs drove from the farm to a supermarket warehouse at 30 mph. He unloaded the eggs and drove back to the farm at 50 mph. The total trip took 2 hours and 15 minutes. How far is the farm from the supermarket warehouse?

9.3 The Ellipse

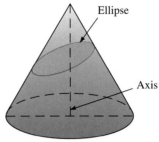

Ellipse

Axis

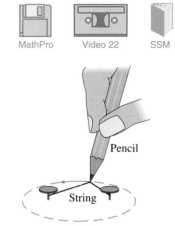

MathPro   Video 22   SSM

Pencil

String

After studying this section, you will be able to:

1. *Graph an ellipse whose center is at the origin.*
2. *Graph an ellipse whose center is at (h, k).*

Suppose a plane cuts a cone at an angle. If the plane is not perpendicular to the axis of the cone, the conic section that is formed is called an ellipse.

We define an **ellipse** as the set of points in a plane such that for each point in the set the *sum* of its distances to two fixed points is constant. The fixed points are called **foci** (plural of focus).

We can use this definition to draw an ellipse using a piece of string tied at each end to a thumbtack. Place a pencil as shown in the drawing and draw the curve, keeping the pencil pushed tightly against the string. The two thumbtacks are the foci of the ellipse that results. Examples of the ellipse can be found in the real world. The orbit of the Earth (and each of the other planets) is approximately an ellipse with the Sun at one focus.

An elliptical surface has a special reflecting property. When sound, light, or some other object originating at one focus reaches the ellipse, it is reflected in such a way that it passes through the other focus. This property can be found in the United States Capitol in a famous room known as the Statuary Hall. If a person whispers at the focus of one end of this elliptically shaped room, he can be easily heard by a person at the other focus.

F

F'

Sun

1 Graphing an Ellipse Whose Center Is at the Origin

The equation of an ellipse is similar to the equation of a circle. The standard form of an ellipse centered at the origin is given below.

Equation of an Ellipse in Standard Form

An ellipse with center at the origin has the equation

$$\frac{x^2}{a^2} + \frac{y^2}{b^2} = 1, \qquad \text{where } a, b > 0$$

To plot the ellipse, we need the x- and y-intercepts.

$$\frac{x^2}{a^2} + \frac{y^2}{b^2} = 1$$

If $x = 0$, then $\dfrac{y^2}{b^2} = 1$

$$y^2 = b^2$$
$$\pm\sqrt{y^2} = \pm\sqrt{b^2}$$
$$\pm y = \pm b \quad or \quad \boxed{y = \pm b}$$

If $y = 0$, then $\dfrac{x^2}{a^2} = 1$

$$x^2 = a^2$$
$$\pm\sqrt{x^2} = \pm\sqrt{a^2}$$
$$\pm x = \pm a \quad or \quad \boxed{x = \pm a}$$

So the x-intercepts are $(a, 0)$ and $(-a, 0)$, and the y-intercepts are $(0, b)$ and $(0, -b)$ for an ellipse of the form $\dfrac{x^2}{a^2} + \dfrac{y^2}{b^2} = 1$.

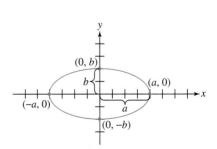

A circle is a special case of an ellipse. If $a = b$, we get

$$\frac{x^2}{a^2} + \frac{y^2}{a^2} = 1$$
$$x^2 + y^2 = a^2$$

which is the equation of a circle of radius a.

EXAMPLE 1 Graph $\dfrac{x^2}{4} + \dfrac{y^2}{9} = 1$. Label the intercepts.

This is the equation of an ellipse with center at the origin; $a^2 = 4$ and $b^2 = 9$, so $a = \pm 2$ and $b = \pm 3$. Therefore, the graph crosses the x-axis at $(-2, 0)$ and $(2, 0)$ and the y-axis at $(0, 3)$ and $(0, -3)$.

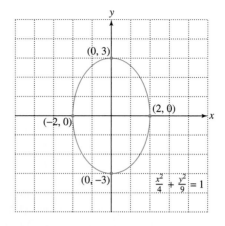

PRACTICE PROBLEM 1

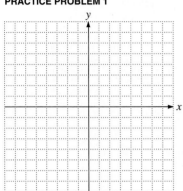

Practice Problem 1 Graph $\dfrac{x^2}{25} + \dfrac{y^2}{16} = 1$. Label the intercepts. ■

EXAMPLE 2 Graph $x^2 + 3y^2 = 12$. Label the intercepts.

Before we can graph this ellipse, we need to put the equation in standard form.

$$\frac{x^2}{12} + \frac{3y^2}{12} = \frac{12}{12} \qquad \textit{Divide each side by } 12.$$

$$\frac{x^2}{12} + \frac{y^2}{4} = 1 \qquad \textit{Simplify.}$$

Thus

$$a^2 = 12 \qquad \text{so} \qquad a = \pm 2\sqrt{3}$$

$$b^2 = 4 \qquad \text{so} \qquad b = \pm 2$$

The x intercepts are $(-2\sqrt{3}, 0)$ and $(2\sqrt{3}, 0)$, and the y-intercepts are $(0, 2)$ and $(0, -2)$. We plot these points and draw the ellipse.

PRACTICE PROBLEM 2

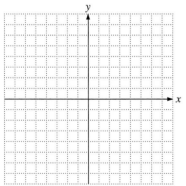

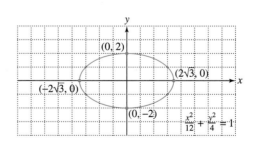

Practice Problem 2 Graph $4x^2 + y^2 = 16$. Label the intercepts. ■

2 Graphing an Ellipse Whose Center Is at (h, k)

If the center of the ellipse is not at the origin, but is at some point whose coordinates are (h, k), then the form of the standard equation is changed.

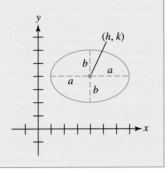

An ellipse with center at (h, k) has the equation

$$\frac{(x - h)^2}{a^2} + \frac{(y - k)^2}{b^2} = 1$$

where $a, b > 0$.

Note that a and b are *not* the x-intercepts now. Why is this? Look at the sketch. You'll see that a is the horizontal distance from the center of the ellipse to a point on the ellipse. Similarly, b is the vertical distance. Hence, when the center of the ellipse is not at the origin, the ellipse may not even cross either axis.

EXAMPLE 3   Graph $\dfrac{(x - 5)^2}{9} + \dfrac{(y - 6)^2}{4} = 1$.

The center of the ellipse is $(5, 6)$, $a = \pm 3$, and $b = \pm 2$. Therefore, we begin at $(5, 6)$. We plot points 3 units to the left, 3 units to the right, 2 units up, and 2 units down from $(5, 6)$.

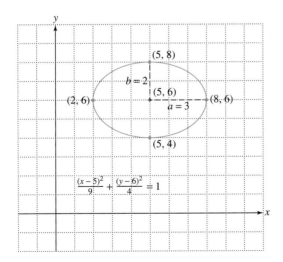

Practice Problem 3   Graph $\dfrac{(x - 2)^2}{16} + \dfrac{(y + 3)^2}{9} = 1$.

GRAPHING CALCULATOR

Graphing Ellipses

In order to graph the ellipse in Example 3 on a graphing calculator we would first need to solve for y.

$$\frac{(y - 6)^2}{4} = 1 - \frac{(x - 5)^2}{9}$$

$$(y - 6)^2 = 4\left[1 - \frac{(x - 5)^2}{9}\right]$$

$$y = 6 \pm 2\sqrt{1 - \frac{(x - 5)^2}{9}}$$

Is it necessary to break up the curve into two "halves" in order to graph the ellipse? Why?

Use the above concepts to graph.

$$\frac{(x - 2)^2}{9} + \frac{(y - 1)^2}{4} = 1$$

Determine from your graph using the trace feature the coordinates of the two x-intercepts and the two y-intercepts. Express your answer to the nearest hundredth.

PRACTICE PROBLEM 3

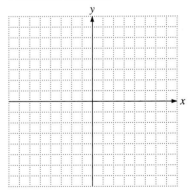

9.3 Exercises

1. What are the x- and y-intercepts of the ellipse $\dfrac{x^2}{9} + \dfrac{y^2}{16} = 1$?

2. What is the center of the ellipse $\dfrac{(x+2)^2}{4} + \dfrac{(y-3)^2}{9} = 1$?

Graph each ellipse. Label the intercepts. You may need to use a scale other than 1 square = 1 unit.

3. $\dfrac{x^2}{9} + \dfrac{y^2}{36} = 1$

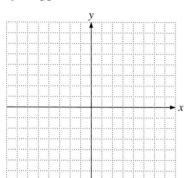

4. $\dfrac{x^2}{4} + \dfrac{y^2}{25} = 1$

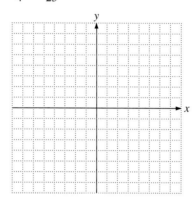

5. $\dfrac{x^2}{49} + \dfrac{y^2}{25} = 1$

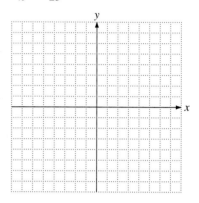

6. $\dfrac{x^2}{36} + \dfrac{y^2}{4} = 1$

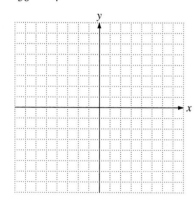

7. $\dfrac{x^2}{121} + \dfrac{y^2}{144} = 1$

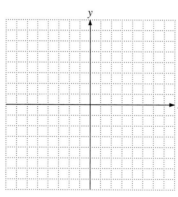

8. $\dfrac{x^2}{81} + \dfrac{y^2}{100} = 1$

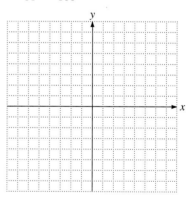

9. $4x^2 + y^2 - 36 = 0$

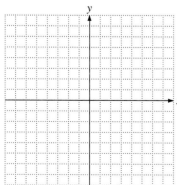

10. $x^2 + 25y^2 - 25 = 0$

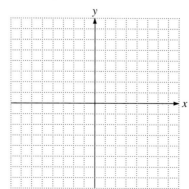

11. $x^2 + 9y^2 = 81$

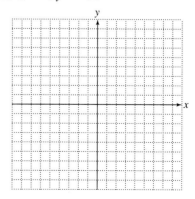

12. $4x^2 + 25y^2 = 100$

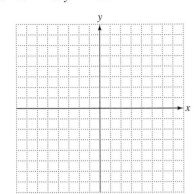

13. $x^2 + 12y^2 = 36$

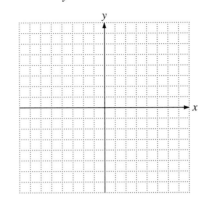

14. $8x^2 + y^2 = 16$

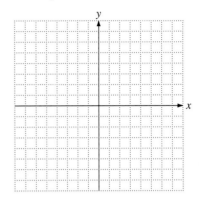

15. $7x^2 + 5y^2 = 35$

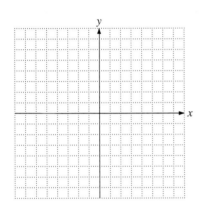

16. $5x^2 + 6y^2 = 30$

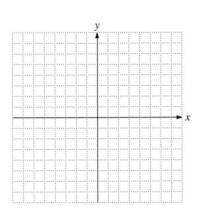

17. $\dfrac{x^2}{\frac{9}{4}} + \dfrac{y^2}{\frac{25}{4}} = 1$

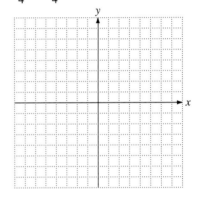

18. $\dfrac{x^2}{\frac{81}{4}} + \dfrac{y^2}{\frac{25}{16}} = 1$

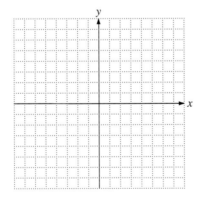

19. Write in standard form the equation of an ellipse with center at the origin, and an x-intercept at $(8, 0)$, and a y-intercept at $(0, 10)$.

20. Write in standard form the equation of an ellipse with center at the origin, and an x-intercept at $(9, 0)$, and a y-intercept at $(0, -2)$.

21. Write in standard form the equation of an ellipse with center at the origin, x-intercept at $(\sqrt{5}, 0)$, and y-intercept at $(0, -6)$.

22. Write in standard form the equation of an ellipse with center at the origin, x-intercept at $(3, 0)$, and y-intercept at $(0, \sqrt{7})$.

Applications

23. The orbit of Venus is an ellipse with the Sun as a focus. If the center of the ellipse is at the origin, an approximate equation for the orbit is

$$\frac{x^2}{5013} + \frac{y^2}{4970} = 1$$

where x and y are measured in millions of miles. Find the largest possible distance across the ellipse. Round your answer to the nearest million miles.

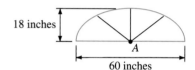

24. The window shown in the following sketch is in the shape of half an ellipse. Find the equation for the ellipse if the center of the ellipse is at point A.

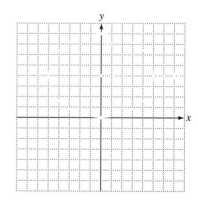

Graph each ellipse. Label the center. Place x and y axes at a convenient place for your graph. You may need to use a scale other than 1 square = 1 unit.

25. $\dfrac{(x-7)^2}{4} + \dfrac{(y-6)^2}{9} = 1$

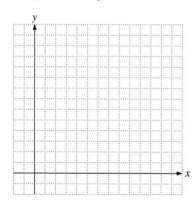

26. $\dfrac{(x-5)^2}{9} + \dfrac{(y-2)^2}{1} = 1$

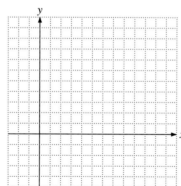

27. $\dfrac{x^2}{25} + \dfrac{(y-4)^2}{16} = 1$

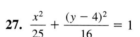

28. $\dfrac{(x+2)^2}{49} + \dfrac{y^2}{25} = 1$

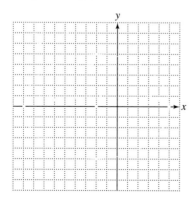

29. $\dfrac{(x+5)^2}{16} + \dfrac{(y+2)^2}{36} = 1$

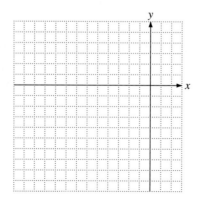

30. $\dfrac{(x+1)^2}{36} + \dfrac{(y+4)^2}{16} = 1$

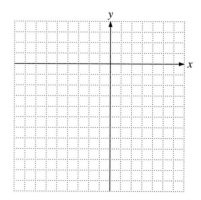

31. Write in standard form the equation of an ellipse whose vertices are $(-3, -2)$, $(5, -2)$, $(1, 1)$, and $(1, -5)$.

32. Write in standard form the equation of an ellipse whose vertices (extreme points or turning points) are $(2, 3)$, $(6, 3)$, $(4, 7)$, and $(4, -1)$.

33. Bob's backyard is a rectangle 40 meters by 60 meters. He drove two posts in the ground and fastened a rope to each post, passing it through the metal ring on his dog's collar. When the dog pulls on the rope while running, its path is an ellipse (see the figure). If the dog can just reach the sides of the rectangle, find the equation of the elliptical path.

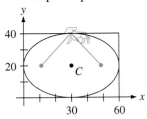

34. For what value of a does the ellipse
$$\frac{(x + 5)^2}{4} + \frac{(y + a)^2}{9} = 1$$
pass through the point $(-4, 4)$?

Optional Graphing Calculator Problems

 Find the four intercepts for each ellipse accurate to four decimal places.

35. $\dfrac{x^2}{12} + \dfrac{y^2}{19} = 1$

36. $\dfrac{(x - 3.6)^2}{14.98} + \dfrac{(y - 5.3)^2}{28.98} = 1$

To Think About

The area enclosed by the ellipse $\dfrac{x^2}{a^2} + \dfrac{y^2}{b^2} = 1$ is given by the equation $A = \pi ab$. Use the value $\pi \approx 3.1416$ to find an approximate value for each of the following problems.

37. An oval mirror is an outer boundary in the shape of an ellipse. The width of the mirror is 20 inches and the length of the mirror is 45 inches. Find the area of the mirror. Round your answer to the nearest tenth.

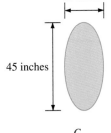

38. In Australia a type of football is played on Aussie Rules fields. These fields are in the shape of an ellipse. Suppose the distance from A to B is 185 meters and the distance from C to D is 154 meters. Find the area of the playing field. Round your answer to the nearest tenth.

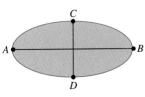

Cumulative Review Problems

Multiply and simplify.

39. $(2\sqrt{3} + 4\sqrt{2})(5\sqrt{6} - \sqrt{2})$

40. $\sqrt{3xy}(\sqrt{2x} + \sqrt{3y} + \sqrt{27})$

Rationalize the denominator.

41. $\dfrac{3}{\sqrt{6xy}}$

42. $\dfrac{5}{\sqrt{2x} - \sqrt{y}}$

9.4 The Hyperbola

MathPro

Video 23

SSM

After studying this section, you will be able to:

1 *Graph a hyperbola whose center is at the origin.*
2 *Graph a hyperbola whose center is at (h, k).*

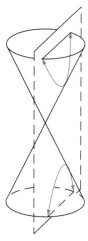

By cutting a cone by a plane as shown in this sketch, we obtain the two branches of a hyperbola. A comet moving with more than enough kinetic energy to escape the Sun's gravitational pull will travel in a hyperbolic path. Similarly, a rocket traveling with more than enough velocity to escape the Earth's gravitational field will follow a hyperbolic path.

We define a **hyperbola** as the set of points in a plane such that for each point in the set the absolute value of the *difference* of its distances to two fixed points (called **foci**) is constant.

1 Graphing a Hyperbola Whose Center Is at the Origin

Notice the similarity to the definition of an ellipse. If we replace the word *difference* by *sum*, we have the definition of an ellipse. Hence, we should expect that the equation of a hyperbola will be that of an ellipse with the plus sign replaced by a minus sign. And it is. If the hyperbola has center at the origin, its equation is

$$\frac{x^2}{a^2} - \frac{y^2}{b^2} = 1 \quad \text{or} \quad \frac{y^2}{b^2} - \frac{x^2}{a^2} = 1$$

The hyperbola has two branches. If the center of the hyperbola is at the origin, and the two branches have two *x*-intercepts but no *y*-intercepts, the hyperbola is a horizontal hyperbola.

The intercepts are also called the **vertices** of the hyperbola. If the center of the hyperbola is at the origin and the two branches have two *y*-intercepts but no *x*-intercepts, the hyperbola is a vertical hyperbola.

Equation of a Hyperbola in Standard Form with Center at Origin

Let a, b be any positive real numbers. A hyperbola with center at the origin and vertices $(-a, 0)$ and $(a, 0)$ has equation

$$\frac{x^2}{a^2} - \frac{y^2}{b^2} = 1$$

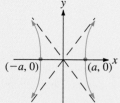

Horizontal Hyperbola

This is called a *horizontal hyperbola.*

A hyperbola with center at the origin and vertices $(0, b)$ and $(0, -b)$ has equation

$$\frac{y^2}{b^2} - \frac{x^2}{a^2} = 1$$

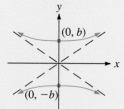

Vertical Hyperbola

This is called a *vertical hyperbola.*

Notice that the two equations are slightly different. Be aware of this difference so that when you look at the equation you will be able to tell whether the hyperbola is horizontal or vertical.

Notice the diagonal lines that we've drawn on the graphs of the hyperbolas. These lines called **asymptotes**. The two branches of the hyperbola come increasingly close to the asymptotes as the value of $|x|$ gets very large. By drawing the asymptotes and plotting the vertices, we can easily graph a hyperbola.

Asymptotes of Hyperbolas

The asymptotes of the hyperbolas $\dfrac{x^2}{a^2} - \dfrac{y^2}{b^2} = 1$ and $\dfrac{y^2}{b^2} - \dfrac{x^2}{a^2} = 1$ are

$$y = \frac{b}{a}x \quad \text{and} \quad y = -\frac{b}{a}x$$

Note that $\dfrac{b}{a}$ and $-\dfrac{b}{a}$ are the slopes of the straight lines.

An easy way to find the asymptotes is to draw extended diagonal lines through the rectangle whose center is at the origin and whose corners are at (a, b), $(a, -b)$, $(-a, b)$, and $(-a, -b)$. (This rectangle is sometimes called the fundamental rectangle.) We draw the fundamental rectangle and the asymptotes with a dashed line because they are not part of the curve.

EXAMPLE 1   Graph $\dfrac{x^2}{25} - \dfrac{y^2}{16} = 1$.

The equation has the form $\dfrac{x^2}{a^2} - \dfrac{y^2}{b^2} = 1$, so it is a horizontal hyperbola; $a^2 = 25$, so $a = 5$; $b^2 = 16$, so $b = 4$; since the hyperbola is horizontal, it has vertices at $(a, 0)$ and $(-a, 0)$ or $(5, 0)$ and $(-5, 0)$.

To draw the asymptotes, we construct a fundamental rectangle with corners at $(5, 4)$, $(5, -4)$, $(-5, 4)$, and $(-5, -4)$. We draw extended diagonal lines through the rectangle as the asymptotes. We construct each branch of the curve passing through a vertex and getting closer to the asymptotes as the graph of the hyperbola gets farther from the origin.

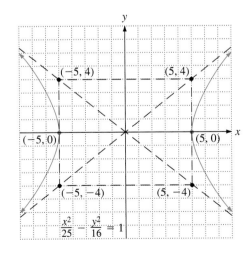

PRACTICE PROBLEM 1

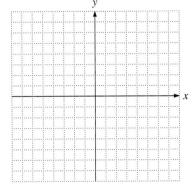

Practice Problem 1   Graph $\dfrac{x^2}{16} - \dfrac{y^2}{25} = 1$. ■

EXAMPLE 2   Graph $4y^2 - 7x^2 = 28$.

To find the vertices and asymptotes, we must put the equation in standard form. Divide each term by 28.

$$\frac{4y^2}{28} - \frac{7x^2}{28} = \frac{28}{28}$$

$$\frac{y^2}{7} - \frac{x^2}{4} = 1$$

We see we have the standard form of a vertical hyperbola with center at the origin. Here $b^2 = 7$, so $b = \sqrt{7}$; $a^2 = 4$, so $a = 2$. We have a vertical hyperbola with vertices at $(0, \sqrt{7})$ and $(0, -\sqrt{7})$. The fundamental rectangle has corners at $(2, \sqrt{7})$, $(2, -\sqrt{7})$, $(-2, \sqrt{7})$, and $(-2, -\sqrt{7})$. To aid us in graphing, we measure the distance $\sqrt{7}$ as approximately 2.6.

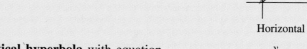

$$\frac{y^2}{7} - \frac{x^2}{4} = 1$$

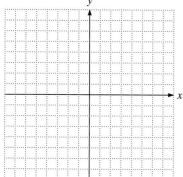

Practice Problem 2   Graph $y^2 - 4x^2 = 4$. ∎

---

2 *Graphing the Hyperbola Whose Center Is at (h, k)*

If the hyperbola does not have its center at the origin, but is shifted h units to the right or left and k units up or down, the equation is as follows.

GRAPHING CALCULATOR

Graphing Hyperbolas

Graph on your graphing calculator the hyperbola in Example 2 using

$$y_1 = \frac{\sqrt{28 + 7x^2}}{2}$$

and

$$y_2 = -\frac{\sqrt{28 + 7x^2}}{2}$$

Do you see how we obtained y_1 and y_2?

Equation of a Hyperbola in Standard Form with Center at (h, k)

A hyperbola with center at (h, k) where $a, b > 0$, is a **horizontal hyperbola** with equation

$$\frac{(x - h)^2}{a^2} - \frac{(y - k)^2}{b^2} = 1$$

and vertices $(h - a, k)$ and $(h + a, k)$.

It is a **vertical hyperbola** with equation

$$\frac{(y - k)^2}{b^2} - \frac{(x - h)^2}{a^2} = 1$$

and vertices $(h, k + b)$ and $(h, k - b)$.

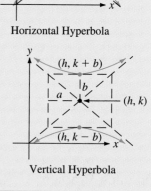

Horizontal Hyperbola

Vertical Hyperbola

EXAMPLE 3 Graph $\dfrac{(x-4)^2}{9} - \dfrac{(y-5)^2}{4} = 1$.

Exploration

Graph the hyperbola in Example 3 by using

$$y_1 = 5 + \sqrt{\dfrac{4(x-4)^2}{9} - 4}$$

and

$$y_2 = 5 - \sqrt{\dfrac{4(x-4)^2}{9} - 4}$$

Do you see how we obtained y_1 and y_2?

The center is (4, 5) and the hyperbola is horizontal. We have $a = 3$ and $b = 2$, so the vertices are $(4 \pm 3, 5)$ or (7, 5) and (1, 5). We can sketch the hyperbola more readily if we can draw a fundamental rectangle. Using (4, 5) as the center, we construct a rectangle $2a$ units wide and $2b$ units high. We then draw and extend the diagonals of the rectangle. The extended diagonals are the asymptotes for the branches of the hyperbola.

In this example, since $a = 3$ and $b = 2$, we draw a rectangle $2a = 6$ units wide and $2b = 4$ units high with a center at (4, 5). We draw extended diagonals through the rectangle. From the vertex at (7, 5), we draw a branch of the hyperobla opening to the right. From the vertex at (1, 5), we draw a branch of the hyperbola opening to the left. The graph of the hyperbola is shown.

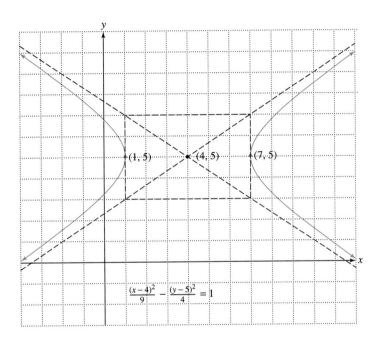

Practice Problem 3 Graph $\dfrac{(y+2)^2}{9} - \dfrac{(x-3)^2}{16} = 1$. ■

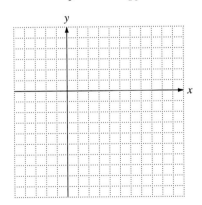

9.4 Exercises

Verbal and Writing Skills

1. What is the standard form of the equation of a horizontal hyperbola centered at the origin?

2. What are the vertices of the hyperbola $\dfrac{y^2}{9} - \dfrac{x^2}{4} = 1$? Is this a horizontal hyperbola or a vertical hyperbola? Why?

3. Explain in your own words how you would draw the graph of the hyperbola $\dfrac{x^2}{16} - \dfrac{y^2}{4} = 1$.

4. What is the center of the hyperbola $\dfrac{(x-2)^2}{4} - \dfrac{(y+3)^2}{25} = 1$?

Find the vertices and graph each hyperbola. If the equation is not in standard form, write it as such.

5. $\dfrac{x^2}{4} - \dfrac{y^2}{25} = 1$

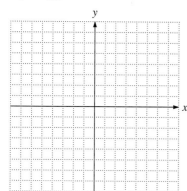

6. $\dfrac{x^2}{9} - \dfrac{y^2}{36} = 1$

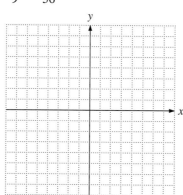

7. $\dfrac{y^2}{36} - \dfrac{x^2}{25} = 1$

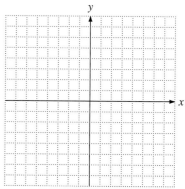

8. $\dfrac{y^2}{49} - \dfrac{x^2}{9} = 1$

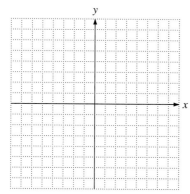

9. $\dfrac{x^2}{16} - \dfrac{y^2}{1} = 1$

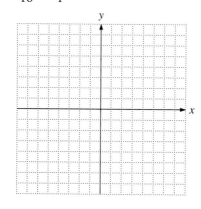

10. $\dfrac{x^2}{9} - \dfrac{y^2}{1} = 1$

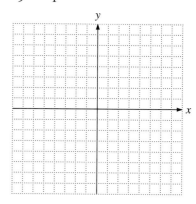

11. $9y^2 - 4x^2 = 36$

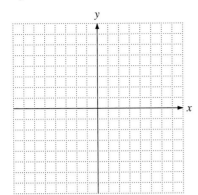

12. $4y^2 - 25x^2 = 100$

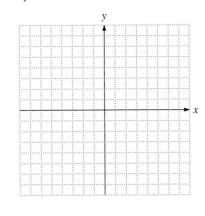

13. $49x^2 - 16y^2 = 196$

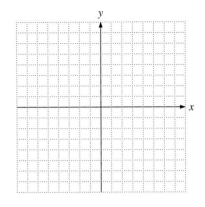

14. $4x^2 - y^2 = 64$

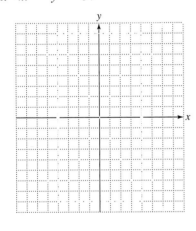

15. $12x^2 - y^2 = 36$

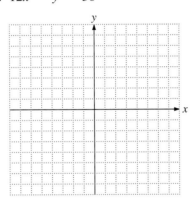

16. $8x^2 - y^2 = 16$

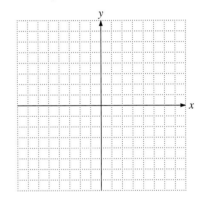

17. $3y^2 - 5x^2 = 15$

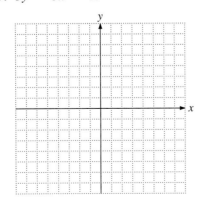

18. $5y^2 - 6x^2 = 30$

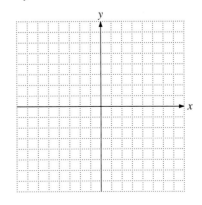

If a = b, the hyperbola is called an equilateral hyperbola. Graph the following equilateral hyperbolas.

19. $\dfrac{x^2}{9} - \dfrac{y^2}{9} = 1$

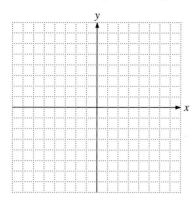

20. $\dfrac{x^2}{16} - \dfrac{y^2}{16} = 1$

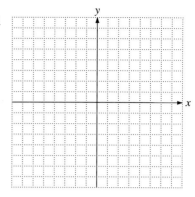

21. $5y^2 - 5x^2 = 10$

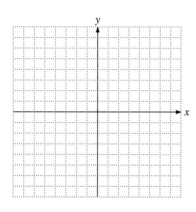

22. $8y^2 - 8x^2 = 24$

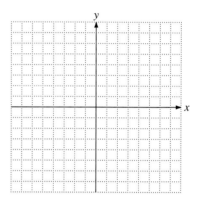

Find the equation of the hyperbola with center at the origin and:

23. Vertices at $(2, 0)$ and $(-2, 0)$; asymptotes $y = \dfrac{3}{2}x$, $y = -\dfrac{3}{2}x$

24. Vertices at $(3, 0)$ and $(-3, 0)$; asymptotes $y = \dfrac{4}{3}x$, $y = -\dfrac{4}{3}x$

25. Vertices at $(0, 7)$ and $(0, -7)$; asymptotes $y = \dfrac{7}{3}x$, $y = -\dfrac{7}{3}x$

26. Vertices at $(0, 6)$ and $(0, -6)$; asymptotes $y = \dfrac{6}{5}x$, $y = -\dfrac{6}{5}x$

Applications

27. A rocket following a hyperbolic path turns rapidly at (4, 0) and follows a path that comes closer and closer to the line $y = \dfrac{2}{3}x$ as the rocket gets farther from the tracking station. Find the equation that describes the path of the rocket if the center of the hyperbola is at (0, 0).

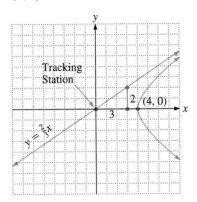

28. Some comets have an orbit that is hyperbolic in shape. The Sun is at the focus of the hyperbola. A comet is heading toward Earth but then veers off. It comes within 120 million miles of Earth. As it travels into the distance it appears to travel a path that becomes closer and closer to the equation $y = 3x$. Find the equation that describes the path of the comet.

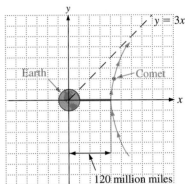

Each square on x-axis = 30 million miles
y-axis = 90 million miles

Find the center and then graph each hyperbola. Draw the axes at a convenient location. You may want to use a scale other than 1 square = 1 unit.

29. $\dfrac{(x-7)^2}{16} - \dfrac{(y-5)^2}{25} = 1$

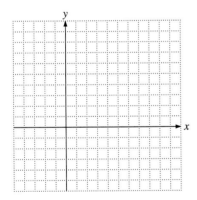

30. $\dfrac{(x-6)^2}{25} - \dfrac{(y-4)^2}{49} = 1$

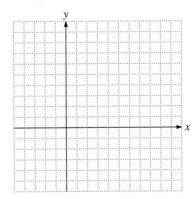

31. $\dfrac{(y+2)^2}{36} - \dfrac{(x+1)^2}{81} = 1$

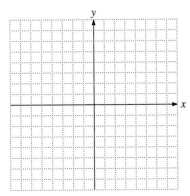

32. $\dfrac{(y+1)^2}{49} - \dfrac{(x+3)^2}{81} = 1$

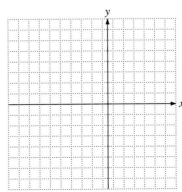

Find the center and the two vertices for the following hyperbolas.

33. $\dfrac{x^2}{5} - \dfrac{(y+3)^2}{10} = 1$

34. $\dfrac{(x+6)^2}{7} - \dfrac{y^2}{3} = 1$

★**35.** A hyperbola center is not at the origin. Its vertices are $(5, 0)$ and $(5, 14)$; one asymptote is $y = \frac{7}{5}x$. Find the equation of the hyperbola.

★**36.** A hyperbola center is not at the origin. Its vertices are $(4, -14)$ and $(4, 0)$. One asymptote is $y = -\frac{7}{4}x$. Find the equation of the hyperbola.

Optional Graphing Calculator Problems

37. In the hyperbola $8x^2 - y^2 = 16$, if $x = 3.5$, what are the two values of y?

38. In the hyperbola $x^2 - 12y^2 = 36$, if $x = 8.2$, what are the two values of y?

Cumulative Review Problems

Factor completely.

39. $2x^3 - 54$

40. $12x^2 + x - 6$

Combine.

41. $\dfrac{3}{x^2 - 5x + 6} + \dfrac{2}{x^2 - 4}$

42. $\dfrac{2x}{5x^2 + 9x - 2} - \dfrac{3}{5x - 1}$

43. The school's hockey team is giving the coach a retirement gift that costs $240. On the day that the captain of the team was collecting money, 4 hockey players were absent. He collected an equal amount from each player present and purchased the gift. Later the team captain said that if everyone had been present to chip in, each person would have contributed $2 less. How many people actually contributed? How many people in total were on the hockey team?

44. A Connecticut FM radio station claims that a minimum of 104,755 songs are played every year.
  (a) Determine the number of songs played daily. (Assume it is not a leap year.)
  (b) How much time would be left over, daily, for advertisements, news, sports, interviews, syndicated shows, and DJ chatter if the average song lasts 4 minutes?
  (c) What percentage of the air time is music?

9.5 Nonlinear Systems of Equations

MathPro   Video 23   SSM

After studying this section, you will be able to:

1 *Solve a nonlinear system by the substitution method.*
2 *Solve a nonlinear system by the addition method.*

1 Solving a Nonlinear System by the Substitution Method

Any equation that is second degree or higher is a nonlinear equation. In other words, the equation is not a straight line (which is what the word *nonlinear* means) and can't be written in the form $y = mx + b$. A **nonlinear system** of equations includes at least one nonlinear equation.

The most frequently used method for solving a nonlinear system is the method of substitution. This method works especially well when one equation of the system is linear. A sketch can often be used to verify the solution(s).

EXAMPLE 1   Solve the nonlinear system,

$$x + y - 1 = 0 \qquad (1)$$

$$y - 1 = x^2 + 2x \quad (2)$$

and verify your answer with a sketch.

We'll use the substitution method.

$$y = -x + 1 \quad (3) \qquad \text{\textit{Solve for y in equation (1).}}$$

$$(-x + 1) - 1 = x^2 + 2x \qquad \text{\textit{Substitute (3) in equation (2).}}$$

$$-x + 1 - 1 = x^2 + 2x$$

$$0 = x^2 + 3x \qquad \text{\textit{Solve the resulting quadratic equation.}}$$

$$0 = x(x + 3)$$

$$x = 0 \quad \text{\textit{or}} \quad x = -3$$

Now substitute the values for x in the equation $y = -x + 1$.

$$\text{For } x = -3: \quad y = -(-3) + 1 = +3 + 1 = 4$$

$$\text{For } x = 0: \quad y = -(0) + 1 = +1 = 1$$

Thus the solutions of the system are $(-3, 4)$ and $(0, 1)$.

To sketch the system, we see that equation (2) describes a parabola. We can rewrite it in the form

$$y = x^2 + 2x + 1$$

$$= (x + 1)^2$$

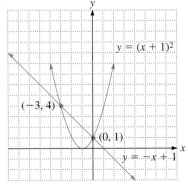

This is a parabola opening upward with a vertex at $(-1, 0)$. Equation (1) can be written as $y = -x + 1$, which is a straight line with slope $= -1$ and y intercept $(0, 1)$.

A sketch shows the two graphs intersecting at $(0, 1)$ and $(-3, 4)$. Thus the solutions are verified.

Practice Problem 1   Solve the system.

$$\frac{x^2}{4} - \frac{y^2}{4} = 1$$

$$x + y + 1 = 0 \quad \blacksquare$$

EXAMPLE 2   Solve the nonlinear system.

$$y - 2x = 0 \quad (1)$$

$$\frac{x^2}{4} + \frac{y^2}{9} = 1 \quad (2)$$

and verify your answer with a sketch.

$$y = 2x \quad (3) \qquad \text{\textit{Solve equation (1) for y.}}$$

$$\frac{x^2}{4} + \frac{(2x)^2}{9} = 1 \qquad \text{\textit{Substitute (3) into equation (2).}}$$

$$\frac{x^2}{4} + \frac{4x^2}{9} = 1 \qquad \text{\textit{Simplify.}}$$

$$36\left(\frac{x^2}{4}\right) + 36\left(\frac{4x^2}{9}\right) = 36(1) \qquad \text{\textit{Clear the fractions.}}$$

$$9x^2 + 16x^2 = 36$$

$$25x^2 = 36$$

$$x^2 = \frac{36}{25}$$

$$x = \pm\sqrt{\frac{36}{25}}$$

$$x = \pm\frac{6}{5} = \pm 1.2$$

For $x = +1.2$: $y = 2(1.2) = 2.4$.

For $x = -1.2$: $y = 2(-1.2) = -2.4$.

Thus the solutions are $(1.2, 2.4)$ and $(-1.2, -2.4)$.

We recognize $\dfrac{x^2}{4} + \dfrac{y^2}{9} = 1$ as an ellipse with center at the origin and vertices $(0, 3)$, $(0, -3)$, $(2, 0)$, and $(-2, 0)$. We recognize $y = 2x$ as a straight line with slope 2 passing through the origin. The sketch below shows that the points of intersection at $(1.2, 2.4)$ and $(-1.2, -2.4)$ seem reasonable.

PRACTICE PROBLEM 2

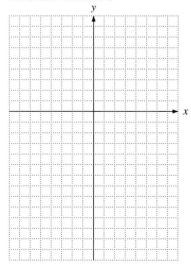

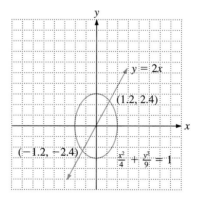

Practice Problem 2   Solve the system. Verify your answer with a sketch.

$$2x - 9 = y$$

$$xy = -4 \quad \blacksquare$$

2 Solving a Nonlinear System by the Addition Method

Sometimes a system may be solved more readily by adding the equations together.

EXAMPLE 3 Solve.

$$4x^2 + y^2 = 1 \quad (1)$$

$$x^2 + 4y^2 = 1 \quad (2)$$

Although we could use the substitution method, it is easier to use the addition method because neither equation is linear.

$$
\begin{array}{l}
-16x^2 - 4y^2 = -4 \\
\underline{x^2 + 4y^2 = 1} \\
-15x^2 = -3
\end{array}
\qquad
\begin{array}{l}
\textit{Multiply equation (1) by } -4 \textit{ and} \\
\textit{add to equation (2).}
\end{array}
$$

$$x^2 = \frac{-3}{-15}$$

$$x^2 = \frac{1}{5}$$

$$x = \pm\sqrt{\frac{1}{5}}$$

If $x = +\sqrt{\dfrac{1}{5}}$, then $x^2 = \dfrac{1}{5}$. Substituting this value into equation (2) gives

$$\frac{1}{5} + 4y^2 = 1$$

$$4y^2 = \frac{4}{5}$$

$$y^2 = \frac{1}{5}$$

$$y = \pm\sqrt{\frac{1}{5}}$$

Similarly, if $x = -\sqrt{\dfrac{1}{5}}$, then $y = \pm\sqrt{\dfrac{1}{5}}$. It is important to determine exactly how many solutions a nonlinear system of equations actually has. In this case, we have four solutions. When x is negative, there are two values for y. When x is positive, there are two values for y. If we rationalize each expression, the four solutions are $\left(\dfrac{\sqrt{5}}{5}, \dfrac{\sqrt{5}}{5}\right)$, $\left(\dfrac{\sqrt{5}}{5}, \dfrac{-\sqrt{5}}{5}\right)$, $\left(\dfrac{-\sqrt{5}}{5}, \dfrac{\sqrt{5}}{5}\right)$, and $\left(\dfrac{-\sqrt{5}}{5}, \dfrac{-\sqrt{5}}{5}\right)$.

Practice Problem 3 Solve the system.

$$x^2 + y^2 = 12$$

$$3x^2 - 4y^2 = 8 \quad \blacksquare$$

Solve each of the following systems by the substitution method. Graph each equation to verify that the answer seems reasonable.

1. $y^2 = 2x$
$y = -2x + 2$

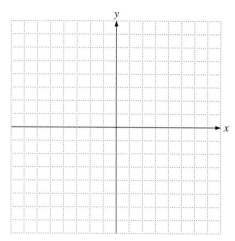

2. $y^2 = 4x$
$y = x + 1$

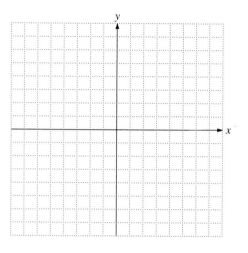

3. $x + 2y = 0$
$x^2 + 4y^2 = 32$

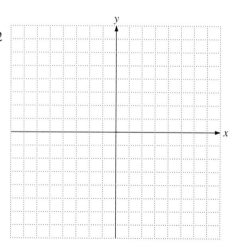

4. $y - 4x = 0$
$4x^2 + y^2 = 20$

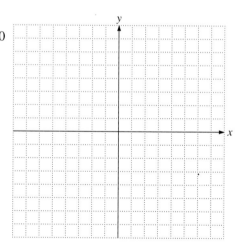

Solve each of the following systems by the substitution method.

5. $\dfrac{x^2}{1} - \dfrac{y^2}{3} = 1$
$x + y = 1$

6. $y = (x + 3)^2 - 3$
$2x - y + 2 = 0$

7. $x^2 + y^2 - 9 = 0$
$2y = 3 - x$

8. $x^2 + y^2 - 25 = 0$
$3y = x + 5$

9. $y = (x - 2)^2$
$x + y - 2 = 0$

10. $y = (x - 3)^2$
$x - 1 - y = 0$

11. $\dfrac{x^2}{4} - \dfrac{y^2}{4} = 1$
$x + y - 4 = 0$

12. $\dfrac{x^2}{3} - \dfrac{y^2}{12} = 1$
$y = -x$

Solve each of the following systems by the addition method. Graph each equation to verify that the answer seems reasonable.

13. $2x^2 + y^2 = 8$
$x^2 + y^2 = 4$

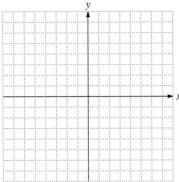

14. $9x^2 + 4y^2 = 36$
$x^2 + y^2 = 9$

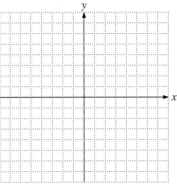

15. $2x^2 - 5y^2 = -2$
$3x^2 + 2y^2 = 35$

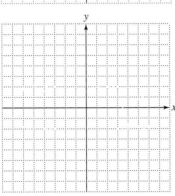

16. $2x^2 - 3y^2 = 5$
$3x^2 + 4y^2 = 16$

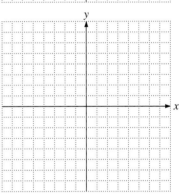

Solve each of the following systems by the addition method.

17. $x^2 + 2y^2 = 12$
$2x^2 + 3y^2 = 21$

18. $2x^2 + 5y^2 = 42$
$3x^2 + 4y^2 = 35$

19. $x^2 + 2y^2 = 8$
$x^2 - y^2 = 1$

20. $x^2 + 4y^2 = 13$
$x^2 - 3y^2 = -8$

Mixed Practice

Solve each of the following systems by any appropriate method. If there is no real number solution, so state.

21. $x^2 + y^2 = 7$
$\dfrac{x^2}{3} - \dfrac{y^2}{9} = 1$

22. $x^2 + (y - 3)^2 = 9$
$x^2 + y^2 = 4$

23. $xy = 3$
$3y = 3x + 6$

24. $xy = 5$
$2y = 2x + 8$

25. $xy = 4$
$x + 2y - 8 = 0$

26. $xy = 1$
$3x + y - 6 = 0$

27. $x + y = 5$
$x^2 + y^2 = 4$

28. $x^2 + y^2 = 0$
$x - y = 6$

Optional Graphing Calculator Problems

Find all solutions for each system. Round your answer to two decimal places.

29. $2y^2 = 5x$
$y = 1.834x - 0.982$

30. $2x^2 + 5.698y^2 = 39.768$
$3x^2 + 4.256y^2 = 34.087$

Applications

31. In an experiment with a laser beam, the path of a particle orbiting a central object is described by the equation $\dfrac{x^2}{49} + \dfrac{y^2}{36} = 1$, where x and y are measured in centimeters from the center of the object. The laser beam follows the path $y = 2x - 6$. Find the coordinates at which the laser will illuminate the particle (that is, when the particle will pass through the beam).

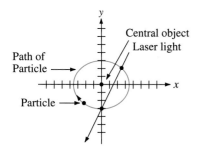

Scale: Each square = 2 units

32. The area of a rectangle is 540 square meters. The diagonal of the rectangle is 39 meters long. Find the dimensions of the rectangle.
Hint: Let x and y represent the length and width and write a system of two nonlinear equations.

Cumulative Review Problems

Simplify.

33. $\dfrac{6x^4 - 24x^3 - 30x^2}{3x^3 - 21x^2 + 30x}$

Divide.

34. $(3x^3 - 8x^2 - 33x - 10) \div (3x + 1)$

35. Ricardo is the staff accountant for a CD-ROM factory. He has determined that his monthly profit factor is given by the expression $11.5n - 290,000$. Here n represents the number of CD-ROMs manufactured each month. The profit this month was $1,187,750. How many CD-ROMs were produced this month?

36. Highway Patrol officers can trap speeders by various methods. Between two certain points in a small town in Georgia on a back country road, a speeding Audi travels 55 miles per hour for 5 seconds. What is the legal speed limit in this town if at the maximum speed, a car driving between those two points requires 11 seconds?

Putting Your Skills to Work

TRACKING THE PATH OF A COMET

On July 23, 1995, the Hale-Bopp comet was discovered independently by Alan Hale of New Mexico and Thomas Bopp of Arizona. This comet was easily viewed and photographed from Earth during March and April 1997. The orbit of this comet (and many of the comets that we can view from Earth is elliptical with the Sun at the focus. The equation of the path of the comet is shown beneath the sketch. In the equation, x and y are measured in astronomical units (AU). One AU is approximately 93 million miles. The path of the comet is an extremely long ellipse. The last time this comet visited our planetary system is estimated to be 4206 years ago!

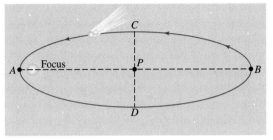

$$\frac{x^2}{33,415.84} + \frac{y^2}{333.32} = 1$$

PROBLEMS FOR INDIVIDUAL INVESTIGATION AND ANALYSIS

1. What is the straight line distance from the point where the comet was farthest from the Sun to the point where the comet was closest to the Sun (point A to point B)?

2. What is the greatest straight line distance from the path of the comet coming toward the Earth to the path of the comet going away from the Sun? (This distance is between points C and D.)

PROBLEMS FOR GROUP INVESTIGATION AND COOPERATIVE GROUP ACTIVITY

Together with some of the other members of your class try to determine the following.

3. Express the equation of the comet in terms of miles instead of astronomical units.

4. The distance from the focus of the ellipse to the center of the elliptical path (point P) is denoted as c. It is obtained from the equation $c^2 = a^2 - b^2$. The eccentricity of the ellipse (the measure of how elongated the ellipse is) is given by the equation $e = c/a$. Orbits that are very close to circular have a value of e close to 0. Orbitals that are very elongated have a value of e close to 1. Determine the value of e of the Hale-Bopp comet. Round your answer to the nearest thousandth.

INTERNET CONNECTION: Go to `http://www.prenhall.com/tobey` to be connected.
Site: Assume a Spherical Comet

This page gives an explanation of comet orbits, including a brief explanation of Kepler's second law.

5. Sketch or print the drawing of a possible comet orbit from the Web page. Label the drawing to show where the comet is moving fastest and where it is moving slowest.

6. The earth's orbit is very nearly circular, with a radius of 1 AU and the sun near the center. Sketch a possible orbit for the earth on your drawing. Assume that the comet is 0.75 AU from the sun at its closest approach.

7. Using Kepler's second law, explain why a comet takes many years to orbit the sun.

Topic	Procedure	Examples
Distance between two points, p. 507.	The distance d between points (x_1, y_1) and (x_2, y_2) is $$d = \sqrt{(x_2 - x_1)^2 + (y_2 - y_1)^2}$$	Find the distance between $(-6, -3)$ and $(5, -2)$. $$\begin{aligned} d &= \sqrt{[5 - (-6)]^2 + [-2 - (-3)]^2} \\ &= \sqrt{(5 + 6)^2 + (-2 + 3)^2} \\ &= \sqrt{121 + 1} \\ &= \sqrt{122} \end{aligned}$$
Standard form of the equation of a circle, p. 508.	The standard form of the equation of a circle with center at (h, k) and radius r is $$(x - h)^2 + (y - k)^2 = r^2$$	Graph $(x - 3)^2 + (y + 4)^2 = 16$. Center at $(h, k) = (3, -4)$, radius $= 4$
Standard form of equation of a vertical parabola, p. 516.	A *vertical parabola* with a vertex at (h, k) can be written in the form $y = a(x - h)^2 + k$. It opens upward if $a > 0$ and downward if $a < 0$.	Graph $y = \dfrac{1}{2}(x - 3)^2 + 5$. $a = \dfrac{1}{2}$, so it opens upward; vertex at $(h, k) = (3, 5)$; if $x = 0$, $y = 9.5$.
Standard form of equation of a horizontal parabola, p. 517.	A *horizontal parabola* with a vertex at (h, k) can be written in the form $x = a(y - k)^2 + h$. It opens to the right if $a > 0$ and to the left if $a < 0$.	Graph $x = \dfrac{1}{3}(y + 2)^2 - 4$. $a = \dfrac{1}{3}$, so it opens to the right; vertex at $(h, k) = (-4, -2)$; if $x = 0$, $y = -2 - 2\sqrt{3}$ and $y = -2 + 2\sqrt{3}$ or $y \approx -5.5$ and $y \approx 1.5$

Topic	Procedure	Examples
Standard form of equation of an ellipse with center at (0, 0), p. 525.	An ellipse with center at the origin has equation $$\frac{x^2}{a^2} + \frac{y^2}{b^2} = 1$$ where $a \neq b$ and $a, b > 0$.	Graph $\dfrac{x^2}{16} + \dfrac{y^2}{4} = 1$. $a^2 = 16$, $a = 4$; $b^2 = 4$, $b = 2$.
Standard form of an ellipse with center at (h, k), p. 527.	An ellipse with center at (h, k) has equation $$\frac{(x - h)^2}{a^2} + \frac{(y - k)^2}{b^2} = 1$$ where $a, b > 0$ and $a \neq b$.	Graph $\dfrac{(x + 2)^2}{9} + \dfrac{(y + 4)^2}{25} = 1$. $(h, k) = (-2, -4)$, $a = 3$, $b = 5$.
Standard form of horizontal hyperbola with center at (0, 0), p. 532.	The horizontal hyperbola with center at the origin and vertices $(a, 0)$ and $(-a, 0)$ has equation $$\frac{x^2}{a^2} - \frac{y^2}{b^2} = 1$$ and asymptotes $y = \pm \dfrac{b}{a} x$.	Graph $\dfrac{x^2}{25} - \dfrac{y^2}{9} = 1$. $a = 5$, $b = 3$.
Standard form of vertical hyperbola with center at (0, 0), p. 532.	The vertical hyperbola with center at the origin and vertices $(0, b)$ and $(0, -b)$ has equation $$\frac{y^2}{b^2} - \frac{x^2}{a^2} = 1$$ and asymptotes $y = \pm \dfrac{b}{a} x$.	Graph $\dfrac{y^2}{9} - \dfrac{x^2}{4} = 1$. $b = 3$, $a = 2$.

Topic	Procedure	Examples
Standard form of horizontal hyperbola with center at (h, k), p. 534.	The horizontal hyperbola with center at (h, k) and vertices $(h - a, k)$ and $(h + a, k)$ has equation $$\frac{(x - h)^2}{a^2} - \frac{(y - k)^2}{b^2} = 1$$	Graph $\dfrac{(x - 2)^2}{4} - \dfrac{(y - 3)^2}{25} = 1$. Center at $(2, 3)$, $a = 2$, $b = 5$.
Standard form of vertical hyperbola with center at (h, k), p. 534.	The vertical hyperbola with center at (h, k) and vertices $(h, k + b)$ and $(h, k - b)$ has equation $$\frac{(y - k)^2}{b^2} - \frac{(x - h)^2}{a^2} = 1$$	Graph: $\dfrac{(y - 5)^2}{9} - \dfrac{(x - 4)^2}{4} = 1$. Center at $(4, 5)$, $b = 3$, $a = 2$.
Nonlinear systems of equations, p. 541.	We can solve a nonlinear system by the substitution method or the addition method. In the addition method, we multiply one or more equations by a numerical value and then add them together so that one variable is eliminated. In the substitution method we solve one equation for one variable and substitute that expression into the other equation.	Solve by substitution. $$2x^2 + y^2 = 18$$ $$xy = 4$$ $$y = \frac{4}{x}$$ $$2x^2 + \left(\frac{4}{x}\right)^2 = 18$$ $$2x^2 + \frac{16}{x^2} = 18$$ $$2x^4 + 16 = 18x^2$$ $$2x^4 - 18x^2 + 16 = 0$$ $$x^4 - 9x^2 + 8 = 0$$ $$(x^2 - 1)(x^2 - 8) = 0$$ $x^2 - 1 = 0 \qquad x^2 - 8 = 0$ $x^2 = 1 \qquad\quad x^2 = 8$ $x = \pm 1 \qquad\quad x = \pm 2\sqrt{2}$ Since $xy = 4$, if $x = 1$ then $y = 4$ if $x = -1$ then $y = -4$ if $x = 2\sqrt{2}$ then $y = \sqrt{2}$ if $x = -2\sqrt{2}$ then $y = -\sqrt{2}$ The solutions are $(1, 4)$, $(-1, -4)$, $(2\sqrt{2}, \sqrt{2})$, and $(-2\sqrt{2}, -\sqrt{2})$.

In Problems 1 and 2, find the distance between the points.

1. $(-7, 3)$ and $(-2, -1)$

2. $(10.5, -6)$ and $(7.5, -4)$

3. Write in standard form the equation of a circle with center at $(0, -7)$ and radius 5.

4. Write in standard form the equation of a circle with center at $(-6, 3)$ and radius $\sqrt{15}$.

Put each equation in standard form. Find the center and the radius of each circle.

5. $x^2 + y^2 - 6x - 8y + 3 = 0$

6. $x^2 + y^2 - 10x + 12y + 52 = 0$

Graph each parabola. Label its vertex and plot at least one intercept.

7. $x = \dfrac{1}{3}y^2$

8. $y = -2(x + 1)^2 - 3$

9. $x = \dfrac{1}{2}(y - 2)^2 + 4$

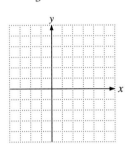

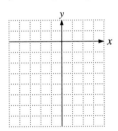

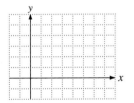

Put each equation in standard form. Find the vertex and determine in which direction the parabola opens.

10. $x^2 + 6x = y - 4$

11. $x + 8y = y^2 + 10$

Graph each ellipse. Label its center and four other points.

12. $\dfrac{x^2}{\dfrac{1}{4}} + \dfrac{y^2}{1} = 1$

13. $16x^2 + y^2 - 32 = 0$

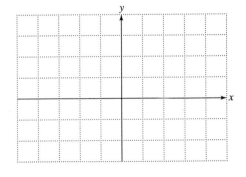

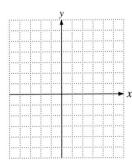

Determine the vertices and the center of each ellipse.

14. $\dfrac{(x + 1)^2}{9} + \dfrac{(y - 2)^2}{16} = 1$

15. $\dfrac{(x + 5)^2}{4} + \dfrac{(y + 3)^2}{25} = 1$

Find the center and vertices of each hyperbola and graph it.

16. $9y^2 - 25x^2 = 225$

17. $x^2 - 4y^2 - 16 = 0$

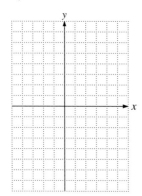

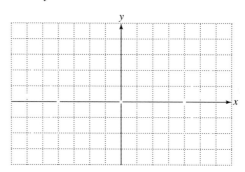

Determine the vertices and the center of each hyperbola.

18. $\dfrac{(x - 2)^2}{4} - \dfrac{(y + 3)^2}{25} = 1$

19. $9(y - 2)^2 - (x + 5)^2 - 9 = 0$

Solve each nonlinear system. If there is no real number solution, so state.

20. $x^2 + y = 9$
    $y - x = 3$

21. $y^2 + x^2 = 3$
    $x - 2y = 1$

22. $2x^2 + y^2 = 17$
    $x^2 + 2y^2 = 22$

23. $xy = -2$
    $x^2 + y^2 = 5$

24. $3x^2 - 4y^2 = 12$
    $7x^2 - y^2 = 8$

25. $y = x^2 + 1$
    $x^2 + y^2 - 8y + 7 = 0$

26. $2x^2 + y^2 = 18$
    $xy = 4$

27. $y^2 - 2x^2 = 2$
    $2y^2 - 3x^2 = 5$

1. Find the distance between $(-6, -8)$ and $(-2, 5)$.

Place the equation in standard form. Find the center or vertex, plot at least one other point, identify the conic, and sketch the curve.

2. $x^2 + y^2 + 6x - 4y + 9 = 0$

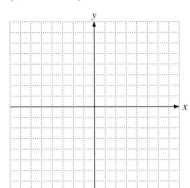

3. $y^2 - 6y - x + 13 = 0$

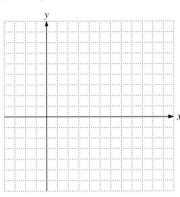

Identify and graph each conic section. Label the center and/or vertex as appropriate.

4. $\dfrac{x^2}{10} - \dfrac{y^2}{9} = 1$

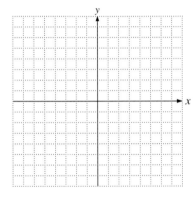

5. $\dfrac{x^2}{25} + \dfrac{y^2}{1} = 1$

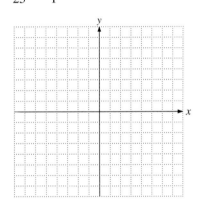

6. $y = -2(x + 3)^2 + 4$

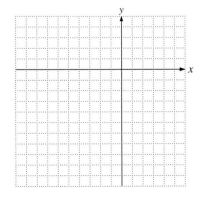

7. $\dfrac{(x + 2)^2}{16} + \dfrac{(y - 5)^2}{4} = 1$

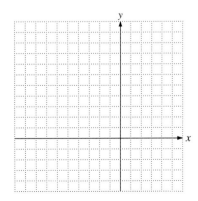

1. _____

2. _____

3. _____

4. _____

5. _____

6. _____

7. _____

8. $7y^2 - 7x^2 = 28$

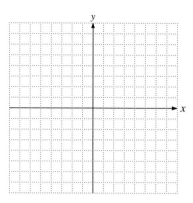

8. _____

9. _____

10. _____

11. _____

12. _____

13. _____

14. _____

15. _____

16. _____

Find the equation in standard form of:

9. a circle of radius $\sqrt{8}$ with a center at $(3, -5)$.

10. a parabola with a vertex at $(-7, 3)$ that opens to the right. This parabola crosses the x-axis at $(2, 0)$. It is of the form $x = (y - k)^2 + h$.

11. an ellipse with a center at $(-4, -2)$ and vertices at $(-4, 1)$, $(-4, -5)$, $(-5, -2)$, and $(-3, -2)$.

12. a hyperbola with a center at $(6, 7)$ that opens upward. This hyperbola has vertices of $(6, 14)$ and $(6, 0)$. The value of a for this hyperbola is 3.

Solve each nonlinear system.

13. $x^2 + y^2 = 9$
   $y = x - 3$

14. $-2x + y = 5$
   $x^2 + y^2 - 25 = 0$

15. $4x^2 + y^2 - 4 = 0$
   $9x^2 - 4y^2 - 9 = 0$

16. $2x^2 + y^2 = 9$
   $xy = -3$

Approximately one-half of this test covers the content of Chapters 1–8. The remainder covers the content of Chapter 9.

1. Identify the property illustrated by the equation $5(-3) = -3(5)$.

2. Evaluate. $3(4 - 6)^3 + \sqrt{25}$

3. Simplify. $2\{x - 3[x - 2(x + 1)]\}$

4. Solve for p. $A = 3bt + prt$

5. Factor. $x^3 + 125$

6. Add. $\dfrac{3}{x - 4} + \dfrac{6}{x^2 - 16}$

7. Solve for x

$$\frac{3}{2x + 3} = \frac{1}{2x - 3} + \frac{2}{4x^2 - 9}$$

8. Solve for (x, y, z).

$$\begin{aligned} 3x - 2y - 9z &= 9 \\ x - y + z &= 8 \\ 2x + 3y - z &= -2 \end{aligned}$$

9. Simplify. $\sqrt{8x} + 3x\sqrt{50} - 4x\sqrt{32}$

10. Multiply and simplify.
$(\sqrt{2} + \sqrt{3})(2\sqrt{6} - \sqrt{3})$

Solve the following inequalities.

11. $2x + (4x - 1) > 6 - x$

12. $\dfrac{6(x - 4)}{5} \geq \dfrac{3(x + 2)}{4}$

13. Find the distance between $(6, -1)$ and $(-3, -4)$.

1. _____

2. _____

3. _____

4. _____

5. _____

6. _____

7. _____

8. _____

9. _____

10. _____

11. _____

12. _____

13. _____

Identify and graph each equation.

14. $y = -\dfrac{1}{2}(x + 2)^2 - 3$

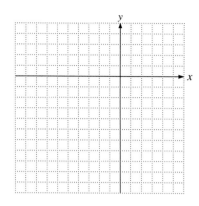

15. $25x^2 + 25y^2 = 125$

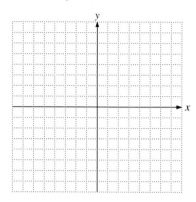

16. $16x^2 - 4y^2 = 64$

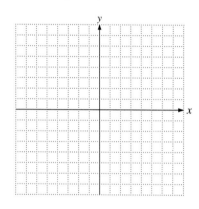

17. $\dfrac{(x - 2)^2}{25} + \dfrac{(y - 3)^2}{16} = 1$

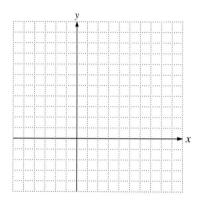

Solve the nonlinear system.

18. $y = 2x^2$
$\quad\,\, y = 2x + 4$

19. $x^2 + 2y^2 = 16$
$\quad\,\, 4x^2 - y^2 = 24$

20. $xy = -15$
$\quad\,\, 4x + 3y = 3$

21. $x^2 + y^2 = 25$
$\quad\,\, x - 2y = -5$

Functions

How is the growth rate of plants influenced by temperature? Are there limits in the tolerance of plants for temperatures that are too hot or too cold for plants to grow? How could the growth rate of plants be predicted if they were genetically altered to survive in higher temperatures and to grow at a faster rate? If these questions interest you, turn to the Putting Your Skills to Work on page 598.

If you are familiar with the topics in this chapter, take this test now. Check your answers with those in the back of the book. If an answer was wrong or you couldn't do a problem, study the appropriate section of the chapter.

If you are not familiar with the topics in this chapter, don't take this test now. Instead, study the examples, work the practice problems, and then take the test.

This test will help you identify those concepts that you have mastered and those that need more study.

Section 10.1

1. For the function $f(x) = 2x - 6$, find
    (a) $f(-3)$ **(b)** $f(a)$ **(c)** $f(2a)$ **(d)** $f(a + 2)$

2. For $f(x) = 5x^2 + 2x - 3$, find
    (a) $f(-2)$ **(b)** $f(a)$ **(c)** $f(a + 1)$

3. For $f(x) = \dfrac{3x}{x + 2}$, find
    (a) $f(a) + f(a - 2)$. Express your answer as one fraction.

    (b) $f(3a) - f(3)$. Express your answer as one fraction.

Section 10.2

Which of these graphs represent functions?

4.

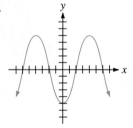

5.

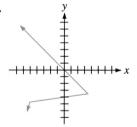

Graph each test item on one coordinate plane.

6. $f(x) = x^2$ and $h(x) = (x + 2)^2 + 3$.

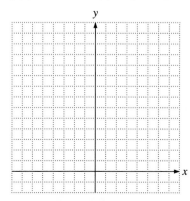

7. $f(x) = |x|$ and $s(x) = |x - 3|$.

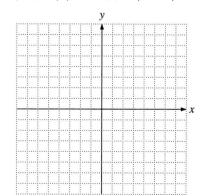

Sidebar answer blanks:

1. (a) _____
   (b) _____
   (c) _____
   (d) _____

2. (a) _____
   (b) _____
   (c) _____

3. (a) _____
   (b) _____

4. _____

5. _____

6. _____

7. _____

8. If $f(x) = 3x - 4$ and $g(x) = -2x^3 - 6x + 3$, find
(a) $(f + g)(x)$ (b) $(f + g)(2)$ (c) $f[g(x)]$

9. If $f(x) = \dfrac{2}{x + 6}$ and $g(x) = -3x + 1$, find
(a) $(fg)(x)$ (b) $(fg)(-4)$ (c) $f[g(x)]$

10. If $f(x) = 6x^2 - 5x - 4$ and $g(x) = 3x - 4$, find
(a) $\left(\dfrac{f}{g}\right)(x)$ (b) $\left(\dfrac{f}{g}\right)(-1)$ (c) $(f \circ g)(x)$ (d) $(g \circ f)(x)$

Section 10.4

Which graphs represent one-to-one functions?

11.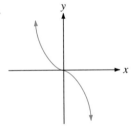

12.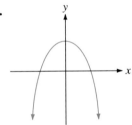

13. Is function A a one-to-one function?
$A = \{(-5, -3), (5, 3), (2, -1), (-2, 1)\}$

14. Determine the inverse of the function $F = \{(7, 1), (6, 3), (2, -1), (-1, 5)\}$.

15. Find the inverse of $g(x) = 3 - 5x$ and graph $g(x)$ and its inverse on the same set of axes.

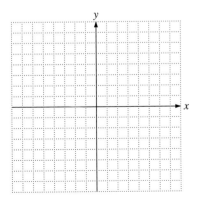

8. (a) _____

(b) _____

(c) _____

9. (a) _____

(b) _____

(c) _____

10. (a) _____

(b) _____

(c) _____

(d) _____

11. _____

12. _____

13. _____

14. _____

15. _____

10.1 Function Notation

After studying this section, you will be able to:

1 *Use function notation to evaluate expressions.*
2 *Use function notation to solve application problems.*

1 *Using Function Notation to Evaluate Expressions*

Function notation is useful in solving a number of interesting problems. Suppose you wanted to sky jump from an airplane. Your instructor tells you that you must wait 20 seconds before you pull the cord to open the parachute. How far will you fall?

The approximate distance an object falls in free fall when there is no initial downward velocity is given by the distance function $d(t) = 16t^2$ where time t is measured in seconds, and distance $d(t)$ is measured in feet.

How far will a person travel during free fall in 20 seconds, if they leave the airplane with no downward velocity? We want to find the distance $d(20)$ where $t = 20$ seconds. To find the distance, we substitute 20 for t in the function $16t^2$.

$$d(20) = 16(20)^2 = 16(400) = 6400$$

Thus a person will fall approximately 6400 feet in 20 seconds during free fall.

Now suppose you wait longer than you should to pull the parachute cord. How will this affect the distance you will fall? Suppose you fall e seconds beyond the 20-second mark.

$$d(20 + e) = 16(20 + e)^2 = 16(20 + e)(20 + e)$$
$$= 16(400 + 40e + e^2)$$
$$= 6400 + 640e + 16e^2$$

Thus, if you waited 5 seconds too long after the 20-second time, you would fall

$$d = 6400 + 640(5) + 16(5)^2$$
$$= 6400 + 3200 + 16(25)$$
$$= 6400 + 3200 + 400$$
$$= 10,000 \text{ feet}$$

This is 3600 feet farther than you would have fallen in 20 seconds. Obviously, an error of 5 seconds could have life or death consequences.

We now turn to finding certain types of function values.

EXAMPLE 1   If $f(x) = 3x - 7$, evaluate

(a) $f(-2)$ **(b)** $f(3)$ **(c)** $f(a)$ **(d)** $f(3a)$

(a) Since $f(x) = 3x - 7$, we replace x by -2 to obtain $f(-2) = 3(-2) - 7 = -6 - 7 = -13$

(b) $f(3) = 3(3) - 7 = 9 - 7 = 2$

(c) $f(a) = 3(a) - 7 = 3a - 7$

(d) $f(3a) = 3(3a) - 7 = 9a - 7$

Practice Problem 1   If $f(x) = 3 - 2x$, find

(a) $f(-3)$ **(b)** $f(2)$ **(c)** $f(a)$ **(d)** $f(2a)$ ∎

EXAMPLE 2 If $g(x) = 5 - 3x$, find

(a) $g(a)$ **(b)** $g(a + 3)$ **(c)** $g(a) + g(3)$

Since $g(x) = 5 - 3x$

(a) $g(a) = 5 - 3a$

(b) $g(a + 3) = 5 - 3(a + 3) = 5 - 3a - 9 = -4 - 3a$

(c) This problem requires us to find each function part separately. Then we add each part together.

$g(a) = 5 - 3a$

$g(3) = 5 - 3(3) = 5 - 9 = -4$

Thus $g(a) + g(3) = (5 - 3a) + (-4)$

$$= 5 - 3a - 4$$

$$= 1 - 3a$$

Notice that $g(a + 3) \neq g(a) + g(3)$.

Practice Problem 2   If $g(x) = \dfrac{1}{2}x - 3$, find

(a) $g(a)$ **(b)** $g(a + 4)$ **(c)** $g(a) + g(4)$ ∎

To Think About Is $g(a + 4) = g(a) + g(4)$? Why or why not?

EXAMPLE 3   If $p(x) = 2x^2 - 3x + 5$, find

(a) $p(-2)$ **(b)** $p(a)$ **(c)** $p(3a)$ **(d)** $p(a - 2)$

Since $p(x) = 2x^2 - 3x + 5$

(a) $p(-2) = 2(-2)^2 - 3(-2) + 5$
$= 2(4) - 3(-2) + 5$
$= 8 + 6 + 5$
$= 19$

(b) $p(a) = 2(a)^2 - 3(a) + 5 = 2a^2 - 3a + 5$

(c) $p(3a) = 2(3a)^2 - 3(3a) + 5$
$= 2(9a^2) - 3(3a) + 5$
$= 18a^2 - 9a + 5$

(d) $p(a - 2) = 2(a - 2)^2 - 3(a - 2) + 5$
$= 2(a - 2)(a - 2) - 3(a - 2) + 5$
$= 2(a^2 - 4a + 4) - 3(a - 2) + 5$
$= 2a^2 - 8a + 8 - 3a + 6 + 5$
$= 2a^2 - 11a + 19$

Practice Problem 3   If $p(x) = -3x^2 + 2x + 4$, find

(a) $p(-3)$ **(b)** $p(a)$ **(c)** $p(2a)$ **(d)** $p(a - 3)$ ∎

EXAMPLE 4   If $f(x) = \sqrt{x + 3}$, find

(a) $f(4)$ **(b)** $f(4b)$ **(c)** $f(a + 2)$ **(d)** $f(a) + f(2)$

Since $f(x) = \sqrt{x + 3}$

(a) $f(4) = \sqrt{4 + 3} = \sqrt{7}$

(b) $f(4b) = \sqrt{4b + 3}$

(c) $f(a + 2) = \sqrt{(a + 2) + 3} = \sqrt{a + 5}$

(d) $f(a) = \sqrt{a + 3}$

$\quad\quad f(2) = \sqrt{2 + 3} = \sqrt{5}$

$\quad\quad f(a) + f(2) = \sqrt{a + 3} + \sqrt{5}$

Practice Problem 4   If $f(x) = \sqrt{3x - 2}$, find

(a) $f(3)$ **(b)** $f(2b)$ **(c)** $f(b - 1)$ **(d)** $f(b) - f(1)$ ∎

EXAMPLE 5   If $r(x) = \dfrac{4}{x + 2}$ find   **(a)** $r(a + 3)$ **(b)** $r(a)$

(c) $r(a + 3) - r(a)$ Express this result as one fraction.

Since $r(x) = \dfrac{4}{x + 2}$

(a) $r(a + 3) = \dfrac{4}{a + 3 + 2} = \dfrac{4}{a + 5}$ **(b)** $r(a) = \dfrac{4}{a + 2}$

(c) $r(a + 3) - r(a) = \dfrac{4}{a + 5} - \dfrac{4}{a + 2}$

To express this as one fraction, we note that the

LCD $= (a + 5)(a + 2)$

$$= \dfrac{4(a + 2)}{(a + 5)(a + 2)} - \dfrac{4(a + 5)}{(a + 2)(a + 5)} = \dfrac{4a + 8}{(a + 5)(a + 2)} - \dfrac{4a + 20}{(a + 5)(a + 2)}$$

$$= \dfrac{4a - 4a + 8 - 20}{(a + 5)(a + 2)} = \dfrac{-12}{(a + 5)(a + 2)}$$

Practice Problem 5   If $r(x) = \dfrac{-3}{x + 1}$, find   **(a)** $r(a + 2)$ **(b)** $r(a)$

(c) $r(a + 2) - r(a)$ Express this result as one fraction.   ∎

EXAMPLE 6 $f(x) = 3x - 7$ Find $\dfrac{f(x + h) - f(x)}{h}$.

First $f(x + h) = 3(x + h) - 7 = 3x + 3h - 7$

and $f(x) = 3x - 7$

so $f(x + h) - f(x) = (3x + 3h - 7) - (3x - 7)$

$\quad\quad\quad\quad\quad\quad\quad\quad = 3x + 3h - 7 - 3x + 7$

$\quad\quad\quad\quad\quad\quad\quad\quad = 3h$

Therefore, $\dfrac{f(x + h) - f(x)}{h} = \dfrac{3h}{h} = 3$

Practice Problem 6 $g(x) = 2 - 5x$ Find $\dfrac{g(x + h) - g(x)}{h}$.   ∎

The surface area of a sphere is given by $S = 4\pi r^2$. If we use $\pi = 3.14$ as an approximation, this becomes $S = 4(3.14)r^2$ or $S = 12.56r^2$.

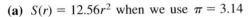

EXAMPLE 7

(a) Write the surface area of a sphere as a function of radius r.

(b) Find the surface area of a sphere with a radius of 3.0 cm.

(c) If the measurement of the radius is made in error, and it is measured to be $3 + e$, find an expression for the surface area as a function of error e.

(d) Evaluate the surface area for $r = 3 + e$ when $e = 0.2$ cm. Round your answer to the nearest hundredth of a centimeter. What is the difference in the surface area due to the error in measurement?

(a) $S(r) = 12.56r^2$ when we use $\pi = 3.14$

(b) $S(3) = 12.56(3)^2 = (12.56)(9) = 113.04$ cm^2

(c) $S(3 + e) = 12.56(3 + e)^2 = 12.56(3 + e)(3 + e)$
$$= 12.56(9 + 6e + e^2)$$
$$= 113.04 + 75.36e + 12.56e^2$$

(d) If an error in measure is made so that the radius is calculated to be $r = 3 + e$ when $e = 0.2$ cm, we can use the function generated in part **(c)**.

$$S = 113.04 + 75.36e + 12.56e^2$$
$$S = 113.04 + 75.36(0.2) + 12.56(0.2)^2$$
$$S = 113.04 + 15.072 + 0.5024$$
$$S = 128.6144$$

Rounding, we have $S = 128.61$ cm^2.

Thus, if the radius of 3 cm were incorrectly measured as 3.2 cm, the surface area would be approximately 15.57 cm^2 too large.

Practice Problem 7

The surface area of a cylinder of height 8 meters and radius r is given by $16\pi r + 2\pi r^2$.

(a) Write the surface area of a cylinder of height 8 meters using $\pi = 3.14$ and radius r as a function of r.

(b) Find the surface area if the radius is 2 meters.

(c) If the measurement of the radius is made in error and it measured to be $2 + e$, find an expression for the surface area as a function of error e.

(d) Evaluate the surface area for $r = 2 + e$ when $e = 0.3$ meter. Round your answer to the nearest hundredth of a meter. What is the difference in the surface area due to the error in measurement? ∎

10.1 Exercises

For the function $f(x) = 3x - 5$, find

1. $f(-3)$ **2.** $f(-4)$ **3.** $f(a)$ **4.** $f(b)$

5. $f(3b)$ **6.** $f(2a)$ **7.** $f(a - 4)$ **8.** $f(b + 3)$

For the function $g(x) = \frac{1}{2}x - 3$, find

9. $g(4) + g(a)$ **10.** $g(6) + g(b)$ **11.** $g(2a)$

12. $g(8b)$ **13.** $g(a - 2)$ **14.** $g(a - 4)$

15. $g(a^2) - g\left(\frac{2}{5}\right)$ **16.** $g(b^2) - g\left(\frac{4}{3}\right)$ **17.** $g(3b + 1)$

18. $g(3b + 5)$ **19.** $g(3b) + g(2a)$ **20.** $g(2b) + g(5)$

If $p(x) = 3x^2 + 4x - 2$, find

21. $p(-2)$ **22.** $p(-3)$ **23.** $p(a)$

24. $p(b)$ **25.** $p(a + 1)$ **26.** $p(b - 1)$

27. $p(4b)$ **28.** $p(3a)$ **29.** $p(2a) - p(a)$

30. $p(3b) - p(b)$ **31.** $p(b + 3) + p(-1)$ **32.** $p(b - 3) - p(3)$

If $h(x) = \sqrt{x + 5}$, find

33. $h(-1)$ **34.** $h(-4)$ **35.** $h(3)$

36. $h(23)$ **37.** $h(a + 2)$ **38.** $h(a - 3)$

39. $h(3a)$ **40.** $h(5a)$ **41.** $h(4a - 1)$

42. $h(4a + 3)$ **43.** $h(a - 5) + h(4)$ **44.** $h(9a + 4) - h(-4)$

If $r(x) = \dfrac{7}{x - 3}$, find the following and write your answer as one fraction.

45. $r(2)$ **46.** $r(-1)$ **47.** $r(5)$

48. $r(3.5)$ **49.** $r(a^2)$ **50.** $r(3b^2)$

51. $r(a + 2)$ **52.** $r(a - 3)$ **53.** $r(a + 2) - r(a)$

54. $r(a - 3) - r(2a)$ **55.** $r(b + 2) + r(2b)$ **56.** $r(b - 1) + r(3b)$

Find $\dfrac{f(x + h) - f(x)}{h}$ if

57. $f(x) = 5 - 2x$ **58.** $f(x) = 2x - 3$

59. $f(x) = 2x^2$ **60.** $f(x) = x^2 - x$

Applications

61. The area of a circle is $A = \pi r^2$. If we use $\pi = 3.14$,
  (a) Write the area of a circle as the function of the radius r.
  (b) Find the area of a circle with a radius of 4.0 feet.
  (c) If the measurement of the radius is made in error and it is measured to be $4 + e$, find an expression for the area as a function of error e.
  (d) Evaluate the area for $r = 4 + e$ when $e = 0.4$ feet. Round your answer to the nearest hundredth.

62. A turbine wind generator produces P kilowatts for wind speed w (measured in miles per hour) by the equation $P = 2.5w^2$.
  (a) Write the number of kilowatts P as a function of w.
  (b) Find the power in kilowatts when $w = 20$ miles per hour.
  (c) If the measurement of the wind is made in error and it is measured to be $20 + e$, find an expression for the power as a function of error e.
  (d) Evaluate the power for $w = 20 + e$ when $e = 2$ miles per hour.

Because of the elimination of lead in automobile gasoline and increased use of emission controls in automobiles and industrial operations, the amount of lead in the air in the United States has shown a marked decrease. The percent of lead in the air $p(x)$ expressed in terms of 1984 levels is defined by the following line graph. The variable x indicates the number of years since 1984. The function $p(x)$ indicates the percent of lead that remains in the air in selected regions of the United States, using the year 1984 as a base of 100.

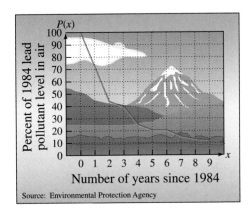

Source: Environmental Protection Agency

63. If $p(x) - 13$ were used instead of $p(x)$ what would happen to the function values? Find $p(3) - 13$.

64. If $p(x + 2)$ were used instead of $p(x)$, what would happen to the function values? Find $p(x + 2)$ when $x = 4$.

If $f(x) = 3x^2 - 4.6x + 1.23$, find each of the following correct to the nearest thousandth.

 65. $f(3.56a)$ **66.** $f(0.026a)$ **67.** $f(a - 0.152)$ **68.** $f(a + 2.23)$

69. A rope 20 feet long is cut into two unequal pieces. Each piece is used to form a square. Write a function $A(x)$ that expresses the total area enclosed by the two squares. Assume that the shorter piece of rope is x feet long. Evaluate $A(2)$, $A(5)$, and $A(8)$.

70. Assume that the smaller piece of rope in Problem 69 is used to form a circle and the longer piece is used to form a square. Write a function $A(x)$ that expresses the total area enclosed by the circle and the square. Evaluate $A(3)$ and $A(9)$. Round answers to the nearest hundredth.

Cumulative Review Problems

Solve for x.

71. $\dfrac{7}{6} + \dfrac{5}{x} = \dfrac{3}{2x}$

72. $3 + \dfrac{2x + 7}{x + 6} = \dfrac{5}{2}$

73. $\dfrac{1}{6} - \dfrac{2}{3x + 6} = \dfrac{1}{2x + 4}$

74. $\dfrac{5}{8} + \dfrac{3}{2x} = \dfrac{1}{4x}$

MathPro

Video 24

SSM

After studying this section, you will be able to:

1 *Determine if a given graph represents a function by the vertical line test.*
2 *Graph a function of the form f(x + h) + k by means of horizontal and vertical shifts from the graph of f(x).*

1 The Vertical Line Test for Functions

Not every graph we observe is a graph of a function. By definition, a function must have no ordered pairs that have the same first coordinates but different second coordinates. A graph that includes the points $(4, 2)$ and $(4, -2)$, for example, would not be the graph of a function. Thus the graph of $x = y^2$ would not be the graph of a function.

If any vertical line crosses a graph of a relation in more than one place, the relation is not a function.

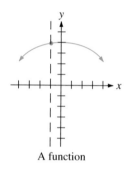

Vertical Line Test

If any vertical line can intersect the graph of a relation more than once, the relation is not a function.

In the sketch below we observe that the dashed vertical line crosses the curve of a function only once. The dashed vertical line crosses the curve of a relation that is *not a* function more than once.

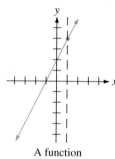

A function

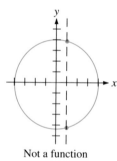

A function

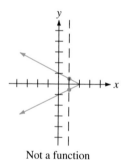

Not a function

Not a function

EXAMPLE 1 Determine if each of the following is a graph of a function.

(a)

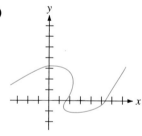

(b)

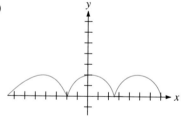

(a) By the vertical line test, this relation is not a function.

(b) By the vertical line test, this relation is a function.

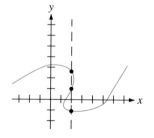

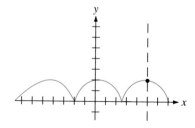

Practice Problem 1   Does this graph represent a function? Why or why not?

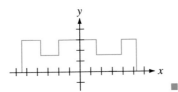

■

2 Transformations of Graphs of Functions

Many functions have graphs that involve a simple vertical shift from the graph of a similar function.

EXAMPLE 2   Graph each function on one coordinate plane.
$f(x) = x^2$ and $h(x) = x^2 + 2$

First we make a table of values for $f(x)$ and for $h(x)$.

$$f(x) = x^2 \qquad h(x) = x^2 + 2$$

x	$f(x)$		x	$h(x)$
-2	4		-2	6
-1	1		-1	3
0	0		0	2
1	1		1	3
2	4		2	6

Now we graph each function on the same coordinate plane.

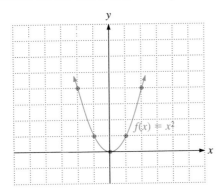

Notice that the graph of $h(x)$ is always 2 units higher than the graph of $f(x)$.

To Think About What would the graph of $j(x) = x^2 - 3$ look like? Verify by making a table of values and drawing a graph of the function.

Practice Problem 2   Graph each function on one coordinate plane.

$$f(x) = x^2 \quad \text{and} \quad h(x) = x^2 - 5 \quad ■$$

We have the following general summary.

PRACTICE PROBLEM 2

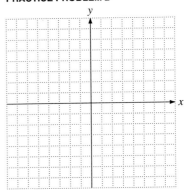

Vertical Shifts

If c is a positive number

1. To obtain the graph of $f(x) + c$, shift the graph of $f(x)$ up c units.
2. To obtain the graph of $f(x) - c$, shift the graph of $f(x)$ down c units.

Exploration

Most graphing calculators have an absolute value function (abs). In this problem use this function to graph $f(x)$ and $h(x)$ on one coordinate plane.

$f(x) = |0.5x|$

$h(x) = |0.5x + 3.5| - 1.75$

Describe how $h(x)$ is shifted from the graph of $f(x)$. Use your calculator to find the approximate coordinates of the point where $f(x)$ and $h(x)$ intersect. Find the answer correct to the nearest hundredth.

Now we turn to the topic of horizontal shifts.

EXAMPLE 3   Graph each function on one coordinate plane.

$f(x) = |x|$ and $p(x) = |x - 3|$

First we make a table of values for $f(x)$ and $p(x)$.

x	$f(x)$	x	$p(x)$
-2	2	-2	5
-1	1	-1	4
0	0	0	3
1	1	1	2
2	2	2	1
3	3	3	0
4	4	4	1

Now we graph each function on the same coordinate plane.

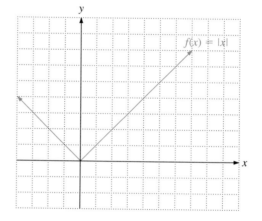

Notice that the graph of $p(x)$ has been shifted 3 units to the right of the graph of $f(x)$.

PRACTICE PROBLEM 3

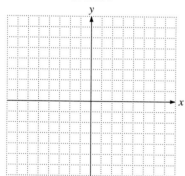

Practice Problem 3   Graph each function on one coordinate plane.

$$f(x) = |x| \text{ and }   p(x) = |x + 2| \quad \blacksquare$$

To Think About What would the graph of $h(x) = (x - 3)^2$ look like? What would the graph of $j(x) = (x + 2)^2$ look like? Verify by making tables of values and drawing the graphs.

Now we can write the following general summary.

Horizontal Shifts

If c is a positive number

1. To obtain the graph of $f(x - c)$, shift the graph of $f(x)$ to the *right* c units.
2. To obtain the graph of $f(x + c)$, shift the graph of $f(x)$ to the *left* c units.

Some graphs will involve both horizontal and vertical shifts.

EXAMPLE 4   Graph each function on one coordinate plane.

$$f(x) = x^3 \quad \text{and} \quad h(x) = (x - 3)^3 - 2$$

First we make a table of values for $f(x)$ and graph the function.

x	$f(x)$
-2	-8
-1	-1
0	0
1	1
2	8

Next we recognize that $h(x)$ will have a similar shape, but the curve will be shifted 3 units to the *right* and 2 units *downward*. We draw the graph of $h(x)$ by using these shifts.

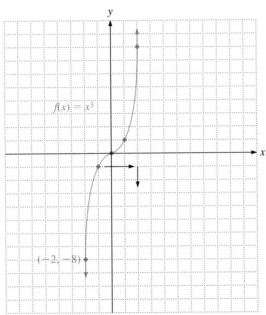

To Think About The point $(-2, -8)$ has been shifted 3 units to the right and 2 units down, that is, $(-2 + 3, -8 + (-2))$. The new coordinates are $(1, -10)$. The point $(-1, -1)$ is a point on $f(x)$. Use the same reasoning to find the image of $(-1, -1)$ on the graph of $h(x)$. Verify by checking the graphs.

Practice Problem 4   Graph each function on one coordinate plane.

$$f(x) = x^3 \quad \text{and} \quad h(x) = (x + 4)^3 + 3 \quad \blacksquare$$

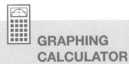

GRAPHING CALCULATOR

Exploration

In this problem graph $f(x)$ and $h(x)$ on one coordinate plane.

$$f(x) = x^3$$

$$h(x) = x^3 - 6x^2 + 12x - 4$$

How many units horizontally and vertically has the graph of $h(x)$ been shifted from the graph of $f(x)$? Use your zoom feature to find the positive value of x where $h(x)$ crosses the x-axis correct to the nearest hundredth.

PRACTICE PROBLEM 4

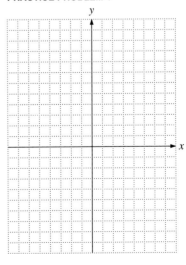

All the functions we have sketched in this section have a domain of all real numbers. Some functions have a restricted domain.

EXAMPLE 5   Graph each function on one coordinate plane. State the domain of each function.

$$f(x) = \frac{4}{x} \quad \text{and} \quad g(x) = \frac{4}{x + 3} + 1$$

First we make a table of values for $f(x)$. The domain of $f(x)$ is all real numbers, $x \neq 0$. Note that $f(x)$ is not defined when $x = 0$ since we cannot divide by 0.

x	$f(x)$
-4	-1
-2	-2
-1	-4
$-\frac{1}{2}$	-8
$\frac{1}{2}$	8
1	4
2	2
4	1

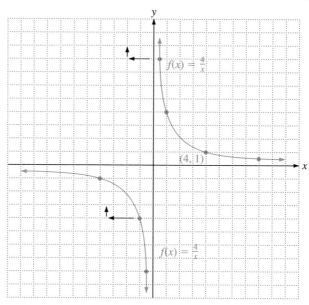

We draw $f(x)$ and note key points. From the equation, we see that the graph of $g(x)$ is 3 units to the left of and 1 unit above $f(x)$. We can find the image of each of the key points on $f(x)$ as a guide in graphing $g(x)$. For example, the image of $(4, 1)$ is $(4 - 3, 1 + 1)$ or $(1, 2)$.

Each point on $f(x)$ is shifted

⟸   3 units left

⇑   1 unit up

to form the graph of $g(x)$.

What is the domain of $g(x)$? Why? $g(x)$ contains a denominator $x + 3$. But $x + 3 \neq 0$. Therefore $x \neq -3$. The domain of $g(x)$ is all real numbers where $x \neq -3$.

PRACTICE PROBLEM 5

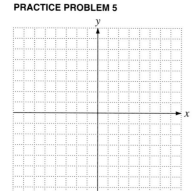

Practice Problem 5   Graph each function on one coordinate plane.

$$f(x) = \frac{2}{x} \quad \text{and} \quad g(x) = \frac{2}{x + 1} - 2 \quad ∎$$

10.2 Exercises

Verbal and Writing Skills

1. Does $f(x + 2) = f(x) + f(2)$? Why or why not? Give an example.

2. Explain what the vertical line test is and why it works.

3. To obtain the graph of $f(x) + c$, shift the graph of $f(x)$ ——————— c units.

4. To obtain the graph of $f(x - c)$, shift the graph of $f(x)$ ——————— c units.

Which of these graphs represent functions?

5.

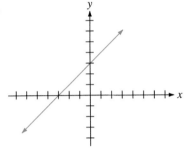

6.

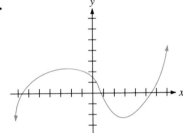

7.

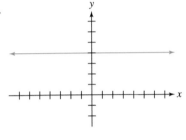

8.

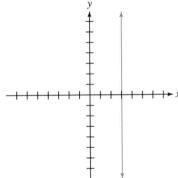

9.

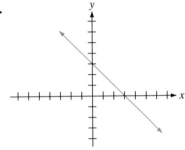

10.

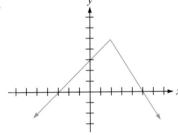

11.

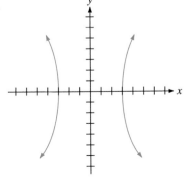

12.

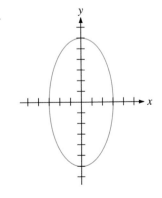

13.

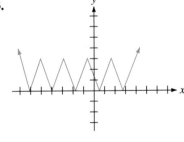

14.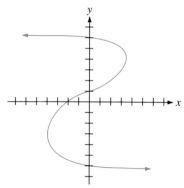

15.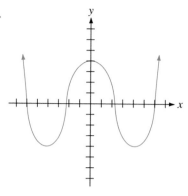

16.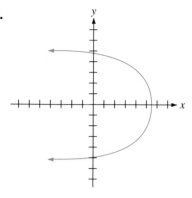

17.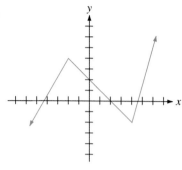

18.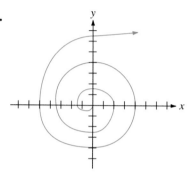

For each of Exercises 19 through 36, graph the two functions on one coordinate plane.

19. $f(x) = x^2$
$h(x) = x^2 - 3$

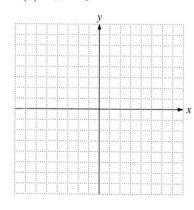

20. $f(x) = x^2$
$h(x) = x^2 + 4$

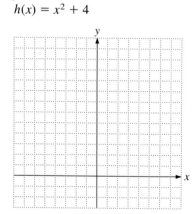

21. $f(x) = x^2$
$p(x) = (x - 2)^2$

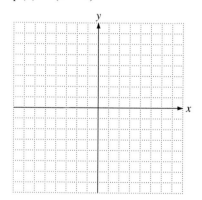

22. $f(x) = x^2$
$p(x) = (x + 1)^2$

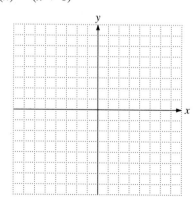

23. $f(x) = x^2$
$g(x) = (x - 2)^2 + 1$

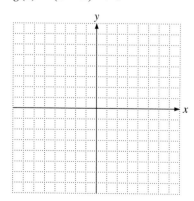

24. $f(x) = x^2$
$g(x) = (x + 1)^2 - 2$

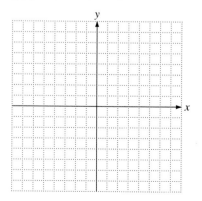

25. $f(x) = |x|$
$r(x) = |x| + 3$

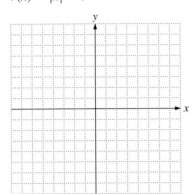

26. $f(x) = |x|$
$r(x) = |x| - 1$

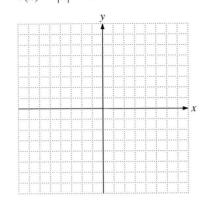

27. $f(x) = |x|$
$s(x) = |x - 2|$

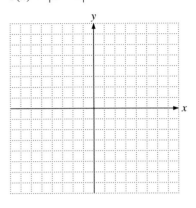

28. $f(x) = |x|$
$s(x) = |x + 4|$

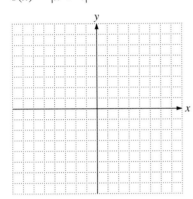

29. $f(x) = |x|$
$t(x) = |x - 3| - 4$

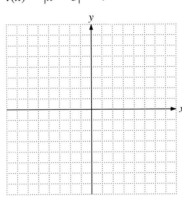

30. $f(x) = |x|$
$t(x) = |x + 1| + 2$

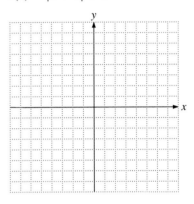

31. $f(x) = x^3$

$j(x) = (x + 3)^3 + 1$

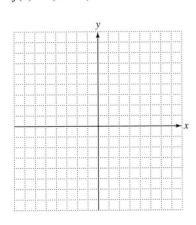

32. $f(x) = x^3$

$j(x) = (x - 3)^3 + 3$

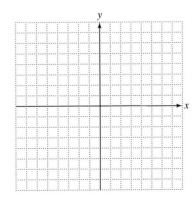

33. $f(x) = \dfrac{2}{x}$

$g(x) = \dfrac{2}{x} + 3$

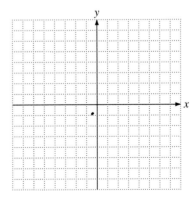

34. $f(x) = \dfrac{3}{x}$

$g(x) = \dfrac{3}{x} - 2$

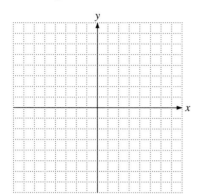

35. $f(x) = \dfrac{4}{x}$

$h(x) = \dfrac{4}{x+1} - 2$

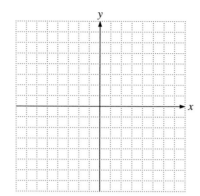

36. $f(x) = \dfrac{4}{x}$

$h(x) = \dfrac{4}{x-2} + 3$

Optional Graphing Calculator Problems

37. Using your graphing calculator, graph $f(x) = x^4$ and $f(x) = (x - 3.2)^4 - 2.6$.

38. Using your graphing calculator, graph $f(x) = x^7$ and $f(x) = (x + 1.3)^7 + 3.3$.

Cumulative Review Problems

Simplify each expression. Assume that all variables are positive.

39. $\sqrt{12} + 3\sqrt{50} - 4\sqrt{27}$

40. $(\sqrt{3x} - \sqrt{y})^2$

41. Rationalize the denominator.

$$\dfrac{2\sqrt{3} + \sqrt{5}}{\sqrt{3} - 2\sqrt{5}}$$

42. Roy is on a diet and can have a maximum of 620 calories for lunch. A tuna fish sandwich on whole wheat bread has 315 calories, and 12 fluid ounces of a fruit juice soft drink has 120 calories. How many French fries can he eat if there are 10 calories in one French fry?

10.3 Algebraic Operations on Functions

After studying this section, you will be able to:

MathPro   Video 25   SSM

1 Find the sum, difference, product, or quotient of two functions.
2 Find the composition of two functions.

1 Finding the Sum, Difference, Product, and Quotient of Functions

When two functions are given, new functions can be formed by combining them. We can add, subtract, multiply, or divide functions just as we add, subtract, multiply or divide numbers. We begin by looking at a real-life application.

Westwood Motors sells cars and trucks. The results of sales at the dealership for the last 5 years are shown in the table below and in the graphs that follow.

d Year	$c(d)$ Number of Cars Sold	$t(d)$ Number of Trucks Sold	$v(d)$ Total Number of Vehicles Sold $v(d) = c(d) + t(d)$
1990	350	200	550
1991	400	150	550
1992	450	100	550
1993	450	200	650
1994	500	100	600
1995	600	200	800

We let d = the year sales are recorded. Also, $c(d)$ is the function that describes the number of cars sold, $t(d)$ is the function that describes the number of trucks sold, and $v(d)$ is the function that describes the total number of vehicles sold. We can display all three functions at one time in graphical form. The number of cars sold $c(d)$ is displayed in red, the number of trucks sold $t(d)$ is displayed in blue, and the total number of vehicles sold $v(d)$ is the total height of the graph.

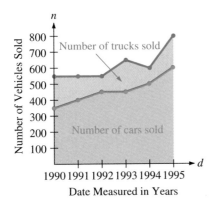

Date Measured in Years

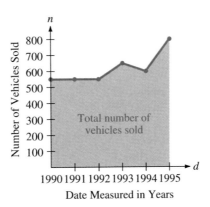

Date Measured in Years

The total number of vehicles sold $v(d)$ is obtained by adding the number of cars sold $c(d)$ to the number of trucks sold $t(d)$. Thus $v(d) = c(d) + t(d)$. This is called a *sum function*. In this section we will study how to combine two functions together to obtain a new function. One such operation is the sum of functions.

When two functions are given, new functions can be formed by combining them, as defined below.

If $f(x)$ represents one function and $g(x)$ represents a second function, then the operations on functions may be performed as follows:

Sum of Functions	$(f + g)(x) = f(x) + g(x)$
Difference of Functions	$(f - g)(x) = f(x) - g(x)$
Product of Functions	$(fg)(x) = f(x) \cdot g(x)$
Quotient of Functions	$\left(\dfrac{f}{g}\right)(x) = \dfrac{f(x)}{g(x)}$
	$g(x) \neq 0$

EXAMPLE 1   Given $f(x) = 3x^2 - 3x + 5$ and $g(x) = 5x - 2$.

(a) Find $(f + g)(x)$.

(b) Find $(f - g)(x)$.

(c) Evaluate $(f + g)(x)$ when $x = 3$.

(d) Evaluate $(f - g)(x)$ when $x = -2$.

(a) $(f + g)(x) = f(x) + g(x)$
$$= (3x^2 - 3x + 5) + (5x - 2)$$
$$= 3x^2 - 3x + 5 + 5x - 2$$
$$= 3x^2 + 2x + 3$$

(b) $(f - g)(x) = f(x) - g(x)$
$$= (3x^2 - 3x + 5) - (5x - 2)$$
$$= 3x^2 - 3x + 5 - 5x + 2$$
$$= 3x^2 - 8x + 7$$

(c) To evaluate $(f + g)(x)$ when $x = 3$, we write $(f + g)(3)$ and we use the formula obtained in (a).

If $(f + g)(x) = 3x^2 + 2x + 3$

then $(f + g)(3) = 3(3)^2 + 2(3) + 3$
$$= 3(9) + 2(3) + 3$$
$$= 27 + 6 + 3 = 36$$

(d) To evaluate $(f - g)(x)$ when $x = -2$, we write $(f - g)(-2)$ and we use the formula obtained in (b).

If $(f - g)(x) = 3x^2 - 8x + 7$

then $(f - g)(-2) = 3(-2)^2 - 8(-2) + 7$
$$= 3(4) - 8(-2) + 7$$
$$= 12 + 16 + 7 = 35$$

Practice Problem 1   Given $f(x) = 4x + 5$ and $g(x) = 2x^2 + 7x - 8$, find

(a) $(f + g)(x)$ **(b)** $(f + g)(4)$ **(c)** $(f - g)(x)$ **(d)** $(f - g)(-3)$ ∎

EXAMPLE 2   Given $f(x) = x^2 - 5x + 6$ and $g(x) = 2x - 1$, find

(a) $(fg)(x)$ **(b)** $(fg)(-4)$

(a) $(fg)(x) = f(x) \cdot g(x)$

$$
\begin{aligned}
&= (x^2 - 5x + 6)(2x - 1) \\
&= 2x^3 - 10x^2 + 12x - x^2 + 5x - 6 \\
&= 2x^3 - 11x^2 + 17x - 6
\end{aligned}
$$

(b) To evaluate $(fg)(x)$ when $x = -4$, we write $(fg)(-4)$.

If $(fg)(x) = 2x^3 - 11x^2 + 17x - 6$

then $(fg)(-4) = 2(-4)^3 - 11(-4)^2 + 17(-4) - 6$
$$
\begin{aligned}
&= 2(-64) - 11(16) + 17(-4) - 6 \\
&= -128 - 176 - 68 - 6 \\
&= -378
\end{aligned}
$$

Practice Problem 2   Given $f(x) = 3x + 2$ and $g(x) = x^2 - 3x - 4$, find

(a) $(fg)(x)$ **(b)** $(fg)(2)$ ▪

When finding the quotient function, we must be careful to avoid division by zero. Thus we always specify a value of x that is eliminated from the domain.

EXAMPLE 3   Given $f(x) = 3x + 1$, $g(x) = 2x - 1$, and $h(x) = 9x^2 + 6x + 1$, find

(a) $\left(\dfrac{f}{g}\right)(x)$ **(b)** $\left(\dfrac{f}{h}\right)(x)$ **(c)** $\left(\dfrac{f}{h}\right)(-2)$

(a) $\left(\dfrac{f}{g}\right)(x) = \dfrac{3x + 1}{2x - 1}$

The denominator of the quotient function can never be zero. Since $2x - 1 \neq 0$, we know that $x \neq \frac{1}{2}$.

(b) $\left(\dfrac{f}{h}\right)(x) = \dfrac{3x + 1}{9x^2 + 6x + 1} = \dfrac{3x + 1}{(3x + 1)(3x + 1)} = \dfrac{1}{3x + 1}$

Since $3x + 1 \neq 0$, we know that $x \neq -\frac{1}{3}$.

(c) $\left(\dfrac{f}{h}\right)(-2)$ means to evaluate $\left(\dfrac{f}{h}\right)(x)$ when $x = -2$.

Since $\left(\dfrac{f}{h}\right)(x) = \dfrac{1}{3x + 1}$

then $\left(\dfrac{f}{h}\right)(-2) = \dfrac{1}{(3)(-2) + 1} = \dfrac{1}{-6 + 1} = -\dfrac{1}{5}$

Practice Problem 3   Given $p(x) = 5x^2 + 6x + 1$, $h(x) = 3x - 2$, and $g(x) = 5x + 1$, find

(a) $\left(\dfrac{g}{h}\right)(x)$ **(b)** $\left(\dfrac{g}{p}\right)(x)$ **(c)** $\left(\dfrac{g}{h}\right)(3)$ ▪

2 Finding the Composition of Functions

Suppose the music section of a department store finds that the number of sales of compact discs (CDs) can be predicted by taking 25% of the people who visit the store in a given day. Thus, if x = the number of people in the store, then sales S can be predicted by $S(x) = 0.25x$.

The average CD in the store sells for $15.00. The store can predict its income from CD sales for x = the number of sales in a given day by $P(x) = 15.00x$.

Thus, if 80 people came into the store

$$S(x) = 0.25x$$

$$S(80) = 0.25(80) = 20$$

we would have 20 people buying a CD.

If 20 people buy a CD and the average price of CDs is $15.00, then

$$P(x) = 15.00x$$

$$P(20) = 15.00(20) = 300$$

The income for sales of CDs for the day is $300.00.

Let us analyze the functions we have described and record a few values of x, $S(x)$, and $P(x)$.

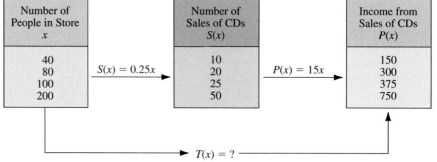

Number of People in Store x		Number of Sales of CDs $S(x)$		Income from Sales of CDs $P(x)$
40		10		150
80	$S(x) = 0.25x$	20	$P(x) = 15x$	300
100		25		375
200		50		750

$$T(x) = ?$$

Is there a function $T(x)$ that we can use to go directly from x, the number of people in the store, to the income from sales of CDs?

The number of sales is

$$S(x) = 0.25x$$

Thus $0.25x$ is the number of sales.

If we replace x in $P(x) = 15x$ by $0.25x$, we have

$$P[S(x)] = P(0.25x) = 15(0.25x) = 3.75x$$

Thus the formula $T(x)$ that goes directly from the number of people in the store to income from sales of CDs is

$$T(x) = 3.75x$$

Is this correct? Let us check by finding $T(200)$. From our table the result should be 750.

If $T(x) = 3.75x$

$$T(200) = 3.75(200) = 750$$

Thus we have verified finding an equation $T(x)$ that is composed by finding $P[S(x)]$.

We now state a definition of forming the composition of one function with the other.

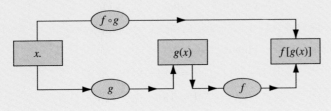

The composition of the functions f and g is $f[g(x)]$. The domain of $f[g(x)]$ is the set of all x values in the domain of g such that $g(x)$ is in the domain of f. $f[g(x)]$ is sometimes written as $(f \circ g)(x)$.

EXAMPLE 4   Given $f(x) = 3x - 2$ and $g(x) = 2x + 5$, find $f[g(x)]$.

$$f[g(x)] = f(2x + 5) \qquad \text{Substitute } g(x) = 2x + 5.$$
$$= 3(2x + 5) - 2 \qquad \text{Apply the formula for } f(x).$$
$$= 6x + 15 - 2 \qquad \text{Remove parentheses.}$$
$$= 6x + 13 \qquad \text{Simplify.}$$

Practice Problem 4   Given $f(x) = 2x - 1$ and $g(x) = 3x - 4$, find $f[g(x)]$. ∎

In most situations $f[g(x)]$ and $g[f(x)]$ are not the same.

EXAMPLE 5   Given $f(x) = \sqrt{x - 4}$ and $g(x) = 3x + 1$, find

(a) $f[g(x)]$ **(b)** $g[f(x)]$

(a) $f[g(x)] = f[3x + 1]$ Substitute $g(x) = 3x + 1.$

      $= \sqrt{(3x + 1) - 4}$ Apply the formula for $f(x).$

      $= \sqrt{3x + 1 - 4}$ Remove parentheses.

      $= \sqrt{3x - 3}$ Simplify.

(b) $g[f(x)] = g\left[\sqrt{x - 4}\right]$ Substitute $f(x) = \sqrt{x - 4}.$

      $= 3(\sqrt{x - 4}) + 1$ Apply the formula for $g(x).$

      $= 3\sqrt{x - 4} + 1$ Remove parentheses.

We note that $g[f(x)] \neq f[g(x)]$.

Practice Problem 5   Given $f(x) = 2x^2 - 3x + 1$ and $g(x) = x + 2$, find

(a) $f[g(x)]$ **(b)** $g[f(x)]$ ∎

EXAMPLE 6   Given $p(x) = 3x + 4$ and $g(x) = 2x^2$, find

(a) $g[p(x)]$ **(b)** $g[p(-3)]$

(a) $g[p(x)] = g[3x + 4]$ *Substitute $p(x) = 3x + 4$.*

$\qquad\qquad\quad = 2(3x + 4)^2$ *Apply the formula for $g(x)$.*

$\qquad\qquad\quad = 2(9x^2 + 24x + 16)$ *Square $(3x + 4)$.*

$\qquad\qquad\quad = 18x^2 + 48x + 32$ *Remove parentheses.*

(b) $g[p(-3)]$ means to evaluate $g[p(x)]$ when $x = -3$.

$\qquad\qquad$ If $\quad g[p(x)] = 18x^2 + 48x + 32$

$\qquad\qquad\qquad g[p(-3)] = 18(-3)^2 + 48(-3) + 32$

$\qquad\qquad\qquad\qquad\qquad = 18(9) + 48(-3) + 32$

$\qquad\qquad\qquad\qquad\qquad = 162 - 144 + 32 = 50$

Practice Problem 6   Given $f(x) = 1 - 2x$ and $g(x) = x^2 + 4x$, find

(a) $f[g(x)]$ **(b)** $f[g(-2)]$ ∎

GRAPHING CALCULATOR

Composition of Functions

You can formulate the composition of functions on most graphing calculators by using the *y*-variable function (Y-VARS). To do example 6 on most graphing calculators you would use:

$$y_1 = 3x + 4$$

$$y_2 = 2(y_1)^2$$

To find the function value you can use the TableSet command to let $x = -3$. Then enter Table and you will see displayed $y_1 = -5$, which represents $p(-3) = -5$, and $y_2 = 50$ which represents $g[p(-3)] = 50$.

Sometimes the notation $(f \circ g)(x)$ is used to denote $f[g(x)]$.

EXAMPLE 7   Given $f(x) = 2x$ and $g(x) = \dfrac{1}{3x - 4}$, $x \neq \dfrac{4}{3}$, find

(a) $(f \circ g)(x)$ **(b)** $(f \circ g)(2)$

(a) $(f \circ g)(x) = f[g(x)] = f\left[\dfrac{1}{3x - 4}\right]$ *Substituting, $g(x) = \dfrac{1}{3x - 4}$.*

$\qquad\qquad\qquad\quad = 2\left(\dfrac{1}{3x - 4}\right)$ *Apply the formula for $f(x)$.*

$\qquad\qquad\qquad\quad = \dfrac{2}{3x - 4}$ *Simplify.*

(b) $(f \circ g)(2) = \dfrac{2}{3(2) - 4} = \dfrac{2}{6 - 4} = \dfrac{2}{2} = 1$

Practice Problem 7   Given $f(x) = 3x + 1$ and $g(x) = \dfrac{2}{x - 3}$, find

(a) $(g \circ f)(x)$ **(b)** $(g \circ f)(-3)$ ∎

10.3 Exercises

For the following functions, find

(a) $(f + g)(x)$ **(b)** $(f - g)(x)$ **(c)** $(f + g)(2)$ **(d)** $(f - g)(-1)$

1. $f(x) = -2x + 3$, $g(x) = 2 + 4x$

2. $f(x) = 3x + 4$, $g(x) = 1 - 2x$

3. $f(x) = 2x^2 - 3x + 1$, $g(x) = 5x + 4$

4. $f(x) = 2 - x$, $g(x) = x^2 + 3x - 1$

5. $f(x) = x^3 - x^2$, $g(x) = x^2 - 3x + 2$

6. $f(x) = 2x^2 + x - 2$, $g(x) = 2x^3 - x$

7. $f(x) = \dfrac{1}{x - 3}$, $g(x) = \dfrac{3}{2x + 1}$

8. $f(x) = \dfrac{3}{x - 3}$, $g(x) = \dfrac{2}{x + 1}$

9. $f(x) = \sqrt{x + 2}$, $g(x) = \sqrt{9x + 18}$

10. $f(x) = \sqrt{4x + 4}$, $g(x) = 3\sqrt{x + 1}$

For the following functions, find   **(a)** $(fg)(x)$ **(b)** $(fg)(-3)$.

11. $f(x) = -2x$, $g(x) = x^2 + 3$

12. $f(x) = 2x^2 - 1$, $g(x) = 3x$

13. $f(x) = x^2 - 3x + 2$, $g(x) = 1 - x$

14. $f(x) = 2x - 3$, $g(x) = -2x^2 - 3x + 1$

15. $f(x) = \dfrac{2}{x + 6}$, $g(x) = 3x + 2$

16. $f(x) = 4x - 1$, $g(x) = \dfrac{3}{x + 5}$

17. $f(x) = \sqrt{-2x + 1}$, $g(x) = -3x$

18. $f(x) = 4x$, $g(x) = \sqrt{3x + 10}$

For the following functions, find   **(a)** $\left(\dfrac{f}{g}\right)x$ **(b)** $\left(\dfrac{f}{g}\right)(2)$

19. $f(x) = 3x$, $g(x) = 4x - 1$

20. $f(x) = x - 6$, $g(x) = 3x$

21. $f(x) = x^2 + 5$, $g(x) = x - 1$

22. $f(x) = x^2 + 4$, $g(x) = 3x + 2$

23. $f(x) = x^2 + 10x + 25$, $g(x) = x + 5$

24. $f(x) = 4x^2 + 4x + 1$, $g(x) = 2x + 1$

25. $f(x) = 4x - 1$, $g(x) = 4x^2 + 7x - 2$

26. $f(x) = 3x + 2$, $g(x) = 3x^2 - x - 2$

27. $f(x) = \dfrac{1}{x + 2}$, $g(x) = \dfrac{3}{x - 4}$

28. $f(x) = \dfrac{2}{x - 7}$, $g(x) = \dfrac{4}{x + 1}$

Let $f(x) = 3x + 2$, $g(x) = x^2 - 2x$, and $h(x) = \dfrac{x - 2}{3}$. Find the following.

29. $(f - g)(x)$

30. $(h + g)(x)$

31. $(f - h)(x)$

32. $(g - f)(x)$

33. $(fg)(x)$

34. $(gh)(x)$

35. $(fg)(-2)$

36. $(gh)(3)$

37. $\left(\dfrac{f}{h}\right)(x)$

38. $\left(\dfrac{g}{f}\right)(x)$

39. $(f - g)\left(\dfrac{1}{2}\right)$

40. $\left(\dfrac{g}{f}\right)(-1)$

Find $f[g(x)]$ for each of the following.

41. $f(x) = 2 - 3x$, $g(x) = 2x + 5$

42. $f(x) = 3x + 2$, $g(x) = 4x - 1$

43. $f(x) = 2x^2$, $g(x) = x - 3$

44. $f(x) = x^2 + 3$, $g(x) = x - 2$

45. $f(x) = 4 - 3x$, $g(x) = 2x^2 - 1$

46. $f(x) = 1 - 2x$, $g(x) = 3x^2 + x - 1$

47. $f(x) = \dfrac{3}{x + 1}$, $g(x) = 2x - 1$

48. $f(x) = \dfrac{4}{x - 3}$, $g(x) = 4x + 1$

49. $f(x) = \sqrt{x - 2}$, $x \geq 2$; $g(x) = 3x + 5$

50. $f(x) = \sqrt{x + 4}$, $x \geq -4$; $g(x) = 2x - 1$

Let $f(x) = x^2 + 2$, $g(x) = 3x + 5$, $h(x) = \dfrac{1}{x}$, and $p(x) = \sqrt{x - 1}$.
Find each of the following.

51. $f[g(x)]$

52. $g[h(x)]$

53. $g[f(x)]$

54. $h[g(x)]$

55. $g[f(3)]$

56. $h[g(2)]$

57. $(p \circ f)(x)$

58. $(f \circ h)(x)$

59. $(g \circ p)(x)$

60. $(f \circ p)(x)$

61. $(p \circ f)(-3)$

62. $(f \circ p)(10)$

63. $(f \circ f)(x)$

64. $(h \circ h)(x)$

65. $p[f(66.52)]$

66. $p[g(126.9)]$

Applications

67. Consider the Celsius function, which converts degrees Fahrenheit to degrees Celsius, as given by $C(F) = \frac{5F - 160}{9}$. A different temperature scale, called the Kelvin scale, is used by many scientists in their research. The Kelvin scale is similar to the Celsius scale but it begins at absolute zero (the coldest possible temperature, which is around $-273°C$). To find the approximate number of Kelvins, we use the function $K(C) = C + 273$. Find $K[C(F)]$, which is the composite function that defines the temperature in Kelvins in terms of the number of degrees Fahrenheit.

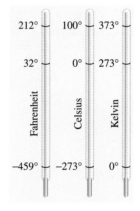

68. Suppose the dollar cost to produce n items in a factory is $c(n) = 5n + 4$. Furthermore, the number of items n produced in x hours is $n(x) = 3x$. Find $c[n(x)]$, which is the composite function that defines the relationship of the cost in terms of the number of hours x involved in the production of the items.

69. An oil tanker with a ruptured hull is leaking oil off the coast of Africa. At the time there is no wind or significant current, so the oil slick is spreading in a circular area with a radius of r feet. This radius is defined by the function $r(t) = 3t$, where t is the time in minutes since the tanker began to leak. The area of the slick for any given radius is approximately determined by the function $a(r) = 3.14r^2$. Find $a[r(t)]$, which is the composite function that defines the area of the oil slick in terms of minutes t since the beginning of the leak. How large is the area after 20 minutes?

70. The volume of polluted water emitted from a discharge pipe from a factory located on the ocean is shaped in a cone. The radius r of the cone of polluted water at the end of each day is given by the equation $r(h) = 3.5h$, where h is the number of hours the factory operates on a given day. The volume function that defines this cone is $v(r) = 31.4r^2$, where r is the radius of the cone measured in feet. Find $v[r(h)]$, which is the composite function that defines the volume of polluted water in terms of the number of hours h the factory is running in a given day. How large is the volume at the end of the day if the factory has been running for 8 hours?

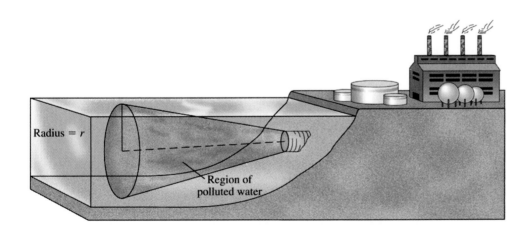

Cumulative Review Problems

Factor each of the following.

71. $25x^4 - 1$ **72.** $36x^2 - 12x + 1$ **73.** $3x^2 - 7x + 2$ **74.** $x^4 - 10x^2 + 9$

10.4 Inverse of a Function

After studying this section, you will be able to:

1. Determine if a function is a one-to-one function.
2. Find the inverse function for a given function.
3. Graph a function and its inverse function.

Americans driving in Canada or Mexico often find a need to quickly convert miles per hour to kilometers per hour, or vice versa.

If I am driving at 55 miles per hour, how fast am I going in kilometers per hour?

Approximate Value in Miles per Hour	Approximate Value in Kilometers per Hour
35	56
40	64
45	72
50	80
55	88
60	96
65	104

A function $f(x)$ that converts from miles per hour to an approximate value in kilometers per hour is $f(x) = 1.6x$.

For example $f(40) = 1.6(40) = 64$

This tells us that 40 miles per hour is approximately equivalent to 64 kilometers per hour.

An inverse function $f^{-1}(x)$ does just the opposite. It converts kilometers per hour to an approximate value in miles per hour. This function is $f^{-1}(x) = 0.625x$.

For example $f^{-1}(64) = 0.625(64) = 40$

This tells us that 64 kilometers per hour is approximately equivalent to 40 miles per hour.

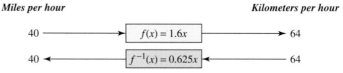

If we made a list of several function values of $f(x)$ and several inverse function values of $f(x)$, we could make up a conversion scale like the one on the left that we could use to help us if we should travel in Mexico or Canada with an American car.

Most American cars have references to kilometers per hour printed in smaller print on the car speedometer. Unfortunately, they are usually hard to read while you are driving a car.

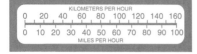

The function we have studied goes from miles per hour to kilometers per hour. The corresponding inverse function goes from kilometers per hour to miles per hour. How do we find inverse functions? Do all functions have inverse functions? These are questions we want to explore in this section.

1 *Determining if a Function Is a One-to-One Function*

First we state that not all functions have inverse functions. To have an inverse that is a function, a function must be one-to-one. This means that for every value of *y* there is only one value of *x*. Or, in the language of ordered pairs, no ordered pairs have the same second coordinate.

Definition of a One-to-One Function

A one-to-one function is a function in which no ordered pairs have the same second coordinate.

To Think About Why must a function be one-to-one in order to have an inverse that is a function?

EXAMPLE 1   Indicate if the following functions are one-to-one.

(a) $M = \{(1, 3), (2, 7), (5, 8), (6, 12)\}$ **(b)** $P = \{(1, 4), (2, 9), (3, 4), (4, 18)\}$

(a) *M* is a function because no ordered pairs have the same first coordinate. *M* is a one-to-one function because no ordered pairs have the same second coordinate.

(b) *P* is a function, *but* it is not one-to-one because the ordered pairs (1, 4) and (3, 4) have the same second coordinates.
Thus the function *M* has an inverse function, but the function *P* does not.

Practice Problem 1

(a) Is the function $A = \{(-2, -6), (-3, -5), (-1, 2), (3, 5)\}$ one-to-one?
(b) Is the function $B = \{(0, 0), (1, 1), (2, 4), (3, 9), (-1, 1)\}$ one-to-one? ■

By examining the graph of a function, we can quickly tell whether it is one-to-one. If any horizontal line crosses the graph of a function in *more* than one place, the function is not one-to-one.

Horizontal Line Test

If any horizontal line can intersect the graph of a function more than once, the function is not one-to-one.

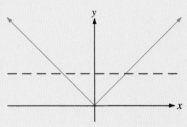

This is *not* a one-to-one function

EXAMPLE 2 Determine whether each of the following is a graph of a one-to-one function.

(a)

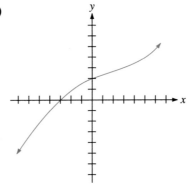

(b)

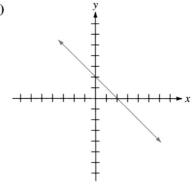

(c)

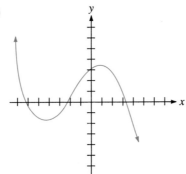

(d)

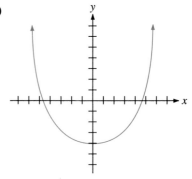

The graphs of **(a)** and **(b)** represent one-to-one functions. A horizontal line crosses the graph only once.

(a)

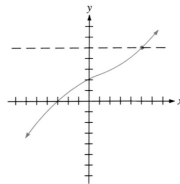

(b)

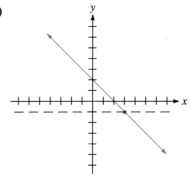

The graphs of **(c)** and **(d)** do not represent one-to-one functions. A horizontal line crosses the graph more than once.

(c)

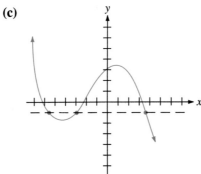

(d)

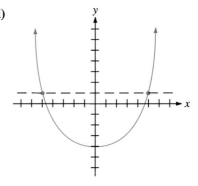

Practice Problem 2   Does the graph represent a one-to-one function? Why or why not?

(a)

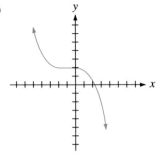

(b)

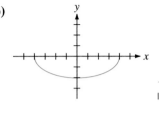

2 Finding the Inverse of a Function

How do we find the inverse of a function? If we have a list of ordered pairs, we simply interchange the ordered pairs. In Example 1, we said *M* has an inverse. What is it?

$$M = \{(1, 3), (2, 7), (5, 8), (6, 12)\}$$

The inverse of *M*, written M^{-1}, is

$$M^{-1} = \{(3, 1), (7, 2), (8, 5), (12, 6)\}$$

Now do you see why a function must be one-to-one to have an inverse function? If not, let's look at the function *P* from Example 1.

$$P = \{(1, 4), (2, 9), (3, 4), (4, 18)\}$$

If *P* had an inverse, it would be

$$P^{-1} = \{(4, 1), (9, 2), (4, 3), (18, 4)\}$$

But we have two ordered pairs with the same first coordinate. Therefore, P^{-1} is not a function (in other words, the *inverse function* does not exist).

A number of real-world situations are described by a function that does have an inverse. Consider the function defined by the ordered pairs (year, U.S. budget in trillions of dollars). Some function values are

$$F = \{(1995, 1.52), (1990, 1.25), (1985, 0.95), (1980, 0.59)\}$$

In this case the inverse of the function is

$$F^{-1} = \{(1.52, 1995), (1.25, 1990), (0.95, 1985), (0.59, 1980)\}$$

By the way, F^{-1} does *not* mean $\dfrac{1}{F}$. Here the -1 simply means "inverse."

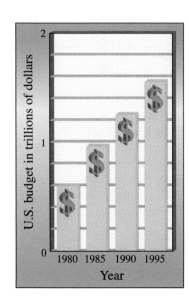

EXAMPLE 3   Determine the inverse function of the function *F* if

$$F = \{(6, 1), (12, 2), (13, 5), (14, 6)\}$$

The inverse function of *F* is denoted by $F^{-1} = \{(1, 6), (2, 12), (5, 13), (6, 14)\}$.

Practice Problem 3   Find the inverse of the one-to-one function $B = \{(1, 2), (7, 8), (8, 7), (10, 12)\}$. ■

Suppose the function is given in the form of an equation. How do we find the inverse? Since, by definition, we interchange the ordered pairs to find the inverse of a function, this means the x values of the function become the y values of the inverse function.

Four steps will help us find the inverse of a one-to-one function when we are given its equation.

Find the Inverse of a One-to-One Function

1. Replace $f(x)$ with y.
2. Interchange x and y.
3. Solve for y in terms of x.
4. Replace y with $f^{-1}(x)$.

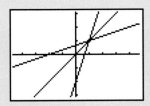

GRAPHING CALCULATOR

Inverse Functions

You can use a graphing calculator to provide a reasonable check if two functions are the inverses of each other. If $f(x)$ and $g(x)$ are inverse functions, then their graphs will be symmetric about the line $y = x$. Use a "square" window setting when graphing. For example, to check the answer in Example 4, graph

$y_1 = 3x - 2$, $y_2 = \dfrac{x + 2}{3}$,

and $y_3 = x$.

Display:

EXAMPLE 4   Find the inverse of $f(x) = 3x - 2$.

Step 1 $y = 3x - 2$ *Replace $f(x)$ with y.*

Step 2 $x = 3y - 2$ *Interchange the variables x and y.*

Step 3 $x + 2 = 3y$ *Solve for y in terms of x.*

$\dfrac{x + 2}{3} = y$

Step 4 $f^{-1}(x) = \dfrac{x + 2}{3}$ *Replace y by $f^{-1}(x)$.*

Practice Problem 4   Find the inverse of the function $g(x) = 4 - 6x$. ∎

Let's see if this technique works on the opening example that converts miles per hour to approximate values in kilometers per hour.

EXAMPLE 5   Find the inverse of $f(x) = 1.6x$.

Step 1 $y = 1.6x$ *Replace $f(x)$ with y.*

Step 2 $x = 1.6y$ *Interchange x and y.*

Step 3 $\dfrac{x}{1.6} = \dfrac{1.6y}{1.6}$ *Solve for y in terms of x.*

$\dfrac{x}{1.6} = y$

Step 4 $f^{-1}(x) = \dfrac{x}{1.6}$ *Replace y by $f^{-1}(x)$*

Is this equivalent to $f^{-1}(x) = 0.625x$? Let's look at our equation.

$$f^{-1}(x) = \dfrac{x}{1.6}$$

This is equivalent to

$$f^{-1}(x) = \dfrac{1x}{1.6}$$

In each of the following cases, graph $f(x)$, $g(x)$, and $y = x$ in an appropriate "square" window to determine if $f(x)$ and $g(x)$ are inverses of each other.

1. $f(x) = (\frac{2}{3})x - 2$
   $g(x) = (\frac{3}{2})x + 3$
2. $f(x) = (\frac{2}{3})x + \frac{5}{3}$
   $g(x) = (\frac{3}{2})x + \frac{5}{2}$
3. $f(x) = (x - 1)^3$
   $g(x) = \sqrt[3]{x} + 1$

Let's rewrite this as a product.

$$f^{-1}(x) = \left(\frac{1}{1.6}\right)(x)$$

If we divide $1 \div 1.6$, we obtain 0.625.

Thus $f^{-1}(x) = 0.625x$

We see that both equations are equivalent. Thus the inverse function of $f(x) = 1.6x$ can be written in either equivalent form $f^{-1}(x) = 0.625x$ or $f^{-1}(x) = \frac{x}{1.6}$.

Practice Problem 5 Find the inverse function of $f(x) = 2(3x + 1) - 4$. ∎

EXAMPLE 6 Find the inverse function of $f(x) = \frac{9}{5}x + 32$, which converts Celsius temperature (x) into equivalent Fahrenheit temperature.

Step 1 $y = \dfrac{9}{5}x + 32$ *Replace $f(x)$ with y.*

Step 2 $x = \dfrac{9}{5}y + 32$ *Interchange x and y.*

Step 3 $5(x) = 5\left(\dfrac{9}{5}\right)y + 5(32)$ *Solve for y in terms of x.*

$$5x = 9y + 160$$

$$5x - 160 = 9y$$

$$\frac{5x - 160}{9} = \frac{9y}{9}$$

$$\frac{5x - 160}{9} = y$$

Step 4 $f^{-1}(x) = \dfrac{5x - 160}{9}$ *Replace y by $f^{-1}(x)$.*

Note: Our inverse function $f^{-1}(x)$ will now convert Fahrenheit temperature to Celsius temperature.

$$f^{-1}(30) = \frac{9}{5}(30) + 32 = (9)(6) + 32 = 86$$

This tells us that a temperature of 86°F corresponds to a temperature of 30°C.

Practice Problem 6 Find the inverse function of $f(x) = 0.75 + 0.55(x - 1)$, which tells you the cost of a telephone call for any call over 1 minute if the telephone company charges 75 cents for the first minute and 55 cents for each minute thereafter. Here $x = $ the number of minutes. ∎

3 Graphing a Function and Its Inverse Function

The graph of a function and its inverse are symmetric about the line $y = x$. Why do you think that this is so?

EXAMPLE 7 Find the inverse function f^{-1} of

$$f = \{(0, 1), (1, 2), (2, 5), (3, 10)\}$$

Graph f and f^{-1} on one coordinate plane.

We interchange the values of (x, y) to (y, x) to obtain

$$f^{-1} = \{(1, 0), (2, 1), (5, 2), (10, 3)\}$$

Now we graph the ordered pairs of f and f^{-1}.

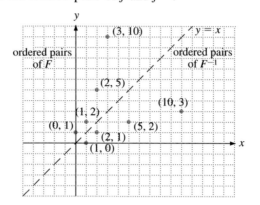

PRACTICE PROBLEM 7

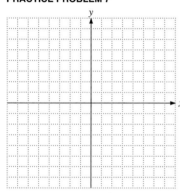

Observe that if we folded the graph paper flat on the line $y = x$ the ordered pairs of f would touch the ordered pairs of f^{-1}.

Practice Problem 7 Find the inverse function f^{-1} if

$$f = \{(3, 8), (2, 4), (1, 2), (0, 1)\}$$

Graph f and f^{-1}. Graph $y = x$ as a reference line. ■

EXAMPLE 8 If $f(x) = 3x - 2$, find $f^{-1}(x)$. Graph $f(x)$ and $f^{-1}(x)$ on the same set of axes. Draw the line $y = x$ as a dashed line for reference.

$f(x) = 3x - 2$ Now we graph each line.

$y = 3x - 2$

$x = 3y - 2$

$x + 2 = 3y$

$\dfrac{x + 2}{3} = y$

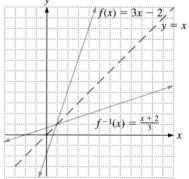

PRACTICE PROBLEM 8

$f^{-1}(x) = \dfrac{x + 2}{3}$

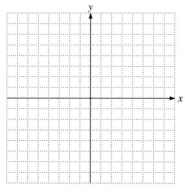

Again we see that the graph of $f(x)$ is symmetric to the graph of $f^{-1}(x)$ about the line $y = x$. If we folded the graph paper flat on the line $y = x$, the graph of $f(x)$ would touch the graph of $f^{-1}(x)$. Try it. Redraw the functions on a separate piece of graph paper. Fold the graph paper on the line $y = x$.

Practice Problem 8 If $f(x) = -\dfrac{1}{4}x + 1$, find $f^{-1}(x)$. Graph $f(x)$ and $f^{-1}(x)$ on the same coordinate plane. Draw the line $y = x$ as a dashed line for reference. ■

10.4 Exercises

Verbal and Writing Skills

Complete the following.

1. A one-to-one function is a function in which no ordered pairs _____.

2. If any horizontal line can intersect the graph of a function more than once, _____.

3. Graph of a function $f(x)$ and its inverse $f^{-1}(x)$ are symmetric about the line _____.

4. Do all functions have inverse functions? Why or why not?

Are the following functions one-to-one?

5. $A = \{(-6, -2), (6, 2), (3, 4)\}$

6. $B = \{(0, 1), (1, 0), (10, 0)\}$

7. $C = \{(12, 3), (-6, 1), (6, 3)\}$

8. $F = \{(\frac{2}{3}, 2), (3, -\frac{4}{5}), (-\frac{2}{3}, -2), (-3, \frac{4}{5})\}$

9. $E = \{(1, 3), (\frac{1}{2}, -5), (-1, -3), (-5, \frac{1}{2})\}$

10. $F = \{(5, 0), (2, -7), (-2, 7), (0, 5)\}$

Which graphs represent one-to-one functions?

11.

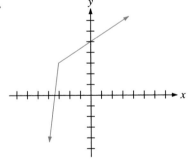

12.

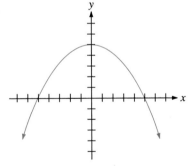

13.

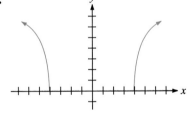

14.

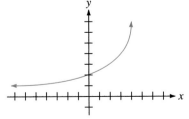

15.

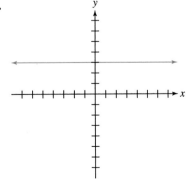

16.

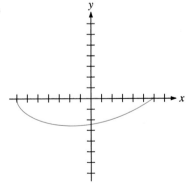

17. Does the graph of a horizontal line represent a function? Why or why not? Does it represent a one-to-one function? Explain.

18. Does the graph of a vertical line represent a function? Why or why not? Does it represent a one-to-one function? Explain.

Find the inverse of each one-to-one function. Graph the function and its inverse on one coordinate plane.

19. $H = \{(2, 7), (-3, -1)\}$

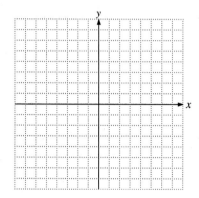

20. $G = \{(-1, 0), (5, 8)\}$

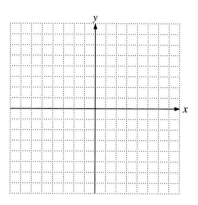

21. $K = \{(-7, 1), (6, 2), (3, -1), (2, 5)\}$

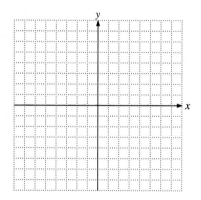

22. $J = \{(8, 2), (1, 1), (0, 0), (-8, -2)\}$

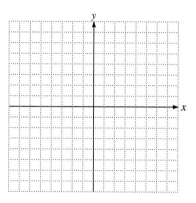

23. $L = \{(1, 4), (2, 8), (3, 6), (-2, -8)\}$

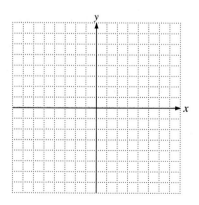

24. $M = \{(0, 0), (-1, 1), (1, -1), (2, 8)\}$

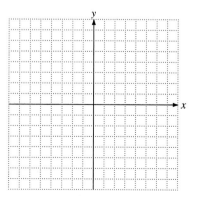

Find the inverse of each function.

25. $f(x) = 3x - 1$

26. $f(x) = \frac{5}{7}x + \frac{1}{2}$

27. $f(x) = \frac{x+4}{5}$

28. $f(x) = \frac{2-4x}{3}$

29. $f(x) = -\frac{4}{x}$

30. $f(x) = \frac{3}{x}$

31. $f(x) = \frac{3}{x-2}$

32. $f(x) = \frac{2}{x+4}$

Verbal and Writing Skills

33. Can you find an inverse function for the function $f(x) = 2x^2 + 3$? Why?

34. Can you find an inverse function for the function $f(x) = |3x + 4|$? Why?

Find the inverse of each function. Graph the function and the inverse on one coordinate plane. Graph the line $y = x$ as a dashed line. What do you notice?

35. $f(x) = 3x + 4$

36. $g(x) = 2x + 5$

37. $p(x) = \frac{2}{3}x - 4$

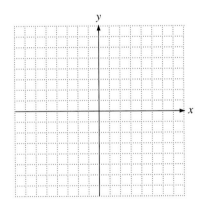

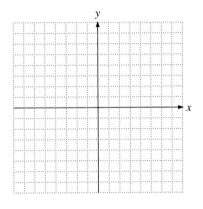

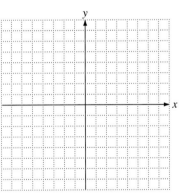

38. $h(x) = \frac{1}{2}x - 2$

39. $k(x) = 3 - 2x$

40. $r(x) = -3x - 1$

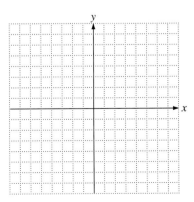

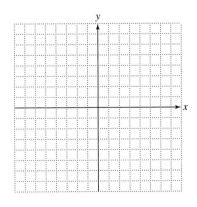

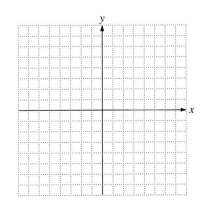

41. $s(x) = 4 - 2x$

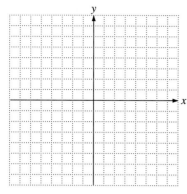

42. $t(x) = 1 - \dfrac{2}{3}x$

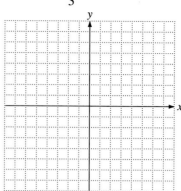

Applications

43. Manuela was in Spain. She brought Spanish pesetas back with her from Madrid and is going to the bank this morning to take advantage of the more favorable rates which the banks offer to convert her leftover Spanish Pesetas to U.S. dollars. The bank teller informs Manuela that today's conversion rate is US$0.0063 for one Spanish Peseta. The bank charges a fee of US$5.00 for each transaction. The function used to convert Spanish Pesetas to U.S. dollars is given by $f(x) = 0.0063x - 5$, where x is the number of Spanish Pesetas. Find the inverse function of $f(x)$. What is the significance of the inverse function? If Manuela wanted to change U.S. dollars to Spanish Pesetas, could she use this inverse function? Why not?

44. Sean was in Ireland. He brought Irish pounds back with him from Dublin and is going to the bank this morning to take advantage of the more favorable rates the banks offer to convert his leftover Irish pounds to U.S. dollars, as opposed to the exchange rate at the airport. The bank teller informs Sean that today's conversion rate is US$1.437 for one Irish pound. The bank charges a fee of US$4.00 for each transaction. The function used to convert Irish pounds to U.S. dollars is given by $f(x) = 1.437x - 4$, where x is the number of Irish pounds. Find the inverse function of $f(x)$. What is the significance of the inverse function? If Sean wanted to change U.S. dollars to Irish pounds, could he use this inverse function? Why not?

To Think About

For every function $f(x)$ and its inverse, $f^{-1}x$, it is true that $f[f^{-1}(x)]$ and $f^{-1}[f(x)] = x$. Show that this is true for each pair of inverse functions.

45. $f(x) = 2x + \dfrac{3}{2}, f^{-1}(x) = \dfrac{1}{2}x - \dfrac{3}{4}$

46. $f(x) = -3x - 10, f^{-1}(x) = \dfrac{-x - 10}{3}$

Cumulative Review Problems

Solve for x.

47. $x^{2/3} + 7x^{1/3} + 12 = 0$

48. $x = \sqrt{15 - 2x}$

49. The average male human has more blood than the average female. In addition, each cubic centimeter of blood is usually richer in red blood cells for males. Each cubic centimeter of blood in men contains from 4.6 million to 6.2 million red blood cells, compared with 4.2 million to 5.4 million for women. Using the lower level, what would be the ratio comparing red blood cells of men to women?

50. Catherine earns $17 per hour for a 40-hour week as an on-call nurse for City Hospital. She earns time and a half for every hour over 40 hours in one week. If Catherine made $1011.50 last week, how many overtime hours did she work?

Topic	Procedure	Examples
Relations, functions, and one-to-one functions, pp. 165, 167, 585.	**Relation** A *relation* is any set of ordered pairs. **Function** A *function* is a relation in which no ordered pairs have the same first coordinate. **One-to-one function** A *one-to-one function* is a function in which no ordered pairs have the same second coordinate.	Is $\{(3, 6), (2, 8), (9, 1), (4, 6)\}$ a one-to-one function? No, since $(3, 6)$ and $(4, 6)$ have the same second coordinate.
Vertical line tests, p. 566.	If any vertical line intersects the graph of a relation more than once, the relation is not a function.	 Does this graph represent a function? No, because a vertical line intersects the curve more than once.
Horizontal line tests, p. 585.	If any horizontal line intersects the graph of a function more than once, the function is not one-to-one.	 Does this graph represent a one-to-one function? Yes, a horizontal line will only cross this function once.
Finding function values, p. 560.	Replace x by the quantity inside the parentheses. Simplify the result.	If $f(x) = 2x^2 + 3x - 4$, then $$f(-2) = 2(-2)^2 + 3(-2) - 4$$ $$= 8 - 6 - 4 = -2$$ $$f(a) = 2a^2 + 3a - 4$$ $$f(a + 2) = 2(a + 2)^2 + 3(a + 2) - 4$$ $$= 2(a^2 + 4a + 4) + 3a + 6 - 4$$ $$= 2a^2 + 8a + 8 + 3a + 6 - 4$$ $$= 2a^2 + 11a + 10$$ $$f(3a) = 2(3a)^2 + 3(3a) - 4$$ $$= 2(9a^2) + 9a - 4$$ $$= 18a^2 + 9a - 4$$

Topic	Procedure	Examples
Vertical shifts of the graph of function $f(x)$, p. 567.	If $c > 0$, **1.** The graph of $f(x) + c$ is shifted c units *upward* from the graph of $f(x)$.	Graph $f(x) = x^2$ and $g(x) = x^2 + 3$.
	2. The graph of $f(x) - c$ is shifted c units *downward* from the graph of $f(x)$.	Graph $f(x) = |x|$ and $g(x) = |x| - 2$.
Horizontal shifts of the graph of function $f(x)$, p. 568.	If $c > 0$, **1.** The graph of $f(x - c)$ is shifted c units to the *right* of the graph of $f(x)$.	Graph $f(x) = x^2$ and $g(x) = (x - 3)^2$.
	2. The graph of $f(x + c)$ is shifted c units to the *left* of the graph of $f(x)$.	Graph $f(x) = x^3$ and $g(x) = (x + 4)^3$.

Chapter Organizer

Topic	Procedure	Examples
Sum, difference, product, and quotient of functions, p. 575.	1. $(f + g)(x) = f(x) + g(x)$ 2. $(f - g)(x) = f(x) - g(x)$ 3. $(f \cdot g)(x) = f(x) \cdot g(x)$ 4. $\left(\dfrac{f}{g}\right)(x) = \dfrac{f(x)}{g(x)}, \; g(x) \neq 0$	If $f(x) = 2x + 3$ and $g(x) = 3x - 4$, then 1. $(f + g)(x) = (2x + 3) + (3x - 4)$ $\qquad\qquad = 5x - 1$ 2. $(f - g)(x) = (2x + 3) - (3x - 4)$ $\qquad\qquad = 2x + 3 - 3x + 4$ $\qquad\qquad = -x + 7$ 3. $(f \cdot g)(x) = (2x + 3)(3x - 4)$ $\qquad\qquad = 6x^2 + x - 12$ 4. $\left(\dfrac{f}{g}\right)(x) = \dfrac{2x + 3}{3x - 4}, \; x \neq \dfrac{4}{3}$
Composition of functions, p. 578.	The composition of functions f and g is $f[g(x)]$ or it may be written as $(f \circ g)(x)$. To find $f[g(x)]$, 1. Replace $g(x)$ by its equation. 2. Apply the $f(x)$ rule to this expression. 3. Simplify the results. Usually, $f[g(x)] \neq g[f(x)]$.	If $f(x) = x^2 - 5$ and $g(x) = -3x + 4$, find $f[g(x)]$ and $g[f(x)]$. $\quad f[g(x)] = f[-3x + 4]$ $\qquad\quad = (-3x + 4)^2 - 5$ $\qquad\quad = 9x^2 - 24x + 16 - 5$ $\qquad\quad = 9x^2 - 24x + 11$ $\quad g[f(x)] = g[x^2 - 5]$ $\qquad\quad = -3(x^2 - 5) + 4$ $\qquad\quad = -3x^2 + 15 + 4$ $\qquad\quad = -3x^2 + 19$
Finding the inverse of a function defined by a set of ordered pairs, p. 587.	Reverse the order of each ordered pair from (a, b) to (b, a).	Find the inverse of $A = \{(5, 6), (7, 8), (9, 10)\}$ $\quad A^{-1} = \{(6, 5), (8, 7), (10, 9)\}$
Finding the inverse of a function defined by an equation, p. 588.	Any one-to-one function has an inverse function. To find the equation of an inverse function $f^{-1}(x)$ when the equation of a one-to-one function $f(x)$ is given: 1. Replace $f(x)$ with y. 2. Interchange x and y. 3. Solve for y in terms of x. 4. Replace y with $f^{-1}(x)$.	Find the inverse of $f(x) = -\dfrac{2}{3}x + 4$. $y = -\dfrac{2}{3}x + 4$ $x = -\dfrac{2}{3}y + 4$ $3x = -2y + 12$ $3x - 12 = -2y$ $\dfrac{3x - 12}{-2} = y$ $-\dfrac{3}{2}x + 6 = y$ $f^{-1}(x) = -\dfrac{3}{2}x + 6$ Find the inverse of $f(x) = (x + 3)^3$. $\quad y = (x + 3)^3$ $\quad x = (y + 3)^3$ $\quad \sqrt[3]{x} = \sqrt[3]{(y + 3)^3}$ $\quad \sqrt[3]{x} = y + 3$ $\quad \sqrt[3]{x} - 3 = y$ $\quad f^{-1}(x) = \sqrt[3]{x} - 3$

Topic	Procedure	Examples
Graphing the inverse of a function, p. 590.	Graph the line $y = x$ as a dashed line for reference. **1.** Graph $f(x)$. **2.** Graph $f^{-1}(x)$. The graph of $f(x)$ is symmetric to the graph of $f^{-1}(x)$ about the line $y = x$.	$f(x) = 2x + 3$ $f^{-1}(x) = \dfrac{x - 3}{2}$ Graph $f(x)$ and $f^{-1}(x)$ on the same set of axes. 

Putting Your Skills to Work

THE TEMPERATURE FUNCTION FOR PLANT LIFE

Most plant life on the earth exists only in the temperature range of 8° to 38° Celsius. The maximum growth rate of many of these plants occurs around 23° Celsius. A function that approximates the growth rate of these plants in terms of the Celsius temperature is given by the equation:

$$G(x) = -0.4(x - 23)^2 + 100$$

A graph of this function is shown below.

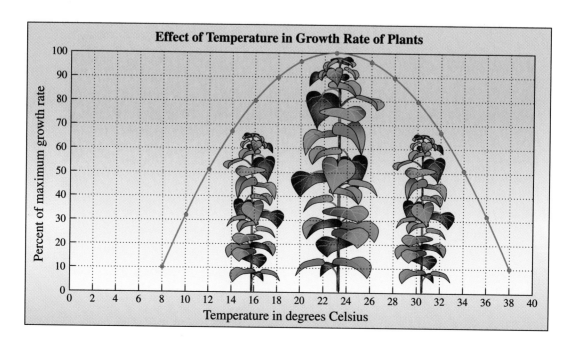

1. The optimal range for temperature for most plants is between 19° and 30° Celsius. If plants grow in that temperature range what is the smallest percent of the maximum growth rate that should be expected for the plants?

2. Biologists are currently seeking to develop "super plants" that achieve a maximum growth rate that is 10% higher than the current maximum. How would this effect the graph of the growth rate function? Write a possible new equation.

3. The new "super plants" may have a maximum growth rate at 25° Celsius. This would allow the plants to be used more readily to provide food crops in desert areas of Africa and other dry and hot regions of the world. What would be a possible new equation that would describe the growth rate function for this as well as the change in Problem 2?

Together with members of your class see if you can do the following.

4. Celsius temperature can be converted for Fahrenheit temperature by the equation $F = 1.8C + 32$. Form a composition of functions that would provide the growth rate in terms of the temperature in degrees Fahrenheit for the original equation.

5. Form a composition of functions that would provide the growth rate in terms of the temperature measured in degrees Kelvin for the original equation.

6. How would the equation for the new "super plants" be modified if in addition to the changes discussed in Problems 2 and 3 the plants also had an increase in the percent of maximum growth rate even at the temperature limits of tolerance so that the smallest growth rate achieved by a plant would be a minimum of 20.4% of the maximum growth rate occurring at 9° and at 41° Celsius?

INTERNET CONNECTION: Go to http://www.prenhall.com/tobey to be connected.

Site: Annual Ryegrass International Fact Sheet

This site gives a detailed description of annual ryegrass *(Lolium multiflorum),* including optimum temperatures for growth.

7. Give a possible growth rate function for annual ryegrass in terms of Fahrenheit temperature. Assume that the range of temperatures given for annual growth is the range in which the growth rate is at least 85% of the maximum growth rate for annual ryegrass.

8. Repeat the preceding problem, using Celsius temperatures.

For the function $f(x) = 7 - 2x$, find

1. $f(4)$

2. $f(-2)$

3. $f(2a)$

4. $f(3b)$

5. $f(2a) + f(4)$

6. $f(3b) + f(-2)$

For the function $f(x) = \dfrac{1}{2}x + 3$, find

7. $f(a - 1)$

8. $f(a + 2)$

9. $f(a - 1) - f(a)$

10. $f(a + 2) - f(a)$

11. $f(2a + 3)$

12. $f(2a - 3)$

For the function $p(x) = -2x^2 + 3x - 1$, find

13. $p(-2)$

14. $p(3)$

15. $p(2a) + p(-2)$

16. $p(3a) + p(3)$

17. $p(a + 2)$

18. $p(a - 3)$

For the function $h(x) = |2x - 1|$, find

19. $h(8a)$

20. $h(7a)$

21. $h(\frac{1}{4}a)$

22. $h(\frac{1}{2}a)$

23. $h(a - 3)$

24. $h(a + 4)$

For the function $r(x) = \dfrac{3x}{x + 4}$, $x \neq -4$, find the following. In each case, write your answer as one fraction, if possible.

25. $r(3)$

26. $r(-2)$

27. $r(a + 3)$

28. $r(a - 2)$

29. $r(3) + r(a)$

30. $r(a) + r(-2)$

Find $\dfrac{f(x + h) - f(x)}{h}$ if

31. $f(x) = 7x - 4$

32. $f(x) = 6x - 5$

33. $f(x) = \dfrac{1}{2}x + 3$

34. $f(x) = \dfrac{1}{4}x + 1$

35. $f(x) = 2x^2 - 5x$

36. $f(x) = 2x - 3x^2$

Examine each of the following graphs. **(a)** *Does the graph represent a function?* **(b)** *Does the graph represent a one-to-one function?*

37.

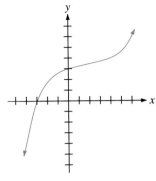

38.

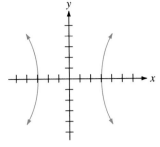

39.

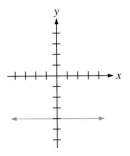

40.

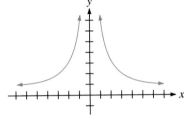

41.

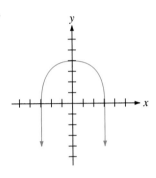

42.

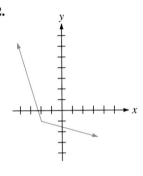

43.

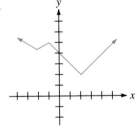

44.

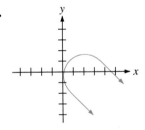

45.

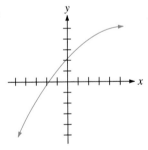

For each of the following pair of functions, graph the two functions on one axis.

46. $f(x) = x^2$
$g(x) = (x + 1)^2 - 3$

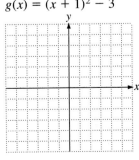

47. $f(x) = x^2$
$g(x) = (x + 2)^2 + 4$

48. $f(x) = |x|$
$g(x) = |x + 3|$

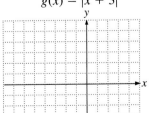

49. $f(x) = |x|$
$g(x) = |x - 4|$

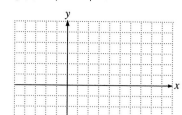

50. $f(x) = |x|$
$h(x) = |x| - 2$

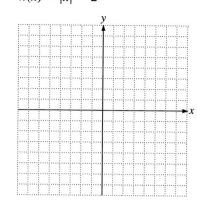

51. $f(x) = |x|$
$h(x) = |x| + 3$

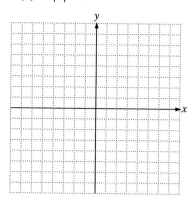

52. $f(x) = x^3$
$r(x) = (x + 3)^3 + 1$

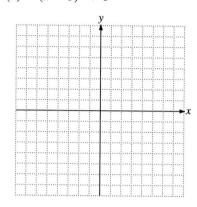

53. $f(x) = x^3$
$r(x) = (x - 1)^3 + 5$

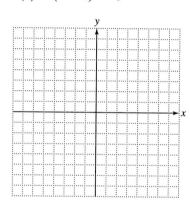

54. $f(x) = \dfrac{4}{x},\ x \neq 0$

$r(x) = \dfrac{4}{x + 2},\ x \neq -2$

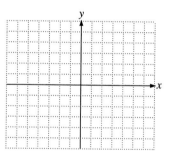

55. $f(x) = \dfrac{2}{x},\ x \neq 0$

$r(x) = \dfrac{2}{x + 3} - 2,\ x \neq -3$

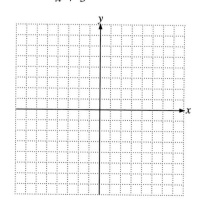

Given the following functions:

$$f(x) = 3x + 5, \qquad g(x) = \frac{2}{x}, \quad x \neq 0, \qquad s(x) = \sqrt{x - 2}, \quad x \geq 2$$

$$h(x) = \frac{x + 1}{x - 4}, \quad x \neq 4, \qquad p(x) = 2x^2 - 3x + 4, \qquad t(x) = -\frac{1}{2}x - 3$$

Find each of the following.

56. $(f + p)(x)$

57. $(f + t)(x)$

58. $(t - f)(x)$

59. $(p - f)(x)$

60. $(t - f)(-3)$

61. $(p - f)(2)$

62. $(fg)(x)$

63. $(tp)(x)$

64. $\left(\dfrac{g}{h}\right)(x)$

65. $\left(\dfrac{g}{f}\right)(x)$

66. $\left(\dfrac{g}{h}\right)(-2)$

67. $\left(\dfrac{g}{f}\right)(-3)$

68. $h[f(x)]$

69. $f[t(x)]$

70. $s[p(x)]$

71. $s[t(x)]$

72. $s[p(2)]$

73. $s[t(-18)]$

74. $(g \circ h)(x)$

75. $(t \circ g)(x)$

76. $f[f(x)]$

77. $t[t(x)]$

78. Show that $f[g(x)] \neq g[f(x)]$.

79. Show that $p[g(x)] \neq g[p(x)]$.

For each set, determine **(a)** *the domain,* **(b)** *the range,* **(c)** *if the set defines a function, and* **(d)** *if the set defines a one-to-one function.*

80. $A = \{(100, 10), (200, 20), (300, 30), (400, 10)\}$

81. $B = \{(3, 7), (7, 3), (0, 8), (0, -8)\}$

82. $C = \{(12, 6), (0, 6), (0, -1), (-6, -12)\}$

83. $D = \left\{\left(\dfrac{1}{2}, 2\right), \left(\dfrac{1}{4}, 4\right), \left(-\dfrac{1}{3}, -3\right), \left(4, \dfrac{1}{4}\right)\right\}$

84. $E = \{(0, 1), (1, 2), (2, 9), (-1, -2)\}$

85. $F = \{(3, 7), (2, 1), (0, -3), (1, 1)\}$

Find the inverse of each of the following functions.

86. $\left\{\left(3, \dfrac{1}{3}\right), \left(-2, -\dfrac{1}{2}\right), \left(-4, -\dfrac{1}{4}\right), \left(5, \dfrac{1}{5}\right)\right\}$

87. $\{(1, 10), (3, 7), (12, 15), (10, 1)\}$

88. $f(x) = -\dfrac{2}{3}x + 4$

89. $g(x) = -5 - 3x$

90. $h(x) = \dfrac{x + 2}{3}$

91. $j(x) = \dfrac{1}{x - 3}$

92. $p(x) = \sqrt[3]{x + 1}$

93. $r(x) = x^3 + 2$

Find the inverse of each function. Graph the function and its inverse on one coordinate plane. Then on that same set of axes graph the line $y = x$ as a dashed line.

94. $f(x) = -\dfrac{3}{4}x + 1$

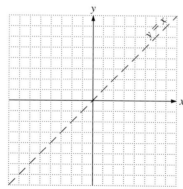

95. $f(x) = \dfrac{-x - 2}{3}$

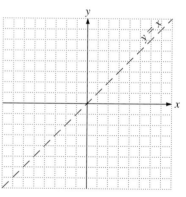

In each of the following cases show that $f\,[f^{-1}(x)] = x$ and that $f^{-1}[f(x)] = x$.

96. $f(x) = \dfrac{1}{2}x - \dfrac{3}{4}$

$f^{-1}(x) = \dfrac{4x + 3}{2}$

97. $f(x) = \dfrac{6 - 3x}{2}$

$f^{-1}(x) = -\dfrac{2}{3}x + 2$

For the function $f(x) = \dfrac{3}{4}x - 2$, *find*

1. $f(-4)$　　　　**2.** $f(2a)$　　　　**3.** $f(a) - f(2)$

For the function $f(x) = 3x^2 - 2x + 4$, *find*

4. $f(3)$　　**5.** $f(a + 1)$　　**6.** $f(a) + f(1)$　　**7.** $f(-2a) - 2$

Look at each graph below. **(a)** *Does the graph represent a function?* **(b)** *Does the graph represent a one-to-one function?*

8.

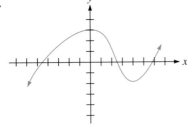

9.

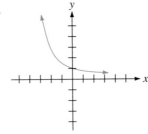

Graph each function on one coordinate plane.

10. $f(x) = x^2$
$g(x) = (x - 1)^2 + 3$

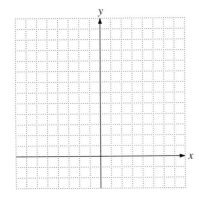

11. $f(x) = |x|$
$g(x) = |x + 1| + 2$

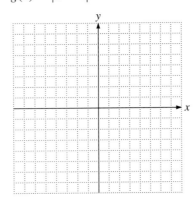

1. _____

2. _____

3. _____

4. _____

5. _____

6. _____

7. _____

8. (a) _____

(b) _____

9. (a) _____

(b) _____

10. _____

11. _____

12. (a) _____

(b) _____

(c) _____

13. (a) _____

(b) _____

(c) _____

14. (a) _____

(b) _____

(c) _____

15. (a) _____

(b) _____

16. (a) _____

(b) _____

17. _____

18. _____

19. _____

12. If $f(x) = 3x^2 - x - 6$ and $g(x) = -2x^2 + 5x + 7$, find
 (a) $(f + g)(x)$ **(b)** $(f - g)(x)$ **(c)** $(f - g)(-2)$

13. If $f(x) = \dfrac{3}{x}$, $x \neq 0$ and $g(x) = 2x - 1$, find

 (a) $(fg)(x)$ **(b)** $\left(\dfrac{f}{g}\right)(x)$ **(c)** $g[f(x)]$

14. If $f(x) = \dfrac{1}{2}x - 3$ and $g(x) = 4x + 5$, find

 (a) $(f \circ g)(x)$ **(b)** $(g \circ f)(x)$ **(c)** $f[f(x)]$

Look at the following functions. **(a)** *Is the function one-to-one?* **(b)** *If so, find the inverse for the function.*

15. $A = \{(1, 5), (2, 1), (4, -7), (0, 7)\}$

16. $B = \{(1, 8), (8, 1), (9, 10), (-10, 9)\}$

17. Determine the inverse function $f^{-1}(x)$ if $f(x) = \dfrac{1}{2}x - \dfrac{1}{5}$.

18. Graph $f(x)$ and the inverse of $f(x)$ on one coordinate plane.
$$f(x) = -3x + 2$$

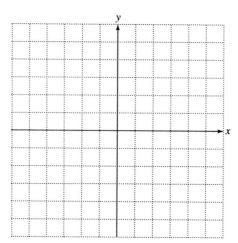

19. Given that $f(x) = \dfrac{3}{7}x + \dfrac{1}{2}$ and that $f^{-1}(x) = \dfrac{14x - 7}{6}$, find $f^{-1}[f(x)]$.

Approximately one-half of this test covers the content of chapters 1–9. The remainder covers the content of Chapter 10.

1. Identify the property illustrated by the equation $5(-3) = -3(5)$.

2. Evaluate. $3(4 - 6)^3 + \sqrt{25}$

3. Simplify. $2\{x - 3[x - 2(x + 1)]\}$

4. Solve for p. $A = 3bt + prt$

5. Factor. $x^3 + 125$

6. Add. $\dfrac{3}{x - 4} + \dfrac{6}{x^2 - 16}$

7. Solve for x.
$$\frac{3}{2x + 3} = \frac{1}{2x - 3} + \frac{2}{4x^2 - 9}$$

8. Solve for (x, y, z).
$$\begin{aligned}
3x - 2y - 9z &= 9 \\
x - y + z &= 8 \\
2x + 3y - z &= -2
\end{aligned}$$

9. Simplify. $\sqrt{8x} + 3x\sqrt{50} - 4x\sqrt{32}$

10. Multiply and simplify.
$(\sqrt{2} + \sqrt{3})(2\sqrt{6} - \sqrt{3})$

11. Find the distance between $(6, -1)$ and $(-3, -4)$.

12. Factor. $12x^2 - 11x + 2$

13. Factor. $81x^4 - 1$

14. Write the equation in standard form for a circle with a radius of 14 and a center at $(-3, 6)$.

15. If $f(x) = 3x^2 - 2x + 1$, find
(a) $f(-2)$
(b) $f(a - 2)$
(c) $f(a) + f(-2)$

16. Graph $f(x) = x^3$ and $g(x) = (x + 2)^3 + 4$ on one coordinate plane.

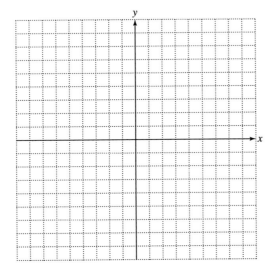

1.	
2.	
3.	
4.	
5.	
6.	
7.	
8.	
9.	
10.	
11.	
12.	
13.	
14.	
15. (a)	
(b)	
(c)	
16.	

17. (a) _____

(b) _____

(c) _____

18. (a) _____

(b) _____

(c) _____

19. _____

20. (a) _____

(b) _____

(c) _____

21. (a) _____

(b) _____

22. _____

17. If $f(x) = 2x^2 - 5x - 6$ and $g(x) = 5x + 3$, find
   (a) $(fg)(x)$
   (b) $\left(\dfrac{f}{g}\right)(x)$
   (c) $f[g(x)]$

18. $A = \{(3, 6), (1, 8), (2, 7), (4, 4)\}$
   (a) Is A a function?
   (b) Is A a one-to-one function?
   (c) Find A^{-1}.

19. Find the inverse function for $f(x) = 7x - 3$.

20. $f(x) = 5x^3 - 3x^2 - 6$
   (a) Find $f(2)$.
   (b) Find $f(-3)$.
   (c) Find $f(2a)$.

21. (a) Find the inverse function for $f(x) = -\dfrac{2}{3}x + 2$.
   (b) Graph $f(x)$ and $f^{-1}(x)$ on one coordinate plane.

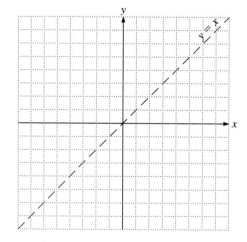

22. Find $f[f^{-1}(x)]$ using your results from question 21.

CHAPTER 11

Logarithmic and Exponential Functions

How long will it be until the spread of an infectious disease will reach another town? How long does it take the population of a certain dangerous bacterium to double? These types of questions are of major concern to medical doctors and research biologists as they seek to study the spread of a disease and the various ways to prevent and treat it. If you would like to see how mathematics assists in this type of research, please turn to the Putting Your Skills to Work on page 656.

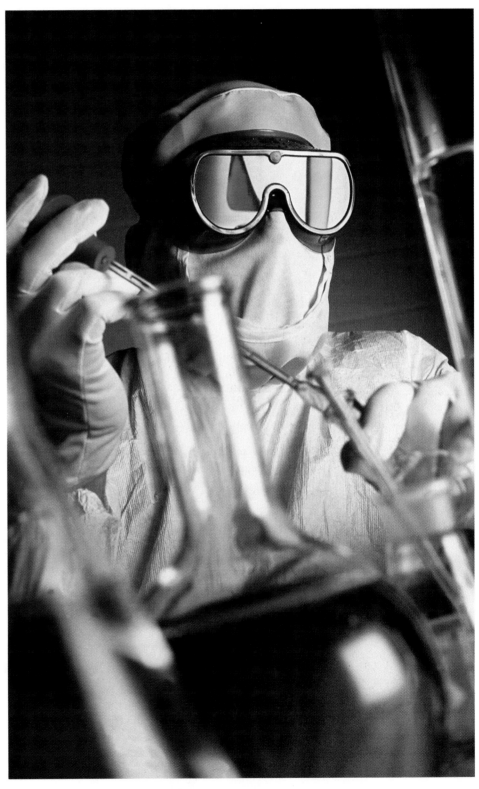

If you are familiar with the topics in this chapter, take this test now. Check your answers with those in the back of the book. If an answer was wrong or you couldn't do a problem, study the appropriate section of the chapter.

If you are not familiar with the topics in this chapter, don't take this test now. Instead, study the examples, work the practice problems, and then take the test.

This test will help you identify those concepts that you have mastered and those that need more study.

Section 11.1

1. Sketch the graph of $f(x) = 2^{-x}$. Plot at least four points.

2. Solve for x. $3^{2x-1} = 27$

3. The amount of money A due after t years when a principal amount P is invested at interest rate r is given by the equation $A = P(1 + r)^t$. How much money will Nancy have in four years if she invests $10,000 in a mutual fund that pays 12% interest compounded annually?

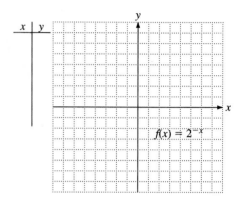

Section 11.2

4. Write in logarithmic form. $\dfrac{1}{36} = 6^{-2}$

5. Solve for x. $\log_4 x = 3$

6. Evaluate. $\log_{10}(10,000)$

Section 11.3

7. Write the logarithm in terms of $\log_5 x$, $\log_5 y$, and $\log_5 z$. $\log_5 \left(\dfrac{x^2 y^5}{z^3} \right)$

8. Express as one logarithm. $\dfrac{1}{2} \log_4 x - 3 \log_4 w$

9. Find x if $\log_3 x + \log_3 2 = 4$.

Use a scientific calculator to evaluate each of the following. Round your answers to the nearest ten-thousandth.

Section 11.4

10. $\log x = 3.9170$

11. $\ln 4.79$

12. $\log_6 5.02$

13. $\log 0.6188$

14. Find the value of x in the equation $\ln x = 22.976$. (Express your answer in scientific notation.)

Section 11.5

15. Solve the following logarithmic equation and check your solution.

$$\log x - \log(x + 3) = -1$$

16. Solve the exponential equation $4^{2x+1} = 9$. (Round your answer to the nearest ten-thousandth.)

17. How long would it take for $2000 to grow to $7000 at 6% annual interest compounded yearly? Use the formula $A = P(1 + r)^t$. Round your answer to the nearest year.

1. _____

2. _____

3. _____

4. _____

5. _____

6. _____

7. _____

8. _____

9. _____

10. _____

11. _____

12. _____

13. _____

14. _____

15. _____

16. _____

17. _____

11.1 The Exponential Function

After studying this section, you will be able to:

1 *Graph an exponential equation.*
2 *Solve an elementary exponential equation.*
3 *Solve an applied problem requiring the use of an exponential equation.*

1 Graphing Exponential Equations

We have defined 2^x for any rational number x. For example,

$$2^{-2} = \frac{1}{4}$$

$$2^{1/2} = \sqrt{2}$$

$$2^{1.7} = 2^{17/10} = \sqrt[10]{2^{17}}$$

We can also define $y = 2^x$ when x is an irrational number, such as π or $\sqrt{2}$. The actual definition is too advanced for this course.

We can define an **exponential function** for all real values of x.

Definition of Exponential Function

The function $f(x) = b^x$, where $b > 0$, $b \neq 1$, and x is a real number, is called the **exponential function.**

The number b is called the **base** of the function. Now let's look at some graphs of exponential functions.

EXAMPLE 1   Graph. $f(x) = 2^x$

We make a table of values for x and evaluate $f(x)$.

$$f(-1) = 2^{-1} = \frac{1}{2}, \qquad f(0) = 2^0 = 1, \qquad f(1) = 2^1 = 2$$

Verify the other values in the table below. We then draw the graph.

x	$f(x)$
-2	$\frac{1}{4}$
-1	$\frac{1}{2}$
0	1
1	2
2	4
3	8

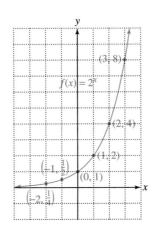

Notice how the curve of $f(x) = 2^x$ comes *very close to* the x-axis but *never* touches it. The x-axis is an asymptote for an exponential function. You should also see that $f(x)$ is always positive, so the range of $f(x)$ includes *all* positive real numbers (the domain includes *all* real numbers). As x increases, $f(x)$ increases faster and faster (the curve gets steeper).

Practice Problem 1   Graph. $f(x) = 3^x$

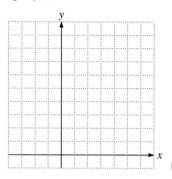

EXAMPLE 2   Graph. $f(x) = \left(\dfrac{1}{2}\right)^x$

We can write $\left(\dfrac{1}{2}\right)^x$ as $f(x) = \left(\dfrac{1}{2}\right)^x = (2^{-1})^x = 2^{-x}$ and evaluate it for a few values of x. We then draw the graph.

x	f(x)
−3	8
−2	4
−1	2
0	1
1	$\dfrac{1}{2}$
2	$\dfrac{1}{4}$

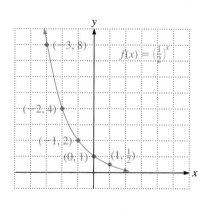

Note that as x increases $f(x)$ decreases.

Practice Problem 2   Graph. $f(x) = \left(\dfrac{1}{3}\right)^x$ ∎

PRACTICE PROBLEM 2

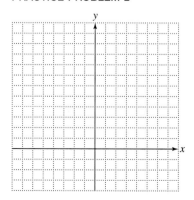

To Think About Look at the graph of $f(x) = 2^x$ in Example 1 and the graph of $f(x) = 2^{-x}$ in Example 2. How are the two graphs related?

EXAMPLE 3   Graph. $f(x) = 3^{x-2}$

We will make a table of values for a few values of x. Then we will graph the function.

$$f(0) = 3^{0-2} = 3^{-2} = \frac{1}{3^2} = \frac{1}{9}$$

$$f(1) = 3^{1-2} = 3^{-1} = \frac{1}{3}$$

$$f(2) = 3^{2-2} = 3^0 = 1$$

$$f(3) = 3^{3-2} = 3^1 = 3$$

$$f(4) = 3^{4-2} = 3^2 = 9$$

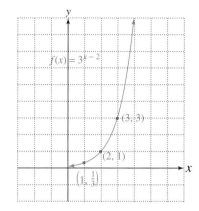

x	f(x)
0	$\frac{1}{9}$
1	$\frac{1}{3}$
2	1
3	3
4	9

We observe in the graph that the curve is that of $f(x) = 3^x$ except that it has been shifted 2 units to the right.

Practice Problem 3   Graph. $f(x) = 3^{x+2}$

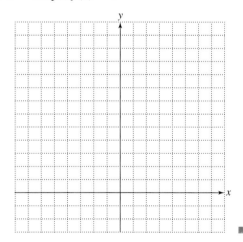

GRAPHING CALCULATOR

Exploration

Using a graphing calculator, graph $f(x) = 2^x$. Then on the same screen, graph $g(x) = 2^{x+3}$. Describe the shift that occurs. What will the graph of $g(x) = 2^{x+5}$ look like? Verify using the graphing calculator.

Using the graphing calculator, graph $f(x) = 2^{x-2}$. Describe the shift that occurs. What will the graph of $g(x) = 2^{x-3}$ look like? Verify using the graphing calculator.

Based on your experience with functions, what would the graph of $f(x) = 2^x + 3$ look like? $f(x) = 2^x - 4$ look like? Verify using the graphing calculator.

To Think About How is the graph of $f(x) = 3^{x+2}$ related to the graph of $f(x) = 3^x$? Without making a table of values, draw the graph of $f(x) = 3^{x+3}$. Draw the graph of $f(x) = 3^{x-3}$.

For the next example we need to discuss a special number that is denoted by the letter e. The letter e is a number like π. It is an irrational number. It occurs in many formulas that describe the world around us, such as the growth of cells or radioactive decay. We need an approximate value for e to use this number in calculations. $e \approx 2.7183$.

An extremely useful function is the exponential function e^x. We usually obtain values for e^x by using a calculator or a computer. If you have a scientific calculator, use the $\boxed{e^x}$ key. (Many scientific calculators require you to press $\boxed{\text{SHIFT}}$ $\boxed{\ln}$ or $\boxed{\text{2nd F}}$ $\boxed{\ln}$ or $\boxed{\text{INV}}$ $\boxed{\ln}$ to obtain the operation e^x.) If you have a calculator that is not a scientific calculator, use $e \approx 2.7183$ as an approximate value. If you don't have any calculator, use Table A–2 in the appendix.

EXAMPLE 4   Graph. $f(x) = e^x$

We evaluate $f(x)$ for some negative and for some positive values of x. We begin with $f(-2)$. To use Table A–2, look for 2 in the x-column. Then locate the value of e^{-x} on the right. Since for e^x, $f(-2) = e^{-2}$ and for e^{-x}, $f(2) = e^{-2}$, we can use e^{-x} to find the value of e^x when x is negative. Thus, we see that for $f(x) = e^x$, $f(-2) = 0.1353$ or 0.14 rounded to the nearest hundredth.

To find e^2 on a scientific calculator, we enter 2 $\boxed{e^x}$ and obtain 7.389056099 as an approximation. (On some scientific calculators you will need to use the keystrokes 2 $\boxed{\text{2nd F}}$ $\boxed{\ln}$ or 2 $\boxed{\text{SHIFT}}$ $\boxed{\ln}$ or 2 $\boxed{\text{INV}}$ $\boxed{\ln}$.) Thus $f(2) = e^2 \approx 7.39$ to the nearest hundredth.

x	$f(x)$
-2	0.14
-1	0.37
0	1
1	2.72
2	7.39

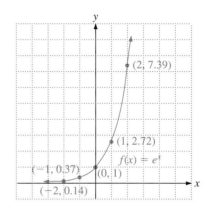

PRACTICE PROBLEM 4

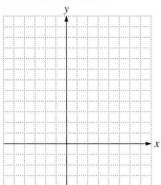

Practice Problem 4   Graph. $f(x) = e^{x-2}$ ∎

To Think About Look at the graphs of $f(x) = e^x$ and $f(x) = e^{x-2}$. Describe the shift that occurs. Without making a table of values, draw the graph of $f(x) = e^{x+3}$.

2 *Solving Elementary Exponential Equations*

All the usual laws of exponents are true for exponential functions. We also have the following important property to help us solve exponential equations.

Property of Exponential Equations

If $b^x = b^y$, then $x = y$ for $b > 0$ and $b \neq 1$.

EXAMPLE 5   Solve. $2^x = \dfrac{1}{16}$

To use the property of exponential equations, we must have the same base on each side of the equation.

$$2^x = \frac{1}{16}$$

$$2^x = \frac{1}{2^4} \qquad \textit{Because } 2^4 = 16.$$

$$2^x = 2^{-4} \qquad \textit{Because } \frac{1}{2^4} = 2^{-4}.$$

$$x = -4 \qquad \textit{Property of exponential equations.}$$

Practice Problem 5   Solve. $2^x = \dfrac{1}{32}$ ∎

3 Solving Applied Problems Using Exponential Functions

An exponential function can be used to solve compound interest problems. If a principal P is invested at an annual interest rate r, the amount of money A accumulated after t years is $A = P(1 + r)^t$.

EXAMPLE 6   If a young married couple invests $5000 in a mutual fund that pays 16% interest compounded annually, how much will they have in 3 years?

Here $P = 5000$, $r = 0.16$, and $t = 3$.

$$A = P(1 + r)^t$$
$$= 5000(1 + 0.16)^3$$
$$= 5000(1.16)^3$$
$$= 5000(1.560896)$$
$$= 7804.48$$

The couple will have $7804.48.

If you have a scientific calculator, you can find the value of $5000(1.16)^3$ immediately by using the $\boxed{\times}$ key and the $\boxed{y^x}$ key. On most scientific calculators you can use the following keystrokes.

$$5000 \; \boxed{\times} \; 1.16 \; \boxed{y^x} \; 3 \; \boxed{=} \; 7804.48$$

Practice Problem 6   If Uncle Jose invests $4000 in a mutual fund that pays 11% interest compounded annually, how much will he have in 2 years? ∎

Interest is often compounded quarterly or monthly or even daily. Therefore, a more useful form of the formula is needed that allows for compounding in these cases. If a principal P is invested at an annual interest rate r that is compounded n times a year, then the amount of money A accumulated after t years is

$$A = P\left(1 + \frac{r}{n}\right)^{nt}$$

EXAMPLE 7 If you invest $8000 in a fund that pays 15% annual interest compounded monthly, how much will you have after 6 years?

In this situation $P = 8000$, $r = 15\% = 0.15$, and $n = 12$. The interest is compounded monthly or 12 times per year. Finally, $t = 6$ since the interest will be compounded for 6 years.

$$A = 8000\left(1 + \frac{0.15}{12}\right)^{(12)(6)}$$
$$= 8000(1 + 0.0125)^{72}$$
$$= 8000(1.0125)^{72}$$
$$\approx 8000(2.445920268)$$
$$\approx 19{,}567.36215$$

Rounding to the nearest cent, you will have $19,567.36. Using a scientific calculator at the second step, you can find $8000(1.0125)^{72}$ directly by using the following keystrokes.

$$8000 \; \boxed{\times} \; 1.0125 \; \boxed{y^x} \; 72 \; \boxed{=} \; 19{,}567.36215$$

Depending on your calculator, your display may read fewer or more digits in the answer.

Practice Problem 7 How much money would Collette have if she invested $1500 for 8 years at 8% annual interest if the interest is compounded quarterly. ■

An exponential function is used to describe radioactive decay. The equation $A = Ce^{kt}$ tells us how much of a radioactive element is left after a specified time.

EXAMPLE 8 The radioactive decay of the chemical element americium 241 can be described by the equation

$$A = Ce^{-0.0016008t}$$

where A is the amount remaining, C the original amount, t the time elapsed in years, and $k = -0.0016008$, the decay constant for americium. If 10 milligrams (mg) is sealed in a laboratory container today, how much americium 241 would theoretically be present in 2000 years? Round your answer to the nearest hundredth.

Here $C = 10$ and $t = 2000$.

$$A = 10e^{-0.0016008(2000)} = 10e^{-3.2016}$$

Using a calculator or Table A–2, we have

$$A \approx 10(0.040697) = 0.40697 \approx 0.41 \text{ mg}$$

The expression $10e^{-3.2016}$ can be found directly on some scientific calculators by the following keystrokes.

$$10 \; \boxed{\times} \; 3.2016 \; \boxed{+/-} \; \boxed{e^x} \; \boxed{=} \; 0.406970366$$

(Scientific calculators with no $\boxed{e^x}$ key will require keystrokes $\boxed{\text{INV}}\boxed{\ln}$ or $\boxed{\text{2nd F}}\boxed{\ln}$ or $\boxed{\text{SHIFT}}\boxed{\ln}$ in place of the $\boxed{e^x}$.)

Practice Problem 8 If 20 milligrams of americium 241 is present now, how much will theoretically be present in 5000 years? Round your answer to the nearest thousandth. ■

11.1 Exercises

Verbal and Writing Skills

1. The exponential function is an equation of the form _____.

2. The irrational number e is a number that is approximately _____. (Give answer to four decimal places.)

Graph each function.

3. $f(x) = 2^x$

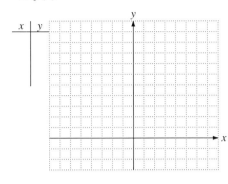

4. $f(x) = 3^x$

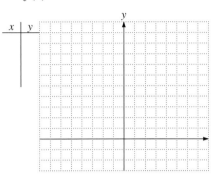

5. $f(x) = 4^x$

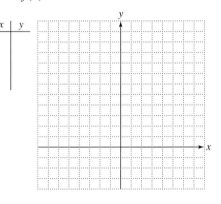

6. $f(x) = 5^x$

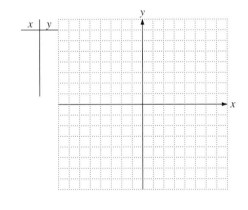

7. $f(x) = \left(\dfrac{1}{4}\right)^x$

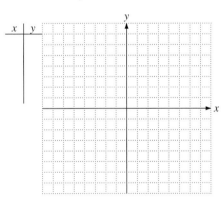

8. $f(x) = \left(\dfrac{1}{3}\right)^x$

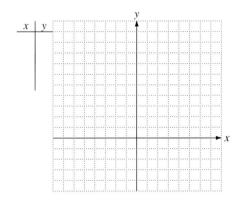

9. $f(x) = 2^{-x}$

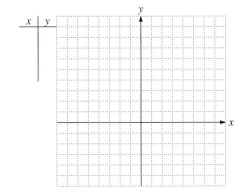

10. $f(x) = 3^{-x}$

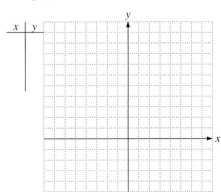

11. $f(x) = 5^{-x}$

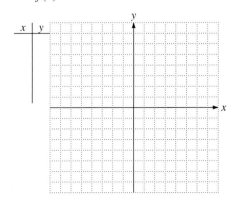

12. $f(x) = 4^{-x}$

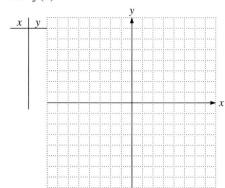

13. $f(x) = 2^{x+1}$

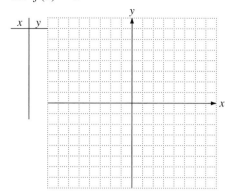

14. $f(x) = 2^{x+5}$

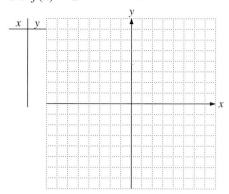

15. $f(x) = 3^{x-1}$

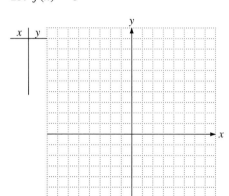

16. $f(x) = 3^{x-4}$

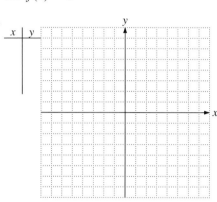

17. $f(x) = \left(\dfrac{3}{4}\right)^x$

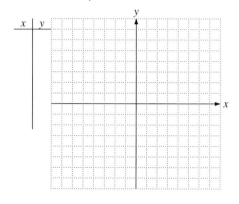

18. $f(x) = \left(\dfrac{2}{3}\right)^x$

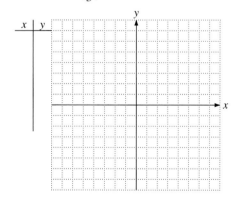

19. $f(x) = 2^x - 2$

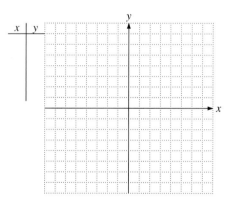

20. $f(x) = 2^x + 2$

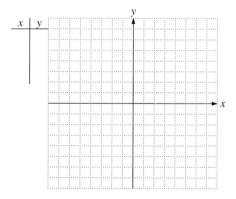

Graph each function. Use a calculator or Table A–2.

21. $f(x) = e^{x+1}$

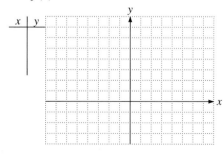

22. $f(x) = e^{x-1}$

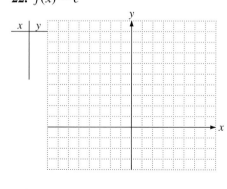

23. $f(x) = 2e^x$

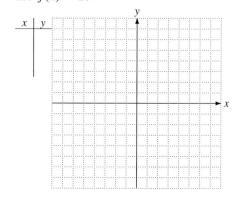

24. $f(x) = 3e^x$

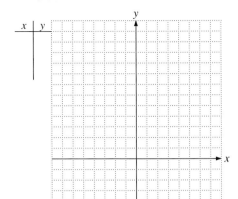

25. $f(x) = e^{2-x}$

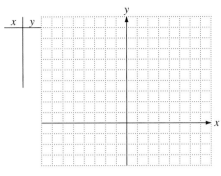

26. $f(x) = e^{1-x}$

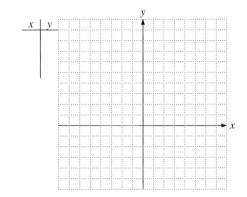

Solve for x.

27. $2^x = 4$

28. $2^x = 8$

29. $2^x = 1$

30. $2^x = 2$

31. $2^x = \dfrac{1}{8}$

32. $2^x = \dfrac{1}{16}$

33. $3^x = 81$

34. $3^x = 27$

35. $3^x = 1$

36. $3^x = 729$

37. $3^{-x} = \dfrac{1}{9}$

38. $3^{-x} = \dfrac{1}{3}$

39. $4^x = 256$

40. $4^{-x} = \dfrac{1}{16}$

41. $5^{x+1} = 25$

42. $6^{2x} = 36$

43. $8^{3x-1} = 64$

44. $5^{2x+3} = 25$

45. $10^{x-1} = 0.001$

46. $10^{x+6} = 0.01$

To solve Problems 47 through 52 use the exponential equation

$A = P\left(1 + \dfrac{r}{n}\right)^{nt}$. *Round your answers to the nearest cent.*

47. Joanna is investing $3000 at an annual rate of 16% interest compounded annually. How much money will Joanna have after 3 years?

48. Enrico is investing $6000 at an annual rate of 15% interest compounded annually. How much money will Enrico have after 4 years?

49. How much money will Suki have in 5 years if she invests $2000 at a 14% annual rate of interest compounded quarterly? How much will she have if it is compounded monthly?

50. How much money will Gina have in 2 years if she invests $8000 at a 17% annual rate if it is compounded quarterly? What would be the rate if it is compounded monthly?

51. How much money will Darci have in 4 years if she invests $5000 at an annual interest rate of 12%, with interest compounded every 2 weeks?

52. How much money will Sally Hanu have in 3 years if she invests $9000 at 8% annual interest if the interest is compounded every 2 weeks?

53. The equation $C(t) = P(1.04)^t$ forecasts tuition cost, where t is time in years and P is the present cost in dollars. If the cost of a college education is increasing 4% per year, how much will a college now charging $3000 for tuition charge in 10 years? How much will a college now charging $12,000 tuition charge in 15 years?

54. The number of bacteria in a culture is given by $B(t) = 4000(2^t)$, where t is the time in hours. How many bacteria will grow in the culture in 3 hours? in 9 hours?

55. U.S. Navy divers off the coast of Nantucket were searching for the wreckage of an old World War II–era submarine. They found that if the water was relatively clear and the surface was calm, the ocean filters out 18% of the sunlight for each 4 feet they descend. How much sunlight is available at a depth of 20 feet? The divers need to use underwater spotlights when the amount of sunlight is less than 10%. Will they need spotlights when working at a depth of 48 feet?

56. Since the city of Manchester put in a municipal sewer, many homeowners have sewer lines connected to their homes. Each year the number of people who use their own private septic tank for their house rather than the public sewer decreases by 8%. What percentage of people will still be using their private septic tanks in 5 years? The city feels that the underground water contamination problem will be solved when the number of homeowners still using septic tanks is less than 10%. Will that goal be achieved in the next 25 years?

Use an exponential equation to solve each problem.

57. The radioactive decay of radium 226 can be described by the equation $A = Ce^{-0.0004279t}$, where A is the amount of radium remaining, C the original amount, and t the elapsed time in years. If 6 milligrams of radium is sealed in a container now, how much radium would be in the container after 1000 years?

58. The radioactive decay of radon 222 can be described by the equation $A = Ce^{-0.1813t}$, where A is the amount of radon remaining, C is the original amount, and t the elapsed time in days. If 1.5 milligrams is in a laboratory container, how much was in there 10 days ago?

Use the following information for problems 59 and 60. The atmospheric pressure measured in pounds per square inch is given by the equation $P = 14.7e^{-0.21d}$, where d is the distance in miles above sea level.

59. What is the pressure in pounds per square inch experienced by a man on a mountain in Colorado that is 2 miles above sea level?

60. What is the pressure in pounds per square inch on an American Airlines jet plane flying 10 miles above sea level?

The population of the world is growing exponentially. By recording a few specific years and the population in billions a pattern of significant increase is seen. A table recording these approximate values and a graph corresponding to these values is shown below.

Year	AD1	1650	1850	1930	1975	1995
Approximate world population in billions	0.2	0.5	1	2	4	5.73

Source: Statistical Division of the United Nations.

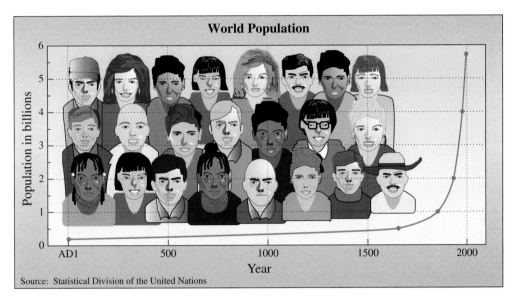

World Population

Source: Statistical Division of the United Nations

61. Based on the graph, what would you estimate was the year when the world population reached 3 billion people?

62. Based on the graph, what would you estimate to be the world population in 1900?

63. The growth rate of the world population during the period 1980–1990 was 1.7% per year. If that rate continued from 1995 to 2005 what would the world population be in 2005?

64. The growth rate of the world population during the period 1990–1997 was 1.4% per year. If that rate was in effect from 1995 to 2010 what would the world population be in 2010?

Optional Graphing Calculator Problems

65. Let $f(x) = \dfrac{e^x + e^{-x}}{2}$. Evaluate $f(x)$ when $x = -1$, -0.5, 0, 0.5, 1, 1.5, and 2. Now use these values to graph the function. [$f(x)$ defines a special function called the hyperbolic cosine. This function is used in advanced mathematics and science to study a variety of technical applications.]

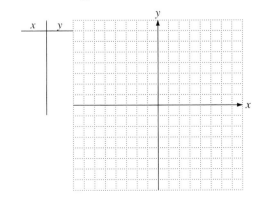

66. Let $g(x) = \dfrac{e^x - e^{-x}}{2}$. Evaluate $g(x)$ when $x = -2$, -1, -0.5, 0, 0.5, 1, 1.5, and 2. Now graph $g(x)$ using these values. [$g(x)$ defines a special function called the hyperbolic sine.]

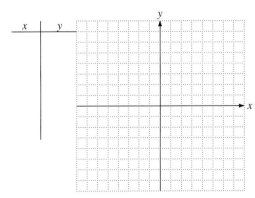

11.2 The Logarithm Function

MathPro

Video 26

SSM

After studying this section, you will be able to:

1. *Write exponential equations in logarithmic form.*
2. *Write logarithmic equations in exponential form.*
3. *Solve elementary logarithmic equations.*
4. *Graph a logarithmic function.*

Logarithms were invented about 400 years ago by the Scottish mathematician John Napier. Napier's amazing invention reduced complicated problems to simple subtraction and addition. Astronomers quickly saw the immense value of logarithms and began using them. The work of Johannes Kepler, Isaac Newton, and others would have been much harder without logarithms.

The most important thing to learn in this chapter is that a logarithm is just an exponent. In Section 11.1 we solved the equation $2^x = 8$. We found the answer was $x = 3$. The question we faced was "To what power do we raise 2 to get 8?" The answer was 3. Mathematicians have to solve this type of problem so often that we have invented a short notation to ask the question. Instead of asking, "To what power do we raise 2 to get 8?" we say instead "What is $\log_2 8$?" Both questions mean the same thing.

Now suppose we had a general equation $x = b^y$ and someone asked, "To what power do we raise b to get x?" We would abbreviate this question by asking "What is $\log_b x$?" Thus we see that $y = \log_b x$ is an equivalent form of the equation $x = b^y$.

The key concept you must remember is that a logarithm is an exponent. We write $\log_b x = y$ to mean that the logarithm of x to the base b is equal to y. y is the exponent.

Definition of Logarithm

The **logarithm** of a *positive* number x is the power (exponent) to which the base b must be raised to produce x. That is, $y = \log_b x$ is the same as $x = b^y$, where $b > 0$, $b \neq 1$.

Often you will need to convert logarithmic expressions to exponential expressions, and vice versa, to solve equations.

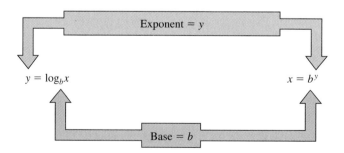

1 Writing Exponential Equations in Logarithmic Form

We begin by converting exponential expressions to logarithmic expressions.

EXAMPLE 1   Write in logarithmic form.

(a) $4 = 2^2$ **(b)** $81 = 3^4$ **(c)** $\dfrac{1}{100} = 10^{-2}$

We use the property that $x = b^y$ is equivalent to $\log_b x = y$.

(a) $4 = 2^2$

Here $x = 4$, $b = 2$, $y = 2$, so $2 = \log_2 4$

(b) $81 = 3^4$

Here $x = 81$, $b = 3$, $y = 4$, so $4 = \log_3 81$

(c) $\dfrac{1}{100} = 10^{-2}$

Here $x = \dfrac{1}{100}$, $b = 10$, $y = -2$, so $-2 = \log_{10}\left(\dfrac{1}{100}\right)$

Practice Problem 1 Write in logarithmic form.

(a) $49 = 7^2$ **(b)** $\dfrac{1}{64} = 4^{-3}$ **(c)** $1000 = 10^3$ ∎

2 Writing Logarithmic Equations in Exponential Form

If we have an equation with a logarithm in it, we can write it in the form of an exponential equation. This is a very important skill. Carefully study the following example.

EXAMPLE 2 Write in exponential form.

(a) $2 = \log_5 25$ **(b)** $3 = \log_{1/4}\left(\dfrac{1}{64}\right)$ **(c)** $-4 = \log_{10}\left(\dfrac{1}{10{,}000}\right)$

(a) $2 = \log_5 25$

Here $y = 2$, $b = 5$, $x = 25$. Thus, since $x = b^y$, $25 = 5^2$

(b) $y = 3$, $b = \dfrac{1}{4}$, $x = \dfrac{1}{64}$, so $\dfrac{1}{64} = \left(\dfrac{1}{4}\right)^3$

(c) $y = -4$, $b = 10$, $x = \dfrac{1}{10{,}000}$, so $\dfrac{1}{10{,}000} = 10^{-4}$

Practice Problem 2 Write in exponential form.

(a) $3 = \log_5 125$ **(b)** $4 = \log_{1/2}\left(\dfrac{1}{16}\right)$ **(c)** $-2 = \log_6\left(\dfrac{1}{36}\right)$ ∎

3 Solving Elementary Logarithmic Equations

Many logarithmic equations are fairly easily solved if we first convert them to an equivalent exponential equation.

EXAMPLE 3 Solve for the variable.

(a) $\log_5 x = -3$ **(b)** $\log_{1/3} 81 = y$ **(c)** $\log_a 16 = 4$

(a) $5^{-3} = x$

$\dfrac{1}{5^3} = x$

$\dfrac{1}{125} = x$

(b) $\left(\dfrac{1}{3}\right)^y = 81$

$(3^{-1})^y = 3^4$

$3^{-y} = 3^4$

$-y = 4$

$y = -4$

(c) $a^4 = 16$

$a^4 = 2^4$

$a = 2$

Practice Problem 3 Solve for the variable.

(a) $\log_6 y = -2$ **(b)** $\log_b 125 = 3$ **(c)** $\log_{1/2} 32 = x$ ∎

With this knowledge we have the ability to solve an additional type of problem.

EXAMPLE 4 Evaluate. $\log_3 81$

Now what exactly is the problem asking for? It is asking, "To what power must we raise 3 to get 81?" Since we do not know the power, we call it x. We have

$$\log_3 81 = x$$

$$81 = 3^x \qquad \textit{Write an equivalent exponential equation.}$$

$$3^4 = 3^x \qquad \textit{Write } 81 \textit{ as } 3^4.$$

$$x = 4 \qquad \textit{If } b^x = b^y, \textit{ then } x = y \textit{ for } b > 0 \textit{ and } b \neq 1.$$

Thus, $\log_3 81 = 4$.

Practice Problem 4 Evaluate. $\log_{10} 0.1$ ■

4 Graphing Logarithmic Functions

We found in Chapter 10 that a function and its inverse when graphed have an interesting property. They are symmetric with respect to the line $y = x$. We also found in Chapter 10 that the procedure for finding the inverse of a function is to interchange the x and y variables. For example, $y = 2x + 3$ and $x = 2y + 3$ are inverse functions. In similar fashion, $y = 2^x$ and $x = 2^y$ are inverse functions. Now another way to write $x = 2^y$ is the logarithmic equation $y = \log_2 x$. The logarithmic function is the **inverse** of the exponential function. If we graph the function $y = 2^x$ and $y = \log_2 x$ on the same set of axes, the graph of one is the reflection of the other about the line $y = x$.

EXAMPLE 5 Graph. $y = \log_2 x$

If we write $y = \log_2 x$ in exponential form, we have $x = 2^y$. We make a table of values and graph the function $x = 2^y$.

In each case, we will pick a value of y as a first step.

$$\text{If } y = -2 \qquad x = 2^y = 2^{-2} = \frac{1}{2^2} = \frac{1}{4}$$

$$\text{If } y = -1 \qquad x = 2^{-1} = \frac{1}{2}$$

$$\text{If } y = 0 \qquad x = 2^0 = 1$$

$$\text{If } y = 1 \qquad x = 2^1 = 2$$

$$\text{If } y = 2 \qquad x = 2^2 = 4$$

x	y
$\frac{1}{4}$	-2
$\frac{1}{2}$	-1
1	0
2	1
4	2

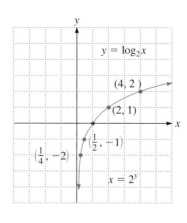

PRACTICE PROBLEM 5

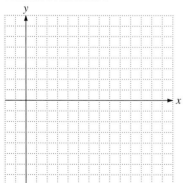

Practice Problem 5 Graph. $y = \log_{1/2} x$ ■

$f(x) = a^x$ and $f(x) = \log_a x$ are inverse functions. As such they have all the properties of inverse functions. We will review a few of these properties as we study the graphs of two inverse functions, $y = 2^x$ and $y = \log_2 x$.

EXAMPLE 6   Graph $y = \log_2 x$ and $y = 2^x$ on the same set of axes.

Make a table of values (ordered pairs) for each equation. Then draw the graph.

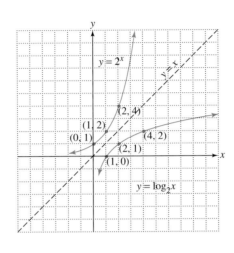

$y = 2^x$

x	y
-2	$\dfrac{1}{4}$
-1	$\dfrac{1}{2}$
0	1
1	2
2	4

$y = \log_2 x$

x	y
$\dfrac{1}{4}$	-2
$\dfrac{1}{2}$	-1
1	0
2	1
4	2

ordered pairs reversed

Note that $y = \log_2 x$ is the inverse of $y = 2^x$ because the ordered pairs (x, y) are reversed. The sketch of the two equations shows the inverse property. If we reflect the graph of $y = 2^x$ about the line $y = x$, it will coincide with the graph of $y = \log_2 x$.

Recall that in function notation $f^{-1}(x)$ means the inverse function of $f(x)$. Thus if we write $f(x) = \log_2 x$, then $f^{-1}(x) = 2^x$.

Practice Problem 6   Graph $y = \log_6 x$ and $y = 6^x$ on the same set of axes.

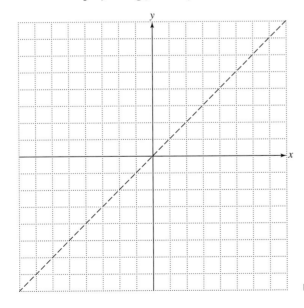

11.2 Exercises

Verbal and Writing Skills

1. A logarithm is really just an _____.
2. In the equation $y = \log_b x$, the value b is called the _____.
3. In the equation $y = \log_b x$, what is the domain (the permitted values of x)?
4. In the equation $y = \log_b x$, what are the permitted values of b?

Write in logarithmic form.

5. $64 = 4^3$
6. $125 = 5^3$
7. $36 = 6^2$
8. $100 = 10^2$
9. $\dfrac{1}{25} = 5^{-2}$

10. $0.01 = 10^{-2}$
11. $\dfrac{1}{32} = 2^{-5}$
12. $\dfrac{1}{64} = 2^{-6}$
13. $y = e^3$
14. $y = e^{-4}$

15. $10^{0.6021} = 4$
16. $10^{0.6990} = 5$
17. $e^{-5} = 0.0067$
18. $e^{-6} = 0.0025$

Write in exponential form.

19. $2 = \log_3 9$
20. $2 = \log_2 4$
21. $0 = \log_5 1$
22. $4 = \log_3 81$

23. $0 = \log_8 1$
24. $2 = \log_{10} 100$
25. $-2 = \log_{10}(0.01)$
26. $-3 = \log_{10}(0.001)$

27. $-4 = \log_3\left(\dfrac{1}{81}\right)$
28. $-5 = \log_2\left(\dfrac{1}{32}\right)$
29. $5 = \log_e x$
30. $6 = \log_e x$

31. $\log_{10} 7 = 0.8451$
32. $\log_{10} 6 = 0.7782$
33. $\log_e 0.3 = -1.2040$
34. $\log_e 0.4 = -0.9163$

Solve.

35. $\log_2 x = 4$
36. $\log_2 x = 6$
37. $\log_{10} x = -3$
38. $\log_{10} x = -2$
39. $\log_4 64 = y$

40. $\log_6 216 = y$
41. $\log_7\left(\dfrac{1}{49}\right) = y$
42. $\log_4\left(\dfrac{1}{64}\right) = y$
43. $\log_a 32 = 5$
44. $\log_a 81 = 4$

45. $\log_a 1000 = 3$
46. $\log_a 100 = 2$
47. $\log_{25} 5 = w$
48. $\log_8 2 = w$
49. $\log_3\left(\dfrac{1}{3}\right) = w$

50. $\log_{12} 1 = w$

51. $\log_{15} w = 0$

52. $\log_{10} w = -3$

53. $\log_w 25 = -2$

54. $\log_w 64 = -6$

55. $\log_e w = 4$

56. $\log_e w = 7$

57. $\log_{25} x = \dfrac{1}{2}$

58. $\log_{81} x = \dfrac{1}{4}$

Evaluate.

59. $\log_{10} 10$

60. $\log_{10} 100$

61. $\log_{10}(0.001)$

62. $\log_{10}(0.0001)$

63. $\log_2 128$

64. $\log_3 27$

65. $\log_8 64$

66. $\log_5 125$

67. $\log_6 \sqrt{6}$

68. $\log_7 \sqrt{7}$

69. $\log_{10} 1$

70. $\log_2 16$

71. $\log_4 \dfrac{1}{16}$

72. $\log_e 1$

Graph.

73. $\log_3 x = y$

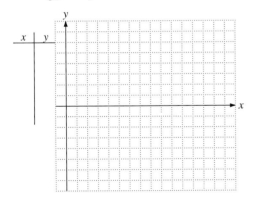

74. $\log_4 x = y$

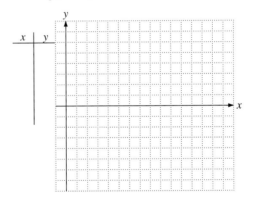

75. $\log_{1/3} x = y$

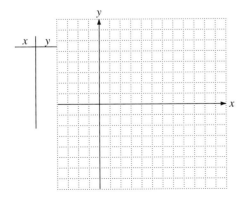

76. $\log_{1/4} x = y$

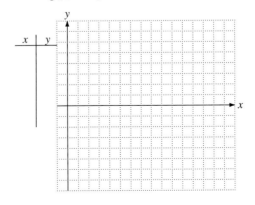

77. $\log_{10} x = y$

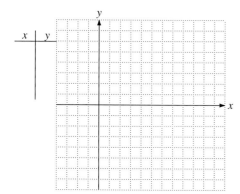

78. $\log_8 x = y$

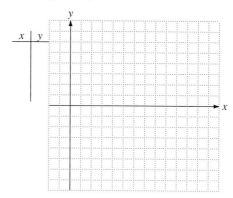

On one coordinate plane, graph the function $f(x)$ and the function $f^{-1}(x)$. Then graph a dashed line for the equation $y = x$.

79. $f(x) = \log_3 x, \quad f^{-1}(x) = 3^x$

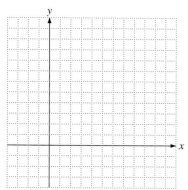

80. $f(x) = \log_4 x, \quad f^{-1}(x) = 4^x$

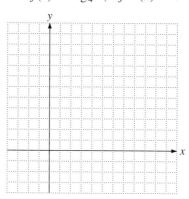

Applications

To determine if a solution is an acid or a base, chemists check the solution's pH. A solution is an acid if its pH is less than 7 and a base if its pH is greater than 7. The pH is defined by pH $= -\log_{10}[H^+]$, where $[H^+]$ is the concentration of the hydrogen ions in the solution.

81. The concentration of hydrogen ions in lemons is approximately 10^{-2}. What would be the pH of the lemons?

82. The concentration of hydrogen ions in cranberries is approximately $10^{-2.5}$. What would be the pH of cranberries?

83. A chemical engineer has mixed a solution with a pH of 8. Find the concentration of hydrogen ions $[H^+]$ in the solution.

84. A highway construction team has created a cement mixture with a pH of 9. Find the concentration of hydrogen ions $[H^+]$ in the solution.

85. A scientist has a batch of experimental industrial solvent with a pH of 9.25. Find the concentration of hydrogen ions $[H^+]$ in the solution. Round your answer to three decimal places.

86. What is the pH of a balsamic vinaigrette salad dressing, where the concentration of hydrogen ions is 1.103×10^{-3}? The logarithm to the base 10 of 0.001103 is approximately -2.957424488. Round your answer to three decimal places.

Maptech is producing map software for people who want to use their home computer to locate any place in the United States and print a detailed map. In the first two years of sales of the software they have found that for N sets of software to be sold they need to invest d dollars for advertising according to the equation

$$N = 1200 + (2500)(\log_{10} d)$$

where d is always a positive number not less than 1.

87. How many sets of software were sold when they spent $10,000 on advertising?

88. How many sets of software were sold when they spent $100,000 on advertising?

89. They have a goal next year of selling 16,200 sets of software. How much should they spend on advertising?

90. They have a goal in two years of selling 18,700 sets of software. How much should they spend on advertising?

91. Evaluate. $5^{\log_5 4}$

92. Evaluate. $\log_2 \sqrt[4]{2}$.

93. Graph. $y = -\dfrac{2}{3}x + 5$

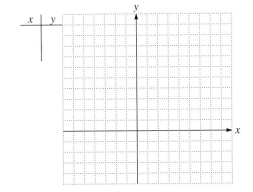

94. Graph. $6x + 3y = -6$

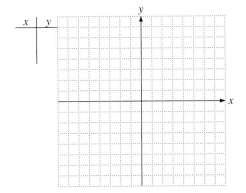

95. Find the slope of a straight line containing $(-6, 3)$ and $(-1, 2)$.

96. Find the equation of a line perpendicular to $y = -\dfrac{2}{3}x + 4$ that contains $(-4, 1)$.

97. The number of viral cells in a laboratory culture in a biology research project is given by $A(t) = 9000(2^t)$, where t is measured in hours.
  (a) How many viral cells will grow in the culture in 2 hours?
  (b) How many viral cells will grow in the culture in 12 hours?

98. The equation $C(t) = P(1.04)^t$ forecasts tuition cost, where t is time in years and P is the present cost in dollars. Assume the cost of a college education is increasing at 4% per year.
  (a) How much will a college now charging $4400 for tuition charge in 5 years?
  (b) How much will a college now charging $16,500 for tuition charge in 10 years?

11.3 Properties of Logarithms

MathPro

Video 26

SSM

After studying this section, you will be able to:

1 Use the property $\log_b MN = \log_b M + \log_b N$.

2 Use the property $\log_b\left(\dfrac{M}{N}\right) = \log_b M - \log_b N$.

3 Use the property $\log_b M^P = P \log_b M$.

4 Solve a simple logarithmic equation.

1 The Logarithm of a Product

We have already said that logarithms reduce complex expressions to addition and subtraction. The following properties show us how to use logarithms this way.

Property 1   The Logarithm of a Product

For any positive real numbers M and N and any positive base $b \neq 1$,

$$\log_b MN = \log_b M + \log_b N$$

To see that this property is true, we let

$$\log_b M = x \quad \text{and} \quad \log_b N = y$$

where x and y are any values. Now we write the expressions in exponential notation:

$$b^x = M \quad \text{and} \quad b^y = N$$

Then

$$MN = b^x b^y = b^{x+y} \qquad \textit{Laws of exponents.}$$

If we convert this equation to logarithmic form, then we must have

$$\log_b MN = x + y \qquad \textit{Definition of logarithm.}$$

$$\log_b MN = \log_b M + \log_b N \qquad \textit{By substitution.}$$

Note that the logarithms must have the same base.

EXAMPLE 1   Write $\log_3 XZ$ as a sum of logarithms.

By property 1, $\log_3 XZ = \log_3 X + \log_3 Z$.

Practice Problem 1   Write as a sum of logarithms. $\log_4 WXY$ ■

EXAMPLE 2   Write as a single logarithm. $\log_{10} 11 + \log_{10} 7$

$$\log_b MN = \log_b M + \log_b N$$

Thus

$$\log_{10} 11 + \log_{10} 7 = \log_{10}(11 \cdot 7)$$

$$\log_{10} 11 + \log_{10} 7 = \log_{10} 77$$

Practice Problem 2   Write as a single logarithm. $\log_9 12 + \log_9 5$ ■

EXAMPLE 3   Write as a single logarithm. $\log_3 16 + \log_3 x + \log_3 y$

If we extend our rule, $\log_b MNP = \log_b M + \log_b N + \log_b P$. Thus

$$\log_3 16 + \log_3 x + \log_3 y = \log_3 16xy$$

Practice Problem 3   Write as a single logarithm. $\log_7 w + \log_7 8 + \log_7 x$ ■

2 The Logarithm of a Quotient

Property 2 is similar to property 1 except that it involves two expressions that are divided, not multiplied.

Property 2   The Logarithm of a Quotient

For any positive real numbers M and N and any positive base $b \neq 1$,

$$\log_b \left(\frac{M}{N} \right) = \log_b M - \log_b N$$

Property 2 can be proved using a similar approach to the one used to prove property 1. The proof will be left as an exercise for you to try.

EXAMPLE 4   Write as the difference of two logarithms. $\log_3 \left(\frac{29}{7} \right)$

$$\log_3 \left(\frac{29}{7} \right) = \log_3 29 - \log_3 7$$

Practice Problem 4   Write as the difference of two logarithms. $\log_3 \left(\frac{17}{5} \right)$ ■

EXAMPLE 5   Express as a single logarithm. $\log_b 36 - \log_b 9$

$$\log_b 36 - \log_b 9 = \log_b \left(\frac{36}{9} \right) = \log_b 4$$

Practice Problem 5   Express as a single logarithm. $\log_b 132 - \log_b 4$ ■

CAUTION Be sure you understand Property 2!

$$\frac{\log_b M}{\log_b N} \neq \log_b M - \log_b N$$

Do you see why?

3 *The Logarithm of a Power*

We now introduce the third property. We will not prove it now, but you will have a chance to verify this property as an exercise.

> ### Property 3   The Logarithm of a Number Raised to a Power
> For any positive real number M, any real number p, and any positive base $b \neq 1$,
> $$\log_b M^p = p \log_b M$$

EXAMPLE 6   Write with no exponent. $\log_3 26^5$
$$\log_3 26^5 = 5 \log_3 26$$

Practice Problem 6   Write with no exponent. $\log_7 16^9$ ■

EXAMPLE 7   Write as a single logarithm. $\dfrac{1}{3} \log_b x + 2 \log_b w - 3 \log_b z$

First, we must eliminate the coefficients of the log terms.
$$\log_b x^{1/3} + \log_b w^2 - \log_b z^3 \qquad \textit{By property 3.}$$

Now we can combine either the sum or difference of the logarithms. We'll do the sum.
$$= \log_b x^{1/3}w^2 - \log_b z^3 \qquad \textit{By property 1.}$$

Now we combine the difference.
$$= \log_b\left(\frac{x^{1/3}w^2}{z^3}\right) \qquad \textit{By property 2.}$$

Practice Problem 7   Write as one logarithm. $\dfrac{1}{3} \log_7 x - 5 \log_7 y$ ■

EXAMPLE 8   Write as three logarithms. $\log_b\left(\dfrac{x^4 y^3}{z^2}\right)$

$$\log_b\left(\frac{x^4 y^3}{z^2}\right) = \log_b x^4 y^3 - \log_b z^2 \qquad \textit{By property 2.}$$
$$= \log_b x^4 + \log_b y^3 - \log_b z^2 \qquad \textit{By property 1.}$$
$$= 4 \log_b x + 3 \log_b y - 2 \log_b z \qquad \textit{By property 3.}$$

Practice Problem 8   Write as three logarithms. $\log_3\left(\dfrac{x^4 y^5}{z}\right)$ ■

4 Solving Logarithmic Equations

If we have a logarithmic equation, these properties are useful in combining separate logarithms together. A major goal in solving many logarithmic equations is to obtain one logarithm on one side of the equation and no logarithm on the other side. In Example 9 we will use property 1 to combine two separate logarithms that are added.

EXAMPLE 9 Find x if $\log_2 x + \log_2 5 = 3$.

$$\log_2 5x = 3 \qquad \textit{Use property 1.}$$

$$5x = 2^3 \qquad \textit{Convert to exponential form.}$$

$$5x = 8 \qquad \textit{Simplify.}$$

$$x = \frac{8}{5} \qquad \textit{Divide both sides by 5.}$$

Practice Problem 9 Find x if $\log_4 x + \log_4 5 = 2$. ∎

In Example 10, two logarithms are subtracted on the left side of the equation. We can use property 2 to combine these two logarithms together. This will allow us to obtain the form of one logarithm on one side of the equation and no logarithm on the other side.

EXAMPLE 10 Solve for x.

$$\log_3(x + 4) - \log_3(x - 4) = 2$$

$$\log_3\left(\frac{x + 4}{x - 4}\right) = 2 \qquad \textit{Property 2.}$$

$$\frac{x + 4}{x - 4} = 3^2 \qquad \textit{Convert to exponential form.}$$

$$x + 4 = 9(x - 4) \qquad \textit{Multiply each side by } (x - 4).$$

$$x + 4 = 9x - 36 \qquad \textit{Simplify.}$$

$$40 = 8x$$

$$5 = x$$

Practice Problem 10 Solve for x. $\log_{10} x - \log_{10}(x + 3) = -1$ ∎

To solve some logarithmic equations, we need a few additional properties of logarithms. We will state these properties now. The proofs of some of them will be left as an exercise for you.

For all positive values of $b \neq 1$ and all positive values of x and y

Property 4 $\log_b b = 1$

Property 5 $\log_b 1 = 0$

Property 6 If $\log_b x = \log_b y$, then $x = y$.

We will now illustrate each property with a sample example.

EXAMPLE 11

(a) Evaluate $\log_7 7$.   **(b)** Evaluate $\log_5 1$.   **(c)** Find x if $\log_3 x = \log_3 17$.

(a) $\log_7 7 = 1$ because $\log_b b = 1$ *Property 4.*

(b) $\log_5 1 = 0$ because $\log_b 1 = 0$ *Property 5.*

(c) If $\log_3 x = \log_3 17$, then $x = 17$.   *Property 6.*

Practice Problem 11   Evaluate.

(a) $\log_7 1$ **(b)** $\log_8 8$ **(c)** Find y if $\log_{12} 13 = \log_{12}(y + 2)$. ■

We now have the mathematical tools to solve a variety of logarithmic equations.

EXAMPLE 12   Find x if $2 \log_7 3 - 4 \log_7 2 = \log_7 x$.

We can use property 3 in two cases.

$$2 \log_7 3 = \log_7 3^2 = \log_7 9$$

$$4 \log_7 2 = \log_7 2^4 = \log_7 16$$

By substituting these results, we now have

$$\log_7 9 - \log_7 16 = \log_7 x$$

$$\log_7 \frac{9}{16} = \log_7 x \qquad \text{\textit{Property 2.}}$$

$$\frac{9}{16} = x \qquad \text{\textit{Property 6.}}$$

Practice Problem 12   Find x if $\log_3 2 - \log_3 5 = \log_3 6 + \log_3 x$. ■

Express as a sum of logarithms.

1. $\log_3 AB$

2. $\log_4 CD$

3. $\log_5(7 \cdot 11)$

4. $\log_6(13 \cdot 5)$

5. $\log_b 9f$

6. $\log_b 5d$

Express as a difference of logarithms.

7. $\log_3\left(\dfrac{13}{5}\right)$

8. $\log_2\left(\dfrac{17}{3}\right)$

9. $\log_a\left(\dfrac{G}{7}\right)$

10. $\log_b\left(\dfrac{H}{10}\right)$

11. $\log_a\left(\dfrac{E}{F}\right)$

12. $\log_6\left(\dfrac{8}{M}\right)$

Express as a product.

13. $\log_3 x^5$

14. $\log_5 y^6$

15. $\log_b A^{-2}$

16. $\log_a B^{-5}$

17. $\log_5 \sqrt{w}$

18. $\log_6 \sqrt{z}$

Mixed Practice

Write each expression as the sum or difference of single logarithms of x, y, and z.

19. $\log_5 \sqrt{x}\,y^3$

20. $\log_3 x^4 \sqrt{y}$

21. $\log_7\left(\dfrac{2x}{y}\right)$

22. $\log_8\left(\dfrac{3y}{x^2}\right)$

23. $\log_2\left(\dfrac{5xy^4}{\sqrt{z}}\right)$

24. $\log_5\left(\dfrac{3x^5\sqrt[3]{y}}{z^4}\right)$

25. $\log_b \sqrt[3]{\dfrac{x}{y^2z}}$

26. $\log_b \sqrt[4]{\dfrac{z}{x^2y^3}}$

27. $\log_e e^5xy^2$

28. $\log_e e^{-4}y^3z$

Write as a single logarithm.

29. $\log_5 x + \log_5 7 + \log_5 3$

30. $\log_5 7 + \log_5 11 + \log_5 y$

31. $5\log_3 x - \log_3 7$

32. $3 \log_8 5 - \log_8 z$

33. $\dfrac{2}{3} \log_b x + \dfrac{1}{2} \log_b y - 3 \log_b z$

34. $\dfrac{3}{4} \log_b x + 2 \log_b y - \dfrac{1}{2} \log_b z$

35. $\dfrac{1}{2} (\log_3 7 - \log_3 x - \log_3 y)$

36. $\dfrac{1}{3} (\log_4 x - \log_4 2 - \log_4 z^2)$

Use the properties of logarithms to simplify each of the following.

37. $\log_3 3$

38. $\log_7 7$

39. $\log_e e$

40. $\log_{10} 10$

41. $\log_9 1$

42. $\log_e 1$

43. $\log_5 5 + \log_5 1$

44. $\log_6 6 + \log_6 1$

Find x in each of the following.

45. $\log_8 x = \log_8 7$

46. $\log_9 x = \log_9 5$

47. $\log_4 2 = \log_4(x - 10)$

48. $\log_5 8 = \log_5(2x + 1)$

49. $\log_3 1 = x$

50. $\log_8 1 = x$

51. $\log_7 7 = x$

52. $\log_5 5 = x$

53. $\log_{10} 10^3 = x$

54. $\log_{10} 10^{-4} = x$

55. $\log_e e^{-4} = x$

56. $\log_e e^8 = x$

57. $\log_{10} x + \log_{10} 25 = 2$

58. $\log_{10} x + \log_{10} 5 = 1$

59. $\log_2 7 = \log_2 x - \log_2 3$

60. $\log_5 1 = \log_5 x - \log_5 8$

61. $3 \log_5 x = \log_5 8$

62. $\dfrac{1}{2} \log_3 x = \log_3 4$

63. $\log_e x - \log_e 2 = 2$

64. $\log_e x + \log_e 3 = 1$

65. $\log_2(x + 3) - \log_2(x - 3) = 3$

66. $\log_4(x + 2) - \log_4(x - 2) = 2$

67. Show that $\log_{10} A = \log_{10}(10{,}000)(1.07^x)$ can be written as $\log_{10} A = 4 + x \log_{10} 1.07$ and that $\dfrac{\log_{10} A - 4}{\log_{10} 1.07} = x.$

68. Show that $\log_{10} A = \log_{10}(1000)(1.12)^x$ can be written as $\log_{10} A = 3 + x \log_{10} 1.12$ and that $\dfrac{\log_{10} A - 3}{\log_{10} 1.12} = x.$

69. It can be shown that $y = b^{\log_b y}$. Use this property to evaluate $5^{\log_5 4} + 3^{\log_3 2}$.

70. It can be shown that $x = \log_b b^x$. Use this property to evaluate $\log_7 \sqrt[4]{7} + \log_6 \sqrt[12]{6}$.

To Think About

71. Prove that $\log_b\left(\dfrac{M}{N}\right) = \log_b M - \log_b N$ by using an argument similar to the proof of property 1.

72. Prove that $\log_b M^p = p \log_b M$ by using an argument similar to the proof of property 1.

73. Prove property 5.

74. Prove property 4.

Cumulative Review Problems

75. Find the area of a circle whose radius is 4 meters.

76. Find the volume of a cylinder with a radius of 2 meters and a height of 5 meters.

77. Solve for (x, y).
$$5x + 3y = 9$$
$$7x - 2y = 25$$

78. Solve for (x, y, z).
$$2x - y + z = 3$$
$$x + 2y + 2z = 1$$
$$4x + y + 2z = 0$$

79. Direct investments in Mexico made by U.S. companies increased from 5.986×10^9 dollars in 1980 to 1.6375×10^{10} dollars in 1993. What was the percent of increase during this 13-year period? What was the average yearly percent of increase?

80. The gross national product of Japan increased from 9.36×10^{11} dollars in 1980 to 2.594×10^{12} dollars in 1994. What was the percent of increase during this 14-year period? If that rate of increase continues, what is an estimated value of the gross national product of Japan in the year 2000?

81. While traveling at 38 feet per second, the driver of a Ford Explorer equipped with antilock brakes hits a patch of sheer ice during the winter in North Dakota. He immediately hits the brakes hard and stops the vehicle in 3 seconds. What is the vehicle's speed in miles per hour 2 seconds after he hits the brake? (Assume that he decelerates at a constant rate.) Was he exceeding the 35 mile per hour speed limit on this back road before he hit the brakes?

82. Eddie has a 1996 Corvette. He rapidly accelerates from a standstill to 45 miles per hour in 5 seconds. If he continues at a constant rate of acceleration how many seconds will it take him to reach the speed limit of 65 miles per hour from the standstill? If he hits the brakes at 65 mph because of an object in the road and comes to a complete stop 14 seconds after starting from a standstill, how many feet did he travel while he was braking?

11.4 Finding Logarithmic Function Values on a Calculator

MathPro   Video 27   SSM

After studying this section, you will be able to:

1 *Find a common logarithm.*
2 *Find the antilogarithm of a common logarithm.*
3 *Find a natural logarithm.*
4 *Find the antilogarithm of a natural logarithm.*
5 *Evaluate a logarithm to a base other than 10 or e.*

1 Finding Common Logarithms on a Scientific Calculator

Although we can find a logarithm of a number to any positive base except 1, the most frequently used bases are 10 and *e*. Logarithms to base 10 are called ***common logarithms*** and are usually written with no subscript.

Definition

The common logarithm of a number x is

$$\log x = \log_{10} x, \qquad \text{for all real numbers } x > 0$$

Before the advent of calculators and computers, people used tables of common logarithms. Now most work with logarithms is done with the aid of a scientific calculator. We will take that approach in this section of the text. To find the common logarithm of a number on a scientific calculator, enter the number and then press the $\boxed{\log x}$ or $\boxed{\log}$ key.

EXAMPLE 1 On a scientific calculator find a decimal approximation for

(a) log 7.32   **(b)** log 73.2   **(c)** log 5632   **(d)** log 0.314

Note: Your calculator may display fewer or more digits in the answer.

(a) 7.32 $\boxed{\log x}$ 0.864511081 ← ⎤
(b) 73.2 $\boxed{\log x}$ 1.864511081 ← ⎥ Note that the only difference in the two answers is the 1 before the decimal point.
(c) 5632 $\boxed{\log x}$ 3.750662646
(d) 0.314 $\boxed{\log x}$ −0.503070352

Practice Problem 1 On a scientific calculator find a decimal approximation for

(a) log 4.36   **(b)** log 436   **(c)** log 1279   **(d)** log 0.2418 ∎

To Think About Why is the difference in the solutions to Example 1 (a) and (b) only 1.00? Can you show why this should be true? Consider the following:

$$
\begin{aligned}
\log 73.2 &= \log(7.32 \times 10^1) && \textit{Use scientific notation.} \\
&= \log 7.32 + \log 10^1 && \textit{By Property 1.} \\
&= \log 7.32 + 1 && \textit{Because } \log_b b = 1. \\
&= 0.864511081 + 1 && \textit{Use a calculator.} \\
&= 1.864511081 && \textit{Add the decimals.}
\end{aligned}
$$

Finding Antilogarithms of a Common Logarithm on a Scientific Calculator

We have previously discussed the function $f(x) = \log x$ and the corresponding inverse function $f^{-1}(x) = 10^x$. The inverse of a logarithmic function is an exponential function. There is another name for this function. It is called an **antilogarithm.** When we find an antilogarithm, we are finding an exponent.

If $f(x) = \log x$ (here the base is understood to be 10), then $f^{-1}(x) = $ antilog $x = 10^x$.

EXAMPLE 2 Find an approximate value for x if $\log x = 4.326$.

Here we are given the value of the logarithm, and we want to find the number that has that logarithm. In other words, we want the **antilogarithm.** We know that $\log_{10} x = 4.326$ is equivalent to $10^{4.326} = x$. So to solve this problem, we want to find a number to a certain power. In this case we want to evaluate 10 to the 4.326 power. Using a calculator, we have

$$4.326 \boxed{10^x} \quad 21183.61135 \qquad \text{Thus } x \approx 21{,}183.61135$$

(If your scientific calculator does not have a $\boxed{10^x}$ key, you can usually use $\boxed{\text{2nd Fn}}$ $\boxed{\log x}$ or $\boxed{\text{INV}}$ $\boxed{\log x}$ or $\boxed{\text{SHIFT}}$ $\boxed{\log x}$ to perform the operation.)

Practice Problem 2 Using a scientific calculator, find an approximate value for x if $\log x = 2.913$. ∎

EXAMPLE 3 Evaluate antilog(-1.6784).

Asking what is antilog(-1.6784) is equivalent to asking what is the value of $10^{-1.6784}$. In determining this on a scientific calculator it will be necessary to enter the digits 1.6784 followed by the $\boxed{+/-}$ key.

$$1.6784 \boxed{+/-} \boxed{10^x} \quad 0.020970076$$

Thus antilog$(-1.6784) \approx 0.020970076$.

Practice Problem 3 Evaluate antilog(-3.0705). ∎

EXAMPLE 4 Using a scientific calculator, find an approximate value for x.

(a) $\log x = 1.156$ **(b)** $\log x = 0.07318$ **(c)** $\log x = -3.1621$

(a) $\log x = 1.156$ is equivalent to $10^{1.156} = x$, and

$$1.156 \boxed{10^x} \quad 14.32187899$$

(b) $\log x = 0.07318$ is equivalent to $10^{0.07318} = x$, and

$$0.07318 \boxed{10^x} \quad 1.183531987$$

(c) $\log x = -3.1621$ is equivalent to $10^{-3.1621} = x$, and

$$3.1621 \boxed{+/-} \boxed{10^x} \quad 0.0006884937465$$

(Some calculators may give the answer in scientific notation as $6.884937465 \times 10^{-4}$. This is often displayed on the calculator screen as $6.884937465 - 4$.)

Practice Problem 4 Using a scientific calculator, find an approximate value for x.

(a) $\log x = 1.823$ **(b)** $\log x = 0.06134$ **(c)** $\log x = -4.6218$ ∎

3 Finding Natural Logarithms on a Scientific Calculator

For most theoretical work in mathematics and other sciences the most useful base for logarithms is e. Logarithms with base e are known as **natural logarithms** and are usually written $\ln x$.

> **Definition**
>
> The natural logarithm of a number x is
>
> $$\ln x = \log_e x, \qquad \text{for all real numbers } x > 0$$

On a scientific calculator we can usually approximate natural logarithms with the $\boxed{\ln x}$ or $\boxed{\ln}$ key.

EXAMPLE 5   On a scientific calculator, approximate the following values.

(a) $\ln 7.21$ **(b)** $\ln 72.1$ **(c)** $\ln 0.0356$

(a) 7.21 $\boxed{\ln x}$ 1.975468951 **(b)** 72.1 $\boxed{\ln x}$ 4.278054044

(c) 0.0356 $\boxed{\ln x}$ -3.335409641

Note that there is no simple relationship between the answers to parts (a) and (b). Do you see why this is different from common logarithms?

Practice Problem 5   On a scientific calculator, approximate the following values.

(a) $\ln 4.82$ **(b)** $\ln 48.2$ **(c)** $\ln 0.0793$ ∎

4 Finding Antilogarithms of a Natural Logarithm on a Scientific Calculator

EXAMPLE 6   On a scientific calculator find an approximate value of x for each equation.

(a) $\ln x = 2.9836$ **(b)** $\ln x = -1.5619$

(a) If $\ln x = 2.9836$, then $e^{2.9836} = x$. On a scientific calculator,

$$2.9836 \ \boxed{e^x} \quad 19.75882051$$

(b) If $\ln x = -1.5619$, then $e^{-1.5619} = x$. On a scientific calculator,

$$1.5619 \ \boxed{+/-} \ \boxed{e^x} \quad 0.209737192$$

Practice Problem 6   On a scientific calculator, find an approximate value of x for each equation.

(a) $\ln x = 3.1628$ **(b)** $\ln x = -2.0573$ ∎

An alternative notation is sometimes used. This is $\text{antilog}_e(x)$.

EXAMPLE 7   Find the value of $\text{antilog}_e(-20.3518)$.

This example is equivalent to asking, "What is x if $\ln x = -20.3518$?" We find x by solving the equation $e^{-20.3518} = x$. On a calculator we enter

$$20.3518 \boxed{+/-} \boxed{e^x} \quad 1.4498583 - 09$$

This type of display is often used on a calculator to indicate scientific notation. This represents 1.4498583×10^{-9}. Thus $\text{antilog}_e(-20.3518) = 1.4498583 \times 10^{-9}$.

Note: Some calculators cannot do this problem directly and require further steps. Be sure you become familiar with the unique characteristics and limitations of your calculator.

Practice Problem 7   Find the value of $\text{antilog}_e(-12.9713)$. ∎

5 *Evaluating a Logarithm to a Base Other Than 10 or e*

A scientific calculator has a separate key to find common logarithms (base 10) and natural logarithms (base e). There is no direct way to find a logarithm with a base 3 or a base 7. What do we do in such cases? The logarithm of a number to a base other than 10 or e can be found from the following formula.

Change of Base Formula

$$\log_b x = \frac{\log_a x}{\log_a b}$$

where $a, b, x > 0$ and $a \neq 1$, $b \neq 1$.

Let's see how this formula would work. If you wanted to use common logarithms to find $\log_3 56$, the value of b in the formula would be 3. You could then write $\log_3 56 = \dfrac{\log_{10} 56}{\log_{10} 3} = \dfrac{\log 56}{\log 3}$. Do you see why?

EXAMPLE 8   Evaluate using common logarithms.

(a) $\log_7 3.67$ **(b)** $\log_3 5.12$

(a) $\log_7 3.67 = \dfrac{\log 3.67}{\log 7}$

On a calculator, we perform

$$3.67 \boxed{\log x} \boxed{\div} 7 \boxed{\log x} \boxed{=} \quad 0.668166340$$

(b) $\log_3 5.12 = \dfrac{\log 5.12}{\log 3}$

On a calculator, we perform

$$5.12 \boxed{\log x} \boxed{\div} 3 \boxed{\log x} \boxed{=} \quad 1.486561234$$

Each of our answers is an approximate value with nine decimal places. Your answer may have more or fewer digits depending on your calculator.

Practice Problem 8   Evaluate using common logarithms.

(a) $\log_6 5.28$ **(b)** $\log_9 3.76$ ∎

If we desire to use base e, then the change of base formula is used with natural logarithms.

EXAMPLE 9 Obtain an approximate value for $\log_4 0.005739$ using natural logarithms.

Using the change of base formula, where $b = 4$ and $x = 0.005739$ while $a = e$, we will have

$$\log_4 0.005739 = \frac{\log_e 0.005739}{\log_e 4} = \frac{\ln 0.005739}{\ln 4}$$

This is done on most scientific calculators as follows.

$$0.005739 \boxed{\ln} \boxed{\div} 4 \boxed{\ln} \boxed{=} \quad -3.722492455$$

Thus we have $\log_4 0.005739 \approx -3.722492455$.

Check: To check our answer we would want to know

$$4^{-3.722492455} \stackrel{?}{=} 0.005739$$

Using a calculator, we can verify this with the $\boxed{y^x}$ key.

$$4 \boxed{y^x} 3.722492455 \boxed{+/-} \boxed{=} 0.005739$$

This is in fact the value we see on the display. The answer does check.

Practice Problem 9 Obtain an approximate value for $\log_8 0.009312$ using natural logarithms. ∎

GRAPHING CALCULATOR

Graphing Logarithmic Functions

You can use the change of base formula to graph logarithmic functions on a graphing calculator.

In Example 10, to graph $y = \log_2 x$ on a graphing calculator, enter the function $y = \dfrac{\log x}{\log 2}$ into the Y = editor of your calculator.

Display:

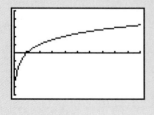

EXAMPLE 10 Using a scientific calculator, graph $y = \log_2 x$.

If we use common logarithms ($\log_{10} x$), then for each value of x we will need to calculate $\log x/\log 2$. Therefore, to find y when $x = 3$, we need to calculate $\log 3/\log 2$. On most scientific calculators we would enter $3 \boxed{\log} \boxed{\div} 2 \boxed{\log} \boxed{=} 1.584962501$. Rounded to the nearest tenth we have when $x = 3$, $y = 1.6$. We can in similar fashion find a number of table values and then graph from them.

x	$y = \log_2 x$
0.5	-1
1	0
2	1
3	1.6
4	2
6	2.6
8	3

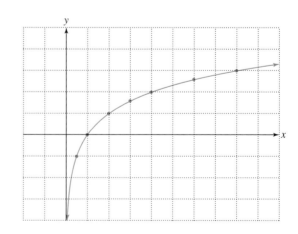

Practice Problem 10 Using a scientific calculator, graph $y = \log_5 x$.

11.4 Exercises

Use a scientific calculator to approximate the following.

1. log 7.36 **2.** log 2.19 **3.** log 25.6 **4.** log 83.8 **5.** log 356 **6.** log 896

7. log 125,000 **8.** log 78,500 **9.** log 0.0123 **10.** log 0.567 **11.** log 0.891 **12.** log 0.00045

13. Try to find $\log(-5.08)$. What happens? Why? **14.** Try to find $\log(-6.63)$. What happens? Why?

Find the approximate value of x using a scientific calculator.

15. $\log x = 0.1614$ **16.** $\log x = 0.2480$ **17.** $\log x = 0.5821$ **18.** $\log x = 0.5922$

19. $\log x = 1.7860$ **20.** $\log x = 1.7896$ **21.** $\log x = 3.9304$ **22.** $\log x = 3.9576$

23. $\log x = 6.4683$ **24.** $\log x = 5.6274$ **25.** $\log x = 0.5353 - 2$ **26.** $\log x = 0.9974 - 3$

27. $\log x = -3.3893$ **28.** $\log x = -4.0458$ **29.** $\log x = 2.0030$ **30.** $\log x = 2.1034$

Approximate the following with a scientific calculator.

31. antilog(7.6215) **32.** antilog(4.3894) **33.** antilog(−4.5113) **34.** antilog(−2.1773)

35. ln 5.62 **36.** ln 8.81 **37.** ln 56.2 **38.** ln 88.1 **39.** ln 136,000

40. ln 129,000 **41.** ln 0.0167 **42.** ln 0.0362 **43.** ln 1.5 **44.** ln 1.01

Find an approximate value of x using a scientific calculator.

45. ln $x = 0.95$ **46.** ln $x = 0.55$ **47.** ln $x = 2.4$ **48.** ln $x = 4.4$ **49.** ln $x = 14$

50. ln $x = 12$ **51.** ln $x = -0.13$ **52.** ln $x = -0.18$ **53.** ln $x = -2.7$ **54.** ln $x = -3.8$

Approximate the following with a scientific calculator.

55. $\text{antilog}_e(4.6782)$ **56.** $\text{antilog}_e(2.4294)$ **57.** $\text{antilog}_e(-2.1298)$ **58.** $\text{antilog}_e(-3.3712)$

*Use a scientific calculator and **common logarithms** to evaluate the following.*

59. $\log_3 9.2$ **60.** $\log_2 6.13$ **61.** $\log_9 8.66$ **62.** $\log_8 7.98$

63. $\log_6 0.127$ **64.** $\log_5 0.173$ **65.** $\log_{14} 156$ **66.** $\log_{15} 243$

*Use a scientific calculator and **natural logarithms** to evaluate the following.*

67. $\log_4 0.07733$ **68.** $\log_7 0.004462$ **69.** $\log_{11} 87,997$ **70.** $\log_{12} 8534$

Mixed Practice

Use a scientific calculator to find an approximate value for the following.

71. $\log 92.81$

72. $\ln 1537$

73. $\log_6 0.5437$

74. $\text{antilog}_e(-1.874)$

Find an approximate value for x if

75. $\log x = 8.5634$

76. $\ln x = 7.9631$

77. $\log_4 x = 0.8645$

78. $\log_3 x = 0.5649$

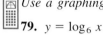 *Use a graphing calculator or a scientific calculator to graph the following.*

79. $y = \log_6 x$

80. $y = \log_4 x$

81. $y = \log_{0.2} x$

Applications: Magnitude of an Earthquake

Suppose we want to measure the magnitude of an earthquake. If an earthquake has a shock wave x times greater than the smallest shock wave that can be measured by a seismograph, then we can measure its magnitude R on the Richter scale by the equation R = log x.

An earthquake that has a shock wave 25,000 times greater than the smallest shock wave that can be detected will have a magnitude of R = log 25,000 ≈ 4.40. (Usually we round the magnitude of an earthquake to the nearest hundredth.)

82. What is the magnitude of an earthquake that has a shock wave that is 56,000 times greater than the smallest shock wave that can be detected?

83. What is the magnitude of an earthquake that has a shock wave that is 184,000 times greater than the smallest shock wave that can be detected?

84. If an earthquake has $R = 6.6$ on the Richter scale, what can you say about the size of its shock wave?

85. If an earthquake has $R = 5.4$ on the Richter scale, what can you say about the size of its shock wave?

Cumulative Review Problems

Solve the quadratic equations. Simplify your answer.

86. $17x^2 - 7x = 0$

87. $3x^2 - 11x - 5 = 0$

88. $2y^2 + 4y - 3 = 0$

89. $3y^2 + 5y + 1 = 0$

11.5 Exponential and Logarithmic Equations

After studying this section, you will be able to:

1 *Solve logarithmic equations.*
2 *Solve exponential equations.*
3 *Solve applied problems using logarithmic or exponential equations.*

MathPro   Video 27   SSM

1 Solving Logarithmic Equations

In general, when solving logarithmic equations we try to obtain all the logarithms on one side of the equation while all the numerical values are on the other side. Then we seek to use the properties of logarithms to obtain a single logarithmic expression on one side.

EXAMPLE 1   Solve $\log_2 2x + \log_2 4 = 8$.

$\log_2 (2x)(4) = 8$	*Property 1.*
$\log_2 (8x) = 8$	*Simplify.*
$8x = 2^8$	*Write the equation in exponential form.*
$8x = 256$	*Evaluate 2^8.*
$x = 32$	*Divide both sides by 8.*

Check: $\log_2[2(32)] + \log_2 4 \overset{?}{=} 8$

$\log_2 64 + \log_2 4 \overset{?}{=} 8$

$6 + 2 \overset{?}{=} 8$

$8 = 8$ ✓

Practice Problem 1   Solve $\log_2 4 + \log_2 x = 3$. ∎

After looking at our results from Example 1 we can describe a general procedure.

Step 1 If an equation contains some logarithms and some terms without logarithms, try to get one logarithm alone on one side and one numerical value on the other.
Step 2 Then convert to an exponential equation using the definition of a logarithm.
Step 3 Solve the equation.

EXAMPLE 2   Solve $\log 5 = 2 - \log(x + 3)$.

$\log 5 + \log(x + 3) = 2$	*Add $\log(x + 3)$ to each side.*
$\log[5(x + 3)] = 2$	*Property 1.*
$\log(5x + 15) = 2$	*Simplify.*
$5x + 15 = 10^2$	*Write the equation in exponential form.*
$5x + 15 = 100$	*Simplify.*
$5x = 85$	*Subtract 15 from each side.*
$x = 17$	*Divide each side by 5.*

Check: $\log 5 \overset{?}{=} 2 - \log(17 + 3)$
$\log 5 \overset{?}{=} 2 - \log 20$

Since these are common logarithms (base 10), probably the easiest way to check is to find decimal approximations for each logarithm on the calculator.

$0.698970004 \overset{?}{=} 2 - 1.301029996$

$0.698970004 = 0.698970004$ ✓

GRAPHING CALCULATOR

Solving Logarithmic Equations

Example 2 could be solved with a graphing calculator in the following way. First write the equation as

$\log 5 + \log(x + 3) - 2 = 0$

and then graph the function

$y = \log 5 + \log(x + 3) - 2$

to find an approximate value for x when $y = 0$. If you set the appropriate window and use your zoom feature, you should be able to obtain $x = 17.0$ if you round your answer to the nearest tenth. Now use your graphing calculator to solve:

1. $\log x + \log(x + 1) = 1$
2. $\log(x - 6) = 2 - \log(x + 15)$
3. $e^{x+2} = 12$

Practice Problem 2   Solve $\log(x + 5) = 2 - \log 5$. ∎

EXAMPLE 3   Solve $\log_3(x + 6) - \log_3(x - 2) = 2$.

$$\log_3\left(\frac{x + 6}{x - 2}\right) = 2 \qquad\qquad \textit{Property 2.}$$

$$\frac{x + 6}{x - 2} = 3^2 \qquad\qquad \textit{Write the equation in exponential form.}$$

$$\frac{x + 6}{x - 2} = 9 \qquad\qquad \textit{Evaluate } 3^2.$$

$$x + 6 = 9(x - 2) \qquad \textit{Multiply each side by } (x - 2).$$

$$x + 6 = 9x - 18 \qquad \textit{Simplify.}$$

$$24 = 8x \qquad\qquad \textit{Add } 18 - x \textit{ to each side.}$$

$$3 = x \qquad\qquad\quad \textit{Divide each side by } 3.$$

$$\textit{Check:} \quad \log_3(3 + 6) - \log_3(3 - 2) \stackrel{?}{=} 2$$

$$\log_3 9 - \log_3 1 \stackrel{?}{=} 2$$

$$2 - 0 \stackrel{?}{=} 2$$

$$2 = 2 \ \checkmark$$

Practice Problem 3   Solve for x. $\log(x + 3) - \log x = 1$ ∎

Some equations consist only of the sum and difference of logarithmic terms. If an equation contains only logarithms, we try to use the properties of logarithms to obtain a single logarithmic expression of each side of the equation. Then we may use the following property. Suppose we know that $b > 0$ and $b \neq 1$ and $M, N > 0$. Then it is true that:

Property 7   If $\log_b M = \log_b N$, then $M = N$.

What if one of our possible solutions gives the logarithm of a negative number? Can we evaluate the logarithm of a negative number? Look again at the graph of $\log x$ on page 624. Note that the domain of $\log x$ is $x > 0$. (The graph is on the positive side of the x-axis.) Therefore, the logarithm of a negative number is *not defined.*

You should be able to see this by using the definition of logarithms. If $\log(-2)$ were valid, we could write

$$y = \log_{10}(-2)$$

$$10^y = -2$$

Obviously, no value for y can make this equation true. Thus we see that **it is not possible to take the logarithm of a negative number.**

Sometimes when we attempt to solve a logarithmic equation we obtain a possible solution that leads to the logarithm of a negative number. We can immediately discard that solution as not valid.

EXAMPLE 4   Solve $\log(x + 6) + \log(x + 2) = \log(x + 20)$.

$$\log(x + 6)(x + 2) = \log(x + 20)$$
$$\log(x^2 + 8x + 12) = \log(x + 20)$$
$$x^2 + 8x + 12 = x + 20$$
$$x^2 + 7x - 8 = 0$$
$$(x + 8)(x - 1) = 0$$
$$x + 8 = 0 \qquad x - 1 = 0$$
$$x = -8 \qquad x = 1$$

Check: $\log(x + 6) + \log(x + 2) = \log(x + 20)$

$x = 1$: $\log(1 + 6) + \log(1 + 2) \overset{?}{=} \log(1 + 20)$

$$\log(7) + \log(3) \overset{?}{=} \log(21)$$
$$\log(7 \cdot 3) \overset{?}{=} \log 21$$
$$\log 21 = \log 21 \quad \checkmark$$

$x = -8$: $\log(-8 + 6) + \log(-8 + 2) \overset{?}{=} \log(-8 + 20)$

$$\log(-2) + \log(-6) \neq \log(12)$$

We can immediately discard any possible solution that leads to taking the logarithm of a negative number. Only $x = 1$ is a solution.

Practice Problem 4   Solve $\log 5 - \log x = \log(6x - 7)$. Check your solution. ■

2 Solving Exponential Equations

You might expect that property 7 could be used in the reverse direction. It would seem logical that if $x = 3$ we should be able to state $\log_4 x = \log_4 3$, for example. This is exactly the case, and we will formally state it as a property.

> **Property 8**   If $M, N > 0$ and $M = N$, then $\log_b M = \log_b N$, where $b > 0$ and $b \neq 1$.

Property 8 is often referred to as "taking the logarithm of each side of the equation." Usually we will take the common logarithm of each side of the equation, although any base can be used.

EXAMPLE 5   Solve $2^x = 7$. Leave your answer in exact form.

$\log 2^x = \log 7$ *Take the logarithm of each side (Property 8).*

$x \log 2 = \log 7$ *Property 3.*

$x = \dfrac{\log 7}{\log 2}$ *Divide each side by log 2.*

Practice Problem 5   Solve $3^x = 5$. ■

In many cases when we solve exponential equations it is useful to find an approximate value for the answer, rather than leave an answer containing several logarithms.

EXAMPLE 6   Solve $3^x = 7^{x-1}$. Approximate your answer to the nearest thousandth.

$$\log 3^x = \log 7^{(x-1)}$$

$$x \log 3 = (x - 1) \log 7$$

$$x \log 3 = x \log 7 - \log 7$$

$$x \log 3 - x \log 7 = -\log 7$$

$$x(\log 3 - \log 7) = -\log 7$$

$$x = \frac{-\log 7}{\log 3 - \log 7}$$

We can approximate the value for x on most scientific calculators by using the following keystrokes.

7 $\boxed{\log}$ $\boxed{+/-}$ $\boxed{\div}$ $\boxed{(}$ 3 $\boxed{\log}$ $\boxed{-}$ 7 $\boxed{\log}$ $\boxed{)}$ $\boxed{=}$ 2.296606943

Rounding to the nearest thousandth, $x \approx 2.297$.

Practice Problem 6   Solve $2^{3x+1} = 9^{x+1}$. Approximate your answer to the nearest thousandth. ■

If the exponential equation involves e raised to a power, it is best to take the *natural logarithm* of each side of the equation.

EXAMPLE 7   Solve $e^{2.5x} = 8.42$. Round your answer to the nearest ten-thousandth.

$$\ln e^{2.5x} = \ln 8.42 \qquad \textit{Take the natural log of each side.}$$

$$(2.5x)(\ln e) = \ln 8.42 \qquad \textit{Property 3.}$$

$$2.5x = \ln 8.42 \qquad \ln e = 1.$$

$$x = \frac{\ln 8.42}{2.5} \qquad \textit{Divide each side by 2.5.}$$

On most scientific calculators the value of x can be approximated with the following keystrokes.

8.42 $\boxed{\ln}$ $\boxed{\div}$ 2.5 $\boxed{=}$ 0.85224393

Rounding to the nearest ten-thousandth, we have $x \approx 0.8522$.

Practice Problem 7   Solve $20.98 = e^{3.6x}$. Round your answer to the nearest ten-thousandth. ■

3 *Solving Applied Problems Using Logarithmic or Exponential Equations*

We will now return to the compound interest formula and consider some other problems that can be solved with its use. For example, perhaps we would like to know how long it takes for a deposit to grow to a specified goal.

EXAMPLE 8 If an amount of money earns interest at 12% compounded annually, the amount available after t years is $A = P(1 + 0.12)^t$, where P is the principal invested. How many years will it take for \$300 to grow to \$1500? Round your answer to the nearest whole year.

$$1500 = 300(1 + 0.12)^t \qquad \text{Substitute } A = 1500, P = 300.$$

$$1500 = 300(1.12)^t \qquad \text{Simplify.}$$

$$\frac{1500}{300} = (1.12)^t \qquad \text{Divide each side by } 300.$$

$$5 = (1.12)^t \qquad \text{Simplify.}$$

$$\log 5 = \log(1.12)^t \qquad \text{Take a common log of each side.}$$

$$\log 5 = t(\log 1.12) \qquad \text{Property 3.}$$

$$\frac{\log 5}{\log 1.12} = t \qquad \text{Divide each side by log } 1.12.$$

On a scientific calculator

$$5 \boxed{\log} \boxed{\div} 1.12 \boxed{\log} \boxed{=} \quad 14.20150519$$

Thus it would take approximately 14 years.

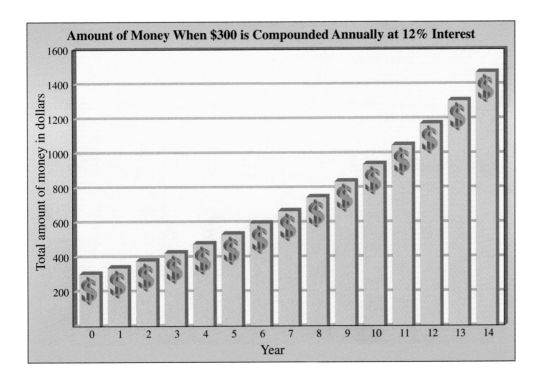

Amount of Money When \$300 is Compounded Annually at 12% Interest

Practice Problem 8 Mon Ling's father has an investment account that earns 8% interest compounded annually. How many years would it take for \$4000 to grow to \$10,000 in that account? Round your answer to the nearest whole year. ■

The growth equation for things that appear to be growing continuously can be modeled by the equation $A = A_0 e^{rt}$, where A is the final amount, A_0 is the original amount, r is the rate that things are growing in a time period, and t is the number of units of that time period.

For example, if cells in a laboratory are reproducing at a rate of 35% per hour and we start with 5000 cells, the number of cells that there will be in 18 hours could be described by the equation

$$A = 5000e^{(0.35)(18)} = 5000e^{6.3} \approx 5000(544.57) = 2,722,850 \text{ cells}$$

EXAMPLE 9   If the population of the world is growing at a rate of 2% per year and the world population is 8 billion people, how many years will it take for the world to have 14 billion people if growth continues at that rate?

$$A = A_0 e^{rt}$$

To make our calculation easier, we will write the values of population in terms of billions. This will allow us to avoid writing numbers like 14,000,000,000 and 8,000,000,000. Do you see why we can do this?

$14 = 8e^{0.02t}$	*Substitute known values.*
$\dfrac{14}{8} = e^{0.02t}$	*Divide each side by 8.*
$\ln\left(\dfrac{14}{8}\right) = \ln e^{0.02t}$	*Take the natural logarithm of each side.*
$\ln(1.75) = (0.02t)\ln e$	*Property 3.*
$\ln 1.75 = 0.02t$	*Since $\ln e = 1$.*
$\dfrac{\ln 1.75}{0.02} = t$	*Divide each side by 0.02.*

Using our calculator, we obtain

$$1.75 \boxed{\ln} \boxed{\div} 0.02 \boxed{=} \quad 27.98078940$$

Rounding to the nearest whole year, we find that the population will grow from 8 billion to 14 billion in about 28 years if the growth continues at the same rate.

Practice Problem 9   The wildlife management team in one region of Alaska has determined that the black bear population is growing at the rate of 4.3% per year. This region currently has approximately 1300 black bears. A food shortage problem will develop if the population reaches 5000. How many years will it take until this food shortage problem occurs if the growth rate remains unchanged? ■

EXAMPLE 10 The magnitude of an earthquake is measured by the formula $R = \log\left(\dfrac{I}{I_0}\right)$, where I is the intensity of the earthquake and I_0 is the minimum measurable intensity. The 1964 earthquake in Anchorage, Alaska, had a magnitude of 8.4. The 1906 earthquake in Taiwan had a magnitude of 7.1. How much more energy was released from the Anchorage earthquake than from the Taiwan earthquake?

Let I_A = intensity of Alaska earthquake. Then

$$8.4 = \log\left(\frac{I_A}{I_0}\right) = \log I_A - \log I_0$$

Solving for $\log I_0$ gives

$$\log I_0 = \log I_A - 8.4$$

Let I_T = intensity of Taiwan earthquake. Then

$$7.1 = \log\left(\frac{I_T}{I_0}\right) = \log I_T - \log I_0$$

Solving for $\log I_0$ gives

$$\log I_0 = \log I_T - 7.1$$

Therefore,

$$\log I_A - 8.4 = \log I_T - 7.1$$

$$\log I_A - \log I_T = 8.4 - 7.1$$

$$\log \frac{I_A}{I_T} = 1.3$$

$$10^{1.3} = \frac{I_A}{I_T}$$

$$19.95262315 = \frac{I_A}{I_T} \qquad \textit{Use a calculator.}$$

$$20 = \frac{I_A}{I_T} \qquad \textit{Round to the nearest whole number.}$$

$$20 I_T = I_A$$

The Alaska earthquake had approximately 20 times the intensity of the Taiwan earthquake.

Practice Problem 10 The 1933 earthquake in Japan had a magnitude of 8.9. The 1989 earthquake in San Francisco had a magnitude of 7.1. How much more energy was released from the Japan earthquake than the San Francisco earthquake? ■

11.5 Exercises

Solve each logarithmic equation and check your solution.

1. $\log_6(x + 3) + \log_6 4 = 2$

2. $\log_4(x + 2) + \log_4 3 = 2$

3. $\log_5(x - 2) + \log_5 3 = 2$

4. $\log_2 4 + \log_2(x - 1) = 5$

5. $\log_3(2x - 1) = 2 - \log_3 4$

6. $\log_5(3x + 1) = 1 - \log_5 2$

7. $1 + \log(4 - x) = \log(3x + 1)$

8. $1 + \log x = \log(9x + 1)$

9. $\log_4(8x) = 2 + \log_4(x - 1)$

10. $\log_2 x = \log_2(x + 5) - 1$

11. $\log(x + 20) - \log x = 2$

12. $\log_3(x + 6) + \log_3 x = 3$

13. $2 \log_3 4 = \log_3 x - \log_3 (x - 1)$

14. $\log_4(x - 3) + \log_4(x + 3) = 2$

15. $1 + \log(x - 3) = \log x$

16. $\log_5(2x) - \log_5(x - 3) = 3 \log_5 2$

17. $\log_2(x + 5) - 2 = \log_2 x$

18. $\log_3(x - 1) - 3 = \log_3(2x + 1)$

19. $2 \log_7 x = \log_7(x + 4) + \log_7 2$

20. $\log x + \log(x - 1) = \log 12$

21. $\log 2 - \log(3x - 5) = \log x$

22. $\log_7(2x - 1) = \log_7(x + 2) + \log_7(x - 2)$

23. $\log(x + 1) + \log(x - 3) = \log(7x - 23)$

24. $\log(x - 3) = \log(3x - 8) - \log x$

Solve each exponential equation. Leave your answer in exact form. Do not approximate.

25. $7^x = 13$

26. $9^x = 11$

27. $5^{x+1} = 9$

28. $4^{x+2} = 7$

29. $2^{3x+4} = 17$

30. $5^{2x-1} = 11$

31. $e^{2x} = 14$

32. $e^{3x} = 55$

Solve each exponential equation. Use your calculator to approximate your solution to the nearest thousandth.

33. $7^{2x+1} = 30$

34. $8^{3x+1} = 26$

35. $5^x = 4^{x+1}$

36. $3^x = 2^{x+3}$

37. $8^x = 9^{x-2}$

38. $9^x = 7^{x+3}$

39. $e^{x-1} = 28$

40. $e^{x+1} = 17$

41. $e^{-x} = 0.12$

42. $e^{-x} = 0.18$

43. $54 = e^{2x-1}$

44. $37 = e^{3x-2}$

Applications

When a principal P earns an annual interest rate r compounded yearly, the amount A after t years is $A = P(1 + r)^t$. Use this information to solve Problems 45 through 48. Round all answers to the nearest whole year.

45. How long will it take $1000 to grow to $4500 at 7% annual interest?

46. How long will it take $1500 to grow to $5000 at 8% annual interest?

47. How long will it take for a principal to double at 5% annual interest?

48. How long will it take for a principal to triple at 6% annual interest?

★ **49.** What interest rate would be necessary to obtain $6500 in 6 years if $5000 is the amount of the original investment and the interest is compounded yearly? (Express the interest rate as a percent rounded to the nearest tenth.)

★ **50.** If $3000 is invested for 3 years with annual interest compounded yearly, what interest rate is needed to achieve an amount of $3600? (Express the interest rate as a percent rounded to the nearest tenth.)

The growth of the world's population can be measured by the equation $A = A_0 e^{rt}$, where t is the number of years, A_0 is the population of the world at time $t = 0$, r is the rate of increase per year, and A is the population at time t. Assume that $r = 2\%$ per year. Use this information to solve Problems 51 through 54. Round your answers to the nearest whole year.

51. How long will it take a population of 7 billion to increase to 12 billion?

52. How long will it take a population of 6 billion to increase to 9 billion?

53. How long will it take for the world's population to double?

54. How long will it take for the world's population to quadruple (become four times as large)?

Use the equation $A = A_0 e^{rt}$ to solve Problems 55 through 62. Round your answers to the nearest whole number.

55. If the population rate of increase was $r = 1\%$, how long would it take the world population to double?

56. If the population rate of increase was $r = 3\%$, how long will it take the world population to double?

57. The population of Melbourne, Australia, is approximately 3 million people. If the growth rate is 3% per year, how many years will it be until there are 3.5 million people?

58. The population of Bethel is 80,000 people and the city is growing at the rate of 1.5% per year. How many years will it take for the city to grow to 120,000 people?

59. The number of new skin cells on a revolutionary skin graft is growing at a rate of 4% per hour. How many hours will it take for 200 cells to become 1800 cells?

60. The work force in Massachusetts is increasing at the rate of 1.5% per year. During the last measured year, the workforce was 3.5 million. How many years will it take to reach 4.5 million if this rate continues?

61. Domestic U.S. deer unfortunately carry ticks that are spreading Lyme disease. The number of people who are exposed to the virus is increasing by 5% every year. If 24,500 people were confirmed with Lyme disease in 1997, how many would be infected by the end of the year 2010?

62. In the city of Scranton, the number of videotape rentals is increasing at 7.5% per year. For the last year that data were available, 1.3 million videos were rented. How many years will it take until 2.0 million videos are rented per year?

To Think About

The magnitude of an earthquake (amount of energy released) is measured by the formula $R = \log\left(\dfrac{I}{I_0}\right)$, where I is the intensity of the earthquake and I_0 is the minimum measurable intensity. Use this formula to solve Problems 63–66.

63. On January 17, 1993, in Northridge, California, Los Angeles area residents experienced an earthquake measuring 6.8 on the Richter Scale that killed 61 people, and undermined supposedly earthquake-proof steel-frame buildings. Exactly one year later, an earthquake measuring 7.2 on the Richter Scale killed more than 5300 people, injured more than 35,000, and destroyed nearly 200,000 homes near Kobe, Japan, in spite of construction codes reputed to be the best in the world. How much more energy was released from the Japan earthquake than from the Northridge earthquake?

64. October 17, 1989, brought tragedy to the San Francisco/Oakland area, when an earthquake measuring 7.1 on the Richter Scale, centered in the Loma Prieta area (Santa Cruz Mountains), collapsed huge sections of freeway and killed 63 people. Almost six years later, an earthquake measuring 8.2 on the Richter Scale killed 190 people in the Kurile Islands of Japan and Russia. How much more energy was released from the Kurile earthquake than the Loma Prieta earthquake?

65. The 1906 earthquake in San Francisco had a magnitude of 8.3. In 1971 an earthquake in Japan measured 6.8. How much more energy was released from the San Francisco earthquake than from the Japan earthquake?

66. The 1933 Japan earthquake had a magnitude of 8.9. In Turkey a 1975 earthquake had a magnitude of 6.7. How much more energy was released from the Japan earthquake than from the Turkey earthquake?

Optional Graphing Calculator Problems

67. Suppose the population of wolves in one region of Alaska is growing according to the equation $y_1 = 34.572x + 850$, where x is the number of years, and the adequate food supply for wolves is growing according to the equation $y_2 = 1000e^{.02x}$, where x is the number of years. In how many years (rounded to the nearest tenth of a year) will the food supply become inadequate for the number of wolves? (When will y_1 be greater than y_2?)

68. In Crystal Lake, north of Amherst, Nova Scotia, the fish population has been out of balance for several years because of an overpopulation of catfish. Environmentalists have taken a number of measures to increase the population of brook trout so that the population of each is at the same level. After several years of dealing with industrial pollution, the environmentalists have succeeded in cleaning the lake sufficiently, so the population of brook trout can reproduce more readily. The growth in the number of brook trout is given by the equation $y = 300e^{0.12x}$, where x is the number of years. The growth in the number of catfish is given by the equation $y = 750 + 100x$ where x is the number of years. How many years will it take until the two fish populations are equal in number? Round your answer to the nearest tenth of a year.

Cumulative Review Problems

Simplify. Assume x, y are positive real numbers.

69. $\sqrt{98x^3y^2}$

70. $\sqrt[3]{81x^6y^9}$

71. $(\sqrt{3} + 2\sqrt{2})(\sqrt{6} - \sqrt{2})$

72. $2\sqrt{50x} + 3\sqrt{72x} - 4\sqrt{128x}$

Putting Your Skills to Work

MEASURING THE DOUBLING TIME

In the course of the investigation of small bacteria, cells, or other microorganisms, it often becomes necessary for scientists to determine the rate of growth of the object of investigation. However, the objects may be so small that they cannot be counted directly. A mathematical measurement called the doubling time is often very useful in such instances. Although a small quantity (the number of bacteria, for example) may be so tiny that it cannot be measured directly, it can often be determined when it has doubled. For example, when the weight of the bacteria has doubled, we assume that the number of bacteria has doubled. Or if the bacteria are spread over a certain area, when the area that the bacteria are contained in has doubled, we assume the number of bacteria has doubled. It is a common practice to evaluate the growth of microorganisms by using the equation $P(t) = P_0 e^{rt}$ where r is the rate of growth per time period and t is the number of time units. P_0 is the initial population size at the time $t = 0$. The value r can be obtained from the equation $r = \dfrac{\ln 2}{T}$ where T is the doubling time.

PROBLEMS FOR INDIVIDUAL INVESTIGATION AND ANALYSIS

Round all answers to four decimal places.

1. A new strain of flesh-eating bacteria has been detected on a patient in a major medical center. The area of infection caused by the bacteria has been observed to double on the patient's skin in 25 minutes. Doctors therefore assume that the number of bacteria has doubled in 25 minutes. Find the value of r in the equation $P(t) = P_0 e^{rt}$, where r is determined by the equation $r = \dfrac{\ln 2}{T}$, where T is the doubling time measured in minutes.

2. Write the equation for the growth of this bacteria using the value of r obtained in Problem 1.

3. If the initial bacteria count for surface area of the skin on the patient when admitted to the hospital is 3000 and it is now 4 hours later, what is the present estimated bacteria count?

4. Graph the growth of the bacteria population using the information obtained in Problems 1, 2, and 3 over the first 4 hours of treatment of the patient.

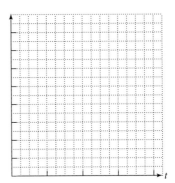

PROBLEMS FOR GROUP INVESTIGATION AND COOPERATIVE GROUP ACTIVITY

Together with some other members of your class see if you can determine the following.

After 4 hours the patient is given an antibiotic that slows the growth of the infection. The new doubling time is 180 minutes.

5. Write the new equation for the growth of the infection that describes the bacteria count for the time after the patient is given the antibiotic.

6. What is the bacteria count for the patient 6 hours after being admitted to the hospital?

7. After 6 hours the patient is given a second antibiotic, which proves to be effective in actually decreasing the bacteria count. The new antibiotic decreases the level of bacteria by 20% during each hour. Write an equation for the new number of bacteria of the patient after he is given the second antibiotic.

8. How many hours after the patient was admitted to the hospital will it be until the bacteria count is reduced down to 3000 again? Round your answer to the nearest hour.

INTERNET CONNECTION: Go to `http://www.prenhall.com/tobey` to be connected.

Site: CSB Rates

The equation $P(t) = P_0 e^{rt}$ can also be used for money in a savings account, provided the interest is compounded *continuously*. In this case, the equation $r =$ can be rewritten in the equivalent form $t =$ and used to find the doubling time—the number of years it would take to double your money in a savings account paying a given rate of interest.

This Web page gives the rates of interest for various accounts at a particular bank. Choose two types of account that pay different rates of continuously compounded interest for a balance of $2500, and answer the following questions.

9. Write the equation for the value of each account after t years if you deposit $2500 today. Then find the value of the account after 5 years.

10. Calculate the doubling time for each account. *Hint:* Be sure to use the interest *rate* (not APY, or *yield*), and remember to convert the rate from a percent to a decimal.

11. For each account, multiply the doubling time by the rate (expressed as a *percent*) and compare your results.

12. What do you think is meant by the "Rule of 70"?

Topic	Procedure	Examples
Exponential function, p. 611.	$f(x) = b^x$, where $b > 0$ and $b \neq 1$ and x is a real number.	Graph. $f(x) = \left(\dfrac{2}{3}\right)^x$
Properties of exponential equations, p. 614.	When $b > 0$ and $b \neq 1$, if $b^x = b^y$, then $x = y$.	Solve for x. $2^x = \dfrac{1}{32}$ $2^x = \dfrac{1}{2^5}$ $2^x = 2^{-5}$ Thus $x = -5$
Definition of logarithm, p. 622.	$y = \log_b x$ if and only if $x = b^y$, where $b > 0$ and $b \neq 1$.	Write in exponential form. $\log_3 17 = 2x$ $3^{2x} = 17$ Write in logarithmic form. $18 = 3^x$ $\log_3 18 = x$ Solve for x. $\log_6\left(\dfrac{1}{36}\right) = x$ $6^x = \dfrac{1}{36}$ $6^x = 6^{-2}$ So $x = -2$
Properties of logarithms, pp. 630–634.	Where $M > 0$, $N > 0$, and $b > 0$, $b \neq 1$: $\log_b MN = \log_b M + \log_b N$ $\log_b\left(\dfrac{M}{N}\right) = \log_b M - \log_b N$ $\log_b M^p = p \log_b M$ $\log_b b = 1$ $\log_b 1 = 0$ If $\log_b M = \log_b N$, then $M = N$. If $M = N$, then $\log_b M = \log_b N$.	Write as separate logarithms of x, y, w. $\log_3\left(\dfrac{x^2 \sqrt[3]{y}}{w}\right)$ $2 \log_3 x + \dfrac{1}{3} \log_3 y - \log_3 w$ Write as one logarithm. $5 \log_6 x - 2 \log_6 w - \dfrac{1}{4} \log_6 z$ $\log_6\left(\dfrac{x^5}{w^2 \sqrt[4]{z}}\right)$ Simplify. $\log 10^5 + \log_3 3 + \log_5 1$ $= 5 \log 10 + \log_3 3 + \log_5 1$ $= 5 + 1 + 0$ $= 6$
Finding logarithms, p. 638.	Use a scientific calculator. $\log x = \log_{10} x$, for all $x > 0$ $\ln x = \log_e x$, for all $x > 0$	Find $\log 3.82$. 3.82 $\boxed{\log}$ $\log 3.82 \approx 0.5820634$ Find $\ln 52.8$. 52.8 $\boxed{\ln}$ $\ln 52.8 \approx 3.9665112$

Topic	Procedure	Examples
Finding antilogarithms, p. 639.	If $\log x = b$, then $10^b = x$. If $\ln x = b$, then $e^b = x$. Use a calculator or a table to evaluate.	Find x if $\log x = 2.1416$. $$10^{2.1416} = x$$ 2.1416 $\boxed{10^x}$ 138.54792 Find x if $\ln x = 0.6218$. $$e^{0.6218} = x$$ 0.6218 $\boxed{e^x}$ 1.8622771
Finding a logarithm to a different base, p. 641.	Change of base formula: $\log_b x = \dfrac{\log_a x}{\log_a b}$, where $a, b, x > 0$ but $a \neq 1$, $b \neq 1$	Evaluate. $\log_7 1.86$ $$\dfrac{\log 1.86}{\log 7}$$ 1.86 $\boxed{\log}$ $\boxed{\div}$ 7 $\boxed{\log}$ $\boxed{=}$ 0.3189132
Solving logarithmic equations, p. 645.	1. If an equation contains some logarithms and some terms without logarithms, try to change the logarithms to one single logarithm on one side and one numerical value on the other. Then convert the equation to exponential form. 2. If an equation contains only logarithms, try to get only one logarithm on each side of the equation. Then use the property that if $\log_b M = \log_b N$ then $M = N$. *Note:* Always check your solutions when solving logarithmic equations.	1. Solve for x. $\log_5 3x - \log_5(x^2 - 1) = \log_5 2$ $$\log_5 3x = \log_5 2 + \log_5(x^2 - 1)$$ $$\log_5 3x = \log_5 2(x^2 - 1)$$ $$3x = 2x^2 - 2$$ $$0 = 2x^2 - 3x - 2$$ $$0 = (2x + 1)(x - 2)$$ $$2x + 1 = 0 \qquad x - 2 = 0$$ Not valid $\quad x = -\dfrac{1}{2} \qquad x = 2$ ✓ valid *Check.* $x = 2$: $\log_5 3(2) - \log_5(2^2 - 1) \overset{?}{=} \log_5 2$ $$\log_5 6 - \log_5 3 \overset{?}{=} \log_5 2$$ $$\log_5\left(\dfrac{6}{3}\right) \overset{?}{=} \log_5 2$$ $$\log_5 2 = \log_5 2 \;\checkmark$$ $x = -\dfrac{1}{2}$: For the expression $\log_5(3x)$ we would obtain $\log_5(-1.5)$. You cannot take a log of a negative number. $x = -\dfrac{1}{2}$ is not a solution.
Solving exponential equations, p. 647.	1. See if each expression can be written so that only one base appears on one side of the equation and the same base appears on the other side. Then use the property that if $b^x = b^y$ then $x = y$. 2. If you can't do step 1, take the logarithm of each side of the equation and use the properties of logarithms to solve for the variable.	1. Solve for x. $2^{x-1} = 7$ $$\log 2^{x-1} = \log 7$$ $$(x - 1)\log 2 = \log 7$$ $$x \log 2 - \log 2 = \log 7$$ $$x \log 2 = \log 7 + \log 2$$ $$x = \dfrac{\log 7 + \log 2}{\log 2}$$ (We can approximate the answer as $x \approx 3.8073549$.)

Graph the function in Problems 1 and 2.

1. $f(x) = 4^{3+x}$

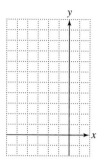

2. $f(x) = e^{x-3}$

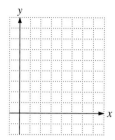

3. Solve. $5^{x+2} = 125$

4. Change to logarithmic form. $\dfrac{1}{32} = 2^{-5}$

5. Write in exponential form. $-3 = \log_{10}(0.001)$

Solve.

6. $\log_3 x = -2$

7. $\log_w 16 = 4$

8. $\log_7 w = -1$

9. $\log_8 x = 0$

10. $\log_w 27 = 3$

11. $\log_{10} w = -3$

12. $\log_{10} 1000 = x$

13. $\log_2 64 = x$

14. $\log_2\left(\dfrac{1}{4}\right) = x$

15. $\log_5 125 = x$

16. Graph the equation. $\log_3 x = y$

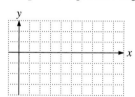

Write each expression as the sum or difference of $\log_2 x$, $\log_2 y$, $\log_2 z$.

17. $\log_2 x^3 \sqrt{y}$

18. $\log_2\left(\dfrac{5x}{\sqrt{w}}\right)$

Write as a single logarithm.

19. $\log_3 x + \log_3 w^{1/2} - \log_3 2$

20. $4 \log_8 w - \dfrac{1}{3} \log_8 z$

21. Evaluate. $\log_e e^6$

Solve.

22. $\log_8 x + \log_8 3 = \log_8 75$

23. $\log_5 100 - \log_5 x = \log_5 4$

Find the value with a scientific calculator.

24. log 23.8   **25.** log 0.0817   **26.** ln 3.92   **27.** ln 803

28. Find *n* if log *n* = 1.1367.   **29.** Find *n* if ln *n* = 1.7.   **30.** $\log_8 2.81$

Solve the equation and check your solution.

31. $\log_7(x + 3) + \log_7(5) = 2$

32. $\log_3(2x + 3) = \log_3(2) - 3$

33. $2\log_3(x + 3) - \log_3(x + 1) = 3\log_3 2$

34. $\log_5(x + 1) - \log_5 8 = \log_5 x$

35. $\log_5(x + 1) + \log_5(x - 3) = 1$

36. $\log_2(x - 2) + \log_2(x + 5) = 3$

37. $\log(2t + 3) + \log(4t - 1) = 2\log 3$

38. $\log(2t + 4) - \log(3t + 1) = \log 6$

Solve each equation. Leave your answer in exact form. Do not approximate.

39. $3^x = 14$

40. $5^x = 4^{x+2}$

41. $e^{2x} = 30.6$

42. $16e^{x+1} = 56$

Solve the equation. Round your answer to the nearest ten-thousandth.

43. $2^{3x+1} = 5^x$ **44.** $3^{x+1} = 7$ **45.** $e^{3x-4} = 20$ **46.** $(1.03)^x = 20$

Use $A = P(1 + r)^t$ for Problems 47 through 50, the formula for problems involving compound interest that is compounded annually.

47. How much money would Chou Lou have after 4 years if he invested $5000 at 6% annual interest?

48. How long will it take Frances to double her money if the annual interest rate is 8%? (Round your answer to nearest year.)

49. Melinda invested $12,000 at 7% annual interest. How many years will it take to amount to $20,000? (Round your answer to the nearest year.)

★ **50.** Robert invested $3500 at 5% annual interest. His brother invested $3500 at 6% annual interest. How many years would it take for Robert's amount to be $500 less than his brother's amount? (Round your answer to the nearest year.)

The growth of the world's population can be measured by the equation $A = A_0 e^{rt}$, where t is the number of years, A_0 is the population of the world at time $t = 0$, r is the rate of population increase per year, and A is the population at time t. Round your answers to the nearest whole year. Use this information to solve Problems 51 through 54.

51. How long will it take a population of 7 billion to increase to 16 billion if $r = 2\%$ per year?

52. How long will it take a population of 6 billion to increase to 10 billion if $r = 2\%$ per year?

53. A town is growing at the rate of 8% per year. How long will it take the town to grow from 40,000 to 95,000 in population?

54. The number of moose in Northern Maine is increasing at a rate of 3% per year. It is estimated in one county that there are now 2000 moose. How many years will it be until there are 2600 moose in that county if the growth rate remains unchanged?

55. An earthquake's magnitude is given by $M = \log\left(\dfrac{I}{I_0}\right)$, where I is the intensity of the earthquake and I_0 is the minimum measurable intensity. The 1964 earthquake in Anchorage, Alaska, had a magnitude of 8.4. The 1975 earthquake in Turkey had a magnitude of 6.7. How much more energy was released from the Alaska earthquake than from the Turkey earthquake?

56. The work W done by a volume of gas expanding at a constant temperature from volume V_0 to volume V_1 is given by $W = p_0 V_0 \ln\left(\dfrac{V_1}{V_0}\right)$, where p_0 is the pressure at volume V_0.
   (a) Find W when $p_0 = 40$ pounds per cubic inch, $V_0 = 15$ cubic inches, and $V_1 = 24$ cubic inches.
   (b) If the amount of work is 100 pound-inch, $V_0 = 8$ cubic inches, and $V_1 = 40$ cubic inches, find p_0.

1. Graph. $f(x) = 3^{4-x}$

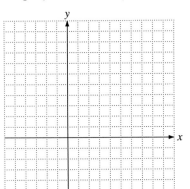

2. Solve. $4^{x+3} = 64$

In Problems 3 and 4, solve for the variable.

3. $\log_8 x = -2$

4. $\log_w 125 = 3$

5. Write as a single logarithm. $\log_8 x + \log_8 w - \dfrac{1}{4}\log_8 3$

Evaluate using a calculator. Round your answer to nearest ten-thousandth.

6. $\log 23.6$

7. $\ln 5.99$

8. $\log_3 1.62$

1. _____

2. _____

3. _____

4. _____

5. _____

6. _____

7. _____

8. _____

Use a scientific calculator to approximate x.

9. $\log x = 3.7284$

10. $\ln x = 0.14$

Solve the equation and check your solution for Problems 11 *and* 12.

11. $\log_8 2x + \log_8 6 = 2$

12. $\log_8(x + 3) - \log_8 2x = \log_8 4$

13. Find the exact value of x.
$29 = 116e^{3x+1}$ Do not approximate.

14. Solve. $5^{3x+6} = 17$
Approximate your answer to the nearest ten-thousandth.

15. How much money will Henry have if he invests \$2000 for 5 years at 8% annual interest compounded annually?

16. How long will it take for Barb to double her money if she invests it at 5% compounded annually? Round to the nearest whole year.

Approximately one-half of this test covers the content of Chapters 1–10. The remainder covers the content of Chapter 11.

1. Evaluate. $2(-3) + 12 \div (-2) + 3\sqrt{36}$

2. Solve for *x*. $H = 3bx - 2ay$

3. Graph. $y = -\dfrac{2}{3}x + 4$

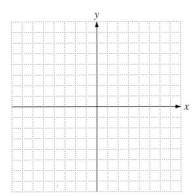

4. Factor. $5ax + 5ay - 7wx - 7wy$

5. Solve for (x, y, z).

$$3x - y + z = 6$$
$$2x - y + 2z = 7$$
$$x + y + z = 2$$

6. Simplify. $(5\sqrt{2} + \sqrt{3})(\sqrt{5} - 2\sqrt{6})$

7. Solve. $x^4 - 5x^2 - 6 = 0$
Express imaginary solutions in *i* notation.

8. Solve for *x* and *y*.

$$2x - y = 4$$
$$4x - y^2 = 0$$

9. Solve. $2x - 3 = \sqrt{7x - 3}$

10. Rationalize the denominator.

$$\frac{5}{\sqrt[3]{2xy^2}}$$

11. Graph. $f(x) = 2^{3-2x}$

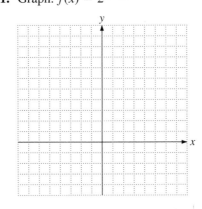

1. _____

2. _____

3. _____

4. _____

5. _____

6. _____

7. _____

8. _____

9. _____

10. _____

11. _____

12. _____	
13. _____	
14. _____	
15. _____	
16. _____	
17. _____	
18. _____	
19. _____	
20. _____	
21. _____	
22. _____	

Solve for the variable.

12. $5^{2x-1} = 25$

13. $\log_x\left(\dfrac{1}{64}\right) = 3$

Evaluate using a calculator or a table.

14. $\log 2.53$

15. Find x if $\log x = 1.8209$.

16. $\log_3 7$

17. Find x if $\ln x = 1.9638$.

Solve the equation.

18. $\log_9 x = 1 - \log_9(x - 8)$

19. $\log_5 x = \log_5 2 + \log_5(x^2 - 3)$

Use a calculator.

20. $3^{x+2} = 5$ Approximate to nearest thousandth.

21. $33 = 66e^{2x}$
Leave your answer as exact. Do not approximate.

22. How much money will Frank and Linda have in 4 years if they invest $3000 at 9% annual interest compounded annually?

Practice Final Examination

Please review the content areas of Chapters 1 through 11. Then try to solve the problems in this Practice Final Examination.

Chapter 1

1. Evaluate. $(4 - 3)^2 + \sqrt{9} \div (-3) + 4$

2. Simplify. $\left(\dfrac{2x^3y^{-2}}{3x^4y^{-3}}\right)^{-2}$

3. Simplify.
$5a - 2ab - 3a^2 - 6a - 8ab + 2a^2$

4. Simplify.
$3[2x - 5(x + y)]$

5. $F = \dfrac{9}{5}C + 32$
Find F when $C = -35$.

Chapter 2

6. Solve for y. $\dfrac{1}{3}y - 4 = \dfrac{1}{2}y + 1$

7. Solve for b. $A = \dfrac{1}{2}a(b + c)$

8. Solve for x. $\left|\dfrac{2}{3}x - 4\right| = 2$

9. A man invested \$4000, part at 12% interest and part at 14% interest. After one year he had earned \$508 in interest. How much was invested at each interest rate?

10. A piece of land is rectangular and has a perimeter of 1760 meters. The length is 200 meters less than twice the width. Find the dimensions of the land.

11. Solve for x. $2x - 3 < x - 2(3x - 2)$

12. Find the value of x that satisfies the given conditions.
$x + 5 \le -4$ or $2 - 7x \le 16$

13. Solve the inequality. $|2x - 5| < 10$

Chapter 3

14. Find the intercepts and then graph the line $7x - 2y = -14$.

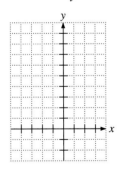

15. Graph the region. $3x - 4y \le 6$

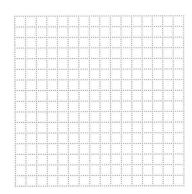

16. Find the slope of the line passing through $(1, 5)$ and $(-2, -3)$.

17. Write the equation of a line in standard form that is parallel to $3x + 2y = 8$ and passes through $(-1, 4)$.

Given the function defined by $f(x) = 3x^2 - 4x - 3$ find the following.

18. $f(3)$

19. $f(-2)$

1. _____

2. _____

3. _____

4. _____

5. _____

6. _____

7. _____

8. _____

9. _____

10. _____

11. _____

12. _____

13. _____

14. _____

15. _____

16. _____

17. _____

18. _____

19. _____

20. Graph the function. $f(x) = |2x - 4|$

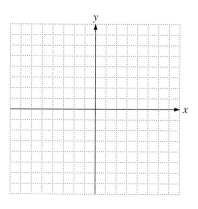

Chapter 4

21. Solve for x and y.

$$\frac{1}{2}x + \frac{2}{3}y = 1$$

$$\frac{1}{3}x + y = -1$$

22. Solve for x and y.

$$4x - 3y = 12$$
$$3x - 4y = 2$$

23. Solve for x, y, and z.

$$2x + 3y - z = 16$$
$$x - y + 3z = -9$$
$$5x + 2y - z = 15$$

24. Solve for x, y, and z.

$$y + z = 2$$
$$x + z = 5$$
$$x + y = 5$$

25. Solve for z only by using Cramer's rule.

$$2x - y + 5z = -2$$
$$x + 3y - z = 6$$
$$4x + y + 3z = -2$$

26. Graph the region. $3y \geq 8x - 12$
$$2x + 3y \leq -6$$

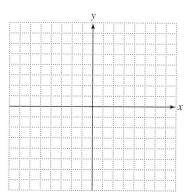

Chapter 5

27. Multiply and simplify your answer.
$(3x - 2)(2x^2 - 4x + 3)$

28. Divide. $(25x^3 + 9x + 2) \div (5x + 1)$

Factor completely the following.

29. $8x^3 - 27$

30. $x^3 + 2x^2 - 4x - 8$

31. $2x^3 + 15x^2 - 8x$

32. Solve for x.
$$x^2 + 15x + 54 = 0$$

Chapter 6

Simplify the following.

33. $\dfrac{9x^3 - x}{3x^2 - 8x - 3}$

34. $\dfrac{x^2 - 9}{2x^2 + 7x + 3} \div \dfrac{x^2 - 3x}{2x^2 + 11x + 5}$

35. $\dfrac{3x}{x + 5} - \dfrac{2}{x^2 + 7x + 10}$

36. $\dfrac{\dfrac{3}{2x + 1} + 2}{1 - \dfrac{2}{4x^2 - 1}}$

20. _____

21. _____

22. _____

23. _____

24. _____

25. _____

26. _____

27. _____

28. _____

29. _____

30. _____

31. _____

32. _____

33. _____

34. _____

35. _____

36. _____

37. Solve for x. $\dfrac{x-1}{x^2-4} = \dfrac{2}{x+2} + \dfrac{4}{x-2}$

Chapter 7

38. Simplify. $\dfrac{5x^{-4}y^{-2}}{15x^{-1/2}y^3}$

39. Simplify. $\sqrt[3]{40x^4y^7}$

40. Combine like terms.
$5\sqrt{2} - 3\sqrt{50} + 4\sqrt{98}$

41. Rationalize the denominator.
$$\dfrac{2\sqrt{3}+1}{3\sqrt{3}-\sqrt{2}}$$

42. Simplify and add together.
$i^3 + \sqrt{-25} + \sqrt{-16}$

43. Solve for x and check your solutions.
$\sqrt{x+7} = x+5$

44. If y varies directly with the square of x and $y = 15$ when $x = 2$, what will y be when $x = 3$?

Chapter 8

45. Solve for x. $5x(x+1) = 1 + 6x$

46. Solve for x. $5x^2 - 9x = -12x$

47. Solve for x. $x^{2/3} + 5x^{1/3} - 14 = 0$

48. The area of a rectangle is 52 square centimeters. The length of the rectangle is 1 meter longer than 3 times its width. Find the dimensions of the rectangle.

49. Graph the quadratic function $f(x) = -x^2 - 4x + 5$. Label the vertex and the intercepts.

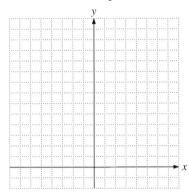

50. Solve. $3x^2 - 11x - 4 \geq 0$

Chapter 9

51. Place the equation of the circle in standard form. Find its center and radius.
$x^2 + y^2 + 6x - 4y = -9$

Identify and graph.

52. $\dfrac{x^2}{16} + \dfrac{y^2}{25} = 1$

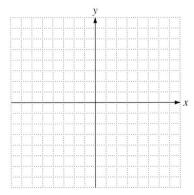

53. $\dfrac{x^2}{4} - \dfrac{y^2}{9} = 1$

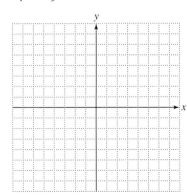

37. _____

38. _____

39. _____

40. _____

41. _____

42. _____

43. _____

44. _____

45. _____

46. _____

47. _____

48. _____

49. _____

50. _____

51. _____

52. _____

53. _____

54. $x = (y - 3)^2 + 5$

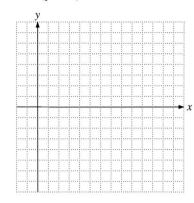

55. Solve the following system of equations.
$$x^2 + y^2 = 16$$
$$x^2 - y = 4$$

Chapter 10

56. Let $f(x) = 3x^2 - 2x + 5$.
  (a) Find $f(-1)$.
  (b) Find $f(a)$.
  (c) Find $f(a + 2)$.

57. If $f(x) = 5x^2 - 3$ and $g(x) = -4x - 2$ find $f[g(x)]$.

58. If $f(x) = \dfrac{1}{2}x - 7$,
  find $f^{-1}(x)$.

59. Graph on one axis. $f(x)$, $f(x + 2)$, and $f(x) - 3$, if $f(x) = |x|$.

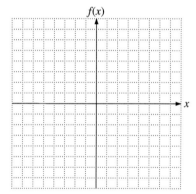

Chapter 11

60. Graph $f(x) = 2^{1-x}$. Plot 3 points.

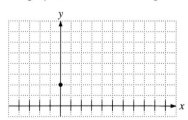

Solve for the variable.

61. $\log_5 x = -4$

62. $\log_4(3x + 1) = 3$

63. $\log_{10} 0.01 = y$

64. $\log_2 6 + \log_2 x = 4 + \log_2(x - 5)$

54.

_____

55.

_____

56.

_____

57.

_____

58.

_____

59.

_____

60.

_____

61.

_____

62.

_____

63.

_____

64.

_____

Glossary

Absolute value inequalities (2.8) Inequalities that contain at least one absolute value expression.

Absolute value of a number (1.2) The distance between the number and 0 on the number line. The absolute value of a number x is written as $|x|$. The absolute value can be determined by

$$|x| = \begin{cases} x, & \text{if } x \geq 0 \\ -x, & \text{if } x < 0 \end{cases}$$

Algebraic expression (1.5) A collection of numerical values, variables, and operation symbols is called an algebraic expression. $3x - 6y + 3yz$ and $\sqrt{5xy}$ are algebraic expressions.

Algebraic fraction (6.1) An expression of the form P/Q, where P and Q are polynomials and Q is not zero. Algebraic fractions are also called rational expressions. For example, $\frac{x+3}{x-4}$ and $\frac{5x^2+1}{6x^3-5x}$ are algebraic fractions.

Approximate value (1.3) A value that is not exact. The approximate value of $\sqrt{3}$ correct to the nearest tenth is 1.7. The symbol $\approx$ is used to indicate "is approximately equal to."

Associative property of addition (1.1) For all real numbers a, b, and c. $a + (b + c) = (a + b) + c$.

Associative property of multiplication (1.1) For all real numbers a, b, c, $a(bc) = (ab)c$.

Asymptote (9.4) A line that a curve continues to approach but never actually touches. Often an asymptote is a helpful reference in making a sketch of a curve, such as a hyperbola.

Augmented matrix (Appendix B) An augmented matrix is a matrix derived from a linear system of equations. It consists of the coefficients of each variable in a linear system and the constants. The augmented matrix of the system $\begin{matrix} -3x + 5y = -22 \\ 2x - y = 10 \end{matrix}$ is the matrix $\left[\begin{array}{cc|c} -3 & 5 & -22 \\ 2 & -1 & 10 \end{array} \right]$. Each row of the augmented matrix represents an equation of the system.

Axis of symmetry of a parabola (9.2) A line passing through the focus and the vertex of a parabola about which the two sides of the parabola are symmetric. See the sketch.

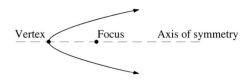

Vertex   Focus   Axis of symmetry

Base (1.3) The number or variable that is raised to a power. In the expression 2^3, the number 2 is the base.

Base of an exponential function (11.1) The number b is called the base of the function $f(x) = b^x$.

Binomial (5.1) A polynomial of two terms. For example, $z^2 - 9$ is a binomial.

Cartesian coordinate system (3.1) Another name for the rectangular coordinate system named after its inventor, Rene Descartes.

Circle (1.6) A geometric figure that consists of a collection of points that are of equal distance from a fixed point called the center.

Center

Circumference of a circle (1.6) The distance around a circle. The circumference of a circle is given by the formulas $C = \pi d$ and $C = 2\pi r$, where d is the diameter of the circle and r is the radius of the circle.

Closure property of addition (1.1) For all real numbers a, b, the sum $a + b$ is a real number.

Closure property of multiplication (1.1) For all real numbers a, b, the product ab is a real number.

Coefficient (1.5) Any factor or group of factors in a term may be called the coefficient of the term. In the term $8xy$, the coefficient of xy is 8. However, the coefficient of x is $8y$. In the term $abcd$, the coefficient of $abcd$ is 1.

Collect like terms (1.5) The process of adding and subtracting like terms. If we collect like terms in the expression $5x - 8y - 7x - 12y$, we obtain $-2x - 20y$.

Combined variation (7.7) If y varies directly with x and z and inversely with d^2, we can write the equation $y = kxz/d^2$, where k is the constant of variation.

Common denominator (1.2) Two fractions that have the same number in the denominator are said to have a common denominator. The fractions $\frac{4}{13}$ and $\frac{7}{13}$ have a common denominator of 13.

Common logarithm (11.4) The common logarithm of a number x is given by $\log x = \log_{10} x$ for all $x > 0$. A common logarithm is a logarithm using base 10.

Commutative property of addition (1.1) For all real numbers a, b, $a + b = b + a$.

Commutative property of multiplication (1.1) For all real numbers a, b, $ab = ba$.

Complex fraction (also called a Complex Rational Expression) (6.3) A fraction made up of polynomials or numerical values in which the numerator or the denominator contains at least one fraction. Examples of complex fractions are

$$\frac{\dfrac{1}{3} + \dfrac{1}{5}}{\dfrac{2}{7}} \quad \text{and} \quad \frac{\dfrac{1}{x} + 3}{2 + \dfrac{5}{x}}$$

Complex number (7.6) A number that can be written in the form $a + bi$, where a and b are real numbers and $i = \sqrt{-1}$.

Compound inequalities (2.7) Two inequality statements connected together by the word "and" or by the word "or."

Conjugate of a binomial with radicals (7.4) The expressions $a\sqrt{x} + b\sqrt{y}$ and $a\sqrt{x} - b\sqrt{y}$ are called conjugates. The conjugate of $2\sqrt{3} + 5\sqrt{2}$ is $2\sqrt{3} - 5\sqrt{2}$. The conjugate of $4 - \sqrt{x}$ is $4 + \sqrt{x}$.

Conjugate of a complex number (7.6) The expressions $a + bi$ and $a - bi$ are called conjugates. The conjugate of $5 + 2i$ is $5 - 2i$. The conjugate of $7 - 3i$ is $7 + 3i$.

Coordinates of a point (3.1) An ordered pair of numbers (x, y) that specifies the location of a point on a rectangular coordinate system.

Counting numbers (1.1) The counting numbers are the natural numbers. They are the numbers in the infinite set $\{1, 2, 3, 4, 5, 6, 7, \ldots \}$.

Critical points of a quadratic inequality (8.6) The critical points of a quadratic inequality of the form $ax^2 + bx + c > 0$ or $ax^2 + bx + c < 0$ are those points where $ax^2 + bx + c = 0$.

The degree of polynomial (5.1) The degree of the highest-degree term in the polynomial. The polynomial $5x^3 + 4x^2 - 3x + 12$ is of degree 3.

The degree of a term (5.1) The sum of the exponents of the term's variables. The term $5x^2y^2$ is of degree 4.

Denominator (6.1) The bottom expression in a fraction. The denominator of $\frac{5}{11}$ is 11. The denominator of $\frac{x-7}{x+8}$ is $x + 8$.

Descending order for a polynomial (5.1) A polynomial is written in descending order if the term of the highest degree is first, the term of the next-to-highest degree is second, and so on, with each succeeding term of less degree. The polynomial $5y^4 - 3y^3 + 7y^2 + 8y - 12$ is in descending order.

Determinant (4.4) A square array of numbers written between vertical lines. For example $\begin{vmatrix} 1 & 5 \\ 2 & 4 \end{vmatrix}$ is a 2×2 determinant. It is also called a second-order determinant. $\begin{vmatrix} 1 & 7 & 8 \\ 2 & -5 & -1 \\ -3 & 6 & 9 \end{vmatrix}$ is a 3×3 determinant. It is also called a third-order determinant.

Different signs (1.2) When one number is positive and one number is negative, the two numbers are said to have different signs. The numbers 5 and -9 have different signs.

Direct variation (7.7) If a variable y varies directly with x, we can write the equation $y = kx$, where k represents some real number that will stay the same over a range of problems. This value k is called the constant of variation.

Discriminant of a quadratic equation (8.2) In the equation $ax^2 + bx + c = 0$ and $a \neq 0$, the expression $b^2 - 4ac$ is called the discriminant. It can be used to determine the nature of the roots of the quadratic equation. If the discriminant is *positive*, there are two rational or irrational roots. The two roots will be rational only if the discriminant is a perfect square. If the discriminant is *zero*, there is only one rational root. If the discriminant is *negative*, there are two complex roots.

Distance between two points (9.1) The distance between point (x_1, y_1) and point (x_2, y_2) is given by the formula $d = \sqrt{(x_2 - x_1)^2 + (y_2 - y_1)^2}$.

Distributive property of multiplication over addition (1.1) For any real numbers $a, b, c, \quad a(b + c) = ab + ac$.

Dividend (5.2) The expression that is being divided by another. In the problem $12 \div 4 = 3$, the dividend is 12. In the problem $x - 5\overline{)5x^2 + 10x - 3}$, the dividend is $5x^2 + 10x - 3$.

Divisor (5.2) The expression that is divided into another. In the problem $12 \div 4 = 3$ the divisor is 4. In the problem $x + 3\overline{)2x^2 - 5x - 14}$, the divisor is $x + 3$.

Domain of a relation or a function (3.5) If all the ordered pairs of the relation or the function are listed, all the different first items of each pair is called the domain.

e (11.1) The number e is an irrational number. It can be approximated by the value $e \approx 2.7183$.

Elements (1.1) The objects that are in a set are called elements.

Ellipse (9.3) The set of points in a plane such that for each point in the set the sum of its distances to two fixed points is constant. Each of the fixed points is called a focus. Each of the following graphs is an ellipse.

Equation (2.1) A mathematical statement that two quantities are equal.

Equations that are quadratic in form (8.3) An equation that can be transformed to an equation that is quadratic. Examples of equations that are quadratic in form are $9x^4 - 25x^2 + 16 = 0$ and $8x^{1/2} + 7x^{1/4} - 1 = 0$.

Equilateral hyperbola (9.4) A hyperbola for which $a = b$ in the equation of the hyperbola.

Equivalent equations (2.1) Equations that have the same solution(s).

Even integers (1.3) Integers that are exactly divisible by 2, such as $\ldots, -4, -2, 0, 2, 4, 6, \ldots$.

Exponent (1.3) The number that indicates the power of a base. If the number is a positive integer, it tells us how many factors of the base occur. In the expression 2^3, the exponent is 3. The number 3 tells us that there are 3 factors each of which is 2 since $2^3 = 2 \cdot 2 \cdot 2$. If an exponent is negative, use the property that $x^{-n} = \frac{1}{x^n}$. If an exponent is zero, use the property that $x^0 = 1$, where $x \neq 0$.

Exponential function (11.1) The exponential function is $f(x) = b^x$, where $b > 0$, $b \neq 1$, and x is any real number.

Expression (1.3) Any combination of mathematical operation symbols with numbers or variables or both is a mathematical expression. Examples of mathematical expressions are $2x + 3y - 6z$ and $\sqrt{7xyz}$.

Extraneous solution to an equation (6.4) A correctly obtained potential solution to an equation, which when substituted back into the original equation does not yield a true statement. For example, $x = 2$ is an extraneous solution to the equation

$$\frac{x}{x - 2} - 4 = \frac{2}{x - 2}$$

An extraneous solution is also called an extraneous root.

Factor (1.5 and 5.4) When two or more numbers, variables, or algebraic expressions are multiplied, each of them is called a factor. In the expression $5st$, the factors are 5, s, and t. In the expression $(x - 6)(x + 2)$, the factors are $(x - 6)$ and $(x + 2)$.

First-degree equation (2.1) A mathematical equation such as $2x - 8 = 4y + 9$ or $7x = 21$ in which each variable has an exponent of 1. It is also called a linear equation.

First-degree equation in one unknown (2.1) An equation such as $x = 5 - 3x$ or $12x - 3(x + 5) = 22$ in which only one kind of variable appears and that variable has an exponent of 1. It is also called a linear equation in one variable.

Focus point of a parabola (9.2) The focus point of a parabola has many properties. For example, the focus point of a parabolic mirror is the point to which all incoming light rays that are parallel to the axis of symmetry will collect. A parabola is a set of points that is the same distance from a fixed line called the directrix and a fixed point. This fixed point is the focus.

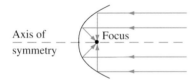

Axis of symmetry   Focus

Formula (1.6) A rule for finding the value of a variable when the values of other variables in the expression are known. For example, the formula for finding the Fahrenheit temperature when the Celsius temperature is known is $F = 1.8C + 32$.

Fractional equation (6.4) An equation that contains a rational expression. Examples of fractional equations are

$$\frac{x}{3} + \frac{x}{4} = 7 \quad \text{and} \quad \frac{2}{3x - 3} + \frac{1}{x - 1} = \frac{-5}{12}$$

Function (3.5) A relation in which no ordered pairs have the same first coordinates.

Graph of a function (3.5, 10.2) A graph in which a vertical line will never cross in more than one place. The following sketches represent the graphs of functions.

Graph of a linear inequality in two variables (3.4) A shaded region in two-dimensional space. It may or may not include the boundary line. If the line is to be included, the sketch will show a solid line. If it is not to be included, the sketch will show a dashed line. Two sketches are shown next.

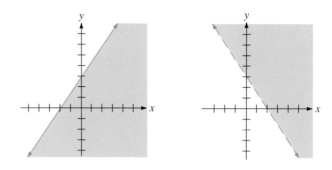

Graph of a one-to-one function (10.4) A graph of a function that has the additional property that a horizontal line will never cross the graph in more than one place. The following sketches represent the graphs of one-to-one functions.

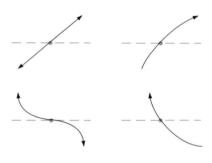

Greater than or equal to symbol (2.6) The symbol $\geq$ means greater than or equal to.

Greater than symbol (1.2, 2.6) The $>$ symbol means greater than. $5 > 3$ is read "5 is greater than 3."

Greatest common factor of a polynomial (5.4) The greatest common factor of a polynomial is a common factor of each term of the polynomial that has the largest possible numerical coefficient and the largest possible exponent for each variable. For example, the greatest common factor of $50x^4y^5 - 25x^3y^4 + 75x^5y^6$ is $25x^3y^4$.

Higher-order equations (8.3) Equations of degree 3 or higher are called higher-order equations. Examples of higher-order equations are

$$x^4 - 29x^2 + 100 = 0, \qquad x^3 + 3x^2 - 4x - 12 = 0$$

Higher-order roots (7.2) Cube roots, fourth roots, and roots with an index greater than 2 are called higher-order roots.

Horizontal line (3.1) A straight line that is parallel to the x-axis. A horizontal line has a slope of zero. The equation of any horizontal line can be written in the form $y = b$, where b is a constant. A sketch is shown.

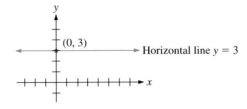

Horizontal line $y = 3$

Horizontal parabolas (9.2) Parabolas that open to the right or to the left. The following graphs represent horizontal parabolas.

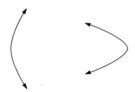

Hyperbola (9.4) The set of points in a plane such that for each point in the set the absolute value of the difference of its distances to two fixed points is constant. Each of these fixed points is called a focus. Each of the following sketches represents the graph of a hyperbola.

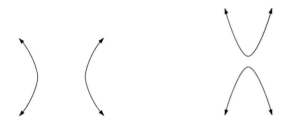

Hypotenuse of a right triangle (8.4) The side opposite the right angle in any right triangle. The hypotenuse is always the longest side of a triangle. In the following sketch the hypotenuse is side c.

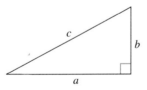

Identity property for addition (1.1) For any real number a, $a + 0 = a = 0 + a$.

Identity property for multiplication (1.1) For any real number a, $a(1) = a = 1(a)$.

Imaginary number (7.6) The imaginary number i is defined as $i = \sqrt{-1}$ and $i^2 = -1$.

Inconsistent system of equations (4.1) A system of equations for which no solution is possible.

Index of a radical (7.2) The index of a radical indicates what type of a root is being taken. The index of a cube root is 3. When we write $\sqrt[3]{x}$, the 3 is called the index of the radical. When we write $\sqrt[4]{y}$, the index is 4. The index of a square root is 2, but the index is not written in the square root symbol, such as $\sqrt{x}$.

Inequality (2.6) A mathematical statement expressing an order relationship. The following are inequalities:

$$x < 3, \qquad x \geq 4.5, \qquad 2x + 3 \leq 5x - 7$$
$$x + 2y < 8, \qquad 2x^2 - 3x > 0$$

Infinite set (1.1) A set that has no end to the numbers of elements that are within it. An infinite set is often indicated by placing an ellipsis (. . .) after listing some of the elements of the set.

Integers (1.1) The integers are the numbers in the infinite set

$$\{ \ldots, -3, -2, -1, 0, 1, 2, 3, \ldots \}$$

Interest (2.5) The charge made for borrowing money or the income received from investing money. Simple interest is calculated by the formula $I = prt$, where p is the principal that is borrowed, r is the rate of interest, and t is the amount of time the money is borrowed.

Inverse function of a one-to-one function (10.4) That function obtained by interchanging the first and second coordinates in each ordered pair of the function.

Inverse property for addition (1.1) For any real number a, $a + (-a) = 0 = (-a) + a$.

Inverse property for multiplication (1.1) For any real number $a \neq 0$, $a\left(\frac{1}{a}\right) = 1 = \left(\frac{1}{a}\right)a$.

Inverse variation (7.7) If a variable y varies inversely with x, we can write the equation $y = \frac{k}{x}$, where k is the constant of variation.

Irrational numbers (1.1) Numbers whose decimal forms are nonterminating and nonrepeating. The numbers π, e, $\sqrt{2}$, and $1.56832574 \ldots$ are irrational numbers.

Joint variation (7.7) If a variable y varies jointly with x and z, we can write the equation $y = kxz$, where k is the constant of variation.

Least common denominator of algebraic fraction (6.2) The least common denominator (LCD) of two or more algebraic fractions is a polynomial that is exactly divisible by each denominator. The LCD is the product of all the *different prime factors*. If a factor occurs more than once in a denominator, you must use the highest power of that factor. For example, the LCD of $\dfrac{5}{2(x + 2)(x - 3)^2}$ and $\dfrac{3}{(x - 3)^4}$ is $2(x + 2)(x - 3)^4$. The LCD of $\dfrac{5}{(x + 2)(x - 3)}$ and $\dfrac{7}{(x - 3)(x + 4)}$ is $(x + 2)(x - 3)(x + 4)$.

Least common denominator of numerical fractions (2.1) The smallest whole number that is exactly divisible by all the denominators of a group of fractions. The least common denominator (LCD) of $\frac{1}{7}, \frac{9}{21}, \frac{3}{14}$ is 42. The number 42 is the smallest number that can be exactly divided by 7, by 21, and by 14. The least common denominator is sometimes called the lowest common denominator.

Leg of a right triangle (8.4) One of the two shortest sides of a right triangle. In the following sketch, sides a and b are the legs of the right triangle.

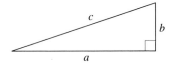

Less than or equal to symbol (2.6) The symbol $\leq$ means less than or equal to.

Less than symbol (1.2) The $<$ symbol means less than. $2 < 8$ is read ''2 is less than 8.''

Like terms (1.5) Terms that have identical variables and also have identical exponents. In the mathematical expression $5x - 8syz + 7x + 15syz$, the terms $5x$ and $7x$ are like terms and the terms $-8syz$ and $15syz$ are like terms.

Linear equation (2.1) A mathematical equation such as $3x + 7 = 5x - 2$ or $5x + 7y = 9$ in which each variable has an exponent of 1.

Linear inequality (2.6) An inequality statement in which each variable has an exponent of 1 and no variables are in the denominator. Some examples of linear inequalities are

$$2x + 3 > 5x - 6, \qquad y < 2x + 1, \qquad x < 8$$

Literal equation (2.2) An equation that has other variables in it besides the variable for which you wish to solve. $I = prt$, $7x + 3y - 6z = 12$, and $P = 2w + 2l$ are examples of literal equations.

Logarithm (11.2) The logarithm of a positive number x is the power to which the base b must be raised to produce x. That is, $y = \log_b x$ is the same as $x = b^y$, where $b > 0$, $b \neq 1$. A logarithm is an exponent.

Logarithmic equation (11.2) An equation that contains at least one logarithm.

Magnitude of an earthquake (11.5) The magnitude of an earthquake is measured by the formula $M = \log\left(\frac{I}{I_0}\right)$, where I is the intensity of the earthquake and I_0 is the minimum measurable intensity.

Matrix (4.4) A matrix is a rectangular array of numbers arranged in rows and columns. We use the symbol [ ] to indicate a matrix. The matrix $\begin{bmatrix} 3 & 4 & 5 \\ 6 & 7 & 8 \end{bmatrix}$ has two rows and three columns and is called a 2×3 matrix.

Minor of an element of a third-order determinant (4.4) The second-order determinant that remains after you delete the row and column in which the element appears. The minor of the element 6 in the determinant $\begin{vmatrix} 1 & 2 & 3 \\ 7 & 6 & 8 \\ -3 & 5 & 9 \end{vmatrix}$ is the second-order determinant $\begin{vmatrix} 1 & 3 \\ -3 & 9 \end{vmatrix}$.

Monomial (5.1) A polynomial of one term. For example, $3a$ is a monomial.

Natural logarithm (11.4) The natural logarithm of a number x is $\ln x = \log_e x$ for all $x > 0$. A natural logarithm is a logarithm using base e.

Negative integers (1.1) The numbers in the infinite set
$$\{-1, -2, -3, -4, -5, -6, -7, \ldots\}.$$

Nonlinear system of equations (9.5) A system of equations in which at least one equation is not a linear equation.

Nonzero (1.4) A nonzero value is a value other than zero. If we say the variable x is nonzero, we mean that x cannot have the value of zero.

Numerator (6.1) The top expression in a fraction. The numerator of $\frac{3}{19}$ is 3. The numerator of $\frac{x + 5}{x^2 + 25}$ is $x + 5$.

Numerical coefficient (1.5) The numerical value multiplied by the variables in a term. The numerical coefficient of $-8xyw$ is -8. The numerical coefficient of abc is 1.

Odd integer (1.3) Integers that are not exactly divisible by 2, such as $\ldots, -3, -1, 1, 3, 5, 7, \ldots$.

One-to-one function (10.4) A function in which no two different ordered pairs have the same second coordinate.

Opposite of a number (1.2) That number with the same absolute value but a different sign. The opposite of -7 is 7. The opposite of 13 is -13.

Ordered pair (3.1) A pair of numbers represented in a specified order. An ordered pair is used to specify the location of a point. Every point on a rectangular coordinate system can be represented by an ordered pair (x, y).

Origin (3.1) The point determined by the intersection of the x-axis and the y-axis. It has the coordinates $(0, 0)$.

Parabola (9.2) The set of points that is the same distance from some fixed line (called the directrix) and some fixed point (called the focus) that is not on the line. The graph of any equation of the form $y = ax^2 + bx + c$ or $x = ay^2 + by + c$, where a, b, c are real numbers and $a \neq 0$, is a parabola. Some examples of the graphs of parabolas are shown below.

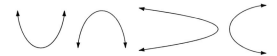

Parallel lines (3.2) Two straight lines are parallel if they never intersect. Parallel lines have the same slope.

Parallelogram (1.6) A four-sided geometric figure with opposite sides parallel. The opposite sides of a parallelogram are equal.

Percent (1.6) Hundredths or ''per one hundred;'' indicated by the % symbol. Thirty-seven hundredths means thirty seven percent. $\frac{37}{100} = 37\%$.

Perfect square (1.3) If x is an integer and a is a positive real number such that $a = x^2$, then x is a square root of a and a is a perfect square. Some numbers that are perfect squares are 1, 4, 9, 16, 25, 36, 49, 64, 81, and 100.

Perfect square trinomials (5.6) Trinomials of the form $a^2 + 2ab + b^2$ or $a^2 - 2ab + b^2$.

Perpendicular lines (3.2) Two straight lines are perpendicular if they meet at a 90-degree angle. If two nonvertical lines have slopes m_1 and m_2 and $m_1, m_2 \neq 0$, then the lines are perpendicular if and only if $m_1 = -\frac{1}{m_2}$.

pH of a solution (11.2) The pH of a solution is defined by the equation $\text{pH} = -\log_{10}(\text{H}^+)$, where H^+ is the concentration of the hydrogen ion in the solution. The solution is an acid if the pH is less than 7 and is a base if the pH is greater than 7.

Pi (1.6) An irrational number, denoted by the symbol π, which is approximately equal to 3.141592654. In most cases, 3.14 can be used as a sufficiently accurate approximation for π.

Point–slope form of the equation of a straight line (3.3) The point–slope form of the equation of a straight line passing through the point (x_1, y_1) and having slope m is $y - y_1 = m(x - x_1)$.

Polynomials (1.5 and 5.1) Variable expressions that contain terms with nonnegative integer exponents. A polynomial must contain no division by a variable. Some examples of polynomials are $5y^2 - 8y + 3$, $-12xy$, $12a - 14b$, and $7x$.

Positive integers (1.1) The numbers in the infinite set $\{1, 2, 3, 4, 5, 6, 7, \ldots\}$. The positive integers are the natural numbers.

Power (1.3) When a number is raised to a power, it means a number with an exponent of that power. Thus, two to the third power means 2^3. The power is the exponent, which is 3. In the expression x^5, we say "x is raised to the fifth power."

Prime factors of a number (6.2) Those factors of a number that are prime. To write the number 40 as a product of prime factors, we would write $40 = 5 \times 2^3$. To write the number 462 as the product of prime factors, we would write $462 = 2 \times 3 \times 7 \times 11$.

Prime factors of a polynomial (6.2) When a polynomial is completely factored, it is written as a product of prime factors. Thus, the prime factors of $x^4 - 81$ are written as $x^4 - 81 = (x^2 + 9)(x - 3)(x + 3)$.

Prime number (6.2) A positive integer that is greater than 1 and that has no factors other than 1 and itself is a prime number. The first ten prime numbers are 2, 3, 5, 7, 11, 13, 17, 19, 23, and 29.

Prime polynomial (5.7) A polynomial that cannot be factored. Examples of prime polynomials are $2x^2 + 100x - 19$, $25x^2 + 9$, and $x^2 - 3x + 5$.

Principal (1.6) In monetary problems, the principal is the original amount of money invested or borrowed.

Principal square root (1.3 and 7.2) The positive square root of a number is called the principal square root. The symbol to find the principal square root is $\sqrt{}$. Thus, $\sqrt{4}$ means to find the principal square root of 4, which is 2.

Proportion (6.5) An equation stating that two ratios are equal. For example, $\frac{a}{b} = \frac{c}{d}$ is a proportion.

Pythagorean theorem (8.4) In any right triangle, if c is the length of the hypotenuse and a and b are the lengths of the two legs, then $c^2 = a^2 + b^2$.

Quadrants (3.1) The four regions into which the x-axis and the y-axis divide the rectangular coordinate system.

Quadratic equation in standard form (5.8, 8.1) An equation of the form $ax^2 + bx + c = 0$, where a, b, c are real numbers and $a \neq 0$, is a quadratic equation in standard form. A quadratic equation is classified as a second-degree equation.

Quadratic formula (8.2) If $ax^2 + bx + c = 0$ and $a \neq 0$, then the roots to the equation are found by the formula $x = \frac{-b \pm \sqrt{b^2 - 4ac}}{2a}$.

Quadratic inequalities (9.5) A quadratic inequality can be written in the form $ax^2 + bx + c > 0$, where $a \neq 0$ and a, b, c are real numbers. The $>$ symbol may be replaced by a $<$, $\geq$, or a $\leq$ symbol.

Quotient (1.4) The result of dividing one number or expression by another. In the problem $12 \div 4 = 3$, the quotient is 3.

Radical equation (7.5) An equation that contains one or more radicals is called a radical equation. The following are examples of radical equations.

$$\sqrt{9x - 20} = x, \qquad 4 = \sqrt{x - 3} + \sqrt{x + 5}$$

Radical sign (1.3, 7.2) The symbol $\sqrt{}$ is used to indicate the root of a number and is called a radical sign.

Radicand (1.3, 7.2) The expression beneath the radical sign is called the radicand. The radicand of $\sqrt{7x}$ is $7x$.

Radius of a circle (1.6) The distance from any point on the circle to the center of the circle.

Range of a relation or a function (3.5) If all the ordered pairs of the relation or the function are listed, all of the different second items of each pair are called the range.

Ratio (6.5) The ratio of two values is the first value divided by the second. The ratio of a to b, where $b \neq 0$, is written as $\frac{a}{b}$, a/b, $a \div b$, or $a:b$.

Rational equation (6.4) An equation that has at least one variable in a denominator. Examples of rational equations are $\frac{x + 6}{3x} = \frac{x + 8}{5}$ and $\frac{x + 3}{x} - \frac{x + 4}{x + 5} = \frac{15}{x^2 + 5x}$.

Rational exponents (7.2) When the exponent is a rational number, we understand this is equivalent to a radical expression in the following way: $x^{m/n} = (\sqrt[n]{x})^m = \sqrt[n]{x^m}$. Thus, $x^{3/7} = (\sqrt[7]{x})^3 = \sqrt[7]{x^3}$.

Rational expressions (6.1) A fraction of the form $\frac{P}{Q}$, where P and Q are polynomials and Q is not zero. Rational expressions are also called algebraic fractions. For example, $\frac{7}{x-8}$ and $\frac{3x-5}{2x^2+1}$ are rational expressions.

Rational numbers (1.1) An infinite set of numbers containing all the integers and all exact quotients of two integers where the denominator is not zero. In set notation, the rational numbers are the set of numbers $\left\{\frac{a}{b} \,\middle|\, a \text{ and } b \text{ are integers but } b \neq 0\right\}$.

Rationalizing the denominator (7.4) The process of transforming a fraction that contains one or more radicals in the denominator to an equivalent fraction that does not contain one. When we rationalize the denominator of $\frac{5}{\sqrt{3}}$, we obtain $\frac{5\sqrt{3}}{3}$. When we rationalize the denominator of $\frac{-2}{\sqrt{11}-\sqrt{7}}$, we obtain $-\frac{\sqrt{11}+\sqrt{7}}{2}$.

Rationalizing the numerator (7.4) The process of transforming a fraction that contains one or more radicals in the numerator to an equivalent fraction that does not contain any radicals in the numerator. When we rationalize the numerator of $\frac{\sqrt{5}}{x}$, we obtain $\frac{5}{x\sqrt{5}}$.

Real number line (1.2) A number line on which all the real numbers are placed. Positive numbers lie to the right of 0 on the number line. Negative numbers lie to the left of 0 on the number line.

Real number line

Real numbers (1.1) The set of numbers containing the rational and irrational numbers.

Reciprocal (1.2) The reciprocal of a number is 1 divided by that number. Therefore, the reciprocal of 12 is $\frac{1}{12}$. The reciprocal of $\frac{3}{4}$ is $\frac{4}{3}$. The reciprocal of $-\frac{5}{8}$ is $-\frac{8}{5}$.

Rectangle (1.6) A four-sided figure with opposite sides parallel and all interior angles measuring 90 degrees. The opposite sides of a rectangle are equal.

Rectangular solid (1.6) A three-dimensional object in which each side is a rectangle. A rectangular solid has the shape of a "box."

Reduce a fraction (6.1) To use the basic rule of fractions to simplify a fraction. The basic rule of fractions is: For any polynomials a, b, c, where b, $c \neq 0$, $\frac{ac}{bc} = \frac{a}{b}$. To reduce the fraction $\frac{x^2-16}{2x+8}$, we have $\frac{(x+4)(x-4)}{2(x+4)} = \frac{x-4}{2}$.

Reduced row echelon form (Appendix B) The reduced row echelon form of an augmented matrix is a matrix with certain properties. All the numbers to the left of the vertical line are 1's along the diagonal from the top left to the bottom right. If there are elements below or above the 1's, these elements are 0's. Two examples of matrices in reduced row echelon form are

$$\begin{bmatrix} 1 & 0 & | & 3 \\ 0 & 1 & | & 4 \end{bmatrix} \quad \text{and} \quad \begin{bmatrix} 1 & 0 & 0 & | & 5 \\ 0 & 1 & 0 & | & 6 \\ 0 & 0 & 1 & | & 7 \end{bmatrix}$$

Relation (3.5) Any set of ordered pairs.

Remainder (5.2) The amount left after the final subtraction when working out a division problem. In the problem

The remainder is 3.

Repeating decimal (1.1) A number that in decimal form has one or more digits that continue to repeat. The numbers $0.33333\ldots$ and $0.128128128\ldots$ are repeating decimals.

Reversing an inequality (2.6) Multiplying or dividing both sides of an inequality by a negative number changes a "greater than" symbol to a "less than" symbol and changes a "less than" symbol to a "greater than" symbol. This is called reversing the direction of an inequality. For example, to solve $-3x < 9$, we divide each side by -3. $\frac{-3x}{-3} > \frac{9}{-3}$, so $x > -3$. The $<$ symbol was reversed to $>$ by dividing both sides by -3.

Rhombus (1.6) A parallelogram with four equal sides and no right angle is called a rhombus.

Right circular cylinder (1.6) A three-dimensional object shaped like a tin can.

Right triangle (8.4) A triangle that contains one right angle. A right angle is an angle that measures exactly 90 degrees. It is indicated by the small rectangle at the corner of the angle.

Root of an equation (2.1) A number that, when substituted into a given equation, yields a true mathematical statement. The root of an equation is also called the solution of an equation.

Scientific notation (1.4) A positive number is written in scientific notation if it is in the form $a \times 10^n$, where $1 \le a < 10$ and n is an integer.

Set (1.1) A collection of objects.

Signed numbers (1.2) Numbers that are either positive, negative, or zero. Positive signed numbers such as 5, 9, or 124 are usually written without a positive sign. Negative signed numbers such as -5, -3.3, or -178 are always written with a negative sign.

Similar radicals (7.3) Two radicals are said to be similar if they are simplified and have the same radicand and the same index. $2\sqrt[3]{7xy^2}$ and $-5\sqrt[3]{7xy^2}$ are similar radicals. Usually similar radicals are referred to as *like radicals*.

Similar triangles (6.5) Two triangles are similar if their corresponding sides are proportional. For example, the following two triangles are proportional.

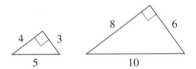

Simplifying a radical (7.3) To simplify a radical when the root cannot be found exactly, we use the product rule for radicals, $\sqrt[n]{ab} = \sqrt[n]{a}\sqrt[n]{b}$ for $a \ge 0$, $b \ge 0$.
To simplify $\sqrt{20}$, we have $= \sqrt{4}\sqrt{5} = 2\sqrt{5}$.
To simplify $\sqrt[3]{16x^4}$, we have $= \sqrt[3]{8x^3}\sqrt[3]{2x} = 2x\sqrt[3]{2x}$.

Simplifying imaginary numbers (7.6) Imaginary numbers are simplified by using the property that, for all positive real numbers a, $\sqrt{-a} = \sqrt{-1}\sqrt{a} = i\sqrt{a}$. Thus, to simplify $\sqrt{-7}$, we have $\sqrt{-7} = \sqrt{-1}\sqrt{7} = i\sqrt{7}$.

Slope–intercept form of the equation of a straight line (3.3) The slope–intercept form of the equation of a straight line with slope m and y-intercept b is $y = mx + b$.

Slope of a straight line (3.2) The slope of a straight line that passes through the points (x_1, y_1) and (x_2, y_2) is

$$\text{Slope} = m = \frac{y_2 - y_1}{x_2 - x_1}, \qquad \text{where } x_1 \ne x_2$$

Solution of an equation (2.1) A number that, when substituted into the equation, yields a true mathematical statement. The solution of an equation is also called the root of an equation.

Sphere (1.6) A perfectly round three-dimensional object shaped like a ball.

Square root (1.3 and 7.2) If x is a real number and a is positive real number such that $a = x^2$, then x is a square root of a. One square root of 16 is 4 since $4^2 = 16$. Another square root of 16 is -4 since $(-4)^2 = 16$.

Standard form of the equation of a circle (9.1) A circle with a center at (h, k) and a radius of r is in standard form if it is written as $(x - h)^2 + (y - k)^2 = r^2$.

Standard form of the equation of an ellipse (9.3) An ellipse with its center at the origin has an equation of the form

$$\frac{x^2}{a^2} + \frac{y^2}{b^2} = 1, \qquad a, b > 0$$

This ellipse has intercepts at $(a, 0)$, $(-a, 0)$, $(0, b)$, and $(0, -b)$.

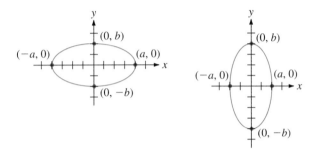

An ellipse with its center at (h, k) has an equation of the form

$$\frac{(x - h)^2}{a^2} + \frac{(y - k)^2}{b^2} = 1, \qquad a, b > 0$$

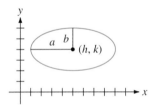

Standard form of the equation of a hyperbola with its center at the origin (9.4) A horizontal hyperbola with its center at the origin has an equation in standard form of $\frac{x^2}{a^2} - \frac{y^2}{b^2} = 1$, where $a, b > 0$. The vertices are at $(-a, 0)$ and $(a, 0)$.

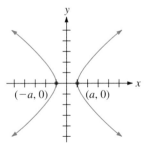

A vertical hyperbola with its center at the origin has an equation in standard form of $\frac{y^2}{b^2} - \frac{x^2}{a^2} = 1$, where $a, b > 0$. The vertices are at $(0, b)$ and $(0, -b)$.

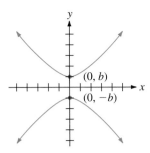

Standard form of the equation of a hyperbola with its center at point (h, k) **(9.4)** A horizontal hyperbola with its center at (h, k) has an equation in standard form of $\frac{(x-h)^2}{a^2} - \frac{(y-k)^2}{b^2} = 1$, where $a, b > 0$. The vertices are $(h - a, k)$ and $(h + a, k)$.

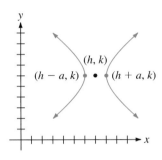

A vertical hyperbola with its center at (h, k) has an equation in standard form of $\frac{(y-k)^2}{b^2} - \frac{(x-h)^2}{a^2} = 1$, where $a, b > 0$. The vertices are at $(h, k + b)$ and $(h, k - b)$.

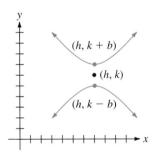

Standard form of the equation of a parabola (9.2) The standard form of the equation of a vertical parabola with its vertex at (h, k) is $y = a(x - h)^2 + k$, where $a \neq 0$. The standard form of the equation of a horizontal parabola with its vertex at (h, k) is $x = a(y - k)^2 + h$, where $a \neq 0$.

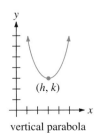 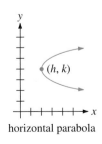

vertical parabola   horizontal parabola

Standard form of the equation of a straight line (3.3) The equation of a straight line is expressed in standard form if it is written in the form $Ax + By = C$, where A, B, and C are real numbers.

Standard form of a quadratic equation (5.8) The standard form of the quadratic equation is $ax^2 + bx + c = 0$, where a, b, c are real numbers and $a \neq 0$. A quadratic equation is classified as a second-degree equation.

Subset (1.1) A set whose elements are all members of another set. For example, the whole numbers are a subset of the integers.

System of dependent equations (4.1) A system of n linear equations in n variables in which some equations are dependent will not have a unique solution. Such a system will have an infinite number of solutions.

System of equations (4.1) A set of two or more equations that must be considered together. The solution to a system of equations is a value for each variable of the system that will satisfy each equation.

$$x + 3y = -7$$
$$4x + 3y = -1$$

is a system of two equations in two unknowns. The solution to this system is $(2, -3)$, or the values $x = 2$, $y = -3$.

System of inequalities (4.6) If two or more inequalities in two variables are considered at one time, we have a system of inequalities. The solution to a system of inequalities in two variables is the region that satisfies every inequality at one time. An example of a system of inequalities is

$$y > 2x + 1$$
$$y < \frac{1}{2}x + 2$$

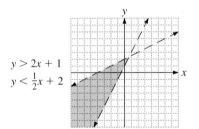

Term (1.5) A real number, a variable, or a product or quotient of numbers and variables. The expression $5xyz$ is one term. The expression $7x + 5y + 6z$ has three terms.

Terminating decimal (1.1) A number in decimal form such as 0.18 or 0.3462. The number of nonzero digits in a terminating decimal is finite.

Trapezoid (1.6) A four-sided geometric figure with two sides parallel. The parallel sides are called the bases of the trapezoid.

Triangle (1.6) A three-sided geometric figure is called a triangle.

Trinomial (5.1) A polynomial of three terms. For example, $2x^2 + 3x - 4$ is a trinomial.

Unknown (2.1) A variable or constant whose value is not known.

Value of a second-order determinant (4.4) The value of the second-order determinant $\begin{vmatrix} a & b \\ c & d \end{vmatrix}$ is defined to be $ad - cb$.

Value of a third-order determinant (4.4) The value of the third-order determinant $\begin{vmatrix} a_1 & b_1 & c_1 \\ a_2 & b_2 & c_2 \\ a_3 & b_3 & c_3 \end{vmatrix}$ is defined to be $a_1 b_2 c_3 + b_1 c_2 a_3 + c_1 a_2 b_3 - a_3 b_2 c_1 - b_3 c_2 a_1 - c_3 a_2 b_1$.

Variable (1.1) When a letter is used to represent a number, it is called a variable.

Vertex of a parabola (9.2) The vertex of a vertical parabola is the lowest point on a parabola opening upward or the highest point on a parabola opening downward.

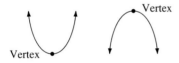

The vertex of a horizontal parabola is the leftmost point on a parabola opening to the right or the rightmost point on a parabola opening to the left.

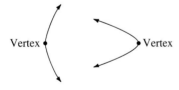

Vertical line (3.1) A straight line that is parallel to the y-axis. The slope of a vertical line is undefined. Therefore, a vertical line has no slope. The equation of a vertical line can be written in the form $x = a$, where a is a constant. A sketch of a vertical line is shown.

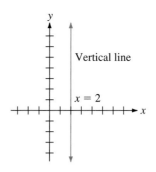

Vertical parabolas (9.2) Parabolas that open up or down. The following graphs represent vertical parabolas.

Whole numbers (1.1) The set of numbers containing the natural numbers as well as the number 0. The whole numbers can be written as the infinite set $\{0, 1, 2, 3, 4, 5, 6, 7, \ldots\}$.

x-Intercept (3.1) The number a is the x-intercept of a line if the line crosses the x-axis at $(a, 0)$. The x-intercept of the line shown below is 5.

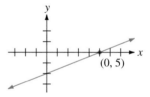

y-Intercept (3.1) The number b is the y-intercept of a line if the line crosses the y-axis at $(0, b)$. The y-intercept of the line shown below is 4.

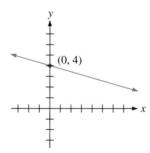

APPENDIX A
Tables

TABLE A-1: TABLE OF SQUARE ROOTS

Square root values ending in 0.000 are exact. All other values are approximate and are rounded to the nearest thousandth.

x	√x	x	√x	x	√x	x	√x	x	√x
1	1.000	41	6.403	81	9.000	121	11.000	161	12.689
2	1.414	42	6.481	82	9.055	122	11.045	162	12.728
3	1.732	43	6.557	83	9.110	123	11.091	163	12.767
4	2.000	44	6.633	84	9.165	124	11.136	164	12.806
5	2.236	45	6.708	85	9.220	125	11.180	165	12.845
6	2.449	46	6.782	86	9.274	126	11.225	166	12.884
7	2.646	47	6.856	87	9.327	127	11.269	167	12.923
8	2.828	48	6.928	88	9.381	128	11.314	168	12.961
9	3.000	49	7.000	89	9.434	129	11.358	169	13.000
10	3.162	50	7.071	90	9.487	130	11.402	170	13.038
11	3.317	51	7.141	91	9.539	131	11.446	171	13.077
12	3.464	52	7.211	92	9.592	132	11.489	172	13.115
13	3.606	53	7.280	93	9.644	133	11.533	173	13.153
14	3.742	54	7.348	94	9.695	134	11.576	174	13.191
15	3.873	55	7.416	95	9.747	135	11.619	175	13.229
16	4.000	56	7.483	96	9.798	136	11.662	176	13.266
17	4.123	57	7.550	97	9.849	137	11.705	177	13.304
18	4.243	58	7.616	98	9.899	138	11.747	178	13.342
19	4.359	59	7.681	99	9.950	139	11.790	179	13.379
20	4.472	60	7.746	100	10.000	140	11.832	180	13.416
21	4.583	61	7.810	101	10.050	141	11.874	181	13.454
22	4.690	62	7.874	102	10.100	142	11.916	182	13.491
23	4.796	63	7.937	103	10.149	143	11.958	183	13.528
24	4.899	64	8.000	104	10.198	144	12.000	184	13.565
25	5.000	65	8.062	105	10.247	145	12.042	185	13.601
26	5.099	66	8.124	106	10.296	146	12.083	186	13.638
27	5.196	67	8.185	107	10.344	147	12.124	187	13.675
28	5.292	68	8.246	108	10.392	148	12.166	188	13.711
29	5.385	69	8.307	109	10.440	149	12.207	189	13.748
30	5.477	70	8.367	110	10.488	150	12.247	190	13.784
31	5.568	71	8.426	111	10.536	151	12.288	191	13.820
32	5.657	72	8.485	112	10.583	152	12.329	192	13.856
33	5.745	73	8.544	113	10.630	153	12.369	193	13.892
34	5.831	74	8.602	114	10.677	154	12.410	194	13.928
35	5.916	75	8.660	115	10.724	155	12.450	195	13.964
36	6.000	76	8.718	116	10.770	156	12.490	196	14.000
37	6.083	77	8.775	117	10.817	157	12.530	197	14.036
38	6.164	78	8.832	118	10.863	158	12.570	198	14.071
39	6.245	79	8.888	119	10.909	159	12.610	199	14.107
40	6.325	80	8.944	120	10.954	160	12.649	200	14.142

TABLE A-2: EXPONENTIAL VALUES

x	e^x	e^{-x}
0.00	1.0000	1.0000
0.01	1.0101	0.9900
0.02	1.0202	0.9802
0.03	1.0305	0.9704
0.04	1.0408	0.9608
0.05	1.0513	0.9512
0.06	1.0618	0.9418
0.07	1.0725	0.9324
0.08	1.0833	0.9231
0.09	1.0942	0.9139
0.10	1.1052	0.9048
0.11	1.1163	0.8958
0.12	1.1275	0.8869
0.13	1.1388	0.8781
0.14	1.1503	0.8694
0.15	1.1618	0.8607
0.16	1.1735	0.8521
0.17	1.1853	0.8437
0.18	1.1972	0.8353
0.19	1.2092	0.8270
0.20	1.2214	0.8187
0.21	1.2337	0.8106
0.22	1.2461	0.8025
0.23	1.2586	0.7945
0.24	1.2712	0.7866
0.25	1.2840	0.7788
0.26	1.2969	0.7711
0.27	1.3100	0.7634
0.28	1.3231	0.7558
0.29	1.3364	0.7483
0.30	1.3499	0.7408
0.35	1.4191	0.7047
0.40	1.4918	0.6703
0.45	1.5683	0.6376
0.50	1.6487	0.6065
0.55	1.7333	0.5769
0.60	1.8221	0.5488
0.65	1.9155	0.5220
0.70	2.0138	0.4966
0.75	2.1170	0.4724
0.80	2.2255	0.4493
0.85	2.3396	0.4274
0.90	2.4596	0.4066
0.95	2.5857	0.3867
1.0	2.7183	0.3679
1.1	3.0042	0.3329
1.2	3.3201	0.3012
1.3	3.6693	0.2725
1.4	4.0552	0.2466
1.5	4.4817	0.2231

x	e^x	e^{-x}
1.6	4.9530	0.2019
1.7	5.4739	0.1827
1.8	6.0496	0.1653
1.9	6.6859	0.1496
2.0	7.3891	0.1353
2.1	8.1662	0.1225
2.2	9.0250	0.1108
2.3	9.9742	0.1003
2.4	11.023	0.0907
2.5	12.182	0.0821
2.6	13.464	0.0743
2.7	14.880	0.0672
2.8	16.445	0.0608
2.9	18.174	0.0550
3.0	20.086	0.0498
3.1	22.198	0.0450
3.2	24.533	0.0408
3.3	27.113	0.0369
3.4	29.964	0.0334
3.5	33.115	0.0302
3.6	36.598	0.0273
3.7	40.447	0.0247
3.8	44.701	0.0224
3.9	49.402	0.0202
4.0	54.598	0.0183
4.1	60.340	0.0166
4.2	66.686	0.0150
4.3	73.700	0.0136
4.4	81.451	0.0123
4.5	90.017	0.0111
4.6	99.484	0.0101
4.7	109.95	0.0091
4.8	121.51	0.0082
4.9	134.29	0.0074
5.0	148.41	0.0067
5.5	244.69	0.0041
6.0	403.43	0.0025
6.5	665.14	0.0015
7.0	1,096.6	0.00091
7.5	1,808.0	0.00055
8.0	2,981.0	0.00034
8.5	4,914.8	0.00020
9.0	8,103.1	0.00012
9.5	13,360	0.000075
10	22,026	0.000045
11	59,874	0.000017
12	162,755	0.0000061
13	442,413	0.0000023
14	1,202,604	0.0000008
15	3,269,017	0.0000003

APPENDIX B
Solving Systems of Linear Equations Using Matrices

After studying this section, you will be able to:

1 *Solve a system of linear equations using matrices.*

1 *Solving a System of Linear Equations Using Matrices*

In Section 4.4 we defined a matrix as any rectangular array of numbers that is arranged in rows and columns.

$$\begin{bmatrix} 2 & 3 \\ 5 & 6 \end{bmatrix}$$ This is a 2 × 2 matrix with two rows and two columns.

$$\begin{bmatrix} 1 & -5 & -6 & 2 \\ 3 & 4 & -8 & -2 \\ 2 & 7 & 9 & -4 \end{bmatrix}$$ This is a 3 × 4 matrix with three rows and four columns.

A matrix that is derived from a linear system of equations is called the **augmented matrix** of the system. This augmented matrix is made up of two smaller matrices separated by a vertical line. The coefficients of each variable in a linear system are placed to the left of the vertical line. The constants are placed to the right of the vertical line.

The augmented matrix for the system of equations

$$-3x + 5y = -22$$
$$2x - y = 10$$

is the 2 × 3 matrix

$$\left[\begin{array}{rr|r} -3 & 5 & -22 \\ 2 & -1 & 10 \end{array}\right]$$

The augmented matrix for the system of equations

$$3x - 5y + 2z = 8$$
$$x + y + z = 3$$
$$3x - 2y + 4z = 10$$

is the 3 × 4 matrix

$$\left[\begin{array}{rrr|r} 3 & -5 & 2 & 8 \\ 1 & 1 & 1 & 3 \\ 3 & -2 & 4 & 10 \end{array}\right]$$

EXAMPLE 1 Write the solution for the system of linear equations represented by the following matrix.

$$\left[\begin{array}{cc|c} 1 & -3 & -7 \\ 0 & 1 & 4 \end{array}\right]$$

This system is represented by the equations

$$x - 3y = -7$$
$$0x + y = 4$$

Since we know that $y = 4$ we can find x by back substitution.

$$x - 3y = -7$$
$$x - 3(4) = -7$$
$$x - 12 = -7$$
$$x = 5$$

Thus the solution to the system is $x = 5$, $y = 4$ or we can write the solution as $(5, 4)$.

Practice Problem 1 Write the solution for the system of linear equations represented by the following matrix.

$$\left[\begin{array}{cc|c} 1 & 9 & 33 \\ 0 & 1 & 3 \end{array}\right]$$

To solve a system of linear equations in matrix form, we use three row operations of the matrix.

Matrix Row Operations

1. Any two rows of a matrix may be interchanged.

2. All the numbers in a row may be multiplied or divided by any nonzero number.

3. All the numbers in any or any multiple of a row may be added to the corresponding numbers of any other row.

To obtain the values for x and y in a system of two linear equations, we use row transformations to obtain an augmented matrix in a form similar to the form of the matrix in Example 1.

The desired form is

$$\left[\begin{array}{cc|c} 1 & a & b \\ 0 & 1 & c \end{array}\right] \quad \text{or} \quad \left[\begin{array}{ccc|c} 1 & a & b & d \\ 0 & 1 & c & e \\ 0 & 0 & 1 & f \end{array}\right]$$

The last row of the matrix will allow us to find the value of one variable. We can then use back substitution to find the other variables.

EXAMPLE 2 Use matrices to solve the system

$$4x - 3y = -13$$
$$x + 2y = 5$$

The augmented matrix for this system of linear equations is

$$\left[\begin{array}{cc|c} 4 & -3 & -13 \\ 1 & 2 & 5 \end{array}\right]$$

First we want to obtain a 1 as the first element in the first row. We can obtain this by interchanging rows one and two.

$$\begin{bmatrix} 1 & 2 & | & 5 \\ 4 & -3 & | & -13 \end{bmatrix} \qquad R_1 \longleftrightarrow R_2$$

Next we wish to obtain a 0 as the first element of the second row. To obtain this we multiply -4 by all the elements of row one and add this to row two.

$$\begin{bmatrix} 1 & 2 & | & 5 \\ 0 & -11 & | & -33 \end{bmatrix} \qquad -R_1 + R_2$$

Next, to obtain a 1 as the second element of the second row we multiply each element of row two by $(-1/11)$.

$$\begin{bmatrix} 1 & 2 & | & 5 \\ 0 & 1 & | & 3 \end{bmatrix} \qquad -\frac{1}{11}R_2$$

This final matrix is in the desired form. It represents the linear system

$$x + 2y = 5$$
$$y = 3$$

Since we know that $y = 3$, we substitute this value into the first equation.

$$x + 2(3) = 5$$
$$x + 6 = 5$$
$$x = -1$$

Thus the solution to the system is $(-1, 3)$.

Practice Problem 2 Use matrices to solve the system

$$3x - 2y = -6$$
$$x - 3y = 5$$

Now we continue with a similar example involving three equations and three unknowns.

EXAMPLE 3 Use matrices to solve the system

$$2x + 3y - z = 11$$
$$x + 2y + z = 12$$
$$3x - y + 2z = 5$$

The augmented matrix that represents this system of linear equations is

$$\begin{bmatrix} 2 & 3 & -1 & | & 11 \\ 1 & 2 & 1 & | & 12 \\ 3 & -1 & 2 & | & 5 \end{bmatrix}$$

To obtain a zero as the first element of the second and third rows we will first need to interchange the first and second rows.

$$\begin{bmatrix} 1 & 2 & 1 & | & 12 \\ 2 & 3 & -1 & | & 11 \\ 3 & -1 & 2 & | & 5 \end{bmatrix} \qquad R_1 \longleftrightarrow R_2$$

Now in order to obtain a 0 as the first element of the second row we multiply row one by -2 and add the result to row two. In order to obtain a 0 as the first element of the third row we multiply row one by -3 and add the result to row three.

$$\begin{bmatrix} 1 & 2 & 1 & | & 12 \\ 0 & -1 & -3 & | & -13 \\ 0 & -7 & -1 & | & -31 \end{bmatrix} \quad \begin{matrix} -2R_1 + R_2 \\ -3R_1 + R_2 \end{matrix}$$

To obtain a 1 as the second element of row two we multiply all the elements of row two by -1.

$$\begin{bmatrix} 1 & 2 & 1 & | & 12 \\ 0 & 1 & 3 & | & 13 \\ 0 & -7 & -1 & | & -31 \end{bmatrix} \quad -1R_2$$

Next, in order to obtain a 0 as the second element of row three we add 7 times row two to row three.

$$\begin{bmatrix} 1 & 2 & 1 & | & 12 \\ 0 & 1 & 3 & | & 13 \\ 0 & 0 & 20 & | & 60 \end{bmatrix} \quad 7R_2 + R_3$$

Finally we multiply all the elements of row three by 1/20. Thus we finally have

$$\begin{bmatrix} 1 & 2 & 1 & | & 12 \\ 0 & 1 & 3 & | & 13 \\ 0 & 0 & 1 & | & 3 \end{bmatrix} \quad \frac{1}{20}R_3$$

From the final line of the matrix we see that $z = 3$. If we substitute this into the values of the second line we have

$$y + 3z = 13$$
$$y + 3(3) = 13$$
$$y + 9 = 13$$
$$y = 4$$

Now we will use the values obtained for y and for z and use the first line of the matrix.

$$x + 2y + z = 12$$
$$x + 2(4) + 3 = 12$$
$$x + 8 + 3 = 12$$
$$x + 11 = 12$$
$$x = 1$$

Thus the solution to this linear system of three equations is $(1, 4, 3)$.

Practice Problem 3 Use matrices to solve the system

$$2x + y - 2z = -15$$
$$4x - 2y + z = 15$$
$$x + 3y + 2z = -5$$

We could continue to use these row operations to obtain an augmented matrix of the form

$$\begin{bmatrix} 1 & 0 & | & a \\ 0 & 1 & | & b \end{bmatrix} \quad \text{or} \quad \begin{bmatrix} 1 & 0 & 0 & | & a \\ 0 & 1 & 0 & | & b \\ 0 & 0 & 1 & | & c \end{bmatrix}$$

This form of the augmented matrix is given a special name. It is known as the **reduced row echelon form.** If the augmented matrix of a system of linear equations is placed in this form, we would immediately know the solution to the system. Thus if a system of linear equations in the variables x, y, z had an augmented matrix that could be placed in the form

$$\begin{bmatrix} 1 & 0 & 0 & | & 7 \\ 0 & 1 & 0 & | & 32 \\ 0 & 0 & 1 & | & 18 \end{bmatrix}$$

we could determine directly that $x = 7$, $y = 32$, and $z = 18$. A similar pattern is obtained for a system of four equations in four unknowns, and so on. Thus, if a system of linear equations in the variables w, x, y, and z had an augmented matrix that could be placed in the form

$$\begin{bmatrix} 1 & 0 & 0 & 0 & | & 23.4 \\ 0 & 1 & 0 & 0 & | & 48.6 \\ 0 & 0 & 1 & 0 & | & 0.73 \\ 0 & 0 & 0 & 1 & | & 5.97 \end{bmatrix}$$

we could directly conclude that $w = 23.4$, $x = 48.6$, $y = 0.73$, and $z = 5.97$. Reducing a matrix to reduced row echelon form is readily done on computers. Many mathematical software packages contain matrix operations that will obtain the reduced row echelon form of an augmented matrix. A number of the newer graphing calculators such as the TI-83 can be used to obtain the reduced row echelon form by using the **rref** command on a given matrix.

GRAPHING CALCULATOR

Obtaining a Reduced Row Echelon Form of the Augmented Matrix

If your graphing calculator has a routine to obtain the **reduced row echelon form** of a matrix (**rref**), then this routine will allow you to quickly obtain the solution of a system of linear equations if one exists. If your calculator has this capability, solve the following system.

$$5w + 2x + 3y + 4z = -8.3$$

$$-4w + 3x + 2y + 7z = -70.1$$

$$6w + x + 4y + 5z = -13.3$$

$$7w + 4x + y + 2z = 14.1$$

Solution:

$$w = 3.1,\ x = 2.2,$$
$$y = 4.6,\ z = -10.5$$

B Exercises

Solve each system of equations by the matrix method. Round answers to the nearest tenth.

1. $2x + 3y = 5$
$5x + y = 19$

2. $3x + 5y = -15$
$2x + 7y = -10$

3. $2x + y = -3$
$5x - y = 24$

4. $x + 5y = -9$
$4x - 3y = -13$

5. $5x + 2y = 6$
$3x + 4y = 12$

6. $-5x + y = 24$
$x + 5y = 10$

7. $3x - 2y + 3 = 5$
$x + 4y - 1 = 9$

8. $3x + y - 4 = 12$
$-2x + 3y + 2 = -5$

9. $-7x + 3y = 2.7$
$6x + 5y = 25.7$

10. $x - 2y - 3z = 4$
$2x + 3y + z = 1$
$-3x + y - 2z = 5$

11. $x + y - z = -2$
$2x - y + 3z = 19$
$4x + 3y - z = 5$

12. $5x - y + 4z = 5$
$6x + y - 5z = 17$
$2x - 3y + z = -11$

13. $x + y - z = -3$
$x + y + z = 3$
$3x - y + z = 7$

14. $2x - y + z = 5$
$x + 2y - z = -2$
$x + y - 2z = -5$

15. $2x - 3y + z = 11$
$x + y + 2z = 8$
$x + 3y - z = -11$

16. $4x + 3y + 5z = 2$
$2y + 7z = 16$
$2x - y = 6$

17. $6x - y + z = 9$
$2x + 3z = 16$
$4x + 7y + 5z = 20$

18. $3x + 2y = 44$
$4y + 3z = 19$
$2x + 3z = -5$

Optional Graphing Calculator Problems

If your graphing calculator has the necessary capability, solve the following problems. Round your answer to the nearest tenth.

19. $5x + 6y + 7z = 45.6$
$1.4x - 3.2y + 1.6z = 3.12$
$9x - 8y + 22z = 70.8$

20. $2x + 12y + 9z = 37.9$
$1.6x + 1.8y - 2.5z = -20.53$
$7x + 8y + 4z = 39.6$

21. $6w + 5x + 3y + 1.5z = 41.7$
$2w + 6.7x - 5y + 7z = -21.92$
$12w + x + 5y - 6z = 58.4$
$3w + 8x - 15y + z = -142.8$

22. $2w + 3x + 11y - 14z = 6.7$
$5w + 8x + 7y + 3z = 25.3$
$-4w + x + 1.5y - 9z = -53.4$
$9w + 7x - 2.5y + 6z = 22.9$

Solutions to Practice Problems

Chapter 1

1.1 Practice Problems

1. (a) 1.26 is a rational number and a real number
   (b) 3 is a natural number, whole number, integer, rational, and a real number
   (c) $\frac{3}{7}$ is a rational number and a real number
   (d) -2 is an integer, a rational number, and a real number
   (e) $5.182671\ldots$ has no repeating pattern. Therefore, it is an irrational number and a real number.

2. (a) $9 + 8 = 8 + 9$ Commutative property of addition
   (b) $17 + 0 = 17$ Identity property of addition
   (c) $-4 + 4 = 0$ Inverse property of addition
   (d) $7 + (3 + x) = (7 + 3) + x$ Associative property of addition

3. (a) $6 \cdot (2 \cdot w) = (6 \cdot 2) \cdot w$ Associative property of multiplication
   (b) $4 \cdot \frac{1}{4} = 1$ Inverse property of multiplication
   (c) $4 \cdot 15 = 15 \cdot 4$ Commutative property of multiplication
   (d) $76 \cdot 1 = 76$ Identity property of multiplication

4. (a) $(20 + 9)3 = (20 \cdot 3) + (9 \cdot 3)$
   $= 60 + 27 = 87$
   (b) $(29 \cdot 9) + (29 \cdot 1) = 29(9 + 1)$
   $= 29 \cdot 10 = 290$

1.2 Practice Problems

1. (a) $|-4| = 4$
   (b) $|3.16| = 3.16$
   (c) $|0| = 0$
   (d) $|12 - 7| = |5| = 5$

2. (a) $3.4 + 2.6 = 6.0$
   (b) $\left(-\frac{3}{4}\right) + \left(-\frac{1}{6}\right) = \left(-\frac{9}{12}\right) + \left(-\frac{2}{12}\right)$
   $= -\frac{11}{12}$
   (c) $(-5) + (-37) = -42$

3. (a) $26 + (-18) = 8$
   (b) $24 + (-30) = -6$
   (c) $-\frac{1}{5} + \frac{2}{3} = \frac{-3}{15} + \frac{10}{15} = \frac{7}{15}$
   (d) $-8.6 + 8.6 = 0$

4. (a) $9 - 2 = 7$
   (b) $-8 - (-3) = -8 + 3 = -5$
   (c) $-5 - (-14) = -5 + 14 = 9$
   (d) $18 - 26 = 18 + (-26) = -8$
   (e) $\frac{1}{2} - \left(-\frac{1}{4}\right) = \frac{2}{4} + \frac{1}{4} = \frac{3}{4}$
   (f) $-0.35 - 0.67 = -0.35 + (-0.67) = -1.02$

5. (a) $(-3)(7) = -21$
   (b) $\left(-\frac{2}{5}\right)\left(\frac{3}{4}\right) = -\frac{6}{20} = -\frac{3}{10}$
   (c) $\frac{150}{-30} = -5$
   (d) $\frac{-4}{\frac{2}{5}} = -4 \div \frac{2}{5} = -4 \times \frac{5}{2} = -10$
   (e) $-36 \div (18) = -2$
   (f) $\frac{0.27}{-0.003} = -90$

6. (a) $(-9)(-3) = 27$
   (b) $(1.23)(0.06) = 0.0738$
   (c) $\frac{-20}{-2} = 10$
   (d) $-60 \div (-5) = 12$
   (e) $\left(\frac{3}{7}\right) \div \left(\frac{1}{5}\right) = \frac{3}{7} \cdot \frac{5}{1} = \frac{15}{7}$

7. $5 + 7(-2) - (-3) + 50 \div (-2)$
   $= 5 + (-14) - (-3) + (-25)$
   $= 5 + (-14) + 3 + (-25)$
   $= -9 + 3 + (-25) = (-6) + (-25) = -31$

8. (a) $3(-5) + 3(-6) - 2(-1) = -15 - 18 + 2$
   $= -33 + 2 = -31$
   (b) $6(-2) + (-20) \div (2)(3) = -12 + (-10)(3)$
   $= -12 + (-30) = -42$
   (c) $\frac{7 + 2 - 12 - (-1)}{(-5)(-6) + 4(-8)} = \frac{-2}{-2} = 1$

1.3 Practice Problems

1. (a) $(-4)(-4)(-4)(-4) = (-4)^4$
   (b) $z \cdot z \cdot z \cdot z \cdot z \cdot z \cdot z = z^7$

2. (a) $(-3)^5 = (-3)(-3)(-3)(-3)(-3) = -243$
   (b) $(-3)^6 = (-3)(-3)(-3)(-3)(-3)(-3) = 729$
   (c) $2^5 = 2 \cdot 2 \cdot 2 \cdot 2 \cdot 2 = 32$
   (d) $(-4)^4 = (-4)(-4)(-4)(-4) = 256$
   (e) $-4^4 = -(4 \cdot 4 \cdot 4 \cdot 4) = -256$
   (f) $\left(\frac{1}{5}\right)^2 = \left(\frac{1}{5}\right)\left(\frac{1}{5}\right) = \frac{1}{25}$

3. Since $(-7)^2 = 49$ and $7^2 = 49$ the square roots of 49 are -7 and 7. The principal square root is 7.

4. (a) $\sqrt{100} = 10$ because $10^2 = 100$
   (b) $\sqrt{1} = 1$ because $1^2 = 1$
   (c) $-\sqrt{100} = -10$ because $(10)^2 = 100$
   (d) $-\sqrt{36} = -6$ because $(6)^2 = 36$
   (e) $\sqrt{0} = 0$

5. (a) $(0.3)^2 = (0.3)(0.3) = 0.09$
   therefore $\sqrt{0.09} = 0.3$
   (b) $\sqrt{\frac{4}{81}} = \frac{\sqrt{4}}{\sqrt{81}} = \frac{2}{9}$

6. (a) $7 - 3(8 + 5 - 6) = 7 - 3(7)$
   $= 7 - 21 = -14$
   (b) $6(12 - 8) + 4 = 6(4) + 4$
   $= 24 + 4 = 28$
   (c) $5[6 - 3(7 - 9)] - 8 = 5[6 - 3(-2)] - 8$
   $= 5[6 + 6] - 8$
   $= 5[12] - 8 = 60 - 8 = 52$

7. $\frac{5 + 2(-3) - 10 + 1}{(1)(-2)(-3)}$
   $= \frac{5 + (-6) - 10 + 1}{6} = \frac{-10}{6} = \frac{-5}{3}$

8. (a) $\sqrt{(-5)^2 + 12^2} = \sqrt{25 + 144} = \sqrt{169} = 13$
   (b) $|-3 - 7 + 2 - (-4)| = |-3 - 7 + 2 + 4|$
   $= |-10 + 6| = |-4| = 4$

9. $-7 - 2(-3) + 4^3 = -7 + 6 + 64$
   $= -1 + 64 = 63$

10. $5 + 6 \cdot 2 - 12 \div (-2) + 3\sqrt{4}$
    $= 5 + 6 \cdot 2 - 12 \div (-2) + 3(2)$
    $= 5 + 6 \cdot 2 + 6 + 3(2)$
    $= 5 + 12 + 6 + 6$
    $= 17 + 6 + 6 = 23 + 6 = 29$

11. $\dfrac{2(3) + 5(-2)}{1 + 2 \cdot 3^2 + 5(-3)} = \dfrac{6 - 10}{1 + 2 \cdot 9 + 5(-3)}$

$\qquad = \dfrac{-4}{1 + 18 + (-15)} = \dfrac{-4}{4} = -1$

1.4 Practice Problems

1. (a) $3^{-2} = \dfrac{1}{3^2} = \dfrac{1}{9}$

  (b) $7^{-1} = \dfrac{1}{7}$

  (c) $z^{-8} = \dfrac{1}{z^8}$

2. (a) $\left(\dfrac{3}{4}\right)^{-2} = \dfrac{1}{\left(\dfrac{3}{4}\right)^2} = \dfrac{1}{\dfrac{9}{16}} = 1 \cdot \dfrac{16}{9} = \dfrac{16}{9}$

  (b) $\left(\dfrac{1}{3}\right)^{-3} = \dfrac{1}{\left(\dfrac{1}{3}\right)^3} = \dfrac{1}{\dfrac{1}{27}} = 1 \cdot \dfrac{27}{1} = 27$

3. (a) $w^5 \cdot w = w^6$

  (b) $2^8 \cdot 2^{15} = 2^{23}$

  (c) $x^2 \cdot x^8 \cdot x^6 = x^{16}$

  (d) $(x + 2y)^4(x + 2y)^{10} = (x + 2y)^{14}$

4. (a) $(7w^3)(2w) = 14w^4$

  (b) $(-5xy)(-2x^2y^3) = 10x^3y^4$

  (c) $(x)(2x)(3xy) = 6x^3y$

5. (a) $(2x^4y^{-5})(3x^2y^{-4}) = 6x^6y^{-9} = \dfrac{6x^6}{y^9}$

  (b) $(7xy^{-2})(2x^{-5}y^{-6}) = 14x^{-4}y^{-8} = \dfrac{14}{x^4y^8}$

6. (a) $\dfrac{w^8}{w^6} = w^{8-6} = w^2$

  (b) $\dfrac{3^7}{3^3} = 3^{7-3} = 3^4$

  (c) $\dfrac{x^5}{x^{16}} = x^{5-16} = x^{-11} = \dfrac{1}{x^{11}}$

  (d) $\dfrac{4^5}{4^8} = 4^{5-8} = 4^{-3} = \dfrac{1}{4^3}$

7. (a) $6y^0 = 6(1) = 6$

  (b) $5x^2y^0 = 5x^2(1) = 5x^2$

  (c) $(3xy)^0 = 1$. Note that the entire expression is raised to the zero power.

  (d) $(5^{-3})(2a)^0 = (5^{-3})(1) = \dfrac{1}{5^3} = \dfrac{1}{125}$

8. (a) $\dfrac{30x^6y^5}{20x^3y^2} = \dfrac{30}{20} \cdot \dfrac{x^6}{x^3} \cdot \dfrac{y^5}{y^2} = \dfrac{3}{2}x^3y^3$ or $\dfrac{3x^3y^3}{2}$

  (b) $\dfrac{-15a^3b^4c^4}{3a^5b^4c^2} = \dfrac{-15}{3} \cdot \dfrac{a^3}{a^5} \cdot \dfrac{b^4}{b^4} \cdot \dfrac{c^4}{c^2} = -5a^{-2}c^2 = -\dfrac{5c^2}{a^2}$

9. (a) $\dfrac{2x^{-3}y}{4x^{-2}y^5} = \dfrac{1}{2}x^{-3-(-2)}y^{1-5} = \dfrac{1}{2}x^{-1}y^{-4} = \dfrac{1}{2xy^4}$

  (b) $\dfrac{5^{-3}xy^{-2}}{5^{-4}x^{-6}y^{-7}} = 5^{-3-(-4)}x^{1-(-6)}y^{-2-(-7)} = 5^1x^7y^5 = 5x^7y^5$

10. (a) $(w^3)^8 = w^{3 \cdot 8} = w^{24}$

  (b) $(5^2)^5 = 5^{2 \cdot 5} = 5^{10}$

  (c) $[(x - 2y)^3]^3 = (x - 2y)^{3 \cdot 3} = (x - 2y)^9$

11. (a) $(4x^3y^4)^2 = 4^2x^6y^8 = 16x^6y^8$

  (b) $\left(\dfrac{4xy}{3x^5y^6}\right)^3 = \dfrac{4^3x^3y^3}{3^3x^{15}y^{18}} = \dfrac{64x^3y^3}{27x^{15}y^{18}} = \dfrac{64}{27x^{12}y^{15}}$

  (c) $(3xy^2)^{-2} = 3^{-2}x^{-2}y^{-4} = \dfrac{1}{9x^2y^4}$

12. (a) $\dfrac{7x^2y^{-4}z^{-3}}{8x^{-5}y^{-6}z^2} = \dfrac{7x^2 \cdot x^5 \cdot y^6}{8 \cdot y^4 \cdot z^3 \cdot z^2} = \dfrac{7x^7y^2}{8z^5}$

  (b) $\left(\dfrac{4x^2y^{-2}}{x^{-4}y^{-3}}\right)^{-3} = \dfrac{4^{-3}x^{-6}y^6}{x^{12}y^9} = \dfrac{y^6}{4^3x^{12} \cdot x^6y^9} = \dfrac{1}{64x^{18}y^3}$

13. $(2x^{-3})^2(-3xy^{-2})^{-3} = (2x^{-6})(-3^{-3}x^{-3}y^6)$

$\qquad = \dfrac{2^2y^6}{-3^3x^6x^3} = -\dfrac{4y^6}{27x^9}$

14. (a) $128{,}320 = 1.2832 \times 10^5$

  (b) $476 = 4.76 \times 10^2$

  (c) $0.0123 = 1.23 \times 10^{-2}$

  (d) $0.007 = 7 \times 10^{-3}$

15. (a) $3 \times 10^4 = 30{,}000$

  (b) $4.62 \times 10^6 = 4{,}620{,}000$

  (c) $1.973 \times 10^{-3} = 0.001973$

  (d) $6 \times 10^{-8} = 0.00000006$

  (e) $4.931 \times 10^{-1} = 0.4931$

16. $\dfrac{(55{,}000)(3{,}000{,}000)}{5{,}500{,}000} = \dfrac{(5.5 \times 10^4)(3.0 \times 10^6)}{5.5 \times 10^6}$

$\qquad = \dfrac{3.0}{1} \times \dfrac{10^{10}}{10^6}$

$\qquad = 3.0 \times 10^{10-6} = 3.0 \times 10^4$

17. $6.0 \times 10^{24} \times 3.4 \times 10^5$

$\qquad = 6.0 \times 3.4 \times 10^{24} \times 10^5$

$\qquad = 20.4 \times 10^{29}$ kilograms

$\qquad = 2.04 \times 10^{30}$ kilograms

1.5 Practice Problems

1. (a) $7x - 2w^3$

  $7x$ is the product of a real number (7) and a variable x, so $7x$ is a term

  $-2w^3$ is the product of a real number (-2) and a variable w^3, so $-2w^3$ is a term

  (b) $5 + 6x + 2y$

  5 is a real number, so it is a term. $6x$ is the product of a real number (6) and a variable x, so $6x$ is a term. $2y$ is the product of a real number (2) and a variable y, so $2y$ is a term.

2. (a) $5x^2y - 3.5w$

  The numerical coefficient of the x^2y term is 5.

  The numerical coefficient of the w term is -3.5.

  (b) $\dfrac{3}{4}x^3 - \dfrac{5}{7}x^2y$

  The numerical coefficient of the x^3 term is $\dfrac{3}{4}$.

  The numerical coefficient of the x^2y term is $-\dfrac{5}{7}$.

  (c) $-5.6abc - 0.34ab + 8.56bc$

  The numerical coefficient of the abc term is -5.6, of the ab term is -0.34, and of the bc term is 8.56.

3. (a) $9x - 12x = (9 - 12)x = -3x$

  (b) $4ab^2c + 15ab^2c = (4 + 15)ab^2c = 19ab^2c$

4. (a) $12x^3 - 5x^2 + 7x - 3x^3 - 8x^2 + x$

  $= 12x^3 - 3x^3 - 8x^2 - 5x^2 + 7x + x$

  $= 9x^3 - 13x^2 + 8x$

  (b) $\dfrac{1}{3}a^2 - \dfrac{1}{5}a - \dfrac{4}{15}a^2 + \dfrac{1}{2}a + 5$

  $= \dfrac{1}{3}a^2 - \dfrac{4}{15}a^2 + \dfrac{1}{2}a - \dfrac{1}{5}a + 5$

  $= \dfrac{1}{15}a^2 + \dfrac{3}{10}a + 5$

  (c) $4.5x^3 - 0.6xy - 9.3x^3 + 0.8xy$

  $= 4.5x^3 - 9.3x^3 - 0.6xy + 0.8xy$

  $= -4.8x^3 + 0.2xy$

5. $-3x^2(2x - 5) = -3x^2[2x + (-5)]$

$= (-3x^2)(2x) + (-3x^2)(-5)$

$= -6x^3 + 15x^2$

6. (a) $-5x(2x^2 - 3x - 1) = -10x^3 + 15x^2 + 5x$

  (b) $3ab(4a^3 + 2b^2 - 6) = 12a^4b + 6ab^3 - 18ab$

7. (a) $(7x^2 - 8) = 1(7x^2 - 8) = 7x^2 - 8$

  (b) $+(5x^2 + 6) = +1(5x^2 + 6) = 5x^2 + 6$

  (c) $-(3x + 2y - 6) = -1(3x + 2y - 6) = -3x - 2y + 6$

  (d) $-5x^2(x + 2xy) = -5x^3 - 10x^3y$

  (e) $\dfrac{3}{4}(8x^2 + 12x - 3) = \dfrac{3}{4}(8x^2) + \dfrac{3}{4}(12x) - \dfrac{3}{4}(3)$

  $= 6x^2 + 9x - \dfrac{9}{4}$

8. $-7(a + b) - 8a(2 - 3b) + 5a$
$= -7a - 7b - 16a + 24ab + 5a$
$= -18a + 24ab - 7b$

9. $-2\{4x - 3[x - 2x(1 + x)]\}$
$= -2\{4x - 3[x - 2x - 2x^2]\}$
$= -2\{4x - 3x + 6x + 6x^2\}$
$= -8x + 6x - 12x - 12x^2 = -14x - 12x^2$

10. $3\{-2a + [b - (3b - a)]\}$
$= 3\{-2a + [b - 3b + a]\}$
$= 3\{-2a + b - 3b + a\}$
$= -6a + 3b - 9b + 3a$
$= -6a + 3b + 3b - 9b = -3a - 6b$

1.6 Practice Problems

1. When $x = -5$
$-6 + 3x = -6 + 3(-5)$
$= -6 - 15 = -21$

2. When $x = -3$
$2x^2 + 3x - 8 = 2(-3)^2 + 3(-3) - 8$
$= 2(9) + 3(-3) - 8$
$= 18 - 9 - 8 = 9 - 8 = 1$

3. When $x = -3$ and $y = 4$
$(x - 3)^2 - 2xy$
$= [(-3) + (-3)]^2 - 2(-3)(4)$
$= (-6)^2 + 24 = 36 + 24 = 60$

4. Evaluate when $x = -4$
   (a) $(-3x)^2 = [-3(-4)]^2$
   $= (12)^2 = 144$
   (b) $-3x^2 = -3(-4)^2$
   $= -3(16) = -48$

5. When $C = 70°$
$F = \dfrac{9}{5}C + 32$
$F = \dfrac{9}{5}(70) + 32$
$F = 9(14) + 32$
$F = 158°$

6. When $L = 128$, $g = 32$
$T = 2\pi\sqrt{\dfrac{L}{g}}$
$T = 2\pi\sqrt{\dfrac{128}{32}}$
$T = 2\pi\sqrt{4}$
$T = 2(3.14)(2)$
$T = 12.56$
The period is about 12.6 seconds.

7. When $p = 600$, $r = 9\%$, and $t = 3$
$A = p(1 + rt)$
$A = 600[1 + (0.09)(3)]$
$A = 600(1 + 0.27)$
$A = 600(1.27)$
The amount is $762.00.

8. When $l = 0.76$, $w = 0.38$
$p = 2l + 2w$
$p = 2(0.76) + 2(0.38)$
$p = 1.52 + 0.76$
The perimeter is 2.28 centimeters.

9. When $a = 12$ and $b = 14$
$A = \dfrac{1}{2}ab$
$A = \dfrac{1}{2}(12)(14)$
$A = 6(14)$
The area is 84 square meters.

10. When $r = 6$, $h = 10$
$V = \pi r^2 h$
$V = \pi(6)^2(10)$
$V = 3.14(36)(10)$
The volume is approximately 1130.4 cubic meters.

Chapter 2

2.1 Practice Problems

1. (a) Is $x = -5$ the root of the equation $2x + 3 = 3x + 8$? Replace x by the value (-5) in the equation.
$$2(-5) + 3 \overset{?}{=} 3(-5) + 8$$
$$-10 + 3 \overset{?}{=} -15 + 8$$
$$-7 = -7 \checkmark$$
Since we obtain a true statement, $x = -5$ is a root of $2x + 3 = 3x + 8$

(b) Is $a = -\dfrac{3}{2}$ a solution of the equation $6a - 5 = -4a + 10$?

Replace a by the value $-\dfrac{3}{2}$ in the equation.
$$6\left(-\dfrac{3}{2}\right) - 5 \overset{?}{=} -4\left(-\dfrac{3}{2}\right) + 10$$
$$-9 - 5 \overset{?}{=} 6 + 10$$
$$-14 \neq 16$$
This last statement is not true. Thus $a = -\dfrac{3}{2}$ is not a solution of $6a - 5 = -4a + 10$.

2. Solve for y.
$y + 5.2 = -2.8$
$y + 5.2 - 5.2 = -2.8 - 5.2$
$y = -8$

Check. $y + 5.2 = -2.8$
$-8 + 5.2 \overset{?}{=} -2.8$
$-2.8 = -2.8$ ✓

3. Solve for w.
$\dfrac{1}{5}w = -6$
$5\left(\dfrac{1}{5}\right)w = -6(5)$
$w = -30$

Check. $\dfrac{1}{5}w = -6$
$\dfrac{1}{5}(-30) \overset{?}{=} -6$
$-6 = -6$ ✓

4. Solve.
$-7x - 2 = 26$
$-7x - 2 + 2 = 26 + 2$
$-7x = 28$
$\dfrac{-7x}{-7} = \dfrac{28}{-7}$
$x = -4$

Check. $-7x - 2 = 26$
$-7(-4) - 2 \overset{?}{=} 26$
$28 - 2 \overset{?}{=} 26$
$26 = 26$ ✓

5. Solve.
$8w - 3 = 2w - 7w + 4$
$8w - 3 = -5w + 4$
$8w + 5w - 3 = -5w + 5w + 4$
$13w - 3 = 4$
$13w - 3 + 3 = 4 + 3$
$13w = 7$
$\dfrac{13w}{13} = \dfrac{7}{13}$
$w = \dfrac{7}{13}$

6. Solve.
$a - 4(2a - 7) = 3(a + 6)$
$a - 8a + 28 = 3a + 18$
$-7a + 28 = 3a + 18$
$-7a - 3a + 28 = 3a - 3a + 18$
$-10a + 28 = 18$
$-10a + 28 - 28 = 18 - 28$
$-10a = -10$
$\dfrac{-10a}{-10} = \dfrac{-10}{-10}$
$a = 1$

Check. $a - 4(2a - 7) = 3(a + 6)$
$1 - 4[2(1) - 7] \overset{?}{=} 3(1 + 6)$
$1 - 4(2 - 7) \overset{?}{=} 3(1 + 6)$
$1 - 4(-5) \overset{?}{=} 3(1 + 6)$
$1 + 20 \overset{?}{=} 3(7)$
$21 = 21$ ✓

7. Solve.
Multiply each term by the LCD = 12.
$$\dfrac{y}{3} + \dfrac{1}{2} = 5 + \dfrac{y - 9}{4}$$
$$12\left(\dfrac{y}{3}\right) + 12\left(\dfrac{1}{2}\right) = 12(5) + 12\left(\dfrac{y}{4}\right) - 12\left(\dfrac{9}{4}\right)$$
$$4y + 6 = 60 + 3y - 27$$
$$4y + 6 = 3y + 33$$
$$4y - 3y + 6 = 3y - 3y + 33$$
$$y + 6 = 33$$
$$y + 6 - 6 = 33 - 6$$
$$y = 27$$

Check.
$$\frac{y}{3} + \frac{1}{2} = 5 + \frac{y-9}{4}$$

$$\frac{27}{3} + \frac{1}{2} \overset{?}{=} 5 + \frac{27-9}{4}$$

$$9 + \frac{1}{2} \overset{?}{=} 5 + \frac{18}{4}$$

$$\frac{18}{2} + \frac{1}{2} \overset{?}{=} \frac{10}{2} + \frac{9}{2}$$

$$\frac{19}{2} = \frac{19}{2} \quad \checkmark$$

8. Solve.
$$4(0.01x + 0.09) - 0.07(x - 8) = 0.83$$
$$0.04x + 0.36 - 0.07x + 0.56 = 0.83$$
Multiply each term by 100.
$$100(0.04x) + 100(0.36) - 100(0.07x) + 100(0.56) = 100(0.83)$$
$$4x + 36 - 7x + 56 = 83$$
$$-3x + 92 = 83$$
$$-3x + 92 - 92 = 83 - 92$$
$$-3x = -9$$
$$\frac{-3x}{-3} = \frac{-9}{-3}$$
$$x = 3$$

Check.
$$4(0.01x + 0.09) - 0.07(x - 8) = 0.83$$
$$4[0.01(3) + 0.09] - 0.07(3 - 8) \overset{?}{=} 0.83$$
$$4(0.03 + 0.09) - 0.07(-5) \overset{?}{=} 0.83$$
$$4(0.12) + 0.35 \overset{?}{=} 0.83$$
$$0.48 + 0.35 \overset{?}{=} 0.83$$
$$0.83 = 0.83 \quad \checkmark$$

2.2 Practice Problems

1. Solve for W.
$$P = 2L + 2W$$
$$P - 2L = 2W$$
$$\frac{P - 2L}{2} = \frac{2W}{2}$$
$$\frac{P - 2L}{2} = W$$

2. Solve for w.
$$8 + 12wx = 18 - 7wx$$
$$8 + 12wx + 7wx = 18 - 7wx + 7wx$$
$$8 + 19wx = 18$$
$$8 - 8 + 19wx = 18 - 8$$
$$19wx = 10$$
$$\frac{19wx}{19x} = \frac{10}{19x}$$
$$w = \frac{10}{19x}$$

3. Solve for a.
$$3(2ax - y) = \frac{1}{2}(ax + 2y)$$
$$6ax - 3y = \frac{1}{2}ax + y$$
$$2(6ax) - 2(3y) = 2\left(\frac{1}{2}ax\right) + 2(y)$$
$$12ax - 6y = ax + 2y$$
$$12ax - ax - 6y = ax - ax + 2y$$
$$11ax - 6y = 2y$$
$$11ax - 6y + 6y = 2y + 6y$$
$$11ax = 8y$$
$$\frac{11ax}{11x} = \frac{8y}{11x}$$
$$a = \frac{8y}{11x}$$

4. Solve for b.
$$-2(ab - 3x) + 2(8 - ab) = 5x + 4ab$$
$$-2ab + 6x + 16 - 2ab = 5x + 4ab$$
$$-4ab + 6x + 16 = 5x + 4ab$$
$$-4ab + 4ab + 6x + 16 = 5x + 4ab + 4ab$$
$$6x + 16 = 5x + 8ab$$
$$6x - 5x + 16 = 5x - 5x + 8ab$$
$$x + 16 = 8ab$$
$$\frac{x + 16}{8a} = b$$

5. (a) Solve for h.
$$A = 2\pi rh + 2\pi r^2$$
$$A - 2\pi r^2 = 2\pi rh + 2\pi r^2 - 2\pi r^2$$
$$A - 2\pi r^2 = 2\pi rh$$
$$\frac{A - 2\pi r^2}{2\pi r} = \frac{2\pi rh}{2\pi r}$$
$$\frac{A - 2\pi r^2}{2\pi r} = h$$

(b) Solve for h when
$A = 100$, $\pi \approx 3.14$, $r = 2.0$
$$\frac{A - 2\pi r^2}{2\pi r} = h$$
$$\frac{100 - 2(3.14)(2.0)^2}{2(3.14)(2.0)} = h$$
$$\frac{100 - (6.28)(4)}{6.28(2.0)} = h$$
$$\frac{100 - 25.12}{12.56} = h$$
$$\frac{74.88}{12.56} = h$$
$$5.96 = h$$

2.3 Practice Problems

1. Solve $|3x - 4| = 23$ and check.
We have two equations.

$3x - 4 = 23$	or	$3x - 4 = -23$
$3x = 27$		$3x = -19$
$x = 9$		$x = -\dfrac{19}{3}$

Check.   if $x = 9$
$$|3x - 4| = 23$$
$$|3(9) - 4| \overset{?}{=} 23$$
$$|27 - 4| \overset{?}{=} 23$$
$$|23| \overset{?}{=} 23$$
$$23 = 23 \quad \checkmark$$

if $x = -\dfrac{19}{3}$
$$|3x - 4| = 23$$
$$\left|3\left(-\frac{19}{3}\right) - 4\right| \overset{?}{=} 23$$
$$|-19 - 4| \overset{?}{=} 23$$
$$|-23| \overset{?}{=} 23$$
$$23 = 23 \quad \checkmark$$

2. Solve. $\left|\dfrac{2}{3}x + 4\right| = 2$
We have two equations.

$\dfrac{2}{3}x + 4 = 2$	or	$\dfrac{2}{3}x + 4 = -2$
$2x + 12 = 6$		$2x + 12 = -6$
$2x = -6$		$2x = -18$
$x = -3$		$x = -9$

Check.   if $x = -3$
$$\left|\frac{2}{3}(-3) + 4\right| \overset{?}{=} 2$$
$$|-2 + 4| \overset{?}{=} 2$$
$$|2| \overset{?}{=} 2$$
$$2 = 2 \quad \checkmark$$

if $x = -9$
$$\left|\frac{2}{3}(-9) + 4\right| \overset{?}{=} 2$$
$$|-6 + 4| \overset{?}{=} 2$$
$$|-2| \overset{?}{=} 2$$
$$2 = 2 \quad \checkmark$$

3. Solve. $|2x + 1| + 3 = 8$
First change the equation so that the absolute value expression alone is on one side of the equation.
$$|2x + 1| + 3 = 8$$
$$|2x + 1| + 3 - 3 = 8 - 3$$
$$|2x + 1| = 5$$
We have the two equations

$2x + 1 = 5$	or	$2x + 1 = -5$
$2x = 4$		$2x = -6$
$x = 2$		$x = -3$

Check.   if $x = 2$
$$|2(2) + 1| + 3 \overset{?}{=} 8$$
$$|5| + 3 \overset{?}{=} 8$$
$$8 = 8 \quad \checkmark$$

if $x = -3$
$$|2(-3) + 1| + 3 \overset{?}{=} 8$$
$$|-5| + 3 \overset{?}{=} 8$$
$$8 = 8 \quad \checkmark$$

4. Solve. $|x - 6| = |5x + 8|$
We write the two possible equations and solve each equation.

$x - 6 = 5x + 8$	or	$x - 6 = -(5x + 8)$
$-4x = 14$		$x - 6 = -5x - 8$
$x = \dfrac{14}{-4} = -\dfrac{7}{2}$		$6x = -2$
		$x = \dfrac{-2}{6} = -\dfrac{1}{3}$

Check.   if $x = -\dfrac{7}{2}$

$$\left|-\dfrac{7}{2} - 6\right| \stackrel{?}{=} \left|5\left(-\dfrac{7}{2}\right) + 8\right|$$

$$\left|-\dfrac{7}{2} - 6\right| \stackrel{?}{=} \left|-\dfrac{35}{2} + 8\right|$$

$$\left|-\dfrac{7}{2} - \dfrac{12}{2}\right| \stackrel{?}{=} \left|-\dfrac{35}{2} + \dfrac{16}{2}\right|$$

$$\left|-\dfrac{19}{2}\right| \stackrel{?}{=} \left|-\dfrac{19}{2}\right|$$

$$\dfrac{19}{2} = \dfrac{19}{2} \checkmark$$

if $x = -\dfrac{1}{3}$

$$\left|-\dfrac{1}{3} - 6\right| \stackrel{?}{=} \left|5\left(-\dfrac{1}{3}\right) + 8\right|$$

$$\left|-\dfrac{1}{3} - 6\right| \stackrel{?}{=} \left|-\dfrac{5}{3} + 8\right|$$

$$\left|-\dfrac{19}{3}\right| \stackrel{?}{=} \left|\dfrac{19}{3}\right|$$

$$\dfrac{19}{3} = \dfrac{19}{3} \checkmark$$

2.4 Practice Problems

1. $\dfrac{3}{4}x + 20 = 110$

   $3x + 80 = 440$

   $3x = 360$

   $x = 120$

2. Let n = number of hours.
   For 12 months per year, we multiply \$400.00 for rent by 12.
   Let $8.00n$ = the cost for using the computer for n hours.

   $$400(12) + 8n = 7680$$
   $$4{,}800 + 8n = 7680$$
   $$8n = 2880$$
   $$n = 360$$

   They used the computer for 360 hours.

3. Let n = the first odd integer.
   Let $n + 2$ = the second odd integer.
   Let $n + 4$ = the third odd integer.

   $$n + n + 2 + n + 4 = 195$$
   $$3n + 6 = 195$$
   $$3n = 189$$
   $$n = 63$$

   Then $n + 2 = 65$ and $n + 4 = 67$
   The three consecutive odd integers are 63, 65, and 67.

4. Let x = the length of the second side.
   Then $2x$ = the length of the first side
   and $3x - 6$ = the length of the third side.

   $$p = a + b + c$$
   $$162 = x + 2x + 3x - 6$$
   $$162 = 6x - 6$$
   $$168 = 6x$$
   $$x = 28$$

   The 1st side is $2x = 2(28) = 56$ meters.
   The 2nd side is $x = 28$ meters.
   The 3rd side is $3x - 6 = 3(28) - 6 = 78$ meters.

2.5 Practice Problems

1. Let x = amount of sales last year
   $0.15x$ = the increase in sales over last year

   $$x + 0.15x = 6900$$
   $$1.15x = 6900$$
   $$x = 6000$$

   6000 computer workstations were sold last year.

2. Let a = the number of cars Alicia sold. Then $43 - a$ = the number of cars Heather sold. If Alicia doubles her sales next month, then she will sell $2a$ cars. If Heather triples hers, she will sell $3(43 - a)$ cars.

   $$2a + 3(43 - a) = 108$$
   $$2a + 129 - 3a = 108$$
   $$-a + 129 = 108$$
   $$-a = -21$$
   $$a = 21$$

   Therefore, Alicia sold 21 cars and Heather sold $43 - 21 = 22$ cars.

3. $I = prt$
   $I = 7000(0.12)(4)$
   The interest is \$3360.00.

4. Let x = the amount invested at 8%, then $5500 - x$ = the amount invested at 12%.

   $$0.08x + 0.12(5500 - x) = 540$$
   $$0.08x + 660 - 0.12x = 540$$
   $$-0.04x = -120$$
   $$x = \$3000.00$$

   Therefore, she invested \$3000.00 at 8% and $5500 - 3000 = \$2500.00$ at 12%.

5.

		A	B	C
		Number of Grams of the Mixture	% Pure Gold	Number of Grams of Pure Gold
68% pure gold source		x	68%	$0.68x$
83% pure gold source		$200 - x$	83%	$0.83(200 - x)$
Final 80% mixture of pure gold		200	80%	$0.80(200)$

Now form an equation from the entries in Column C.

$$0.68x + 0.83(200 - x) = 0.80(200)$$
$$0.68x + 166 - 0.83x = 160$$
$$-0.15x = -6$$
$$x = 40$$

if $x = 40$, then $200 - x = 200 - 40 = 160$.
Thus, we have 40 grams of 68% pure gold and 160 grams of 83% pure gold.

6.

	r	t	d
Steady speed	x	4	$4x$
Slower speed	$x - 10$	2	$2(x - 10)$
Entire trip	Not given	6	352

$$4x + 2(x - 10) = 352$$
$$4x + 2x - 20 = 352$$
$$6x - 20 = 352$$
$$6x = 372$$
$$x = 62$$

Thus, Wally drove 62 miles per hour for 4 hours and $x - 10 = 52$ miles per hour for 2 hours.

2.6 Practice Problems

1. (a) $-1 > -2$

   (b) $\dfrac{2}{3} < \dfrac{3}{4}$

   (c) $-0.561 < 0.5555$

2. **(a)** $(-8 - 2) \; ? \; (-3 - 12)$
$\qquad (-10) > (-15)$
(b) $|-15 + 8| \; ? \; |7 - 13|$
$\qquad |-7| \; ? \; |-6|$
$\qquad\quad 7 > 6$
(c) $(x - 3) > (x - 4)$
Since x represents the same number in both expressions, $x - 3$ will be greater than $x - 4$.

3. **(a)** $x > 3.5$

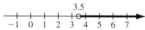

(b) $x \leq -10$

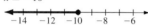

(c) $x \geq -2$

(d) $-4 > x$

4. **(a)** $x + 2 > -12 \qquad$ To check, we choose $x = -13$
$\qquad\quad x > -14 \qquad$ Is $-13 + 2 > -12$
$\qquad\qquad\qquad\qquad\qquad -11 > -12$ ✓ True

(b) $12x \leq 11x + 5$
$\qquad\quad x \leq 5$

To check we choose $x = 5$ and $x = 4$.
Is $12(5) \leq 11(5) + 5 \qquad$ Is $12(4) \leq 11(4) + 5$
$\quad 60 \leq 55 + 5 \qquad\qquad\quad 48 \leq 44 + 5$
$\quad 60 \leq 60$ ✓ True $\qquad\quad 48 \leq 49$ ✓ True

5. $8x - 8 \geq 5x + 1$
$\quad 3x \geq 9$
$\quad\; x \geq 3$

6. $2 - 12x > 7(1 - x)$
$\;\; 2 - 12x > 7 - 7x$
$\qquad -5x > 5$
$\qquad\quad x < -1$

7. $0.5x - 0.4 \leq -0.8x + 0.9$
$\qquad\quad 1.3x \leq 1.3$
$\qquad\qquad x \leq 1$

8. $\dfrac{1}{5}(x - 6) < \dfrac{1}{3}(x - 2)$

$\dfrac{1}{5}x - \dfrac{6}{5} < \dfrac{1}{3}x - \dfrac{2}{3}$

$3x - 18 < 5x - 10$
$\qquad -2x < 8$
$\qquad\quad x > -4$

9. $\dfrac{1}{5} - \dfrac{1}{10}x \leq \dfrac{3}{10}x + 1$

$2 - x \leq 3x + 10$
$\quad -4x \leq 8$
$\qquad x \geq -2$

2.7 Practice Problems

1. $-8 < x < -2$

2. $-1 \leq x \leq 5$

3. $-10 \leq x \leq -5.5$

4. $200 \leq x \leq 950$

5. $x < 8 \quad$ or $\quad x > 12$

6. $x \leq -6 \quad$ or $\quad x > 3$

7.

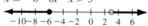

8. $3x - 4 < -1 \quad$ or $\quad 2x + 3 > 13$
$\qquad 3x < 3 \qquad\qquad\qquad 2x > 10$
$\qquad\; x < 1 \qquad$ or $\qquad\; x > 5$

9. $3x + 6 > -6 \quad$ and $\quad 4x + 5 < 1$
$\qquad 3x > -12 \qquad\qquad\qquad 4x < -4$
$\qquad\; x > -4 \quad$ and $\qquad x < -1$
$\qquad\qquad\quad -4 < x < -1$

10. $-2x + 3 < -7 \quad$ and $\quad 7x - 1 > -15$
$\qquad\; -2x < -10 \qquad\qquad\quad 7x > -14$
$\qquad\qquad x > 5 \quad$ and $\qquad\; x > -2$
$x > 5$ and at the same time $x > -2$. Thus $x > 5$ is the solution to the compound inequality.

11. $-3x - 11 < -26 \quad$ and $\quad 5x + 4 < 14$
$\qquad\quad -3x < -15 \qquad\qquad\quad 5x < 10$
$\qquad\qquad x > 5 \quad$ and $\qquad\; x < 2$
Now clearly it is impossible for one number to be greater than 5 and at the same time less than 2. There is no solution.

2.8 Practice Problems

1. $|x| < 2$
$\quad -2 < x < 2$

$-2 < x < 2$

2. $|x - 6| < 15$

$-15 < x - 6 < 15$

$-15 + 6 < x - 6 + 6 < 15 + 6$

$-9 < x < 21$

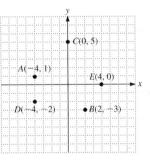 (number line from -12 to 24, open circles at -9 and 21)

$-9 < x < 21$

3. $\left|x + \dfrac{3}{4}\right| \le \dfrac{7}{6}$

$-\dfrac{7}{6} \le x + \dfrac{3}{4} \le \dfrac{7}{6}$

$-14 \le 12x + 9 \le 14$

$-14 - 9 \le 12x + 9 - 9 \le 14 - 9$

$-23 \le 12x \le 5$

$\dfrac{-23}{12} \le \dfrac{12x}{12} \le \dfrac{5}{12}$

$-1\dfrac{11}{12} \le x \le \dfrac{5}{12}$

(number line from -4 to 4 with closed circles at $-1\frac{11}{12}$ and $\frac{5}{12}$)

$-1\frac{11}{12} \le x \le \frac{5}{12}$

4. $|2 + 3(x - 1)| < 20$

$|2 + 3x - 3| < 20$

$|-1 + 3x| < 20$

$-20 < -1 + 3x < 20$

$-20 + 1 < 1 - 1 + 3x < 20 + 1$

$-19 < 3x < 21$

$\dfrac{-19}{3} < \dfrac{3x}{3} < \dfrac{21}{3}$

$\dfrac{-19}{3} < x < 7$

$-6\dfrac{1}{3} < x < 7$

(number line from -7 to 7 with open circles at $-6\frac{1}{3}$ and 7)

$-6\frac{1}{3} < x < 7$

5. $|x| > 2.5$

$x > 2.5$ or $x < -2.5$

(number line from -3 to 3 with open circles at -2.5 and 2.5)

$x < -2.5 \text{ or } x > 2.5$

6. $|x + 6| > 2$

$x + 6 > 2$ or $x + 6 < -2$

$x > -4$ or $x < -8$

(number line from -10 to -2 with open circles at -8 and -4)

$x < -8 \text{ or } x > -4$

7. $|-5x - 2| > 13$

$-5x - 2 > 13$ or $-5x - 2 < -13$

$-5x > 15$ or $-5x < -11$

$x < -3$ or $x > \dfrac{11}{5}$

$x > 2\dfrac{1}{5}$

(number line from -4 to 4 with open circles at -3 and $2\frac{1}{5}$)

8. $\left|4 - \dfrac{3}{4}x\right| \ge 5$

$4 - \dfrac{3}{4}x \ge 5$ or $4 - \dfrac{3}{4}x \le -5$

$16 - 3x \ge 20$ $\qquad$ $16 - 3x \le -20$

$-3x \ge 4$ $\qquad\qquad$ $-3x \le -36$

$x \le \dfrac{-4}{3}$ $\qquad\qquad$ $x \ge 12$

$x \le -1\dfrac{1}{3}$

(number line with closed circles at $-1\frac{1}{3}$ and 12)

9.
$$|d - s| \le 0.37$$
$$|d - 276.53| \le 0.37$$
$$-0.37 \le d - 276.53 \le 0.37$$
$$-0.37 + 276.53 \le d - 276.53 + 276.53 \le 0.37 + 276.53$$
$$276.16 \le d \le 276.90$$

Thus the diameter of the transmission must be at least 276.16 millimeters, but not greater than 276.90 millimeters.

Chapter 3

3.1 Practice Problems

1. $A\ (-4, 1)$
$B\ (2, -3)$
$C\ (0, 5)$
$D\ (-4, -2)$
$E\ (4, 0)$

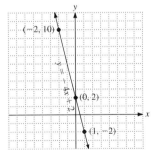

2. $y = -4x + 2$

x	y
-2	10
0	2
1	-2

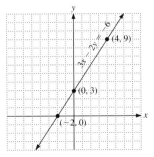

3. $3x - 2y = -6$

Let $x = 0$

$3(0) - 2y = -6$

$-2y = -6$

$y = 3$

The y-intercept is $(0, 3)$

Now let $y = 0$

$3x - 2(0) = -6$

$3x = -6$

$x = -2$

The x-intercept is $(-2, 0)$.

4. $5x + 8 = 9x$

$-4x = -8$

$x = 2$

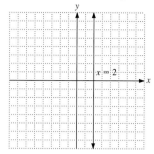

The graph $x = 2$ is a vertical line 2 units to the right of the y-axis.

5. Let $n = 0, 1,000$, and $2,000$. We then have

n	C
0	300
1000	450
2000	600

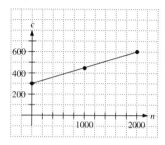

3.2 Practice Problems

1. Slope $= m = \dfrac{-2 - 1}{-5 - (-6)} = \dfrac{-3}{1} = -3$

2. **(a)** $m = \dfrac{-3.4 - (-6.2)}{-2.2 - (1.8)} = \dfrac{2.8}{-4} = -0.7$

(b) $m = \dfrac{-\dfrac{3}{4} - \left(-\dfrac{1}{2}\right)}{\dfrac{4}{15} - \left(\dfrac{1}{5}\right)} = \dfrac{-\dfrac{1}{4}}{\dfrac{1}{15}} = -\dfrac{15}{4}$

(c) $m = \dfrac{-14 - (12)}{-3 - (-3)} = -\dfrac{26}{0} =$ undefined

(d) $m = \dfrac{-6 - (-6)}{-5 - 7} = -\dfrac{0}{12} = 0$

3. $m = \dfrac{25.95}{1296} = 0.02$ or $\dfrac{1}{50}$

4. $m_k = \dfrac{-3 - (-2)}{-5 - 7} = \dfrac{-1}{-12} = \dfrac{1}{12}$

The slope of a line parallel to k is the same.

5. $m_l = \dfrac{-2 - 0}{6 - (5)} = \dfrac{-2}{1} = -2$

$m_h = \dfrac{1}{2}$

6. $m_{ab} = \dfrac{1 - 5}{-1 - 1} = \dfrac{-4}{-2} = 2$

$m_{bc} = \dfrac{-1 - 1}{-2 - (-1)} = \dfrac{-2}{-1} = 2$

Since the slopes are the same and have a point in common, the points lie on the same line.

7. $m_1 = \dfrac{4850 - 3650}{4.5 - 3.5} = \dfrac{1200}{1} = 1200$

$m_2 = \dfrac{3650 - 1010}{3.5 - 1.3} = \dfrac{2640}{2.2} = 1200$

The slopes of the two portions of the flight are the same. The two portions of the flight have one point in common. Therefore the jet is descending in a straight line.

3.3 Practice Problems

1. $y = mx + b$

$m = 4$ and $b = -\dfrac{3}{2}$

$y = 4x - \dfrac{3}{2}$

2. y-intercept is at $(0, -3)$, thus $b = -3$.
Another point on the line is $(5, 0)$. Thus

$$m = \dfrac{(-3) - 0}{5 - 0} = \dfrac{3}{5}$$

$$y = \dfrac{3}{5}x - 3$$

3. $3x - 4y = -8$

$-4y = -3x - 8$

$y = \dfrac{3}{4}x + 2$

The slope is $\dfrac{3}{4}$ and
the y-intercept is 2.

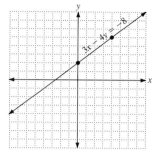

4. $y - y_1 = m(x - x_1)$

$y - (-2) = \dfrac{3}{4}(x - 5)$

$y + 2 = \dfrac{3}{4}x - \dfrac{15}{4}$

$4y + 8 = 3x - 15$

$4y = 3x - 23$

$y = \dfrac{3}{4}x - \dfrac{23}{4}$

5. $(-4, 1)$ and $(-2, -3)$

$$m = \dfrac{-3 - 1}{-2 - (-4)} = \dfrac{-4}{2} = -2$$

Substituting $m = -2$ and $(x_1, y_1) = (-4, 1)$

$y - 1 = -2(x + 4)$

$y - 1 = -2x - 8$

$2x + y = -7$

6. First, we need to find the slope of the line $5x - 3y = 10$. We do this by writing the equation in slope–intercept form

$$5x - 3y = 10$$

$$-3y = -5x + 10$$

$$y = \dfrac{5}{3}x - \dfrac{10}{3}$$

The slope is $\frac{5}{3}$. A line parallel to this passing through $(4, -5)$ would have an equation

$$y - (-5) = \dfrac{5}{3}(x - 4)$$

$$y + 5 = \dfrac{5}{3}x - \dfrac{20}{3}$$

$$3y + 15 = 5x - 20$$

$$-5x + 3y = -35$$

$$5x - 3y = 35$$

7. Find the slope of the line

$$6x + 3y = 7$$

$$3y = -6x + 7$$

$$y = -2x + \dfrac{7}{3}$$

The slope is -2. A line perpendicular to this passing through $(-4, 3)$ would have a slope of $\frac{1}{2}$.

$$y - 3 = \dfrac{1}{2}(x + 4)$$

$$y - 3 = \dfrac{1}{2}x + 2$$

$$2y - 6 = x + 4$$

$$-x + 2y = 10$$

$$x - 2y = -10$$

3.4 Practice Problems

1. The boundary line is $y = 3x + 1$. Graph the boundary line with a dashed line since the inequality contains $>$. Substituting $(0, 0)$ into the inequity $y > 3x + 1$ gives $0 > 1$, which is false. Thus we shade on the opposite side of the boundary from $(0, 0)$. The solution is the shaded region above the dashed line.

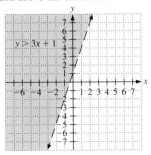

2. The boundary line is $-4x + 5y = -10$. Graph the boundary line with a solid line because the inequality contains $\leq$. Substituting $(0, 0)$ into the inequality $-4x + 5y \leq -10$ gives $0 \leq -10$ which is false. Therefore shade the region opposite the sides of the boundary from $(0, 0)$. The solution is the solid line and the shaded region below the solid line.

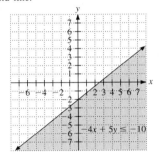

3. The boundary line is $3y + x = 0$. We use a dashed line since the inequality contains $<$. We cannot use $(0, 0)$ to test the inequality, let's pick $(-2, -3)$. Substituting $(-2, -3)$ into $3y + x < 0$ gives $-11 < 0$ which is true. Therefore shade on the same side of the boundary line as $(-2, -3)$.

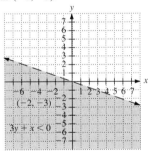

4. Graph the vertical line $x = -2$. We use a dashed line since the inequality contains $>$. The region we want to shade is the region to the right of the dashed line.

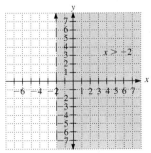

5. Simplify the original inequality by dividing each side by 6, which gives $y \leq 3$. We use a solid line since the inequality contains $\leq$. The region we want to shade is the region below the horizontal line. The solution is the line $y = 3$ and the region below that line.

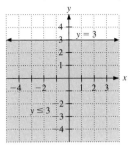

3.5 Practice Problems

1. $\{(1936, 11.5), (1948, 11.9), (1960, 11.0), (1972, 11.07), (1984, 10.97), (1996, 10.94)\}$

Year	1936	1948	1960	1972	1984	1996
Time	11.5	11.9	11.0	11.07	10.97	10.94

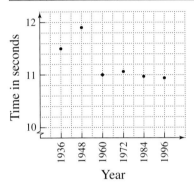

2. (a) Domain of $h = \{2, 3, 4, 6\}$
Range of $h = \{1\}$
h is a function.
  (b) Domain of $p = \{158, 160, 161, 163\}$
Range of $p = \{89.2, 94.9, 98.6, 101.4, 102.3\}$
p is not a function.
  (c) Domain is all real numbers less than 6.
Range of $x = \{0, 1, 2, 3, 4, 5\}$
x is a function.

3. If $f(x) = 2x^2 - 8$ find
  (a) $f(-3) = 2(-3)^2 - 8 = 10$
  (b) $f(4) = 2(4)^2 - 8 = 24$
  (c) $f\left(\dfrac{1}{2}\right) = 2\left(\dfrac{1}{2}\right)^2 - 8 = -7\dfrac{1}{2}$

3.6 Practice Problems

1. $f(x) = -3x + 2$

x	$f(x)$
-1	5
0	2
1	-1

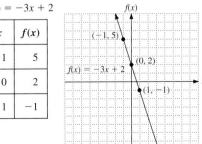

2. Income = 10,000 + 15% of total sales

Let x = amount of total sales in dollars.

$$d(x) = 10,000 + 0.15(x)$$
$$d(0) = 10,000$$
$$d(40,000) = 10,000 + 0.15(40,000) = 16,000$$
$$d(80,000) = 10,000 + 0.15(80,000) = 22,000$$

Modify the table by recording x and $d(x)$ in thousands to make the task of graphing easier.

x	d(x)
0	10
40	16
80	22

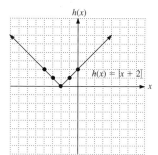

3. $h(x) = |x + 2|$

x	h(x)
−4	2
−3	1
−2	0
−1	1
0	2

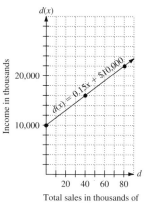

4. (a) $r(x) = (x − 2)^2$

x	r(x)
0	4
1	1
2	0
3	1
4	4

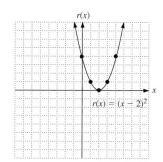

(b) $s(x) = x^2 + 2$

x	s(x)
−2	6
−1	3
0	2
1	3
2	6

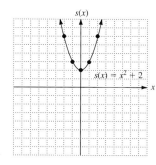

5. $f(x) = x^3 − 2$

x	f(x)
−1	−3
0	−2
1	−1
2	6

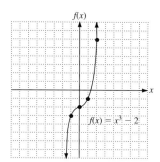

6. $q(x) = -\dfrac{4}{x}$

x	q(x)
8	$-\dfrac{1}{2}$
4	−1
2	−2
1	−4
$\dfrac{1}{2}$	−8
$-\dfrac{1}{2}$	8
−1	4
−2	2
−4	1
−8	$\dfrac{1}{2}$

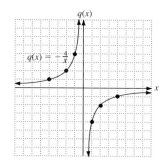

7. (a)

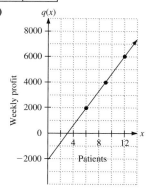

(b) As the number of patients per hour increases so does the weekly profit.

(c) 0 profit

(d) Loss of $2,000.00

(e) In terms of mathematics, the function will increase indefinitely; however, is it reasonable to think that a doctor could see an infinite number of patients?

8. (a)

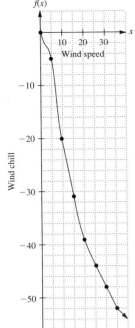

(b) −49°F
(c) −54°F
(d) Slightly more than 11 mph
(e) In theory based on our graph, the domain is all nonnegative real numbers. The range is all nonpositive.

9.

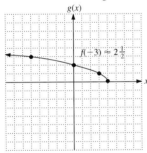

$f(-3) \approx 2\frac{1}{2}$

(a) Approximately 2.5
(b) Approximately 1.6
(c) The curve is moving upward more slowly.

Chapter 4

4.1 Practice Problems

1. Substitute $(-3, 4)$ into the first equation to see if the ordered pair is a solution.

$$2x + 3y = 6$$
$$2(-3) + 3(4) \overset{?}{=} 6$$
$$-6 + 12 \overset{?}{=} 6$$
$$6 = 6 \checkmark$$

Likewise, we will determine if $(-3, 4)$ is a solution to the second equation.

$$3x - 4y = 7$$
$$3(-3) - 4(4) \overset{?}{=} 7$$
$$-9 - 16 \overset{?}{=} 7$$
$$-25 \neq 7$$

Since $(-3, 4)$ is not a solution to each equation in the system, it is not a solution to the system itself.

2. You can use any method we developed in Chapter 3 to graph each line. We will change each equation to slope–intercept form to graph.

$$3x + 2y = 10$$
$$2y = -3x + 10$$
$$y = -\frac{3}{2}x + 5$$

$$x - y = 5$$
$$y = x - 5$$

The lines intersect at the point $(4, -1)$. Thus $(4, -1)$ is a solution. We verify this by substituting $x = 4$ and $y = -1$ into the system of equations.

$$3x + 2y = 10 \qquad\qquad x - y = 5$$
$$3(4) + 2(-1) \overset{?}{=} 10 \qquad 4 - (-1) \overset{?}{=} 5$$
$$12 - 2 \overset{?}{=} 10 \qquad\qquad 5 = 5 \checkmark$$
$$10 = 10 \checkmark$$

3. $2x - y = 7$ [1]
$3x + 4y = -6$ [2]
Solve equation [1] for y.
$$-y = 7 - 2x$$
$$y = -7 + 2x \quad [3]$$
Substitute $-7 + 2x$ for y in equation [2].
$$3x + 4(-7 + 2x) = -6$$
$$3x - 28 + 8x = -6$$
$$11x - 28 = -6$$
$$11x = 22$$
$$x = 2$$
Substitute $x = 2$ into equation [3].
$$y = -7 + 2(2)$$
$$y = -7 + 4$$
$$y = -3$$
The solution is $x = 2$, $y = -3$, or $(2, -3)$.

4. $\dfrac{x}{2} + \dfrac{2y}{3} = 1$

$\dfrac{x}{3} + y = -1$

First we clear both equations of fractions.
$$3x + 4y = 6 \quad [1]$$
$$x + 3y = -3 \quad [2]$$
Step 1 Solve for x in equation [2].
$$x + 3y = -3$$
$$x = -3y - 3$$
Step 2 Substitute this expression for x into equation [1] and solve for y.
$$3(-3y - 3) + 4y = 6$$
$$-9y - 9 + 4y = 6$$
$$-5y = 15$$
$$y = -3$$
Step 3 Substitute $y = -3$ into equation [1] or [2].
$$x + 3y = -3$$
$$x + 3(-3) = -3$$
$$x - 9 = -3$$
$$x = 6$$
So our solution is $x = 6$ and $y = -3$, or $(6, -3)$.

5. $-3x + y = 5$ [1]

$\quad\;\; 2x + 3y = 4$ [2]

Multiply equation [1] by -3 and add to equation [2].

$$9x - 3y = -15$$
$$\underline{2x + 3y = \quad 4}$$
$$11x \quad\quad = -11$$
$$x = -1$$

Now substitute $x = -1$ into equation [1].

$$-3(-1) + y = 5$$
$$3 + y = 5$$
$$y = 2$$

The solution is $x = -1$, $y = 2$, or $(-1, 2)$.

6. $5x + 4y = 23$ [1]

$\;\; 7x - 3y = 15$ [2]

Multiply equation [1] by 3 and equation [2] by 4.

$$15x + 12y = \quad 69$$
$$\underline{28x - 12y = \quad 60}$$
$$43x \quad\quad\;\; = 129$$
$$x = 3$$

Now substitute $x = 3$ into equation [1] or [2].

$$5(3) + 4y = 23$$
$$15 + 4y = 23$$
$$4y = 8$$
$$y = 2$$

The solution is $x = 3$, $y = 2$, or $(3, 2)$.

7. $3x + 12y = 25$ [1]

$\;\; 2x - 6y = 12$ [2]

Multiply equation [2] by 2 and add to equation [1].

$$3x + 12y = 25$$
$$\underline{4x - 12y = 24}$$
$$7x \quad\quad = 49$$
$$x = 7$$

Now substitute $x = 7$ into equation [2].

$$2(7) - 6y = 12$$
$$14 - 6y = 12$$
$$-6y = -2$$
$$y = \frac{1}{3}$$

The solution is $x = 7$, $y = \frac{1}{3}$, or $\left(7, \frac{1}{3}\right)$.

8. $\;\; 4x - 2y = 6$ [1]

$\;\; -6x + 3y = 9$ [2]

Multiply equation [1] by 3 and equation [2] by 2 and add together.

$$12x - 6y = 18$$
$$\underline{-12x + 6y = 18}$$
$$0 = 36$$

This statement is of course false. Thus, we conclude that this system of equations is inconsistent, so there is **no solution.**

9. $\;\; 0.3x - 0.9y = \quad 1.8$ [3]

$\;\; -0.4x + 1.2y = -2.4$ [2]

Multiply both equations by 10 to obtain a more convenient form.

$$3x - 9y = \quad 18$$ [3]
$$-4x + 12y = -24$$ [4]

Multiply equation [3] by 4 and equation [4] by 3.

$$12x - 36y = \quad 72$$
$$\underline{-12x + 36y = -72}$$
$$0 = 0$$

This statement is always true. Hence these are dependent equations. There are an infinite number of solutions.

4.2 Practice Problems

1. Substitute $x = 3$, $y = -2$, $z = 2$ into each equation.

$$2(3) + 4(-2) + (2) \overset{?}{=} 0$$
$$6 - 8 + 2 \overset{?}{=} 0$$
$$-2 + 2 \overset{?}{=} 0$$
$$0 = 0 \;\; \checkmark$$

$$(3) - 2(-2) + 5(2) \overset{?}{=} 17$$
$$3 + 4 + 10 \overset{?}{=} 17$$
$$17 = 17 \;\; \checkmark$$
$$3(3) - 4(-2) + 2 \overset{?}{=} 19$$
$$9 + 8 + 2 \overset{?}{=} 19$$
$$19 = 19 \;\; \checkmark$$

Since we obtained three true statements, the ordered triple $(3, -2, 2)$ is a solution to the system.

2. $\;\; x + 2y + 3z = \quad 4$ [1]

$\;\; 2x + \;\; y - 2z = \quad 3$ [2]

$\;\; 3x + 3y + 4z = 10$ [3]

We eliminate x by multiplying equation [1] by -2 and adding it to equation [2].

$$-2x - 4y - 6z = -8$$ [4]
$$\underline{2x + \;\; y - 2z = \quad 3}$$ [2]
$$-3y - 8z = -5$$ [5]

Now we eliminate x by multiplying equation [1] by -3 and adding it to equation [3].

$$-3x - 6y - 9z = -12$$ [6]
$$\underline{3x + 3y + 4z = \quad 10}$$ [3]
$$-3y - 5z = -2$$ [7]

We now eliminate y and solve for z in the system formed by equation [5] and equation [7].

$$-3y - 8z = -5$$ [5]
$$-3y - 5z = -2$$ [7]

To do this we multiply equation [5] by -1 and add it to equation [7].

$$3y + 8z = \quad 5$$ [8]
$$\underline{-3y - 5z = -2}$$ [7]
$$3z = \quad 3$$
$$z = 1$$

Substitute $z = 1$ into equation [8] and solve for y.

$$3y + 8(1) = 5$$
$$3y + 8 = 5$$
$$3y = -3$$
$$y = -1$$

Substitute $z = 1$, $y = -1$ into equation [1] and solve for x.

$$x + 2(-1) + 3(1) = 4$$
$$x + 1 = 4$$
$$x = 3$$

The solution is $x = 3$, $y = -1$, $z = 1$ or $(3, -1, 1)$.

3. First we must transform each equation to an equivalent equation in the proper form.

$$3x - \;\; y - 3z = 1$$ [1]
$$x - 5y + 5z = 3$$ [2]
$$2x - 3y + \;\; z = 2$$ [3]

Let's eliminate the variable x. Multiply each term in equation [2] by -3 and add the result to equation [1].

$$3x - \;\; y - \;\; 3z = \quad 1$$ [1]
$$\underline{-3x + 15y - 15z = -9}$$ [4]
$$14y - 18z = -8$$ [5]

Now let's eliminate x by working with a different pair of equations. We will use equations [2] and [3].

Multiply equation [2] by -2 and add the result to equation [3].

$$-2x + 10y - 10z = -6$$ [6]
$$\underline{2x - \;\; 3y + \;\; z = \quad 2}$$ [3]
$$7y - \;\; 9z = -4$$ [7]

We will now solve the system formed by equation [5] and equation [7].

$$14y - 18z = -8$$ [5]
$$7y - \;\; 9z = -4$$ [7]

Multiply equation [7] by -2 and add the result to equation [5].

$$14y - 18z = -8$$ [5]
$$\underline{-14y + 18z = \quad 8}$$ [8]
$$0 = 0$$

This statement is true. Thus we conclude that this system of equations is dependent. There are an infinite number of solutions.

4. $2x + y + z = 11$ [1]
$\qquad 4y + 3z = -8$ [2]
$\quad x - 5y \quad = 2$ [3]

Multiply equation [1] by -3 and add the results to equation [2], thus eliminating the z terms.

$$-6x - 3y - 3z = -33 \quad [4]$$
$$\underline{\qquad\quad 4y + 3z = -8} \quad [2]$$
$$-6x + y \qquad = -41 \quad [5]$$

We can solve the system formed by equation [3] and equation [5].

$$x - 5y = 2 \quad [3]$$
$$-6x + y = -41 \quad [5]$$

Multiply equation [3] by 6 and add the results to equation [5].

$$6x - 30y = 12 \quad [6]$$
$$\underline{-6x + y = -41} \quad [5]$$
$$-29y = -29$$
$$y = 1$$

Now substitute $y = 1$ into equation [2] and solve for z.

$$4(1) + 3z = -8$$
$$4 + 3z = -8$$
$$3z = -12$$
$$z = -4$$

Now substitute $y = 1$, $z = -4$ into equation [1] and solve for x.

$$2x + 1 + (-4) = 11$$
$$2x - 3 = 11$$
$$2x = 14$$
$$x = 7$$

The solution is $x = 7$, $y = 1$, $z = -4$, or $(7, 1, -4)$.

4.3 Practice Problems

1. Let x = the number of baseballs purchased
$\quad y$ = the number of bats purchased
Last week: $6x + 21y = 318$ [1]
This week: $5x + 17y = 259$ [2]
Multiply equation [1] by 5 and equation [2] by -6.

$$30x + 105y = 1590 \quad [3]$$
$$\underline{-30x - 102y = -1554} \quad [4]$$
$$3y = 36 \quad \text{Add equations [3] and [4].}$$
$$y = 12$$

Substitute $y = 12$ into equation [2].

$$5x + 17(12) = 259$$
$$5x + 204 = 259$$
$$5x = 55$$
$$x = 11$$

Thus 11 baseballs and 12 bats were purchased.

2. Let x = The number of small chairs
$\quad y$ = The number of large chairs
(HINT: change all hours to minutes)

$$30x + 40y = 1560 \quad [1]$$
$$75x + 80y = 3420 \quad [2]$$

Multiply equation [1] by -2 and add the results to equation [2].

$$-60x - 80y = -3120 \quad [3]$$
$$\underline{75x + 80y = 3420} \quad [2]$$
$$15x = 300$$
$$x = 20$$

Substitute $x = 20$ in either equation [1] or [2] and solve for y.

$$30(20) + 40y = 1560$$
$$600 + 40y = 1560$$
$$40y = 960$$
$$y = 24$$

Therefore, the company can make 20 small chairs and 24 large chairs each day.

3. Let a = the speed of the airplane in still air in kilometers per hour
$\quad w$ = the speed of the wind in kilometers per hour

	R	$\cdot$ T	$=$ D
Against the wind	$a - w$	3	1950
With the wind	$a + w$	2	1600

We obtain a system of equations from the chart.

$$(a - w)3 = 1950$$
$$(a + w)2 = 1600$$

We remove the parentheses.

$$3a - 3w = 1950 \quad [1]$$
$$2a + 2w = 1600 \quad [2]$$

Multiply equation [1] by 2 and equation [2] by 3 and add the resulting equations.

$$6a - 6w = 3900$$
$$\underline{6a + 6w = 4800}$$
$$12a = 8700$$
$$a = 725$$

Substituting $a = 725$ into equation [2] we have

$$2(725) + 2w = 1600$$
$$1450 + 2w = 1600$$
$$2w = 150$$
$$w = 75$$

Thus, the speed of the plane in still air is 725 kilometers per hour and the speed of the wind is 75 kilometers per hour.

4. $\quad A + B + C = 260$ [1]
$3A + 2B \qquad = 390$ [2]
$\qquad 3B + 4C = 655$ [3]

Multiply equation [1] by -3 and add it to equation [2].

$$-3A - 3B - 3C = -780 \quad [4]$$
$$\underline{3A + 2B \qquad = 390} \quad [2]$$
$$-B - 3C = -390 \quad [5]$$

Now multiply equation [5] by 3 and add it to equation [3].

$$-3B - 9C = -1170 \quad [6]$$
$$\underline{3B + 4C = 655} \quad [3]$$
$$-5C = -515$$
$$C = 103$$

Substitute $C = 103$ into equation [3] and solve for B.

$$3B + 4(103) = 655$$
$$3B + 412 = 655$$
$$3B = 243$$
$$B = 81$$

Now substitute $B = 81$ into equation [2] and solve for A.

$$3A + 2(81) = 390$$
$$3A + 162 = 390$$
$$3A = 228$$
$$A = 76$$

Machine A wraps 76 boxes per hour, machine B wraps 81 boxes per hour, and machine C wraps 103 boxes per hour.

5. Let x = the radius of wire A
$\quad y$ = the radius of wire B
$\quad z$ = the radius of wire C
The distance from A to B is 13 millimeters. Thus $x + y = 13$. The distance from B to C is 22 millimeters. Thus $y + z = 22$. The distance between the outside edges is 52 millimeters. Thus, $2x + 2y + 2z = 52$. We can divide each term by 2 to obtain $x + y + z = 26$. Our system is

$$x + y \qquad = 13 \quad [1]$$
$$y + z = 22 \quad [2]$$
$$x + y + z = 26 \quad [3]$$

Multiply equation [2] by -1 and add it to equation [3].

$$-y - z = -22 \quad [3]$$
$$\underline{x + y + z = 26} \quad [4]$$
$$x = 4$$

Now substitute $x = 4$ into equation [1] and solve for y.

$$4 + y = 13$$
$$y = 9$$

Now substitute $x = 4$ and $y = 9$ into equation [3] and solve for z.

$$4 + 9 + z = 26$$
$$13 + z = 26$$
$$z = 13$$

Thus, the radius of wire A is 4 millimeters, of wire B is 9 millimeters, and of wire C is 13 millimeters.

4.4 Practice Problems

1. (a) $\begin{vmatrix} 2 & -5 \\ 3 & 1 \end{vmatrix} = (2)(1) - (3)(-5)$
$= 2 + 15 = 17$

(b) $\begin{vmatrix} -6 & -4 \\ -5 & 2 \end{vmatrix} = (-6)(2) - (-5)(-4)$
$= -12 - (20) = -32$

2. (a) $\begin{vmatrix} -7 & 3 \\ -4 & -2 \end{vmatrix} = (-7)(-2) - (-4)(3)$
$= 14 + 12 = 26$

(b) $\begin{vmatrix} 5 & 6 \\ 0 & -5 \end{vmatrix} = (5)(-5) - (0)(6)$
$= -25 - 0 = -25$

3. (a) $\begin{vmatrix} -0.6 & 1.3 \\ 0.5 & 5.1 \end{vmatrix} = (-0.6)(5.1) - (0.5)(1.3)$
$= (-3.06) - (0.65) = -3.71$

(b) $\begin{vmatrix} \frac{1}{6} & \frac{7}{12} \\ \frac{-3}{4} & \frac{1}{4} \end{vmatrix} = \left(\frac{1}{6}\right)\left(\frac{1}{4}\right) - \left(\frac{-3}{4}\right)\left(\frac{7}{12}\right)$
$= \frac{1}{24} + \frac{7}{16} = \frac{2}{48} + \frac{21}{48} = \frac{23}{48}$

4. Solve for x. $\begin{vmatrix} x+2 & 5 \\ x & -4 \end{vmatrix} = 46$

The equation states that the determinant is equal to 46. We write the value of the determinant and solve for x.
$$(x+2)(-4) - (x)(5) = 46$$
$$-4x - 8 - 5x = 46$$
$$-9x - 8 = 46$$
$$-9x = 54$$
$$x = -6$$

5. Find **(a)** the minor of 3 and
(b) the minor of -6 in the determinant
$$\begin{vmatrix} 1 & 2 & 7 \\ -4 & -5 & -6 \\ 3 & 4 & -9 \end{vmatrix}$$

(a) $\begin{vmatrix} 2 & 7 \\ -5 & -6 \end{vmatrix}$ **(b)** $\begin{vmatrix} 1 & 2 \\ 3 & 4 \end{vmatrix}$

6. $\begin{vmatrix} 1 & 2 & -3 \\ 2 & -1 & 2 \\ 3 & 1 & 4 \end{vmatrix} = 1\begin{vmatrix} -1 & 2 \\ 1 & 4 \end{vmatrix} - 2\begin{vmatrix} 2 & -3 \\ 1 & 4 \end{vmatrix} + 3\begin{vmatrix} 2 & -3 \\ -1 & 2 \end{vmatrix}$
$= 1[(-1)(4) - (1)(2)] - 2[(2)(4) - (1)(-3)]$
$\qquad\qquad\qquad\qquad\qquad + 3[(2)(2) - (-1)(-3)]$
$= 1(-4 - 2) - 2(8 + 3) + 3(4 - 3)$
$= -6 - 22 + 3 = -28 + 3 = -25$

7. $\begin{vmatrix} 5 & 2 & 1 \\ 3 & 0 & -2 \\ -4 & -1 & 2 \end{vmatrix} = -3\begin{vmatrix} 2 & 1 \\ -1 & 2 \end{vmatrix} + 0\begin{vmatrix} 5 & 1 \\ -4 & 2 \end{vmatrix} - (-2)\begin{vmatrix} 5 & 2 \\ -4 & -1 \end{vmatrix}$
$= -3[(2)(2) - (-1)(1)] + 0 + 2[(5)(-1) - (-4)(2)]$
$= -3(4 + 1) + 2(-5 + 8)$
$= -3(5) + 2(3) = -15 + 6 = -9$

8. $\begin{vmatrix} 3 & 6 & -2 \\ -4 & 0 & -1 \\ 5 & 0 & 8 \end{vmatrix}$ Expanding by the second column gives
$-6\begin{vmatrix} -4 & -1 \\ 5 & 8 \end{vmatrix} + 0\begin{vmatrix} 3 & -2 \\ 5 & 8 \end{vmatrix} - 0\begin{vmatrix} 3 & -2 \\ -4 & -1 \end{vmatrix}$
$= -6[(-4)(8) - (5)(-1)] + 0 - 0$
$= -6(-32 + 5) = -6(-27) = 162$

4.5 Practice Problems

1. $-2x + y = 8$
$3x + 4y = -1$

$D = \begin{vmatrix} -2 & 1 \\ 3 & 4 \end{vmatrix} = (-2)(4) - (3)(1) = -8 - 3 = -11$

$D_x = \begin{vmatrix} 8 & 1 \\ -1 & 4 \end{vmatrix} = (8)(4) - (-1)(1) = 32 + 1 = 33$

$D_y = \begin{vmatrix} -2 & 8 \\ 3 & -1 \end{vmatrix} = (-2)(-1) - (3)(8) = 2 - 24 = -22$

$x = \dfrac{D_x}{D} = \dfrac{33}{-11} = -3 \quad y = \dfrac{D_y}{D} = \dfrac{-22}{-11} = 2$

The solution is $x = -3$, $y = 2$, or $(-3, 2)$.

2. $5x + 3y = 17$
$2x - 5y = 13$

$D = \begin{vmatrix} 5 & 3 \\ 2 & -5 \end{vmatrix} = (5)(-5) - (2)(3) = -25 - 6 = -31$

$D_x = \begin{vmatrix} 17 & 3 \\ 13 & -5 \end{vmatrix} = (17)(-5) - (13)(3) = -85 - 39 = -124$

$D_y = \begin{vmatrix} 5 & 17 \\ 2 & 13 \end{vmatrix} = (5)(13) - (2)(17) = 65 - 34 = 31$

$x = \dfrac{D_x}{D} = \dfrac{-124}{-31} = 4 \quad y = \dfrac{D_y}{D} = \dfrac{31}{-31} = -1$

The solution is $x = 4$, $y = -1$, or $(4, -1)$.

3. $2x + 3y - z = -1$
$3x + 5y - 2z = -3$
$x + 2y + 3z = 2$

We will expand each determinant by the first row.

$D = \begin{vmatrix} 2 & 3 & -1 \\ 3 & 5 & -2 \\ 1 & 2 & 3 \end{vmatrix} = 2\begin{vmatrix} 5 & -2 \\ 2 & 3 \end{vmatrix} - 3\begin{vmatrix} 3 & -2 \\ 1 & 3 \end{vmatrix} - 1\begin{vmatrix} 3 & 5 \\ 1 & 2 \end{vmatrix}$
$= +38 - 33 - 1 = 4$

$D_x = \begin{vmatrix} -1 & 3 & -1 \\ -3 & 5 & -2 \\ 2 & 2 & 3 \end{vmatrix} = -1\begin{vmatrix} 5 & -2 \\ 2 & 3 \end{vmatrix} - 3\begin{vmatrix} -3 & -2 \\ 2 & 3 \end{vmatrix} - 1\begin{vmatrix} -3 & 5 \\ 2 & 2 \end{vmatrix}$
$= -19 + 15 + 16 = 12$

$D_y = \begin{vmatrix} 2 & -1 & -1 \\ 3 & -3 & -2 \\ 1 & 2 & 3 \end{vmatrix} = 2\begin{vmatrix} -3 & -2 \\ 2 & 3 \end{vmatrix} + 1\begin{vmatrix} 3 & -2 \\ 1 & 3 \end{vmatrix} - 1\begin{vmatrix} 3 & -3 \\ 1 & 2 \end{vmatrix}$
$= -10 + 11 - 9 = -8$

$D_z = \begin{vmatrix} 2 & 3 & -1 \\ 3 & 5 & -3 \\ 1 & 2 & 2 \end{vmatrix} = 2\begin{vmatrix} 5 & -3 \\ 2 & 2 \end{vmatrix} - 3\begin{vmatrix} 3 & -3 \\ 1 & 2 \end{vmatrix} - 1\begin{vmatrix} 3 & 5 \\ 1 & 2 \end{vmatrix}$
$= 32 - 27 - 1 = 4$

$x = \dfrac{D_x}{D} = \dfrac{12}{4} = 3 \quad y = \dfrac{D_y}{D} = \dfrac{-8}{4} = -2 \quad z = \dfrac{D_z}{D} = \dfrac{4}{4} = 1$

The solution is $x = 3$, $y = -2$, $z = 1$, or $(3, -2, 1)$.

4. First we write the complete form showing coefficients of 0.
$$2x - 3y + 2z = -1$$
$$x + 2y + 0z = 14$$
$$-x + 0y + 3z = 5$$

Then we use Cramer's rule to find the value of x.

$D = \begin{vmatrix} 2 & -3 & 2 \\ 1 & 2 & 0 \\ -1 & 0 & 3 \end{vmatrix} = 25 \quad D_x = \begin{vmatrix} -1 & -3 & 2 \\ 14 & 2 & 0 \\ 5 & 0 & 3 \end{vmatrix} = 100$

$x = \dfrac{D_x}{D} = \dfrac{100}{25} = 4$

4.6 Practice Problems

1.

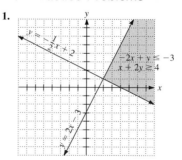

2.

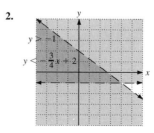

3.

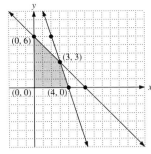

Boundary lines intersect at $(3, 3)$, $(0, 6)$, $(4, 0)$, and $(0, 0)$.

Chapter 5

5.1 Practice Problems

1. **(a)** $p = 3x^5 - 6x^4 + x^2$ **(b)** $g = 5x^2 + 2$
p is a trinomial of degree 5 g is a binomial of degree 2
(c) $h = 3ab + 5a^2b^2 - 6a^4b$ **(d)** $f = 16x^4y^6$
h is a trinomial of degree 5 f is a monomial of degree 10

2. $(-7x^2 + 5x - 9) + (2x^2 - 3x + 5)$
We remove the parentheses and combine like terms.
$= -7x^2 + 2x^2 + 5x - 3x - 9 + 5$
$= -5x^2 + 2x - 4$

3. $(2x^2 - 14x + 9) - (-3x^2 + 10x + 7)$
We add the opposite of the second polynomial to the first polynomial.
$= (2x^2 - 14x + 9) + (3x^2 - 10x - 7)$
$= 2x^2 + 3x^2 - 14x - 10x + 9 - 7 = 5x^2 - 24x + 2$

4. **(a)** $(2x^3 - 6x^2 + 5x - 2) + (-7x^3 - 5x^2 + 3x + 4)$
$= 2x^3 - 7x^3 - 6x^2 - 5x^2 + 5x + 3x - 2 + 4$
$= -5x^3 - 11x^2 + 8x + 2$
(b) $(2x^3 - 6x^2 + 5x - 2) - (-7x^3 - 5x^2 + 3x + 4)$
$= 2x^3 + 7x^3 - 6x^2 + 5x^2 + 5x - 3x - 2 - 4$
$= 9x^3 - x^2 + 2x - 6$

5.
$$\begin{array}{r} 2x^2 - 3x + 1 \\ x^2 - 5x \\ \hline -10x^3 + 15x^2 - 5x \\ 2x^4 - 3x^3 + x^2 \\ \hline 2x^4 - 13x^3 + 16x^2 - 5x \end{array}$$

5. (REVISITED)
$(2x^2 - 3x + 1)(x^2 - 5x)$
$= (2x^2 - 3x + 1)(x^2) + (2x^2 - 3x + 1)(-5x)$
$= 2x^4 - 3x^3 + x^2 - 10x^3 + 15x^2 - 5x$
$= 2x^4 - 3x^3 - 10x^3 + x^2 + 15x^2 - 5x$
$= 2x^4 - 13x^3 + 16x^2 - 5x$

6. $(7x + 3)(2x - 5)$

$14x^2 - 35x + 6x - 15$
$14x^2 - 29x - 15$

7. $(5a - 2b)(3c - 4d)$

$15ac - 20ad - 6bc + 8bd$

8. **(a)** $(7x)^2 - (2y)^2$
$= 49x^2 - 4y^2$
(b) $(5a^2)^2 - (8b^2)^2$
$= 25a^4 - 64b^4$

9. **(a)** $(3x - 7y)^2 = (3x)^2 - 2(3x)(7y) + (7y)^2 = 9x^2 - 42xy + 49y^2$
(b) $(4u + 5v)^2 = (4u)^2 + 2(4u)(5v) + (5v)^2 = 16u^2 + 40uv + 25v^2$
(c) $(9a + 2b)^2 = (9a)^2 + 2(9a)(2b) + (2b)^2 = 81a^2 + 36ab + 4b^2$
(d) $(7x^2 - 3y^2)^2 = (7x^2)^2 - 2(7x^2)(3y^2) + (3y^2)^2$
$= 49x^4 - 42x^2y^2 + 9y^4$

5.2 Practice Problems

1. $\dfrac{16x^3 - 8x^2 + 3x}{2x} = \dfrac{16x^3}{2x} - \dfrac{8x^2}{2x} + \dfrac{3x}{2x}$

$= 8x^2 - 4x + \dfrac{3}{2}$

2. $\dfrac{-16x^4 + 16x^3 + 8x^2 + 64x}{8x}$

$= \dfrac{-16x^4}{8x} + \dfrac{16x^3}{8x} + \dfrac{8x^2}{8x} + \dfrac{64x}{8x}$

$= -2x^3 + 2x^2 + x + 8$

3.
$$\begin{array}{r} 2x + 5 \\ 4x - 3\overline{\smash{)}8x^2 + 14x - 14} \\ \underline{8x^2 - 6x} \\ 20x - 14 \\ \underline{20x - 15} \\ 1 \end{array}$$

The answer is $2x + 5$ remainder 1 or $2x + 5 + \dfrac{1}{4x - 3}$.

4.
$$\begin{array}{r} 4x^2 - 6x + 9 \\ 2x + 3\overline{\smash{)}8x^3 + 0x^2 + 0x + 27} \\ \underline{8x^3 + 12x^2} \\ -12x^2 + 0x \\ \underline{-12x^2 - 18x} \\ 18x + 27 \\ \underline{18x + 27} \\ 0 \end{array}$$

The answer is $4x^2 - 6x + 9$.

5.
$$\begin{array}{r} x^2 - 3x + 1 \\ x^2 + 0x - 1\overline{\smash{)}x^4 - 3x^3 + 0x^2 + 3x + 4} \\ \underline{x^4 - 0x^3 - x^2} \\ -3x^3 + x^2 + 3x \\ \underline{-3x^3 - 0x^2 + 3x} \\ x^2 + 0x + 4 \\ \underline{x^2 + 0x - 1} \\ 5 \end{array}$$

The answer is $x^2 - 3x + 1 + \dfrac{5}{x^2 - 1}$.

5.3 Practice Problems

1.
$$\begin{array}{r|rrrr} -3 & 1 & -3 & 4 & -5 \\ & & -3 & +18 & -66 \\ \hline & 1 & -6 & 22 & -71 \end{array}$$

The quotient is $x^2 - 6x + 22 + \dfrac{-71}{x + 3}$.

2.
$$\begin{array}{r|rrrrr} 3 & 2 & 0 & -1 & 5 & -12 \\ & & 6 & 18 & 51 & 168 \\ \hline & 2 & 6 & 17 & 56 & 156 \end{array}$$

The quotient is $2x^3 + 6x^2 + 17x + 56 + \dfrac{156}{x - 3}$.

3.
$$\begin{array}{r|rrrrr} 3 & 2 & -9 & 5 & 13 & -3 \\ & & 6 & -9 & -12 & 3 \\ \hline & 2 & -3 & -4 & 1 & 0 \end{array}$$

The quotient is $2x^3 - 3x^2 - 4x + 1$.

5.4 Practice Problems

1. **(a)** $19x^3 - 38x^2 = 19x^2(x - 2)$
(b) $100a^4 - 50a^2 = 50a^2(2a^2 - 1)$
2. **(a)** $3x(7x^2 - 6xy + 8y^2)$
(b) $2xy(6y - 7x + 10xy + 18x^2)$
3. **(a)** $8a^3b^4(2a^2 - 4 - 3a^3)$
(b) $25xyz^2(2 - 3xy + 4x^2y)$
4. $3a^2(3a - 4b^2 - 5a^2)$
To check, we multiply.
$3a^2(3a - 4b^2 - 5a^2) = 9a^3 - 12a^2b^2 - 15a^4$
This is the original polynomial. It checks.

5. $(x + 2y)(7x - 8y - 1)$

6. $b(x + 5y) + 2w(x + 5y)$
$(x + 5y)(b + 2w)$

7. To factor $5x^2 - 12y + 4xy - 15x$, rearrange the terms. Then factor.
$5x^2 - 15x + 4xy - 12y$
$5x(x - 3) + 4y(x - 3)$
$(x - 3)(5x + 4y)$

8. To factor $xy - 12 - 4x + 3y$, rearrange the terms. Then factor.
$xy - 4x + 3y - 12$
$x(y - 4) + 3(y - 4)$
$(y - 4)(x + 3)$

9. To factor $2x^3 - 15 - 10x + 3x^2$, rearrange the terms. Then factor.
$2x^3 - 10x + 3x^2 - 15$
$2x(x^2 - 5) + 3(x^2 - 5)$
$(x^2 - 5)(2x + 3)$

5.5 Practice Problems

1. $x^2 + 14x + 48 = (x + 6)(x + 8)$

2. $x^2 - 10x + 21 = (x - 7)(x - 3)$

3. $x^2 - 13x - 48 = (x - 16)(x + 3)$

4. $x^4 + 9x^2 + 8 = (x^2 + 8)(x^2 + 1)$

5. (a) $x^2 + 15x + 50 = (x + 5)(x + 10)$
   (b) $x^2 - 12x + 35 = (x - 5)(x - 7)$
   (c) $a^2 + 2a - 48 = (a + 8)(a - 6)$
   (d) $x^4 + 2x^2 - 15 = (x^2 + 5)(x^2 - 3)$

6. (a) $x^2 - 16xy + 15y^2 = (x - 15y)(x - y)$
   (b) $x^2 + xy - 42y^2 = (x + 7y)(x - 6y)$

7. $4x^2 - 44x + 72 = 4(x^2 - 11x + 18) = 4(x - 9)(x - 2)$

8. Factor. $3x^2 + 2x - 8$
The grouping number is -24. Two numbers whose product is -24 and whose sum is $+2$ are $+6$ and -4.
   $3x^2 + 6x - 4x - 8$
$= 3x(x + 2) - 4(x + 2)$
$= (x + 2)(3x - 4)$

9. Factor. $10x^2 - 9x + 2$
The grouping number is 20. Two numbers whose product is 20 and whose sum is -9 are -5 and -4.
   $10x^2 - 5x - 4x + 2$
$= 5x(2x - 1) - 2(2x - 1)$
$= (2x - 1)(5x - 2)$

10. $9x^3 - 15x^2 - 6x = 3x(3x^2 - 5x - 2)$
                             $= 3x(3x^2 - 6x + x - 2)$
                             $= 3x[3x(x - 2) + 1(x - 2)]$
                             $= 3x(x - 2)(3x + 1)$

11. $3x^2 + 31x + 10 = (3x + 1)(x + 10)$

12. $8x^2 - 6x - 5 = (4x - 5)(2x + 1)$

13. $6x^4 + 13x^2 - 5 = (2x^2 + 5)(3x^2 - 1)$

5.6 Practice Problems

1. $x^2 - 9 = (x + 3)(x - 3)$

2. $64x^2 - 121y^2 = (8x + 11y)(8x - 11y)$

3. (a) $49x^2 - 25y^4 = (7x + 5y^2)(7x - 5y^2)$
   (b) $100x^6 - 1 = (10x^3 + 1)(10x^3 - 1)$

4. $7x^2 - 28 = 7(x^2 - 4) = 7(x + 2)(x - 2)$

5. $9x^2 - 30x + 25 = (3x - 5)^2$

6. $25x^2 - 70x + 49 = (5x - 7)^2$

7. $242x^2 + 88x + 8 = 2(121x^2 + 44x + 4)$
                          $= 2(11x + 2)^2$

8. (a) $49x^4 + 28x^2 + 4 = (7x^2 + 2)^2$
   (b) $36x^4 + 84x^2y^2 + 49y^4 = (6x^2 + 7y^2)^2$

9. $27x^3 + 64$
   $(3x)^3 = 27x^3$ $(4)^3 = 64$
   Thus, $27x^3 + 64 = (3x)^3 + (4)^3$
                      $= (3x + 4)(9x^2 - 12x + 16)$

10. $8x^3 + 125y^3 = (2x + 5y)(4x^2 - 10xy + 25y^2)$

11. $64x^3 - 125y^3 = (4x - 5y)(16x^2 + 20xy + 25y^2)$

---

12. $27w^3 - 125z^6 = (3w - 5z^2)(9w^2 + 15wz^2 + 25z^4)$

13. $54x^3 - 16 = 2(27x^3 - 8)$
            $= 2(3x - 2)(9x^2 + 6x + 4)$

14. $64a^6 - 1$
Use the difference of two squares first.
$(8a^3 + 1)(8a^3 - 1)$
Now use the formula for the sum and difference of two cubes.
$(2a + 1)(4a^2 - 2a + 1)(2a - 1)(4a^2 + 2a + 1)$

5.7 Practice Problems

1. (a) $7x^5 + 56x^2 = 7x^2(x^3 + 8)$
                  $= 7x^2(x + 2)(x^2 - 2x + 4)$
   (b) $125x^2 + 50xy + 5y^2$
        $= 5(25x^2 + 10xy + y^2)$
        $= 5(5x + y)^2$
   (c) $12x^2 - 75 = 3(4x^2 - 25)$
                 $= 3(2x + 5)(2x - 5)$
   (d) $3x^2 - 39x + 126$
        $= 3(x^2 - 13x + 42)$
        $= 3(x - 7)(x - 6)$
   (e) $6ax + 6ay + 18bx + 18by$
        $= 6(ax + ay + 3bx + 3by)$
        $= 6[a(x + y) + 3b(x + y)]$
        $= 6(x + y)(a + 3b)$
   (f) $6x^3 - x^2 - 12x$
        $= x(6x^2 - x - 12)$
        $= x(6x^2 - 9x + 8x - 12)$
        $= x[3x(2x - 3) + 4(2x - 3)]$
        $= x(2x - 3)(3x + 4)$

2. $x^2 - 3x - 20$
It has no common factor. There are no factors of -20 whose sum is -3; therefore, it is prime.

3. $3x^2 - 10x + 4$
Prime. There are no factors of 12 whose sum is -10.

4. $16x^2 + 81$
Prime. Binomials of the form $a^2 + b^2$ cannot be factored.

5.8 Practice Problems

1. $x^2 + x = 56$
          $x^2 + x - 56 = 0$
        $(x + 8)(x - 7) = 0$
       $x + 8 = 0$ $x - 7 = 0$
          $x = -8$ $x = 7$

2. $12x^2 - 11x + 2 = 0$
     $(4x - 1)(3x - 2) = 0$
     $4x - 1 = 0$ $3x - 2 = 0$
        $x = \dfrac{1}{4}$ $x = \dfrac{2}{3}$

3. $7x^2 - 14x = 0$
      $7x(x - 2) = 0$
    $7x = 0$ $x - 2 = 0$
     $x = 0$ $x = 2$

4. $5x^2 - 7x = 2x(x - 3)$
   $5x^2 - 7x = 2x^2 - 6x$
     $3x^2 - x = 0$
   $x(3x - 1) = 0$
   $x = 0$ $3x - 1 = 0$
              $x = \dfrac{1}{3}$

5. $16x(x - 2) = 8x - 25$
        $16x^2 - 32x = 8x - 25$
   $16x^2 - 32x - 8x + 25 = 0$
      $16x^2 - 40x + 25 = 0$
         $(4x - 5)^2 = 0$
   $4x - 5 = 0$ $4x - 5 = 0$
     $x = \dfrac{5}{4}$ is a double root

6.
$$3x^3 + 6x^2 = 45x$$
$$3x^3 + 6x^2 - 45x = 0$$
$$3x(x^2 + 2x - 15) = 0$$
$$3x(x + 5)(x - 3) = 0$$

$3x = 0 \qquad x + 5 = 0 \qquad x - 3 = 0$
$x = 0 \qquad\quad x = -5 \qquad\quad x = 3$

7. $A = \dfrac{1}{2}ab$

Let base $= b$
  altitude $= b + 5$
$$52 = \frac{1}{2}(b + 5)(b)$$
$$104 = (b + 5)(b)$$
$$104 = b^2 + 5b$$
$$0 = b^2 + 5b - 104$$
$$0 = b^2 - 8b + 13b - 104$$
$$0 = (b - 8)(b + 13)$$

$b - 8 = 0 \qquad b + 13 = 0$
$b = 8 \qquad\quad b = -13$

The base of a triangle must be a positive number, so we disregard -13. Thus,
base $= b = 8$ feet
altitude $= b + 5 = 13$ feet

8. Let $x =$ length in square feet of last year's garden
Let $3x + 2 =$ length in square feet of this year's garden
$$(3x + 2)^2 = 112 + x^2$$
$$9x^2 + 12x + 4 = 112 + x^2$$
$$8x^2 + 12x - 108 = 0$$
$$4(2x^2 + 3x - 27) = 0$$
$$4(x - 3)(2x + 9) = 0$$

$x - 3 = 0 \qquad 2x + 9 = 0$

$x = 3 \qquad\qquad x = -\dfrac{9}{2}$

A garden cannot measure $-\frac{9}{2}$ in length, so we reject the negative answer. We use $x = 3$. Last year's garden is a square with each side measuring 3 feet. This year's garden measures $3x + 2 = 3(3) + 2 = 11$ feet on each side.

Chapter 6

6.1 Practice Problems

1. Solve the equation $x^2 - 9x - 22 = 0$.
$$(x + 2)(x - 11) = 0$$
$x + 2 = 0 \qquad x - 11 = 0$
$x = -2 \qquad\quad x = 11$
The domain is all real numbers except -2 and 11.

2. $\dfrac{15x^2 - 18x}{6x^2 + 3x} = \dfrac{3x(5x - 6)}{3x(2x + 1)} = \dfrac{5x - 6}{2x + 1}$

3. $\dfrac{x^2 - 36y^2}{x^2 - 3xy - 18y^2} = \dfrac{(x + 6y)(x - 6y)}{(x + 3y)(x - 6y)} = \dfrac{x + 6y}{x + 3y}$

4. $\dfrac{9x^2y}{3xy^2 + 6x^2y} = \dfrac{\overset{3x}{9x^2y}}{3xy(y + 2x)} = \dfrac{3x}{y + 2x}$

5. $\dfrac{2x^2 - 8x - 10}{2x^2 - 20x + 50} = \dfrac{2(x - 5)(x + 1)}{2(x - 5)(x - 5)} = \dfrac{x + 1}{x - 5}$

6. (a) $\dfrac{-3x + 6y}{x^2 - 7xy + 10y^2} = \dfrac{-3(x - 2y)}{(x - 2y)(x - 5y)} = \dfrac{-3}{x - 5y}$

(b) $\dfrac{36 - x^2}{x^2 - 10x + 24} = \dfrac{(6 + x)(6 - x)}{(x - 4)(x - 6)} = -\dfrac{6 + x}{x - 4}$

7. $\dfrac{7a^2 - 23ab + 6b^2}{4b^2 - 49a^2} = \dfrac{(7a - 2b)(a - 3b)}{(2b - 7a)(2b + 7a)} = -\dfrac{a - 3b}{7a + 2b}$

8. $\dfrac{2x^2 + 5xy + 2y^2}{4x^2 - y^2} \cdot \dfrac{2x^2 + xy - y^2}{x^2 + xy - 2y^2}$

$= \dfrac{(2x + y)(x + 2y)}{(2x + y)(2x - y)} \cdot \dfrac{(2x - y)(x + y)}{(x + 2y)(x - y)} = \dfrac{x + y}{x - y}$

9. $\dfrac{2x^3 - 3x^2}{3x^2 + 3x} \cdot \dfrac{9x + 36}{10x^2 - 15x}$

$= \dfrac{x^2(2x - 3)}{3x(x + 1)} \cdot \dfrac{\overset{3}{9}(x + 4)}{5x(2x - 3)} = \dfrac{3(x + 4)}{5(x + 1)}$

10. $\dfrac{8x^3 + 27}{64x^3 - 1} \div \dfrac{4x^2 - 9}{16x^2 + 4x + 1}$

$= \dfrac{8x^3 + 27}{64x^3 - 1} \cdot \dfrac{16x^2 + 4x + 1}{4x^2 - 9}$

$= \dfrac{(2x + 3)(4x^2 - 6x + 9)}{(4x - 1)(16x^2 + 4x + 1)} \cdot \dfrac{16x^2 + 4x + 1}{(2x + 3)(2x - 3)}$

$= \dfrac{4x^2 - 6x + 9}{(4x - 1)(2x - 3)}$ or $\dfrac{4x^2 - 6x + 9}{8x^2 - 14x + 3}$

11. $\dfrac{4x^2 - 9}{2x^2 + 11x + 12} \div (-6x + 9)$

$= \dfrac{4x^2 - 9}{2x^2 + 11x + 12} \cdot \dfrac{1}{-6x + 9}$

$= \dfrac{(2x + 3)(2x - 3)}{(2x + 3)(x + 4)} \cdot \dfrac{1}{-3(2x - 3)} = -\dfrac{1}{3(x + 4)}$

6.2 Practice Problems

1. Find the LCD. $\dfrac{8}{x^2 - x - 12}, \dfrac{3}{x - 4}$

Factor each denominator completely.
$x^2 - x - 12 = (x - 4)(x - 3)$
$x - 4$ cannot be factored.
The LCD is the product of all the different factors.
LCD $= (x - 4)(x - 3)$

2. Find the LCD. $\dfrac{2}{15x^3y^2}, \dfrac{13}{25xy^3}$

Factor each denominator.
$15x^3y^2 = 3 \cdot 5 \cdot x \cdot x \cdot x \cdot y \cdot y$
$25xy^3 = \quad 5 \cdot 5 \cdot x \cdot y \cdot y \cdot y$
LCD $= 3 \cdot 5 \cdot 5 \cdot x \cdot x \cdot x \cdot y \cdot y \cdot y$
LCD $= 75x^3y^3$

3. $\dfrac{7}{3a^2b} + \dfrac{-2 + a}{3a^2b} = \dfrac{5 + a}{3a^2b}$

4. $\dfrac{4x}{(x + 6)(2x - 1)} - \dfrac{3x + 1}{(x + 6)(2x - 1)} = \dfrac{x - 1}{(x + 6)(2x - 1)}$

5. $\dfrac{8}{(x - 4)(x + 3)} + \dfrac{3}{x - 4}$

LCD $= (x - 4)(x + 3)$

$= \dfrac{8}{(x - 4)(x + 3)} + \dfrac{3}{(x - 4)} \cdot \dfrac{(x + 3)}{(x + 3)}$

$= \dfrac{8 + 3x + 9}{(x - 4)(x + 3)} = \dfrac{3x + 17}{(x - 4)(x + 3)}$

6. $\dfrac{5}{x + 4} + \dfrac{3}{4x}$

LCD $= 4x(x + 4)$

$= \dfrac{5}{x + 4} \cdot \dfrac{4x}{4x} + \dfrac{3}{4x} \cdot \dfrac{(x + 4)}{(x + 4)}$

$= \dfrac{20x + 3x + 12}{4x(x + 4)} = \dfrac{23x + 12}{4x(x + 4)}$

7. $\dfrac{7}{4ab^3} + \dfrac{1}{3a^3b^2}$

LCD $= 12a^3b^3$

$= \dfrac{7}{4ab^3} \cdot \dfrac{(3a^2)}{(3a^2)} + \dfrac{1}{3a^3b^2} \cdot \dfrac{(4b)}{(4b)} = \dfrac{21a^2 + 4b}{12a^3b^3}$

8. $\dfrac{7}{x+4}+\dfrac{-5x}{x^2-x-20}+\dfrac{2}{x-5}$

$=\dfrac{7}{x+4}+\dfrac{-5x}{(x+4)(x-5)}+\dfrac{2}{x-5}$

$\text{LCD}=(x+4)(x-5)$

$=\dfrac{7}{(x+4)}\cdot\dfrac{(x-5)}{(x-5)}+\dfrac{-5x}{(x+4)(x-5)}+\dfrac{2}{(x-5)}\cdot\dfrac{(x+4)}{(x+4)}$

$=\dfrac{7x-35-5x+2x+8}{(x+4)(x-5)}=\dfrac{4x-27}{(x+4)(x-5)}$

9. $\dfrac{4x+2}{x^2+x-12}-\dfrac{3x+8}{x^2+6x+8}$

$=\dfrac{4x+2}{(x-3)(x+4)}-\dfrac{3x+8}{(x+2)(x+4)}$

$\text{LCD}=(x+2)(x-3)(x+4)$

$=\dfrac{4x+2}{(x+4)(x-3)}\cdot\dfrac{(x+2)}{(x+2)}-\dfrac{3x+8}{(x+2)(x+4)}\cdot\dfrac{(x-3)}{(x-3)}$

$=\dfrac{4x^2+10x+4}{(x+4)(x-3)(x+2)}-\dfrac{3x^2-x-24}{(x+4)(x-3)(x+2)}$

$=\dfrac{x^2+11x+28}{(x+4)(x-3)(x+2)}=\dfrac{(x+4)(x+7)}{(x+4)(x-3)(x+2)}$

$=\dfrac{x+7}{(x-3)(x+2)}$

10. $\dfrac{7x-3}{4x^2+20x+25}-\dfrac{3x}{4x+10}$

$=\dfrac{7x-3}{(2x+5)(2x+5)}-\dfrac{3x}{2(2x+5)}$

$\text{LCD}=2(2x+5)(2x+5)$

$=\dfrac{7x-3}{(2x+5)(2x+5)}\cdot\dfrac{2}{2}-\dfrac{3x}{2(2x+5)}\cdot\dfrac{(2x+5)}{(2x+5)}$

$=\dfrac{14x-6-6x^2-15x}{2(2x+5)(2x+5)}=\dfrac{-6x^2-x-6}{2(2x+5)^2}$

6.3 Practice Problems

1. You can use either method 1 or method 2 to simplify a complex fraction.

$$\dfrac{y+\dfrac{3}{y}}{\dfrac{2}{y^2}+\dfrac{5}{y}}$$

METHOD 1

Simplify the numerator.

$y+\dfrac{3}{y}=\dfrac{y^2}{y}+\dfrac{3}{y}=\dfrac{y^2+3}{y}$

Simplify the denominator.

$\dfrac{2}{y^2}+\dfrac{5}{y}=\dfrac{2}{y^2}+\dfrac{5y}{y^2}=\dfrac{2+5y}{y^2}$

Divide the numerator by the denominator.

$\dfrac{\dfrac{y^2+3}{y}}{\dfrac{2+5y}{y^2}}$

$=\dfrac{(y^2+3)}{y}\cdot\dfrac{y^2}{2+5y}$

$=\dfrac{y(y^2+3)}{2+5y}$

METHOD 2

Find the LCD.

$\text{LCD}=y^2$

Multiply the numerator and denominator by the LCD.

$\dfrac{\left(y+\dfrac{3}{y}\right)\cdot y^2}{\left(\dfrac{2}{y^2}+\dfrac{5}{y}\right)\cdot y^2}$

$=\dfrac{y(y^2)+\dfrac{3}{y}(y^2)}{\dfrac{2}{y^2}(y^2)+\dfrac{5}{y}(y^2)}$

$=\dfrac{y^3+3y}{2+5y}$

$=\dfrac{y(y^2+3)}{2+5y}$

2. Simplify.

$$\dfrac{\dfrac{4}{16x^2-1}+\dfrac{3}{4x+1}}{\dfrac{x}{4x-1}+\dfrac{5}{4x+1}}$$

METHOD 1

Simplify the numerator.

$\dfrac{4}{(4x+1)(4x-1)}+\dfrac{3(4x-1)}{(4x+1)(4x-1)}=\dfrac{12x+1}{(4x+1)(4x-1)}$

Simplify the denominator.

$\dfrac{x(4x+1)}{(4x-1)(4x+1)}+\dfrac{5(4x-1)}{(4x+1)(4x-1)}=\dfrac{4x^2+21x-5}{(4x-1)(4x+1)}$

To divide the numerator by the denominator, we multiply the numerator by the reciprocal of the denominator.

$\dfrac{12x+1}{(4x+1)(4x-1)}\cdot\dfrac{(4x+1)(4x-1)}{4x^2+21x-5}=\dfrac{12x+1}{4x^2+21x-5}$

METHOD 2

Multiply the numerator and denominator by the LCD.

$\text{LCD}=(4x+1)(4x-1)$. Notice we factored $16x^2-1$.

$$\dfrac{\left[\dfrac{4}{(4x+1)(4x-1)}+\dfrac{3}{(4x+1)}\right](4x+1)(4x-1)}{\left[\dfrac{x}{(4x-1)}+\dfrac{5}{(4x+1)}\right](4x+1)(4x-1)}$$

$=\dfrac{\dfrac{4(4x+1)(4x-1)}{(4x+1)(4x-1)}+\dfrac{3(4x+1)(4x-1)}{4x+1}}{\dfrac{x}{4x-1}\cdot(4x+1)(4x-1)+\dfrac{5}{4x+1}\cdot(4x+1)(4x-1)}$

$=\dfrac{4+3(4x-1)}{x(4x+1)+5(4x-1)}=\dfrac{4+12x-3}{4x^2+x+20x-5}=\dfrac{12x+1}{4x^2+21x-5}$

3. Simplify by **METHOD 1.**

$\dfrac{4+\dfrac{1}{x+3}}{\dfrac{2}{x^2+4x+3}}=\dfrac{4+\dfrac{1}{x+3}}{\dfrac{2}{(x+1)(x+3)}}$

$=\dfrac{4\cdot\dfrac{(x+3)}{(x+3)}+\dfrac{1}{x+3}}{\dfrac{2}{(x+1)(x+3)}}=\dfrac{\dfrac{4x+12+1}{x+3}}{\dfrac{2}{(x+1)(x+3)}}$

$=\dfrac{4x+13}{x+3}\cdot\dfrac{(x+1)(x+3)}{2}$

$=\dfrac{(4x+13)(x+1)}{2}\quad\text{or}\quad\dfrac{4x^2+17x+13}{2}$

4. Simplify by **METHOD 2.**

$\text{LCD}=y(y+3)$

$\dfrac{\dfrac{7}{y+3}-\dfrac{3}{y}}{\dfrac{2}{y}+\dfrac{5}{y+3}}\cdot\dfrac{y(y+3)}{y(y+3)}$

$=\dfrac{\dfrac{7}{y+3}\cdot y(y+3)-\dfrac{3}{y}\cdot y(y+3)}{\dfrac{2}{y}\cdot y(y+3)+\dfrac{5}{y+3}\cdot y(y+3)}$

$=\dfrac{7y-3(y+3)}{2(y+3)+5(y)}=\dfrac{7y-3y-9}{2y+6+5y}=\dfrac{4y-9}{7y+6}$

6.4 Practice Problems

1.
$\dfrac{4}{3x}+\dfrac{x+1}{x}=\dfrac{1}{2}$

$6x\left[\dfrac{4}{3x}\right]+6x\left[\dfrac{x+1}{x}\right]=6x\left[\dfrac{1}{2}\right]$

$8+6(x+1)=3x$

$8+6x+6=3x$

$3x=-14$

$x=-\dfrac{14}{3}$

Check.
$$\frac{4}{3\left(\frac{-14}{3}\right)} + \frac{\frac{-14}{3} + 1}{-\frac{14}{3}} \overset{?}{=} \frac{1}{2}$$

$$\frac{4}{-14} + \frac{\frac{-11}{3}}{\frac{-14}{3}} \overset{?}{=} \frac{1}{2}$$

$$-\frac{4}{14} + \frac{11}{14} \overset{?}{=} \frac{1}{2}$$

$$\frac{7}{14} \overset{?}{=} \frac{1}{2}$$

$$\frac{1}{2} = \frac{1}{2} \checkmark$$

2.
$$\frac{1}{3x - 9} = \frac{1}{2x - 6} - \frac{5}{6}$$

$$\frac{1}{3(x - 3)} = \frac{1}{2(x - 3)} - \frac{5}{6}$$

$$6(x - 3)\left[\frac{1}{3(x - 3)}\right] = 6(x - 3)\left[\frac{1}{2(x - 3)}\right] - 6(x - 3)\left[\frac{5}{6}\right]$$

$$2(1) = 3(1) - (x - 3)(5)$$

$$2 = 3 - 5x + 15$$

$$2 = 18 - 5x$$

$$-16 = -5x$$

$$\frac{16}{5} = x$$

Check.
$$\frac{1}{3\left(\frac{16}{5}\right) - 9} \overset{?}{=} \frac{1}{2\left(\frac{16}{5}\right) - 6} - \frac{5}{6}$$

$$\frac{1}{\frac{48}{5} - 9} \overset{?}{=} \frac{1}{\frac{32}{5} - 6} - \frac{5}{6}$$

$$\frac{1}{\frac{3}{5}} \overset{?}{=} \frac{1}{\frac{2}{5}} - \frac{5}{6}$$

$$\frac{5}{3} = \frac{5}{3} \checkmark$$

3.
$$\frac{y^2 + 4y - 2}{y^2 - 2y - 8} = 1 + \frac{4}{y - 4}$$

$$\frac{y^2 + 4y - 2}{(y + 2)(y - 4)} = 1 + \frac{4}{y - 4}$$

$$(y + 2)(y - 4)\left[\frac{y^2 + 4y - 2}{(y + 2)(y - 4)}\right] = (y + 2)(y - 4) + (y + 2)(y - 4)\left[\frac{4}{(y - 4)}\right]$$

$$y^2 + 4y - 2 = y^2 - 2y - 8 + 4y + 8$$

$$4y - 2 = 2y$$

$$2y = 2$$

$$y = 1$$

4.
$$\frac{2x - 1}{x^2 - 7x + 10} + \frac{3}{x - 5} = \frac{5}{x - 2}$$

$$\frac{2x - 1}{(x - 2)(x - 5)} + \frac{3}{x - 5} = \frac{5}{x - 2}$$

$$(x - 2)(x - 5)\left[\frac{2x - 1}{(x - 2)(x - 5)}\right] + (x - 2)(x - 5)\left[\frac{3}{x - 5}\right] = (x - 2)(x - 5)\left[\frac{5}{x - 2}\right]$$

$$2x - 1 + (x - 2)3 = (x - 5)5$$

$$2x - 1 + 3x - 6 = 5x - 25$$

$$5x - 7 = 5x - 25$$

$$0x = -18$$

Of course $0 \neq -18$. Therefore, no values of x makes the original equation true. Hence the equation has **no solution.**

5.
$$\frac{y}{y - 2} - 3 = 1 + \frac{2}{y - 2}$$

$$(y - 2)\left[\frac{y}{y - 2}\right] - (y - 2)[3] = (y - 2)[1] + (y - 2)\left[\frac{2}{y - 2}\right]$$

$$y - 3(y - 2) = 1(y - 2) + 2$$

$$y - 3y + 6 = y - 2 + 2$$

$$-2y + 6 = y$$

$$6 = 3y$$

$$2 = y$$

Check.
$$\frac{2}{2 - 2} - 3 \overset{?}{=} 1 + \frac{2}{2 - 2}$$

$$\frac{2}{0} - 3 \overset{?}{=} 1 + \frac{2}{0}$$

Division by zero is not defined. The value $y = 2$ is therefore not a solution to the original equation. There is **no solution.**

1. Solve for r.
$$P = \frac{A}{1+r}$$
$$(1+r)[P] = (1+r)\left[\frac{A}{1+r}\right]$$
$$P + Pr = A$$
$$Pr = A - P$$
$$r = \frac{A-P}{P}$$

2. Solve for t.
$$\frac{1}{t} = \frac{1}{c} + \frac{1}{d}$$
$$cdt\left[\frac{1}{t}\right] = cdt\left[\frac{1}{c}\right] + cdt\left[\frac{1}{d}\right]$$
$$cd = dt + ct$$
$$cd = t(d+c)$$
$$\frac{cd}{d+c} = t$$

3. Solve for p_1.
$$C = \frac{Bp_1p_2}{d^2}$$
$$Cd^2 = Bp_1p_2$$
$$\frac{Cd^2}{Bp_2} = p_1$$

4. $\dfrac{\text{perfect parts}}{\text{defective parts}} = \dfrac{414}{36} = \dfrac{23}{2}$ or $\quad 23:2$

5. We set up a proportion and solve.
$$\text{acid 1} \rightarrow \frac{13}{37} = \frac{x}{296} \leftarrow \text{acid 2}$$
$$\text{solution 1} \rightarrow \qquad\qquad \leftarrow \text{solution 2}$$
$$37x = 3848$$
$$x = 104$$
Therefore there is 104 milliliters of acid in 2196 milliliters of solution.
We could have set up the proportion
$$\text{acid 1} \rightarrow \frac{13}{x} = \frac{37}{296} \leftarrow \text{solution 1}$$
$$\text{acid 2} \rightarrow \qquad\qquad \leftarrow \text{solution 2}$$
The answer will be the same.

6. $\dfrac{21}{2} = \dfrac{x}{168}$
$$2x = 3528$$
$$x = 1764$$
The number of students enrolled should be 1764 to maintain that ratio.

7. We will draw the picture as two similar triangles.

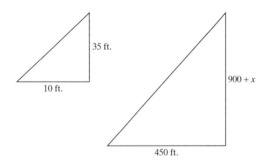

35 ft.

10 ft.

900 + x

450 ft.

We write a proportion and solve.
$$\frac{10}{35} = \frac{450}{900+x}$$
$$\text{LCD} = 35(900+x)$$
$$35(900+x)\left[\frac{10}{35}\right] = 35(900+x)\left[\frac{450}{900+x}\right]$$
$$10(900+x) = 35(450)$$
$$9000 + 10x = 15{,}750$$
$$10x = 6750$$
$$x = 675$$
The helicopter is about 675 feet over the building.

8. We construct a table for the data.

	Rate	·	Time	=	Amount
	Rate of Work per Hour		Time Worked in Hours		Amount of Task Done
Alfred	$\dfrac{1}{4}$		x		$\dfrac{x}{4}$
Son	$\dfrac{1}{5}$		x		$\dfrac{x}{5}$

The amount of work done by Alfred plus the amount of work done by his son equals 1 completed job.
Write an equation and solve.
$$\frac{x}{4} + \frac{x}{5} = 1$$
$$5x + 4x = 20$$
$$9x = 20$$
$$x = 2.2\overline{2}$$
Alfred and his son can mow the lawn working together in approximately 2.2 hours.

Chapter 7

7.1 Practice Problems

1. $\left(\dfrac{3x^{-2}y^4}{2x^{-5}y^2}\right)^{-3} = \dfrac{(3x^{-2}y^4)^{-3}}{(2x^{-5}y^2)^{-3}}$
$$= \frac{3^{-3}x^6y^{-12}}{2^{-3}x^{15}y^{-6}}$$
$$= \frac{3^{-3}}{2^{-3}} \cdot \frac{x^6}{x^{15}} \cdot \frac{y^{-12}}{y^{-6}}$$
$$= \frac{2^3}{3^3} \cdot x^{6-15} \cdot y^{-12-(-6)}$$
$$= \frac{8}{27}x^{-9}y^{-6}$$

2. (a) $(x^4)^{3/8} = x^{4 \cdot 3/8} = x^{3/2}$
   (b) $\dfrac{x^{3/7}}{x^{2/7}} = x^{3/7-2/7} = x^{1/7}$
   (c) $x^{-7/5} \cdot x^{4/5} = x^{-7/5+4/5} = x^{-3/5}$

3. (a) $\dfrac{x^4}{4x^{1/2}} = \dfrac{x^{8/2-1/2}}{4} = \dfrac{x^{7/2}}{4}$
   (b) $(-3x^{1/4})(2x^{1/2}) = -6x^{3/4}$
   (c) $\dfrac{13x^{1/12}y^{-1/4}}{26x^{-1/3}y^{1/2}} = \dfrac{x^{1/12-(-1/3)}y^{-1/4-1/2}}{2}$
$$= \frac{x^{5/12}y^{-3/4}}{2}$$
$$= \frac{x^{5/12}}{2y^{3/4}}$$

4. $-3x^{1/2}(2x^{1/4} + 3x^{-1/2}) = -6x^{3/4} - 9$

5. (a) $(4)^{5/2} = (2^2)^{5/2} = 2^5 = 32$
   (b) $(27)^{4/3} = (3^3)^{4/3} = 3^4 = 81$

6. $3x^{1/3} + x^{-1/3} = 3x^{1/3} + \dfrac{1}{x^{1/3}}$
$$= \frac{x^{1/3}}{x^{1/3}}(3x^{1/3}) + \frac{1}{x^{1/3}}$$
$$= \frac{3x^{2/3} + 1}{x^{1/3}}$$

7. $4y^{3/2} - 8y^{5/2} = 4y(y^{1/2} - 2y^{3/2})$

8. $x^{5/2} - 2x^{-3/2} = x^{-3/2}(x^{8/2} - 2)$
$$= x^{-3/2}(x^4 - 2)$$

7.2 Practice Problems

1. (a) $\sqrt[3]{216} = \sqrt[3]{(6)^3} = 6$
   (b) $\sqrt[5]{32} = \sqrt[5]{(2)^5} = 2$

(c) $\sqrt[3]{-8} = \sqrt[3]{(-2)^3} = -2$

(d) $\sqrt[4]{-81}$ is not a real number.

2. (a) $\sqrt[3]{x^3} = (x^3)^{1/3} = x$

(b) $\sqrt[4]{y^4} = (y^4)^{1/4} = y$

(c) $\sqrt[5]{(8)^5} = [(8)^5]^{1/5} = 8$

(d) $\sqrt[6]{x^{12}} = (x^{12})^{1/6} = x^{12/6} = x^2$

3. (a) $\sqrt[4]{x^3} = x^{3/4}$

(b) $\sqrt[5]{(xy)^7} = (xy)^{7/5}$

(c) $(\sqrt[5]{z})^4 = z^{4/5}$

4. (a) $\sqrt[4]{81x^{12}} = (3^4 x^{12})^{1/4} = 3x^3$

(b) $\sqrt[3]{-27x^6} = [(-3)^3 x^6]^{1/3} = -3x^2$

(c) $(32x^5)^{3/5} = (2^5 x^5)^{3/5} = 2^3 x^3 = 8x^3$

(d) $16(x^6 y^8)^{7/2} = 16x^{21} y^{28}$

5. (a) $x^{3/4} = \sqrt[4]{x^3}$

(b) $y^{-1/3} = \dfrac{1}{y^{1/3}} = \dfrac{1}{\sqrt[3]{y}}$

(c) $(2x)^{4/5} = \sqrt[5]{(2x)^4} = \sqrt[5]{16x^4}$

(d) $2x^{4/5} = 2\sqrt[5]{x^4}$

6. (a) $8^{2/3} = (\sqrt[3]{8})^2 = 2^2 = 4$

(b) $4^{3/2} = (\sqrt{4})^3 = 2^3 = 8$

(c) $(-8)^{4/3} = (\sqrt[3]{-8})^4 = (-2)^4 = 16$

(d) $(-25)^{3/2} = \sqrt{(-25)^3} =$ This is not a real number.

(e) $100^{-3/2} = \dfrac{1}{100^{3/2}} = \dfrac{1}{(\sqrt{100})^3} = \dfrac{1}{10^3} = \dfrac{1}{1000}$

(f) $32^{-3/5} = \dfrac{1}{32^{3/5}} = \dfrac{1}{(\sqrt[5]{32})^3} = \dfrac{1}{2^3} = \dfrac{1}{8}$

7. (a) $\sqrt[5]{(-3)^5} = -3$

(b) $\sqrt[4]{(-5)^4} = |-5| = 5$

(c) $\sqrt[4]{w^4} = |w|$

(d) $\sqrt[7]{y^7} = y$

8. (a) $\sqrt{36x^2} = 6x$

(b) $\sqrt[3]{125x^3 y^6} = (5^3 x^3 y^6)^{1/3} = 5xy^2$

(c) $\sqrt[4]{16y^8} = (2^4 y^8)^{1/4} = 2y^2$

7.3 Practice Problems

1. $\sqrt{20} = \sqrt{4} \cdot \sqrt{5} = 2\sqrt{5}$

2. (a) $\sqrt{75} = \sqrt{25} \cdot \sqrt{3} = 5\sqrt{3}$

(b) $\sqrt{27} = \sqrt{9} \cdot \sqrt{3} = 3\sqrt{3}$

3. (a) $\sqrt[3]{24} = \sqrt[3]{8} \cdot \sqrt[3]{3} = 2\sqrt[3]{3}$

(b) $\sqrt[3]{-108} = \sqrt[3]{-27} \cdot \sqrt[3]{4} = -3\sqrt[3]{4}$

4. $\sqrt[4]{64} = \sqrt[4]{16} \cdot \sqrt[4]{4} = 2\sqrt[4]{4}$

5. (a) $\sqrt{45x^6 y^7} = \sqrt{9x^6 y^6} \cdot \sqrt{5y}$
$$= 3x^3 y^3 \sqrt{5y}$$

(b) $\sqrt{81a^4 b^0 c^7} = \sqrt{81a^4 c^6} \cdot \sqrt{c}$
$$= 9a^2 c^3 \sqrt{c}$$

(c) $\sqrt{27a^7 b^8 c^9} = \sqrt{9a^6 b^8 c^8} \cdot \sqrt{3ac}$
$$= 3a^3 b^4 c^4 \sqrt{3ac}$$

(d) $\sqrt{256a^8 b^{11} c^{17}} = \sqrt{256a^8 b^{10} c^{16}} \sqrt{bc}$
$$= 16a^4 b^5 c^8 \sqrt{bc}$$

6. (a) $5\sqrt[3]{7} - 12\sqrt[3]{7} - 6\sqrt[3]{7} = (5 - 12 - 6)\sqrt[3]{7} = -13\sqrt[3]{7}$

(b) $19\sqrt{xy} + 5\sqrt{xy} - 10\sqrt{xy} = (19 + 5 - 10)\sqrt{xy} = 14\sqrt{xy}$

7. $\quad 4\sqrt{2} - 5\sqrt{50} - 3\sqrt{98}$
$= 4\sqrt{2} - 5\sqrt{25} \cdot \sqrt{2} - 3\sqrt{49} \cdot \sqrt{2}$
$= 4\sqrt{2} - 5(5)\sqrt{2} - 3(7)\sqrt{2}$
$= 4\sqrt{2} - 25\sqrt{2} - 21\sqrt{2}$
$= (4 - 25 - 21)\sqrt{2}$
$= -42\sqrt{2}$

8. $\quad 4\sqrt{2x} + \sqrt{18x} - 2\sqrt{125x} - 6\sqrt{20x}$
$= 4\sqrt{2x} + \sqrt{9} \cdot \sqrt{2x} - 2\sqrt{25} \cdot \sqrt{5x} - 6\sqrt{4} \cdot \sqrt{5x}$
$= 4\sqrt{2x} + 3\sqrt{2x} - 2(5)\sqrt{5x} - 6(2)\sqrt{5x}$
$= 4\sqrt{2x} + 3\sqrt{2x} - 10\sqrt{5x} - 12\sqrt{5x}$
$= 7\sqrt{2x} - 22\sqrt{5x}$

9. $\quad 3x\sqrt[3]{54x^4} - 3\sqrt[3]{16x^7}$
$= 3x\sqrt[3]{27x^3} \cdot \sqrt[3]{2x} - 3\sqrt[3]{8x^6} \cdot \sqrt[3]{2x}$
$= 3x(3x)\sqrt[3]{2x} - 3(2x^2)\sqrt[3]{2x}$
$= 9x^2 \sqrt[3]{2x} - 6x^2 \sqrt[3]{2x}$
$= 3x^2 \sqrt[3]{2x}$

10. $\sqrt{300}$

Using a calculator, we have

$300 \;\boxed{\sqrt{}}\; 17.320508$

Thus we conclude that $\sqrt{300} \approx 17.320508$

If we use Table A-1, we will need to do an extra step since the table only goes to 200. We have
$$\sqrt{300} = \sqrt{100}\sqrt{3} = 10\sqrt{3}$$
$$\approx 10(1.732)$$
$$\approx 17.32$$

7.4 Practice Problems

1. (a) $\sqrt{7}\sqrt{5} = \sqrt{35}$

(b) $(-4\sqrt{2})(-3\sqrt{13x}) = 12\sqrt{26x}$

2. (a) $\sqrt{3}(\sqrt{6} - 5\sqrt{7}) = \sqrt{18} - 5\sqrt{21}$
$$= \sqrt{9}\sqrt{2} - 5\sqrt{21}$$
$$= 3\sqrt{2} - 5\sqrt{21}$$

(b) $\quad \sqrt{2x}(\sqrt{5} + 2\sqrt{3x} + \sqrt{8})$
$= \sqrt{10x} + 2\sqrt{6x^2} + \sqrt{16x}$
$= \sqrt{10x} + 2x\sqrt{6} + 4\sqrt{x}$

3. $\quad (\sqrt{7} + 4\sqrt{2})(2\sqrt{7} - 3\sqrt{2})$
$= 2\sqrt{49} - 3\sqrt{14} + 8\sqrt{14} - 12\sqrt{4}$
$= 2(7) + 5\sqrt{14} - 12(2)$
$= 14 + 5\sqrt{14} - 24$
$= -10 + 5\sqrt{14}$

4. $(2 - 5\sqrt{5})(3 - 2\sqrt{2}) = 6 - 4\sqrt{2} - 15\sqrt{5} + 10\sqrt{10}$

5. $(\sqrt{5x} + \sqrt{10})^2 = (\sqrt{5x} + \sqrt{10})(\sqrt{5x} + \sqrt{10})$
$$= \sqrt{25x^2} + \sqrt{50x} + \sqrt{50x} + \sqrt{100}$$
$$= 5x + 2\sqrt{25}\sqrt{2x} + 10$$
$$= 5x + 2(5)\sqrt{2x} + 10$$
$$= 5x + 10\sqrt{2x} + 10$$

6. $(\sqrt{4x-1} + 3)^2 = (\sqrt{4x-1} + 3)(\sqrt{4x-1} + 3)$
$$= (\sqrt{4x-1})^2 + 3\sqrt{4x-1} + 3\sqrt{4x-1} + 9$$
$$= 4x - 1 + 6\sqrt{4x-1} + 9$$
$$= 4x + 8 + 6\sqrt{4x-1}$$

7. (a) $\quad \sqrt[3]{2x}(\sqrt[3]{4x^2} + 3\sqrt[3]{y})$
$= \sqrt[3]{8x^3} + 3\sqrt[3]{2xy}$
$= 2x + 3\sqrt[3]{2xy}$

(b) $\quad (\sqrt[3]{7} + \sqrt[3]{x^2})(2\sqrt[3]{49} - \sqrt[3]{x})$
$= 2\sqrt[3]{343} - \sqrt[3]{7x} + 2\sqrt[3]{49x^2} - \sqrt[3]{x^3}$
$= 2\sqrt[3]{7^3} - \sqrt[3]{7x} + 2\sqrt[3]{49x^2} - x$
$= 2(7) - \sqrt[3]{7x} + 2\sqrt[3]{49x^2} - x$
$= 14 - \sqrt[3]{7x} + 2\sqrt[3]{49x^2} - x$

8. (a) $\sqrt{\dfrac{144}{16}} = \dfrac{\sqrt{144}}{\sqrt{16}} = \dfrac{12}{4} = 3$

(b) $\dfrac{\sqrt{75}}{\sqrt{3}} = \sqrt{\dfrac{75}{3}} = \sqrt{25} = 5$

(c) $\sqrt[3]{\dfrac{27}{64}} = \dfrac{\sqrt[3]{27}}{\sqrt[3]{64}} = \dfrac{3}{4}$

(d) $\dfrac{\sqrt{54a^3 b^7}}{\sqrt{6b^5}} = \sqrt{\dfrac{54a^3 b^7}{6b^5}} = \sqrt{9a^3 b^2} = 3ab\sqrt{a}$

9. $\dfrac{7}{\sqrt{3}} \cdot \dfrac{\sqrt{3}}{\sqrt{3}} = \dfrac{7\sqrt{3}}{3}$

10. $\dfrac{8}{\sqrt{20x}} = \dfrac{8}{2\sqrt{5x}} \cdot \dfrac{\sqrt{5x}}{\sqrt{5x}} = \dfrac{8\sqrt{5x}}{10x} = \dfrac{4\sqrt{5x}}{5x}$

11. $\sqrt[3]{\dfrac{6}{5x}} = \dfrac{\sqrt[3]{6}}{\sqrt[3]{5x}} \cdot \dfrac{\sqrt[3]{25x^2}}{\sqrt[3]{25x^2}} = \dfrac{\sqrt[3]{150x^2}}{\sqrt[3]{125x^3}} = \dfrac{\sqrt[3]{150x^2}}{5x}$

12. $\dfrac{4}{2 + \sqrt{5}} = \dfrac{4}{2 + \sqrt{5}} \cdot \dfrac{2 - \sqrt{5}}{2 - \sqrt{5}}$
$$= \dfrac{4(2 - \sqrt{5})}{4 + 2\sqrt{5} - 2\sqrt{5} - \sqrt{25}}$$
$$= \dfrac{4(2 - \sqrt{5})}{4 - 5}$$
$$= \dfrac{4(2 - \sqrt{5})}{-1}$$
$$= -(8 - 4\sqrt{5})$$

13.

$$\frac{\sqrt{11} + \sqrt{2}}{\sqrt{11} - \sqrt{2}} \cdot \frac{\sqrt{11} + \sqrt{2}}{\sqrt{11} + \sqrt{2}}$$

$$= \frac{\sqrt{121} + \sqrt{22} + \sqrt{22} + \sqrt{4}}{\sqrt{121} + \sqrt{22} - \sqrt{22} - \sqrt{4}}$$

$$= \frac{11 + 2\sqrt{22} + 2}{11 - 2} = \frac{13 + 2\sqrt{22}}{9}$$

7.5   Practice Problems

1. $\sqrt{3x + 1} = 5$

$(\sqrt{3x + 1})^2 = (5)^2$

$3x + 1 = 25$

$3x = 24$

$x = 8$

Check. $\sqrt{3(8) + 1} \overset{?}{=} 5$

$\sqrt{25} \overset{?}{=} 5$

$5 = 5$ ✓

2. $\sqrt{3x - 8} + 2 = x$

$\sqrt{3x - 8} = x - 2$

$(\sqrt{3x - 8})^2 = (x - 2)^2$

$3x - 8 = x^2 - 4x + 4$

$0 = x^2 - 7x + 12$

$0 = (x - 3)(x - 4)$

$x = 3$ or $x = 4$

Check.   for $x = 3$ $\sqrt{3(3) - 8} + 2 \overset{?}{=} 3$

$\sqrt{1} + 2 \overset{?}{=} 3$

$3 = 3$ ✓

for $x = 4$ $\sqrt{3(4) - 8} + 2 \overset{?}{=} 4$

$\sqrt{4} + 2 \overset{?}{=} 4$

$4 = 4$ ✓

3. $\sqrt{x + 4} = x + 4$

$(\sqrt{x + 4})^2 = (x + 4)^2$

$x + 4 = x^2 + 8x + 16$

$= x^2 + 7x + 12$

$= (x + 3)(x + 4)$

$x = -3$ or $x = -4$

Check.   for $x = -3$ $\sqrt{-3 + 4} \overset{?}{=} -3 + 4$

$\sqrt{1} \overset{?}{=} 1$

$1 = 1$ ✓

for $x = -4$ $\sqrt{-4 + 4} \overset{?}{=} -4 + 4$

$\sqrt{0} \overset{?}{=} 0$

$0 = 0$ ✓

4. $\sqrt{2x + 5} - 2\sqrt{2x} = 1$

$\sqrt{2x + 5} = 2\sqrt{2x} + 1$

$(\sqrt{2x + 5})^2 = (2\sqrt{2x} + 1)^2$

$2x + 5 = 8x + 4\sqrt{2x} + 1$

$-6x + 4 = 4\sqrt{2x}$

$-3x + 2 = 2\sqrt{2x}$

$(-3x + 2)^2 = (2\sqrt{2x})^2$

$9x^2 - 12x + 4 = 8x$

$9x^2 - 20x + 4 = 0$

$(9x - 2)(x - 2) = 0$

$x = \dfrac{2}{9}$ or $x = 2$

Check.   for $x = \dfrac{2}{9}$

$\sqrt{2\left(\dfrac{2}{9}\right) + 5} - 2\sqrt{2\left(\dfrac{2}{9}\right)} \overset{?}{=} 1$

$\sqrt{\dfrac{4}{9} + 5} - 2\sqrt{\dfrac{4}{9}} \overset{?}{=} 1$

$\sqrt{\dfrac{49}{9}} - 2\sqrt{\dfrac{4}{9}} \overset{?}{=} 1$

$\dfrac{7}{3} - \dfrac{4}{3} \overset{?}{=} 1$

$\dfrac{3}{3} \overset{?}{=} 1$

$1 = 1$ ✓

$x = \dfrac{2}{9}$ is a valid solution.

for $x = 2$

$\sqrt{2(2) + 5} - 2\sqrt{2(2)} \overset{?}{=} 1$

$\sqrt{9} - 2\sqrt{4} \overset{?}{=} 1$

$3 - 4 \overset{?}{=} 1$

$-1 \neq 1$

$x = 2$ is not a valid solution.

5. $\sqrt{y - 1} + \sqrt{y - 4} = \sqrt{4y - 11}$

$(\sqrt{y - 1} + \sqrt{y - 4})^2 = (\sqrt{4y - 11})^2$

$y - 1 + 2(\sqrt{y - 1})(\sqrt{y - 4}) + y - 4 = 4y - 11$

$2y - 5 + 2(\sqrt{y - 1})(\sqrt{y - 4}) = 4y - 11$

$2(\sqrt{y - 1})(\sqrt{y - 4}) = 2y - 6$

$(\sqrt{y - 1})(\sqrt{y - 4}) = y - 3$

$[(\sqrt{y - 1})(\sqrt{y - 4})]^2 = (y - 3)^2$

$y^2 - 5y + 4 = y^2 - 6y + 9$

$y - 5 = 0$

$y = 5$

Check. $\sqrt{5 - 1} + \sqrt{5 - 4} \overset{?}{=} \sqrt{4(5) - 11}$

$2 + 1 \overset{?}{=} 3$

$3 = 3$ ✓

7.6   Practice Problems

1.   **(a)** $\sqrt{-49} = \sqrt{49}\sqrt{-1} = 7i$

(b) $\sqrt{-31} = \sqrt{-1}\sqrt{31} = i\sqrt{31}$

2.   **(a)** $\sqrt{-98} = \sqrt{-1}\sqrt{49}\sqrt{2} = 7i\sqrt{2}$

(b) $\sqrt{-150} = \sqrt{-1}\sqrt{25}\sqrt{6} = 5i\sqrt{6}$

3. $\sqrt{-8} \cdot \sqrt{-2} = \sqrt{-1}\sqrt{8} \cdot \sqrt{-1}\sqrt{2}$

$= i\sqrt{8} \cdot i\sqrt{2}$

$= i^2\sqrt{16}$

$= -1(4) = -4$

4. $-7 + 2yi\sqrt{3} = x + 6i\sqrt{3}$

$x = -7$ $2yi\sqrt{3} = 6i\sqrt{3}$

$y = 3$

5.   **(a)** $(-6 - 2i) + (5 - 8i)$

$= (-6 + 5) + (-2 - 8)i$

$= -1 - 10i$

(b) $(3 - 4i) - (-2 - 18i)$

$= [3 - (-2)] + [-4 - (-18)]i$

$= 5 + 14i$

6. $(4 - 2i)(3 - 7i)$

$= 12 - 28i - 6i + 14i^2$

$= 12 - 28i - 6i + 14(-1)$

$= 12 - 28i - 6i - 14 = -2 - 34i$

7. $-2i(5 + 6i)$

$= -10i - 12i^2$

$= -10i - 12(-1) = 12 - 10i$

8.   **(a)** $\sqrt{-50} \cdot \sqrt{-4}$

$= \sqrt{-1}\sqrt{50} \cdot \sqrt{-1}\sqrt{4}$

$= i\sqrt{50} \cdot i\sqrt{4}$

$= i^2\sqrt{200} = 10i^2\sqrt{2} = -10\sqrt{2}$

(b) $-3i(2 - 3i + \sqrt{-4})$

$= -6i + 9i^2 - 3i\sqrt{-4}$

$= -6i + 9(-1) - 3i\sqrt{-1}\sqrt{4}$

$= -6i - 9 - 3i^2(2)$

$= -6i - 9 - 6(-1)$

$= -6i - 9 + 6 = -6i - 3$

9.   **(a)** $i^{42} = (i^6)^7 = (-1)^7 = -1$

(b) $i^{53} = i^{50} \cdot i^3 = (i^{10})^5 \cdot i^3 = (-1)^5 \cdot (-i) = i$

(c) $i^{103} = i^{53} \cdot i^{50} = i \cdot (i^{10})^5 = i \cdot (-1)^5 = -i$

10. $\dfrac{4 + 2i}{3 + 4i} \cdot \dfrac{3 - 4i}{3 - 4i} = \dfrac{12 - 16i + 6i - 8i^2}{9 - 16i^2}$

$= \dfrac{12 - 10i - 8(-1)}{9 - 16(-1)}$

$= \dfrac{12 - 10i + 8}{9 + 16} = \dfrac{20 - 10i}{25}$

$= \dfrac{5(4 - 2i)}{25} = \dfrac{4 - 2i}{5}$

11. $\dfrac{5-6i}{-2i} \cdot \dfrac{2i}{2i}$

$= \dfrac{10i - 12i^2}{-4i^2} = \dfrac{10i - 12(-1)}{-4(-1)}$

$= \dfrac{10i + 12}{4} = \dfrac{2(5i + 6)}{4} = \dfrac{6 + 5i}{2}$

7.7 Practice Problems

1. $y = kx$

$9 = k\left(\dfrac{1}{2}\right)$

$18 = k$

Now use the equation with the k value known.

We want to find y when $x = \dfrac{2}{3}$

$y = 18x$

$y = 18\left(\dfrac{2}{3}\right)$

$y = 12$

2. Let s = speed

h = horsepower

$s = k\sqrt{h}$

Substitute $s = 128$ mph and $h = 256$

$128 = k\sqrt{256}$

$128 = 16k$

$8 = k$

Now we know the value of k so

$s = 8\sqrt{h}$

when $h = 225$

$s = 8(\sqrt{225})$

$s = 120$ mph

3. $y = \dfrac{k}{x}$

$45 = \dfrac{k}{16}$

$720 = k$

Find the value of y when $x = 36$.

$y = \dfrac{720}{x}$

$y = \dfrac{720}{36}$

$y = 20$

4. Let r = resistance

c = amount of current

$r = \dfrac{k}{c^2}$

Find the value of k when $r = 800$ ohms and $c = 0.01$ amps.

$800 = \dfrac{k}{(0.01)^2}$

$0.08 = k$

Now substitute $k = 0.08$ and $c = 0.04$ and solve for r.

$r = \dfrac{0.08}{(0.04)^2}$

$r = 50$ ohms

5. $y = \dfrac{kzw^2}{x}$

To find the value of k substitute $y = 20$, $z = 3$, $w = 5$ and $x = 4$. Solve for k.

$20 = \dfrac{k(3)(5)^2}{4}$

$20 = \dfrac{75k}{4}$

$\dfrac{80}{75} = k$

$\dfrac{16}{15} = k$

We now substitute $\dfrac{16}{15}$ for k.

$y = \dfrac{16zw^2}{15x}$

We use this equation to find y when $z = 4$, $w = 6$, and $x = 2$.

$y = \dfrac{16(4)(6)^2}{15(2)}$

$y = \dfrac{2304}{30} = \dfrac{384}{5}$

Chapter 8

8.1 Practice Problems

1. $x^2 = 29$ Check. $(\sqrt{29})^2 \overset{?}{=} 29$ $(-\sqrt{29})^2 \overset{?}{=} 29$

$x = \pm\sqrt{29}$ $29 = 29$ ✓ $29 = 29$ ✓

2. $x^2 - 121 = 0$

$x^2 = 121$

$x = \pm 11$

Check. $(11)^2 - 121 \overset{?}{=} 0$ $(-11)^2 - 121 \overset{?}{=} 0$

       $121 - 121 \overset{?}{=} 0$ $121 - 121 \overset{?}{=} 0$

            $0 = 0$ ✓ $0 = 0$ ✓

3. $x^2 = 18$ **4.** $5x^2 + 1 = 46$

$x = \pm\sqrt{18}$ $5x^2 = 45$

$x = \pm 3\sqrt{2}$ $x = \pm\sqrt{9}$

                    $x = \pm 3$

5. $3x^2 = -27$

$x = \pm\sqrt{-9}$

$x = \pm 3i$

Check. $3(3i)^2 \overset{?}{=} -27$ $3(-3i)^2 \overset{?}{=} -27$

      $3(9)(-1) \overset{?}{=} -27$ $3(9)(-1) \overset{?}{=} -27$

        $-27 = -27$ ✓ $-27 = -27$ ✓

6. $(2x + 3)^2 = 7$ **7.** $x^2 + 8x + 3 = 0$

$(2x + 3) = \pm\sqrt{7}$ $x^2 + 8x = -3$

$2x + 3 = \pm\sqrt{7}$ $x^2 + 8x + (4)^2 = -3 + (4)^2$

$2x = -3 \pm \sqrt{7}$ $(x + 4)^2 = 13$

$x = \dfrac{-3 \pm \sqrt{7}}{2}$ $x + 4 = \pm\sqrt{13}$

                   $x = -4 \pm \sqrt{13}$

8. $x^2 + 3x - 18 = 0$

$x^2 + 3x = 18$

$x^2 + 3x + \left(\dfrac{3}{2}\right)^2 = 18 + \dfrac{9}{4}$

$\left(x + \dfrac{3}{2}\right)^2 = \dfrac{81}{4}$

$\left(x + \dfrac{3}{2}\right) = \pm\dfrac{\sqrt{81}}{\sqrt{4}}$

$x + \dfrac{3}{2} = \pm\dfrac{9}{2}$

$x = -\dfrac{3}{2} \pm \dfrac{9}{2}$

$x = -\dfrac{3}{2} + \dfrac{9}{2} = 3$ or $x = -\dfrac{3}{2} - \dfrac{9}{2} = -6$

9. $2x^2 + 4x + 1 = 0$

$x^2 + 2x = \dfrac{-1}{2}$

$x^2 + 2x + (1)^2 = \dfrac{-1}{2} + 1$

$(x + 1)^2 = \dfrac{1}{2}$

$(x + 1) = \pm \sqrt{\dfrac{1}{2}}$

$x + 1 = \pm\dfrac{1}{\sqrt{2}}$

$x = -1 \pm \dfrac{\sqrt{2}}{2}$ or $\dfrac{-2 \pm \sqrt{2}}{2}$

8.2 Practice Problems

1. $3x^2 + 4x - 5 = 0$

$a = 3, b = 4, c = -5$

$x = \dfrac{-4 \pm \sqrt{4^2 - 4(3)(-5)}}{2(3)}$

$x = \dfrac{-4 \pm \sqrt{76}}{6}$

$x = \dfrac{-4 \pm 2\sqrt{19}}{6}$

$x = \dfrac{-2 \pm \sqrt{19}}{3}$

2. $x^2 + 5x = -1 + 2x$

$x^2 + 3x + 1 = 0$

$a = 1, b = 3, c = 1$

$x = \dfrac{-3 \pm \sqrt{3^2 - 4(1)(1)}}{2(1)}$

$x = \dfrac{-3 \pm \sqrt{5}}{2}$

3. $2x^2 + 7x + 6 = 0$

$a = 2, b = 7, c = 6$

$x = \dfrac{-7 \pm \sqrt{7^2 - 4(2)(6)}}{2(2)}$

$x = \dfrac{-7 \pm \sqrt{49 - 48}}{4}$

$x = \dfrac{-7 \pm \sqrt{1}}{4}$

$x = \dfrac{-7 + 1}{4}$ or $x = \dfrac{-7 - 1}{4}$

$x = -\dfrac{6}{4} = -\dfrac{3}{2}$ $x = -2$

4. $2x^2 - 26 = 0$

$a = 2, b = 0, c = -26$

$x = \dfrac{-0 \pm \sqrt{0^2 - 4(2)(-26)}}{2(2)}$

$x = \dfrac{\pm\sqrt{208}}{4} = \dfrac{\pm 4\sqrt{13}}{4} = \pm\sqrt{13}$

5.

$\dfrac{1}{x} + \dfrac{1}{x-1} = \dfrac{5}{6}$ LCD is $6x(x - 1)$

$6x(x-1)\left[\dfrac{1}{x}\right] + 6x(x-1)\left[\dfrac{1}{x-1}\right] = 6x(x-1)\left[\dfrac{5}{6}\right]$

$6(x - 1) + 6x = 5(x^2 - x)$

$6x - 6 + 6x = 5x^2 - 5x$

$0 = 5x^2 - 17x + 6$

$a = 5, b = -17, c = 6$

$x = \dfrac{-(-17) \pm \sqrt{(-17)^2 - 4(5)(6)}}{2(5)}$

$x = \dfrac{17 \pm \sqrt{289 - 120}}{10}$

$x = \dfrac{17 \pm \sqrt{169}}{10}$

$x = \dfrac{17 \pm 13}{10}$

$x = \dfrac{17 + 13}{10} = \dfrac{30}{10} = 3$ $x = \dfrac{17 - 13}{10} = \dfrac{2}{5}$

6. $2x^2 - 4x + 5 = 0$

$a = 2, b = -4, c = 5$

$x = \dfrac{-(-4) \pm \sqrt{(-4)^2 - 4(2)(5)}}{2(2)}$

$x = \dfrac{4 \pm \sqrt{-24}}{4}$

$x = \dfrac{4 \pm 2i\sqrt{6}}{4} = \dfrac{2 \pm i\sqrt{6}}{2}$

7. $9x^2 + 12x + 4 = 0$

$a = 9, b = 12, c = 4$

$b^2 - 4ac = 12^2 - 4(9)(4) = 144 - 144 = 0$

Since the discriminant is 0 there is one rational solution.

8. (a) $x^2 - 4x + 13 = 0$

$a = 1, b = -4, c = 13$

$b^2 - 4ac = (-4)^2 - 4(1)(13) = 16 - 52 = -36$

Since the discriminant is negative there are two complex solutions containing i.

(b) $9x^2 + 6x + 7 = 0$

$a = 9, b = 6, c = 7$

$b^2 - 4ac = 6^2 - 4(9)(7)$

$= 36 - 252 = -216$

Since the discriminant is negative there are two complex solutions containing i.

9. $x = -10$ $x = -6$

$x + 10 = 0$ $x + 6 = 0$

$(x + 10)(x + 6) = 0$

$x^2 + 6x + 10x + 60 = 0$

$x^2 + 16x + 60 = 0$

10. $x = 2i\sqrt{3}$ $x = -2i\sqrt{3}$

$x - 2i\sqrt{3} = 0$ $x + 2i\sqrt{3} = 0$

$(x - 2i\sqrt{3})(x + 2i\sqrt{3}) = 0$

$x^2 - 4i^2(\sqrt{9}) = 0$

$x^2 - 4(-1)(3) = 0$

$x^2 + 12 = 0$

11. $x = \dfrac{4}{5}$ $x = \dfrac{3}{7}$

$5x - 4 = 0$ $7x - 3 = 0$

$(5x - 4)(7x - 3) = 0$

$35x^2 - 15x - 28x + 12 = 0$

$35x^2 - 43x + 12 = 0$

8.3 Practice Problems

1. $x^4 - 5x^2 - 36 = 0$

Let $y = x^2$. Then $y^2 = x^4$

Thus our new equation is

$y^2 - 5y - 36 = 0$

$(y - 9)(y + 4) = 0$

$y - 9 = 0$ $y + 4 = 0$

$y = 9$ $y = -4$

$x^2 = 9$ $x^2 = -4$

$x = \pm\sqrt{9}$ $x = \pm\sqrt{-4}$

$x = \pm 3$ $x = \pm 2i$

2. $x^6 - 5x^3 + 4 = 0$

Let $y = x^3$, then $y^2 = x^6$

$y^2 - 5y + 4 = 0$

$(y - 1)(y - 4) = 0$

$y - 1 = 0$ $y - 4 = 0$

$y = 1$ $y = 4$

$x^3 = 1$ $x^3 = 4$

$x = 1$ $x = \sqrt[3]{4}$

3. $3x^{4/3} - 5x^{2/3} + 2 = 0$

Let $y = x^{2/3}$ and $y^2 = x^{4/3}$

$3y^2 - 5y + 2 = 0$

$(3y - 2)(y - 1) = 0$

$3y - 2 = 0$ $y - 1 = 0$

$y = \dfrac{2}{3}$ $y = 1$

$x^{2/3} = \dfrac{2}{3}$ $x^{2/3} = 1$

$(x^{2/3})^3 = \left(\dfrac{2}{3}\right)^3$ $(x^{2/3})^3 = 1^3$

$x^2 = \dfrac{8}{27}$ $x^2 = 1$

$x = \pm\sqrt{\dfrac{8}{27}}$ $x = \pm\sqrt{1}$

$x = \pm\dfrac{2\sqrt{2}}{3\sqrt{3}}$ $x = \pm 1$

$x = \pm\dfrac{2\sqrt{2}}{3\sqrt{3}} \cdot \dfrac{\sqrt{3}}{\sqrt{3}}$

$x = \pm\dfrac{2\sqrt{6}}{9}$

Check.   for $x = \dfrac{2\sqrt{6}}{9}$

$$3\left(\dfrac{2\sqrt{6}}{9}\right)^{4/3} - 5\left(\dfrac{2\sqrt{6}}{9}\right)^{2/3} + 2 \overset{?}{=} 0$$

$$3\left(\dfrac{4}{9}\right) - 5\left(\dfrac{2}{3}\right) + 2 \overset{?}{=} 0$$

$$\dfrac{4}{3} - \dfrac{10}{3} + 2 \overset{?}{=} 0$$

$$0 = 0 \checkmark$$

for $x = -\dfrac{2\sqrt{6}}{9}$

$$3\left(-\dfrac{2\sqrt{6}}{9}\right)^{4/3} - 5\left(-\dfrac{2\sqrt{6}}{9}\right)^{2/3} + 2 \overset{?}{=} 0$$

$$3\left(\dfrac{4}{9}\right) - 5\left(\dfrac{2}{3}\right) + 2 \overset{?}{=} 0$$

$$\dfrac{4}{3} - \dfrac{10}{3} + 2 \overset{?}{=} 0$$

$$0 = 0 \checkmark$$

for $x = 1$

$$3(1)^{4/3} - 5(1)^{2/3} + 2 \overset{?}{=} 0$$
$$3(1) - 5(1) + 2 \overset{?}{=} 0$$
$$0 = 0 \checkmark$$

for $x = -1$

$$3(-1)^{4/3} - 5(-1)^{2/3} + 2 \overset{?}{=} 0$$
$$3(1) - 5(1) + 2 \overset{?}{=} 0$$
$$0 = 0 \checkmark$$

4.
$$3x^{1/2} = 8x^{1/4} - 4$$
$$3x^{1/2} - 8x^{1/4} + 4 = 0$$
Let $y = x^{1/4}$ and $y^2 = x^{1/2}$
$$3y^2 - 8y + 4 = 0$$
$$(3y - 2)(y - 2) = 0$$
$$3y - 2 = 0 \qquad y - 2 = 0$$
$$y = \dfrac{2}{3} \qquad\qquad y = 2$$
$$x^{1/4} = \dfrac{2}{3} \qquad\qquad x^{1/4} = 2$$
$$(x^{1/4})^4 = \left(\dfrac{2}{3}\right)^4 \qquad (x^{1/4})^4 = (2)^4$$
$$x = \dfrac{16}{81} \qquad\qquad x = 16$$

Check.   for $x = \dfrac{16}{81}$

$$3\left(\dfrac{16}{81}\right)^{1/2} \overset{?}{=} 8\left(\dfrac{16}{81}\right)^{1/4} - 4$$

$$3\left(\dfrac{4}{9}\right) \overset{?}{=} \dfrac{16}{3} - 4$$

$$\dfrac{4}{3} = \dfrac{4}{3} \checkmark$$

for $x = 16$

$$3(16)^{1/2} \overset{?}{=} 8(16)^{1/4} - 4$$
$$3(4) \overset{?}{=} 8(2) - 4$$
$$12 = 12 \checkmark$$

8.4 Practice Problems

1. $V = \dfrac{1}{3}\pi r^2 h$ Solve for r.

$$\dfrac{3V}{\pi h} = r^2$$

$$\pm\sqrt{\dfrac{3V}{\pi h}} = r$$

2. $5x^2 - 8 = 3x^2 + 2w$ Solve for x.
$$2x^2 = 2w + 8$$
$$x^2 = w + 4$$
$$x = \pm\sqrt{w + 4}$$

3. $2y^2 + 9wy + 7w^2 = 0$ Solve for y.
$$(2y + 7w)(y + w) = 0$$
$$2y + 7w = 0 \qquad y + w = 0$$
$$2y = -7w \qquad\qquad y = -w$$
$$y = -\dfrac{7}{2}w$$

4. $3y^2 + 2yf - 7g = 0$ Solve for y.
Use the quadratic formula.
$$a = 3, \ b = 2f, \ c = -7g$$
$$y = \dfrac{-2f \pm \sqrt{(2f)^2 - 4(3)(-7g)}}{2(3)}$$
$$y = \dfrac{-2f \pm \sqrt{4f^2 + 84g}}{6}$$
$$y = \dfrac{-2f \pm \sqrt{4(f^2 + 21g)}}{6}$$
$$y = \dfrac{-2f + 2\sqrt{(f^2 + 21g)}}{6}$$
$$y = \dfrac{-f \pm \sqrt{f^2 + 21g}}{3}$$

5. $d = \dfrac{n^2 - 3n}{2}$ Solve for n.

Multiply each term by 2
$$2d = n^2 - 3n$$
$$0 = n^2 - 3n - 2d$$
Use the Quadratic Formula
$$a = 1, \ b = -3, \ c = -2d$$
$$n = \dfrac{-(-3) \pm \sqrt{(-3)^2 - 4(1)(-2d)}}{2}$$
$$n = \dfrac{3 \pm \sqrt{9 + 8d}}{2}$$

6. (a) $a^2 + b^2 = c^2$ Solve for b.
$$b^2 = c^2 - a^2$$
$$b = \sqrt{c^2 - a^2}$$
(b) $b = \sqrt{c^2 - a^2}$
$$b = \sqrt{(26)^2 - (24)^2}$$
$$b = \sqrt{676 - 576}$$
$$b = \sqrt{100}$$
$$b = 10$$

7. $x + x - 7 + c = 30$
$$2x - 7 + c = 30$$
$$c = -2x + 37$$
$$a = x, \ b = x - 7, \ c = -2x + 37$$
By the Pythagorean theorem,
$$x^2 + (x - 7)^2 = (-2x + 37)^2$$
$$x^2 + x^2 - 14x + 49 = 4x^2 - 148x + 1369$$
$$-2x^2 + 134x - 1320 = 0$$
$$x^2 - 67x + 660 = 0$$
By the quadratic formula,
$$a = 1, \ b = -67, \ c = 660$$
$$x = \dfrac{67 \pm \sqrt{(67)^2 - 4(1)(660)}}{2}$$
$$x = \dfrac{67 \pm \sqrt{4489 - 2640}}{2}$$
$$x = \dfrac{67 \pm \sqrt{1849}}{2}$$
$$x = \dfrac{67 \pm 43}{2}$$
$$x = \dfrac{67 + 43}{2} = 55 \quad \text{or} \quad x = \dfrac{67 - 43}{2} = 12$$

The only answer that makes sense is $x = 12$, therefore
$$x = 12$$
$$x - 7 = 5$$
$$-2x + 37 = 13$$
The legs are 5 miles and 12 miles long. The hypotenuse of the triangle is 13 miles long.

8. $A = \pi r^2$
$$A = \pi(6)^2$$
$$= 36\pi$$
Let $x =$ the radius of the new pipe.
(Area of new pipe) minus (Area of old pipe) $= 45\pi$
$$\pi x^2 - 36\pi = 45\pi$$
$$\pi x^2 = 45\pi + 36\pi$$
$$x^2 = 81$$
$$x = \pm 9$$
Since the radius must be positive, we select $x = 9$. The radius of the new pipe is 9 inches. The radius of the new pipe has been increased by 3 inches.

9. Let $w =$ width; then $2w - 3 =$ the length.
$$w(2w - 3) = 54$$
$$2w^2 - 3w = 54$$
$$2w^2 - 3w - 54 = 0$$
$$(2w + 9)(w - 6) = 0$$
$$2w + 9 = 0 \qquad w - 6 = 6$$
$$w = \frac{-9}{2} \qquad w = 6$$
We do not use the negative value.
Thus
$$\text{width} = 6 \text{ feet}$$
$$\text{length} = 2w - 3 = 2(6) - 3 = 9 \text{ feet}$$

10.

	Distance	Rate	Time
Secondary Road	150	x	$\dfrac{150}{x}$
Better Road	240	$x + 10$	$\dfrac{240}{x + 10}$
TOTAL	390	(not used)	7

$$\frac{150}{x} + \frac{240}{x + 10} = 7$$
The LCD of this equation is $x(x + 10)$. Multiply each term by the LCD.
$$x(x + 10)\left[\frac{150}{x}\right] + x(x + 10)\left[\frac{240}{x + 10}\right] = x(x + 10)[7]$$
$$150(x + 10) + 240x = 7x(x + 10)$$
$$150x + 1500 + 240x = 7x^2 + 70x$$
$$7x^2 - 320x - 1500 = 0$$
$$(x - 50)(7x + 30) = 0$$
$$x - 50 = 0 \qquad 7x + 30 = 0$$
$$x = 50 \qquad\qquad x = \frac{-30}{7}$$

We disregard the negative answer. Thus $x = 50$ mph, so Carlos drove 50 mph on the secondary road and 60 mph on the better road.

11.

Driver	Distance	Rate	Time
Ms. McSwiggin	100	$r + 25$	$\dfrac{100}{r + 25}$
Mr. McSwiggin	100	r	$\dfrac{100}{r}$

$$\frac{100}{r + 25} = \frac{100}{r} - 2$$
The LCD of this equation is $r(r + 25)$. Multiply each term by the LCD
$$r(r + 25)\left[\frac{100}{r + 25}\right] = r(r + 25)\left[\frac{100}{r}\right] - r(r + 25)[2]$$
$$100r = 100r + 2500 - 2r^2 - 50r$$
$$0 = -2r^2 - 50r + 2500$$
$$0 = r^2 + 25r - 1250$$
$$0 = (r + 50)(r - 25)$$

$$r + 50 = 0 \qquad\qquad r - 25 = 0$$
$$r = -50 \qquad\qquad r = 25$$
The negative speed is not used. Thus Mr. McSwiggin traveled at 25 mph. Ms. McSwiggin traveled at 50 mph.

8.5 Practice Problems

1. $f(x) = x^2 - 6x + 5$
$a = 1$, $b = -6$, $c = 15$

Step 1 The vertex occurs at $x = \dfrac{-b}{2a}$ thus
$$x = \frac{-(-6)}{2(1)} = 3$$
The vertex has an x-coordinate of 3.
To find the y-coordinate, we evaluate $f(3)$.
$$f(3) = 3^2 - 6(3) + 5$$
$$= 9 - 18 + 5$$
$$= -4$$
Thus the vertex is $(3, -4)$

Step 2 The y-intercept is at $f(0)$.
$$f(0) = 0^2 - 6(0) + 5$$
$$= 5$$
The y-intercept is $(0, 5)$.

Step 3 The x-intercept is at $f(x) = 0$
$$x^2 - 6x + 5 = 0$$
$$(x - 5)(x - 1) = 0$$
$$x - 5 = 0 \qquad x - 1 = 0$$
$$x = 5 \qquad\quad x = 1$$
Thus the x-intercepts are $(5, 0)$ and $(1, 0)$.

2. $g(x) = x^2 - 2x - 2$
$a = 1$, $b = -2$, $c = -2$

Step 1 vertex occurs at
$$x = \frac{-b}{2a}$$
$$x = \frac{-(-2)}{2(1)} = \frac{2}{2} = 1$$
The vertex has an x-coordinate of 1. To find the y-coordinate we evaluate $f(1)$.
$$g(1) = 1^2 - 2(1) - 2$$
$$= 1 - 2 - 2$$
$$= -3$$
Thus the vertex is $(1, -3)$.

Step 2 The y-intercept is at $g(0)$.
$$g(0) = 0^2 - 2(0) - 2$$
$$= -2$$
The y-intercept is $(0, -2)$.

Step 3 The x-intercepts occur when $g(x) = 0$.
We set $x^2 - 2x - 2 = 0$ and solve for x. The equation does not factor so we use the quadratic formula.
$$x = \frac{-(-2) \pm \sqrt{12}}{2} = \frac{2 \pm 2\sqrt{3}}{2} = 1 \pm \sqrt{3}$$
The x-intercepts are approximately $(2.7, 0)$ and $(-0.7, 0)$.

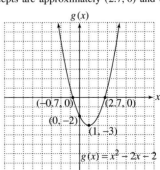

3. $g(x) = -2x^2 - 8x - 6$

$a = -2$, $b = -8$, $c = -6$

Since $a < 0$, the parabola opens downward.

The vertex occurs at

$$x = \frac{-b}{2a} = \frac{-(-8)}{2(-2)} = -2$$

To find the y-coordinate evaluate $g(-2)$

$$g(-2) = -2(-2)^2 - 8(-2) - 6$$
$$= -8 + 16 - 6$$
$$= 2$$

Thus the vertex is $(-2, 2)$.

The y-intercept is at $g(0)$.

$$g(0) = -2(0)^2 - 8(0) - 6$$
$$= -6$$

The y-intercept is $(0, -6)$

The x-intercepts occur when $g(x) = 0$

Using the quadratic formula

$$x = \frac{-(-8) \pm \sqrt{64 - 4(-2)(-6)}}{2(-2)}$$

$$= \frac{8 \pm \sqrt{16}}{-4} = -2 \pm 1$$

$x = -3$, $x = -1$

The x-intercepts are $(-3, 0)$ and $(1, 0)$.

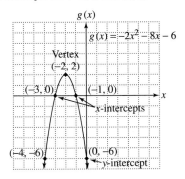

8.6 Practice Problems

1. $x^2 - 2x - 8 < 0$

Replace the inequality by an equal sign and solve.

$(x - 4)(x + 2) = 0$

$x = 4$ $x = -2$

Region I	$x < -2$	$(-3)^2 - 2(-3) - 8$
	$x = -3$	$= 9 + 6 - 8 = 7 > 0$
Region II	$-2 < x < 4$	$0^2 - 2(0) - 8 = -8 < 0$
	$x = 0$	
Region III	$x > 4$	$(5)^2 - 2(5) - 8 = 7 > 0$
	$x = 5$	

Thus, $x^2 - 2x - 8 < 0$ when $-2 < x < 4$

$$\xleftarrow{\quad} \overset{-3\ -2\ -1\ \ 0\ \ 1\ \ 2\ \ 3\ \ 4\ \ 5}{+\ \circ\ +\ +\ +\ +\ +\ \circ\ +} \xrightarrow{\quad}$$

2. $3x^2 - x - 2 \geq 0$

$(3x + 2)(x - 1) = 0$

$3x + 2 = 0$ $x - 1 = 0$

$x = -2/3$ $x = 1$

Region I	$x < -\dfrac{2}{3}$	
	$x = -1$	$3(-1)^2 + 1 - 2 = 2 > 0$
Region II	$-\dfrac{2}{3} \leq x \leq 1$	
	$x = 0$	$3(0) - 0 - 2 = -2 < 0$
Region III	$x > 1$	
	$x = 2$	$3(2)^2 - 2 - 2 = 8 > 0$

Thus $3x^2 - x - 2 \geq 0$ when $x \leq -\dfrac{2}{3}$ or when $x \geq 1$

$$\xleftarrow{\quad} \overset{-3\ \ -2\ \ -1\ \ 0\ \ 1\ \ 2\ \ 3}{\quad\ \bullet\quad\ \ \ \bullet\quad} \xrightarrow{\quad}$$

3.
$$x^2 + 2x < 7$$
$$x^2 + 2x - 7 < 0$$
$$x^2 + 2x - 7 = 0$$
$$x = \frac{-2 \pm \sqrt{4 + 28}}{2}$$
$$x = -1 \pm 2\sqrt{2}$$

$x \approx 1.8$ and $x \approx -3.8$

Region I	$x < -3.8$	
	$x = -5$	$(-5)^2 + 2(-5) - 7 = 8 > 0$
Region II	$-3.8 < x < 1.8$	
	$x = 0$	$(0)^2 + 2(0) - 7 = -7 < 0$
Region III	$x > 1.8$	
	$x = 3$	$(3)^2 + 2(3) - 7 = 8 > 0$

Thus $x^2 + 2x - 7 < 0$ when $-1 - 2\sqrt{2} < x < -1 + 2\sqrt{2}$

Approximately $-3.8 < x < 1.8$

$$\overset{\text{about } -3.8 \qquad\qquad \text{about } 1.8}{\xleftarrow{\quad} \underset{-4\ -3\ -2\ -1\ \ 0\ \ 1\ \ 2\ \ 3\ \ 4}{\circ\!-\!+\!-\!+\!-\!+\!-\!+\!-\!+\!-\!\circ\!-\!+\!-\!+} \xrightarrow{\quad}}$$

Chapter 9

9.1 Practice Problems

1. The distance between $(-6, -2)$ and $(3, 1)$ is
$$d = \sqrt{[3 - (-6)]^2 + [1 - (-2)]^2}$$
$$= \sqrt{(3 + 6)^2 + (1 + 2)^2}$$
$$= \sqrt{81 + 9} = \sqrt{90} = 3\sqrt{10}$$

2. $\sqrt{10} = \sqrt{(1 - 2)^2 + (7 - y)^2}$

$\sqrt{10} = \sqrt{1 + y^2 - 14y + 49}$

$\sqrt{10} = \sqrt{y^2 - 14y + 50}$

$10 = y^2 - 14y + 50$

$0 = y^2 - 14y + 40$

$0 = (y - 4)(y - 10)$

$\quad y = 4$ and $y = 10$

The points $(2, 4)$ and $(2, 10)$ are $\sqrt{10}$ units from $(1, 7)$

3. $(x + 1)^2 + (y + 2)^2 = 9$

If we compare this to $(x - h)^2 + (y - k)^2 = r^2$ we can write it in the form

$$[x - (-1)]^2 + [y - (-2)]^2 = 3^2$$

Thus we see the center $(h, k) = (-1, -2)$ and the radius $r = 3$.

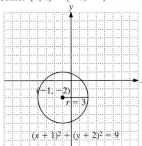

4. We are given that $(h, k) = (-5, 0)$ and $r = \sqrt{3}$. Thus,

$(x - h)^2 + (y - k)^2 = r^2$ becomes

$$[x - (-5)]^2 + [y - 0]^2 = (\sqrt{3})^2$$
$$(x + 5)^2 + y^2 = 3$$

5. To put $x^2 + 4x + y^2 + 2y - 20 = 0$ in standard form, we complete the square.

$$x^2 + 4x + \underline{\quad\quad} + y^2 + 2y + \underline{\quad\quad} = 20$$
$$x^2 + 4x + 4 + y^2 + 2y + 1 = 20 + 4 + 1$$
$$(x + 2)^2 + (y + 1)^2 = 25$$

The circle has its center at $(-2, -1)$ and a radius of 5.

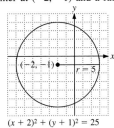

9.2 Practice Problems

1. $y = 3 - x^2$

x	y
-2	-1
-1	2
0	3
1	2
2	-1

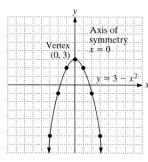

The vertex is $(0, 3)$ and the axis of symmetry is the y-axis.

2. $y = -(x + 3)^2$

x	y
-5	-4
-4	-1
-3	0
-2	-1
-1	-4
0	-9

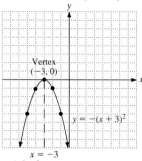

The vertex is $(-3, 0)$ and the axis of symmetry is the line $x = -3$.

3. $y = (x - 6)^2 + 4$

The vertex is $(6, 4)$ and the parabola opens upward. If $x = 0$, $y = (0 - 6)^2 + 4 = 36 + 4 = 40$, so the y-intercept is $(0, 40)$.

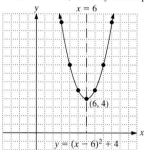

4. $y = \dfrac{1}{4}(x - 2)^2 + 3$

The vertex is $(2, 3)$ and the parabola opens upward. If $x = 0$, then $y = 4$ so the y intercept is $(0, 4)$. The axis of symmetry is $x = 2$.

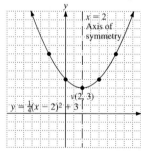

5. $x = -2y^2 + 4$

x	y
-4	2
2	1
4	0
2	-1
-4	-2

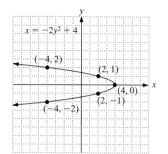

6. $x = -(y + 1)^2 - 3$

The vertex is $(-3, -1)$ and the parabola opens to the left. If $y = 0$ then $x = -4$, so the x intercept is $(-4, 0)$. The axis of symmetry is $y = -1$.

x	y
-12	2
-7	1
-4	0
-3	-1
-7	-3

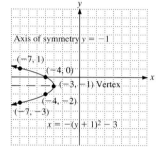

7. Since the y term is squared, we have a horizontal parabola. The standard form is $x = a(y - k)^2 + h$.

$$x = y^2 - 6y + 13$$
$$x = y^2 - 6y + 9 - 9 + 13 \quad \text{Complete the square.}$$
$$x = (y^2 - 6y + 9) + 4$$
$$x = (y - 3)^2 + 4$$

Therefore, we know that $a = 1$, $k = 3$, and $h = 4$. The vertex is at $(4, 3)$. If $y = 0$, $x = (-3)^2 + 4 = 9 + 4 = 13$. So the x-intercept is $(13, 0)$.

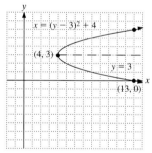

8. $y = 2x^2 + 8x + 9$

Since the x term is squared, we have a vertical parabola. The standard form is

$$y = a(x - h)^2 + k$$
$$y = 2x^2 + 8x + 9$$
$$= 2(x^2 + 4x + \underline{\quad}) - \underline{\quad} + 9 \quad \text{Complete the square.}$$
$$= 2(x^2 + 4x + 4) - 2(4) + 9$$
$$= 2(x + 2)^2 + 1$$

The parabola opens up. The vertex is $(-2, 1)$ and the y intercept is $(0, 9)$. The axis of symmetry is $x = -2$

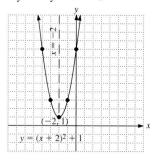

9.3 Practice Problems

1. This is the equation of an ellipse with center at origin. $a = 5$ and $b = 4$. Therefore the graph crosses the x axis at $(-5, 0)$ and $(5, 0)$ and the y axis at $(0, 4)$ and $(0, -4)$

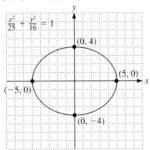

2. $4x^2 + y^2 = 16$

$$\frac{4x^2}{16} + \frac{y^2}{16} = 1$$

$$\frac{x^2}{4} + \frac{y^2}{16} = 1$$

This is the equation of an ellipse with center at origin. $a = 2$ and $b = 4$. Therefore the graph crosses the x axis at $(2, 0)$ and $(-2, 0)$ and the y axis at $(0, 4)$ and $(0, -4)$.

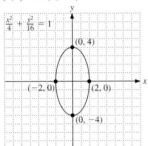

3. $\dfrac{(x - 2)^2}{16} + \dfrac{(y + 3)^2}{9} = 1$

The center is $(h, k) = (2, -3)$, $a = 4$, and $b = 3$. We thus start at $(2, -3)$ and measure to the right and to the left 4 units, and up and down 3 units.

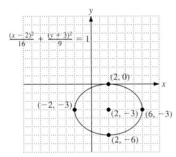

9.4 Practice Problems

1. $\dfrac{x^2}{16} - \dfrac{y^2}{25} = 1$

This is the equation of a horizontal hyperbola with center $(0, 0)$, where $a = 4$ and $b = 5$. The vertices are $(-4, 0)$ and $(4, 0)$.

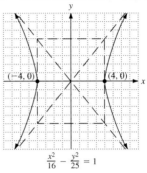

2. $y^2 - 4x^2 = 4$

$$\frac{y^2}{4} - \frac{x^2}{1} = 1$$

This is the equation of a vertical hyperbola with center $(0, 0)$ where $a = 1$ and $b = 2$. The vertices are $(0, 2)$ and $(0, -2)$

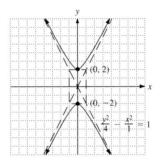

3. $\dfrac{(y + 2)^2}{9} - \dfrac{(x - 3)^2}{16} = 1$

This is a vertical hyperbola with center at $(3, -2)$, $a = 4$, and $b = 3$. The vertices are $(3, 1)$ and $(3, -5)$.

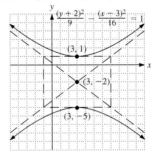

9.5 Practice Problems

1. $\dfrac{x^2}{4} - \dfrac{y^2}{4} = 1 \qquad x + y + 1 = 0$

$x^2 - y^2 = 4$ (1) $\qquad\qquad x = -1 - y$ (2)

Substituting (2) into (1).

$(-1 - y)^2 - y^2 = 4$

$1 + 2y + y^2 - y^2 = 4$

$1 + 2y = 4$

$2y = 3$

$y = \dfrac{3}{2} = 1.5$

If $y = 1.5$, $x = -1 - y = -1.0 - 1.5 = -2.5$

The solution is $(-2.5, 1.5)$.

2. $2x - 9 = y$ (1)

$xy = -4$ (2) $\qquad$ Solve equation (2) for y

$y = \dfrac{-4}{x}$ (3)

Substitute equation (3) into (1) and solve for x.

$2x - 9 = \dfrac{-4}{x}$

$2x^2 - 9x + 4 = 0$

$(2x - 1)(x - 4) = 0$

$2x - 1 = 0 \qquad x - 4 = 0$

$x = \dfrac{1}{2} \qquad\qquad x = 4$

$x = \dfrac{1}{2}$ and $x = 4$

if $x = \dfrac{1}{2}$ then $y = -8$

if $x = 4$ then $y = -1$

The solution is $(4, -1)$ and $\left(\dfrac{1}{2}, -8\right)$.

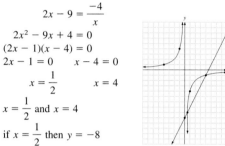

3. (1) $x^2 + y^2 = 12 \qquad 4x^2 + 4y^2 = 48 \qquad$ Multiply (1) by 4.

(2) $3x^2 - 4y^2 = 8 \qquad \dfrac{3x^2 - 4y^2 = 8}{7x^2 \qquad\quad = 56} \qquad$ Add the equations.

$x^2 = 8$

$x = \pm\sqrt{8}$

$x = \pm 2\sqrt{2}$

If $x = \pm 2\sqrt{2}$, then $x^2 = 8$. Substituting this value into equation (1) gives

$8 + y^2 = 12$

$y^2 = 4$

$y = \pm\sqrt{4}$

$y = \pm 2$

Thus the four solutions are $(2\sqrt{2}, 2)$, $(2\sqrt{2}, -2)$, $(-2\sqrt{2}, 2)$, $(-2\sqrt{2}, -2)$.

Chapter 10

10.1 Practice Problems

1. (a) $f(-3) = 3 - 2(-3) = 3 + 6 = 9$

(b) $f(2) = 3 - 2(2) = 3 - 4 = -1$

(c) $f(a) = 3 - 2a$

(d) $f(2a) = 3 - 2(2a) = 3 - 4a$

2. (a) $g(a) = \dfrac{1}{2}a - 3$

(b) $g(a + 4) = \dfrac{1}{2}(a + 4) - 3$

$= \dfrac{1}{2}a + 2 - 3$

$= \dfrac{1}{2}a - 1$

(c) $g(a) + g(4) = \dfrac{1}{2}a - 3 + \dfrac{1}{2}(4) - 3$

$= \dfrac{1}{2}a - 3 + 2 - 3$

$= \dfrac{1}{2}a - 4$

3. (a) $p(-3) = -3(-3)^2 + 2(-3) + 4$

$= -3(9) + (-6) + 4$

$= -27 - 6 + 4$

$= -29$

(b) $p(a) = -3(a)^2 + 2a + 4$

$= -3a^2 + 2a + 4$

(c) $p(2a) = -3(2a)^2 + 2(2a) + 4$

$= -3(4a^2) + 4a + 4$

$= -12a^2 + 4a + 4$

(d) $p(a - 3) = -3(a - 3)^2 + 2(a - 3) + 4$

$= -3(a^2 - 6a + 9) + 2a - 6 + 4$

$= -3a^2 + 18a - 27 + 2a - 2$

$= -3a^2 + 20a - 29$

4. (a) $f(3) = \sqrt{3(3) - 2} = \sqrt{9 - 2} = \sqrt{7}$

(b) $f(2b) = \sqrt{3(2b) - 2} = \sqrt{6b - 2}$

(c) $f(b - 1) = \sqrt{3(b - 1) - 2}$

$= \sqrt{3b - 3 - 2}$

$= \sqrt{3b - 5}$

(d) $f(b) - f(1) = \sqrt{3b - 2} - \sqrt{3(1) - 2}$

$= \sqrt{3b - 2} - \sqrt{1}$

$= \sqrt{3b - 2} - 1$

5. (a) $r(a + 2) = \dfrac{-3}{(a + 2) + 1}$

$= \dfrac{-3}{a + 3}$

(b) $r(a) = \dfrac{-3}{a + 1}$

(c) $r(a + 2) - r(a) = \dfrac{-3}{a + 3} - \left(\dfrac{-3}{a + 1}\right)$

$= \dfrac{-3}{a + 3} + \dfrac{3}{a + 1}$

$= \dfrac{(a + 1)(-3)}{(a + 1)(a + 3)} + \dfrac{3(a + 3)}{(a + 1)(a + 3)}$

$= \dfrac{-3a - 3 + 3a + 9}{(a + 1)(a + 3)}$

$= \dfrac{6}{(a + 1)(a + 3)}$

6. $g(x + h) = 2 - 5(x + h) = 2 - 5x - 5h$

$g(x) = 2 - 5x$

$g(x + h) - g(x) = 2 - 5x - 5h - (2 - 5x)$

$= 2 - 5x - 5h - 2 + 5x$

$= -5h$

Therefore $\dfrac{g(x + h) - g(x)}{h} = \dfrac{-5h}{h} = -5$

7. (a) $S(r) = 50.24r + 6.28r^2$

(b) $S(2) = 50.24(2) + 6.28(4)$

$= 125.6 \text{ m}^2$

(c) $S(2 + e) = 50.24(2 + e) + 6.28(2 + e)^2$

$= 125.6 + 75.36e + 6.28e^2$

(d) $S = 125.6 + 75.36(0.3) + 6.28(0.3)^2$

$= 125.6 + 22.608 + 0.5652$

$= 148.77 \text{ m}^2$

Thus, if the radius of 2 m were incorrectly measured as 2.3 m, the surface area would be approximately 23.17 m² too large.

10.2 Practice Problems

1. By the vertical line test, this relation is not a function.

2. $f(x) = x^2 \qquad\qquad h(x) = x^2 - 5$

x	$f(x)$
-2	4
-1	1
0	0
1	1
2	4

x	$h(x)$
-2	-1
-1	-4
0	-5
1	-4
2	-1

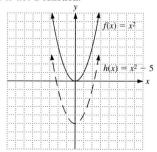

3. $f(x) = |x|$ $\qquad$ $p(x) = |x + 2|$

x	$f(x)$
-2	2
-1	1
0	0
1	1
2	2

x	$p(x)$
-4	2
-3	1
-2	0
-1	1
0	2
1	3
2	4

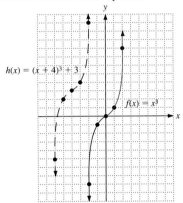

$p(x) = |x + 2|$ $\qquad$ $f(x) = |x|$

4. $f(x) = x^3$

x	$f(x)$
-2	-8
-1	-1
0	0
1	1
2	8

We recognize $h(x)$ will have a similar shape, but the curve will be shifted 4 units to the left and 3 units upward.

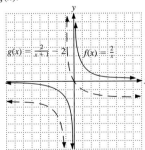

$h(x) = (x + 4)^3 + 3$

$f(x) = x^3$

5. $f(x) = \dfrac{2}{x}$

x	$f(x)$
-2	-1
-1	-2
0	undefined
1	2
2	1

The graph of $g(x)$ is 1 unit to the left and 2 units below $f(x)$. We use each point on $f(x)$ to guide us in graphing $g(x)$. Note: $x = -1$ is undefined on $g(x)$.

$g(x) = \dfrac{2}{x+1} - 2$ $\qquad$ $f(x) = \dfrac{2}{x}$

10.3 Practice Problems

1. (a) $(f + g)(x) = (4x + 5) + (2x^2 + 7x - 8)$
$\qquad = 2x^2 + 11x - 3$

(b) Using the answer found in **(a)**
$(f + g)(x) = 2x^2 + 11x - 3$
$(f + g)(4) = 2(4)^2 + 11(4) - 3$
$\qquad = 2(16) + 44 - 3$
$\qquad = 32 + 44 - 3$
$\qquad = 73$

(c) $(f - g)(x) = (4x + 5) - (2x^2 + 7x - 8)$
$\qquad = -2x^2 - 3x + 13$

(d) Using the answer found in **(c)**
$(f - g)(x) = -2x^2 - 3x + 13$
$(f - g)(-3) = -2(-3)^2 - 3(-3) + 13$
$\qquad = -2(9) + 9 + 13$
$\qquad = -18 + 9 + 13$
$\qquad = 4$

2. (a) $(fg)(x) = (3x + 2)(x^2 - 3x - 4)$
$\qquad = 3x^3 - 9x^2 - 12x + 2x^2 - 6x - 8$
$\qquad = 3x^3 - 7x^2 - 18x - 8$

(b) Using the answer found in **(a)**
$(fg)(x) = 3x^3 - 7x^2 - 18x - 8$
$(fg)(2) = 3(2)^3 - 7(2)^2 - 18(2) - 8$
$\qquad = 24 - 28 - 36 - 8$
$\qquad = -48$

3. (a) $\left(\dfrac{g}{h}\right)(x) = \dfrac{5x + 1}{3x - 2}$

(b) $\left(\dfrac{g}{p}\right)(x) = \dfrac{5x + 1}{5x^2 + 6x + 1} = \dfrac{5x + 1}{(5x + 1)(x + 1)} = \dfrac{1}{(x + 1)}$

(c) $\left(\dfrac{g}{h}\right)(3) = \dfrac{5(3) + 1}{3(3) - 2} = \dfrac{16}{7}$

4. $f[g(x)] = f(3x - 4)$
$\qquad = 2(3x - 4) - 1$
$\qquad = 6x - 8 - 1$
$\qquad = 6x - 9$

5. (a) $f[g(x)] = f(x + 2)$
$\qquad = 2(x + 2)^2 - 3(x + 2) + 1$
$\qquad = 2(x^2 + 4x + 4) - 3x - 6 + 1$
$\qquad = 2x^2 + 8x + 8 - 3x - 6 + 1$
$\qquad = 2x^2 + 5x + 3$

(b) $g[f(x)] = g(2x^2 - 3x + 1)$
$\qquad = 2x^2 - 3x + 1 + 2$
$\qquad = 2x^2 - 3x + 3$

6. (a) $f[g(x)] = f(x^2 + 4x)$
$\qquad = 1 - 2(x^2 + 4x)$
$\qquad = 1 - 2x^2 - 8x$
$\qquad = -2x^2 - 8x + 1$

(b) $f[g(-2)] = -2(-2)^2 - 8(-2) + 1$
$\qquad = -2(4) + 16 + 1$
$\qquad = -8 + 16 + 1$
$\qquad = 9$

7. (a) $(g \circ f)(x) = g[f(x)]$
$\qquad = g(3x + 1)$
$\qquad = \dfrac{2}{(3x + 1) - 3} = \dfrac{2}{3x - 2}$

(b) $(g \circ f)(-3) = \dfrac{2}{3(-3) - 2} = -\dfrac{2}{11}$

10.4 Practice Problems

1. (a) A is a function. No two pairs have the same second coordinate. Thus it is a one-to-one function.

(b) B is a function. The pair $(1, 1)$ and $(-1, 1)$ share a common second coordinate. Therefore, B is not a one-to-one function.

2. A horizontal line through the graphs of (a) and (b) intersects the graphs more than once. The graphs do not represent one-to-one functions.

(a)

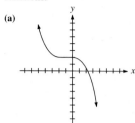

(b)

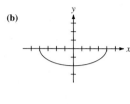

3. The inverse of B is obtained by interchanging x and y values for each ordered pair.

$$B^{-1} = \{(2, 1), (8, 7), (7, 8), (12, 10)\}$$

4. $g(x) = 4 - 6x$

$$
\begin{array}{ll}
y = 4 - 6x & \text{Replace } g(x) \text{ with } y. \\
x = 4 - 6y & \text{Interchange the variables } x \text{ and } y. \\
x - 4 = -6y & \text{Solve for } y \text{ in terms of } x. \\
-x + 4 = 6y & \\
\dfrac{-x + 4}{6} = y & \\
g^{-1}(x) = \dfrac{-x + 4}{6} &
\end{array}
$$

5. $f(x) = 2(3x + 1) - 4$

$$
\begin{array}{ll}
y = 2(3x + 1) - 4 & \text{Replace } f(x) \text{ with } y. \\
x = 2(3y + 1) - 4 & \text{Interchange the variables } x \text{ and } y. \\
x = 6y + 2 - 4 & \text{Solve for } y \text{ in terms of } x. \\
x = 6y - 2 & \\
\dfrac{x + 2}{6} = y & \\
f^{-1}(x) = \dfrac{x + 2}{6} &
\end{array}
$$

6. $f(x) = 0.75 + 0.55(x - 1)$

$$
\begin{aligned}
y &= 0.75 + 0.55(x - 1) \\
x &= 0.75 + 0.55(y - 1) \\
x &= 0.75 + 0.55y - 0.55 \\
x &= 0.55y + 0.2 \\
x - 0.2 &= 0.55y \\
\dfrac{x - 0.2}{0.55} &= y \\
f^{-1}(x) &= \dfrac{x - 0.2}{0.55}
\end{aligned}
$$

7. $f^{-1} = \{(8, 3), (4, 2), (2, 1), (1, 0)\}$

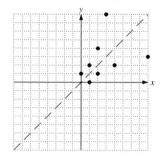

8. $f(x) = -\dfrac{1}{4}x + 1$

$$
\begin{aligned}
y &= -\dfrac{1}{4}x + 1 \\
x &= -\dfrac{1}{4}y + 1 \\
x - 1 &= -\dfrac{1}{4}y \\
-4x + 4 &= y \\
f^{-1}(x) &= -4x + 4
\end{aligned}
$$

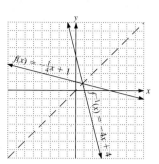

Chapter 11

11.1 Practice Problems

1. Graph. $f(x) = 3^x$

$f(x) = 3^x$

$f(-2) = 3^{-2} = \left(\dfrac{1}{3}\right)^2 = \dfrac{1}{9}$

$f(-1) = 3^{-1} = \left(\dfrac{1}{3}\right) = \dfrac{1}{3}$

$f(0) = 3^0 = 1$

$f(1) = 3^1 = 3$

$f(2) = 3^2 = 9$

x	$f(x)$
-2	$\dfrac{1}{9}$
-1	$\dfrac{1}{3}$
0	1
1	3
2	9

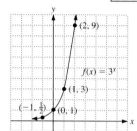

2. Graph. $f(x) = \left(\dfrac{1}{3}\right)^x$

$f(x) = \left(\dfrac{1}{3}\right)^x$

$f(-2) = \left(\dfrac{1}{3}\right)^{-2} = 3^2 = 9$

$f(-1) = \left(\dfrac{1}{3}\right)^{-1} = 3^1 = 3$

$f(0) = \left(\dfrac{1}{3}\right)^0 = 1$

$f(1) = \left(\dfrac{1}{3}\right)^1 = \dfrac{1}{3}$

$f(2) = \left(\dfrac{1}{3}\right)^2 = \dfrac{1}{9}$

x	$f(x)$
-2	9
-1	3
0	1
1	$\dfrac{1}{3}$
2	$\dfrac{1}{9}$

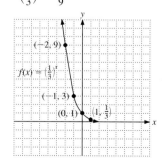

3. Graph.

$f(x) = 3^{x+2}$

$f(-4) = 3^{-4+2} = 3^{-2} = \dfrac{1}{3^2} = \dfrac{1}{9}$

$f(-3) = 3^{-3+2} = 3^{-1} = \dfrac{1}{3}$

$f(-2) = 3^{-2+2} = 3^0 = 1$

$f(-1) = 3^{-1+2} = 3$

$f(0) = 3^{0+2} = 3^2 = 9$

x	$f(x)$
-4	$\dfrac{1}{9}$
-3	$\dfrac{1}{3}$
-2	1
-1	3
0	9

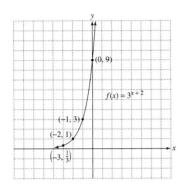

4. Graph. $f(x) = e^{x-2}$

$f(x) = e^{x-2}$ (values rounded to nearest hundredth)

$f(4) = e^{4-2} = e^2 = 7.39$

$f(3) = e^{3-2} = e^1 = 2.72$

$f(2) = e^{2-2} = e^0 = 1$

$f(1) = e^{1-2} = e^{-1} = 0.37$

$f(0) = e^{0-2} = e^{-2} = 0.14$

x	$f(x)$
4	7.39
3	2.72
2	1
1	0.37
0	0.14

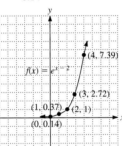

5. Solve. $2^x = \dfrac{1}{32}$

$2^x = \dfrac{1}{32}$

$2^x = \dfrac{1}{2^5}$ Because $2^5 = 32$

$2^x = 2^{-5}$ Because $\dfrac{1}{2^5} = 2^{-5}$

$x = -5$ Property of Exponential Equations

6. If Uncle Jose invests $4000 in a mutual fund that pays 11% interest compounded annually, how much will he have in 2 years?

Here $P = 4000$, $r = 0.11$, $t = 2$

$A = P(1 + r)^t$

$A = 4000(1 + 0.11)^2$

$A = 4000(1.11)^2$

$A = 4000(1.2321)$

$A = 4928.4$

He will have $4928.40.

7. How much money would Collette have if she invested $1500 for 8 years at 8% annual interest if the interest is compounded quarterly (four times a year)?

Here $P = 1500$, $r = 0.08$, $t = 8$, $n = 4$

$A = P\left(1 + \dfrac{r}{n}\right)^{nt}$

$A = 1500\left(1 + \dfrac{.08}{4}\right)^{[4(8)]}$

$A = 1500(1 + .02)^{32}$

$A = 1500(1.02)^{32}$

$A \approx 1500(1.88454)$

Collette will have approximately $2826.81.

8. If 20 milligrams of americum 241 is present now, how much will theoretically be present in 5000 years? Round your answer to the nearest thousandth.

$A = Ce^{-0.0016008t}$ where $C = 20$ and $t = 5000$.

$A = 20e^{-0.0016008(5000)}$

$A = 20e^{-8.004}$

$A \approx 20(0.0003341) \approx 0.006682$

The amount that will be present will be approximately 0.007 mg.

11.2 Practice Problems

1. Write in logarithmic form.

(a) $49 = 7^2$ **(b)** $\dfrac{1}{64} = 4^{-3}$ **(c)** $1000 = 10^3$

(a) If $49 = 7^2$, then $2 = \log_7 49$

(b) If $\dfrac{1}{64} = 4^{-3}$ then $-3 = \log_4 \left(\dfrac{1}{64}\right)$

(c) If $1000 = 10^3$, then $3 = \log_{10} 1000$

2. Write in exponential form.

(a) $3 = \log_5 125$ **(b)** $4 = \log_{1/2}\left(\dfrac{1}{16}\right)$

(c) $-2 = \log_6\left(\dfrac{1}{36}\right)$

(a) $125 = 5^3$

(b) $\dfrac{1}{16} = \left(\dfrac{1}{2}\right)^4$

(c) $\dfrac{1}{36} = 6^{-2}$

3. (a) $\log_6 y = -2$; then $y = 6^{-2}$ **(b)** $\log_b 125 = 3$; then $125 = b^3$

$\qquad\qquad\qquad\qquad y = \dfrac{1}{6^2} \qquad\qquad\qquad\qquad\qquad\qquad 5^3 = b^3$

$\qquad\qquad\qquad\qquad y = \dfrac{1}{36} \qquad\qquad\qquad\qquad\qquad\qquad b = 5$

(c) $\log_{1/2} 32 = x$; then $32 = \left(\dfrac{1}{2}\right)^x$

$\qquad\qquad\qquad 2^5 = \left(\dfrac{1}{2}\right)^x$

$\qquad\qquad\qquad \dfrac{1}{2^{-5}} = \left(\dfrac{1}{2}\right)^x$

$\qquad\qquad\qquad \left(\dfrac{1}{2}\right)^{-5} = \left(\dfrac{1}{2}\right)^x$

$\qquad\qquad\qquad\qquad x = -5$

4. Evaluate. $\log_{10} 0.1$

$\log_{10} 0.1 = x$

$\qquad 0.1 = 10^x$

$\qquad 10^{-1} = 10^x$

$\qquad\quad -1 = x$

$\log_{10} 0.1 = -1$

5. Graph. $y = \log_{1/2} x$

To graph $y = \log_{1/2} x$, we first write $x = \left(\dfrac{1}{2}\right)^y$. We make a table of values.

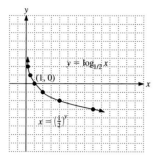

x	y
$\dfrac{1}{2}$	1
1	0
2	-1
4	-2

6. Graph $y = \log_6 x$ and $y = 6^x$ on the same set of axes.

$y = \log_6 x$

x	y
$\dfrac{1}{6}$	-1
1	0
6	1
36	2

$y = 6^x$

x	y
-1	$\dfrac{1}{6}$
0	1
1	6
2	36

Ordered pairs reversed

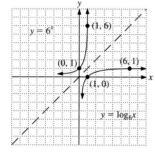

11.3 Practice Problems

1. Property 1 can be extended to three logarithms
$$\log_b MNP = \log_b M + \log_b N + \log_b P$$
Thus
$$\log_4 WXY = \log_4 W + \log_4 X + \log_4 Y$$

2. By property 1,
$$\log_b M + \log_b N = \log_b MN$$
Thus
$$\log_9 12 + \log_9 5 = \log_9 (12 \cdot 5) = \log_9 (60)$$

3. $\log_b M + \log_b N + \text{lob}_b P = \log_b MNP$
Therefore,
$$\log_7 w + \log_7 8 + \log_7 x = \log_7 (w \cdot 8 \cdot x) = \log_7 8wx$$

4. By property 2,
$$\log_b \left(\dfrac{M}{N}\right) = \log_b M - \log_b N$$
Thus
$$\log_3 \left(\dfrac{17}{5}\right) = \log_3 17 - \log_3 5$$

5. $\log_b 132 - \log_b 4 = \log_b \left(\dfrac{132}{4}\right) = \log_b 33$

6. By Property 3
$$\log_b M^p = p \log_b M$$
Thus
$$\log_7 16^9 = 9 \log_7 16$$

7. $\dfrac{1}{3} \log_7 x - 5 \log_7 y = \log_7 x^{1/3} - \log_7 y^5$ By property 3.
$$= \log_7 \left(\dfrac{x^{1/3}}{y^5}\right) \qquad \text{By property 2.}$$

8. $\log_3 \left(\dfrac{x^4 y^5}{z}\right) = \log_3 x^4 y^5 - \log_3 z \qquad$ By property 2.
$$= \log_3 x^4 + \log_3 y^5 - \log_3 z \qquad \text{By property 1.}$$
$$= 4 \log_3 x + 5 \log_3 y - \log_3 z \qquad \text{By property 3.}$$

9. $\log_4 x + \log_4 5 = 2$
$$\log_4 5x = 2 \qquad \text{By property 1.}$$
Converting to exponent form, we have
$$4^2 = 5x$$
$$16 = 5x$$
$$\dfrac{16}{5} = x$$

10. $\log_{10} x - \log_{10} (x + 3) = -1$
$$\log_{10} \left(\dfrac{x}{(x + 3)}\right) = -1 \qquad \text{By property 2.}$$
$$\left(\dfrac{x}{x + 3}\right) = 10^{-1} \qquad \text{Convert to exponential form.}$$
$$x = \dfrac{1}{10}(x + 3) \qquad \text{Multiply each side by } x + 3.$$
$$x = \dfrac{1}{10}x + \dfrac{3}{10} \qquad \text{Simplify.}$$
$$\dfrac{9}{10}x = \dfrac{3}{10}$$
$$x = \dfrac{1}{3}$$

11. (a) $\log_7 1 = 0 \qquad$ By property 5.
(b) $\log_8 8 = 1 \qquad$ By property 4.
(c) $\log_{12} 13 = \log_{12}(y + 2)$
$$13 = y + 2 \qquad \text{By property 6.}$$
$$y = 11$$

12. $\log_3 2 - \log_3 5 = \log_3 6 + \log_3 x$
$$\log_3 \left(\dfrac{2}{5}\right) = \log_3 6x \qquad \text{By property 2}$$
$$\dfrac{2}{5} = 6x \qquad \text{By property 6.}$$
$$x = \dfrac{1}{15}$$

11.4 Practice Problems

1. On a scientific calculator find a decimal approximation for
(a) log 4.36   (b) log 436   (c) log 1279   (d) log 0.2418
(a) 4.36 $\boxed{\log x}$ 0.639486489
(b) 436 $\boxed{\log x}$ 2.639486489
(c) 1279 $\boxed{\log x}$ 3.106870544
(d) 0.2418 $\boxed{\log x}$ -0.616543703

2. We know that $\log x = 2.913$ is equivalent to $10^{2.913} = x$. Using a calculator we have,
2.913 $\boxed{10^x}$ 818.46479
$$x \approx 818.46479$$

3. To evaluate antilog(-3.0705) using a scientific calculator, we have
3.0705 $\boxed{+/-}$ $\boxed{10^x}$ 8.5015869×10^{-4}
$$\text{antilog}(-3.0705) \approx .00085015869$$

4. Using a scientific calculator, find an approximate value for x.
(a) $\log x = 1.823$ (b) $\log x = 0.06134$
(c) $\log x = -4.6218$
(a) 1.823 $\boxed{10^x}$ 66.527316
$$x \approx 66.527316$$
(b) .06134 $\boxed{10^x}$ 1.1517017
$$x \approx 1.1517017$$
(c) 4.6218 $\boxed{+/-}$ $\boxed{10^x}$ 2.3889112×10^{-5}
$$x \approx .000023889112$$

5. On a scientific calculator, approximate the following values.
(a) ln 4.82   (b) ln 48.2   (c) ln 0.0793
(a) 4.82 $\boxed{\ln x}$ 1.572773928
(b) 48.2 $\boxed{\ln x}$ 3.875359021
(c) 0.0793 $\boxed{\ln x}$ -2.53451715

6. On a scientific calculator, find an approximate value of x.
(a) $\ln x = 3.1628$ (b) $\ln x = -2.0573$
(a) 3.1628 $\boxed{e^x}$ 23.636686
$$x \approx 23.636686$$
(b) 2.0573 $\boxed{+/-}$ $\boxed{e^x}$ 0.1277986
$$x \approx 0.1277986$$

7. To find antilog$_e$ (-12.9713) using a scientific calculator, we have

12.9713 $\boxed{+/-}$ $\boxed{e^x}$ 2.3261407 $\times$ 10^{-6}

antilog$_e$ $(-12.9713) \approx 2.3261407 \times 10^{-6}$

8. **(a)** To evaluate $\log_6 5.28$, we use the change of base formula.

$$\log_6 5.28 = \frac{\log 5.28}{\log 6}$$

On a calculator, we perform

5.28 $\boxed{\log x}$ $\boxed{\div}$ 6 $\boxed{\log x}$ $\boxed{=}$ 0.9286548

(b) To evaluate $\log_9 3.76$, we use the change of base formula.

$$\log_9 3.76 = \frac{\log 3.76}{\log 9}$$

On a calculator, we perform

3.76 $\boxed{\log x}$ $\boxed{\div}$ 9 $\boxed{\log x}$ $\boxed{=}$ 0.602769

9. By the change of base formula,

$$\log_8 0.009312 = \frac{\log_e 0.009312}{\log_e 8} = \frac{\ln 0.009312}{\ln 8}$$

On a calculator, perform

0.009312 $\boxed{\ln}$ $\boxed{\div}$ 8 $\boxed{\ln}$ $\boxed{=}$ -2.24889774

$\log_8 0.009312 \approx -2.24889774$

11.5 Practice Problems

1. $\log_2 4 + \log_2 x = 3$

$$\log_2 4x = 3$$
$$4x = 2^3$$
$$4x = 8$$
$$x = 2$$

2. $\qquad \log(x + 5) = 2 - \log 5$

$$\log(x + 5) + \log 5 = 2$$
$$\log 5(x + 5) = 2$$
$$\log (5x + 25) = 2$$
$$5x + 25 = 10^2$$
$$5x = 75$$
$$x = 15$$

3. $\log(x + 3) - \log x = 1$

$$\log\left(\frac{x + 3}{x}\right) = 1$$
$$\frac{x + 3}{x} = 10^1$$
$$x + 3 = 10x$$
$$3 = 9x$$
$$\frac{1}{3} = x$$

4. $\log 5 - \log x = \log(6x - 7)$

$$\log\left(\frac{5}{x}\right) = \log(6x - 7)$$
$$\frac{5}{x} = 6x - 7$$
$$5 = 6x^2 - 7x$$
$$0 = 6x^2 - 7x - 5$$
$$0 = (3x - 5)(2x + 1)$$

$3x - 5 = 0 \quad$ or $\quad 2x + 1 = 0$

$\quad 3x = 5 \qquad\qquad\quad 2x = -1$

$\quad x = \dfrac{5}{3} \qquad\qquad\quad x = -\dfrac{1}{2}$

Check. $\qquad \log 5 - \log x = \log(6x - 7)$

$x = \dfrac{5}{3}:\ \log 5 - \log\left(\dfrac{5}{3}\right) \overset{?}{=} \log\left[6\left(\dfrac{5}{3}\right) - 7\right]$

$$\log 5 - \log\frac{5}{3} \overset{?}{=} \log 3$$

$$\log\left[\frac{5}{\frac{5}{3}}\right] \overset{?}{=} \log 3$$

$$\log 3 = \log 3 \quad ✔$$

$x = -\dfrac{1}{2}:\ \log 5 - \log\left(-\dfrac{1}{2}\right) \overset{?}{=} \log\left[6\left(-\dfrac{1}{2}\right) - 7\right]$

Since logarithms of negative numbers do not exist, $x = -\dfrac{1}{2}$ is not a valid solution.

The only solution is $x = \dfrac{5}{3}$.

5. Solve. $3^x = 5$

Take the logarithm of each side.

$$\log 3^x = \log 5$$
$$x \log 3 = \log 5$$
$$x = \frac{\log 5}{\log 3}$$

6. Solve for x. $2^{3x+1} = 9^{x+1}$

Take the logarithm of each side.

$$\log 2^{3x+1} = \log 9^{x+1}$$
$$(3x + 1) \log 2 = (x + 1) \log 9$$
$$3x \log 2 + \log 2 = x \log 9 + \log 9$$
$$3x \log 2 - x \log 9 = \log 9 - \log 2$$
$$x(3 \log 2 - \log 9) = \log 9 - \log 2$$
$$x = \frac{\log 9 - \log 2}{3 \log 2 - \log 9}$$
$$x \approx -12.76989838.$$
$$x \approx -12.770$$

7. Solve $20.98 = e^{3.6x}$. Round answer to nearest ten-thousandth.

Take the natural logarithm of each side.

$$\ln 20.98 = \ln e^{3.6x}$$
$$\ln 20.98 = (3.6x)(\ln e)$$
$$\ln 20.98 = 3.6x$$
$$\frac{\ln 20.98}{3.6} = x$$

On a scientific calculator, perform

20.98 $\boxed{\ln}$ $\boxed{\div}$ 3.6 $\boxed{=}$ 0.8454

8. We use the formula $A = P(1 + r)^t$ where

$A = \$10,000;\ P = \$4000;\ r = 8\%$

$$10,000 = 4000(1 + .08)^t$$
$$10,000 = 4000(1.08)^t$$
$$\frac{10,000}{4000} = (1.08)^t$$
$$2.5 = (1.08)^t$$
$$\log(2.5) = \log(1.08)^t$$
$$\log 2.5 = t[\log(1.08)]$$
$$\frac{\log 2.5}{\log 1.08} = t$$

On a scientific calculator

2.5 $\boxed{\log}$ $\boxed{\div}$ 1.08 $\boxed{\log}$ $\boxed{=}$ 11.905904

Thus, it would take about 12 years.

9.
$$A = A_0 e^{rt}$$

$5000 = 1300e^{0.043t} \qquad$ Substitute known values.

$\dfrac{5000}{1300} = e^{0.043t} \qquad$ Divide each side by 1300.

$\ln\left(\dfrac{5000}{1300}\right) = \ln e^{0.043t} \qquad$ Take the natural logarithm of each side.

$$\ln\left(\frac{50}{13}\right) = 0.043t \ln e$$
$$\ln 50 - \ln 13 = 0.043t$$
$$\frac{\ln 50 - \ln 13}{0.043} = t$$

Using a scientific calculator, we obtain $t \approx 31.32729414$. It will take about 31 years.

10. Let I_J = intensity of Japan earthquake.
Let I_S = intensity of San Francisco earthquake.

$8.9 = \log\left(\dfrac{I_J}{I_0}\right) = \log I_J - \log I_0$

Solving for $\log I_0$ gives $\log I_0 = \log I_J - 8.9$

$7.1 = \log\left(\dfrac{I_S}{I_0}\right) = \log I_S - \log I_0$

Solving for $\log I_0$ gives $\log I_0 = \log I_S - 7.1$

$\log I_J - 8.9 = \log I_S - 7.1$

$\log I_J - \log I_S = 1.8$

$\log\left(\dfrac{I_J}{I_S}\right) = 1.8$

$\dfrac{I_J}{I_S} = 10^{1.8}$

$I_J \approx 63 I_S$

The earthquake in Japan is about 63 times as intense as the San Francisco earthquake.

Selected Answers

Chapter 1

Pretest Chapter 1

1. $\pi, \sqrt{7}$ **2.** $\sqrt{9}, 3, \frac{6}{2}, 0$ **3.** Associative property of addition **4.** 8 **5.** $\frac{1}{6}$ **6.** 24 **7.** -54 **8.** $-\frac{14b^4c^2}{a}$ **9.** $\frac{27y^{10}}{16x^7}$
10. $\frac{xy^5}{-4}$ **11.** $\frac{9}{4a^6b^6}$ **12.** $\frac{y^4}{4x^6}$ **13.** 5.8×10^{-5} **14.** 89,500,000 **15.** $2x^3 - 5x^2 - 4x$ **16.** $-6a^3b^2 + 9a^2b^4 - 3ab^2$
17. $28x + 24y$ **18.** 23 **19.** 50.24 square meters **20.** 25.12 seconds

1.1 Exercises

5. natural, whole, integers, rational, real **7.** integers, rational, real **9.** rational, real **11.** rational, real **13.** irrational, real
15. rational, real **17.** irrational, real **19.** $-\frac{28}{7}, -25, -\frac{18}{5}, -0.763, -0.333\ldots, 0, \frac{1}{10}, \frac{2}{7}, 9, \frac{283}{5}, 52.8$ **21.** 9

23. $-\frac{28}{7}, -25, -\frac{18}{5}, -\pi, -0.763, -0.333\ldots$ **25.** 9 **27.** $-0.763, 52.8$ **29.** $\ldots -4, -2, 0, 2, 4\ldots$ **31.** 0, 1, 2, 3, 4, 5, 6, 7, 8, 9
33. True **35.** False **37.** Commutative property of addition **39.** Distributive property of multiplication over addition **41.** Inverse
property of addition **43.** Identity property of multiplication **45.** Associative property of multiplication **47.** Identity property of addition
49. Associative property of addition **51.** Distributive property of multiplication over addition **53.** Commutative property of multiplication
55. Identity property of addition **57.** Identity property of multiplication **59.** Distributive property of multiplication over addition
61. The distance from base to peak is 31,784 feet for Mauna Kea and 17,022 feet for Mount Everest.

1.2 Exercises

1. 8 **3.** 9 **5.** 8.3 **7.** 2 **9.** a **11.** -3 **13.** -5 **15.** -3 **17.** $-\frac{5}{3}$ or $-1\frac{2}{3}$ **19.** 5 **21.** -9.07 **23.** 5.6

25. -3.1 **27.** $\frac{25}{28}$ **29.** $-\frac{7}{6}$ or $-1\frac{1}{6}$ **31.** -42 **33.** -10 **35.** 0 **37.** -10 **39.** -1 **41.** 0 **43.** Not possible
45. 0 **47.** 0 **49.** 1 **51.** 7 **53.** 0 **55.** 7 **57.** 55 **59.** 1 **61.** 4 **63.** -0.678197169 **65.** One or three of the
quantities a, b, c must be negative. In other words, when multiplying an odd number of negative numbers, the answer will be negative. **66.** Either
two quantities are negative or all are positive. In other words, when multiplying an even number of negative numbers, the answer will be positive.
When all the quantities being multiplied are positive, the answer will be positive. **67.** Commutative property of addition **68.** Associative
property of multiplication **69.** $\left\{-\frac{1}{2}\pi, \sqrt{3}\right\}$ **70.** $\left\{-16, 0, \frac{19}{2}, 9.36, 10.\overline{5}\right\}$ **71.** 9,783,000,000,000,000,000,000,000,000 atoms

1.3 Exercises

7. 2^8 **9.** x^6 **11.** $(-6)^5$ **13.** 243 **15.** -64 **17.** $-(8)^2 = -64$ **19.** -1 **21.** -1 **23.** $\frac{4}{9}$ **25.** $-\frac{1}{125}$ **27.** 0.36

29. 0.04 **31.** 1.4641 **33.** 5 **35.** 9 **37.** -4 **39.** $\frac{5}{7}$ **41.** 0.3 **43.** 0.02 **45.** 70 **47.** 4 **49.** 1 **51.** Does not
exist **53.** -5 **55.** -20 **57.** 6 **59.** 40 **61.** -35 **63.** -14 **65.** 18 **67.** 0 **69.** 2 **71.** 3 **73.** 4 **75.** 0
77. 7685.702373 **79.** 4789 **81.** Associative property of addition **82.** Distributive property **83.** 700% increase **84.** Three times
greater **86.** Inverse property for addition

1.4 Exercises

1. $\frac{1}{9}$ **3.** $\frac{1}{x^5}$ **5.** $\frac{64}{27}$ **7.** -9 **9.** x^2 **11.** x^{12} **13.** b^{11} **15.** 3^{17} **17.** $-6x^6$ **19.** $36x^8y^3$ **21.** $-10xy^5z^9$
23. $\frac{20}{a^5b}$ **25.** $-\frac{12}{y^{11}}$ **27.** x^5 **29.** $\frac{1}{a^5}$ **31.** 8 **33.** $\frac{2}{x^5}$ **35.** $-5b^3c$ **37.** $\frac{2}{3a^{15}b}$ **39.** $-\frac{7}{4}a^3b^{10}$ **41.** x^{16}

43. $81a^{20}b^4$ **45.** $\dfrac{x^{12}y^{18}}{z^6}$ **47.** $-8y^6z^3$ **49.** $\dfrac{9a^2c^6}{16b^{12}}$ **51.** $\dfrac{1}{8x^{12}y^{18}}$ **53.** $4x^4y^2$ **55.** $\dfrac{-256}{a^3b^{17}}$ **57.** $\dfrac{2a^4}{b^7}$ **59.** $\dfrac{y^4}{9x^8}$ **61.** $\dfrac{x^6}{y^8}$

63. $\dfrac{b^5}{a^3}$ **65.** $\dfrac{1}{4x^2}$ **67.** $\dfrac{7}{5}$ **69.** $-\dfrac{6x}{y}$ **71.** $72a^{14}b^{13}$ **73.** $-9460.906704x^{21}y^{28}$ **75.** 204,283 cubic meters per second

77. 4.7×10^2 **79.** 1.73×10^6 **81.** 1.7×10^{-2} **83.** 8.346×10^{-6} **85.** 713,000 **87.** 0.01863 **89.** 0.000000901

91. 36×10^{-8} **93.** 3×10^1 **95.** 1.02×10^7 cycles **97.** 1.06×10^{-18} gram **99.** 24 **100.** -3 **101.** -15 **102.** -1

1.5 Exercises

5. 1, 2, 3 **7.** 5, -3, 1 **9.** 6.5, -0.02, 3.05 **11.** $11ab$ **13.** $6x - 8y$ **15.** $-3x^2 - x$ **17.** $-ab$ **19.** $5y^2 - 9y + 6$

21. $-0.4x^2 + 3x$ **23.** $\dfrac{a}{6}$ **25.** $-\dfrac{3a^2}{10} + 2b$ **27.** $-7.7x^2 - 3.6x$ **29.** $-3a^2b - 14ab - 5ab^2$ **31.** $8x^2 + 4xy$ **33.** $x^4 - 3x^3 - 5x^2$

35. $-9a^2 + 3a - 21$ **37.** $2x^3y - 6x^2y^2 + 8xy^3$ **39.** $\dfrac{2}{3}x + 2xy - 4$ **41.** $\dfrac{x^3}{6} + \dfrac{5x^2}{6} - \dfrac{3x}{2}$ **43.** $2a^6b - 6a^5b + 2a^3b - 4a^2b$

45. $1.5x^3 - 3x^2y + 4.5x^2y^2$ **47.** $-y - 12z$ **49.** $2x^2 + x + 4$ **51.** $4x - 5y$ **53.** $18x + 16$ **55.** $5a^2 + 6a - 22$ **57.** 10

58. 26 **59.** -1.8 **60.** -249 **61.** Approximately 7.4547×10^{10} organisms **62.** Approximately 27.3%

1.6 Exercises

1. -2 **3.** 8 **5.** -12 **7.** -2 **9.** -55 **11.** 6 **13.** 14 **15.** 8 **17.** 48.41439 **19.** $-76°$ **21.** 50°C

23. 6.28 seconds **25.** \$5664 **27.** \$1596 **29.** 576 feet **31.** 256 feet **33.** $\dfrac{18}{5}$ **35.** 114 **37.** 0.785 sq. in.

39. 84 sq. meters **41.** 40 sq. yards **43.** 0.035 sq. meter **45.** 1.52 meters **47.** (a) 197.82 cu. ft (b) 188.4 sq ft **49.** Approximately 18.16937 meters **51.** Approximately 113.04 square centimeters; approximately 37.68 centimeters **53.** (a) 51,960 watts (b) 45,032 calories

55. $-7a + 5b$ **56.** $x^2 - 2xy$ **57.** $\dfrac{25x^4}{4y^6}$ **58.** $20x + 22$

Putting Your Skills to Work

1. 5.99905×10^{24} kg **2.** 3.30008×10^{23} kg The mass of Mercury is approximately 5.5% of the mass of Earth.

Chapter 1 Review Problems

1. Integers, rational, real **2.** Rational, real **3.** Natural, whole, integers, rational, real **4.** Rational, real **5.** Irrational, real
6. Commutative property of addition **7.** Associative property of multiplication **8.** Yes **9.** -5 **10.** -6.8 **11.** 12 **12.** -4
13. $-\dfrac{2}{5}$ **14.** $\dfrac{13}{7}$ or $1\dfrac{6}{7}$ **15.** 120 **16.** 4 **17.** 5.4 **18.** 0 **19.** Does not exist **20.** 0 **21.** 5 **22.** 1 **23.** 18

24. 0 **25.** -9 **26.** -5 **27.** $\dfrac{2}{3}$ **28.** 4 **29.** -0.064 **30.** $-24x^4y^6$ **31.** $\dfrac{2b^7}{a^4}$ **32.** $\dfrac{25x^2}{y^2}$ **33.** $\dfrac{b^2c}{3a^4}$ **34.** $\dfrac{81y^4}{16x^4z^8}$

35. $-\dfrac{72c^2}{a^{17}b^8}$ **36.** $\dfrac{a^6}{8b^{12}}$ **37.** $\dfrac{x^9}{27y}$ **38.** $\dfrac{9x^5}{4y^4}$ **39.** $\dfrac{3a^3}{8b^2}$ **40.** $27a^{30}b^{18}$ **41.** $\dfrac{1}{125x^9y^{12}}$ **42.** 7.21×10^{-3} **43.** 1.06×10^{16}

44. $-ab - 2a^2b - 6b^2 + 5b^3$ **45.** $-5a^4b^2 - 10a^3b^3 + 15ab^3 + 20ab^2$ **46.** $-3a^2 - 36a$ **47.** $4x^2 - 5x - 2$ **48.** 28
49. 56.52 cubic meters

Chapter 1 Test

1. $\pi, 2\sqrt{5}$ **2.** $-2, 12, \frac{9}{3}, \frac{25}{25}, 0, \sqrt{4}$ **3.** Commutative property of multiplication **4.** 26 **5.** 2 **6.** $\dfrac{30x}{y^4}$ **7.** $\dfrac{4a^2b^4}{5c^3}$ **8.** $\dfrac{16x^{10}}{y^6}$
9. $-7y^2 + 3y$ **10.** $-9a^2 - 6a + 4b$ **11.** $12x^2y^2 - 9xy^3 + 6x^3y^2$ **12.** 2.186×10^{-6} **13.** 2,158,000,000 **14.** 1.52×10^{-6}
15. 3×10^6 **16.** $14x + 60$ **17.** 38 **18.** 9 **19.** 78 sq. m **20.** 113.04 sq. m **21.** \$9200

Chapter 2

Pretest Chapter 2

1. -3.7 **2.** -18 **3.** -2.4 **4.** 2 **5.** $\dfrac{15 - 5x}{-8}$ or $\dfrac{5x - 15}{8}$ or $-\dfrac{15 - 5x}{8}$ **6.** $\dfrac{b + 24}{11b}$ **7.** (a) $\dfrac{A - P}{Pt}$ (b) $\frac{3}{50} = 0.06$

8. 3; $-\frac{5}{3}$ **9.** 4; 12 **10.** 2.5; -5.5 **11.** 5; 0.75 **12.** 9 cm $\times$ 23 cm **13.** 26 **14.** 80 grams at 77% pure copper; 20 grams at 92% pure copper **15.** \$25,500 at 10%, \$14,500 at 15% **16.** $x > 6$ **17.** $x \le -6$

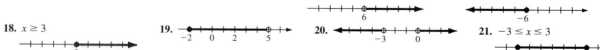

18. $x \ge 3$ **19.** **20.** **21.** $-3 \le x \le 3$

22. $x < -4$ or $x > 3$ **23.** $-\frac{10}{3} < x < 2$ **24.** $-\frac{15}{4} \le x \le \frac{21}{4}$ **25.** $x < -3$ or $x > \frac{11}{5}$

2.1 Exercises

7. 59 **9.** −7 **11.** 4 **13.** −10 **15.** −1 **17.** −5 **19.** −2 **21.** $-\dfrac{1}{3}$ **23.** $\dfrac{2}{3}$ **25.** −1 **27.** $\dfrac{2}{3}$ **29.** $-\dfrac{18}{5}$ or

$-3\dfrac{3}{5}$ **31.** $\dfrac{18}{5}$ or $3\dfrac{3}{5}$ **33.** 5 **35.** 3 **37.** 0 **39.** 6 **41.** 19 **43.** −20 **45.** $\dfrac{1}{2}$ **47.** −4.8 **49.** 8 **51.** $\dfrac{9}{8}$

53. 4 **55.** $\dfrac{1}{2}$ **57.** 2 **59.** $\dfrac{29}{2}$ **61.** −0.8901 **63.** (a) Approximately 3% (b) 25,689 **65.** −5 **66.** $\dfrac{27y^3}{8x^3}$ **67.** 16

68. $\dfrac{x^{-6}y^{-2}}{4} = \dfrac{1}{4x^6y^2}$

2.2 Exercises

1. $\dfrac{8+3y}{5}$ **3.** $\dfrac{y}{3}$ **5.** $\dfrac{4y}{ab}$ **7.** $\dfrac{8-3y}{5}$ **9.** $\dfrac{3y+12}{2}$ **11.** $\dfrac{3a-3b}{2}$ **13.** $\dfrac{d}{r}$ **15.** $\dfrac{2A-hb}{h}$ **17.** $\dfrac{x}{2}$ **19.** 0.5a

21. $\dfrac{5-3b}{15}$ **23.** $\dfrac{A+3966.9228}{3318.7328}$ **25.** (a) $\dfrac{2A}{b}$ (b) 16 **27.** (a) $\dfrac{A-a+d}{d}$ (b) $2\frac{2}{3}$ **29.** (a) $\dfrac{2D-2Vt}{t^2}$ (b) 3 **31.** (a) 1.15k

(b) 33.35 knots **33.** (a) $\dfrac{C-5.8263}{0.6547}$ (b) $5.7 billion **35.** $\dfrac{x^6}{4y^2}$ **36.** $\dfrac{y^{15}}{125x^{18}}$ **37.** $9610 **38.** 19.8 miles per gallon

2.3 Exercises

1. 14; −14 **3.** 5; −9 **5.** 9; −4 **7.** $-\dfrac{3}{2}$; 4 **9.** 10; 2 **11.** −5; $\dfrac{25}{3}$ **13.** 6; −10 **15.** 7; −10 **17.** 4; $\dfrac{-3}{2}$

19. $\dfrac{-8}{3}$; $\dfrac{16}{3}$ **21.** 9; −1 **23.** 5; −1 **25.** −4; 4 **27.** 3; 1 **29.** −2; −7 **31.** −18.41; −3.92 **33.** 1.15; 0.29

35. 6.5; −2.5 **37.** −12 **39.** No solution **41.** $\dfrac{1}{6}$; $-\dfrac{5}{6}$ **43.** 6; $-\dfrac{6}{5}$ **45.** 11; −5 **46.** −3 **47.** $\dfrac{3}{2}$ **48.** $\dfrac{8y^3z^3}{x^6}$

49. Each beaker was $30. Each Bunsen burner was $75.

Putting Your Skills to Work

1. From 1985 to 1990 **2.** 23.8 miles per gallon

2.4 Exercises

1. −128 **3.** 1.75 million people **5.** 21 round trips **7.** 72 weeks **9.** 84; 86; 88 **11.** Melissa drives 38 miles, Marcia drives 19 miles, and John drives 55 miles. **13.** The width is 14 feet. The length is 68 feet. **15.** The first side is 69 cm, the second side is 48 cm, and the third side is 36 cm. **17.** $31,500 at Golden West College; $60,000 at Westmont College **19.** (a) There are 120 ounces in the large box, 40 ounces in the small box, and 68 ounces in the medium box. (b) The medium box is the best buy. **21.** The total number of copies was 14,451. 13,051 copies were made in excess of the suggested amount of 1400 copies. **23.** Identity property of addition **24.** Associative property of addition

2.5 Exercises

1. Approximately 191.6 million people **3.** $24,107 per year **5.** 295 penguins **7.** The first number is 66. The second number is 11. **9.** The shorter piece of board is 7 linear feet. The longer piece of board is 9 linear feet. **11.** $288 **13.** $150 **15.** She invested $3900 at 5% and $2500 at 8%. **17.** They invested $12,000 at 12% and $18,000 at 15%. **19.** She should mix 12 grams of cheese that contains 45% fat and 18 grams of cheese that contains 20% fat. **21.** He should mix 36 lb of the tea that is $6.00 per pound with 108 lb of the tea that is $8.00 per pound. **23.** He should mix 200 lb of imported candy. **25.** Her speed on secondary roads was 35 mph. **27.** They each rode for $\frac{1}{2}$ hour on the stationary bike. **29.** The population was 50,000 in 1950. **31.** The profit was $18,000,000. **33.** The area of the parallelogram is 32 cm². **35.** 15 **36.** 7 **37.** 18 **38.** −13

2.6 Exercises

7. > **9.** < **11.** > **13.** > **15.** < **17.** < **19.** **21.**

23. **25.** **27.** $x \le 5$ **29.** $x > -1$

31. $x > -1$ **33.** $x \ge 2$ **35.** $x > \dfrac{3}{2}$ **37.** $x < -4$ **39.** $x \le 4$ **41.** $x < 7.5$ **43.** $x \le -2.5$ **45.** $x \ge 7$

47. $x \ge 4$ **49.** $x > 0$ **51.** $x > -4.54$ **53.** More than 8 customers **55.** At least 236 time shares per month **57.** At least 52 weeks

59. $\dfrac{-y^2}{6x}$ **60.** $\dfrac{4x^4}{y^6}$ **61.** $\dfrac{27x^3w^{12}}{8y^6}$ **62.** $\dfrac{x^4z^{12}}{16y^8}$

2.7 Exercises

1. **3.** **5.** **7.**

9. [number line with open circles at 3 and 5]

11. [number line: −2.5, marks −3 −2 −1 0 ... 4]

13. [number line with marks at −1 and 4]

15. $-5 \le x < -1$ [number line: −5 to −1]

17. [number line: −5, 1.5 at 2]

19. No Solution

21. $s < 10 \ or \ s > 12$

23. $490 \le c \le 2000$

25. $46.4° \le F \le 73.4°$ **27.** $\$143.53 \le d \le \$259.81$ **29.** $-2 < x < 2$ **31.** $x = 2$ **33.** $x < -3 \ or \ x \ge 1$ **35.** $x < 2$

37. All real numbers **39.** $x \ge 6$ **41.** No Solution **43.** $-2.5 < x \le 1$ **45.** $-4 < x < 6$ **47.** $x \le 3$ **49.** $-4.85 < x < 4.78$

51. $3 \le x \le 8$ **52.** No Solution **53.** $\dfrac{2}{3} < x \le \dfrac{7}{3}$ **54.** All real numbers **55.** $x = \dfrac{3y - 8}{5}$ **56.** $y = \dfrac{-7x - 12}{6}$

57. $a = \dfrac{3d - 5x}{6}$ **58.** $p = \dfrac{I}{rt}$

2.8 Exercises

1. $-8 \le x \le 8$ **3.** $x > 5 \ or \ x < -5$ **5.** $-9.5 < x < 0.5$ **7.** $-2 \le x \le 14$ **9.** $-10 \le x \le 7$

[number line: −8 −4 0 4 8] [number line: −5 −2 0 2 5] [number line: −9.5 ... 0.5; −10 −8 −6 −4 −2 0 2]

11. $\dfrac{4}{3} \le x \le 2$ **13.** $-2 < x < 3$ **15.** $-32 < x < 16$ **17.** $-\dfrac{7}{3} \le x \le 1$ **19.** $-9 < x < 11$ **21.** $-6 < x < 0$ **23.** $x > 3 \ or \ x < -7$

25. $x \ge 5 \ or \ x \le -1$ **27.** $x \ge 5 \ or \ x \le \dfrac{1}{3}$ **29.** $x < -8 \ or \ x > 16$ **31.** $x < -\dfrac{19}{2} \ or \ x > \dfrac{21}{2}$ **33.** $x \le -\dfrac{4}{7} \ or \ x \ge 2$

35. $-13 < x < 17$ **37.** $18.53 \le m \le 18.77$ **39.** $9.63 \le n \le 9.73$ **41.** Statement says $12 < -12$, which is a false statement.

42. Step 2: $2 < -3x < 15$; Next step: Divide by -3, *but* remember to reverse the inequality symbol. Next step should read:

Step 3: $\dfrac{2}{-3} > \dfrac{-3x}{-3} > \dfrac{15}{-3}$ **43.** -11 **44.** 2 **45.** 9.94 seconds **46.** 7.85 seconds

Chapter 2 Review Problems

1. -2 **2.** -10 **3.** $\dfrac{10}{3}$ **4.** $\dfrac{14}{5}$ **5.** 15 **6.** -3 **7.** -4 **8.** 1 **9.** $\dfrac{4x - 5}{8}$ **10.** $-\dfrac{2y}{3x}$ **11.** (a) $\dfrac{P - 2L}{2}$ (b) 29.5 m

12. (a) $\dfrac{9C + 160}{5}$ (b) 50° **13.** $7; -9$ **14.** $15; -9$ **15.** $3; -\dfrac{1}{2}$ **16.** $6; -\dfrac{22}{3}$ **17.** $12; -\dfrac{4}{3}$ **18.** $2; \dfrac{8}{3}$ **19.** $44; -20$

20. $-3; \dfrac{29}{7}$ **21.** $\dfrac{13}{2}; \dfrac{3}{2}$ **22.** 464 **23.** 120 men, 160 women **24.** 350 miles **25.** 134, 136, 138 **26.** Width = 9 mm,

length = 35 mm **27.** Retirement = $10, state tax = $23, federal tax = $69 **28.** 2650 **29.** 65,000 two-door sedans, 195,000 four-door sedans

30. $4500 at 12%, $2500 at 8% **31.** 8 liters at 2%, 16 liters at 5% **32.** 12 lb at $4.25, 18 lb at $4.50 **33.** 500 full-time students,

390 part-time students **34.** $x < -1$ **35.** $x < -1$ **36.** $x \le -2$ **37.** $x \ge 2$ **38.** $x < 7$ **39.** $x < -2$ **40.** $x \ge 2$ **41.** $x \le 5$

42. $x > 1$ **43.** $x < 3$ **44.** $x \le -1$ **45.** $x \le 6$ **46.** $x < 3$ **47.** $x > 4$ **48.** $x \le 1$ **49.** [number line: −3 ... 2]

50. [number line: −4 ... 5] **51.** [number line: −8 −4] **52.** [number line: −9 −6] **53.** [number line: −2 5]

54. [number line: −3 6] **55.** [number line: −5 −1] **56.** [number line: −8 −3] **57.** [number line: 4 5]

58. $x > 9 \ or \ x < -1$ **59.** No Solution **60.** $-6 < x < 3$ **61.** $-4 \le x \le \dfrac{5}{3}$ **62.** $\dfrac{9}{5} \le x < 5$ **63.** All real numbers

64. $-13 < x < 7$ **65.** $-7 < x < -1$ **66.** $-7 \le x \le 8$ **67.** $-\dfrac{17}{3} \le x \le 7$ **68.** $-\dfrac{15}{2} < x < -\dfrac{1}{2}$ **69.** $-26 < x < -4$

70. $x \ge 5 \ or \ x \le -4$ **71.** $x \ge 1 \ or \ x \le -\dfrac{1}{3}$ **72.** $x \ge \dfrac{8}{3} \ or \ x \le -\dfrac{2}{3}$ **73.** $x \ge 5 \ or \ x \le 1$ **74.** $x < -14 \ or \ x > 6$

75. $x < -8 \ or \ x > 6$

Chapter 2 Test

1. -5 **2.** $\frac{1}{2}$ **3.** 1 **4.** $1\frac{3}{14}$ **5.** $\dfrac{L - a + d}{d}$ **6.** $\dfrac{5F - 160}{9}$ **7.** $-40°$ **8.** $\dfrac{4H - 12b + 1}{2}$ **9.** $-7; \frac{39}{5}$ **10.** $6; -18$

11. 1st side = 16 meters; 2nd side = 32 meters; 3rd side = 21 meters **12.** 900 hr **13.** 2.5 gal at 90%, 7.5 gal at 50%

14. $1800 at 6%, $3200 at 10% **15.** $x > -2$ **16.** $x \le -1$ **17.** $-10 < x < -5$

[number line: −4 −3 −2 −1 0 1 2] [number line: −5 −4 −3 −2 −1 0 1 2 3]

18. $x \le -2 \ or \ x \ge 1$ **19.** $-\frac{15}{7} \le x \le 3$ **20.** $x < -\frac{8}{3} \ or \ x > 2$

Cumulative Test for Chapters 1–2

1. $-12, -3, 0, \frac{1}{4}, 2.16, 2.333\ldots, -\frac{5}{8}, 3$ **2.** Associative property of addition **3.** 58 **4.** $\frac{x^6}{4y^8}$ **5.** $-\frac{b^5}{2a^4}$ **6.** 54 cm

7. 153.86 square inches **8.** $30x^3 - 4x^2 - 8x$ **9.** 3 **10.** $b = \dfrac{3h - 2d}{2}$ **11.** 1st side = 25 meters; 2nd side = 35 meters;

3rd side = 45 meters **12.** 340 miles **13.** \$2000 at 12%; \$4500 at 10% **14.** 6 gallons at 80%; 3 gallons at 50%

15. $x > -10$ **16.** $x \le 4$ **17.** $-1 < x < 2$ **18.** $x \le -9 \ or \ x \ge -2$ **19.** $-20 \le x \le 12$

20. $x < -\frac{7}{3} \ or \ x > 5$

Chapter 3

Pretest Chapter 3

1. $a = -\frac{24}{5}$ **2.**

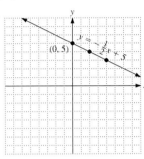

3.

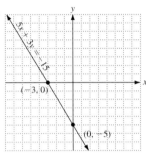

4. $m = \frac{14}{3}$ **5.** $m\perp = -\frac{9}{2}$
6. $2x + y = 11$ or $y = -2x + 11$
7. $2x - y = 0$ or $y = 2x$

8.

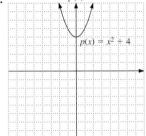

9.

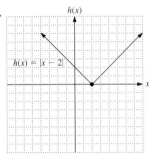

10. Domain = {0, 3, 4, 5}; range = {1, 2, 4, 6, 7}; not a function
11. Function **12.** Not a function **13.** $f(-3) = -24$
14. $g(-1) = 5$

15.

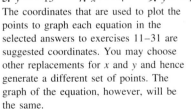

16.

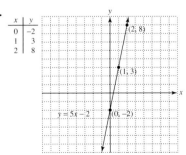

17.

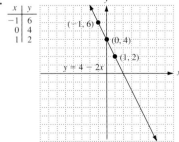

3.1 Exercises

5. $y = -13$ **7.** $x = -4$ **9.** $y = -2$ **11.**

The coordinates that are used to plot the points to graph each equation in the selected answers to exercises 11–31 are suggested coordinates. You may choose other replacements for x and y and hence generate a different set of points. The graph of the equation, however, will be the same.

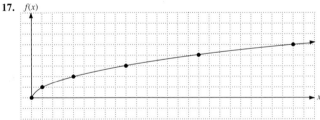

13.

15.

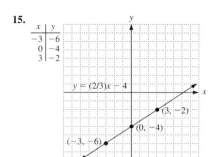

x	y
−3	−6
0	−4
3	−2

$y = (2/3)x - 4$

(3, −2)
(0, −4)
(−3, −6)

17.

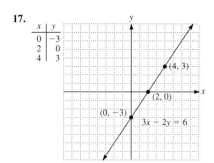

x	y
0	−3
2	0
4	3

(4, 3)
(2, 0)
(0, −3)
$3x - 2y = 6$

19.

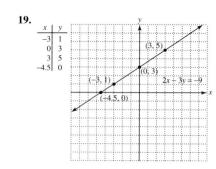

x	y
−3	1
0	3
3	5
−4.5	0

(3, 5)
(0, 3)
(−3, 1)
$2x - 3y = -9$
(−4.5, 0)

21.

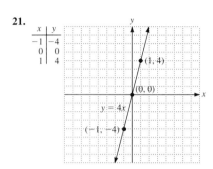

x	y
−1	−4
0	0
1	4

(1, 4)
(0, 0)
$y = 4x$
(−1, −4)

23.

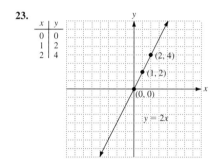

x	y
0	0
1	2
2	4

(2, 4)
(1, 2)
(0, 0)
$y = 2x$

25.

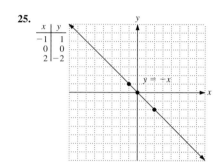

x	y
−1	1
0	0
2	−2

$y = -x$

27.

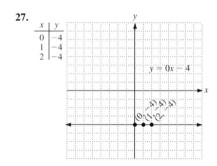

x	y
0	−4
1	−4
2	−4

$y = 0x - 4$

(0, −4)(1, −4)(2, −4)

29.

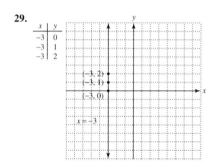

x	y
−3	0
−3	1
−3	2

(−3, 2)
(−3, 1)
(−3, 0)
$x = -3$

31.

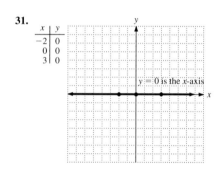

x	y
−2	0
0	0
3	0

$y = 0$ is the x-axis

33.

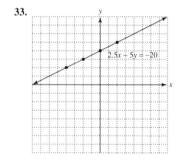

$2.5x - 5y = -20$

35.

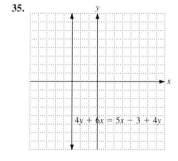

$4y + 6x = 5x - 3 + 4y$

37.

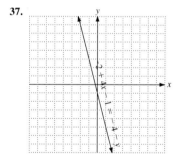

$2 + 4x - 1 = -4 - y$

39.

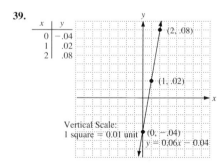

x	y
0	−.04
1	.02
2	.08

(2, .08)
(1, .02)

Vertical Scale:
1 square = 0.01 unit
(0, −.04)
$y = 0.06x - 0.04$

41.

a.

m	G
0	900
10	750
20	600
30	450
60	0

b.

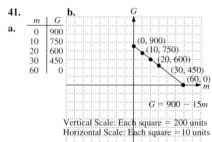

(0, 900)
(10, 750)
(20, 600)
(30, 450)
(60, 0)
$G = 900 - 15m$

Vertical Scale: Each square = 200 units
Horizontal Scale: Each square = 10 units

c. This would have no meaning since the
tank is empty after $m = 60$ minutes.

43.

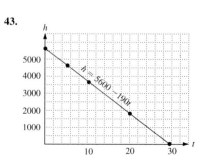

$h = 5600 - 190t$

45.   **47.**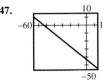

3.2 Exercises

7. $-\dfrac{3}{5}$ **9.** $-\dfrac{1}{2}$ **11.** 4   **13.** 0   **15.** $-\dfrac{1}{2}$ **17.** no slope   **19.** $\dfrac{3}{5}$ or 0.6   **21.** $\dfrac{1}{4}$ or 0.25   **23.** 80 ft   **25.** -2 **27.** $-\dfrac{4}{3}$

29. 0   **31.** $-\dfrac{5}{3}$ **33.** $\dfrac{1}{3}$ **35.** $-\dfrac{4}{3}$ **37.** $m_{AD} = m_{BC} = -\dfrac{1}{6}$; $m_{AB} = m_{CD} = 1$ **39.** **(a)** $\dfrac{1}{12}$ **(b)** 2 feet **(c)** 20.4 feet   **41.** -56

42. 2   **43.** $-\dfrac{10x^3}{y^{12}}$ **44.** $\dfrac{5x^{10}}{y^3}$

3.3 Exercises

3. $y = -3x + 6$ **11.** 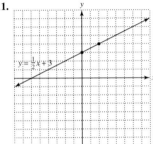   **13.** 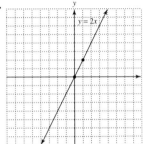   **15.**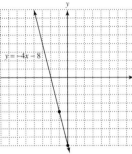
5. $y = \frac{3}{4}x - 9$
7. $5x + y = -8$
9. $3x - 4y = -2$

17. $y = -6x + 7$, $m = -6$, $b = 7$ **25.** 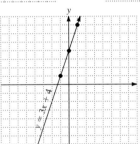   **27.** 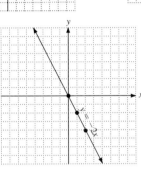   **29.**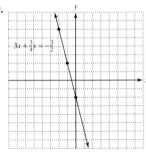
19. $y = \dfrac{2}{3}x + \dfrac{8}{3}$, $m = \dfrac{2}{3}$, $b = \dfrac{8}{3}$

21. $y = -\dfrac{1}{8}x + \dfrac{5}{4}$

   $m = -\dfrac{1}{8}$, $b = \dfrac{5}{4}$

23. $y = 2x - 5$
   $m = 2$, $b = -5$

31. $y = 3x - 9$ **33.** $y = 5x + 33$ **35.** $y = -\dfrac{2}{3}x - \dfrac{8}{3}$ **37.** $2x + y = -4$ **39.** $x - 8y = 23$ **41.** $13x + 2y = 12$

43. $3x - y = -3$ **45.** $x + 2y = -13$ **47.** $x + 5y = -6$ **49.** $4x + y = 11$ **51.** $x = 5$ (vertical line)   **53.** Perpendicular
54. Neither   **55.** Neither   **56.** Parallel   **57.** Neither   **58.** Perpendicular   **59.** Yes, they are parallel.

Putting Your Skills to Work

1. $y = -0.8x + 160$

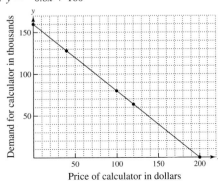

2. The slope is -0.8. The negative slope indicates a decrease in demand for each increase in price. The y-intercept is 160. This is theoretically the number of calculators in thousands that would be sold for a price of \$0.   **3.** The demand is increased by 8000 calculators when the price is lowered by \$10. The demand is increased by 80,000 calculators when the price is lowered by \$100.

3.4 Exercises

1.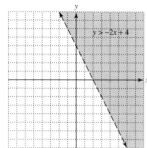

3.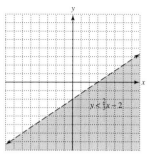

5.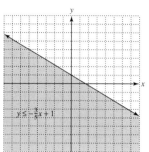

7.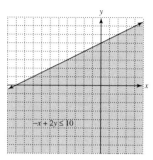

9.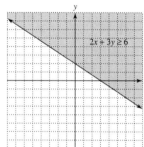

11.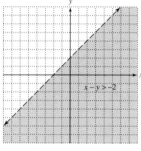

13.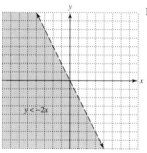

15.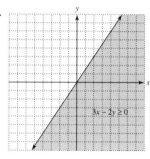

17. The point $(0, 0)$ is on the line. We obtain a false statement.

19.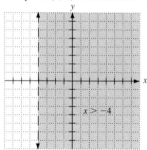

21.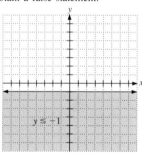

23.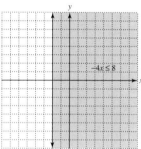

25.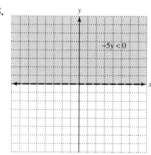

27.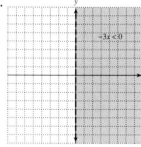

29.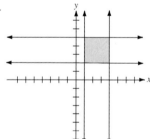

31. -2 **32.** 24

3.5 Exercises

5. Domain = {10, 12, 14, 16}; range = {1, 2, 3, 4}; relation is a function.   **7.** Domain = {$\frac{1}{2}$, -3, $\frac{3}{2}$}; range = {5, 7, -1}; relation is *not* a function.
9. Domain = {3, 4, 5, 8}; range = {-6, -2, 7}; relation is a function.   **11.** Domain = {1, 3, 5, 7}; range = {8}; relation is a function.
13. Domain = {6, 8, 10, 12, 14}; range = {38, 40, 42, 44, 46}; relation is a function.   **15.** Domain = {Nile, Amazon, Chang Jiang, Ob-Irtysh, Huang He, Congo}; range = {2900, 2903, 3362, 3964, 4000, 4160}; relation is a function.   **17.** Domain = {Chicago, New York}; range = {1046, 1127, 1136, 1250, 1350, 1454}; relation is *not* a function.   **19.** Domain = {1, 2, 3, 4, 5}; range = {1.61, 3.22, 4.83, 6.44, 8.05}; relation is a function.   **21.** Function   **23.** Function   **25.** Function   **27.** Not a Function   **29.** 19   **31.** -5 **33.** -11

35. $-4\frac{1}{3}$ **37.** 0   **39.** $2\frac{2}{3}$ **41.** $-\frac{1}{2}$ **43.** 17   **45.** -3 **47.** -9 **49.** Range = {1, 2, 3, 4, 5}   **50.** Range = {3, 4, 7}
51. Domain = {6, 9, 11, 12}   **52.** Domain = {12, 11, 9, -4}   **53.** $-7x - 6y + 15$ **54.** $9x + 36$ **55.** She should deposit $433.93.
56. 3140 tons of raw material

3.6 Exercises

1.

x	$f(x)$
0	1
1	3
-1	-1

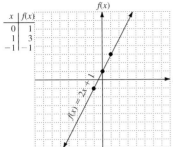

3.

x	$f(x)$
0	-4
3	-2
-3	-6

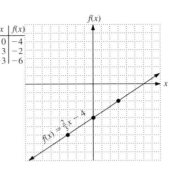

5.

x	$f(x)$
0	6
3	5
6	4

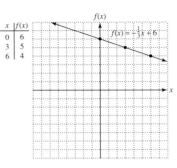

7.

x	$f(x)$
0	$\frac{1}{2}$
2	$-\frac{1}{2}$
-2	$1\frac{1}{2}$

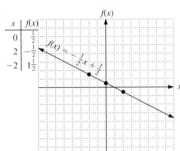

9.

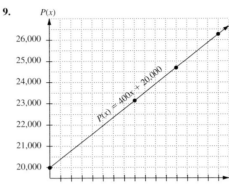

$p(x) = 400x + 20{,}000;$ $p(0) = 20{,}000$
$p(8) = 23{,}200;$ $p(12) = 24{,}800;$
$p(16) = 26{,}400$

11.

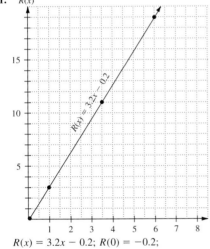

$R(x) = 3.2x - 0.2;$ $R(0) = -0.2;$
$R(1) = 3;$ $R(3.5) = 11;$ $R(6) = 19$

13.

x	$f(x)$
0	1
1	0
-1	2
2	1
3	2

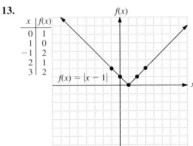

15.

x	$f(x)$
0	2
1	3
2	4
-1	3
-2	4

$f(x) = |x| + 2$

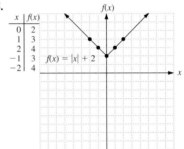

17.

x	$f(x)$
0	1
1	2
2	5
-1	2
-2	5

$f(x) = x^2 + 1$

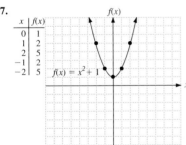

19.

x	$f(x)$
1	4
2	1
3	0
4	1
5	4

$f(x) = (x-3)^2$

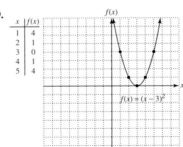

21.

x	$f(x)$
0	2
1	3
2	10
-1	1
-2	-6

$f(x) = x^3 + 2$

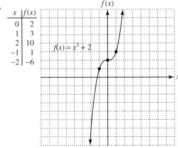

23.

x	$p(x)$
-1	1
-2	0
-3	-1
0	8
-4	-8

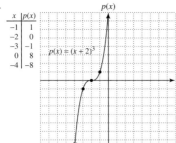
$p(x) = (x + 2)^3$

25.

x	$f(x)$
1	2
2	1
4	$\frac{1}{2}$
$\frac{1}{2}$	4
$\frac{1}{4}$	8
-1	-2
-2	-1
-4	$-\frac{1}{2}$
$-\frac{1}{2}$	-4
$-\frac{1}{4}$	-8

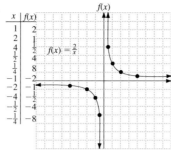
$f(x) = \frac{2}{x}$

27.

x	$h(x)$
-3	2
-2	3
2	-3
3	-2
6	-1

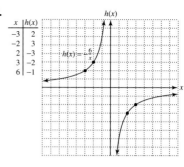
$h(x) = \frac{-6}{x}$

29.

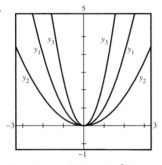

When the coefficient of x^2 is greater than 1, the curve is closer to the y-axis. When the coefficient is less than 1, the curve is farther away from the y-axis.

31.

x	$f(x)$
-5	-2
3	2
5	3

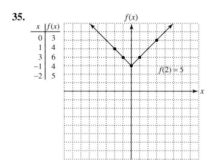
$f(2) = 1\frac{1}{2}$

33.

x	$f(x)$
-2	0.25
-1	0.5
0	1
2.5	5.7
3	8

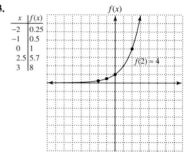
$f(2) = 4$

35.

x	$f(x)$
0	3
1	4
3	6
-1	4
-2	5

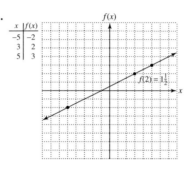
$f(2) = 5$

37. (a)

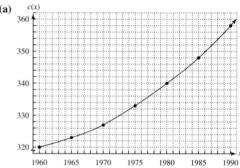

(b) About 329 parts per million **(c)** About 370 parts per million **(d)** It increased at a greater rate from 1975 to 1990 than it did from 1960 to 1975. **(e)** Between 1972 and 1973

39. $x = \dfrac{8y + 15}{a}$ **40.** $x = 17y$ **41.** $x = \dfrac{6y - 3}{5}$ **42.** $x = 7.25$ **43.** 8.698×10^{13} square feet **44.** It will breathe 21,840 times.

Chapter 3 Review Problems

1. $a = -1$ **2.** $b = -9$

3.

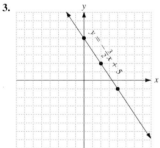
$y = -\frac{3}{2}x + 5$

4.

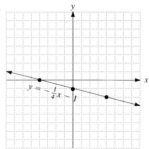
$y = -\frac{1}{4}x - 1$

5.

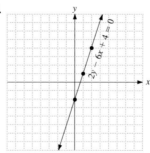
$2y - 6x + 4 = 0$

6.

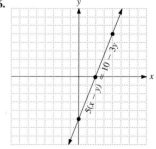
$5(x - y) = 10 - 3y$

7.

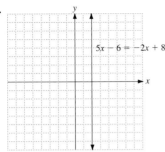

$5x - 6 = -2x + 8$

8.

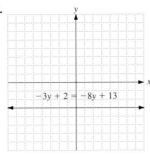

$-3y + 2 = -8y + 13$

9. 15 or more computers   **10.** $m = \dfrac{1}{10}$ **11.** $-\dfrac{1}{2}$

12. No slope   **13.** $m = -2$ **14.** No.

Slope 1st two pts $= -6$. Slope 2nd two pts $= -\dfrac{9}{4}$

15. $y = x - 3$, $m = 1$, $b = -3$ **16.** $2x - 3y = 12$

17. $4x + y = 0$ **18.** $y = 1$ **19.** $13x - 12y = -7$

20. $x = -6$ **21.** $3x - 2y = 13$ **22.** $8x - 7y = -51$

23.

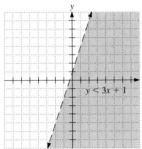

$y < 3x + 1$

24.

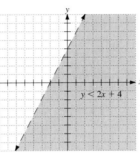

$y < 2x + 4$

25.

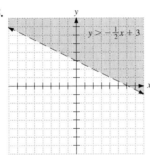

$y > -\frac{1}{2}x + 3$

26.

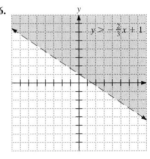

$y > -\frac{2}{3}x + 1$

27.

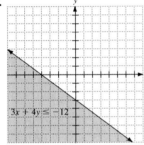

$3x + 4y \leq -12$

28.

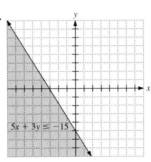

$5x + 3y \leq -15$

29.

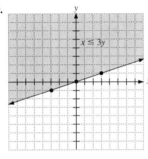

$x \leq 3y$

30.

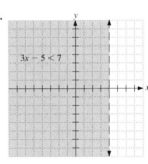

$3x - 5 < 7$

31.

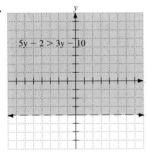

$5y - 2 > 3y - 10$

32. Domain $= \{0, 1, 2, 3\}$;
range $= \{0, 1, 4, 9, 16\}$;
relation is not a function.

33. Domain $= \{-20, -18, -16, -12\}$;
range $= \{14, 16, 18\}$;
relation is a function.

34.

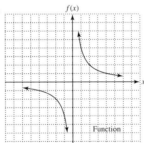

$f(x)$

Function

35.

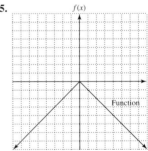

$f(x)$

Function

36.

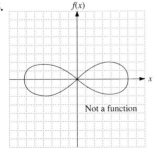

$f(x)$

Not a function

37. $f(-2) = -14$

38. $g(-3) = 22$

39. $h(-1) = 14$

40. $p(3) = 21$

41.

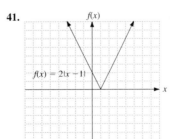

$f(x) = 2|x - 1|$

42.

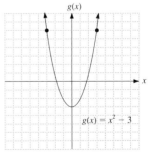

$g(x)$

$g(x) = x^2 - 3$

43.

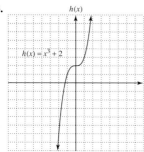

44.

x	$f(x)$
-1	5
-3	-3
-4	-4
-5	-3
-6	0
-7	5

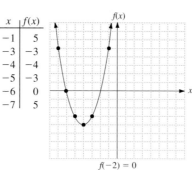

$f(-2) = 0$

45.

x	$f(x)$
-3	4
-1	2
0	1
1	0
2	-1
3	0
4	1

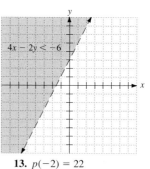

$f(-2) = 3$

46.

x	$f(x)$
-5	7
0	3
10	-5

47.

x	$f(x)$
-3	31
0	4
4	24

48.

x	$f(x)$
-1	-7
0	-4
2	20

49.

x	$f(x)$
-2	-7
0	$\frac{7}{3}$
2	1

Chapter 3 Test

1.

x	y
-2	-2
1	-4
4	-6

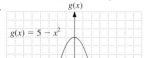

$(-2, -2)$ $2x + 3y = -10$
$(1, -4)$
$(4, -6)$

2. 2   **3.** $-\frac{3}{8}$

4. $x + 8y = -11$

5. $7x + 6y = -12$

6. $y = 2$

7.

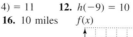

$y \geq -4x$

8.

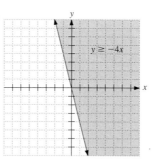

$4x - 2y < -6$

9. Domain = $\{0, 1, 2\}$; Range = $\{-4, -1, 0, 1, 4\}$ **10.** $f(\frac{3}{4}) = -1\frac{1}{2}$ **11.** $g(-4) = 11$ **12.** $h(-9) = 10$ **13.** $p(-2) = 22$

14.

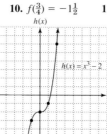

$g(x) = 5 - x^2$

15.

$h(x) = x^3 - 2$

16. 10 miles $f(x)$

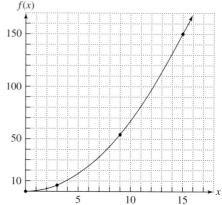

Cumulative Test for Chapters 1–3

1. Inverse property of addition   **2.** 12   **3.** $\dfrac{y^{12}}{81x^8}$ **4.** $7x^2 - 15xy - 12$ **5.** $x = -2$, $x = \frac{10}{3}$ **6.** $x < -2$ or $x > 4$

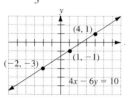

7. $x = \dfrac{6b - 2y}{3}$ **8.** Width = 15 cm; Length = 31 cm   **9.** \$1700 at 5%; \$1300 at 8%   **10.** 14.13 sq in.

11.

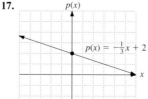

$(4, 1)$
$(-2, -3)$ $(1, -1)$
$4x - 6y = 10$

12. $\frac{1}{2}$ **13.** $2x + y = 11$

14. $3x + 2y = -12$

15. $D = \{\frac{1}{2}, 2, 3, 5\}$;
$R = \{-1, 2, 7, 8\}$;
Yes, it is a function.

16. $f(-3) = -5$

17.

$p(x)$

$p(x) = -\frac{1}{3}x + 2$

18.

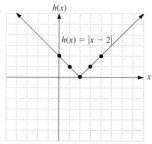

$h(x) = |x - 2|$

19.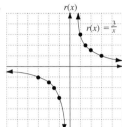

20.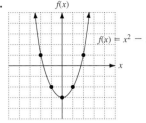

21.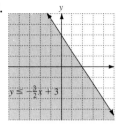

22.

x	$f(x)$
-2	20
0	4
3	-50

Chapter 4

Pretest Chapter 4

1. $x = 5$, $y = 2$ **2.** $x = 3$, $y = -2$ **3.** Infinite number of solutions; dependent equations   **4.** $x = 1$, $y = 3$, $z = 0$
5. $x = -2$, $y = 3$, $z = 4$ **6.** \$12 shirts, \$17 pants
7. Packet $A = 3$; packet $B = 2$; packet $C = 4$ **8.** 2
9. -1 **10.** 0   **11.** 7   **12.** $x = 2$, $y = -2$
13. $x = -4$

14.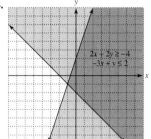

15.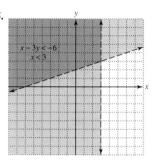

4.1 Exercises

1. $(-2, -3)$ is a solution to the system.   **3.** $(13, -10)$ is not a solution to the system.   **5.** $(\frac{1}{3}, -2)$ is a solution to the system.

7.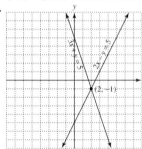

9.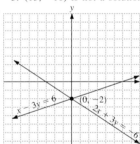

11.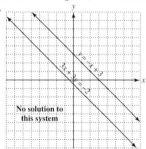

13. $x = 1$, $y = 2$ **15.** $x = 23$, $y = -43$ **17.** $x = 2$, $y = -2$ **19.** $a = -3$, $b = -2$
21. $x = 15$, $y = 4$ **23.** $x = 3$, $y = -2$ **25.** $x = 0$, $y = 1$ **27.** $a = 1$, $b = -1$ **29.** $s = 1$,
$t = \dfrac{5}{3}$ **31.** $x = 1$, $y = -3$ **33.** $x = 5$, $y = 1$ **35.** $x = -1$, $y = 3$ **37.** $x = 6$, $y = -8$
39. No solution; inconsistent system of equations   **41.** Infinite number of solutions; dependent equations
43. $x = 5$, $y = -3$ **45.** No solution; inconsistent system of equations   **47.** $x = 4$, $y = -1$
49. $x = 1.6867$, $y = -3.7664$ **51.** 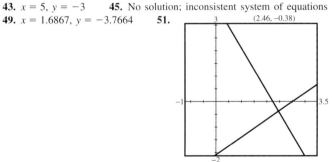   **53.**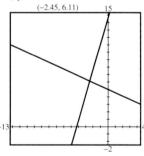

55. Approximately \$0.01 per pound   **56.** 341,889 cars

Putting Your Skills to Work

1. A minimum of 162 bricks will be needed.   **2.** $N = 4.5(A - 0.10A)$ or $N = 4.05A$. A minimum of 329 bricks will be needed.

4.2 Exercises

1. $(2, 1, -4)$ is a solution.   **3.** $x = 1$, $y = 3$, $z = -2$ **5.** $x = 3$, $y = -1$, $z = 4$ **7.** $x = 1$, $y = -1$, $z = -1$ **9.** $x = 1$, $y = -1$, $z = 2$
11. $x = 2$, $y = 1$, $z = -4$ **13.** $a = 4$, $b = 0$, $c = 2$ **15.** $a = -1$, $b = 2$, $c = -3$ **17.** $x = 1$, $y = 2$, $z = -3$ **19.** $x = 1.10551$,

$y = 2.93991$, $z = 1.73307$ **21.** $x = 3$, $y = -2$, $z = 1$ **23.** $x = \dfrac{1}{2}$, $y = \dfrac{2}{3}$, $z = \dfrac{5}{6}$ **25.** $x = 1$, $y = 3$, $z = 5$ **27.** $a = 2$, $b = 4$, $c = 4$

29. $x = 4$, $y = -2$, $z = 1$ **31.** $x = \dfrac{1}{2}$, $y = \dfrac{1}{3}$, $z = -1$ **33.** Infinite number of solutions; dependent equations   **35.** No solution;

inconsistent system of equations   **37.** $m = \dfrac{4 - 3}{1 + 2} = \dfrac{1}{3}$, $y - 4 = \dfrac{1}{3}(x - 1)$, $x - 3y = -11$ **38.** $m(\perp \text{ line}) = \dfrac{3}{2}$, $y - 2 = \dfrac{3}{2}(x + 4)$,

$3x - 2y = -16$ **39. (a)** He will buy 57 sheep and 22 cattle. **(b)** After the purchase he will have 346 horses, 602 sheep, and 623 cattle.
40. 12.5 miles per hour

4.3 Exercises

1. 16 heavy equipment operators; 19 general laborers   **3.** 51 tickets for regular coach seats; 47 tickets for sleeper car seats   **5.** 300 full-time
students; 220 part-time students   **7.** 15 new salespeople; 12 experienced salespeople   **9.** 9 packets of mixture 1; 12 packets of mixture 2
11. One doughnut costs $0.45; one large coffee costs $0.89   **13.** Speed of plane in still air 216 mph; speed of wind 36 mph   **15.** Average
speed of the canoe in still water 4.5 mph; rate of the current $= 1.5$ mph   **17.** He scored 10 2-point baskets and 6 3-point baskets.   **19.** She
drove 315 miles on the highway and 105 miles in city driving.   **21.** The lodge has 2 large vans, 4 Dodge Minivans, and 8 Ford Explorers.
23. A total of 80 adults, 170 high school students, and 50 children not yet in high school attended.   **25.** A total of 800 senior citizens, 10,000
adults, and 1200 children under 12 ride during the rush hour.   **27.** Pump A pumps 10,000 gallons per hour. Pump B pumps 12,000 gallons per
hour. Pump C pumps 15,000 gallons per hour.   **29.** She can prepare 2 of Box A, 3 of Box B, and 4 of Box C.   **31.** The department pays
$10,258 for a car and $17,300 for a truck.   **33.** The scientist should use 2 of packet A, 1 of packet B, 5 of packet C, and 3 of packet D.
34. $x = \frac{26}{7}$ **35.** $x = \frac{7}{18}$ **36.** $y = \frac{5}{3}$

4.4 Exercises

1. -7 **3.** 15   **5.** 2   **7.** 47   **9.** 18   **11.** 0   **13.** 0   **15.** -0.6 **17.** $-7a - 4b$ **19.** $\dfrac{11}{84}$ **21.** $-\dfrac{11}{4}$ **23.** -7

25. $-\dfrac{2}{5}$ **27.** $\begin{vmatrix} 6 & 10 \\ -5 & 9 \end{vmatrix}$ **29.** $\begin{vmatrix} 3 & -4 \\ 1 & -5 \end{vmatrix}$ **31.** -7 **33.** -26 **35.** 11   **37.** -27 **39.** -8 **41.** 0   **43.** -3.179

45. 18,553   **47.** 10   **49.** 21   **51.** -2 **53.** $-1.4679086 \times 10^{10}$ **55.** $-25x^3 - 10x^2$ **56.** $\dfrac{1}{6}a + 2y + 3z$ **57.** $-\dfrac{24y^7z^2}{x}$

58. $\dfrac{64y^2}{81x^{17}}$ **59.** $2660 for Eurovision, $1000 from Traveland, and $1660 from Unitech   **60.** The rental rate was $35 per day for the car, $8 per

4.5 Exercises

1. $x = 2$, $y = 3$ **3.** $x = -2$, $y = 5$ **5.** $x = 10$, $y = 2$ **7.** $x = \dfrac{1}{2}$, $y = -2$ **9.** $x = -1$, $y = -1$ **11.** $x = 4$, $y = -2$

13. $x = 1.5795$, $y = -0.0902$ **14.** $x = 1.2117$, $y = 0.3436$ **15.** $x = 7$, $y = -8$ **17.** $x = -4$, $y = 2$ **19.** $x = 2$, $y = 6$ **21.** $x = 1$,

$y = 1$, $z = 1$ **23.** $x = -\dfrac{1}{2}$, $y = \dfrac{1}{2}$, $z = 2$ **25.** $x = 4$, $y = -2$, $z = 1$ **27.** $x = -3$, $y = -2$, $z = 2$ **29.** $x = -0.219$, $y = 1.893$, $z = -3.768$

31. $z = -\dfrac{7}{8}$ **33.** $y = -7$ **35.** No solution; inconsistent system of equations   **37.** Infinite number of solutions; dependent equations

39. $z = \dfrac{-18}{6} = -3$ **41.** Width is $15\dfrac{1}{3}$ meters; length is $34\dfrac{2}{3}$ meters.   **42.** Number of miles driven is 190 miles.   **43.** She invested $2600
at 8% and $1400 at 6%.   **44.** $a = 3$; $b = -8$; the equation is $y = 3x - 8$.

4.6 Exercises

1.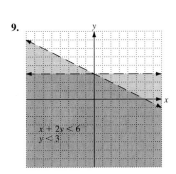
$y \geq 2x - 1$
$x + y \leq 6$

3.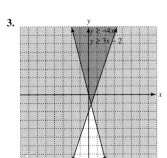
$y \geq -4x$
$y \geq 3x - 2$

5.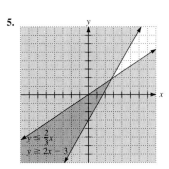
$y \leq \frac{2}{3}x$
$y \geq 2x - 3$

7.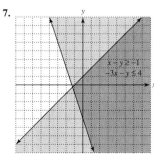
$x - y \geq -1$
$-3x - y \leq 4$

9.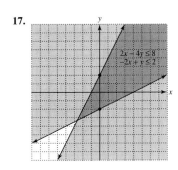
$x + 2y < 6$
$y < 3$

11.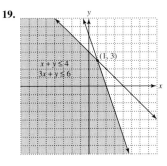
$3x + y > 6$
$x > 0$

13.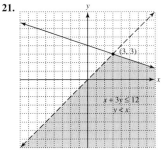
$y < 4$
$x > -2$

15.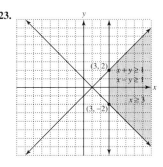
$x + 2y > 6$
$x + 2y < 6$

17.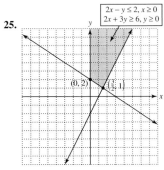
$2x - 4y \leq 8$
$-2x + y \leq 2$

19.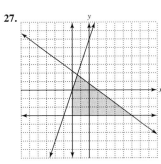
$(1, 3)$
$x + y \leq 4$
$3x + y \leq 6$

21.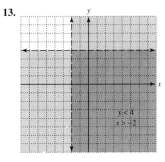
$(3, 3)$
$x + 3y \leq 12$
$y \leq x$

23.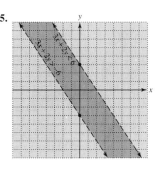
$(3, 2)$
$x + y \geq 1$
$x - y \leq 1$
$x \leq 3$
$(3, -2)$

25.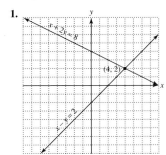
$2x - y \leq 2, x \geq 0$
$2x + 3y \geq 6, y \geq 0$
$(0, 2)$ $\left(\frac{3}{2}, 1\right)$

27.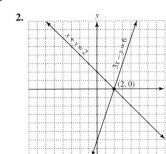

29. $-\dfrac{1}{2}$

30. Slope $= -\dfrac{3}{4}$;
y-intercept $= -2$

31. $5x - 4y = -14$

32. $y = -2x + 10$

33. The Cinema takes in $4200 on a rainy day and $300 on a sunny day. **34.** The volunteers establish 43 ft of bicycle trails each day. The professionals establish 65 ft of bicycle trails each day. **35.** They had each worked for 12 weeks and had each sold $100,000 worth of goods. **36.** One roast beef sandwich costs $2.50. One order of french fries costs $1.75. One soda costs $0.95.

Chapter 4 Review Problems

1.
$x + 2x = 8$
$(4, 2)$
$x - y = 2$

2.
$x + y \leq 2$
$3x - y = 6$
$(2, 0)$

3.
$3x + 4y = 4$
$2x + y = 6$
$(4, -2)$

4. $x = -1, y = 3$ **5.** $x = 1, y = -7$ **6.** $x = 1, y = 2$ **7.** $x = 1, y = 3$ **8.** $x = 1, y = -2$ **9.** $x = 3, y = -2$
10. $x = 2, y = 3$ **11.** $x = 2, y = -3$ **12.** No solution; inconsistent system of equations   **13.** Infinite number of solutions; dependent equations   **14.** $x = 2, y = -4$ **15.** $x = -\dfrac{1}{3}, y = \dfrac{1}{2}$ **16.** $x = 0, y = 3$ **17.** $x = -2, y = 6$ **18.** $a = \dfrac{4}{3}, b = -\dfrac{1}{2}$
19. $a = \dfrac{17}{2}, b = -\dfrac{7}{2}$ **20.** $x = 0, y = \dfrac{2}{3}$ **21.** No solution; inconsistent system of equations   **22.** No solution; inconsistent system of equations   **23.** $x = 5, y = 2$ **24.** $x = 1, y = 1, z = -2$ **25.** $x = 1, y = -2, z = 3$ **26.** $x = 5, y = -3, z = 8$ **27.** $x = 7, y = \dfrac{1}{2}$, $z = -3$ **28.** $x = 3, y = 0, z = -2$ **29.** $x = 1, y = -2, z = 3$ **30.** $x = 1, y = 2, z = -4$ **31.** $x = -2, y = -4, z = -8$
32. Speed of plane in still air = 264 mph; speed of wind = 24 mph   **33.** New employees = 7; laid-off employees = 10   **34.** Laborers = 15; mechanics = 10   **35.** Children's tickets = 340; adult tickets = 250   **36.** Hats = \$3; shirts = \$15; pants = \$12   **37.** $A = 2; B = 3; C = 4$
38. One jar of jelly = \$0.70; one jar of peanut butter = \$1.00; one jar of honey = \$0.80   **39.** Buses = 2; station wagons = 4; sedans = 3
40. -1 **41.** $\dfrac{1}{2}$ **42.** -34 **43.** 96   **44.** -0.1 **45.** $\dfrac{27}{40}$ **46.** 52   **47.** 33   **48.** 16   **49.** $x = 0, z = 1$ **50.** $x = \dfrac{20}{3}$, $y = \dfrac{10}{3}$ **51.** $x = -3, y = 2$ **52.** $x = 1, y = 0, z = -1$ **53.** $x = 3, y = -1, z = -2$ **54.** $y = -5$ **55.** $x = \dfrac{10}{13}$

56.   **57.**   **58.**   **59.**

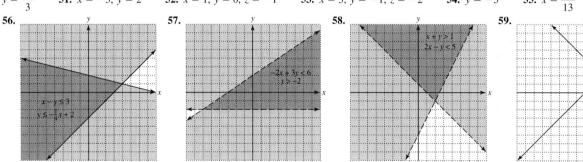

Chapter 4 Test

1. $x = -2, y = 1$ **2.** $x = \frac{2}{3}, y = 3$ **3.** $a = \frac{1}{2}, b = \frac{3}{2}$ **4.** No solution; inconsistent system of equations   **5.** $x = 2, y = -1, z = 3$ **6.** $x = -2, y = 3, z = 5$ **7.** Speed of plane in still air = 450 mph; speed of wind = 50 mph   **8.** Station wagons = 4, 2-door sedans = 6, 4-door sedans = 6   **9.** $x = 3, y = 4$ **10.** $x = 1, y = 2$ **11.** $z = -1$

12.

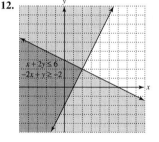

Cumulative Test for Chapters 1–4

1. Identity property of addition   **2.** 2   **3.** $15x^{-6}y^2$ or $\dfrac{15y^2}{x^6}$ **4.** $22x + 12$ **5.** $P = \dfrac{A}{3 + 4rt}$ **6.** $x = 68$

7.

x	y
0	-1.25
4	0.75
7	2.25

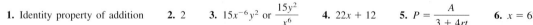

8. $m = \dfrac{1}{10}$ **9.** $x > -10$ **10.** $5 \le x \le 11$

11. $6x - 5y = 27$ **12.** 1st side = 17 m; 2nd side = 24 m; 3rd side = 28 m   **13.** \$1500 at 7%; \$4500 at 9%   **14.** $x = 2, y = -4$ **15.** $x = 2, y = -1, z = -1$ **16.** Shirts = \$21; slacks = \$30   **17.** $x = 5, y = 3$ **18.** $z = -2$ **19.** Infinite number of solutions; dependent equations   **20.**

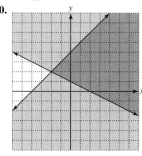

Chapter 5

Pretest Chapter 5

1. $x^2 - 11x + 4$ **2.** $2x^3 - 9x^2 + x + 12$ **3.** $5a^2 - 43a + 56$ **4.** $4y^2 - 9$ **5.** $9x^4 + 24x^2 + 16$ **6.** $5x - 6y - 10$

7. $3y^2 + y + 4 + \dfrac{7}{y - 2}$ **8.** $x^3 + 2x^2 - x - 2$ **9.** $12a^3b^2(2 + 3a - 5b)$ **10.** $(4x - 3y)(3x - 2)$ **11.** $(5w + 3z)(2x - 5y)$

12. $(x - 5y)(x - 2y)$ **13.** $(2y - 5)(2y + 3)$ **14.** $(7x - 3y)(4x - y)$ **15.** $(6x - 5y)^2$ **16.** $(11x - 1)(11x + 1)$

17. $(2x - y)(4x^2 + 2xy + y^2)$ **18.** $(4x + 3)(16x^2 - 12x + 9)$ **19.** $y^3(x - 3y)(x^2 + 3xy + 9y^2)$ **20.** $2x(x - 4)(x + 3)$ **21.** prime

22. $a(9a + 7y)^2$ **23.** $x = \frac{2}{3}, x = -\frac{3}{4}$ **24.** $x = 0, x = \frac{7}{4}$ **25.** $x = 4, x = -4$ **26.** $W = 4$ m; $L = 13$ m

5.1 Exercises

1. Binomial, 4th degree   **3.** Monomial, 9th degree   **5.** Trinomial, 8th degree   **7.** $-2x^2 - 2x + 8$ **9.** $-3x - 6$

11. $2x^3 - 9x^2 + 2x + 13.8$ **13.** $5a^3 - a^2 + 12$ **15.** $\frac{5}{6}x^2 - 6\frac{3}{4}x$ **17.** $-3.2x^3 + 1.8x^2 - 4$ **19.** $6x^3 - 10x^2 + 2x$

21. $-\frac{2}{3}x^2y + 2xy^2 - 5xy$ **23.** $x^2 - x - 30$ **25.** $6x^2 + 7x + 2$ **27.** $28x^2 - 45x + 18$ **29.** $15aw + 6ad - 20bw - 8bd$

31. $-12x^2 + 11xy - 2y^2$ **33.** $10r^2 - 8rs^2 - 18s^4$ **35.** $2x^3 - 5x^2 + 5x - 3$ **37.** $6x^3 - 7x^2y - 10xy^2 + 6y^3$

39. $2x^4 - 17x^3 + 34x^2 - 17x + 2$ **41.** $5a^4 - 18a^3 + 11a^2 - 10a + 12$ **43.** $3r^4 + 5r^3s - 20r^2s^2 + 2rs^3 + 4s^4$ **45.** $25x^2 - 64y^2$

47. $25a^2 - 20ab + 4b^2$ **49.** $49m^2 - 14m + 1$ **51.** $4x^4 - 1$ **53.** $4a^4b^4 - 12a^2b^2 + 9$ **55.** $49x^2 - 9y^4$ **57.** $2x^3 - 7x^2 - 7x + 30$

59. $3a^3 - a^2 - 22a + 24$ **61.** $10.732x^2 - 10.729x$ **63.** $6x^3 + 25x^2 + 49x + 40$ cm³   **65.** $-10x - 12y$ **66.** $2x^3 - 4x^2 - 8x + 6$

67. 77.73 parts per million   **68.** 51 parts per million   **69.** 3 parts per million   **70.** Approximately 0.28 parts per million

5.2 Exercises

1. $3x^2 + 4x - 6$ **3.** $3x^3 - x^2 + 7x$ **5.** $2x^2 - x + 2.5$ **7.** $9a^2 + 6a - 2$ **9.** $3x + 4$ **11.** $7x - 2$ **13.** $2a - 7 + \dfrac{5}{3a + 5}$

15. $x^2 - 2x + 13 - \dfrac{14}{x + 1}$ **17.** $2x^2 + 3x + 6 + \dfrac{5}{x - 2}$ **19.** $x^2 + 5x - 3 + \dfrac{3}{2x + 3}$ **21.** $x^2 - 3x + 9$ **23.** $x^3 + 8x + 4$

25. $5a^2 + a + 3 + \dfrac{a + 7}{3a^2 - 1}$ **27.** $2t^2 - 3t + 2$ **29.** $58.9x^2 - 9.87x + 258.1$

31. The graph of $y_1 = \dfrac{2x^2 - x - 10}{2x - 5}$ and $y_2 = x + 2$ should coincide   **33.** $x = 2$ **34.** $x = 10$ **35.** $x = 6$ **36.** $\dfrac{24}{5} = x$

5.3 Exercises

1. $x + 9$ **3.** $2x + 1 + \dfrac{-2}{x - 6}$ **5.** $3x^2 - 2x + 1 + \dfrac{3}{x + 1}$ **7.** $x^2 + 4x + 5$ **9.** $x^2 + 6x + 11 + \dfrac{27}{x - 2}$ **11.** $x^2 - 4x + 8 - \dfrac{8}{x + 2}$

13. $6x^3 + 3x^2 - 6x - 16 + \dfrac{26}{x + 2}$ **15.** $x^3 - 7x^2 + 8x - 8 - \dfrac{1}{x + 1}$ **17.** $3x^4 - 3x^3 + 3x^2 - 3x + 4 - \dfrac{5}{x + 1}$

19. $2x^4 + 11x^3 + 31x^2 + 95x + 283 + \dfrac{852}{x - 3}$ **21.** $x^5 - x^4 + x^3 - x^2 + x - 1 - \dfrac{3}{x + 1}$ **23.** $x^2 + 3.7x + 0.84$ remainder 6.408

25. $b = -\dfrac{13}{3}$ **27.** $x^2 - 2x + 4$ **28.** $2x^2 - 6x + 9 - \dfrac{21}{2x + 3}$

29. We are using the basic property of fractions that for any nonzero polynomials a, b, and c, $\dfrac{ac}{bc} = \dfrac{a}{b}$.   **30.** To have an equivalent fraction, you must divide both the numerator and the denominator by the same value. However, in the case of the remainder, the denominator has not changed. When we write the remainder in fraction form, the denominator is still $ax + b$.   **31.** 268,000 cubic feet   **32.** 2.1 feet deep

5.4 Exercises

1. $15(2 - y)$ **3.** $xy(1 - 3x)$ **5.** $b(bx^2 + x + 1)$
7. $2x(x^2 - 4x + 6)$ **9.** $9ab(ab - 4 + 5b)$
11. $4ab^2c^2(-3b^2c - 2a^2 + 1)$ **13.** There is no common factor.   **15.** $12xy^2(y - 2x^2 + 3xy^2 - 5x^3y)$
17. $(x + y)(3x - 2)$ **19.** $(a - 3b)(5b + 8)$
21. $(a + 5b)(3x + 1)$ **23.** $(3x - y)(2a^2 - 5b^3)$
25. $(5x + y)(3x - 8y - 1)$ **27.** $(x + 5)(x^2 + 3)$
29. $(x + 1)(4 - 3w)$ **31.** $(a - y)(a - 3)$
33. $(x - 3y)(5a - 2b)$ **35.** $(t^2 - 5)(y + 5)$
37. $(7x + 2y^2)(4x + 3w)$ **39.** $(4a - c)(3a^2 - c^2)$

41. $7.37(2x - 3y + 8)$ **45.** $m = \dfrac{3 + 1}{2 - 6} = -1$

43.

x	y
-2	0
-1	3
0	6

$6x - 2y = -12$; points $(0, 6)$, $(-1, 3)$, $(-2, 0)$

44.

x	y
-3	-4
0	-2
3	0

$x = \frac{2}{3}x - 2$; points $(3, 0)$, $(0, -2)$, $(-3, -4)$

46. $m = -3, b = -\dfrac{3}{2}$ **47.** Assuming he only wants to swim complete laps, he would need to swim 17 laps.   **48.** He answered 8 multiple choice questions correctly.

5.5 Exercises

1. $(x + 1)(x + 8)$ **3.** $(x + 9)(x - 2)$ **5.** $(x + 6)(x - 5)$ **7.** $(x + 6)(x + 2)$
9. $(a - 10)(a - 3)$ **11.** $(a + 9)(a - 5)$ **13.** $(x - 4y)(x - 5y)$ **15.** $(x + 7y)(x - 2y)$
17. $(x^2 - 8)(x^2 + 5)$ **19.** $(x^2 + 7y^2)(x^2 + 9y^2)$ **21.** $2(x + 11)(x + 2)$ **23.** $x(x + 12)(x - 3)$
25. $(2x - 1)(x - 3)$ **27.** $(6x + 1)(5x - 1)$ **29.** $(3x - 5)(2x + 1)$ **31.** $(3a - 5)(a - 1)$
33. $(4a + 9)(2a - 1)$ **35.** $(3x - 2)(2x - 3)$ **37.** $(2x + 3)(x + 5)$ **39.** $(3x^2 + 1)(x^2 - 3)$
41. $(3x + y)(2x + 11y)$ **43.** $(2x - 3y)(x - 5y)$ **45.** $x(2x + 5)(2x - 3)$
47. $3x(3x + 2)(2x + 1)$ **49.** $(x - 9)(x + 7)$ **51.** $(3x + 2)(2x - 1)$ **53.** $(x - 17)(x - 3)$
55. $(5x + 2)(3x - 1)$ **57.** $2(x + 8)(x - 6)$ **59.** $3(3x + 2)(2x + 1)$ **61.** $4x(x - 6)(x + 2)$
63. $2x(3x - 2)(x + 5)$ **65.** $(3x^2 - 5)(x^2 + 1)$ **67.** $(7x - y)(x - 3y)$ **69.** $(x^3 - 13)(x^3 + 3)$

71. $x(3x + 4)(2x - 1)$ **73.** 28.26 in.^{2} **74.** $\dfrac{2A - 2a}{5} = b$ **75.** $x + 3y = -6$

77. They should stock 45 bike racks and 75 helmets.   **78.** She should add exactly 0.195 ounces of milk.

76.

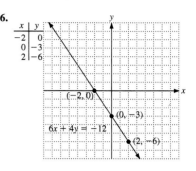

5.6 Exercises

1. $(x - 5)(x + 5)$ **3.** $(7x - 2)(7x + 2)$ **5.** $(8x - 1)(8x + 1)$ **7.** $(t^2 + 1)(t + 1)(t - 1)$ **9.** $(9x^2 + 1)(3x + 1)(3x - 1)$
11. $(7m - 3n)(7m + 3n)$ **13.** $(1 - 9xy)(1 + 9xy)$ **15.** $(9x - 11)(9x + 11)$ **17.** $2(4x + 3)(4x - 3)$ **19.** $5x(1 + 2x)(1 - 2x)$
21. $(7x - 1)^2$ **23.** $(w - 3)^2$ **25.** $(6x - 1)^2$ **27.** $(z + 8)^2$ **29.** $(9w + 2t)^2$ **31.** $(5x - 4y)^2$ **33.** $2(2x + 3)^2$ **35.** $3x(x - 4)^2$
37. $(2x + 3)(4x^2 - 6x + 9)$ **39.** $(x + 5)(x^2 - 5x + 25)$ **41.** $(4x - 1)(16x^2 + 4x + 1)$ **43.** $(5x - 2)(25x^2 + 10x + 4)$
45. $(1 - 3x)(1 + 3x + 9x^2)$ **47.** $(4x + 5)(16x^2 - 20x + 25)$ **49.** $(4s^2 + t^2)(16s^4 - 4s^2t^2 + t^4)$ **51.** $6(y - 1)(y^2 + y + 1)$
53. $x(4x + 3)(16x^2 - 12x + 9)$ **55.** $x^2(x - 2y)(x^2 + 2xy + 4y^2)$ **57.** $(5w^3 - 1)(5w^3 + 1)$ **59.** $2(2w^3 + 1)^2$
61. $(2a - 3b)(4a^2 + 6ab + 9b^2)$ **63.** $(5m + 2n)(25m^2 - 10mn + 4n^2)$ **65.** $(3x - 10y)(3x + 10y)$ **67.** $(2w - 5z)^2$
69. $9(2a - 3b)(2a + 3b)$ **71.** $x^2(4x + yz)(16x^2 - 4xyz + y^2z^2)$ **73.** $(9x^2 - 2)^2$ **75.** $(4x^2 + 9y^2)(2x - 3y)(2x + 3y)$
77. $(5x + 4)(5x + 1)$ **79.** $(4x - 3)(x - 3)$ **81.** $(9x^8 + 16)(3x^4 - 4)(3x^4 + 4)$ **82.** $(2x^5 + 7y^7)(4x^{10} - 14x^5y^7 + 49y^{14})$ **83.** $(11x^8 - 5y^{10})^2$
84. $(m - 2n)(m^2 + 2mn + 4n^2)(m + 2n)(m^2 - 2mn + 4n^2)$ **85.** The hot water tap runs at 12 gallons per minute. The cold water tap runs at 20
gallons per minute.   **86.** \$1400 at 14%; \$2600 at 11%   **87.** 1st side is 20 cm; 2nd side is 30 cm; 3rd side is 16 cm   **88.** Melinda paid
\$388; Hector paid \$278; Alice paid \$192

5.7 Exercises

3. $3y(x - 2z)$ **5.** $(2x - 1)(x + 3)$ **7.** $(2x - 5y)(4x^2 + 10xy + 25y^2)$ **9.** $x(x + 2y - z)$ **11.** $2a(b + 5)(b - 5)$
13. $3x(x - 1)(x + 2)$ **15.** $(6x + 1)(x - 4)$ **17.** $(x - 1)(x - y)$ **19.** $2x(2x - 1)(4x^2 + 2x + 1)$ **21.** $2x(x - 3)(x + 3)(x^2 + 1)$
23. $3x^2(x^2 + 9)$ **25.** $3x(2x - 5)(x + 1)$ **27.** $2(2x^2 - 4x - 3)$ **29.** Prime   **31.** $2(a - 7)(a - 5)$ **33.** $a(a - 3)(2y + x)$
35. Prime   **37.** $(2x^2 - 5)(x^2 + 1)$ **39.** Prime   **41.** $2y^6(3x^2 - 1)(9x^4 + 3x^2 + 1)$ **43.** $(2x^2 + 5y^4)^2$ **45.** $b = 2$ **46.** $a = -130$

47. $x \le -9$ **48.** $-\dfrac{1}{5} < x < \dfrac{3}{5}$ **49.** $x < -\dfrac{7}{4}$ or $x > \dfrac{17}{4}$ **50.** $x \ge 11$ or $x \le 4$ **51.** Approximately \$238.3 million

52. Approximately 49.3%   **53.** Approximately \$213.5 million   **54.** Approximately \$237.8 million

5.8 Exercises

1. $x = -10, x = -1$ **3.** $x = 9, x = 4$ **5.** $x = 0, x = \dfrac{6}{5}$ **7.** $x = \dfrac{5}{2}, x = 1$ **9.** $x = -\dfrac{4}{3}, x = 2$ **11.** $x = \dfrac{2}{3}, x = -\dfrac{5}{2}$ **13.** $x = \dfrac{3}{8},$

$x = 1$ **15.** $x = 9, x = -3$ **17.** $x = -7, x = 3$ **19.** $x = 0, x = -1$ **21.** $x = 0, x = \dfrac{1}{8}$ **23.** $x = 0, x = -3, x = -2$ **25.** $x = 0,$

$x = 5, x = -4$ **27.** $x = 3, x = -3, x = 0$ **29.** $x = -7, x = 2, x = 0$ **31.** $x = \dfrac{3}{4}, x = -\dfrac{5}{2}$ **33.** $x = -\dfrac{3}{2}, x = 4$ **35.** $x = 0, x = \dfrac{1}{6}$

37. $x = 0, x = 4$ **39.** $x = -\dfrac{2}{3}, x = -\dfrac{3}{2}$ **41.** $b = 1, x = 3$ **42.** $c = -2, x = 2$ **43.** Altitude is 11 cm; base is 44 cm.

45. Altitude is 7 yards; base is 34 yards.   **47.** (a) Width is 20 cm; length is 24 cm. (b) Width is 200 mm; length is 240 mm.
49. Each side of the rug is 12 feet.   **51.** Width is 4 feet; length is 13 feet.   **53.** Width is 5 mi; length is 17 mi   **55.** Old side is 7 cm; new

side is 16 cm   **57.** 20 units   **59.** 6 units   **61.** $200x^{11}y^{10}$ **62.** $\dfrac{a^4}{2b^2}$ **63.** $\dfrac{x^4}{9y^8z^2}$ **64.** $\dfrac{125x^{12}}{8y^9}$

Putting Your Skills to Work

1. 20 cannon balls with 4 rows; 35 cannon balls with 5 rows   **2.** $x\left(\dfrac{1}{2}x + 1\right)\left(\dfrac{1}{3}x + \dfrac{1}{3}\right)$

Chapter 5 Review Problems

1. $-x^2 - 10x + 13$ **2.** $x^2y - 5xy - 8y$ **3.** $-11x^2 + 10xy + 6y^2$ **4.** $-11x^2 + 15x - 15$ **5.** $2x + 5$ **6.** $-x^3 - 7x^2 + 3x + 3$
7. $3x^3y - 3x^2y^2 + 3xy^3$ **8.** $6x^3 - 3x^2 + 2x - 1$ **9.** $25x^4 + 30x^2 + 9$ **10.** $2x^3 - 7x^2 - 7x + 30$ **11.** $-2x^4 + 7x^3 - 7x^2 + 7x - 2$
12. $9x^3 - 9x^2 - 22x + 20$ **13.** $-5x^2 + 3x + 20$ **14.** $4x - 8 + \dfrac{12}{3x + 2}$ **15.** $2x^2 - 3x - 4$ **16.** $3y^2 + 9y + 25 + \dfrac{80}{y - 3}$

17. $5a^2 - a + 3 + \dfrac{-a+7}{3a^2-1}$ **18.** $x^2 + 2x + 1$ **19.** $2x^3 - 3x^2 + x - 4$ **20.** $3x^3 - x^2 + x - 1$ **21.** $4x^3 - x + 2 - \dfrac{4}{x+3}$

22. $2x^3 - 3x^2 - 4x + 1$ **23.** $(x+3)(x+12)$ **24.** $(5x-1)(x-2)$ **25.** $(3x-11)(3x+11)$ **26.** Prime   **27.** $(x+4w)(x-2y)$

28. $x(x+6)(x+2)$ **29.** Prime   **30.** $(x+9y)(x-3y)$ **31.** $x(3x-1)(9x^2+3x+1)$ **32.** $(7a+2b)(3a+2b)$

33. $-a^2b^3(3a-2b+1)$ **34.** $a^2b^4(a+3)(a-2)$ **35.** $(3x^2+1)(x^2-2)$ **36.** $2(x^2+12)(x^2-2)$ **37.** $b(3a+7)(3a-2)$

38. Prime   **39.** $3(2x+1)^2$ **40.** $y^2(4y-9)(y-1)$ **41.** $y^2(y+7)(y-5)$ **42.** $4x^2(y^2-3y-2)$ **43.** $(3x^2+2)(x^2-3)$

44. $(a+b^3)(a+4b^3)$ **45.** $(2x+3)(x+2)(x-2)$ **46.** $2(x-3)(x+3)(x^2+3)$ **47.** $4(2-x)(a+b)$ **48.** $(4x^2+3y^2)(2x^2+7y^2)$

49. $2x(2x-1)(x+3)$ **50.** $x(2a-1)(a-7)$ **51.** $(4x^2y-7)^2$ **52.** $2xy(8x-1)(8x+1)$ **53.** $13xy(2x^2-y^2+4xy^3)$

54. $(5x+4y)(b-7)$ **55.** $3ab(3c-2)(3c+2)$ **56.** $5a^3(a+2b)(a^2-2ab+4b^2)$ **57.** $2x^2(25x^2-50x+32)$

58. $5(6x-y)(2x-3y)$ **59.** $x = -\dfrac{1}{5}, x = 2$ **60.** $x = \dfrac{3}{2}, x = 4$ **61.** $x = \dfrac{5}{6}, x = 3$ **62.** $x = 0, x = 3$ **63.** $x = 0, x = -3, x = -4$

64. $x = 0, x = -\dfrac{10}{3}$ **65.** Base is 11 meters; altitude is 14 meters.   **66.** Width is 4 miles; length is 10 miles.   **67.** Old side is 1 yard; new side is 5 yards.   **68.** 5 calculators

Chapter 5 Test

1. $-4x^2y - 1$ **2.** $5a^2 - 9a - 2$ **3.** $2x^3 - 6x^2y$ **4.** $4x^2 - 12xy^2 + 9y^4$ **5.** $x^4 + 3x^3 - 24x^2 - 18x + 8$ **6.** $5x^2 + 4x - 7$

7. $x^3 - x^2 + x - 2$ **8.** $x^2 - 3x + 1$ **9.** $x^3 - 1 - \dfrac{2}{x+1}$ **10.** $2x^4 + x^3 + 4x^2 + x + 3 + \dfrac{17}{x-4}$ **11.** $(11x - 5y)(11x + 5y)$

12. $(3x + 5y)^2$ **13.** $x(x-2)(x-24)$ **14.** $2(4x-1)(3x+2)$ **15.** $4x^2y(x+2y+1)$ **16.** $(x+3y)(x-2w)$ **17.** Prime

18. $3x^2(x+2)(x+10)$ **19.** $3(6x-5)(x+1)$ **20.** $y^4(5x-4)(5x+4)$ **21.** $2a(3a-2)(9a^2+6a+4)$ **22.** $x(3x^2-y)^2$

23. $(3x^2+2)(x^2+5)$ **24.** $(x-6y)(x-2y)$ **25.** $(x+2y)(3-5a)$ **26.** $(4x^2+1)(2x+1)(2x-1)$ **27.** $x = -2, x = 7$

28. $x = -\frac{1}{3}, x = 4$ **29.** $x = 0, x = \frac{2}{7}$ **30.** Base is 14 in., altitude is 10 in.

Cumulative Test for Chapters 1–5

1. Associative property of multiplication   **2.** 2   **7.**

3. 29   **4.** $x = \dfrac{2-7y}{5}$ **5.** $x = 0$ **6.** $m = \frac{8}{3}$

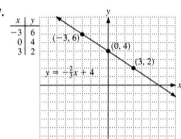

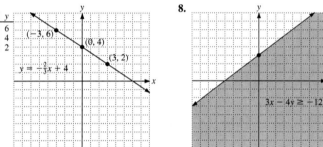

8.

9. $x > \frac{1}{3}$ **10.** Width is 6 m; length is 17 m.   **11.** $5a^2 - 9ab + 12b^2$ **12.** $-6x^2y^2 - 9xy^3 + 15x^2y^3$ **13.** $10x^3 - 19x^2 - 14x + 8$

14. $-3x^2 + 2x - 4$ **15.** $2x^2 + x + 5 + \dfrac{6}{x-2}$ **16.** $2x^2(x-5)$ **17.** $(8x+7)(8x-7)$ **18.** $x(3x-4)^2$ **19.** $(5x+6)^2$

20. $3(x-7)(x+2)$ **21.** $2(x+2)(x+10)$ **22.** Prime   **23.** $x(3x+1)(2x+3)$ **24.** $x(3x+4)(9x^2-12x+16)$

25. $(2m+1)(7n+5p)$ **26.** $x = -\frac{2}{3}, x = 2$ **27.** $x = 2, x = -13$ **28.** Base is 8 meters; altitude is 17 meters.

Chapter 6

Pretest Chapter 6

1. $\dfrac{7x+3y}{x+y}$ **2.** $\dfrac{2x^2+3x-1}{1-5x-6x^2}$ **3.** $2(a-5)$ **4.** $\dfrac{15(x-5y)}{2y^4(x+5y)}$ **5.** $\dfrac{x^2-4x+8}{3x(x-2)}$ **6.** $\dfrac{12x+5}{(x+5)(x-5)}$ **7.** $\dfrac{10y-6}{(y+3)(y-3)(y+4)}$

8. $\dfrac{21x}{8}$ **9.** $\dfrac{x}{(2x-1)(6x+1)}$ **10.** $y = \dfrac{25}{2}$ or 12.5   **11.** $y = 5$ **12.** $d_2 = \dfrac{d_1 w_2}{w_1}$ **13.** $n = \dfrac{IR}{E-Ir}$ **14.** 31.5 ft

15. 8 ft by $13\frac{1}{3}$ ft

6.1 Exercises

1. All real numbers except 3   **3.** All real numbers except -4 and 9   **5.** $\dfrac{3x}{2y}$ **7.** $\dfrac{x-8}{x+4}$ **9.** $\dfrac{2x}{3x-4}$ **11.** $-3y+5$ **13.** $\dfrac{a+2}{a+4}$

15. $\dfrac{1}{x-5}$ **17.** $\dfrac{7}{5}$ **19.** $\dfrac{-2}{x^2}$ **21.** $y-2$ **23.** $-\dfrac{x+6}{x+4}$ **25.** $\dfrac{2(y+3)}{y+4}$ **27.** $-\dfrac{2y+5}{3+y}$ **29.** $\dfrac{2}{5xy^6}$ **31.** $a(a-2)$

33. $\dfrac{1}{2}$ **35.** $(x-8y)(x+7y)$ **37.** $\dfrac{y+3}{y+4}$ **39.** $\dfrac{y(x-5)}{x^2}$ **41.** $\dfrac{y^2}{2(y-2)}$ **43.** $\dfrac{(4a+5)(a-4)}{3a+2}$ **45.** $\dfrac{x-3y}{x(x^2+2)}$ **47.** $-\dfrac{5}{2}$

49. $\dfrac{a+3b}{a-3b}$ **51.** $\dfrac{1}{3x^2y^2}$ **53.** $\dfrac{x}{2}$ **55.** $\dfrac{3(x+7)}{xy}$ **57.** $-\dfrac{x-4}{2x(x+2)}$ **59.** Cannot be simplified   **61.** $\dfrac{7(x+1)^2}{2x^2}$

63. All real numbers except $x \approx -1.4$ and $x \approx 0.9$

65.

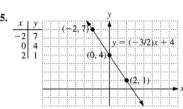

x	y
-2	7
0	4
2	1

$y = (-3/2)x + 4$

$(-2, 7)$ $(0, 4)$ $(2, 1)$

66.

x	y
-1	2
0	4
1	6

$(1, 6)$ $(0, 4)$ $(-1, 2)$ $6x - 3y = -12$

67. $4x - y = -5$ **68.** $4x - 3y = -28$ **69.** 3960 inquiries **70.** No

6.2 Exercises

1. $7x^2y$ **3.** $2x(x + 2)$ **5.** $(x + 3y)(x + 4y)$ **7.** $(x + 6)(2x + 5)^3$ **9.** $x(5x - 7)(x - 1)$ **11.** $(3y + 7)(3y - 7)^2$

13. $24x^2(x + 3)(x - 8)$ **15.** $\dfrac{x - 6}{2x(x + 2)}$ **17.** $\dfrac{12y + 2x}{5x^2y}$ **19.** $\dfrac{8x - 15}{x(x - 4)(x - 3)}$ **21.** $\dfrac{28x + 15y}{(7x - 5y)(7x + 5y)}$ **23.** $\dfrac{3y - 4}{(y + 2)(y - 3)}$

25. $\dfrac{2a^2 + 5a - 10}{2a(a + 2)(a - 2)}$ **27.** $\dfrac{y^2 - 3y + 18}{(2y - 3)(y + 3)}$ **29.** $\dfrac{x + 5}{x - 6}$ **31.** $\dfrac{x^2 - x}{(x + 5)(x - 2)(x + 3)}$ **33.** $\dfrac{2x^2 - 6x - 11}{(x + 3)(x + 2)(x - 2)}$ **35.** $\dfrac{4}{y + 4}$

37. $\dfrac{3a^2 + 4a - 13}{3a - 5}$ **39.** $\dfrac{3x + y}{(x + 2)(x + y)(x - 2)}$ **41.** $\dfrac{27x^3 - 37x^2 + 26x - 8}{(3x - 2)^2}$ **43.** Assuming people were working 24 hours per day, it

would take 1460 hours, or approximately 60.83 days. **45.** Tony's car cost \$6000, Melissa's car cost \$13,000, Alreda's car cost \$7500.
47. Speed of boat in still water is 20 km/hr; speed of the current is 5 km/hr.

6.3 Exercises

1. $\dfrac{xy(x + y)}{2}$ **3.** $\dfrac{3x + 2}{5x - 6}$ **5.** $\dfrac{y - 5}{y}$ **7.** $\dfrac{y^2 - 3}{3}$ **9.** $\dfrac{2}{(y - 3)(y + 6)}$ **11.** $\dfrac{4(7 - 2x)}{x(x^2 + 1)}$ **13.** $\dfrac{3(x - 4)}{x^2 - 4x - 5}$ **15.** $\dfrac{x(2x + 5)}{5(2x + 3)}$

17. $\dfrac{2ab(5a + b)}{(a + b)(a - b)(3a + b)}$ **19.** $-\dfrac{1}{x(x + a)}$ **21.** $\dfrac{y^2 - 3y + 2}{y(y - 4)}$ **23.** $-\dfrac{1}{y - 3}$ **24.** $\dfrac{2x^2}{x + 1}$ **25.** $x = -\dfrac{2}{3}$ or $x = 2$ **26.** $x = -5$

or $x = 15$ **27.** $-\dfrac{3}{5} < x < \dfrac{9}{5}$ **28.** $x \geq 1.5$ or $x \leq -2.5$ **29.** (a) \$253,440,000 was spent per mile in the 1970s. (b) Starting in the year

2001 a total of 6.5 miles can be built with that budget limit. **30.** She should ship 1750 kilograms by airfreight and 3850 kilograms by ocean
freighter.

6.4 Exercises

1. $3 = x$ **3.** $x = \dfrac{3}{4}$ **5.** $-1 = x$ **7.** $\dfrac{4}{3} = x$ **9.** $-1 = y$ **11.** No solution **13.** $x = 5$ **15.** $x = -2$ **17.** $x = 4$

19. $y = -10$ **21.** $x = 0$ **23.** $x = -9$ **25.** $y = 1$ **27.** $z = \dfrac{34}{3}$ **29.** $y = 7$ **31.** No solution **33.** $x = -3$

35. No solution **39.** $x \approx 70.6$ **41.** $7(x + 3)(x - 3)$ **42.** $2(x + 5)^2$ **43.** $(4x - 3y)(16x^2 + 12xy + 9y^2)$ **44.** $(3x - 7)(x - 2)$

6.5 Exercises

1. $r = \dfrac{d}{t}$ **3.** $t = \dfrac{A - P}{Pr}$ **5.** $b = \dfrac{af}{a - f}$ **7.** $x = \dfrac{ab}{3R}$ **9.** $\dfrac{2F}{y + z} = x$ **11.** $n = \dfrac{IR}{E - Ir}$ **13.** $\dfrac{A}{2\pi(b + h)} = r$ **15.** $\dfrac{Er}{R + r} = e$

17. $T_1 = \dfrac{P_1V_1T_2}{P_2V_2}$ **19.** $d^2 = \dfrac{Gm_1m_2}{F}$ **21.** $T_1 = \dfrac{ET_2}{T_2 - 1}$ **23.** $x_1 = \dfrac{mx_2 - y_2 + y_1}{m}$ **25.** $t = \dfrac{2S}{V_1 + V_2}$ **27.** $t_2 = \dfrac{kAt_1 - QL}{kA}$

29. $T_2 = \dfrac{-qT_1}{W - q}$ **31.** $\dfrac{VM}{v - V} = m$ **33.** $T = T_0 + 0.1702\left(\dfrac{V}{V_0}\right) - 0.1656$ **35.** Approximately 5.63 kilometers **37.** 68.18 mph

39. (a) 328 balls will pass inspection. (b) 56 balls will be defective. **41.** 80 grizzly bears **43.** 189.73 km **45.** 21.6 in. **47.** 125 kg
49. 48 station wagons, 176 sedans **51.** He should walk 49 miles. He should run 14 miles. **53.** 6.4 feet high **55.** 196 times
57. 3.6 hours **59.** 2.73 hours **61.** 16.07 minutes **63.** 12 hours **65.** 282 ft **66.** 784 ft **67.** $(4x - 5)(2x + 1)$
68. $(x - 12)(x + 10)$ **69.** $(5x - 9y)^2$ **70.** $5y(x + 2)(x - 2)$

Putting Your Skills to Work

1. 3 watts per square yard. The intensity is $\frac{1}{4}$ as strong. **2.** Approximately 5.2 miles

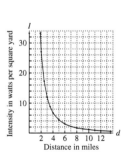

Chapter 6 Review Problems

1. $\dfrac{x}{2}$ **2.** $\dfrac{4x^2}{x^3 - 5}$ **3.** $\dfrac{4b}{5a^3}$ **4.** $\dfrac{a + 4}{a + 3}$ **5.** $-\dfrac{y + 7}{y - 7}$ **6.** $\dfrac{a - b}{3(x - 2)}$ **7.** $\dfrac{(a + 1)(a - 1)}{a^2 + 2}$ **8.** $\dfrac{2(x + 3)}{x - 3}$ **9.** $\dfrac{x(2x - 1)}{x - 2}$

10. $\dfrac{1}{8xy}$ **11.** $\dfrac{y - 5}{y - 4}$ **12.** $\dfrac{x^2(x + 5)}{2y(x + 4)}$ **13.** $\dfrac{x + 6}{3(x + 2)}$ **14.** $\dfrac{2a^2}{x^2 + a^2}$ **15.** $\dfrac{(y + 1)^2(y + 3)}{y + 5}$ **16.** $\dfrac{5}{2}$ **17.** $\dfrac{2y + 5}{2y - 5}$

18. $\dfrac{2a+5}{2a(2a^2-7a-13)}$ **19.** $\dfrac{4x+3y-2}{x^2y^2}$ **20.** $\dfrac{17x-1}{(x-3)(3x+1)}$ **21.** $\dfrac{-7x+20}{4x(x+4)}$ **22.** $-\dfrac{x^2+10x-9}{(2x+1)(x-2)}$ **23.** $\dfrac{7y-18}{(y+5)(y-5)}$

24. $\dfrac{-6y-11}{36y}$ **25.** $-\dfrac{y+2}{y+3}$ **26.** $\dfrac{4y^2-y+3}{(y+1)^2(y-1)}$ **27.** $\dfrac{a-2}{a+3}$ **28.** $\dfrac{4a+20}{(a+1)(a+2)(a+3)}$ **29.** $\dfrac{4a^2+17a+11}{a+4}$

30. $\dfrac{9a^2+6a+4}{3a}$ **31.** $\dfrac{2y+3x}{7}$ **32.** $\dfrac{2(2y+3x)}{15x}$ **33.** $\dfrac{x}{x-5}$ **34.** $\dfrac{4(x-2)}{2x+5}$ **35.** $\dfrac{y^2+y+1}{y^2-y-1}$ **36.** $\dfrac{5}{2}$ **37.** $\dfrac{2(2x-1)}{3x}$

38. $-\dfrac{1}{x+y}$ **39.** $\dfrac{y^2}{2y-1}$ **40.** $-\dfrac{(x+3)(x+4)}{x(x-4)}$ **41.** $x=6$ **42.** $x=-1$ **43.** $x=-1$ **44.** $x=5$ **45.** $a=-6$ **46.** $a=4$

47. $y=\dfrac{3}{4}$ **48.** $y=\dfrac{1}{7}$ **49.** $a=-\dfrac{1}{2}$ **50.** $a=-\dfrac{1}{2}$ **51.** $x=\dfrac{3}{4}$ **52.** $y=-1$ **53.** $M=\dfrac{mV}{N}-N$ **54.** $x=\dfrac{y-y_0+mx_0}{m}$

55. $T_1=\dfrac{P_1V_1T_2}{P_2V_2}$ **56.** $a=\dfrac{bf}{b-f}$ **57.** $t=\dfrac{2S}{V_1+V_2}$ **58.** $p=\dfrac{12I}{A(1+3r)}$ **59.** $R_2=\dfrac{dR_1}{L-d}$ **60.** $r=\dfrac{S-P}{Pt}$ **61.** 57

62. 65 kilograms **63.** 5.6 hr **64.** 500 rabbits **65.** 15 cm **66.** 28 officers **67.** 12.25 nautical miles **68.** 182 feet tall

69. 3.6 hours **70.** 2.1 hours **71.** 6 minutes **72.** 7.2 hours

Chapter 6 Test

1. $\dfrac{x+2}{x-3}$ **2.** $\dfrac{y-2}{y^2-2y+4}$ **3.** $\dfrac{2(2y-1)}{3y+5}$ **4.** $-\dfrac{6}{(2x-1)(x+1)}$ **5.** $\dfrac{-2x-17}{(x+5)(x-5)}$ **6.** $\dfrac{3x^2+8x+6}{(x+3)^2(x+2)}$ **7.** $-\dfrac{2}{5}$

8. $-x+1$ **9.** $x=3$ **10.** $x=-2$ **11.** $y=5$ **12.** $x=\tfrac{7}{2}$ **13.** $W=\dfrac{S-2Lh}{2h+2L}$ **14.** $b=\dfrac{3ac}{4c-2a}$ **15.** 52.5 minutes

16. 75 miles per hour

Cumulative Test for Chapters 1–6

1. $\dfrac{x^4z^8}{9y^6}$ **2.** $x=\dfrac{55}{24}$ **3.**

x	y
0	−6
1	−3
2	0

$-6x+2y=-12$

(2, 0)
(1, −3)
(0, −6)

4. $5x-6y=13$ **5.** \$700 at 5% interest; \$6300 at 8% interest

6. $x<-1$ **7.** -22

$-3\;-2\;-1\;\;0\;\;1\;\;2\;\;3$

8. $-2\le x\le\dfrac{14}{3}$ **9.** $(3x+5y)(9x^2-15xy+25y^2)$ **10.** $x(9x-5y)^2$ **11.** $x=-10,\ x=-2$ **12.** $x=4,\ x=-\tfrac{1}{3}$ **13.** $\dfrac{7(x-2)}{x+4}$

14. $\dfrac{x(x+1)}{2x+5}$ **15.** $\dfrac{(x^2-3x+9)(2x+5)}{(x-3)^2}$ **16.** $-\dfrac{x+25}{2(x-4)(x-5)}$ **17.** $\dfrac{2(x+1)(2x-1)}{16x^2-7}$ **18.** $\dfrac{7x-12}{(x-6)(x+4)}$ **19.** $x=-4$

20. $x=-2$ **21.** $b=\dfrac{5H-2x}{3+4H}$ **22.** 8250 cars

Chapter 7

Pretest Chapter 7

1. $\dfrac{6y^{5/6}}{x^{1/4}}$ **2.** $\dfrac{9x^4}{y^6}$ **3.** $6x^3y^{5/3}$ **4.** $-\dfrac{64y}{x^{3/4}}$ **5.** $\dfrac{1}{27}$ **6.** 9 **7.** $3a^4b^2c^5$ **8.** $2x^2y^3\sqrt[4]{2y^3}$ **9.** $2y\sqrt{3y}+5\sqrt[3]{2}$

10. $108-57\sqrt{2}$ **11.** $\dfrac{2\sqrt[3]{3x^2}}{x}$ **12.** $-5-2\sqrt{6}$ **13.** $x=2,\ x=3$ **14.** $x=3,\ x=11$ **15.** $4-5i$ **16.** $4i$ **17.** $-16+30i$

18. $\dfrac{12-5i}{13}$ **19.** 50 **20.** 7.2

7.1 Exercises

1. $\dfrac{16x^2}{y^6}$ **3.** $-\dfrac{1}{8a^3b^3}$ **5.** $x^{3/2}$ **7.** y^2 **9.** $x^{1/2}$ **11.** $x^{5/2}$ **13.** $x^{4/7}$ **15.** $y^{1/2}$ **17.** 3 **19.** $5^{7/10}$ **21.** $\dfrac{1}{x^{3/4}}$

23. $\dfrac{b^{1/3}}{a^{5/6}}$ **25.** $\dfrac{1}{6^{1/2}}$ **27.** $\dfrac{2}{a^{1/4}}$ **29.** $x^{5/6}y$ **31.** $-14x^{7/12}y^{1/12}$ **33.** $\dfrac{1}{5^{1/2}}$ **35.** $2x^{7/10}$ **37.** $-\dfrac{4x^{5/2}}{y^{6/5}}$ **39.** $2ab$ **41.** $\dfrac{a^3}{27b^6}$

43. $49x^{4/3}y^{1/2}z^3$ **45.** $25x^2yz^3$ **47.** $x^2-x^{13/15}$ **49.** $a^{1/3}+3a^{31/30}$ **51.** $2^{1/4}a^{1/20}b^{1/6}$ **53.** $x^{41/24}$ **55.** 5 **57.** 8 **59.** 73

61. $\dfrac{-17.92628036x^{3/2}}{y^2}$ **63.** $\dfrac{3y+1}{y^{1/2}}$ **65.** $\dfrac{1+4^{1/3}y^{2/3}}{y^{2/3}}$ **67.** $3x^{1/2}(-4x+2x^2)$ **69.** $2a(5a^{1/4}-2a^{3/5})$ **71.** $a=-\dfrac{3}{8}$

73. radius 1.86 m **75.** $-\dfrac{3}{2}=x$ **76.** $\dfrac{2A-ah}{h}=b$ **77.** Approximately 147 milligrams **78.** Approximately 5 years

7.2 Exercises

5. 8 **7.** 12 **9.** $-\dfrac{1}{3}$ **11.** 3 **13.** -5 **15.** 5 **17.** -2 **19.** -2 **21.** Not a real number **23.** 8 **25.** 5 **27.** $-\dfrac{1}{4}$

29. $9^{1/3}$ **31.** $(2x)^{1/5}$ **33.** $(a+b)^{3/7}$ **35.** $x^{1/6}$ **37.** $(3x)^{5/6}$ **39.** x^2 **41.** 17 **43.** xy^2 **45.** y^5 **47.** $6x^4y^2$ **49.** $2a^2b^5$

51. $-5x^{10}$ **53.** $(\sqrt[4]{x})^3$ **55.** $\dfrac{1}{(\sqrt[3]{7})^2}$ **57.** $(\sqrt[7]{2a+b})^5$ **59.** $(\sqrt[5]{-x})^3$ **61.** $\sqrt[5]{8x^3y^3}$ **63.** 8 **65.** -2 **67.** $\dfrac{8}{27}$ **69.** $\dfrac{1}{5x^2}$

71. $3a^2b^3$ **73.** $11x^2$ **75.** $10a^4b^5$ **77.** $6x^3y^4z^5$ **79.** $6ab^3c^4$ **81.** $2x^2y^3 + 3x^2y^3 = 5x^2y^3$ **83.** 23 **85.** -7 **87.** $5|x|$
89. $-2x^2$ **91.** x^2y^4 **93.** $|a^3b|$ **95.** $2x^4y^2$ **97.** \$1215 per day **99.** 4.485×10^{16} Btu **100.** 22.75%

7.3 Exercises

1. $5\sqrt{2}$ **3.** $3\sqrt{2}$ **5.** $2\sqrt{30}$ **7.** 9 **9.** $3x\sqrt{x}$ **11.** $2a^2b^2\sqrt{15b}$ **13.** $7x^2y^3\sqrt{2xz}$ **15.** 2 **17.** $3\sqrt[3]{4}$ **19.** $2\sqrt[3]{7y}$
21. $2ab^2\sqrt[3]{b^2}$ **23.** $2x^3y^4\sqrt[3]{7x}$ **25.** $-2a^2bc^4\sqrt[3]{2}$ **27.** $2x^2y^3\sqrt[4]{3y}$ **29.** $3p^5\sqrt[4]{kp^3}$ **31.** $-2xy\sqrt[5]{y}$ **33.** $xy^2\sqrt[6]{8}$ or $xy^2\sqrt{2}$
35. $a = 4$ **36.** $b = 8$ **37.** 17 **39.** $6\sqrt{3}$ **41.** $3\sqrt{3}$ **43.** $-12\sqrt{2}$ **45.** 0 **47.** $2\sqrt{11} - \sqrt{7x}$ **49.** $-13x\sqrt{5}$
51. $11\sqrt[3]{2}$ **53.** $-10xy\sqrt[3]{y} + 6xy^2$ **55.** $a\sqrt[3]{2ab} - 5\sqrt{5a}$ **57.** 2.236 **59.** 12.490 **61.** 22.361 **63.** 18.547
65. $20.78460969 = 20.78460969$ ✓ **66.** $28.28427125 = 28.28427125$ ✓ **67.** 7.071 amps **69.** $y(9x + 5)(9x - 5)$ **70.** $x(4x - 7y)^2$
71. 5 small servings of scallops (30 scallops); 4 small servings of skim milk (4 cups)

7.4 Exercises

1. $\sqrt{15}$ **3.** $-12\sqrt{3}$ **5.** $4x\sqrt{10}$ **7.** $2x\sqrt{2y}$ **9.** $30x^2y^2\sqrt{3}$ **11.** $5\sqrt{2} + 30$ **13.** $6x - 8\sqrt{5x}$ **15.** $-3\sqrt{2}$ **17.** $4 - 6\sqrt{6}$
19. $14 + 11\sqrt{35x} + 60x$ **21.** $x - 6\sqrt{x} + 8$ **23.** $\sqrt{15} + 2\sqrt{10} + 3 + 2\sqrt{6}$ **25.** $-11x$ **27.** $29 - 4\sqrt{30}$
29. $3x + 13 + 6\sqrt{3x + 4}$ **31.** $4a - 20\sqrt{ab} + 25b$ **33.** $6x^2\sqrt[3]{3}$ **35.** $3\sqrt[3]{4x^3} - \sqrt[3]{x^7} = 3x\sqrt[3]{4} - 4x^2\sqrt[3]{x}$
37. $3\sqrt[3]{x^2} + 3x\sqrt[3]{2} - \sqrt[3]{4x^2} - 2x$ **39.** In the form of $(a + b)(a^2 - ab + b^2)$, which equals $a^3 + b^3$ we have $(\sqrt[3]{6})^3 + (\sqrt[3]{5})^3 = 6 + 5 = 11.$

40. In the form of $(a - b)(a^2 + ab + b^2)$, which equals $a^3 - b^3$ we have $(\sqrt[3]{9})^3 - (\sqrt[3]{4})^3 = 9 - 4 = 5.$ **41.** $\dfrac{7}{5}$ **43.** $\dfrac{4}{5}$ **45.** $\dfrac{2x\sqrt{x}}{9y^2}$

47. $\dfrac{2xy^2\sqrt[3]{x^2}}{3}$ **49.** $5x$ **51.** $2xy$ **53.** $\dfrac{3\sqrt{2}}{2}$ **55.** $\dfrac{\sqrt{2x}}{4}$ **57.** $\dfrac{4\sqrt{5}}{5}$ **59.** $\dfrac{\sqrt{2ab}}{b}$ **61.** $\dfrac{\sqrt{5y}}{5y}$ **63.** $\dfrac{3\sqrt{6x}}{4x}$ **65.** $\dfrac{\sqrt{35y}}{7x}$

67. $\dfrac{2(\sqrt{6} - \sqrt{3})}{3}$ **69.** $\sqrt{15} + 2\sqrt{3}$ **71.** $\dfrac{x\sqrt{3} - \sqrt{2x}}{3x - 2}$ **73.** $4 + \sqrt{15}$ **75.** $\dfrac{3x - 3\sqrt{3xy} + 2y}{3x - y}$ **77.** $\dfrac{3\sqrt{6} + 4}{2}$

79. $5x - 2x\sqrt{5} + \sqrt{5} - 2$ **81.** $\dfrac{40 + 5\sqrt{6}}{58}$ **83.** $\dfrac{3\sqrt[4]{8x}}{2x}$ **85.** $\dfrac{\sqrt[3]{49}}{7}$ **87.** $x\sqrt[5]{8x^3y}$ **89.** 1.194938299, 1.194938299; yes

90. 0.563508327, 0.563508327; yes **91.** $\dfrac{5}{7\sqrt{5}}$ **93.** $\dfrac{-25}{8(\sqrt{3} - 2\sqrt{7})}$ **95.** Cost = \$2.92 **97.** $x = 2, y = 3$ **98.** $x = 1, y = -5, z = 3$

99. On January 11 she will reach her goal if she continues this pattern. **100.** On January 1 Carlos had 5 cups of coffee and 6 cups of tea. If he continues the pattern he will reach his goal on January 22.

7.5 Exercises

3. $x = 51$ **5.** $x = 3$ **7.** $x = 7$ **9.** $x = 0, x = 3$ **11.** $y = 7$ **13.** $y = 0, y = -1$ **15.** $x = 3, x = 7$ **17.** $x = 9, x = -1$

19. $x = \dfrac{5}{2}$ **21.** $x = 2, x = 4$ **23.** $x = -9$ **25.** $x = 12$ **27.** $x = \dfrac{1}{4}$ **29.** No solution **31.** $x = 5, x = 13$ **33.** $x = 0, x = 8$

35. $x = -1$ **37.** $x = 9$ **39.** $x = 1$ **41.** $y = 3$ **43.** $x = 4.9232, 0.4028$ **45. (a)** $\dfrac{V^2}{12} = S$ **(b)** 27 feet **47.** $c = 9$

48. $b = -12$ **49.** $16x^4$ **50.** $\dfrac{1}{2x^2}$ **51.** $-6x^2y^3$ **52.** $2x^3y^4\sqrt[4]{4}$ **53.** The current flows at 3 miles per hour.

54. She skis at 15 miles per hour.

7.6 Exercises

5. $7i$ **7.** $i\sqrt{19}$ **9.** $5i\sqrt{2}$ **11.** $3i\sqrt{7}$ **13.** $-9i$ **15.** $2 + i\sqrt{3}$ **17.** $-3 + 2i\sqrt{6}$ **19.** $7i$ **21.** $x = 5; y = -3$

23. $x = -8; y = 2$ **25.** $x = -6, y = 3$ **27.** $x = -3, y = -\dfrac{7}{2}$ **29.** $-5 - 12i$ **31.** $1 - i$ **33.** $7 - 6i$ **35.** $0.6 - 0.3i$

37. $7 + 4i$ **39.** $-\dfrac{3}{4} + \dfrac{13}{12}i$ **41.** $34 - 6i$ **43.** $-10 - 12i$ **45.** $27 - 36i$ **47.** $-13 + 84i$ **49.** $-\sqrt{21}$ **51.** $-\sqrt{6}$

53. -12 **55.** $-3\sqrt{3} + 12i$ **57.** $30 + 5\sqrt{7} - 25i$ **59.** $12 - \sqrt{6} + 2i\sqrt{2} + 6i\sqrt{3}$ **61.** i **63.** 1 **65.** -1 **67.** 0

69. $\dfrac{1 + i}{2}$ **71.** $\dfrac{3 + 6i}{10}$ **73.** $-\dfrac{6 + 8i}{3}$ **75.** $\dfrac{35 + 42i}{61}$ **77.** $-2299.95 + 3293.32i$ **79.** $0.7676145826 - 3.136283633i$

81. $\dfrac{2 - 3i}{3}$

7.7 Exercises

5. 1.6 **7.** 12.5 **9.** 71.4 psi **11.** 576 feet **13.** 1.2 **15.** 16.9 **17.** Approx. 1333.3 Btu per hour **19.** 30 miles per hour

21. 21 **23.** 2.6 **25.** Approximately 62.7 mph **27.** 2.2 oersteds **29.** $x = 8, x = 7$ **30.** $x = \dfrac{1}{2}, x = -3$ **31.** $x = \dfrac{2}{3}, x = 2$

32. $x = 1, x = -8$ **33.** \$460 **34.** 55 gallons of paint

Putting Your Skills to Work

1. Approximately 12.3%   **2.** Approximately 10.5%

Chapter 7 Review Problems

1. $\dfrac{15x^3}{y^{5/2}}$ **2.** $-28x^{4/3}y$ **3.** $-4x^{7/2}$ **4.** $9x^2$ **5.** $\dfrac{4x^{4/3}}{y^2}$ **6.** $5a^{3/2}b^2$ **7.** $5^{3/4}$ **8.** $-6a^{5/6}b^{3/4}$ **9.** $\dfrac{x^{1/2}y^{3/10}}{2}$ **10.** $\dfrac{x}{32y^{1/2}z^4}$

11. $7a^5b$ **12.** $2a^{19/20} - 3a^{3/10}$ **13.** $x^{1/2}y^{1/5}$ **14.** $3x^{n+1}$ **15.** $6^{5/7}$ **16.** $\dfrac{2x+1}{x^{2/3}}$ **17.** $3x(2x^{1/2} - 3x^{-1/2})$ **18.** $(2x)^{1/10}$

19. $(\sqrt[9]{2x+3y})^4$ **20.** 8 **21.** $\sqrt[6]{-64}$ is not a real no.; $-\sqrt[6]{64} = -2$ **22.** $\dfrac{1}{81}$ **23.** $\dfrac{8}{27}$ **24.** $3xy^3z^5\sqrt{11x}$ **25.** $-2a^2b^3c^4\sqrt[3]{7a^2b}$

26. $2x^2z^2\sqrt[4]{y^3z^3}$ **27.** $xy^2z^4\sqrt[5]{x^2z^3}$ **28.** $12x^5y^6$ **29.** $5a^3b^2c^{100}$ **30.** y **31.** $|y|$ **32.** $|xy|$ **33.** x^2 **34.** x^7 **35.** x^4

36. $11\sqrt{2}$ **37.** $13\sqrt{7}$ **38.** $2 - 6\sqrt[3]{2}$ **39.** $7y\sqrt[3]{2y}$ **40.** $11\sqrt{2x} - x\sqrt{2}$ **41.** $-8x\sqrt[3]{5y} + x\sqrt[3]{3y}$ **42.** $90\sqrt{2}$

43. $12x\sqrt{2} - 36\sqrt{3x}$ **44.** $4 - 9\sqrt{6}$ **45.** $34 - 14\sqrt{3}$ **46.** $74 - 12\sqrt{30}$ **47.** $2x + 2\sqrt[3]{3x^2} - \sqrt[3]{2xy} - \sqrt[3]{6y}$ **48.** 11.832

49. 16.432 **50.** 9.568 **51.** $\dfrac{x\sqrt{3y}}{y}$ **52.** $\dfrac{2\sqrt{3y}}{3y}$ **53.** $\sqrt{3}$ **54.** $2\sqrt{6} + 2\sqrt{5}$ **55.** $\dfrac{3x - \sqrt{xy}}{9x - y}$ **56.** $\dfrac{-(\sqrt{35} + 3\sqrt{5})}{2}$

57. $\dfrac{2 + 3\sqrt{2}}{7}$ **58.** $\dfrac{10\sqrt{3} - 3\sqrt{2} + 5\sqrt{6} - 3}{3}$ **59.** $\dfrac{3x + 4\sqrt{xy} + y}{x - y}$ **60.** $\dfrac{\sqrt[3]{4x^2y}}{2y}$ **61.** $\dfrac{5\sqrt{6}}{6}$ **62.** $\dfrac{\sqrt{35}}{5}$ **63.** $4i + 3i\sqrt{5}$

64. $x = \dfrac{-7 + \sqrt{6}}{2}$; $y = -3$ **65.** $-9 - 11i$ **66.** $-10 + 2i$ **67.** $29 - 29i$ **68.** $48 - 64i$ **69.** $-8 + 6i$ **70.** $-5 - 4i$ **71.** -1

72. i **73.** $\dfrac{13 - 34i}{25}$ **74.** $\dfrac{11 + 13i}{10}$ **75.** $\dfrac{-(4i + 3)}{5}$ **76.** $\dfrac{18 + 30i}{17}$ **77.** $-2i$ **78.** $x = 7$ **79.** $x = 2$ **80.** $x = 4$

81. $x = -1$ **82.** $x = 4$ **83.** $x = 5$ **84.** $x = 5, 1$ **85.** $x = 1, \dfrac{3}{2}$ **86.** $y = 9.6$ **87.** $y = 12.5$ **88.** 168 feet

89. 3.5 seconds   **90.** $y = 0.5$ **91.** 16.8 pounds   **92.** $y = \dfrac{4}{3}$ **93.** 160 cubic centimeters

Chapter 7 Test

1. $-6x^{5/6}y^{1/2}$ **2.** $\dfrac{7x^{9/4}}{4}$ **3.** $16\sqrt{2x}$ or $16(2x)^{1/2}$ **4.** $6^{4/5}$ **5.** $\dfrac{1}{4}$ **6.** 512 **7.** $5xy^2\sqrt[3]{2x}$ **8.** $8x^3y^2\sqrt{y}$

9. $5a^2b^4\sqrt{3b}$ **10.** $18\sqrt{3} + x\sqrt[3]{2x^2}$ **11.** $10\sqrt{2}$ **12.** $18\sqrt{2} - 10\sqrt{6}$ **13.** $12 + 39\sqrt{2}$ **14.** $\dfrac{4\sqrt{5x}}{5x}$ **15.** $\dfrac{\sqrt{5x}}{x}$ **16.** $2 + \sqrt{3}$

17. $x = 2, x = 1$ **18.** $x = 10$ **19.** $x = 6$ **20.** $2 + 14i$ **21.** $-1 + 4i$ **22.** $18 + i$ **23.** $\dfrac{-13 + 11i}{10}$ **24.** $27 + 36i$

25. $-i$ **26.** $y = 3$ **27.** $y = \dfrac{5}{6}$ **28.** 83.3 feet

Cumulative Test for Chapters 1–7

1. Associative property of addition   **2.** $6a^4 - 3a^3 + 15a^2 - 8a$ **3.** -64 **4.** $x = \dfrac{3y + 24}{2}$ **5.**

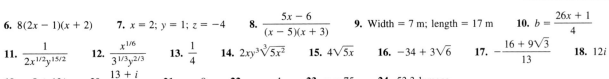

x	y
0	-3
5	0
10	3

$3x - 5y = 15$

6. $8(2x - 1)(x + 2)$ **7.** $x = 2; y = 1; z = -4$ **8.** $\dfrac{5x - 6}{(x - 5)(x + 3)}$ **9.** Width = 7 m; length = 17 m   **10.** $b = \dfrac{26x + 1}{4}$

11. $\dfrac{1}{2x^{1/2}y^{15/2}}$ **12.** $\dfrac{x^{1/6}}{3^{1/3}y^{2/3}}$ **13.** $\dfrac{1}{4}$ **14.** $2xy^3\sqrt[3]{5x^2}$ **15.** $4\sqrt{5x}$ **16.** $-34 + 3\sqrt{6}$ **17.** $-\dfrac{16 + 9\sqrt{3}}{13}$ **18.** $12i$

19. $-5 + 12i$ **20.** $\dfrac{13 + i}{10}$ **21.** $x = 8$ **22.** $x = -1$ **23.** $y = 75$ **24.** 53.3 lumens

Chapter 8

Pretest Chapter 8

1. $x = \pm 2\sqrt{3}$ **2.** $x = \dfrac{2 \pm \sqrt{10}}{2}$ **3.** $x = \dfrac{1 \pm 2\sqrt{2}}{2}$ **4.** $x = \dfrac{1 \pm \sqrt{21}}{10}$ **5.** $x = \dfrac{-3 \pm 2i\sqrt{2}}{2}$ **6.** $x = \frac{2}{3}, -\frac{3}{4}$ **7.** $x = 0, \frac{7}{4}$

8. $x = -\frac{2}{3}, 3$ **9.** $x = 2, -1$ **10.** $w = \pm 8, \pm 2\sqrt{2}$ **11.** $x = \dfrac{-w \pm \sqrt{w^2 - 24w}}{3}$ **12.** $W = 4$ m; $L = 13$ m

13. Vertex $(-1, -12)$; y-intercept $(0, -9)$; x-intercepts $(1, 0)$, $(-3, 0)$ **14.** **15.** $x < -2$ or $x > 3$

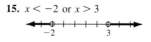

16. $-3 \le x \le -\dfrac{3}{2}$ **17.** $-1.8 < x < 1.1$

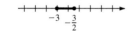

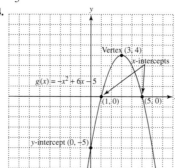

8.1 Exercises

1. $x = \pm 12$ **3.** $x = \pm 2\sqrt{5}$ **5.** $x = \pm\sqrt{13}$ **7.** $x = \pm\sqrt{5}$ **9.** $x = \pm 6$ **11.** $x = \pm 5i$ **13.** $x = \pm i\sqrt{2}$ **15.** $x = 3 \pm 2\sqrt{3}$

17. $x = \dfrac{-1 \pm \sqrt{7}}{2}$ **19.** $x = \dfrac{9}{4}, -\dfrac{3}{4}$ **21.** $x = \dfrac{4 \pm 2\sqrt{2}}{3}$ **23.** $x = \dfrac{-1 \pm \sqrt{7}}{2}$ **25.** $x = 2 \pm \sqrt{15}$ **27.** $x = -5 \pm 2\sqrt{5}$

29. $x = 4 \pm \sqrt{33}$ **31.** $x = 3 \pm \sqrt{5}$ **33.** $x = \dfrac{-5 \pm \sqrt{41}}{2}$ **35.** $y = \dfrac{-5 \pm \sqrt{3}}{2}$ **37.** $x = \dfrac{3}{2}, x = -\dfrac{1}{2}$ **39.** $y = 2, y = -\dfrac{3}{2}$

41. $x = \dfrac{1 \pm i\sqrt{3}}{2}$ **43.** $x = \dfrac{-4 \pm \sqrt{13}}{3}$ **45.** $x = \dfrac{4 \pm i\sqrt{5}}{3}$ **47.** $(-1 + \sqrt{6})^2 + 2(-1 + \sqrt{6}) - 5 \overset{?}{=} 0$;

$1 - 2\sqrt{6} + 6 - 2 + 2\sqrt{6} - 5 \overset{?}{=} 0$; $0 = 0$ ✓ **48.** $(2 + \sqrt{3})^2 - 4(2 + \sqrt{3}) + 1 \overset{?}{=} 0$; $4 + 4\sqrt{3} + 3 - 8 - 4\sqrt{3} + 1 \overset{?}{=} 0$; $0 = 0$ ✓

49. Approximately 0.88 second **51.** 15 seconds **53.** 8 **54.** 7 **55.** 40 **56.** 4

8.2 Exercises

5. $x = \dfrac{-5 \pm \sqrt{65}}{2}$ **7.** $x = \dfrac{-1 \pm \sqrt{33}}{4}$ **9.** $x = 0, x = \dfrac{2}{3}$ **11.** $x = -4 \pm \sqrt{3}$ **13.** $x = \dfrac{1}{2}, x = -\dfrac{1}{3}$ **15.** $x = \dfrac{-3 \pm \sqrt{41}}{8}$

17. $x = \dfrac{\pm\sqrt{21}}{3}$ **19.** $x = \dfrac{4 \pm \sqrt{6}}{2}$ **21.** $x = \pm 1$ **23.** $x = \dfrac{1 \pm \sqrt{29}}{2}$ **25.** $x = \dfrac{-7 \pm \sqrt{17}}{4}$ **27.** $x = \dfrac{5 \pm \sqrt{73}}{2}$

29. $y = \dfrac{16 \pm 2\sqrt{34}}{5}$ **31.** $y = 9, y = 5$ **33.** $x = 2 \pm 2i$ **35.** $x = \dfrac{\pm i\sqrt{30}}{2}$ **37.** $x = \dfrac{4 \pm i\sqrt{5}}{3}$ **39.** $x = \dfrac{-1 \pm i\sqrt{5}}{2}$

41. 2 nonreal complex roots **43.** 2 irrational roots **45.** 2 rational roots **47.** 1 rational root **49.** $x^2 + 6x - 55 = 0$
51. $x^2 + 16x + 60 = 0$ **53.** $x^2 + 36 = 0$ **55.** $x^2 + 20 = 0$ **57.** $2x^2 - x - 15 = 0$ **59.** $x = -2.7554, x = 1.0888$ **61.** $x = 2.8515$,
$x = 0.7116$ **63.** $a = 2$ **64.** $c = -2$ **65.** $7 + 4\sqrt{3} = 7 + 4\sqrt{3}$ **66.** $11 + 6\sqrt{2} = 11 + 6\sqrt{2}$ **67.** $-3x^2 - 10x + 11$
68. $-y^2 + 3y$ **69.** The suits cost \$95 and the goggles cost \$29 last year. **70.** The width is 9 feet and length is 16 feet.

8.3 Exercises

1. $x = \pm\sqrt{5}, x = \pm 2$ **3.** $x = \dfrac{\pm\sqrt{6}}{2}, x = \pm 2i$ **5.** $x = \pm\dfrac{i\sqrt{6}}{3}, x = \pm 2$ **7.** $x = 2, x = -1$ **9.** $x = -\dfrac{\sqrt[3]{4}}{2}, x = \sqrt[3]{4}$

11. $x = \pm\sqrt[4]{2}, x = \pm 1$ **13.** $x = \pm\dfrac{\sqrt[4]{54}}{3}$; these are the only real roots. **15.** $x = 8; x = -64$ **17.** $x = -27; x = 1$ **19.** $x = 81$

21. $x = 16$ **23.** $x = 1; x = -32$ **25.** $x = -2, x = 1, x = \dfrac{-1 \pm \sqrt{13}}{2}$ **27.** $x = 49$ **29.** $x = -5; x = -2$ **31.** $x = \dfrac{1}{2}, x = 3$

33. The only two real roots are $x \approx \pm 3.66$ **35.** $x = \dfrac{5}{6}, x = \dfrac{3}{2}$ **36.** $x = \sqrt[3]{\dfrac{-13 \pm \sqrt{41}}{2}}$ **37.** $-15\sqrt{2x}$ **38.** 0 **39.** $3\sqrt{10} - 12\sqrt{3}$

41. The salaries for men are 29.8% greater for college and University teachers, 13.7% greater for elementary teachers, 34.0% greater for physicians, and 3.6% greater for secretaries and typists. **42.** She would have to work an additional 8.9 years.

8.4 Exercises

1. $t = \pm\dfrac{\sqrt{S}}{4}$ **3.** $r = \pm\sqrt{\dfrac{V}{\pi h}}$ **5.** $x = \pm\sqrt{\dfrac{6H}{a}}$ **7.** $y = \dfrac{\pm\sqrt{2A - 5}}{2}$ **9.** $V = \pm\sqrt{\dfrac{Fr}{kb}}$ **11.** $r = \pm\sqrt{\dfrac{V - \pi R^2 h}{\pi h}}$

13. $h = \pm\sqrt{\dfrac{kd^4}{L}} = \pm d^2\sqrt{\dfrac{k}{L}}$ **15.** $x = 2b$ **17.** $a = -\dfrac{b}{2}, a = 3b$ **19.** $I = \dfrac{E \pm \sqrt{E^2 - 4RP}}{2R}$ **21.** $w = \dfrac{-b \pm \sqrt{b^2 + 40}}{5}$

23. $r = \dfrac{-\pi h \pm \sqrt{\pi^2 h^2 + S\pi}}{\pi}$ **25.** $x = \dfrac{-5 \pm \sqrt{25 - 8aw - 8w}}{2a + 2}$ **27.** $c = \sqrt{13}$ **29.** $12 = a$ **31.** $4 = b$

33. $b = \dfrac{24\sqrt{5}}{5}$, $a = \dfrac{12\sqrt{5}}{5}$ **35.** The hypotenuse is 15 miles long. The shorter leg is 9 miles long.   **37.** The shorter leg is approx. 6.13 miles long. The final leg is approx. 9.13 miles long.   **39.** Width is 7 feet; length is 18 feet.   **41.** Base is 8 cm; altitude is 18 cm.   **43.** The rate on the old road was 40 mph; the rate on the new road was 45 mph.   **45.** 105 miles   **47.** $w = \dfrac{-7b - 21 \pm \sqrt{529b^2 + 294b + 441}}{10}$

48. Interest rate is 4% per year.   **49.** $\dfrac{4\sqrt{3x}}{3x}$ **50.** $\dfrac{\sqrt{30}}{2}$ **51.** $\dfrac{3(\sqrt{x} - \sqrt{y})}{x - y}$ **52.** $-2 - 2\sqrt{2}$ **53.** $\dfrac{3\sqrt[3]{a^2b}}{2}$

Putting Your Skills to Work

1.

n	0	40	80	86	120	140	160	171
M	0	1052	1464	1471	1236	882	368	17

2.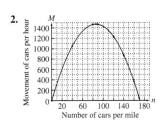

The maximum movement of cars occurs when there are 86 cars per mile.

8.5 Exercises

1. $V(4, -36)$ $I(0, -20)$; $(10, 0)$; $(-2, 0)$ **3.** $V(-2, 16)$ $I(0, 12)$; $(-6, 0)$; $(2, 0)$ **5.** $V(-2, -9)$ $I(0, 3)$; $(-0.3, 0)$; $(-3.7, 0)$

7. $V\left(-\dfrac{1}{3}, -\dfrac{17}{3}\right)$; no x-intercepts; y-intercept $(0, -6)$ **9.** $V\left(\dfrac{5}{4}, \dfrac{57}{8}\right)$ $I(0, 4)$; $(3.1, 0)$; $(-0.6, 0)$ **11.** $V\left(-\dfrac{1}{4}, -\dfrac{121}{8}\right)$ $I(0, -15)$; $(-3, 0)$; $(2.5, 0)$

13. $V(-1, 6)$ $I(0, 11)$;
No x-intercepts

15.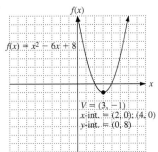

17.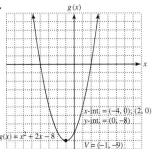

19.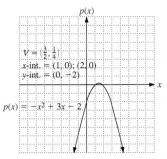

21.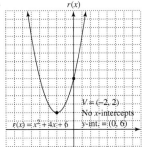

23.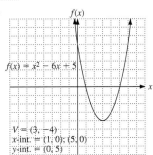

25.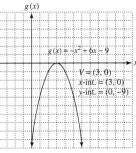

27. The maximum height is 56 feet. It will take about 2.9 seconds.   **31.** x-intercepts $(-0.3, 0)$ and $(2.6, 0)$
29. Vertex $(2.2, 2.8)$; y-intercept $(0, 7.59)$; no x-intercepts

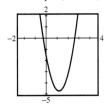

8.6 Exercises

3. $-4 < x < 3$
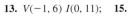

5. $x < 0$ or $x > \dfrac{5}{2}$

7. $-\dfrac{3}{2} < x < 1$

9. $x \le -2$ or $x \ge 2$

11. $-\dfrac{1}{2} \le x \le 6$ **13.** $-5 < x < 4$ **15.** $x < -\dfrac{2}{3}$ or $x > \dfrac{3}{2}$ **17.** $-4 \le x \le \dfrac{1}{2}$ **19.** $-9 \le x \le 4$ **21.** $x = 2$
23. Approx. $x < -1.2$ or $x > 3.2$ **25.** Approx. $1.6 < x < 4.4$ **27.** Approx. $x \le 0.6$ or $x \ge 2.4$ **29.** No real numbers satisfy this inequality.   **31.** For time t greater than 30 seconds or less than 10 seconds   **33.** **(a)** Approx. $15.1 < x < 264.9$ **(b)** \$187,500 **(c)** \$230,000
35. $x > 0.3$ or $x < -0.7$ **37.** She must score a combined total of 167 points on the two tests. Any two test scores that total 167 will be sufficient to participate in synchronized swimming.   **38.** 52 ounces of potato chips, 122 ounces of peanuts, 62 ounces of popcorn, and 124 ounces of pretzels.

Chapter 8 Review Problems

1. $x = \pm 3\sqrt{2}$ **2.** $x = -7, 3$ **3.** $x = -4 \pm \sqrt{3}$ **4.** $x = \dfrac{2 \pm \sqrt{3}}{2}$ or $1 \pm \dfrac{\sqrt{3}}{2}$ **5.** $x = \dfrac{5 \pm \sqrt{7}}{3}$ **6.** $x = 3 \pm \sqrt{13}$

7. $x = \dfrac{4 \pm i\sqrt{2}}{3}$ **8.** $x = \dfrac{3}{2}$ **9.** $x = 0, \dfrac{9}{7}$ **10.** $x = \dfrac{3}{4}, \dfrac{5}{2}$ **11.** $x = \pm i\sqrt{5}$ **12.** $x = \pm i\sqrt{3}$ **13.** $x = -8, 4$ **14.** $x = 0, \dfrac{2}{5}$

15. $x = \dfrac{5}{2}, -3$ **16.** $x = -1 \pm \sqrt{5}$ **17.** $x = \dfrac{3 \pm i\sqrt{23}}{8}$ **18.** $x = \dfrac{-5 \pm \sqrt{13}}{6}$ **19.** $x = -\dfrac{2}{3}, \dfrac{1}{3}$ **20.** $x = \dfrac{11 \pm \sqrt{21}}{10}$ **21.** $x = -7$

22. $x = -8$ **23.** $x = \dfrac{3 \pm i}{5}$ **24.** $x = -\dfrac{1}{4}, -1$ **25.** $y = -\dfrac{5}{6}, -2$ **26.** $y = -\dfrac{5}{2}, \dfrac{3}{5}$ **27.** $y = -5, 3$ **28.** $y = 9$

29. $y = -3, 1$ **30.** $y = 0, -\dfrac{5}{2}$ **31.** $x = -2, 3$ **32.** $x = -3, 2$ **33.** Two rational solutions **34.** Two irrational solutions

35. Two complex solutions **36.** One rational solution **37.** $x^2 - 25 = 0$ **38.** $x^2 + 9 = 0$ **39.** $x^2 - 32 = 0$ **40.** $8x^2 + 10x + 3 = 0$

41. $x = -0.5, x = -2.5$ **42.** $x = 1.6, x = -1.2$ **43.** $x = \pm 2, \pm\sqrt{2}$ **44.** $x = \dfrac{-\sqrt[3]{4}}{2}, \sqrt[3]{3}$ **45.** $x = 256$ **46.** $x = -512, -1$

47. $x = 1, 2$ **48.** $x = \pm 1, \pm\sqrt{2}$ **49.** $B = \pm\sqrt{\dfrac{3AH}{2C}}$ **50.** $b = \pm\sqrt{\dfrac{2H}{3g} - a^2}$ **51.** $d = \dfrac{x}{4}, \dfrac{-x}{5}$ **52.** $x = \dfrac{3 \pm \sqrt{9 + 28y}}{2y}$

53. $y = \dfrac{2a \pm \sqrt{4a^2 - 6a}}{3}$ **54.** $x = \dfrac{-1 \pm \sqrt{1 - 15y^2 + 5PV}}{5}$ **55.** $c = \sqrt{22}$ **56.** $a = 4\sqrt{15}$ **57.** The car is approximately 3.3 miles

from the observer. **58.** Width is 7 m; length is 29 m **59.** Base is 7 cm; Alt is 20 cm **60.** 50 mph during 1st part, 45 mph during rain
61. 20 mph cruising; 5 mph trolling **62.** The walkway should be approximately 2.4 feet wide. **63.** The walkway should be 2 feet wide.
64. Vertex $(-5, 0)$; one x-intercept $(-5, 0)$; y-intercept $(0, 25)$ **65.** Vertex $(3, -2)$; y-intercept $(0, -11)$; no x-intercepts
66.

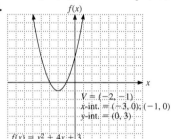

$V = (-2, -1)$
x-int. $= (-3, 0); (-1, 0)$
y-int. $= (0, 3)$
$f(x) = x^2 + 4x + 3$

67.

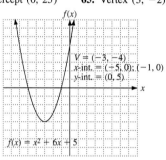

$V = (-3, -4)$
x-int. $= (-5, 0); (-1, 0)$
y-int. $= (0, 5)$
$f(x) = x^2 + 6x + 5$

68.

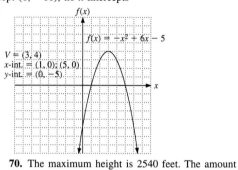

$f(x) = -x^2 + 6x - 5$
$V = (3, 4)$
x-int. $= (1, 0); (5, 0)$
y-int. $= (0, -5)$

69. $R(x) = x(1200 - x)$; The maximum revenue will occur if the price is \$600 for each unit. **70.** The maximum height is 2540 feet. The amount of time for the complete flight is 25.1 seconds. **71.** $-9 < x < 2$

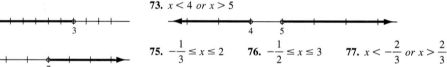

72. $-7 < x < 3$

73. $x < 4$ or $x > 5$

74. $x < 4$ or $x > 7$

75. $-\dfrac{1}{3} \le x \le 2$ **76.** $-\dfrac{1}{2} \le x \le 3$ **77.** $x < -\dfrac{2}{3}$ or $x > \dfrac{2}{3}$

78. $x < -\dfrac{5}{4}$ or $x > \dfrac{5}{4}$ **79.** $x < (1 - \sqrt{5})$ or $x > (1 + \sqrt{5})$; approximately $x < -1.2$ or $x > 3.2$ **80.** $x \le -6$ or $x \ge -2$

81. $x < -8$ or $x > 2$ **82.** $x < 1.4$ or $x > 2.6$ **83.** No real solution **84.** No real solution **85.** $x < -4$ or $2 < x < 3$
86. $-4 < x < -1$ or $x > 2$

Chapter 8 Test

1. $x = 0, \frac{7}{5}$ **2.** $x = \frac{1}{3}, -2$ **3.** $x = 2, -\frac{2}{9}$ **4.** $x = 8$ **5.** $x = \pm\sqrt{10}$ **6.** $x = \frac{7}{2}, -1$ **7.** $x = \dfrac{3 \pm \sqrt{3}}{2}$ **8.** $x = \dfrac{3 \pm i}{2}$

9. $x = \pm\sqrt{7}, \pm\sqrt{2}$ **10.** $x = \frac{1}{5}, -\frac{3}{4}$ **11.** $x = (1 \pm \sqrt{13})^3$ **12.** $z = \pm\sqrt{\dfrac{xyw}{B}}$ **13.** $y = \dfrac{-b \pm \sqrt{b^2 - 30w}}{5}$ **14.** $c = 4\sqrt{3}$

15. Width is 5 miles; length is 16 miles. **17.** $V = (-3, 4)$; y-int. $(0, -5)$; x-int. $(-5, 0)$; $(-1, 0)$
16. 2 mph during 1st part; 3 mph after lunch
18. $-\frac{2}{3} \le x \le 4$ **19.** $x \le -\frac{9}{2}$ or $x \ge 3$
20. $x < -4.5$ or $x > 1.5$

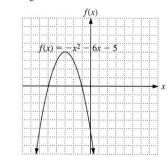

$f(x) = -x^2 - 6x - 5$

Cumulative Test for Chapters 1–8

1. $\dfrac{81y^{12}}{x^8}$ **2.** $\frac{1}{4}a^3 - a^2 - 3a$ **3.** $y = \dfrac{ab + 4}{a}$ **4.**

x	y
0	4
−2	0

5. $x + 2y = 4$ **6.** $\dfrac{32\pi}{3}$ cubic inches

7. $(5x - 3y)(25x^2 + 15xy + 9y^2)$

8. $4x^2y^2\sqrt{3y}$ **9.** $4\sqrt{6} + 5\sqrt{3}$ **10.** $\dfrac{5\sqrt{7}}{7}$ **11.** $x = 0, \frac{3}{5}$ **12.** $x = \frac{1}{3}, -5$ **13.** $x = \dfrac{3 \pm 2\sqrt{3}}{2}$ **14.** $x = \dfrac{2 \pm i\sqrt{11}}{3}$

15. $x = 16$ **16.** $x = -27, -216$ **17.** $y = \dfrac{-5w \pm \sqrt{25w^2 + 56z}}{4}$ **18.** $y = \pm\sqrt{\dfrac{5w - 16z^2}{3}}$ **19.** $\sqrt{15}$

20. Base is 5 meters; altitude is 18 meters.   **21.** Vertex $(4, 4)$; y-intercept $(0, -12)$; x-intercepts $(2, 0)$, $(6, 0)$

22.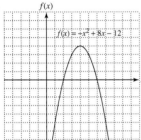

23. $x < -5$ or $x > 3$ **24.** $-\frac{1}{2} \le x \le \frac{2}{3}$

Chapter 9

Pretest Chapter 9

1. $d = 3\sqrt{5}$

2. $(x - 8)^2 + (y + 2)^2 = 7$

3. Center $= (1, 2)$; radius $= 2$; $(x - 1)^2 + (y - 2)^2 = 4$

4. $x = (y + 1)^2 + 2$

5. $y = (x + 2)^2 - 3$

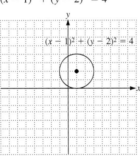

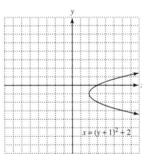

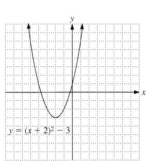

6. $4x^2 + y^2 - 36 = 0$ **7.** $\dfrac{(x + 3)^2}{25} + \dfrac{(y - 1)^2}{16} = 1$ **8.** $\dfrac{y^2}{9} - \dfrac{x^2}{25} = 1$ **9.** $\dfrac{(x - 2)^2}{4} - \dfrac{(y + 1)^2}{9} = 1$

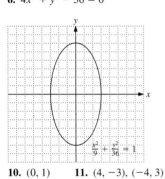

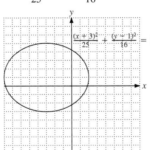

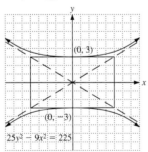

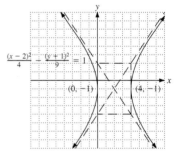

10. $(0, 1)$ **11.** $(4, -3)$, $(-4, 3)$

9.1 Exercises

5. $\sqrt{5}$ **7.** $\sqrt{74}$ **9.** $\sqrt{29}$ **11.** 5   **13.** 13   **15.** $4\sqrt{2}$ **17.** $\dfrac{2\sqrt{26}}{5}$ **19.** $5\sqrt{2}$ **21.** 2   **23.** $d \approx 5.228852647$

25. $y = 10, y = -6$ **27.** $y = 0, y = 4$ **29.** $x = 6, x = 8$ **31.** 9.5 miles   **33.** $(x - 7)^2 + (y - 12)^2 = 225$ **35.** $(x - 7)^2 + (y + 4)^2 = 4$

37. $(x + 1)^2 + (y + 7)^2 = 5$ **39.** $(x + 3.5)^2 + y^2 = 36$ **41.** $x^2 + y^2 = 144$ **43.** $(x - 26.8)^2 + (y - 29.2)^2 = 2165.0409$

45.

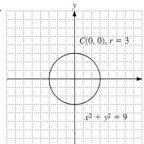

$C(0, 0), r = 3$

$x^2 + y^2 = 9$

47.

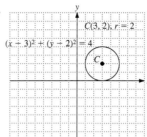

$C(3, 2), r = 2$

$(x - 3)^2 + (y - 2)^2 = 4$

49.

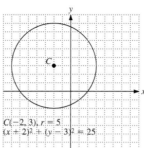

$C(-2, 3), r = 5$

$(x + 2)^2 + (y - 3)^2 = 25$

51.

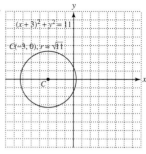

$(x + 3)^2 + y^2 = 11$

$C(-3, 0), r = \sqrt{11}$

53. $(x + 3)^2 + (y - 2)^2 = 16; C(-3, 2), r = 4$ **55.** $(x - 6)^2 + (y + 1)^2 = 49; C(6, -1), r = 7$ **57.** $(x + 6)^2 + y^2 = 26; C(-6, 0), r = \sqrt{26}$

59. $(x - 42.7)^2 + (y - 29.7)^2 = 630.01$ **61.**

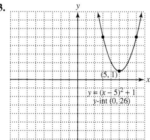

63. $x = \dfrac{2}{3}, x = 1$ **64.** $x = -\dfrac{2}{3}, x = \dfrac{1}{3}$

65. $x = \dfrac{-1 \pm \sqrt{5}}{4}$ **66.** $x = \dfrac{3 \pm 2\sqrt{11}}{5}$

9.2 Exercises

5.

x	y
-1	0.5
0	0
1	0.5
4	8
-4	8

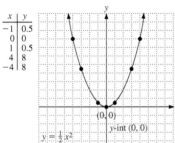

$(0, 0)$

y-int $(0, 0)$

$y = \frac{1}{2}x^2$

7.

x	y
-1	-3
0	0
1	-3
2	-12
-2	-12

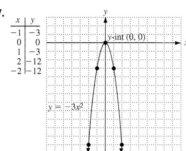

y-int $(0, 0)$

$y = -3x^2$

9.

x	y
-2	-2
0	-6
2	-2
3	3
-3	3

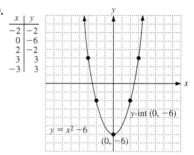

y-int $(0, -6)$

$y = x^2 - 6$

$(0, -6)$

11.

x	y
-2	-4
0	4
2	-4

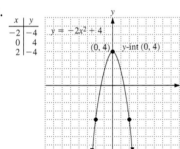

$y = -2x^2 + 4$

$(0, 4)$ y-int $(0, 4)$

13.

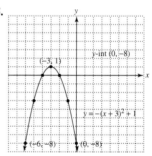

$(5, 1)$

$y = (x - 5)^2 + 1$
y-int $(0, 26)$

15.

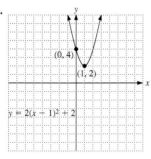

y-int $(0, -8)$

$(-3, 1)$

$y = -(x + 3)^2 + 1$

$(-6, -8)$ $(0, -8)$

17.

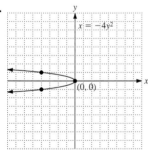

$(0, 4)$

$(1, 2)$

$y = 2(x - 1)^2 + 2$

19.

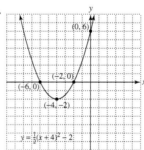

$(0, 6)$

$(-2, 0)$

$(-6, 0)$

$(-4, -2)$

$y = \frac{1}{2}(x + 4)^2 - 2$

21.

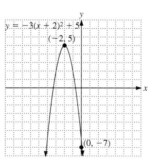

$y = -3(x + 2)^2 + 5$

$(-2, 5)$

$(0, -7)$

23.

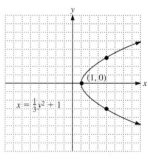

$(1, 0)$

$x = \frac{1}{3}y^2 + 1$

25.

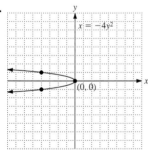

$x = -4y^2$

$(0, 0)$

27.

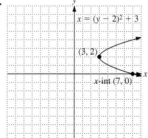

$x = (y - 2)^2 + 3$

$(3, 2)$

x-int $(7, 0)$

29.

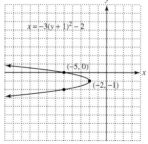

$x = -3(y + 1)^2 - 2$

$(-5, 0)$

$(-2, -1)$

31.

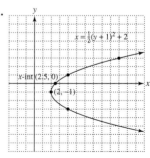

$x = \frac{1}{2}(y + 1)^2 + 2$

x-int $(2.5, 0)$

$(2, -1)$

SA–28   *Selected Answers*

33.

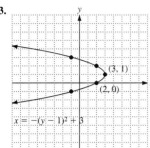

$x = -(y - 1)^2 + 3$

35. $y = (x + 3)^2 + 1$ **(a)** Vertical **(b)** Opens up **(c)** $v(-3, 1)$ **37.** $y = -2(x - 1)^2 - 1$ **(a)** Vertical **(b)** Opens down **(c)** $v(1, -1)$ **39.** $x = (y + 5)^2 - 2$ **(a)** Horizontal **(b)** Opens right **(c)** $v(-2, -5)$

41. $y = \dfrac{1}{36}x^2$ **43.** $p = 9$. The distance from $(0, 0)$ to the focus point is 9 inches.

45. Vertex $(-1.62, -5.38)$; y-int -0.1312; x-intercepts 0.020121947 and -3.260121947

47. Maximum profit is \$36,000; number of items needed is 40. **48.** Maximum profit is \$52,000; number of items needed is 50. **49.** Maximum yield is 202,500; number of trees per acre is 450. **50.** Maximum sensitivity is 52,812.5; dosage is 162.5 milligrams. **51.** $5x\sqrt{2x}$ **52.** $2xy\sqrt[3]{5y}$

53. $2x\sqrt{2} - 8\sqrt{2x}$ **54.** $2x\sqrt[3]{2x} - 20x\sqrt[3]{2}$ **55.** $27\dfrac{1}{3}$ miles **56.** Approximately 42.2 miles

9.3 Exercises

3.

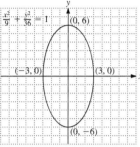

5.

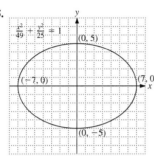

7.

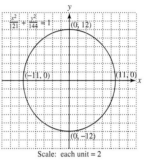

Scale: each unit = 2

9.

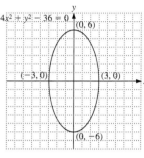

11.

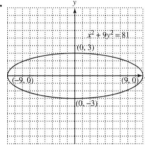

13.

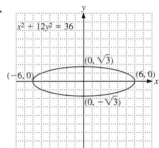

15.

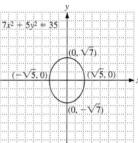

17.

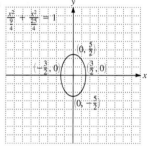

19. $\dfrac{x^2}{64} + \dfrac{y^2}{100} = 1$ **21.** $\dfrac{x^2}{5} + \dfrac{y^2}{36} = 1$ **23.** 142 million miles

25.

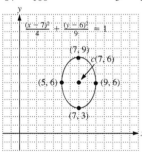

27.

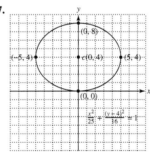

29. $\dfrac{(x + 5)^2}{16} + \dfrac{(y + 2)^2}{36} = 1$

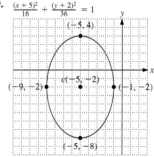

31. $\dfrac{(x - 1)^2}{16} + \dfrac{(y + 2)^2}{9} = 1$ **33.** $\dfrac{(x - 30)^2}{900} + \dfrac{(y - 20)^2}{400} = 1$ **35.** $(\pm 3.4641, 0)$; $(0, \pm 4.3589)$ **37.** 706.9 square inches

38. 22,376.0 square meters **39.** $30\sqrt{2} + 40\sqrt{3} - 2\sqrt{6} - 8$ **40.** $x\sqrt{6y} + 3y\sqrt{x} + 9\sqrt{xy}$ **41.** $\dfrac{\sqrt{6xy}}{2xy}$ **42.** $\dfrac{5(\sqrt{2x} + \sqrt{y})}{2x - y}$

9.4 Exercises

5.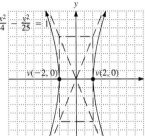

7.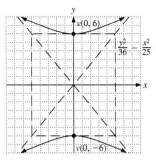

9.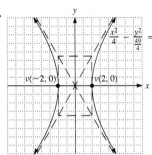

11.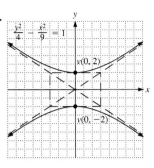

13.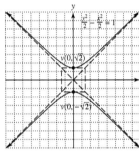

15.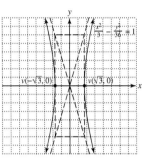

17.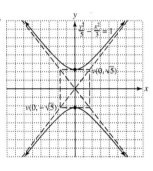

19.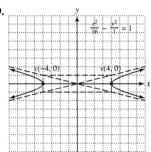

21.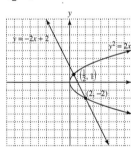

23. $\dfrac{x^2}{4} - \dfrac{y^2}{9} = 1$

25. $\dfrac{y^2}{49} - \dfrac{x^2}{9} = 1$

27. $\dfrac{x^2}{16} - \dfrac{y^2}{\frac{64}{9}} = 1$

29.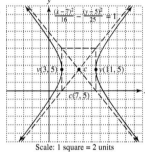

Scale: 1 square = 2 units

31.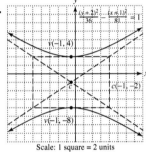

Scale: 1 square = 2 units

33. Center $(0, -3)$; vertices $(\sqrt{5}, -3)$, $(-\sqrt{5}, -3)$ **35.** $\dfrac{(y-7)^2}{49} - \dfrac{(x-5)^2}{25} = 1$ **37.** $y = \pm 9.055385138$

39. $2(x-3)(x^2 + 3x + 9)$ **40.** $(4x+3)(3x-2)$ **41.** $\dfrac{5x}{(x-3)(x-2)(x+2)}$ **42.** $\dfrac{-x-6}{(5x-1)(x+2)}$ or $-\dfrac{x+6}{(5x-1)(x+2)}$

43. 20 people contributed. There were 24 people on the hockey team. **44. (a)** 287 songs per day **(b)** 292 minutes per day
(c) 79.7% of the air time is music.

9.5 Exercises

1. $\left(\dfrac{1}{2}, 1\right)$, $(2, -2)$

3. $(-4, 2)$, $(4, -2)$

5. $(-2, 3)$, $(1, 0)$ **7.** $(3, 0)$, $\left(-\dfrac{9}{5}, \dfrac{12}{5}\right)$

9. $(2, 0)$, $(1, 1)$ **11.** $\left(\dfrac{5}{2}, \dfrac{3}{2}\right)$

13. $(2, 0)$, $(-2, 0)$

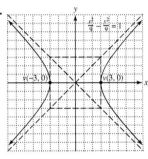

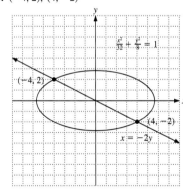

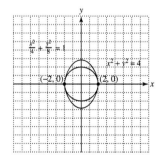

15. $(3, \pm 2)$, $(-3, \pm 2)$

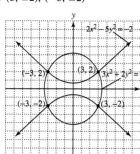

17. $(\sqrt{6}, \pm\sqrt{3})$, $(-\sqrt{6}, \pm\sqrt{3})$ **19.** $\left(\dfrac{\sqrt{30}}{3}, \pm\dfrac{\sqrt{21}}{3}\right)\left(-\dfrac{\sqrt{30}}{3}, \pm\dfrac{\sqrt{21}}{3}\right)$

21. $(2, \pm\sqrt{3})$, $(-2, \pm\sqrt{3})$ **23.** $(1, 3)$, $(-3, -1)$

25. $(4 - 2\sqrt{2}, 2 + \sqrt{2})$; $(4 + 2\sqrt{2}, 2 - 2\sqrt{2})$

27. No real solution   **29.** $(1.64, 2.02)$, $(0.17, -0.66)$ **31.** $(0, -6)$, $\left(\dfrac{147}{29}, \dfrac{120}{29}\right)$

33. $\dfrac{2x(x + 1)}{x - 2}$ **34.** $x^2 - 3x - 10$ **35.** 128,500 CD-ROMs were produced.

36. 25 miles per hour

Putting Your Skills to Work

1. 365.6 AU   **2.** 36.5 AU

Chapter 9 Review Problems

1. $\sqrt{41}$ **2.** $\sqrt{13}$ **3.** $x^2 + (y + 7)^2 = 25$ **4.** $(x + 6)^2 + (y - 3)^2 = 15$ **5.** $(x - 3)^2 + (y - 4)^2 = 22$; $C(3, 4)$, $r = \sqrt{22}$

6. $(x - 5)^2 + (y + 6)^2 = 9$; $C(5, -6)$, $r = 3$ **7.**

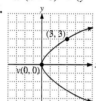

8.

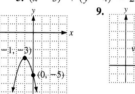

9.

10. $y = (x + 3)^2 - 5$; vertex at $(-3, -5)$; parabola opens upward.   **11.** $x = (y - 4)^2 - 6$; vertex at $(-6, 4)$; parabola opens to the right.

12.

13.

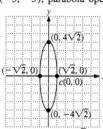

Scale: each unit is $\sqrt{2}$ units

14. $C = (-1, 2)$; $v = (2, 2)$, $(-4, 2)$, $(-1, 6)$, $(-1, -2)$

15. $C = (-5, -3)$; $v = (-3, -3)$, $(-7, -3)$, $(-5, 2)$, $(-5, -8)$ **16.**

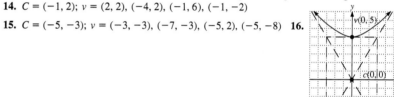

17.

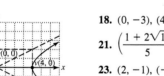

18. $(0, -3)$, $(4, -3)$; $C = (2, -3)$ **19.** $(-5, 3)$, $(-5, 1)$; $C = (-5, 2)$ **20.** $(-3, 0)$, $(2, 5)$

21. $\left(\dfrac{1 + 2\sqrt{14}}{5}, \dfrac{-2 + \sqrt{14}}{5}\right)$, $\left(\dfrac{1 - 2\sqrt{14}}{5}, \dfrac{-2 - \sqrt{14}}{5}\right)$ **22.** $(2, \pm 3)$, $(-2, \pm 3)$

23. $(2, -1)$, $(-2, 1)$, $(1, -2)$, $(-1, 2)$ **24.** No real solution   **25.** $(0, 1)$; $(\sqrt{5}, 6)$; $(-\sqrt{5}, 6)$

26. $(1, 4)$; $(-1, -4)$; $(2\sqrt{2}, \sqrt{2})$; $(-2\sqrt{2}, -\sqrt{2})$ **27.** $(1, \pm 2)$; $(-1, \pm 2)$

Chapter 9 Test

1. $\sqrt{185}$ **2.** Circle $C(-3, 2)$; $r = 2$; $(x + 3)^2 + (y - 2)^2 = 4$

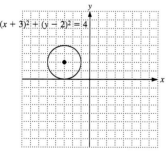

3. Parabola; vertex $(4, 3)$; $x = (y - 3)^2 + 4$

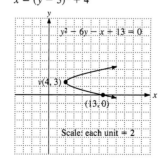

Scale: each unit = 2

4. Hyperbola; $C(0, 0)$

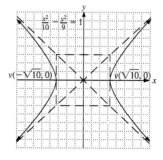

5. Ellipse; $C(0, 0)$

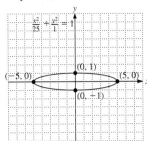

$\frac{x^2}{25} + \frac{y^2}{1} = 1$

$(-5, 0)$ $(0, 1)$ $(5, 0)$ $(0, -1)$

6. Parabola; $V(-3, 4)$

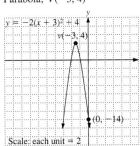

$y = -2(x + 3)^2 + 4$

$v(-3, 4)$

$(0, -14)$

Scale: each unit $= 2$

7. Ellipse; $C(-2, 5)$

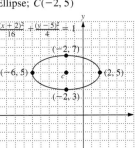

$\frac{(x + 2)^2}{16} + \frac{(y - 5)^2}{4} = 1$

$(-2, 7)$

$(-6, 5)$ $(2, 5)$

$(-2, 3)$

8. Hyperbola; $C(0, 0)$

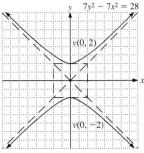

$7y^2 - 7x^2 = 28$

$v(0, 2)$

$v(0, -2)$

9. $(x - 3)^2 + (y + 5)^2 = 8$ **10.** $x = (y - 3)^2 - 7$ **11.** $\frac{(x + 4)^2}{1} + \frac{(y + 2)^2}{9} = 1$ **12.** $\frac{(y - 7)^2}{49} - \frac{(x - 6)^2}{9} = 1$

13. $(0, -3), (3, 0)$ **14.** $(0, 5), (-4, -3)$ **15.** $(1, 0), (-1, 0)$ **16.** $(\sqrt{3}, -\sqrt{3}), (-\sqrt{3}, \sqrt{3}); \left(\frac{\sqrt{6}}{2}, -\sqrt{6}\right), \left(-\frac{\sqrt{6}}{2}, \sqrt{6}\right)$

Cumulative Test for Chapters 1–9

1. Commutative property of multiplication **2.** -19 **3.** $8x + 12$ **4.** $p = \frac{A - 3bt}{rt}$ **5.** $(x + 5)(x^2 - 5x + 25)$ **6.** $\frac{3x + 18}{(x + 4)(x - 4)}$

7. $x = \frac{7}{2}$ **8.** $x = 4, y = -3, z = 1$ **9.** $2\sqrt{2}x - x\sqrt{2}$ **10.** $4\sqrt{3} + 6\sqrt{2} - \sqrt{6} - 3$ **11.** $x > 1$ **12.** $x \geq 14$ **13.** $3\sqrt{10}$

14. Parabola; $V(-2, -3)$

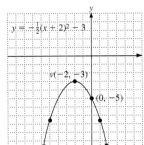

$y = -\frac{1}{2}(x + 2)^2 - 3$

$v(-2, -3)$ $(0, -5)$

15. Circle; $C(0, 0)$; radius $= \sqrt{5}$

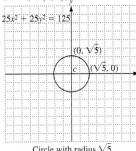

$25x^2 + 25y^2 = 125$

$(0, \sqrt{5})$

c $(\sqrt{5}, 0)$

Circle with radius $\sqrt{5}$

16. Hyperbola; $C(0, 0)$

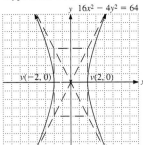

y $16x^2 - 4y^2 = 64$

$v(-2, 0)$ $v(2, 0)$

17. Ellipse; $C(2, 3)$

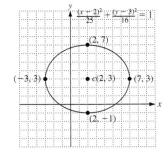

$\frac{(x - 2)^2}{25} + \frac{(y - 3)^2}{16} = 1$

$(2, 7)$

$(-3, 3)$ $c(2, 3)$ $(7, 3)$

$(2, -1)$

18. $(2, 8), (-1, 2)$ **19.** $\left(\frac{8}{3}, \pm\frac{2\sqrt{10}}{3}\right), \left(-\frac{8}{3}, \pm\frac{2\sqrt{10}}{3}\right)$ **20.** $(\frac{15}{4}, -4), (-3, 5)$ **21.** $(-5, 0), (3, 4)$

Chapter 10

Pretest Chapter 10

1. (a) -12 **(b)** $2a - 6$ **(c)** $4a - 6$ **(d)** $2a - 2$

2. (a) 13 **(b)** $5a^2 + 2a - 3$ **(c)** $5a^2 + 12a + 4$

3. (a) $\frac{6(a^2 - 2)}{a(a + 2)}$ **(b)** $\frac{18(a - 1)}{5(3a + 2)}$

4. Function **5.** Not a function

6.

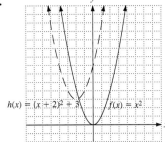

$h(x) = (x + 2)^2 + 3$ $f(x) = x^2$

7.

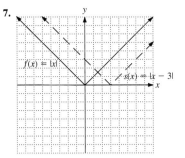

$f(x) = |x|$ $s(x) = |x - 3|$

8. (a) $-2x^3 - 3x - 1$ **(b)** -23 **(c)** $-6x^3 - 18x + 5$

9. (a) $\frac{-6x + 2}{x + 6}$ **(b)** 13 **(c)** $\frac{2}{-3x + 7}$

10. (a) $2x + 1, x \neq \frac{4}{3}$ **(b)** -1 **(c)** $54x^2 - 159x + 112$ **(d)** $18x^2 - 15x - 16$ **11.** One-to-one function **12.** Not a one-to-one function **13.** Yes **14.** $F^{-1} = \{(1, 7), (3, 6), (-1, 2)(5, -1)\}$

15. $g^{-1}(x) = \frac{3 - x}{5}$

15.

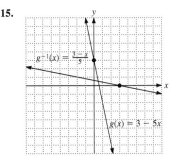

$g^{-1}(x) = \frac{3 - x}{5}$

$g(x) = 3 - 5x$

10.1 Exercises

1. -14 **3.** $3a - 5$ **5.** $9b - 5$ **7.** $3a - 17$ **9.** $\frac{1}{2}a - 4$ **11.** $a - 3$ **13.** $\frac{1}{2}a - 4$ **15.** $\frac{1}{2}a^2 - \frac{1}{5}$ **17.** $\frac{3}{2}b - \frac{5}{2}$

19. $\frac{3}{2}b + a - 6$ **21.** 2 **23.** $3a^2 + 4a - 2$ **25.** $3a^2 + 10a + 5$ **27.** $48b^2 + 16b - 2$ **29.** $9a^2 + 4a$ **31.** $3b^2 + 22b + 34$

33. 2 **35.** $2\sqrt{2}$ **37.** $\sqrt{a + 7}$ **39.** $\sqrt{3a + 5}$ **41.** $2\sqrt{a + 1}$ **43.** $\sqrt{a} + 3$ **45.** -7 **47.** $\frac{7}{2}$ **49.** $\frac{7}{a^2 - 3}$ **51.** $\frac{7}{a - 1}$

53. $\frac{-14}{(a - 1)(a - 3)}$ **55.** $\frac{7(3b - 4)}{(b - 1)(2b - 3)}$ **57.** -2 **59.** $4x + 2h$ **61.** (*a*) $A(r) = 3.14r^2$ (**b**) 50.24 square feet

(**c**) $A(e) = 3.14e^2 + 25.12e + 50.24$ (**d**) The area is 60.79 square feet.   **63.** They would decrease by 13. $p(3) - 13 \approx 26$

65. $38.021a^2 - 16.376a + 1.23$ **67.** $3a^2 - 5.512a + 1.999$ **69.** $A(x) = \dfrac{x^2 - 20x + 200}{8}$; $A(2) = 20.5$; $A(5) = 15.625$; $A(8) = 13$

71. $x = -3$ **72.** $x = -4$ **73.** $x = 5$ **74.** $x = -2$

10.2 Exercises

5. Function   **7.** Function   **9.** Function   **11.** Not a function   **13.** Function   **15.** Function   **17.** Function

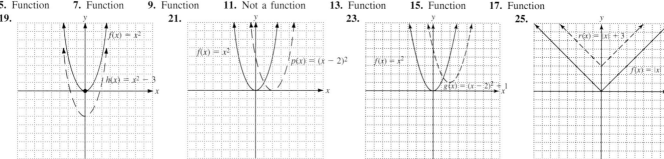

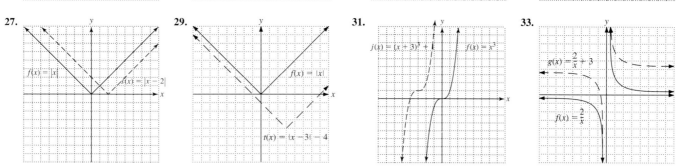

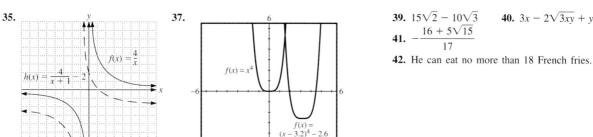

39. $15\sqrt{2} - 10\sqrt{3}$ **40.** $3x - 2\sqrt{3xy} + y$

41. $-\dfrac{16 + 5\sqrt{15}}{17}$

42. He can eat no more than 18 French fries.

10.3 Exercises

1. (**a**) $2x + 5$ (**b**) $-6x + 1$ (**c**) 9 (**d**) 7 **3.** (**a**) $2x^2 + 2x + 5$ (**b**) $2x^2 - 8x - 3$ (**c**) 17 (**d**) 7 **5.** (**a**) $x^3 - 3x + 2$ (**b**) $x^3 - 2x^2 + 3x - 2$

(**c**) 4 (**d**) -8 **7.** (**a**) $\dfrac{5x - 8}{(x - 3)(2x + 1)}$ (**b**) $\dfrac{-x + 10}{(x - 3)(2x + 1)}$ (**c**) $-\frac{2}{5}$ (**d**) $\frac{11}{3}$ **9.** (**a**) $4\sqrt{x + 2}$ (**b**) $-2\sqrt{x + 2}$ (**c**) 8 (**d**) -2

11. (**a**) $-2x^3 - 6x$ (**b**) 72 **13.** (**a**) $-x^3 + 4x^2 - 5x + 2$ (**b**) 80 **15.** (**a**) $\dfrac{6x + 4}{x + 6}$ (**b**) $-\frac{14}{3}$ **17.** (**a**) $-3x\sqrt{-2x + 1}$ (**b**) $9\sqrt{7}$

19. (**a**) $\dfrac{3x}{4x - 1}$, $x \neq \frac{1}{4}$ (**b**) $\frac{6}{7}$ **21.** (**a**) $\dfrac{x^2 + 5}{x - 1}$, $x \neq 1$ (**b**) 9 **23.** (**a**) $x + 5$, $x \neq -5$ (**b**) 7 **25.** (**a**) $\dfrac{1}{x + 2}$, $x \neq -2$, $x \neq \frac{1}{4}$ (**b**) $\frac{1}{4}$

27. (**a**) $\dfrac{x - 4}{3x + 6}$, $x \neq -2$, $x \neq 4$ (**b**) $-\frac{1}{6}$ **29.** $-x^2 + 5x + 2$ **31.** $\frac{8}{3}x + \frac{8}{3}$ or $\dfrac{8x + 8}{3}$ **33.** $3x^3 - 4x^2 - 4x$ **35.** -32

37. $\dfrac{9x + 6}{x - 2}, x \neq 2$ **39.** $\frac{17}{4}$ **41.** $-6x - 13$ **43.** $2x^2 - 12x + 18$ **45.** $7 - 6x^2$ **47.** $\dfrac{3}{2x}, x \neq 0$ **49.** $\sqrt{3x + 3}, x \geq -1$

51. $9x^2 + 30x + 27$ **53.** $3x^2 + 11$ **55.** 38   **57.** $\sqrt{x^2 + 1}$ **59.** $3\sqrt{x - 1} + 5$ **61.** $\sqrt{10}$ **63.** $x^4 + 4x^2 + 6$

65. Approximately 66.52751611   **67.** $K[C(F)] = \dfrac{5F + 2297}{9}$ **69.** $a[r(t)] = 28.26t^2$; The area is 11,304 square feet after 20 minutes.

71. $(5x^2 + 1)(5x^2 - 1)$ **72.** $(6x - 1)^2$ **73.** $(3x - 1)(x - 2)$ **74.** $(x + 3)(x - 3)(x + 1)(x - 1)$

10.4 Exercises

5. Yes   **7.** No   **9.** Yes   **11.** Yes   **13.** No   **15.** No   **17.** Yes, it passes the vertical line test. No, it does not pass the horizontal line test.   **19.** $H^{-1} = \{(7, 2), (-1, -3)\}$ **21.** $K^{-1} = \{(1, -7), (2, 6), (-1, 3), (5, 2)\}$ **23.** $L^{-1} = \{(4, 1), (8, 2), (6, 3), (-8, -2)\}$

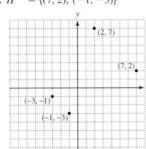
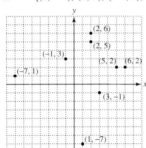
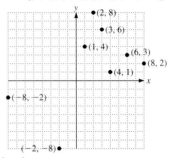

25. $f^{-1}(x) = \dfrac{x + 1}{3}$ or $\dfrac{1}{3}x + \dfrac{1}{3}$ **27.** $f^{-1}(x) = 5x - 4$ **29.** $f^{-1}(x) = -\dfrac{4}{x}$ **31.** $f^{-1}(x) = \dfrac{3}{x} + 2$ or $\dfrac{3 + 2x}{x}$

35.   **37.**   **39.**   **41.**

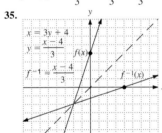
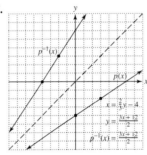
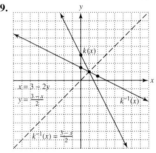
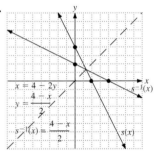

43. $f^{-1}(x) = \dfrac{x + 5}{0.0063}$; The inverse function would tell you how many Spanish pesetas were given by the bank for x dollars. No. Because of the bank fee, it would not work for Manuela's transaction.   **45.** $f[f^{-1}(x)] = 2\left[\dfrac{1}{2}x - \dfrac{3}{4}\right] + \dfrac{3}{2} = x; f^{-1}[f(x)] = \dfrac{1}{2}\left[2x + \dfrac{3}{2}\right] - \dfrac{3}{4} = x$

47. $x = -64; x = -27$ **48.** $x = 3$ is the only solution   **49.** The ratio would be $23 : 21$.   **50.** She worked 13 overtime hours.

Putting Your Skills to Work

1. 80.4% of the maximum growth rate.   **2.** It would raise the graph 10 units higher. $G(x) = -0.4(x - 23)^2 + 110$

3. $G(x) = -0.4(x - 25)^2 + 110$

Chapter 10 Review Problems

1. -1 **2.** 11   **3.** $7 - 4a$ **4.** $7 - 6b$ **5.** $6 - 4a$ **6.** $18 - 6b$ **7.** $\dfrac{1}{2}a + \dfrac{5}{2}$ **8.** $\dfrac{1}{2}a + 4$ **9.** $-\dfrac{1}{2}$ **10.** 1

11. $a + \dfrac{9}{2}$ **12.** $a + \dfrac{3}{2}$ **13.** -15 **14.** -10 **15.** $-8a^2 + 6a - 16$ **16.** $-18a^2 + 9a - 11$ **17.** $-2a^2 - 5a - 3$

18. $-2a^2 + 15a - 28$ **19.** $|16a - 1|$ **20.** $|14a - 1|$ **21.** $\left|\dfrac{1}{2}a - 1\right|$ **22.** $|a - 1|$ **23.** $|2a - 7|$ **24.** $|2a + 7|$ **25.** $\dfrac{9}{7}$

26. -3 **27.** $\dfrac{3a + 9}{a + 7}$ **28.** $\dfrac{3a - 6}{a + 2}$ **29.** $\dfrac{30a + 36}{7a + 28}$ **30.** $\dfrac{-12}{a + 4}$ **31.** 7   **32.** 6   **33.** $\dfrac{1}{2}$ **34.** $\dfrac{1}{4}$ **35.** $4x + 2h - 5$

36. $-6x - 3h + 2$ **37.** (a) Yes (b) Yes   **38.** (a) No (b) No   **39.** (a) Yes (b) No   **40.** (a) Yes (b) No   **41.** (a) No (b) No

42. (a) Yes (b) Yes   **43.** (a) Yes (b) No   **44.** (a) No (b) No   **45.** (a) Yes (b) Yes

46.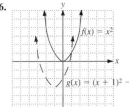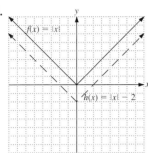

47. 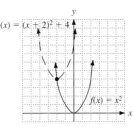 $g(x) = (x + 2)^2 + 4$ 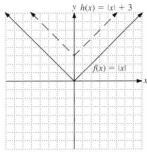 $f(x) = x^2$

48. 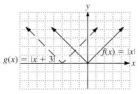 $g(x) = |x + 3|$ $f(x) = |x|$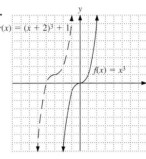

49. 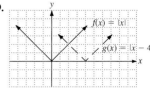 $f(x) = |x|$ $g(x) = |x - 4|$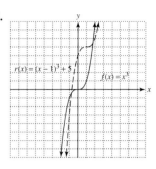

50. $f(x) = |x|$ $h(x) = |x| - 2$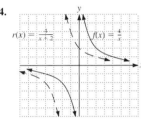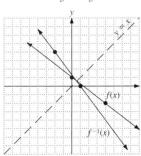

51. y $h(x) = |x| + 3$ $f(x) = |x|$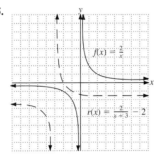

52. $r(x) = (x + 2)^3 + 1$ $f(x) = x^3$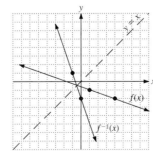

53. $r(x) = (x - 1)^3 + 5$ $f(x) = x^3$

54. $r(x) = \frac{-4}{x + 2}$ $f(x) = \frac{4}{x}$

55. $f(x) = \frac{2}{x}$ $r(x) = \frac{-2}{x + 3} - 2$

56. $2x^2 + 9$ **57.** $\frac{5}{2}x + 2$ **58.** $-\frac{7}{2}x - 8$ **59.** $2x^2 - 6x - 1$

60. $\frac{5}{2}$ **61.** -5 **62.** $\frac{6x + 10}{x}$, $x \neq 0$ **63.** $-x^3 - \frac{9}{2}x^2 + 7x - 12$

64. $\frac{2x - 8}{x^2 + x}$, $x \neq 0$, $x \neq -1$, $x \neq 4$ **65.** $\frac{2}{3x^2 + 5x}$, $x \neq 0$, $x \neq -\frac{5}{3}$

66. -6 **67.** $\frac{1}{6}$ **68.** $\frac{3x + 6}{3x + 1}$, $x \neq -\frac{1}{3}$ **69.** $-\frac{3}{2}x - 4$

70. $\sqrt{2x^2 - 3x + 2}$ **71.** $\sqrt{-\frac{1}{2}x - 5}$, $x \leq -10$ **72.** 2 **73.** 2 **74.** $\frac{2x - 8}{x + 1}$, $x \neq 4$, $x \neq -1$ **75.** $-\frac{1}{x} - 3$, $x \neq 0$ **76.** $9x + 20$

77. $\frac{1}{4}x - \frac{3}{2}$ **78.** $f[g(x)] = \frac{6}{x} + 5 = \frac{6 + 5x}{x}$; $g[f(x)] = \frac{2}{3x + 5}$; $f[g(x)] \neq g[f(x)]$ **79.** $p[g(x)] = \frac{8}{x^2} - \frac{6}{x} + 4 = \frac{8 - 6x + 4x^2}{x^2}$;

$g[p(x)] = \frac{2}{2x^2 - 3x + 4}$; $p[g(x)] \neq g[p(x)]$ **80. (a)** $D = \{100, 200, 300, 400\}$ **(b)** $R = \{10, 20, 30\}$ **(c)** Yes **(d)** No **81. (a)** $D = \{0, 3, 7\}$

(b) $R = \{-8, 3, 7, 8\}$ **(c)** No **(d)** No **82. (a)** $D = \{12, 0, -6\}$ **(b)** $R = \{-1, -12, 6\}$ **(c)** No **(d)** No **83. (a)** $D = \left\{\frac{1}{2}, \frac{1}{4}, -\frac{1}{3}, 4\right\}$

(b) $R = \left\{2, 4, -3, \frac{1}{4}\right\}$ **(c)** Yes **(d)** Yes **84. (a)** $D = \{0, 1, 2, -1\}$ **(b)** $R = \{1, 2, 9, -2\}$ **(c)** Yes **(d)** Yes **85. (a)** $D = \{0, 1, 2, 3\}$

(b) $R = \{-3, 1, 7\}$ **(c)** Yes **(d)** No **86.** $\left\{\left(\frac{1}{3}, 3\right), \left(-\frac{1}{2}, -2\right), \left(-\frac{1}{4}, -4\right), \left(\frac{1}{5}, 5\right)\right\}$ **87.** $\{(10, 1), (7, 3), (15, 12), (1, 10)\}$ **88.** $f^{-1}(x) = -\frac{3}{2}x + 6$

89. $g^{-1}(x) = -\frac{1}{3}x - \frac{5}{3}$ **90.** $h^{-1}(x) = 3x - 2$ **91.** $j^{-1}(x) = \frac{1}{x} + 3$ **92.** $p^{-1}(x) = x^3 - 1$ **93.** $r^{-1}(x) = \sqrt[3]{x - 2}$

94. $f^{-1}(x) = -\frac{4}{3}x + \frac{4}{3}$

95. $f^{-1}(x) = -3x - 2$

96. $f^{-1}[f(x)] = \dfrac{4\left[\frac{1}{2}x - \frac{3}{4}\right] + 3}{2} = \dfrac{2x - 3 + 3}{2} = x$;

$f[f^{-1}(x)] = \frac{1}{2}\left[\frac{4x + 3}{2}\right] - \frac{3}{4} = \frac{4x + 3}{4} - \frac{3}{4} = x$

97. $f^{-1}[f(x)] = -\frac{2}{3}\left[\frac{6 - 3x}{2}\right] + 2 = \dfrac{-6 + 3x}{3} + 2 = x$;

$f[f^{-1}(x)] = \dfrac{6 - 3\left(-\frac{2}{3}x + 2\right)}{2} = \dfrac{6 + 2x - 6}{2} = x$

Chapter 10 Test

1. -5 **2.** $\frac{3}{2}a - 2$ **3.** $\frac{3}{4}a - \frac{3}{2}$ **4.** 25 **5.** $3a^2 + 4a + 5$ **6.** $3a^2 - 2a + 9$ **7.** $12a^2 + 4a + 2$ **8. (a)** Yes **(b)** No
9. (a) Yes **(b)** Yes

10.

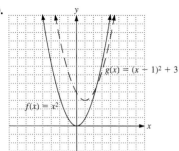

$g(x) = (x - 1)^2 + 3$

$f(x) = x^2$

11.

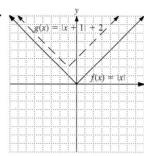

$g(x) = |x + 1| + 2$

$f(x) = |x|$

12. (a) $x^2 + 4x + 1$ **(b)** $5x^2 - 6x - 13$ **(c)** 19

13. (a) $\dfrac{6x - 3}{x}$, $x \neq 0$ **(b)** $\dfrac{3}{2x^2 - x}$, $x \neq 0$, $x \neq \frac{1}{2}$ **(c)** $\dfrac{6}{x} - 1$,

$x \neq 0$ **14. (a)** $2x - \frac{1}{2}$ **(b)** $2x - 7$ **(c)** $\frac{1}{4}x - \frac{9}{2}$

15. (a) Yes **(b)** $A^{-1} = \{(5, 1), (1, 2), (-7, 4), (7, 0)\}$

16. (a) Yes **(b)** $B^{-1} = \{(8, 1), (1, 8), (10, 9), (9, -10)\}$

17. $f^{-1}(x) = 2x + \frac{2}{5}$

18. $f^{-1}(x) = -\frac{1}{3}x + \frac{2}{3}$ **19.** $f^{-1}[f(x)] = x$

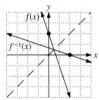

$f(x)$

$f^{-1}(x)$

Cumulative Test for Chapters 1–10

1. Commutative property of multiplication   **2.** -19 **3.** $8x + 12$ **4.** $p = \dfrac{A - 3bt}{rt}$ **5.** $(x + 5)(x^2 - 5x + 25)$ **6.** $\dfrac{3x + 18}{(x + 4)(x - 4)}$

7. $x = \frac{7}{2}$ **8.** $x = 4$, $y = -3$, $z = 1$ **9.** $2\sqrt{2x} - x\sqrt{2}$ **10.** $4\sqrt{3} + 6\sqrt{2} - \sqrt{6} - 3$ **11.** $3\sqrt{10}$ **12.** $(4x - 1)(3x - 2)$

13. $(9x^2 + 1)(3x + 1)(3x - 1)$ **14.** $(x + 3)^2 + (y - 6)^2 = 196$ **15. (a)** 17 **(b)** $3a^2 - 14a + 17$ **(c)** $3a^2 - 2a + 18$

16.

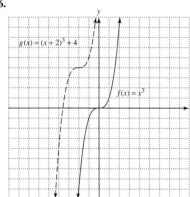

$g(x) = (x + 2)^3 + 4$

$f(x) = x^3$

17. (a) $10x^3 - 19x^2 - 45x - 18$

(b) $\dfrac{2x^2 - 5x - 6}{5x + 3}$, $x \neq -\frac{3}{5}$ **(c)** $50x^2 + 35x - 3$

18. (a) Yes **(b)** Yes **(c)** $A^{-1} = \{(6, 3), (8, 1),$

$(7, 2), (4, 4)\}$ **19.** $f^{-1}(x) = \dfrac{x + 3}{7}$

20. (a) 22 **(b)** -168 **(c)** $40a^3 - 12a^2 - 6$

21. (a) $f^{-1}(x) = -\dfrac{3}{2}x + 3$

(b)

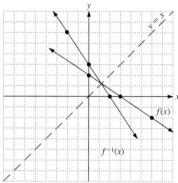

$y = x$

$f(x)$

$f^{-1}(x)$

22. $f[f^{-1}(x)] = x$

Chapter 11

Pretest Chapter 11

1.

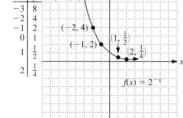

x	y
-3	8
-2	4
-1	2
0	1
1	$\frac{1}{2}$
2	$\frac{1}{4}$

$(-3, 8)$

$(-2, 4)$

$(-1, 2)$

$(1, \frac{1}{2})$

$(2, \frac{1}{4})$

$f(x) = 2^{-x}$

2. $x = 2$ **3.** $\$10,000(1.12)^4$ or $\$15,735.19$ **4.** $\log_6 \frac{1}{36} = -2$ **5.** $x = 64$ **6.** 4

7. $2 \log_5 x + 5 \log_5 y - 3 \log_5 z$ **8.** $\log_4 \dfrac{\sqrt{x}}{w^3}$ **9.** $x = \frac{81}{2}$ **10.** $x = 8260.3795$ **11.** 1.5665

12. 0.9005   **13.** -0.2084 **14.** 9.5137×10^9 **15.** $x = \frac{1}{3}$ **16.** $x \approx 0.2925$ **17.** 21 yr

11.1 Exercises

3.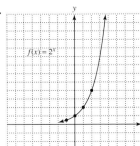

5.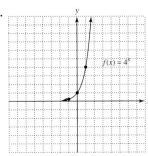

7.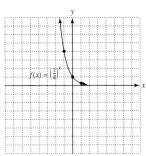

9.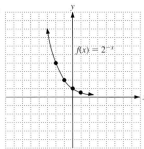

11.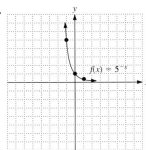

13.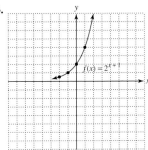

15.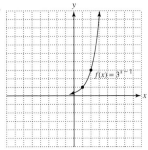

17.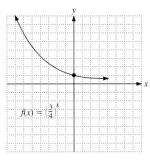

19.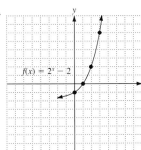

21.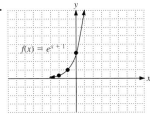

23.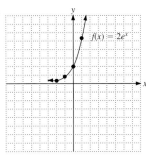

25.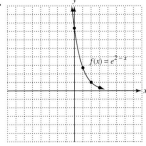

27. $x = 2$ **29.** $x = 0$ **31.** $x = -3$ **33.** $x = 4$ **35.** $x = 0$ **37.** $x = 2$ **39.** $x = 4$ **41.** $x = 1$ **43.** $x = 1$ **45.** $x = -2$
47. $4682.69   **49.** $3979.58; $4011.22   **51.** $8071.45   **53.** $4440.73; $21,611.32   **55.** At a depth of 20 feet, they will have 37% of available sunlight. Yes, they will need spotlights at a depth of 48 feet.   **57.** Approximately 3.91 mg   **59.** The pressure is approximately 9.66 pounds per square inch.   **61.** 1955   **63.** 6.8 billion people   **65.**

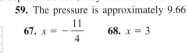

67. $x = -\dfrac{11}{4}$ **68.** $x = 3$

69. $x = 1$ **70.** $x = \dfrac{2}{3}$

11.2 Exercises

5. $\log_4 64 = 3$ **7.** $\log_6 36 = 2$ **9.** $\log_5\left(\dfrac{1}{25}\right) = -2$ **11.** $\log_2\left(\dfrac{1}{32}\right) = -5$ **13.** $\log_e y = 3$ **15.** $\log_{10} 4 = 0.6021$

17. $\log_e 0.0067 = -5$ **19.** $3^2 = 9$ **21.** $5^0 = 1$ **23.** $8^0 = 1$ **25.** $10^{-2} = 0.01$ **27.** $3^{-4} = \dfrac{1}{81}$ **29.** $e^5 = x$ **31.** $10^{0.8451} = 7$

33. $e^{-1.2040} = 0.3$ **35.** $16 = x$ **37.** $\dfrac{1}{1000} = x$ **39.** $y = 3$ **41.** $y = -2$ **43.** $a = 2$ **45.** $a = 10$ **47.** $w = \dfrac{1}{2}$

49. $w = -1$ **51.** $w = 1$ **53.** $w = \dfrac{1}{5}$ **55.** $w = e^4$ **57.** $x = 5$ **59.** $x = 1$ **61.** $x = -3$ **63.** $x = 7$

65. $x = 2$ **73.**
67. $x = \dfrac{1}{2}$
69. $x = 0$
71. $x = -2$

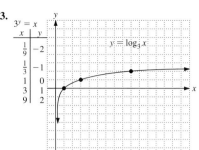

x	y
$\frac{1}{9}$	-2
$\frac{1}{3}$	-1
1	0
3	1
9	2

75.

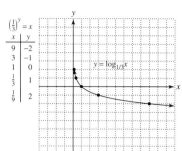

x	y
9	-2
3	-1
1	0
$\frac{1}{3}$	1
$\frac{1}{9}$	2

77.

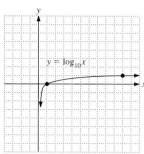

79.

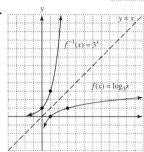

81. pH $= 2$
83. $[\text{H}^+] = 10^{-8}$
85. $[\text{H}^+] \approx 5.623 \times 10^{-10}$
87. 11,200 sets of software
89. They should spend $1,000,000 on advertising.
91. $y = 4$

93.

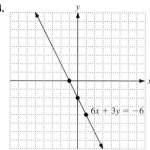

94.

95. $m = -\dfrac{1}{5}$ **96.** $y = \dfrac{3}{2}x + 7$ **97.** (a) 36,000 cells (b) 36,864,000 cells **98.** (a) $5353.27 (b) $24,424.03

11.3 Exercises

1. $\log_3 A + \log_3 B$ **3.** $\log_5 7 + \log_5 11$ **5.** $\log_b 9 + \log_b f$ **7.** $\log_3 13 - \log_3 5$ **9.** $\log_a G - \log_a 7$ **11.** $\log_a E - \log_a F$

13. $5\log_3 x$ **15.** $-2\log_b A$ **17.** $\dfrac{1}{2}\log_5 w$ **19.** $\dfrac{1}{2}\log_5 x + 3\log_5 y$ **21.** $\log_7 2 + \log_7 x - \log_7 y$

23. $\log_2 5 + \log_2 x + 4\log_2 y - \dfrac{1}{2}\log_2 z$ **25.** $\dfrac{1}{3}\log_b x - \dfrac{2}{3}\log_b y - \dfrac{1}{3}\log_b z$ **27.** $5 + \log_e x + 2\log_e y$ **29.** $\log_5 21x$

31. $\log_3\left(\dfrac{x^5}{7}\right)$ **33.** $\log_b \dfrac{\sqrt[3]{x^2}\sqrt{y}}{z^3}$ **35.** $\log_3 \sqrt{\dfrac{7}{xy}}$ **37.** 1 **39.** 1 **41.** 0 **43.** 1 **45.** $x = 7$ **47.** $x = 12$ **49.** $x = 0$

51. $x = 1$ **53.** $x = 3$ **55.** $x = -4$ **57.** $x = 4$ **59.** $x = 21$ **61.** $x = 2$ **63.** $x = 2e^2$ **65.** $x = \dfrac{27}{7}$ **67.** $\dfrac{\log_{10} A - 4}{\log_{10} 1.07} = x$

69. 6 **73.** Use the equation $1 = b^y$ **74.** Use the equation $\log_b b = y$ **75.** Approximately 50.27 square meters
76. Approximately 62.83 cubic meters **77.** $(3, -2)$ **78.** $(-1, -2, 3)$ **79.** There was a 173.6% increase for the 13-year period. This is a 13.4% average yearly increase. **81.** The speed is $12\frac{2}{3}$ feet per second (or approximately 8.64 miles per hour) after 2 seconds. No, he was not.

11.4 Exercises

1. 0.866877814 **3.** 1.408239965 **5.** 2.551449998 **7.** 5.096910013 **9.** -1.910094889 **11.** -0.050122296 **13.** Error. You cannot take the log of a negative number. **15.** 1.450106836 **17.** 3.820322269 **19.** 61.09420249 **21.** 8519.223264
23. 2939679.609 **25.** 0.034300464 **27.** 0.000408037 **29.** 100.6931669 **31.** 41831168.87 **33.** 0.000030811 **35.** 1.726331664
37. 4.028916757 **39.** 11.82041016 **41.** -4.09234656 **43.** 0.405465108 **45.** 2.585709659 **47.** 11.02317638 **49.** 1202604.284
51. 0.878095431 **53.** 0.067205513 **55.** 107.5762609 **57.** 0.1188610637 **59.** 2.020006063 **61.** 0.98247341 **63.** -1.151699337
65. 1.913507505 **67.** -1.846414 **69.** 4.747938 **71.** 1.9675948 **73.** -0.34008898 **75.** 3.6593167×10^8 **77.** 3.3149796
79.

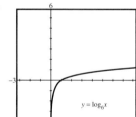

81.

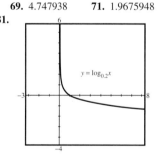

83. $R \approx 5.26$ **85.** The shock wave is about 251,000 times greater than the smallest detectable shock wave. **86.** $x = 0, x = \dfrac{7}{17}$

87. $x = \dfrac{11 \pm \sqrt{181}}{6}$ **88.** $y = \dfrac{-2 \pm \sqrt{10}}{2}$ **89.** $y = \dfrac{-5 \pm \sqrt{13}}{6}$

11.5 Exercises

1. $x = 6$ **3.** $x = \frac{31}{3}$ **5.** $x = \frac{13}{8}$ **7.** $x = 3$ **9.** $x = 2$ **11.** $x = \dfrac{20}{99}$ **13.** $x = \dfrac{16}{15}$ **15.** $x = \frac{10}{3}$ **17.** $x = \frac{5}{3}$ **19.** $x = 4$

21. $x = 2$ **23.** $x = 4, x = 5$ **25.** $x = \dfrac{\log 13}{\log 7}$ **27.** $x = \dfrac{\log 9}{\log 5} - 1$ or $x = \dfrac{\log 9 - \log 5}{\log 5}$ **29.** $x = \dfrac{\log 17 - 4\log 2}{3\log 2}$ **31.** $x = \dfrac{\ln 14}{2}$

33. $x \approx 0.374$ **35.** $x \approx 6.213$ **37.** $x \approx 37.310$ **39.** $x \approx 4.332$ **41.** $x \approx 2.120$ **43.** $2.494 \approx x$ **45.** $t \approx 22$ yr **47.** $t \approx 14$ yr
49. rate $\approx 4.5\%$ **51.** $t \approx 27$ yr **53.** $t \approx 35$ yr **55.** 69 yr **57.** approx. 6 years **59.** approx. 55 hours **61.** Approximately 46,931 people would be infected. **63.** 2.5 times greater **65.** It was 32 times greater. **67.** 12.7 years **69.** $7xy\sqrt{2x}$ **70.** $3x^2y^3\sqrt[3]{3}$
71. $3\sqrt{2} + 4\sqrt{3} - \sqrt{6} - 4$ **72.** $-4\sqrt{2x}$

Putting Your Skills to Work

1. $r \approx 0.0277$ **2.** $P = P_0 e^{0.0277t}$ **3.** Approximately 2.3137×10^6 or 2,313,700 **4.**

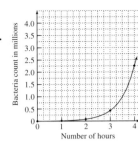
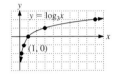

Chapter 11 Review Problems

1. 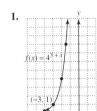 **2.** 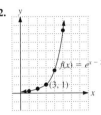 **3.** $x = 1$ **4.** $\log_2\left(\frac{1}{32}\right) = -5$ **5.** $10^{-3} = 0.001$

6. $\frac{1}{9}$ **7.** 2 **8.** $\frac{1}{7}$ **9.** 1 **10.** 3 **11.** 0.001

12. 3 **13.** 6 **14.** -2 **15.** 3

$y = \log_3 x$, (1, 0)

17. $3 \log_2 x + \frac{1}{2} \log_2 y$ **18.** $\log_2 5 + \log_2 x - \frac{1}{2} \log_2 w$ **19.** $\log_3 \frac{x\sqrt{w}}{2}$ **20.** $\log_8 \frac{w^4}{\sqrt[3]{z}}$ **21.** 6 **22.** 25 **23.** 25

24. 1.376576957 **25.** -1.087777943 **26.** 1.366091654 **27.** 6.688354714 **28.** 13.69935122 **29.** 5.473947392 **30.** 0.49685671

31. $x = \frac{34}{5}$ **32.** $x = \frac{-79}{54}$ **33.** $x = 1$ **34.** $x = \frac{1}{7}$ **35.** $x = 4$ **36.** $x = 3$ **37.** $t = \frac{3}{4}$ **38.** $t = -\frac{1}{8}$ **39.** $x = \frac{\log 14}{\log 3}$

40. $x = \frac{2 \log 4}{\log 5 - \log 4}$ **41.** $x = \frac{\ln 30.6}{2}$ **42.** $x = -1 + \ln 3.5$ **43.** $x \approx -1.4748$ **44.** $x \approx 0.7712$ **45.** $x \approx 2.3319$

46. $x \approx 101.3482$ **47.** \$6312.38 **48.** 9 yr **49.** Approx. 8 years **50.** Approx. 9 years **51.** 41 yr **52.** 26 yr **53.** 11 yr

54. 9 yr **55.** 50.1 times as much **56. (a)** 282 **(b)** 7.77

Chapter 11 Test

1. 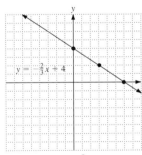 **2.** $x = 0$ **3.** $x = \frac{1}{64}$ **4.** $w = 5$ **5.** $\log_8 \frac{xw}{\sqrt[4]{3}}$ **6.** 1.3729 **7.** 1.7901 **8.** 0.4391

9. $x = 5350.569382$ **10.** $x = 1.150273799$ **11.** $x = \frac{16}{3}$ **12.** $x = \frac{3}{7}$ **13.** $x = \frac{-1 + \ln 0.25}{3}$

14. $x \approx -1.4132$ **15.** \$2938.66 **16.** 14 yr

Cumulative Test for Chapters 1–11

1. 6 **2.** $x = \frac{H + 2ay}{3b}$ **3.** 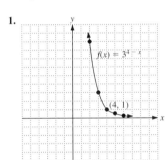 $y = -\frac{2}{3}x + 4$

4. $(x + y)(5a - 7w)$ **5.** $x = 1, y = -1, z = 2$

6. $5\sqrt{10} + \sqrt{15} - 20\sqrt{3} - 6\sqrt{2}$ **7.** $x = \pm\sqrt{6}; x = \pm i$

8. $x = 4, y = 4; \quad x = 1, y = -2$

9. $x = 4$ is only valid solution **10.** $\frac{5\sqrt[3]{4x^2y}}{2xy}$

11. 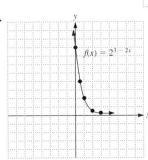 $f(x) = 2^{3 - 2x}$

12. $x = \frac{3}{2}$ **13.** $x = \frac{1}{4}$ **14.** 0.403120521 **15.** $x = 66.20640403$ **16.** 1.771243749

17. $x \approx 7.1263558$ **18.** $x = 9$ **19.** $x = 2$ **20.** $x = -0.535$ **21.** $x = \frac{\ln 0.5}{2}$

22. \$4234.74

Practice Final Examination

1. 4   **2.** $\dfrac{9x^2}{4y^2}$ **3.** $-a^2 - 10ab - a$ **4.** $-9x - 15y$ **5.** $F = -31$ **6.** $y = -30$ **7.** $b = \dfrac{2A - ac}{a}$ **8.** $x = 3$, $x = 9$

9. $2600 at 12%; $1400 at 14%   **10.** Length = 520 m; width = 360 m   **11.** $x < 1$ **12.** $x \le -9$ or $x \ge -2$ **13.** $-\frac{5}{2} < x < \frac{15}{2}$

14. x-intercept $(-2, 0)$;   **15.**   **16.** $m = \frac{8}{3}$ **17.** $3x + 2y = 5$ **20.**
     y-intercept $(0, 7)$ **18.** $f(3) = 12$ **19.** $f(-2) = 17$

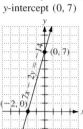

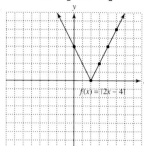

21. $x = 6$; $y = -3$ **22.** $x = 6$; $y = 4$ **23.** $x = 1$; $y = 4$; $z = -2$ **24.** $x = 4$; $y = 1$; $z = 1$ **25.** $z = 1$

26.   **27.** $6x^3 - 16x^2 + 17x - 6$ **28.** $5x^2 - x + 2$ **29.** $(2x - 3)(4x^2 + 6x + 9)$

30. $(x + 2)(x + 2)(x - 2)$ **31.** $x(2x - 1)(x + 8)$ **32.** $x = -6$, $x = -9$ **33.** $\dfrac{x(3x - 1)}{x - 3}$

34. $\dfrac{x + 5}{x}$ **35.** $\dfrac{3x^2 + 6x - 2}{(x + 5)(x + 2)}$ **36.** $\dfrac{8x^2 + 6x - 5}{4x^2 - 3}$ **37.** $x = -1$ **38.** $\dfrac{1}{3x^{7/2}y^5}$

39. $2xy^2\sqrt[3]{5xy}$ **40.** $18\sqrt{2}$ **41.** $\dfrac{18 + 2\sqrt{6} + 3\sqrt{3} + \sqrt{2}}{25}$ **42.** $8i$ **43.** $x = -3$

44. $y = 33.75$ **45.** $x = \dfrac{1 \pm \sqrt{21}}{10}$ **46.** $x = 0$, $x = -\frac{3}{5}$ **47.** $x = 8$, $x = -343$

48. Width = 4 cm; length = 13 cm

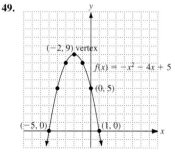

49.   **50.** $x \le -\frac{1}{3}$ or $x \ge 4$ **52.** $\dfrac{x^2}{16} + \dfrac{y^2}{25} = 1$; ellipse   **53.** $\dfrac{x^2}{4} - \dfrac{y^2}{9} = 1$; hyperbola

51. $(x + 3)^2 + (y - 2)^2 = 4$;
     center at $(-3, 2)$;
     radius = 2

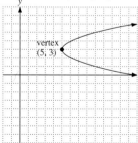

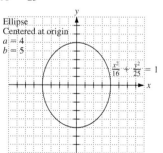

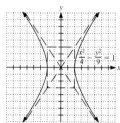

54. Parabola opening right   **55.** $(0, -4)$, $(\sqrt{7}, 3)$, $(-\sqrt{7}, 3)$ **56.** $f(-1) = 10$;   **59.**
     $f(a) = 3a^2 - 2a + 5$; $f(a + 2) = 3a^2 + 10a + 13$
     57. $f[g(x)] = 80x^2 + 80x + 17$ **58.** $f^{-1}(x) = 2x + 14$

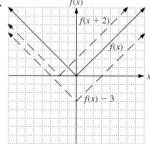

60. $f(x) = 2^{1-x}$

x	0	1	-2	4	-1
y	2	1	8	$\frac{1}{8}$	4

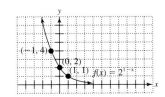

61. $0.0016 = x$ or $\frac{1}{625} = x$ **62.** $21 = x$ **63.** $y = -2$ **64.** $x = 8$

Appendix B Exercises

1. $(4, -1)$ **3.** $(3, -9)$ **5.** $(0, 3)$ **7.** $(2, 2)$ **9.** $(1.2, 3.7)$ **11.** $(3, -1, 4)$ **13.** $(1, -1, 3)$ **15.** $(0, -2, 5)$ **17.** $(0.5, -1, 5)$
19. $(3.6, 1.8, 2.4)$ **21.** $(4.2, -3.6, 8.8, 5.4)$

Photo Credits

P-1